NATURAL PRODUCTS CHEMISTRY

Vol. 2

Edited by

KOJI NAKANISHI
DEPT. OF CHEMISTRY
COLUMBIA UNIVERSITY, NEW YORK

TOSHIO GOTO
DEPT. OF AGRICULTURAL CHEMISTRY
NAGOYA UNIVERSITY, NAGOYA

SHÔ ITÔ
DEPT. OF CHEMISTRY
TOHOKU UNIVERSITY, SENDAI

SHINSAKU NATORI
NATIONAL INSTITUTE OF HYGIENIC SCIENCES, TOKYO

SHIGEO NOZOE
CHEMISTRY DIVISION
THE INSTITUTE OF APPLIED MICROBIOLOGY,
UNIVERSITY OF TOKYO, TOKYO

KODANSHA LTD. ACADEMIC PRESS, INC. New York and London
Tokyo *A Subsidiary of Harcourt Brace Jovanovich, Publishers*

CHEMISTRY

KODANSHA SCIENTIFIC BOOKS

Co-published by

KODANSHA LTD.
12-21 Otowa 2-chome, Bunkyo-ku, Tokyo 112
and

ACADEMIC PRESS, INC.
111 Fifth Avenue, New York, New York 10003

United Kingdom Edition published by
ACADEMIC PRESS, INC. (LONDON) LTD.
24/28 Oval Road, London NW 1

INTERNATIONAL STANDARD BOOKS NUMBER: 0-12-513902-0
LIBRARY OF CONGRESS CATALOG CARD NUMBER: 74-6431

PRINTED IN JAPAN

NATURAL PRODUCTS CHEMISTRY

Vol. 2

LIST OF CONTRIBUTORS

*: major contributors

Chapter 7

Jun FURUKAWA,   Faculty of Pharmaceutical Sciences, University of Tokyo, Tokyo
Yoshimasa MACHIDA,   Institute of Applied Microbiology, University of Tokyo, Tokyo
Naoki MIYATA,   Faculty of Pharmaceutical Sciences, University of Tokyo, Tokyo
Shigeo NOZOE*,   Institute of Applied Microbiology, University of Tokyo, Tokyo
Takashi YATSUNAMI,   Faculty of Pharmaceutical Sciences, University of Tokyo, Tokyo

Chapters 8 and 11

Toshio GOTO*,   Department of Agricultural Chemistry, Nagoya University, Nagoya
Yoshito KISHI*,   Department of Chemistry, Harvard University, Cambridge, Mass.
Yasuaki OKUMURA,   Department of Chemistry, Shizuoka University, Shizuoka
Nobutaka SUZUKI,   Department of Applied Chemistry, Mie University, Tsu
Tadao KONDO,   Department of Agricultural Chemistry, Nagoya University, Nagoya
Minoru ISOBE,   Department of Agricultural Chemistry, Nagoya University, Nagoya
Hideo TANINO,   Department of Agricultural Chemistry, Nagoya University, Nagoya
Shin-ichi NAKATSUKA,   Department of Agricultural Chemistry, Nagoya University, Nagoya
Fumiaki NAKATSUBO,   Wood Research Institute, Kyoto University, Uji

Chapter 9

Masamichi FUKUOKA,   National Institute of Hygienic Sciences, Tokyo
Masanori KUROYANAGI,   National Institute of Hygienic Sciences, Tokyo
Shinsaku NATORI*,   National Institute of Hygienic Sciences, Tokyo
Shigeo NOZOE,   Institute of Applied Microbiology, University of Tokyo, Tokyo
Tamotsu SAITOH*,   Faculty of Pharmaceutical Sciences, University of Tokyo, Tokyo
Ushio SANKAWA*,   Faculty of Pharmaceutical Sciences, University of Tokyo, Tokyo
Setsuko SEKITA,   National Institute of Hygienic Sciences, Tokyo
Michiko TOHYAMA,   Better Living Information Center, Tokyo
Mikio YAMAZAKI*,   Research Institute for Chemobiodynamics, Chiba University, Chiba

Akiko YOKOYAMA, National Institute of Hygienic Sciences, Tokyo

Kunitoshi YOSHIHIRA, National Institute of Hygienic Sciences, Tokyo

Chapter 10

Norio AIMI*, Faculty of Pharmaceutical Sciences, Chiba University, Chiba

Toshio GOTO*, Department of Agricultural Chemistry, Nagoya University, Nagoya

Hiroshi IRIE*, Faculty of Pharmaceutical Sciences, Kyoto University, Kyoto

Hisashi ISHII*, Faculty of Pharmaceutical Sciences, Chiba University, Chiba

Shô ITÔ*, Department of Chemistry, Tohoku University, Sendai

Mitsuaki KODAMA*, Department of Chemistry, Tohoku University, Sendai

Shinsaku NATORI*, National Institute of Hygienic Sciences, Tokyo

Takeshi OISHI*, Faculty of Pharmaceutical Sciences, Hokkaido University, Sapporo

Shinichiro SAKAI*, Faculty of Pharmaceutical Sciences, Chiba University, Chiba

Mikio YAMAZAKI*, Research Institute for Chemobiodynamics, Chiba University, Chiba

Chapter 12

Koji NAKANISHI, Department of Chemistry, Columbia University, New York

Hideo SATO*, Research Laboratories (Tokyo), Fuji Photo-Film Co., Asaka

ACKNOWLEDGEMENTS

We are most grateful to the numerous contributors (see List of Contributors) who undertook the painstaking job of literature survey, writing, and condensing of material. We would also like to thank Mr. W. R. S. Steele and staff of Kodansha for their invaluable assistance in the preparation of the English manuscripts comprising this book.

April, 1975   EDITORS

CONTENTS

CONTENTS OF VOLUME 1

DATA CONVENTIONS USED IN THIS BOOK

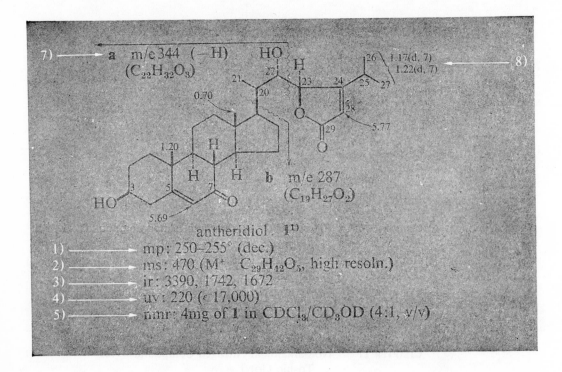

7) ⟶ **a** m/e 344 (−H)
$(C_{22}H_{32}O_3)$

8)

26 \ 1.17(d, 7)
1.22(d, 7)

b m/e 287
$(C_{19}H_{27}O_2)$

antheridiol **1**[1]

1) ⟶ mp: 250–255° (dec.)
2) ⟶ ms: 470 (M⁺ $C_{29}H_{42}O_5$, high resoln.)
3) ⟶ ir: 3390, 1742, 1672
4) ⟶ uv: 220 (ϵ 17,000)
5) ⟶ nmr: 4mg of **1** in $CDCl_3/CD_3OD$ (4:1, v/v)

1) Melting point.
2) Mass spectroscopic data. See also number 7 below.
3) Infrared data: state of measurement, if given, is shown in brackets. e.g. ir $(CHCl_3)$: Values (cm^{-1}) are given only for pertinent bands.
4) Ultraviolet/visible spectral data: the solvent, if given, is shown in brackets, e.g. uv (EtOH): The wavelength is given in nanometers with the intensity in brackets (as log ϵ, unless otherwise stated).
5) Nuclear magnetic resonance data: the solvent is usually given in brackets, e.g. nmr $(CDCl_3)$: Values are given as ppm from TMS. 1H, 2H, etc. indicate the intensity. s: singlet, d: doublet, t: triplet, m: multiplet. J values are given in Hz, and 14-H indicates a proton attached to C-14. See also number 8 below.
6) Rotation data are given as follows: "α_D (EtOH): +65" indicates the specific rotation at the D line in EtOH. "cd (MeOH): 215 ($\Delta\epsilon$ +13.17)" indicates the circular dichroism extrema in MeOH, 215 nanometers maximum (or minimum) with a $\Delta\epsilon$ value of +13.17 (or −13.17). Rotation data are not given in the above example.
7) Mass spectroscopic data: fragmentation (a) with loss of · H gives m/e 344 $(C_{22}H_{32}O_3)$ fragment arising from the C-1 to C-22 portion of the molecule.
8) Nuclear magnetic resonance data: the isopropyl methyls appear at ppm values of 1.17 and 1.22 as doublets with $J=7$ Hz.

CHAPTER 7

Fatty Acid Derivatives and Related Compounds

7.1 INTRODUCTION

A number of natural products are biogenetically derived through the acetate-malonate pathway. They include fatty acids and a variety of substances originating from fatty acid or poly-β-ketomethylene intermediates. Although the structural diversity of this class of compounds is not as remarkable as that of other classes of natural products (e.g. isoprenoids, alkaloids, etc.), this class includes many important compounds having essential functions within living systems or having important physiological activities.

Biosynthetic mechanisms of fatty acids have been identified in considerable detail by experiments using cell-free systems or purified enzymes. The processes involved in the biosynthesis of fatty acids are outlined below.

[ACP=acyl carrier protein; Enz=enzyme]

1) The carboxylation of acetyl coenzyme A is mediated by biotin, giving malonyl coenzyme A **2**. A malonyl residue is then transferred to acyl carrier protein. (**1**⟶**2**⟶**3**).
2) An acetyl residue is also transferred from coenzyme A to acyl carrier protein (**1**⟶**4**) and then to an enzyme (**4**⟶**5**).
3) Condensation of malonyl ACP with an enzyme-bound acetyl residue, followed by decarboxylation, gives acetoacetyl ACP (**5**+**3**⟶**6**).

---

1) F. Lynen, *Pure Appl. Chem.*, **14**, 137 (1967).

4) Acetoacetyl ACP undergoes, successively, reduction (**6**——>**7**), dehydration (**7**——>**8**), and hydrogenation (**8**——>**9**) to yield butyryl ACP.

5) An enzyme-bound butyryl residue is then converted into **11** by a sequence analogous to **5**——>**6**——>**7**——>**8**——>**9**. This process of chain elongation is repeated until C_{16} or C_{18} fatty acid is formed.

The fatty acids are then subjected to a variety of secondary modification processes such as dehydrogenation, chain shortening, alkylation, oxygenation, cyclization, etc., yielding diverse structural types. The compounds belonging to this category and discussed in this chapter are C_{11}–C_{15} marine products (e.g. dictyopterene, laurencin, etc.) C_{12}–C_{18} insect pheromones, fatty acid derivatives containing a cyclopentane ring (e.g. jasmone, prostaglandin, etc.), compounds with a cyclopropane ring or rings (e.g. lactobacillic acid, mycolic acids, etc.), and polyacetylene derivatives. A further group of compounds whose biosynthesis involves fatty acid intermediates is also included in this chapter.

The macrolide antibiotics which are biogenetically derived from poly-β-ketomethylene intermediates (see Chapter 8) are also collected at end of this chapter. In the biosynthesis of macrolide antibiotics and polyether antibiotics, methyl malonate is frequently used instead of malonate for chain building.

7.2   STRUCTURE OF DICTYOPTERENES

5.22 (m, 11, 7.5)   5.89 (tt, 11, 11, .1.3)

0.7 (m)

2.08   H   H   5.04 (q, 15, 8)

H H   1.3 (m)

H   4.96 (q, 17, 2.6)

11   9   8

10   7   4   H

6

H   5   3   2   H

0.92 (t, 7)   H   1

6.34 (q, 11, 15)   H   4.82 (q, 9.5, 2.6)

1.30 (m)   H   H

5.30 (m, 7.5, 9.5, 17)

α_D: $-43°$
ut(EtOH): 247
nmr in C_6D_6

(−)-dictyopterene-B **1**[1]

1) The presence of a *trans, cis* conjugated double bond in **1** is indicated by the nmr coupling constants i.e., $J_{6,7}=15$ Hz (*trans* coupling), $J_{7,8}=11$ Hz, and $J_{8,9}=11$ Hz (*cis* coupling).[1]

2) The formation of (+)-*trans*-cyclopropane-1,2-dicarboxylic acid **2** on oxidation established the absolute configuration of the cyclopropane moiety.[1]

3) On heating in benzene, the cyclic diene **3** is formed by concerted Cope rearrangement.[1,3] *Cis*-dictyopterene-A **4** (synthetic material) undergoes Cope rearrangement at lower temperature (15°C) yielding the corresponding cycloheptadiene **5**.[2]

1) J. A. Pettus Jr., R. E. Moore, *Chem. Commun.*, 1093 (1970).
2) G. Ohloff, W. Pickenhagen, *Helv. Chim. Acta*, **52**, 880 (1969).
3) J. A. Pettus Jr., R. E. Moore, *J. Am. Chem. Soc.*, **93**, 3087 (1971).
4) W. Pickenhagen, F. Näf, G. Ohloff, P. Müller, J-C. Perlberger, *Helv, Chim. Acta*, **56**, 1868 (1973).

4) Irradiation of the *trans*-divinylcyclopropane derivative (e.g. **6**) in benzene at 40° gave *cis*-divinylcyclopropane **7** and cycloheptadiene **5**.[4]

Synthesis

1 and **9** were synthesized from **8** by Wittig reaction.[5]

Remarks

Dictyopterene-B **1** and -A **9**[6] are major constituents of the essential oil of an odoriferous seaweed, *Dictyopteris*. The compounds **10–13** are also found in this seaweed as minor constituents.[3] Dictyopterene-C′ **14** and -D′ **15** obtained from same source have an absolute configuration opposite to that of the Cope rearrangement products of **9** and **1** respectively.[3] The compound **15** has been isolated as a sex attractant secreted by female gametes of the marine brown algae *Ectocarpus siliculosus*.[7,9]

The cyclopropane rings of **1** and **9** might be formed by di-π-methane rearrangement of the acyclic polyenes.[8]

dictyopterene-A **9**   **10**   **11**

12   **13**   dictyopterene-C′ **14** dictyopterene-D′ **15**

5) A. Ali, D. Sarantakis, B. Weinstein, *Chem. Commun.*, 940 (1971); K. C. Das, B. Weinstein, *Tetr. Lett.*, 3459 (1969).
6) R. E. Moore, J. A. Pettus Jr., M. S. Doty, *Tetr. Lett.*, 4787 (1968); R. E. Moore, J. A. Pettus Jr., J. Mistysyn, *J. Org. Chem.*, **39**, 2210 (1974).
7) D. G. Müller, L. Jaenicke, T. Akintori, *Science*, **171**, 815 (1971).
8) cf. H. E. Zimmerman, P. S. Mariano, *J. Am. Chem. Soc.*, **91**, 1718 (1969).
9) cf. L. Jaenicke, D. G. Müller, R. E. Moore, *J. Am. Chem. Soc.*, **96**, 3324 (1974).

7.3   SULFUR CONTAINING LIPIDS FROM *Dictyopteris*

Sulfur-containing compounds, e.g. **2–7**, have been isolated from odoriferous seaweeds along with the enone **1**. These compounds have a structural as well as biogenetic relationship to the C_{11} hydrocarbon dictyopterenes, which were obtained from same source. The natural substances **1–6** were chemically interrelated by the reactions shown below.[1,2]

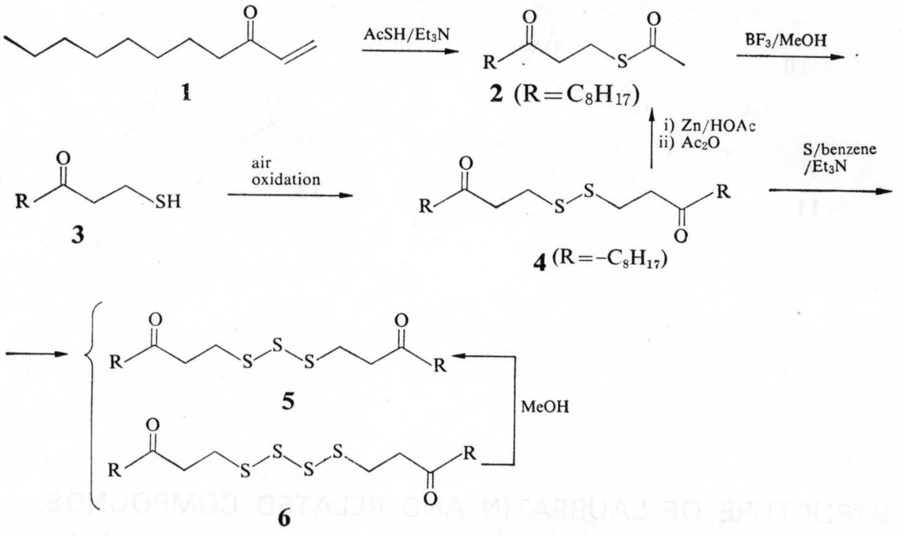

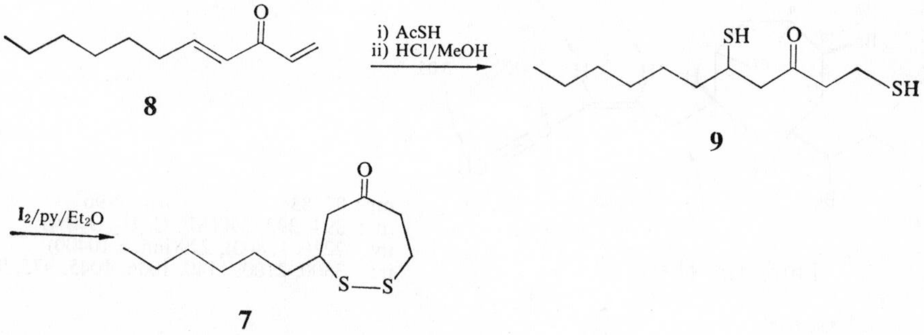

1) Michael-type addition of AcSH to the enone **1** afforded **2**.

2) The disulfide **4** was readily formed by air oxidation of **3** obtained by acid-catalyzed hydrolysis of **2**.

3) The naturally occurring polysulfides **5** and **6** were shown to be identical with the products obtained by Et_3N-catalyzed reaction of **4** and sulfur.

1) P. Roller, K. Au, R. E. Moore, *Chem. Commun.*, 503 (1971).
2) R. E. Moore, *Chem. Commun.*, 1168 (1971).
3) A. E. Asato, R. E. Moore, *Tetr. Lett.*, 4941 (1973).
4) R. E. Moore, J. Mistysyn, J. A. Pettus, Jr., *Chem. Commun.*, 326 (1972).

4) On prolonged standing, **6** decomposed, giving **5** and sulfur.

5) The cyclic disulfide **7** was synthesized from **8**.[3]

Remarks

The compounds **1–7** were isolated from *Dictyopteris plagiogramma* along with the compounds **10, 11**[4] and **12**.[1]

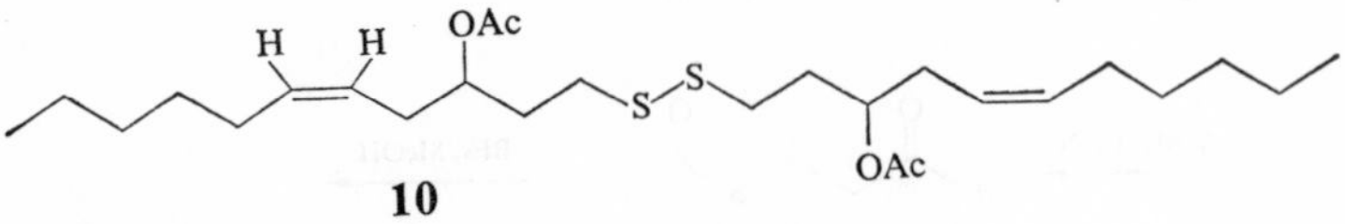

10

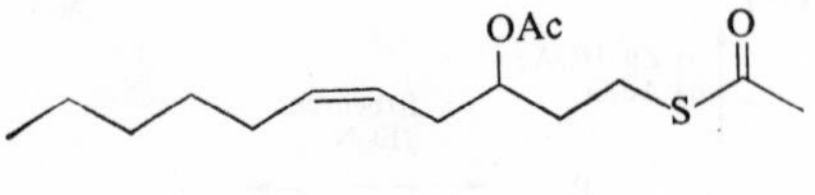

11

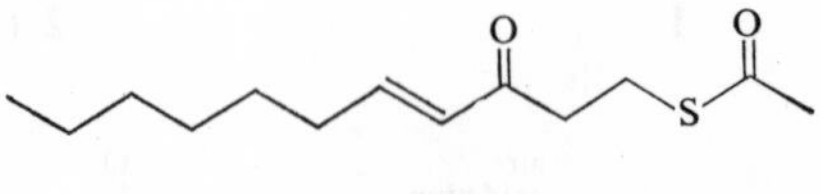

12

7.4 STRUCTURE OF LAUREATIN AND RELATED COMPOUNDS

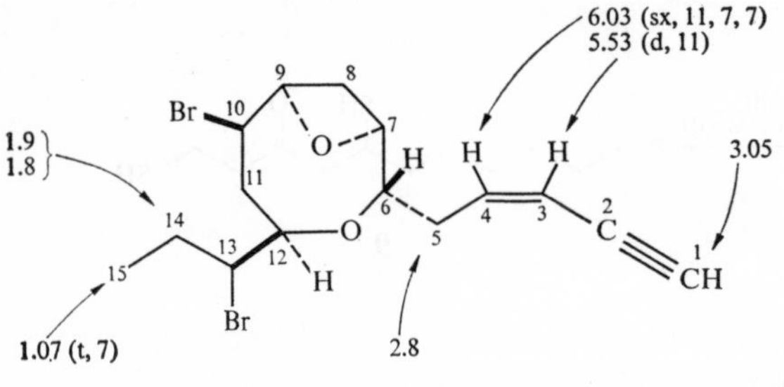

laureatin **1**[1,2]

mp: 82–83° α_D: +96°
ms: 394, 392, 390(M$^+$ C$_{15}$H$_{20}$O$_2$Br$_2$)
uv: 223(ε 12800), 229(infl. ε 10400)
ir: 3300, 2100, 1140, 1086, 1045, 975, 965, 758

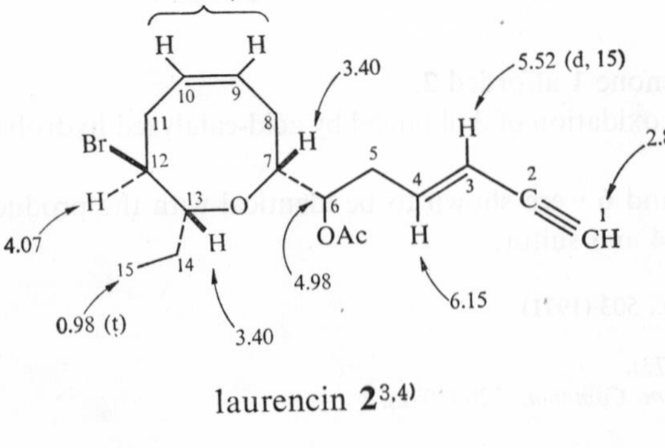

laurencin **2**[3,4]

mp: 73–74° α_D: +70.2°
uv: 224(ε 16400), 234(ε 11000)
ir: 3285, 2100, 1168, 1080, 3040, 950, 750

The absolute configuration of **1** was established as 6S, 7S, 9R, 10R, 12R and 13S from the following evidence.[2]

1) The configuration at C-6 and C-7 was determined from the optical rotation of the glycol diacetate **3** obtained from **1**. The antipodal diacetate **4** was obtained from laurencin **2**.

2) The *trans* relationship of the substituents at C-9 and C-10 was deduced from the coupling constant ($J = 2.5$ Hz) and also from the facile formation of **6** from **5** on treatment with zinc.

3) Considerable downfield shift of the 12-H signal indicates the proximity of this to the oxetane oxygen.

4) C-12 and C-13 are assumed to be *erythro* from a biogenetic point of view.

5) On acid treatment, **1** was converted into isolaureatin **8**. The rearrangement might be due to oxetane ring strain in **1**.[5]

6) The complete structure of **8** was confirmed by an x-ray diffraction study.[2]

$$\textbf{1} \xrightarrow[\substack{\text{i) Zn/HOAc} \\ \text{ii) H}_2/\text{PtO}_2 \\ \text{iii) Ac}_2\text{O}}]{} \quad \text{Me-(CH}_2)_7\text{-CH}\underset{7}{-}\text{CH}\underset{6}{-}\text{(CH}_2)_4\text{-Me}$$

QAc   OAc

3 (6S, 7S)
4 (6R, 7R)

$$\textbf{1} \xrightarrow{\text{H}_2/\text{Pd}} \textbf{5} \xrightarrow{\text{Zn/HOAc}} \textbf{6}$$

5

6

$$\textbf{1} \xrightarrow{\text{ZnCl}_2/\text{AcOH}} [\ \textbf{7}\ ] \longrightarrow \text{isolaureatin } \textbf{8}$$

7

isolaureatin **8**

Remarks

The compounds **1**, and **8–12** were isolated from the essential oil of red marine algae, *Laurencia nipponica*, and **2** from *L. glandulifera*.

Chondriol **13** and rhodophytin, a halogenated vinyl peroxide have been isolated from marine algae *Chondria oppositiclada*[9] and Laurencia species[10] respectively.

1) T. Irie, M. Izawa, E. Kurosawa, *Tetr.*, **26**, 851 (1970); *Tetr. Lett.*, 2091, 2735 (1968).
2) E. Kurosawa, A. Furusaki, M. Izawa, A. Fukuzawa, T. Irie, *Tetr. Lett.*, 3857 (1973).
3) T. Irie, M. Suzuki, T. Masamune, *Tetr.*, **24**, 4193 (1968); *Tetr. Lett.*, 1091 (1965).
4) A. F. Cameron, K. K. Cheung, G. Furguson, J. M. Robertson, *J. Chem. Soc.*, B, 559 (1969); *Chem. Commun.*, 638 (1965).
5) A. Fukuzawa, E. Kurosawa, T. Irie, *J. Org. Chem.*, **37**, 680 (1972).

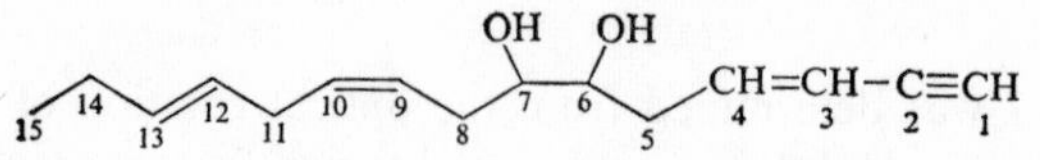

trans-laurediol **9** (C-3/C-4 trans)[6]
cis-laurediol **10** (C-3/C-4 cis)[6]

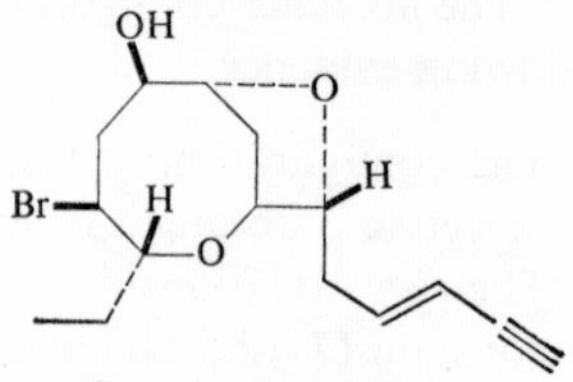

laurefucin **11**[7]

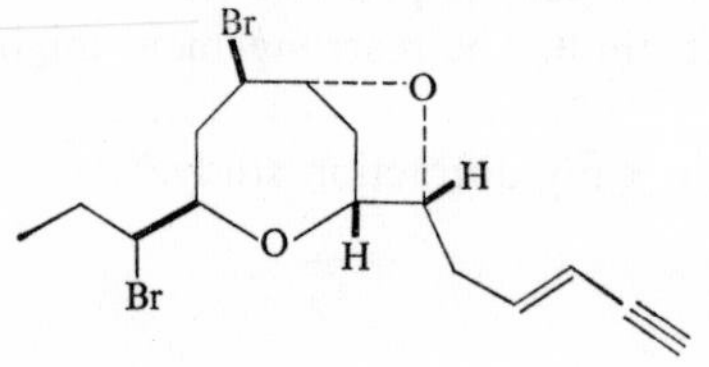

isoprelaurefucin **12**[8]

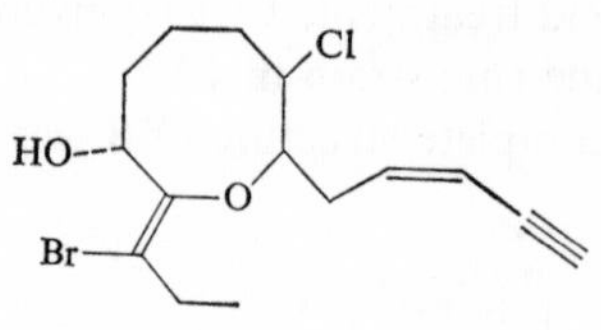

chondriol **13**[9]

6) E. Kurosawa, A. Fukuzawa, T. Irie, *Tetr. Lett.*, 2121 (1972).
7) A. Furusaki E. Kurosawa, A. Fukuzawa, T. Irie, *Tetr. Lett.*, 4579 (1973).
8) E. Kurosawa, A. Fukuzawa, T. Irie, *Tetr. Lett.*, 4135 (1973).
9) W. Fenical, K. B. Gifkins, J. Clardy, *Tetr. Lett.*, 1507 (1974); cf. W. Fenical, J. J. Sim, P. Radlick, *ibid.*, 313 (1973).
10) W. Fenical, *J. Am. Chem. Soc.*, **96**, 5580 (1974).

7.5   INSECT SEX PHEROMONES

There has recently been increasing activity in insect pheromone chemistry because of their chemical and biological interest as well as their potential use as aids in the control of specific injurious insects. Insect pheromones can usually be classified, from the type of activity, as sex pheromones (attractant), alarm pheromones, aggregating pheromones, and trail-marking hormones, etc. Most of them are structurally simple compounds of low molecular weight, and are biogenetically fatty acid- and terpene-derived substances. The methods of structural determination differ somewhat from those of other natural products, since extremely small amounts of the biologically active materials are available. General features of structural work are as follows.

1) The gross structure of biologically active compounds is assumed on the basis of the spectral properties (ir, uv, etc.) as well as the retention time on gc analysis and mobilities on tlc.

2) Observations of loss of biological activity due to simple chemical treatment (e.g. hydrolysis, $KMnO_4$, etc.) supply information as to functional groups or unsaturation in the molecule.

3) More detailed information is obtainable by mass spectrometric analysis or more conveniently by the use of combined gc-ms analysis of the pheromone itself or chemically transformed products obtained by micro-scale reactions (e.g. ozonolysis, hydrogenation, etc.).

4) The structure will then be confirmed by chemical synthesis and the geometry of double bond(s) can be assigned by biological tests of the synthetic isomers.

5) That the compound is a real pheromone is established by field tests of the sample.

6) Combined use of electroantennography and gas-chromatography (gc-eag) is convenient for the identification of pheromones.

The following are the chemical structures and sources of well-characterized insect sex pheromones of fatty acid origin.[1] Most of these have a C_{12}–C_{18} aliphatic carbon chain with an acetoxyl group at the terminal position and contain one or two olefinic bond.

C_{12} group

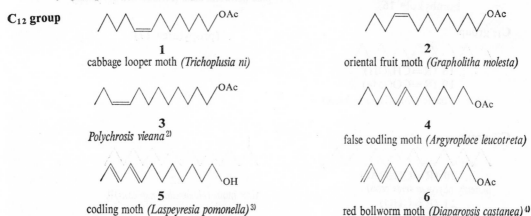

1
cabbage looper moth *(Trichoplusia ni)*

2
oriental fruit moth *(Grapholitha molesta)*

3
Polychrosis vieana [2]

4
false codling moth *(Argyroploce leucotreta)*

5
codling moth *(Laspeyresia pomonella)* [3]

6
red bollworm moth *(Diaparopsis castanea)* [4]

1) J. H. Law, F. E. Regnier, *Ann. Rev. Biochem.*, 533 (1971).
2) J. A. Kuhn, T. A. Brindley, *J. Econ. Entomol.*, **63**, 779 (1970).
3) W. Roelofs, A. Comeau, A. Hill, G. Milicevic, *Science*, **174**, 297 (1971).
4) B. F. Nesbitt, P. S. Beevor, R. A. Cole, R. Lester, R. G. Poppi, *Nature New Biol.*, **244**, 208 (1973).
5) W. L. Roelofs, A. Comeau, *Pesticide Chemistry* (ed. A. S. Tobotu) Vol. 8, p. 91–112, Cordon of Beach, 1971.
6) Y. Kuwahara, H. Hara, S. Ishii, H. Fukami, *Science*, **171**, 801 (1971); U. E. Brady, J. H. Tumlinson, R. G. Brownlee, R. M. Silverstein, *ibid.*, **171**, 802 (1971)
7) W. L. Roelofs, R. T. Carde, *Science*, **171**, 684 (1971).
8) K. H. Dahm, D. Meyer, W. E. Finn, V. Reinhold, H. Röller, *Naturwiss.*, **58**, 265 (1971).

C₁₄group

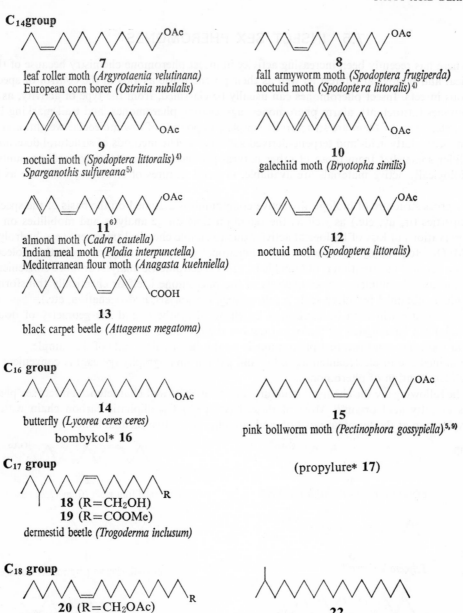

7

leaf roller moth *(Argyrotaenia velutinana)*
European corn borer *(Ostrinia nubilalis)*

8

fall armyworm moth *(Spodoptera frugiperda)*
noctuid moth *(Spodoptera littoralis)*[4]

9

noctuid moth *(Spodoptera littoralis)*[4]
Sparganothis sulfureana[5]

10

galechiid moth *(Bryotopha similis)*

11[6]

almond moth *(Cadra cautella)*
Indian meal moth *(Plodia interpunctella)*
Mediterranean flour moth *(Anagasta kuehniella)*

12

noctuid moth *(Spodoptera littoralis)*

13

black carpet beetle *(Attagenus megatoma)*

C₁₆ group

14

butterfly *(Lycorea ceres ceres)*

bombykol* **16**

15

pink bollworm moth *(Pectinophora gossypiella)*[5,9]

C₁₇ group

18 (R=CH₂OH)
19 (R=COOMe)

dermestid beetle *(Trogoderma inclusum)*

(propylure* **17**)

C₁₈ group

20 (R=CH₂OAc)
butterfly *(Lycorea ceres ceres)*
21 (R=CHO)
Achroia grisella[8]

22

tiger moth *(Holomelina nigricans)*[7]

C₁₉ group   disparlure* **23**

C₂₃ group   muscalure* **24**

The compounds asterisked are described in more detail later in this chapter. For the compounds without reference numbers, see ref. 1.

9) The sex pheromone of the pink bollworm moth has recently been assigned as a 1:1 mixture of (7Z, 11Z)- and (7Z, 11E)-7,11-hexadecadien-1-yl acetate.
   H. E. Hummel, L. K. Gaston, H. H. Shorey, R. S. Kaae, K. J. Byrne, R. M. Silverstein, *Science*, **181**, 873 (1973); B. A. Bierl, M. Beroza, R. T. Staten, P. E. Sonnet, V. E. Adler, *J. Econ. Entomol*, **87**, 211 (1974).

7.6   STRUCTURE AND SYNTHESIS OF BOMBYKOL

bombykol 1[1)]

ir: 9.5–9.75μ (−OH), 10.18, 10.56μ,
   (conjugated *cis, trans* double bonds),
   13.89μ ((CH_2)$_n$, $n > 4$)
uv(for NAB ester): 230 (conjugated diene),
              331 (azo chromophore)

1) **1** forms a *p*-nitrophenylazobenzoate (NAB) ester **2** on esterification.
2) Hydrogenation of **1** (2H_2) yields cetyl alcohol, Me(CH_2)$_{14}$$CH_2$OH.
3) The position of the double bond was determined by oxidative cleavage, which yielded **3**, **4** and **5**.
4) The geometry of the two double bonds was determined by synthesis of all four possible isomers. Comparison of the chemical and biological properties of the synthetic specimens with those of the natural substance established that bombykol **1** is 10-*trans*, 12-*cis*-hexadecadien-1-ol.[1,2)]

Remarks

Bombykol **1** is a sex attractant secreted by the male moth *Bombyx mori*, and it is physiologically active in concentrations of 10^{-10} μg/ml. Twelve mg of pure bombykol was obtained from 500,000 male moths as a crystalline derivative.

Synthesis[2,3)]

1) A. Butenandt, R. Beckmann, D. Stamm, E. Hecker, *Z. Naturforsch.*, **14b**, 283 (1959).
2) A. Butenandt, E. Hecker, M. Hopp, W. Koch, *Ann. Chem.*, **658**, 39 (1962).
3) A. Butenandt, E. Hecker, *Angew. Chem.*, **73**, 349 (1961).

7.7 STRUCTURE OF DISPARLURE

disparlure **1**[1)]

The structure of disparlure was established indirectly by determining the structure of the olefin precursor **2**, from which the biologically active substance was formed by epoxidation with *m*-Cl-perbenzoic acid.[1)]

$$\text{olefin precursor } \mathbf{2} \xrightarrow{\;m\text{–Cl–PBA}\;} \text{disparlure } \mathbf{1}$$

1) The position of the double bond in **2** was determined by the formation of **3** and **4** on ozonolysis.

2) The position of the methyl substituent was determined from the fragmentation of **5** in mass spectrum.

3) The structure of **2** was confirmed by synthesis, **6** ⟶ **7** ⟶ **2**.

$$\mathbf{2} \xrightarrow{\;O_3\;} CH_3\text{-}(CH_2)_9\text{-CHO} \qquad + \qquad C_7H_{15}CHO \text{ (branched alkyl)}$$

3   **4**

$$\mathbf{2} \xrightarrow{\;H_2\;} CH_3(CH_2)_{15}\!\!-\!\!CH\!\!-\!\!CH_3$$

$$\underset{CH_3}{|}$$

m/e 225 (M-43)

ms : 268 (M$^+$C$_{19}$H$_{40}$)

m/e 253 (M-15)   **5**

$$\underset{\text{Me}}{\overset{|}{\text{Me-CH}}}\text{-(CH}_2)_5\text{-Br} \xrightarrow{\;PPh_3/MeCN\;} \underset{\text{Me}}{\overset{|}{\text{Me-CH}}}\text{-(CH}_2)_5\text{-P}^+Ph_3Br^- \longrightarrow$$

6   **7**

i) *n*-BuLi/DMSO
ii) undecanal ⟶

1.98   0.86

0.86   *cis*   + *trans* isomer

1.25   5.26   1.25   1.6

2

Remarks

Disparlure **1** is a potent sex attractant produced by the female gypsy moth, *Porthetria dispar* (Lymantriidae).[1)] As little as 2×10^{-12} g of the synthetic epoxide was active in the laboratory bioassay. The *cis* epoxide was about ten times as active as the *trans* isomer. *cis*-7-Hexadecene-1,10-diol (gyptol) had been identified as a gypsy moth sex attractant,[2)] but it was later found that this compound was inactive.[3,4)] **1** has been synthesized in optically active form from (S)(+)-glutamic acid.[5)]

1) B. A. Bierl, M. Beroza, C. W. Collier, *Science*, **170**, 87 (1970).
2) M. Jacobson, M. Beroza, W. A. Jones, *Science*, **132**, 1011 (1960).
3) K. Eiter, E. Truscheit, M. Boness, *Ann. Chem.*, **709**, 29 (1967).
4) M. Jacobson, R. M. Waters, M. Schwarz, *J. Econ. Entomol.*, **63**, 943 (1970).
5) S. Iwaki, S. Marumo, T. Saito, M. Yamada, K. Katagiri, *J. Am. Chem. Soc.*, **96**, 7842 (1974).

7.8   STRUCTURE AND SYNTHESIS OF MUSCALURE

$$CH_3 - (CH_2)_6 - CH_2 - \overset{H}{C} = \overset{H}{C} - CH_2 - (CH_2)_{11} - CH_3$$

0.88 (t)   5.28   0.88

1.27   1.99   1.99   1.27

muscalure **1**[1]

ms: 322 (M$^+$ C$_{23}$H$_{46}$)
nmr (for synthetic material) in CDCl$_3$

1) Micro-scale ozonolysis of **1** (10 μg sample) followed by gc analysis indicated the formation of nonanal and tetradecanal.
2) Hydrogenation of **1** yielded *n*-tricosane.
3) Condensation of nonanal with Wittig reagent prepared from 1-bromotetradecane gave 9-tricosene containing 85% *cis* and 15% *trans* isomer. Synthetic *cis* isomer was found to be identical with muscalure.

Synthesis

Route I[2]

The *cis* olefinic linkage in **1** was stereospecifically constructed by hydrogenation of the acetylenic intermediate **3**.

$$Me-(CH_2)_{12}-C \equiv CH \xrightarrow[ii) Me(CH_2)_7Br]{i) n\text{-}BuLi} Me-(CH_2)_{12}C \equiv C-(CH_2)_7-Me \xrightarrow{H_2/Lindlar} \mathbf{1}$$

2   **3**

Route II[3]

1 was synthesized from the readily available erucic acid **4** by two simple reaction steps in an overall yield of 85%.

$$Me-(CH_2)_7-\overset{H}{C}=\overset{H}{C}-(CH_2)_{11}-COOH \xrightarrow{MeLi} Me-(CH_2)_7\overset{H}{C}=\overset{H}{C}-(CH_2)_{11}-\overset{O}{\overset{\|}{C}}-Me \xrightarrow{Huang-Minlon} \mathbf{1}$$

4   **5**

Route III[4]

Mixed Kolbe electrolysis of oleic acid and heptanoic acid afforded **1** in 14% yield.

$$Me-(CH_2)_7\overset{H}{C}=\overset{H}{C}-(CH_2)_7COOH + Me-(CH_2)_5COOH \xrightarrow[(-2CO_2)]{electrolysis/NaOMe/MeOH} \mathbf{1}$$

6   **7**

Remarks

Muscalure **1** is a sex pheromone isolated from the cuticle and feces of the female house fly, *Musca domestica*.

1) D. A. Carlson, M. S. Mayer, D. L. Silhacek, J. D. James, M. Beroza, B. A. Bierl, *Science*, **174**, 76 (1971).
2) K. Eiter, *Naturwiss.*, **59**, 468 (1972).
3) R. L. Cargill, M. G. Rosenblum, *J. Org. Chem.*, **37**, 3971 (1972).
4) G. W. Gribble, J. K. Sanstead, J. W. Sullivan, *Chem. Commun.*, 735 (1973).

7.9 SYNTHESIS OF BREVICOMIN

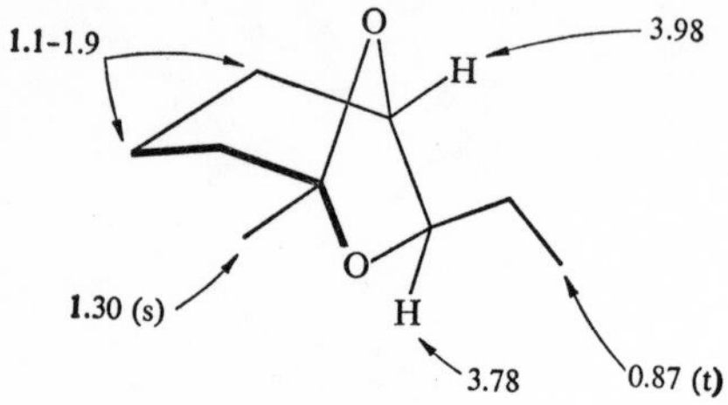

brevicomin **1**[1]

ms: 156.11572 (M+ $C_9H_{16}O_2$)
ir: no OH or C=O absorption
uv: no absorption
nmr in CCl_4

Route I[1,2]

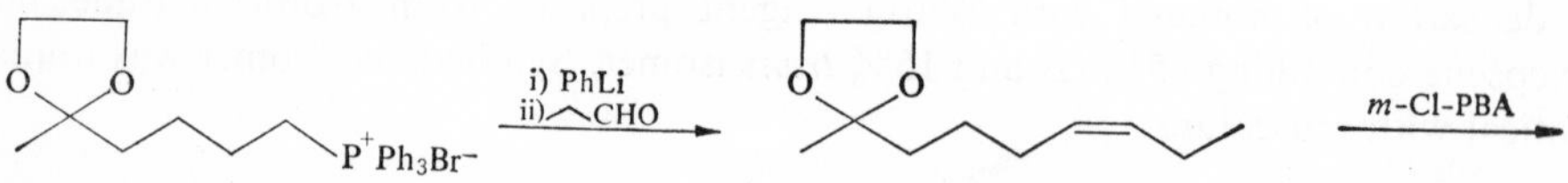

2

i) PhLi
ii) ⌐CHO

3 (*cis, trans* mixture)

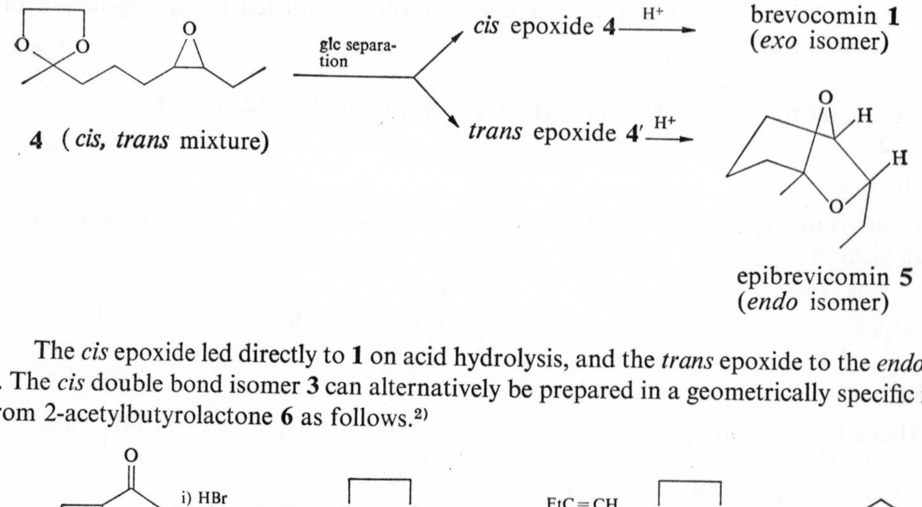

m-Cl-PBA

4 (*cis, trans* mixture)

glc separation

cis epoxide **4** $\xrightarrow{H^+}$ brevocomin **1** (*exo* isomer)

trans epoxide **4'** $\xrightarrow{H^+}$

epibrevicomin **5** (*endo* isomer)

The *cis* epoxide led directly to **1** on acid hydrolysis, and the *trans* epoxide to the *endo* isomer **5**. The *cis* double bond isomer **3** can alternatively be prepared in a geometrically specific manner from 2-acetylbutyrolactone **6** as follows.[2]

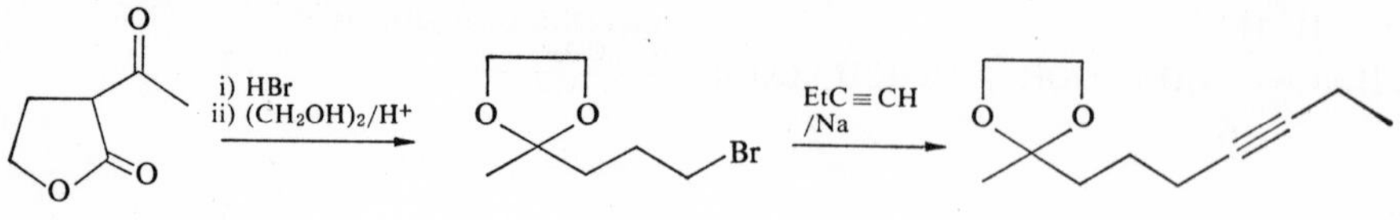

6

i) HBr
ii) (CH₂OH)₂/H⁺

7

EtC≡CH
/Na

8

$\xrightarrow{H_2/Ni\,(OAc)_2}$ *cis*-**3**

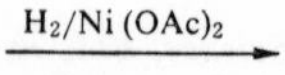

1) R. M. Silverstein, R. G. Brownlee, T. E. Bellas, D. L. Wood, L. E. Browne, *Science*, **159**, 889 (1968).
2) T. E. Bellas, R. G. Brownlee, R. M. Silverstein, *Tetr.*, **25**, 5149 (1969).

Route II[3]

Brevicomin 1 was also synthesized by thermal rearrangement of the δ,ϵ-epoxy ketone 10.

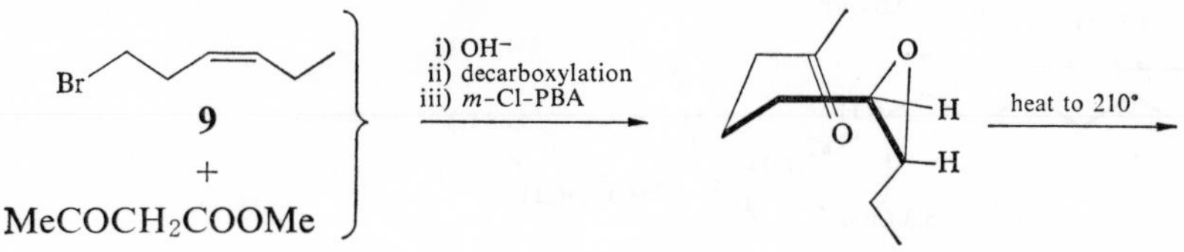

i) OH⁻
ii) decarboxylation
iii) m-Cl-PBA

heat to 210°

9

+

MeCOCH₂COOMe

10

brevicomin **1** + epibrevicomin **5**

90%   10%

In a like manner, the corresponding *trans* epoxide is transformed to **5** and **1** in a ratio of 91:9.

Remarks

Brevicomin 1 has been identified as an aggregating pheromone produced by the female western pine beetle *Dendroctonus brevicomis* boring in ponderosa pine. Epibrevicomin 5 is also beetle *D. frontalis* has been found to possess the structure 11.[4] 1 was synthesized in optically active form from (2S:3S)-D-(−)-tartaric acid.[5]

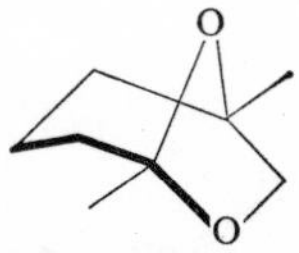

frontalin **11**

3) H. H. Wasserman, E. H. Barber, *J. Am. Chem. Soc.*, **91**, 3674 (1969).
4) G. W. Kinzer, A. F. Fentiman Jr., T. F. Page Jr., R. L. Foltz, J. P. Vité, G. B. Pitman, *Nature*, **221**, 477 (1969).
5) K. Mori, *Tetr.*, **30**, 4223 (1974).

7.10   ALLENIC ESTER FROM MALE DRIED BEAN BEETLE

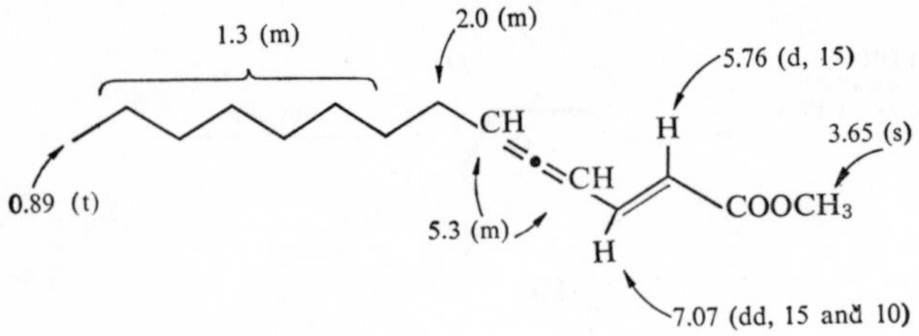

allenic ester from *Acanthoscelides obtectus* **1**[1]

ms: 138 (M$^+$ C$_{15}$H$_{24}$O$_2$)
uv: 254 (ε 16,000)
ir: 1940 (–C=C=C–), 1721, 1630 (–C=C–C=O),
     981 (*trans*–C=C–)

The structure of the allenic ester **1** was confirmed by the following synthesis.

Route I[2]

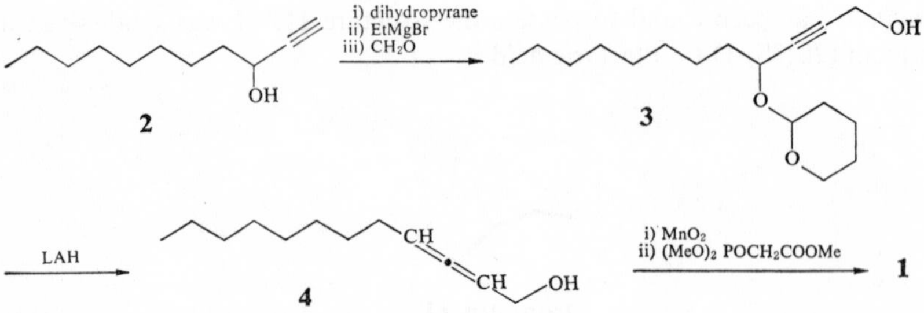

3 ⟶ 4: Reductive elimination of the tetrahydropyranyloxy group in **3** converts the acetylenic linkage into an allenic linkage.

Route II[3]

Another synthesis of **1**, employing the reaction of lithium dialkylcuprate with propargylic allylic acetate producing the allene-ene system, has been reported.

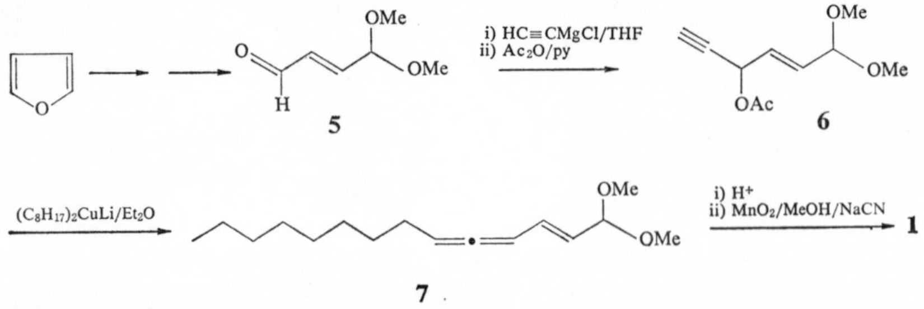

1) D F.. Horler, *J. Chem. Soc.*, C, 859 (1970).
2) P. D. Landor, S. R. Landor, S. Mukasa, *Chem. Commun.*, 1638 (1971).
3) C. Descoins, C. A. Henrick, J. B. Siddall, *Tetr. Lett.*, 3777 (1972).
4) R. M. Silverstein, J. O. Rodin, W. E. Burkholder, J. E. Gorman, *Science*, **175**, 85 (1967).

Remarks

The C_{14} allenic ester **1** was isolated from the male dried bean beetle, *Acanthoscelides obtectus*. Although crude fractions containing this material attract female beetles, the amounts of **1** produced are far greater than is usual for a pheromone. Thus there is still a question as to whether this compound is the sex pheromone of the beetles.

The C_{14} unsaturated fatty acid ester **8**, which is closely related to **1** has been obtained as the principal component of the sex attractant of the black carpet beetle, *Attagenus megatoma*.[4]

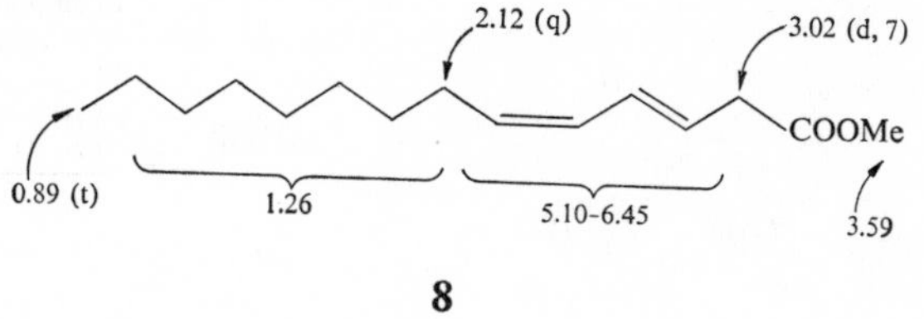

8

7.11   SYNTHESIS OF PROPYLURE

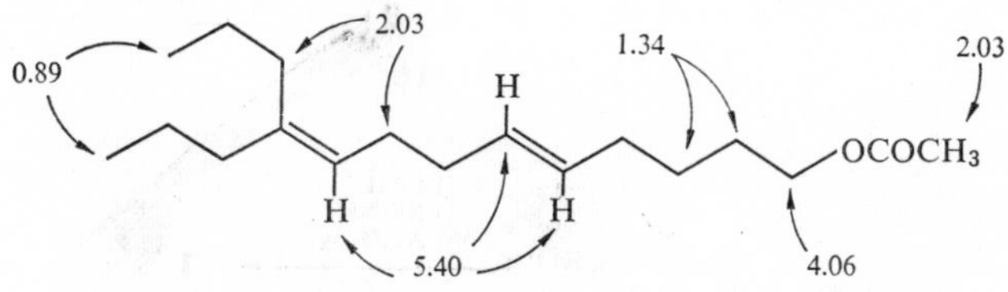

propylure **1**[1]

ms: 280 (M⁺ $C_{18}H_{32}O_2$)
ir: 1755, 1235, 1038(acetyl), 1660, 965 (*trans* double bond), 723 (methylenes)
uv: end absorption
nmr in $CDCl_3$:
    allylic methylene at 2.03,
    other methylenes at 1.34

The structure of propylure **1** was confirmed by chemical synthesis. Here, two routes for the synthesis of **1** are described, both using acetylenic intermediates for the geometrically specific construction of a *trans* double bond.

---

1) W. A. Jones, M. Jacobson, D. F. Martin, *Science*, **152**, 1516 (1966).

Route I[2)]

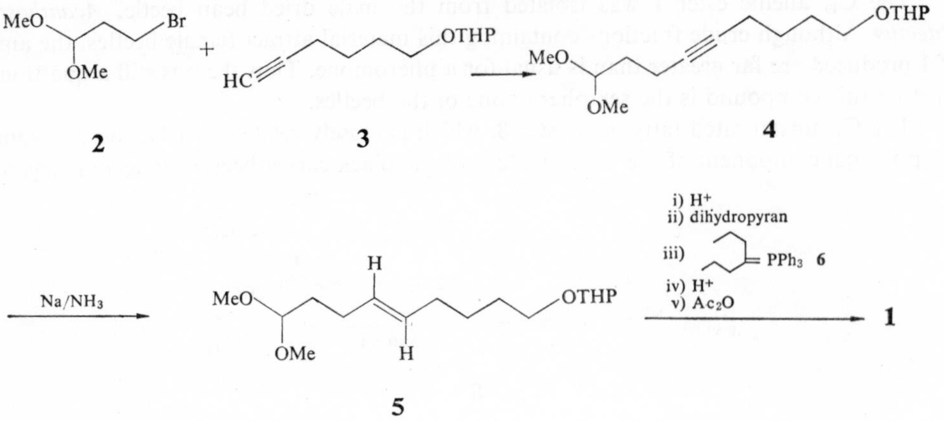

2   **3**   **4**

5

Route II[3)]

Propylure **1** was synthesized by an alternative method involving fragmentation of the tosyl-hydrazone of **10** as a key step, giving rise to the acetylenic aldehyde **11**.

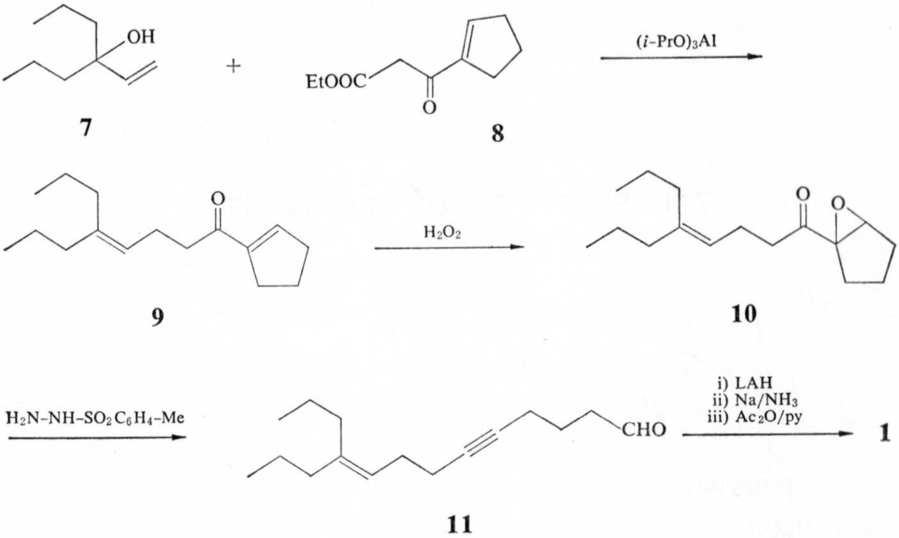

7   **8**

9   **10**

11

Route III[4)]

By means of a novel method for aldehyde synthesis utilizing the dihydro-1,3-oxazine deriva-tive **13**, propylure was synthesized in an overall yield of 31%. Wittig condensation of **17** yielded a *cis/trans* mixture of products which isomerized to **1** by heating with Se.

2) G. Pattenden, *J. Chem. Soc.* C, 2385 (1968).
3) M. Stoll, I. Flament, *Helv. Chim. Acta*, **52**, 1996 (1969).
4) A. I. Meyers, E. W. Collington, *Tetr.*, **27**, 5979 (1971).

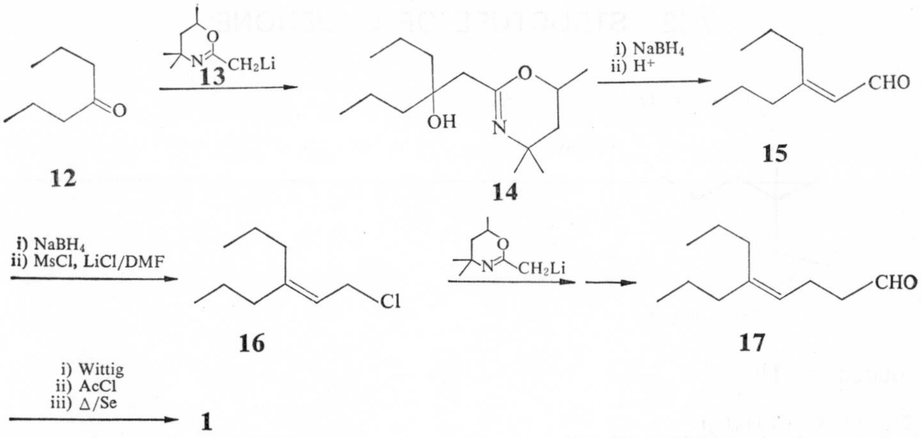

Route IV[5)]

1 was synthesized by the following route involving photochemical ring opening of the cyclopentanone **20** as a key step (Norrish type I reaction). The diene-aldehyde **21** formed in this reaction was found to be a mixture of *trans* and *cis* isomers in the ratio of 2 : 1.

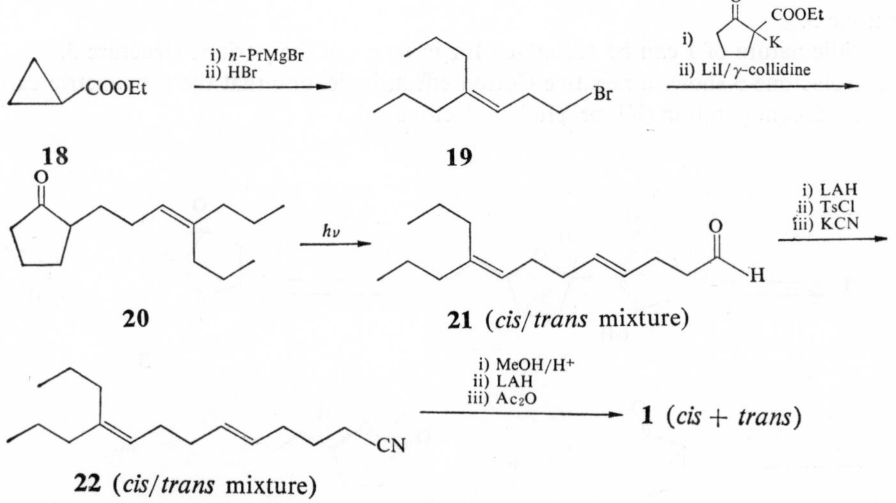

20

21 (*cis/trans* mixture)

22 (*cis/trans* mixture)

1 (*cis* + *trans*)

For other syntheses of **1** see refs. 6 and 7.

Remarks

Propylure **1** was reported to be a sex attractant produced by the female pink bollworm moth, *Pectinophora gossypiella*, which is a destructive pest of cotton.[1)] However, doubt has been raised as to the structure of this pheromone, since a synthetic material showed no biological activity.[6,8)]

5) J. Kossanyi, B. Furth, J-P. Morizur, *Tetr. Lett.*, 4559 (1973).
6) K. Eiter, E. Truscheit, M. Boness, *Ann. Chem.*, **709**, 29 (1967).
7) J. C. Stowell, *J. Org. Chem.*, **35**, 244 (1970).
8) For gossyplure, a recently identified pheromone of pink bollworm moth, see ref. 9 of page 10 and R. J. Anderson, C. A. Henrick, *J. Am. Chem. Soc.*, **97**, 4327 (1975).

7.12   STRUCTURE OF OUDENONE

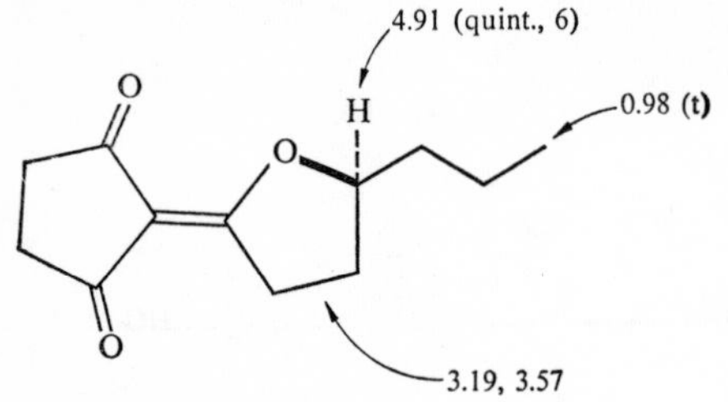

4.91 (quint., 6)

—0.98 (t)

—3.19, 3.57

oudenone **1**[1]

ms: 208.112 (M$^+$ C$_{12}$H$_{16}$O$_3$)
uv (MeOH): 285 (ε 19,000), 220 (ε 12,000)
uv (NaOH): 247 (ϵ 23,000), 230 (sh)
ir: 1662, 1563
pK_a: 4.1

1) Acid treatment of **1** affords two known compounds, **4** and **5**, which might be formed via the triketone **3**.

2) The acidic nature of **1** can be accounted for in terms of the γ-trione structure **3**.

3) γ-Butyrolactone **5** shows a negative Cotton effect, indicating that the asymmetric carbon in **1** has an S configuration (Klyne–Hudson lactone rule).

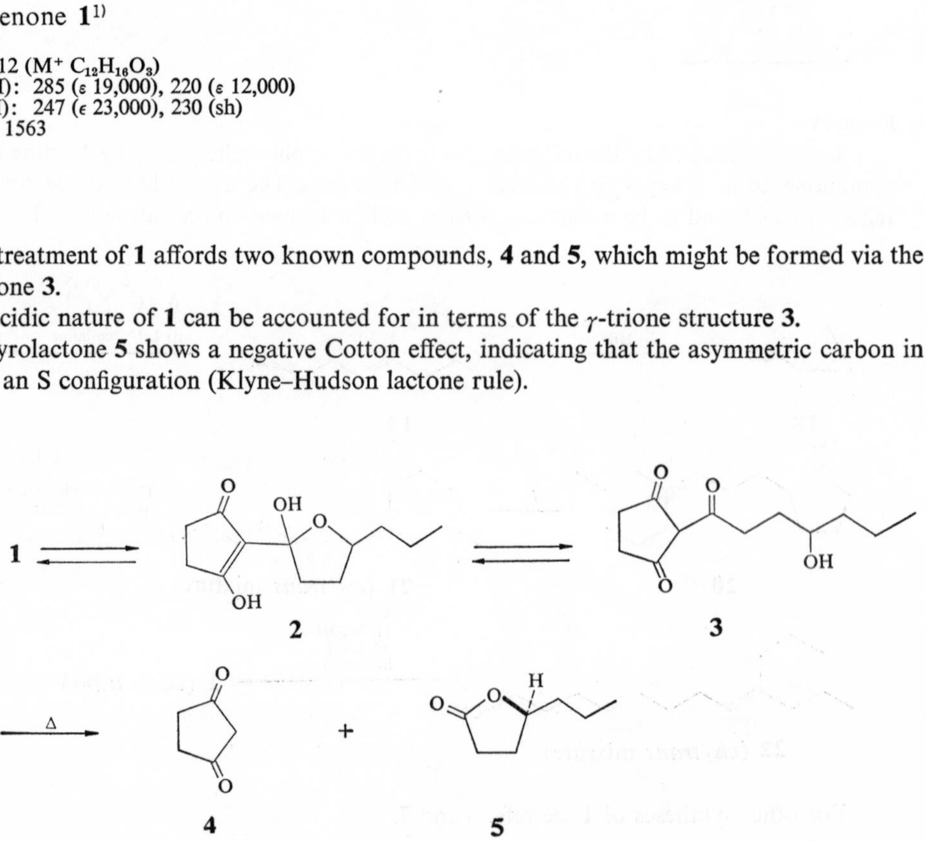

1

2   **3**

Δ   +

4   **5**

Synthesis

The structure was confirmed by the following synthesis. The synthetic *dl*-compound was optically resolved as its brucine salt.

1) M. Ohno, M. Okamoto, N. Kawabe, H. Umezawa, T. Takeuchi, H. Imura, S. Takahashi, *J. Am. Chem. Soc.*, **92**, 1285 (1971).

2) H. Umezawa, T. Takeuchi, H. Imura, K. Suzuki, H. Ito, M. Matsuzaki, T. Nagata, O. Tanabe, *J. Antibiotics*, **23**, 514 (1970).

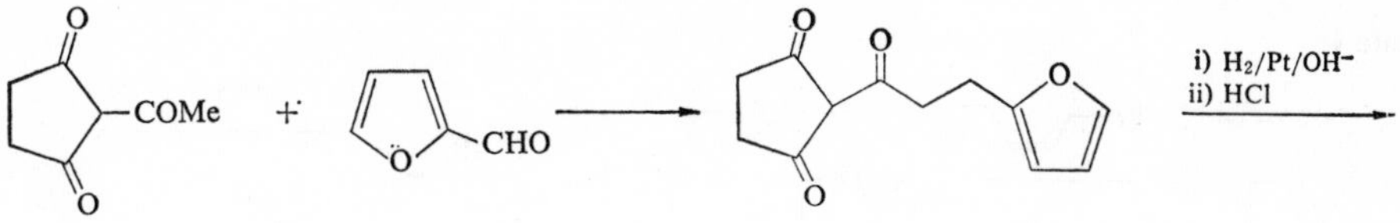

6   **7**   **8**

1   **9**

Remarks

Oudenone **1** was isolated from *Oudemansiella radicata*. It shows strong inhibitory activity towards a specific enzyme, tyrosine hydroxylase, and also has the effect of reducing blood pressure.[2]

7.13 SYNTHESIS OF *cis*-JASMONE

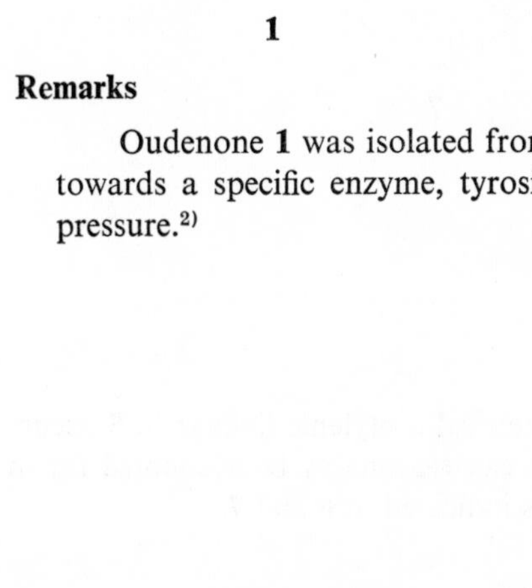

cis-jasmone **1**

Cis-jasmone **1**, an important substance in the perfume industry, has been synthesized by a route involving the base-catalyzed cyclization of 1,4-diketones or alklylation of preformed cyclopentenones.

Various methods have been reported for preparing the 1,4-diketones **2** and **17**, which are valuable synthetic intermediates for *cis*-jasmone **1**. Several examples of synthetic routes to **1** are given overleaf.[1]

---

1) For previous syntheses of **1**, see L. Crombie, S. H. Harper, *J. Chem. Soc.*, 869 (1952); S. H. Harper, R. J. D. Smith, *ibid.*, 1512 (1955); J. H. Amin, R. K. Razden, S. C. Bhattachary, *Perfum. Essent. Oil Rec.*, **49**, 502 (1958); K. Sisido, S. Torii, M. Kawasaki, *J. Org. Chem.*, **29**, 2290 (1964).

Route I[2]

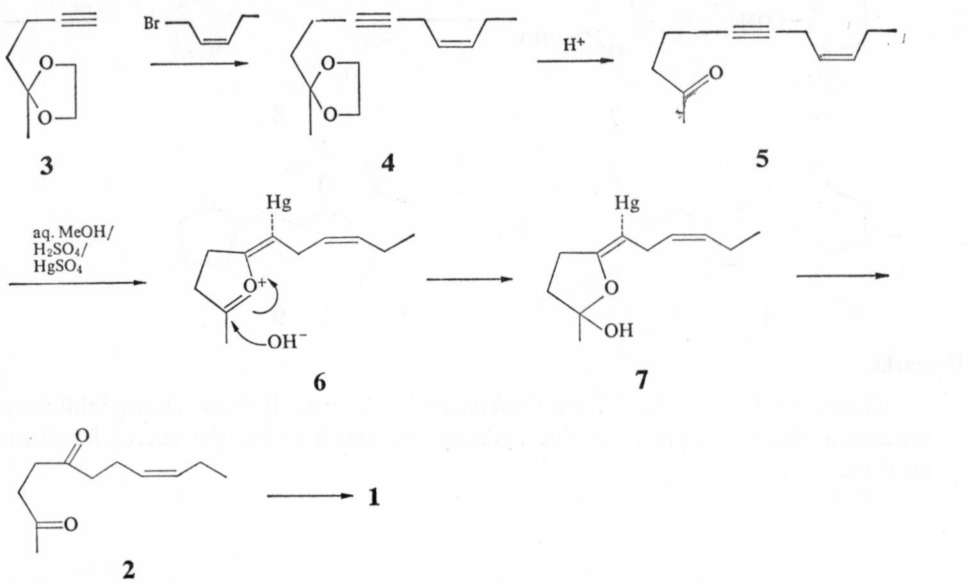

3   4   5

6   7

5 ⟶ 2: Mercury ion-catalyzed hydration of the unsymmetrical acetylenic linkage in **5** occurs selectively to yield only the 1,4-diketone **2**. This can reasonably be accounted for in terms of participation of the carbonyl oxygen, as indicated in **6** and **7**.

Route II[3]

By simple alkylation of furan, which is a potential form of 1,4-dicarbonyl compounds, the desired compound **2** was obtained after hydrolysis.

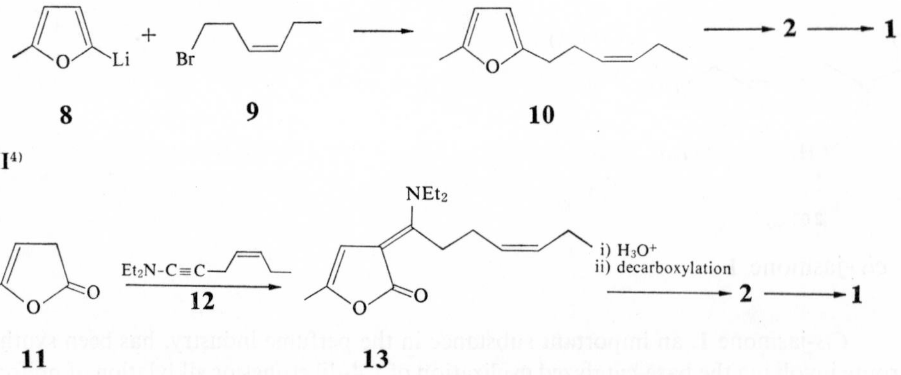

8   9   10

Route III[4]

11   13

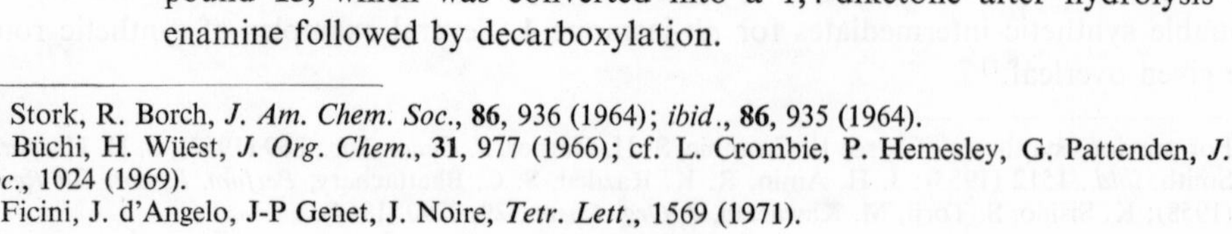

11 + 12 ⟶ 13: An ynamine, **12**, was with condensed with angelicalactone **11** to afford the compound **13**, which was converted into a 1,4-diketone after hydrolysis of the enamine followed by decarboxylation.

2) G. Stork, R. Borch, *J. Am. Chem. Soc.*, **86**, 936 (1964); *ibid.*, **86**, 935 (1964).
3) G. Büchi, H. Wüest, *J. Org. Chem.*, **31**, 977 (1966); cf. L. Crombie, P. Hemesley, G. Pattenden, *J. Chem. Soc.*, 1024 (1969).
4) J. Ficini, J. d'Angelo, J-P Genet, J. Noire, *Tetr. Lett.*, 1569 (1971).

Route IV[5)]

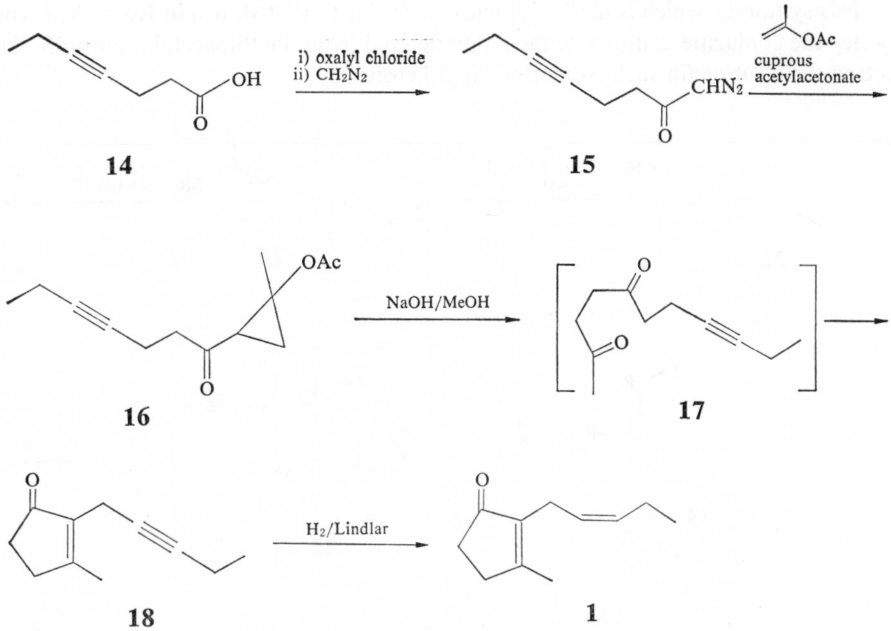

14   **15**

16   **17**

18   **1**

15 ⟶ 16: The intermolecular insertion of a ketocarbene derived from the diazoketone **15** into isopropenylacetate affords the cyclopropyl ketone **16**, which can easily be cleaved into the 1,4-diketone **17** under basic conditions.

Route V[6)]

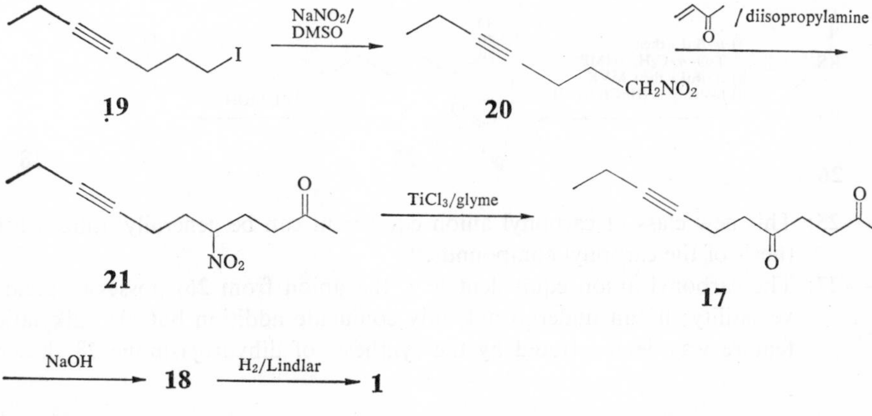

19   **20**

21   **17**

20 ⟶ 21: Nitro-stabilized carbanions undergo conjugate addition with enones to yield γ-nitro-ketones.

21 ⟶ 17: Transformation of the nitro group into carbonyl under mild conditions can be achieved with TiCl$_3$, and this can be utilized for the preparation of the 1,4-diketone.

5) J. E. McMurry T. E. Glass, *Tetr. Lett.*, 2575 (1971); cf. E. Wenkert, R. A. Mueller, E. J. Reardon, Jr., S. S. Sathe, D. J. Scharf, G. Tosi, *J. Am. Chem. Soc.*, **92**, 7428 (1970).
6) J. E. McMurry, J. Melton, *J. Am. Chem. Soc.*, **93**, 5309 (1971).

Route VI[7]

This synthesis, which is methodologically similar to that shown in Route V, involves as the key step the conjugate addition of the anion derived from the thioacetal monosulfoxide **24** to an electron-deficient olefin such as methyl vinyl ketone.

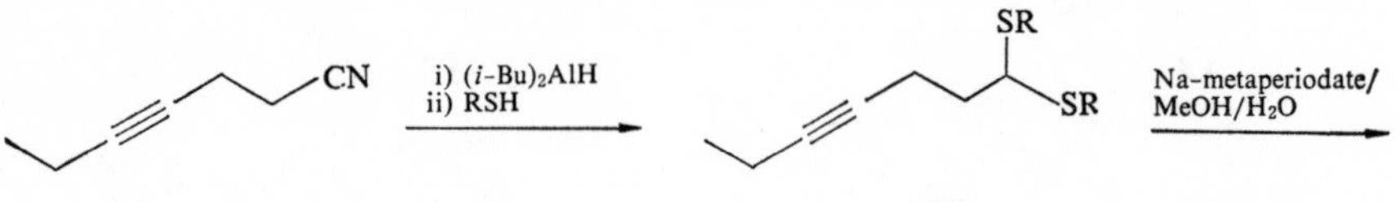

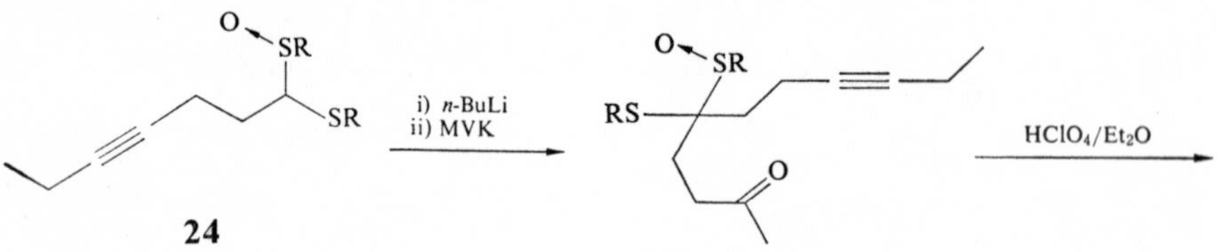

24 ⟶ 25: This new class of carbonyl anion equivalent can be generally utilized for the synthesis of the carbonyl compounds.[8]

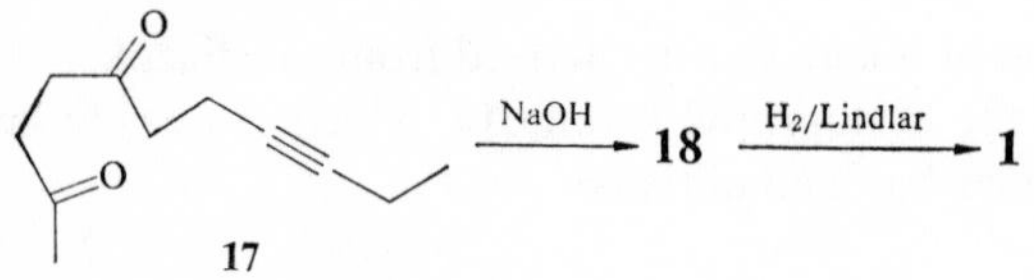

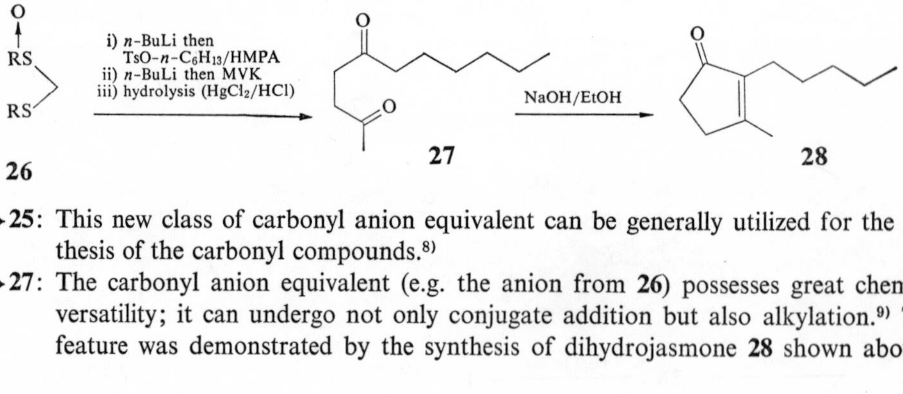

26 ⟶ 27: The carbonyl anion equivalent (e.g. the anion from **26**) possesses great chemical versatility; it can undergo not only conjugate addition but also alkylation.[9] This feature was demonstrated by the synthesis of dihydrojasmone **28** shown above.[7]

7) J. L. Herrmann, J. E. Richman, R H. Schlessinger, *Tetr. Lett.*, 3275 (1973).
8) J. L. Herrmann, J. E. Richman, R. H. Schlessinger, *Tetr. Lett.*, 3271 (1973).
9) J. E Richman, J. L. Herrmann, R. H. Schlessinger, *Tetr. Lett.*, 3267 (1973).

Route VII[10)]

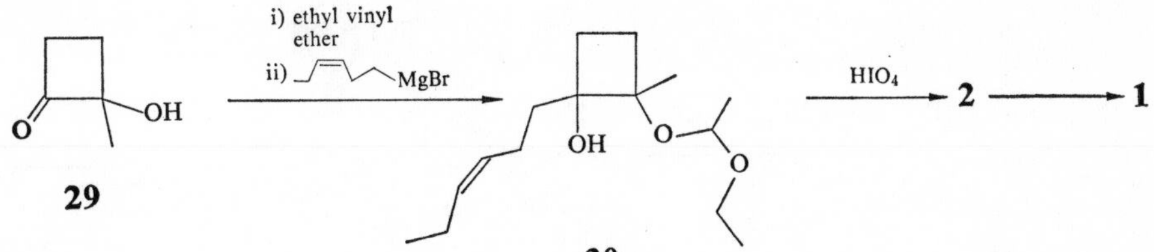

29   **30**

The 1,4-diketone **2** was obtained by oxidative cleavage of the cyclobutanediol derivative **30**.

Route VIII[11)]

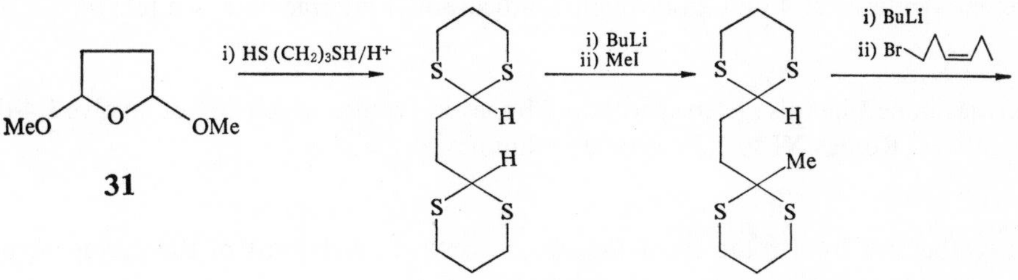

31   **32**   **33**

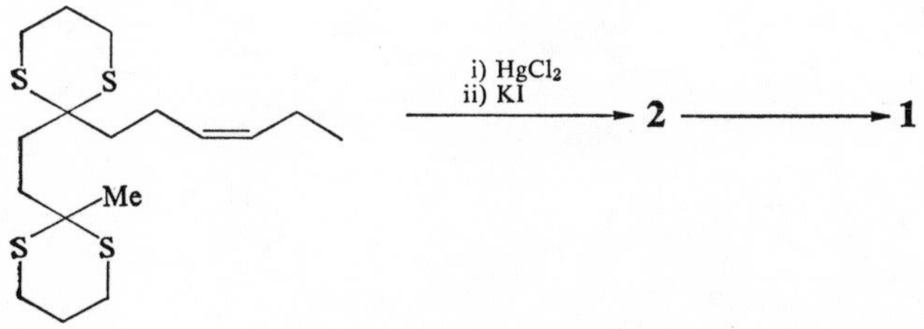

34

Two successive alkylations of the di-1,3-dithiane derivative **32**, followed by hydrolysis, afforded the 1,4-diketone **2**.

Route IX[12)]

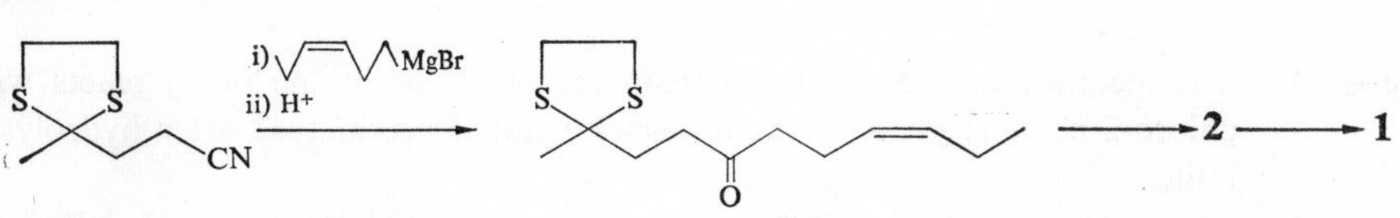

35   **36**

The 1,4-diketone **2** was also obtained by a practical route involving three simple reactions from the levulonitrile thioketal **35**.

10) S. M. Weinreb, R. J. Cretovich, *Tetr. Lett.*, 1233 (1972).
11) R. A. Ellison, W. D. Woesher, *Chem. Commun.*, 529 (1972).
12) H. C. Ho, T. -L, Ho, C. M. Wong, *Can. J. Chem.*, **50**, 2718 (1972).

Route X[13)]

37   **38**

Grignard reagent reacts selectively with the ester carbonyl in the 8-acyloxyquinoline derivative **37** giving rise to the 1,4-diketone **2**. This might proceed through an Mg complex such as **38**.
For recent synthesis of **1** and **28** through 1,4-dicarbonyl intermediate, see ref. 14.

Cis-jasmone **1** has also been synthesized by several routes which do not involve 1,4-diketone
intermediates. Routes XI to XIV provide examples.[19)]

Route XI[15)]

1 was synthesized by a route involving two successive alkylations of the cyclopentenone **39**,
which is readily available from cyclopentadiene.

39   **40**   **41**

42

40 ⟶ 41: The metalloenamine generated from the cyclohexylimine of **40** reacts with *cis*-1-
chloro-2-pentene yielding the α-monoalkylated product **41** after hydrolysis of the
imine.

41 ⟶ 42: Thermal cracking of the adduct **41** (retro Diels-Alder) affords **42**, which could be
isomerized to **1**.

13) T. Sakan, Y. Mori, T. Yamazaki, *Chem. Lett.*, 713 (1973); T. Sakan, Y. Mori, *ibid.*, 793 (1972).
14) T. Mukaiya, K. Narasaka, M. Furusato, *J. Am. Chem. Soc.*, **94**, 8641 (1972); K. Oshima, H. Yamamoto,
    H. Nozaki, *ibid.*, **95**, 4446 (1973); T. Mukaiyama, M. Arai, H. Takei, *ibid.*, **95**, 4763 (1973); T. Wakamatsu,
    K. Akasaka, Y. Ban, *Tetr. Lett.*, 3883 (1974); Th. Cuvigny, M. Larcheveque, H. Normant, *ibid.*, 1237 (1974).
15) G. Stork, G. L. Nelson, F. Rouessac, O. Gringore, *J. Am. Chem. Soc.*, **93**, 3091 (1971).
16) G. Büchi, B. Egger, *J. Org. Chem.*, **36**, 2021 (1971).

Route XII[16)]

1 and methyl jasmonate were synthesized from the common intermediate **46** which was obtained from dihydroresorcinol. Here, the synthesis of **1** is shown.

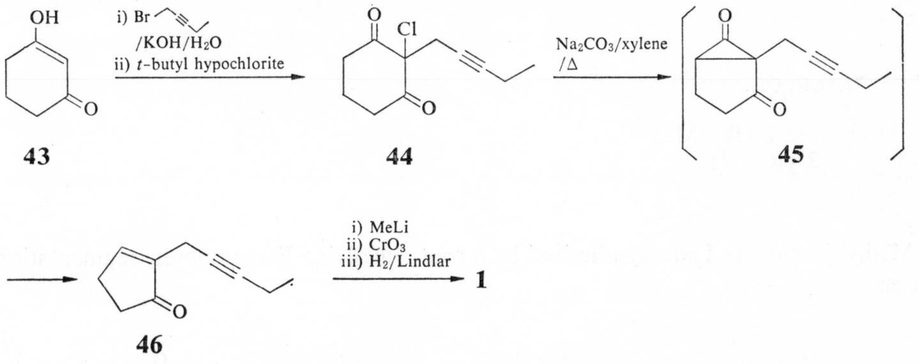

43   **44**   **45**

46

44 ⟶ 46: Dehydrochlorination of **44** gave the cyclopentenone **46** directly. This can be interpreted in terms of intermediary formation of the cyclopropanone **45** followed by thermal decarbonylation.

Route XIII[17)]

This route involves thermal rearrangement of cyclopropyl ketones into cyclopentenones. **1** was obtained from **47** in 32% overall yield.

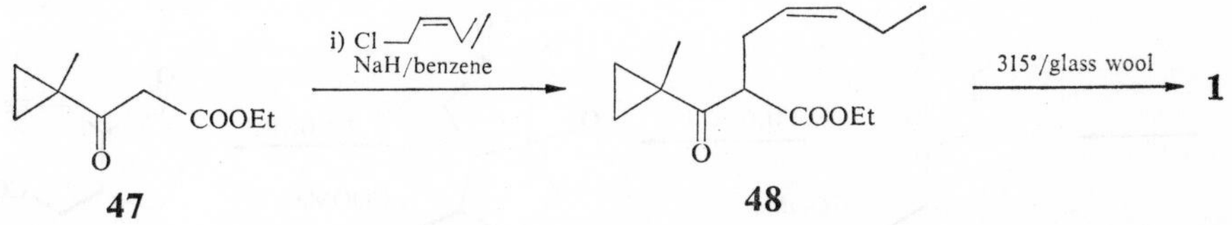

47   **48**

Route XIV[18)]

Cyclopentadiene was converted to *cis*-jasmone in a seven-step reaction with an overall yield of 40%.

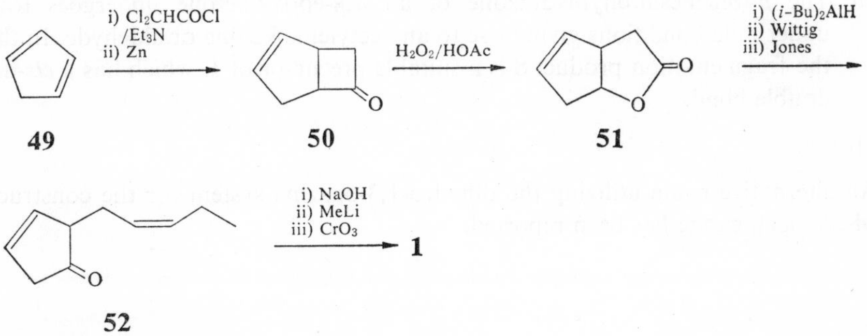

49   **50**   **51**

52

17) W. F. Berkowitz, *J. Org. Chem.*, **37**, 341 (1972).

18) P. A. Grieco, *J. Org. Chem.*, **37**, 2363 (1972).

19) See also P. M. McCurry Jr., K. Abe, *Tetr. Lett.*, 1387 (1974); I. Kawamoto, S. Muramatsu, Y. Yura, *ibid.*, 4223 (1974).

7.14   SYNTHESIS OF METHYL JASMONATE

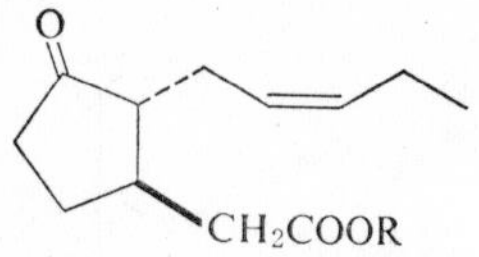

methyl jasmonate **1** (R=Me)[1,2]
jasmonic acid **2** (R=H)[8]

Route I[3]

Methyl jasmonate **1** was synthesized by a route involving Eschenmoser fragmentation as a key step.

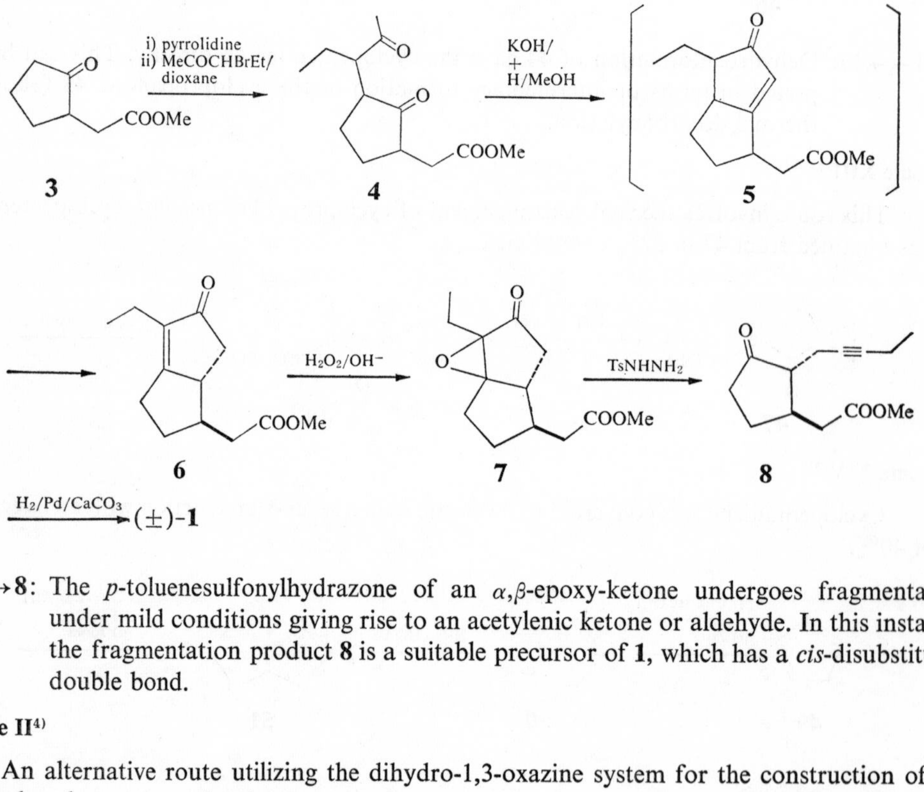

7⟶8: The *p*-toluenesulfonylhydrazone of an α,β-epoxy-ketone undergoes fragmentation under mild conditions giving rise to an acetylenic ketone or aldehyde. In this instance, the fragmentation product **8** is a suitable precursor of **1**, which has a *cis*-disubstituted double bond.

Route II[4]

An alternative route utilizing the dihydro-1,3-oxazine system for the construction of the 2-alkyl-cyclopentenone has been reported.

1) E. Demole, E. Lederer, D. Mercier, *Helv. Chim. Acta,* **45**, 675 (1962).
2) R. K. Hill, A. G. Edwards, *Tetr.*, **21**, 1501 (1965) (absolute configuration).
3) K. Shishido, S. Koizumi, K. Utimoto, *J. Org. Chem.*, **34**, 2661 (1969).
4) A. I. Meyers, N. Nazarenko, *J. Org. Chem.*, **38**, 175 (1973).

12 ⟶ 14: The ketene *N,O*-acetal **13**, a highly reactive nucleophile, undergoes Michael addition to the cyclopentenone **12** giving **14** after hydrolysis. This provides a general method for the introduction of a –CH₂COOR group into an electron-deficient olefin.

For other syntheses, see reference 5.

Remarks

Methyl jasmonate **1**, an essential constituent for the characteristic odor of jasmine, was isolated from *Jasminum grandiflorum*[1] and *Rosmarinus officinalis*[6]. From jasmine oil, jasmone and the keto-lactone **15** were obtained.[7] Jasmonic acid **2** was isolated from *Lasiodiplodia theobromae* as a plant-growth inhibitor[8]. *N*-dihydrojasmonyl-isoleucine and *N*-jasmonyl-isoleucine have recently been isolated from *Gibberella fujikuroi*.[9]

15

5) E. Demole, M. Stoll, *Helv. Chim. Acta*, **45**, 692 (1962); G. Büchi, B. Egger, *J. Org. Chem.*, **36**, 2021 (1970); P. Ducos, F. Rouessac, *Tetr.*, **29**, 3233 (1973).

6) L. Crabalona, *Compt. Rend.*, **264**, 2074 (1967).

7) E. Demole, B. Willhalm, M. Stoll, *Helv. Chim. Acta*, **47**, 1152 (1964).

8) D. C. Aldridge, S. Galt, D. Giles, W. B. Turner, *J. Chem. Soc.* C, 1623 (1971).

9) B. E. Cross, G. R. B. Webster, *J. Chem. Soc.* C, 1839 (1970).

7.15   SYNTHESIS OF RETHROLONES

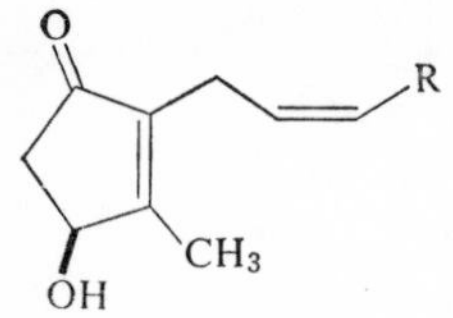

cinerolone **1** (R=CH$_3$)
jasmololone **2** (R=C$_2$H$_5$)
allethrolone **4** (R=H)

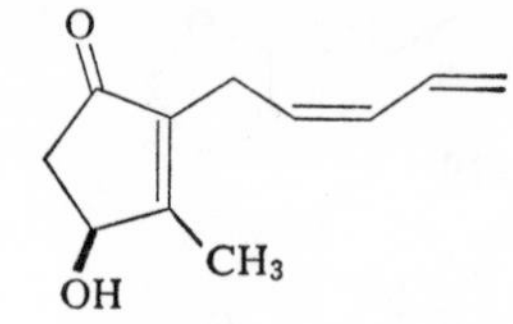

pyrethrolone **3**

Route I[1]

Allethrolone **4** and cinerolone **1** were synthesized starting from acetylacetone. **4** was obtained with 36% overall yield. The synthesis of **1** is shown below.

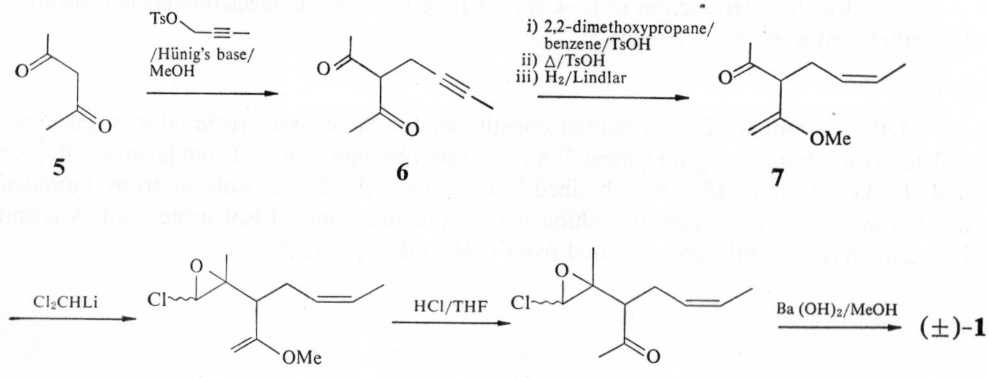

7⟶8: Dichloromethyllithium is an efficient reagent for one-carbon homologation of ketones to give α,β-unsaturated aldehydes or their equivalents.[2]

Route II[3]

Pyrethrolone **3** was synthesized from 5-bromopentan-2-one via the following route in an overall yield of 21%. By an analogous method, **1** and **2** were also synthesized.

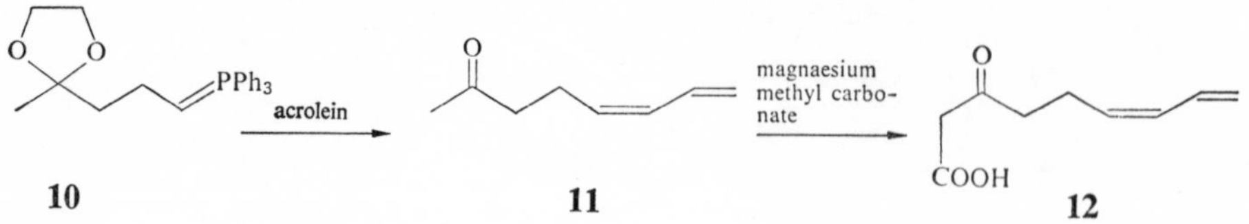

1) G. Büchi, D. Minster, J. C. F. Young, *J. Am. Chem. Soc.*, **93**, 4319 (1971).
2) See also H. Taguchi, S. Tanaka, H. Yamamoto, H. Nozaki, *Tetr. Lett.*, 2465 (1973).
3) L. Crombie, P. Hemesley, G. Pattenden, *J. Chem. Soc.*, C, 1016 (1969); *ibid.*, 1024 (1969).
4) L. Crombie, M. Elliott, *Progr. Chem. Org. Nat. Prod.*, **19**, 121 (1961).
5) For alternative synthesis of rethrolones, see R. F. Romanet, R. H. Schlessinger, *J. Am. Chem. Soc.*, **96**, 3701 (1974); cf. G. Pattenden, R. Storer, *J. Chem. Soc.*, **P-1**, 1603 (1974) and references therein.

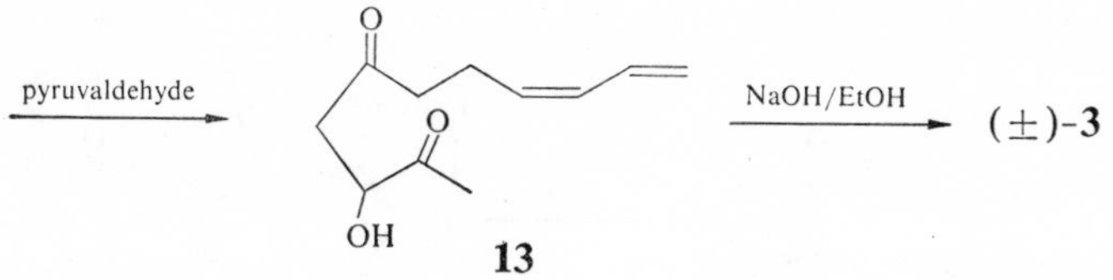

pyruvaldehyde ⟶ [structure 13] NaOH/EtOH ⟶ (±)-3

13

10 ⟶ 11: Wittig condensation of **10** with vaporized acraldehyde, under salt-free conditions gave almost exclusively *cis* diene **11**.

11 ⟶ 12: Magnesium methylcarbonate reacts with a methyl ketone yielding a free carboxylic acid derivative (e.g. **12**) directly.

Remarks

The compounds **1–3** (rethrolones) are the alcohol components of cinerin, jasmolin, and pyrethrin, which are esters of chrysanthemic acid **14** and pyrethric acid **15**. These esters are contained in the flower heads of *Chrysanthemum cinerariaefolium*[4] and are important insecticides.

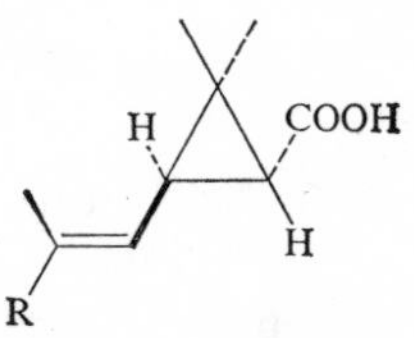

14 (R=Me)
15 (R=COOMe)

7.16   STRUCTURE OF BREFELDIN-A

$$J_{am} = 1.8\,\text{Hz}$$
$$J_{ax} = 3.5\,\text{Hz}$$
$$J_{mx} = 15.5\,\text{Hz}$$

1.25(d, 6.5)

4.75(m)

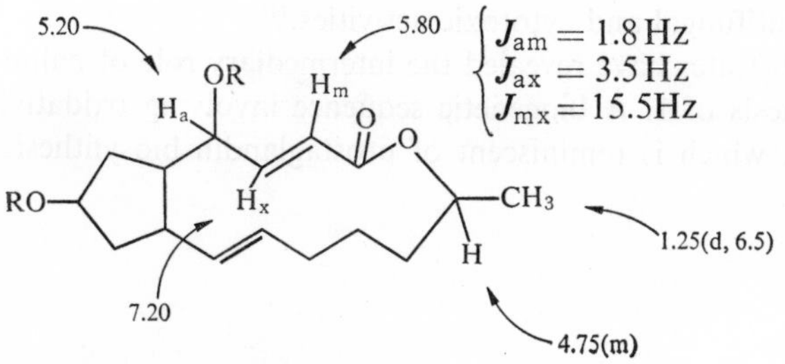

brefeldin-A **1** (R=H)

mp: 204–205° α_D: +96°
ms: 280 (M⁺ $C_{16}H_{24}O_4$)
uv: 215 (4.05)
ir: 3370 (OH), 1711, 1260 (lactone), 1646 (C=C)
nmr for the diacetate (R=Ac)

The structure of brefeldin-A **1** was determined on the basis of the chemical and spectroscopic results shown below. The coupling constant $J_{mx} = 15.5$ Hz indicates a *trans* geometry for the double bond.

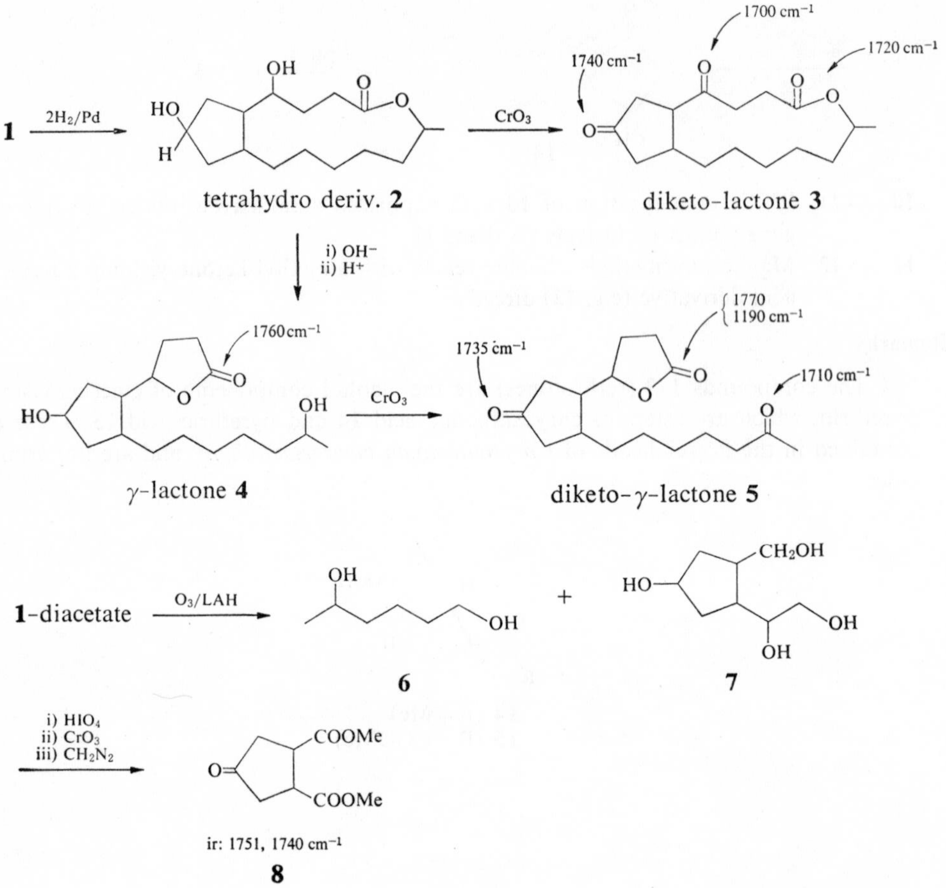

tetrahydro deriv. **2**

diketo-lactone **3**

γ-lactone **4**

diketo-γ-lactone **5**

1-diacetate

6   +   **7**

ir: 1751, 1740 cm^{-1}

8

Remarks

Brefeldin-A **1** was isolated from *Penicillium brefeldianum, Curvularia lunata, Penicillium cyaneum*, etc. as an antibiotic. It shows antifungal and cytotoxic activities.[1]

Feeding experiments with labeled substrates have revealed the intermediary role of palmitate[3] as well as acetate[2] in the biosynthesis of **1**. A biogenetic sequence involving oxidative cyclization of the unsaturated fatty acid, which is reminiscent of prostaglandin biosynthesis, was proposed.[3]

1) H. P. Sigg, *Helv. Chim. Acta*, **47**, 1401 (1964); E. Harri, W. Loeffler, H. P. Sigg, H. Stähelin, Ch. Tamm, *ibid.*, **46**, 1235 (1963).
2) R. G. Coombe, P. S. Foss, T. R. Watson, *Chem. Commun.*, 1229 (1967).
3) J. D. Bu'Lock, P. T. Clay, *Chem. Commun.*, 237 (1969).

7.17 BIOSYNTHESIS OF PROSTAGLANDINS

The prostaglandins consist of a group of monocyclic hydroxy-fatty acids of widespread occurrence in mammalian tissues and of potent physiological activities.[1] Six prostaglandins of the E series (structures 1–3) and the F series (structures 4–6) are designated as primary prostaglandins.[1]

8, 11, 14-eicosatri-
enoic acid 7

5, 8, 11, 14-eicosatetra-
enoic acid (arachidoic
acid) 8

5, 8, 11, 14, 17-eicosa-
pentanoic acid 9

prostaglandin E₁ 1

prostaglandin E₂ 2

prostaglandin E₃ 3

prostaglandin F₁α 4

prostaglandin F₂α 5

prostaglandin F₃α 6

That the prostaglandins are derived biosynthetically from polyunsaturated fatty acids has been shown by the enzymatic conversion of 7, 8 and 9 into prostaglandins-E (e.g. 1–3) and -F (e.g. 5) by incubation with homogenates from sheep vesicular glands and lung tissue, respectively.[2-4]

Possible biosynthetic schemes for these compounds have been proposed on the basis of various experimental results, as outlined below.

7

10

1) For reviews, see S. Bergström, *Science*, **157**, 382 (1967); S. Bergström, L. A. Carlson, J. R. Weeks, *Pharmacol. Rev.*, **20**, 1 (1968).
2) S. Bergström, B. Samuelsson, *The Prostaglandings, Nobel Symposium II*, Almqvist and Wiksell, 1967.
3) S. Bergström, H. Danielsson, B. Samuelsson, *Biochim. Biophys. Acta*, **90**, 207 (1964); S. Bergström, H Danielsson, D. Klenberg, B. Samuelsson, *J. Biol. Chem.*, **239**, PC 4006 (1964); D. A. van Dorp, D. A. R. K Beerthuis, D. H. Nugteren, H. Vonkeman, *Biochim. Biophys, Acta*, **90**, 204 (1964); *Nature*, 839 (1964).
4) M. Hamberg, B. Samuelsson, *J. Biol. Chem.*, **242**, 5336 (1967).

11

1) The initial step of biosynthesis involves stereospecific elimination of the pro-S hydrogen at C-13 and oxygenation at C-11, giving rise to the peroxide **10**. This was shown by tracer experiments using stereospecifically ^{3}H-labeled **7**.[4]

2) The peroxide **10** then cyclizes into the endoperoxide **11** by a concerted oxygenation process as shown in **10**. The evidence for the formation of the endoperoxide **11** is outlined in 3) to 5) below.

3) Mass spectrometric analysis of PGE$_1$ **1** biosynthesized in a mixture of ^{18}O–^{18}O and ^{16}O–^{16}O indicated that the three oxygens at C-9, C-11, and C-15 are derived from molecular oxygen and the former two originate from the same molecule of oxygen.[5]

4) In some enzyme preparations, fragmentation occurs yielding C$_{17}$ acid and malonaldehyde.[6]

5) Selective inhibition of PG synthetase by a bicyclo[2,2,1]heptane derivative structurally reminiscent of **11** was observed.[7]

6) PGE and PGF are not interconvertible in tissue homogenates but are formed by independent steps.

7) Non-enzymatic formation of PGE$_1$ **1** was achieved in a yield of 0.1% by autooxidaion of **7**.[8]

Remarks[1]

Prostaglandin was first isolated from human semen and then from sheep sperm or vesicular glands of male sheep. Some of the primary prostaglandins were found to exist in various tissues, e.g. lungs, iris, brain, thymus, pancreas, kidney etc., albeit in low concentrations. Recently, PGE, and other derivatives have been isolated from soft coral, *Plexaura homomalla*.[9]

Prostaglandins have potent physiological activities, stimulating smooth muscle and depressing blood pressure.

5) B. Samuelsson, *J. Am. Chem. Soc.*, **87**, 3011 (1965).
6) M. Hamberg, B. Samuelsson, *J. Am. Chem. Soc.*, **88**, 2349 (1966); *J. Biol. Chem.*, **242**, 5344 (1967).
7) P. Wlodawer, B. Samuelsson, S. M. Albonico, E. J. Corey, *J. Am. Chem. Soc.*, **93**, 2815 (1971).
8) D. H. Nugteren, H. Vonkeman, D. A. van Dorp, *Rec. Trav. Chim.*, **86**, 1 (1967).
9) See refs. in "Prostaglandins from coral".

7.18 PROSTAGLANDIN DERIVATIVES IN CORAL

Esterified derivatives of prostaglandin, such as 15(R)-PGA$_2$ **1**[1], 15(S)-PGA$_2$ **2**[2] 15(S)-PGE$_2$ **3**,[2] and 15(S)-Δ^5-*trans*-PGA$_2$ **4**[3] have recently been isolated from a soft coral, the gorgonian, *Plexaura homomalla*. **2** and **3** are identical with prostaglandins derived from mammalian sources.

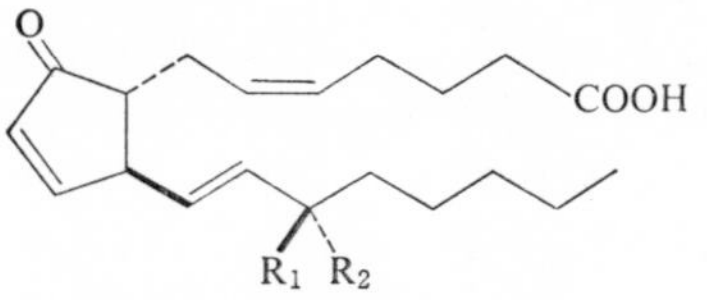

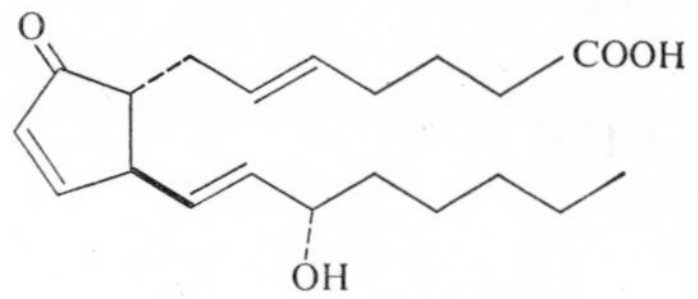

1 (R$_1$=OH; R$_2$=H)
2 (R$_1$=H; R$_2$=OH)

4

15(S)-PGA$_2$ ester-acetate **5** (R$_1$=Ac; R$_2$=Me) was chemically transformed into PGE$_2$ and PGF$_2\alpha$ by the reactions shown below.[4]

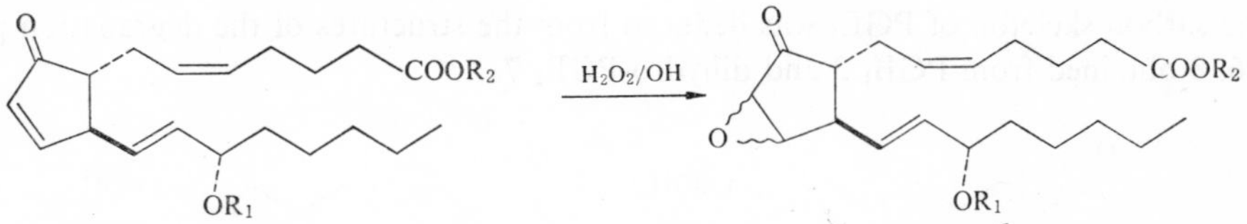

5 $\xrightarrow{\text{H}_2\text{O}_2/\text{OH}}$ **6**

$\xrightarrow[\text{or Al-Hg}]{\text{chromous acetate/HOAc}}$ COOMe $\xrightarrow[\text{esterase}]{\text{acetone-insoluble}}$

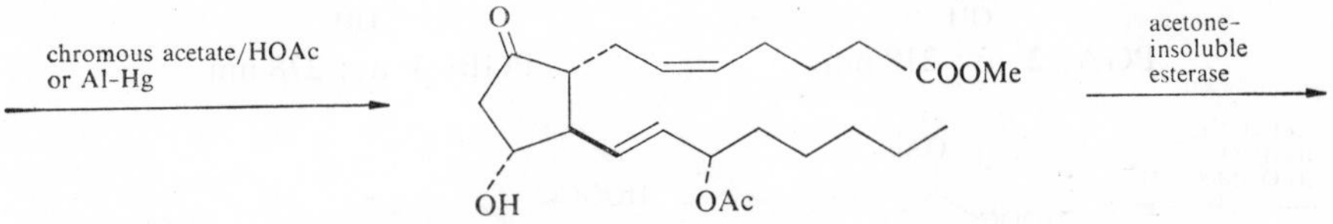

7

COOH $\xrightarrow{\text{NaBH}_4}$ PGF$_2\alpha$

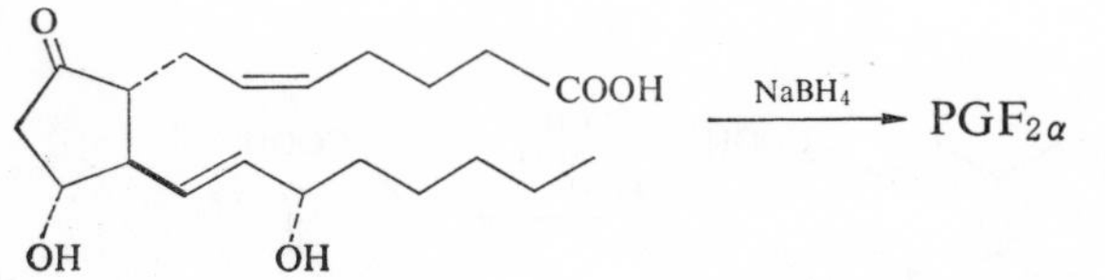

PGE$_2$ **3**

5——→**6**: Epoxidation of **5** (R$_1$=Ac; R$_2$=Me) gave a mixture of two epoxides moderately favoring the α-isomer. Introduction of a bulky substituent at C-15 improves the α:β ratio.[5] The 15-tri-*p*-xylylsilyl ether **5** [R$_1$=Si–(CH$_2$–Ph–Me)$_3$; R$_2$=H] afforded the α-oxide almost exclusively (α : β=94 : 6) upon epoxidation.[6]

1) A. J. Weinheimer, R. L. Spraggins, *Tetr. Lett.*, 5183 (1969).
2) W. P. Schneider, R. D. Hamilton, L. E. Rhuland, *J. Am. Chem. Soc.*, **94**, 2122 (1972).
3) G. L. Bundy, E. G. Daniels, F. H. Lincoln, J. E. Pike, *J. Am. Chem. Soc.*, **94**, 2124 (1972).
4) G. L. Bundy, W. P. Scheneider, F. H. Lincoln, J. E. Pike, *J. Am. Chem. Soc.*, **94**, 2123 (1972).
5) W. P. Schneider, G. L. Bundy, F. H. Lincoln, *Chem. Commun.*, 254 (1973).
6) E. J. Corey, H. E. Ensley, *J. Org. Chem.*, **38**, 3187 (1973).

7.19   STRUCTURE OF PROSTAGLANDIN-E₁

prostaglandin E_1 (PGE₁) **1**[1,2)]

mp: 115–117°
ms: 354 (M⁺ $C_{20}H_{34}O_5$)
ir: 1733 (5-membered ring ketone), 971 (C=C)

1) Treatment of PGE₁ with weak base or weak acid affords the dehydration product **2** (PGA₁), which shows α,β-unsaturated carbonyl absorption. Further alkaline treatment of **2** yields the conjugated dienone **3** (PGB₁).

2) The carbon skeleton of PGE₁ was deduced from the structures of the degradation products **4–6, 8** obtained from PGB₁ **3** and dihydro-PGB₁ **7**.

3) The partial structure was confirmed by chemical synthesis of the compound **12**, which was found to be identical with the naturally derived substance.

4) The structure, stereochemistry and absolute configuration were conclusively determined by an x-ray diffraction study of the tris-*p*-bromobenzoate of PGF₁β.

1) S. Bergstrom, R. Ryhage, B. Samuelsson, J. Sjovall, *Acta. Chim. Scand.*, **16**, 501, 6969 (1962); *ibid.*, **17**, 2217 (1963).
2) B. Samuelsson, *Angew. Chem. Intern. Ed.*, **4**, 410 (1965).

7.20 SYNTHESIS OF PROSTAGLANDINS

Route I

A general and stereocontrolled synthesis of all the primary prostaglandins **1–6** (PGE$_1$, PGE$_2$, PGE$_3$, PGF$_{1\alpha}$, PGF$_{2\alpha}$ and PGF$_{3\alpha}$) in the naturally occurring forms from common intermediates such as **7**, **8**, and **9** has been developed by Corey.[1-10] The key intermediates were synthesized by an efficient route from the bicyclo[2,2,2]heptene system (*vide infra*).

Synthesis of PGE$_1$, PGE$_2$, PGF$_{1\alpha}$, and PGF$_{2\alpha}$[1-5]

7 (R=Ac)
8 (R=p-C$_6$H$_5$C$_6$H$_4$CO)

10 (R=Ac)
11 (R=p-C$_6$H$_5$C$_6$H$_4$CO)
12 (R=p-C$_6$H$_5$C$_6$H$_4$NHCO)

13 (R=Ac)
14 (R=p-C$_6$H$_5$C$_6$H$_4$CO)
15 (R=p-C$_6$H$_5$C$_6$H$_4$NHCO)

1) E. J. Corey, T. K. Schaaf, W. Huber, U. Koelliker, N. M. Weinshenker, *J. Am. Chem. Soc.*, **92**, 397 (1970); E. J. Corey, N. M. Weinshenker, T. K. Schaaf, W. Huber, *ibid.*, **91**, 5675 (1969).
2) E. J. Corey, S. M. Albonico, U. Koelliker, T. K. Schaaf, R. K. Varma, *J. Am. Chem. Soc.*, **93**, 1491 (1971).
3) E. J. Corey, K. B. Becker, R. K. Varma, *J. Am. Chem. Soc.*, **94**, 8616 (1972).
4) E. J. Corey, R. Noyori, T. K. Shcaaf, *J. Am. Chem. Soc.*, **92**, 2586 (1970).
5) E. J. Corey, R. K. Varma, *J. Am. Chem. Soc.*, **93**, 7319 (1971).
6) E. J. Corey, H. Shirahama, H. Yamamoto, A. Venkateswarlu, T. K. Schaaf, *J. Am. Chem. Soc.*, **93**, 1490 (1971).
7) E. J. Corey, U. Koelliker, J. Neuffer *J. Am. Chem. Soc.*, **93**, 1489 (1971).
8) E. J. Corey, T. Ravindranathan, S. Terashima, *J. Am. Chem. Soc.*, **93**, 4326 (1971).
9) See also E. J. Corey, G. Moinet, *J. Am. Chem. Soc.*, **95**, 6831 (1973); J. S. Bindra, A. Grodski, T. K. Schaaf, E. J. Corey, *ibid.*, **95**, 7522 (1973).
10) E. J. Corey, Z. Arnold, J. Hutton, *Tetr. Lett.*, 307 (1970).

16

17 (R=THP)
18 (R=dimethylisopropylsilyl)

7 ⟶ 10: Reaction of the aldehyde derived from **2** with the sodio derivative of dimethyl 2-oxoheptylphosphonate afforded the *trans* enone lactone **10** stereospecifically.[1]

10 ⟶ 13; 11 ⟶ 14; 12 ⟶ 15: Reduction of the enone **10** with zinc borohydride gave a mixture of the 15α and 15β hydroxyl derivatives in a ratio of 1 : 1.[1] When a diisopinocamphenylborane/methyllithium/hexamethylphosphoramide system is used for reduction of the *p*-phenylbenzoate **11**, the ratio of the resulting 15α and 15β isomers is improved to 66 : 31.[2] A trialkylborane derivative **19** prepared from limonene and thexylborane was found to be a more satisfactory reagent for the reduction of **11** to **14** ($\alpha : \beta = 4.5 : 1$).[2] The most outstanding result was obtained when the *p*-phenylphenylurethane **12** was reduced by **19** ($\alpha : \beta = 92 : 8$).[3]

16 ⟶ 17: Condensation of the aldehyde derived from **16** with the Wittig reagent prepared from 5-triphenylphosphoniovalerate ion yielded the desired Δ^5-*cis* product **17** exclusively.[1]

17 ⟶ 4; 17 ⟶ 1: Catalytic hydrogenation of **17** afforded a dihydro derivative saturated selectively at the Δ^5 double bond.[4] Selective hydrogenation was also performed more effectively by the use of the bis-dimethylisopropylsilyl derivative **18**.[5] This might be due to steric screening of the Δ^{13} double bond by the bulky protecting group.

2 ⟶ 5; 1 ⟶ 4: Stereospecific reduction of PGE's to PGF's was achieved with the bulky trialkyl borohydride reagent **19** or **20**.[5]

19   **20**

Synthesis of PGE$_3$ and PGF$_{3\alpha}$[6]

9 $\longrightarrow$ 22: The optically active aldehyde derived from **9** was converted stereospecifically to the unsaturated alcohol **22** by the β-oxido ylide reagent derived from the hydroxy phosphonium salt (S)(+)-**21**.

Synthesis of the key intermediates[1,7-9]

The key intermediates **7–9** in the above prostaglandin syntheses was prepared by a stereo-controlled route through the bicyclo[2,2,2]heptenone derivative **27**, as outlined below.

31 (R′=Ac, or THP or p-C$_6$H$_5$C$_6$H$_4$CO)

24 ⟶ 25: Alkylation of the thallous cyclopentadienide **24** is preferable to the use of sodium or lithium salts.[7]

25 ⟶ 27: Diels-Alder addition of **25** with 2-chloroacrylonitrile followed by hydrolysis afforded the *anti*-bicyclic ketone **27**.[7] Reaction of **25** with 2-chloroacryloylchloride also gave **27** more efficiently.[8]

29: The optical resolution of **29** was achieved via the (+)-amphetamine or (+)-ephedrine salt. The primary prostaglandins were synthesized as the naturally occuring forms starting with optically active **7–9**.

The aldehyde **35** was synthesized by the alternative route shown below.[10]

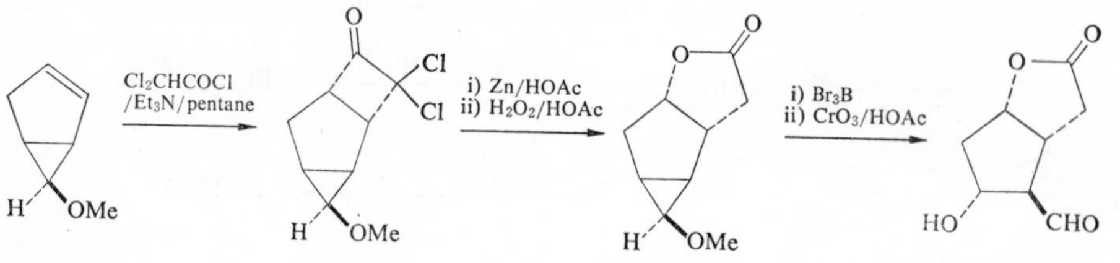

32   **33**   **34**   **35**

Route II[11,12]

Two routes leading to the compound **44**, an important synthetic intermediate for primary prostaglandins, by a semi-pinacolic ring contraction have been reported.

In the first route, the dihydroxy amino derivative **43** having substituents with the correct stereochemistry was synthesized from **36** by a stereochemically controlled process as outlined below.[11]

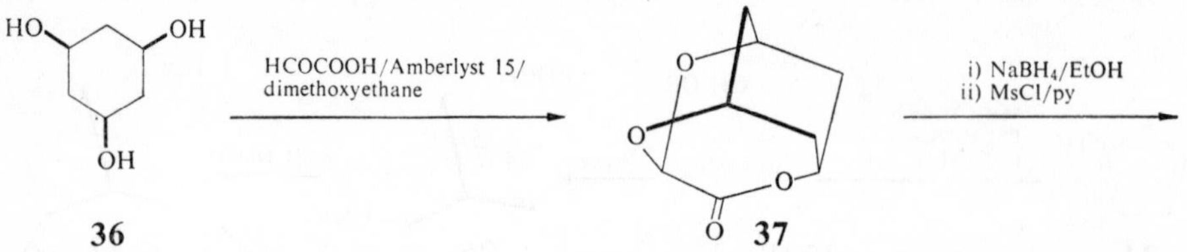

36   **37**

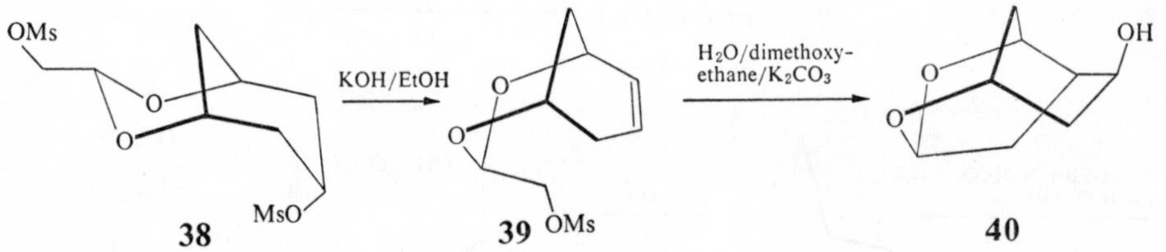

38   **39**   **40**

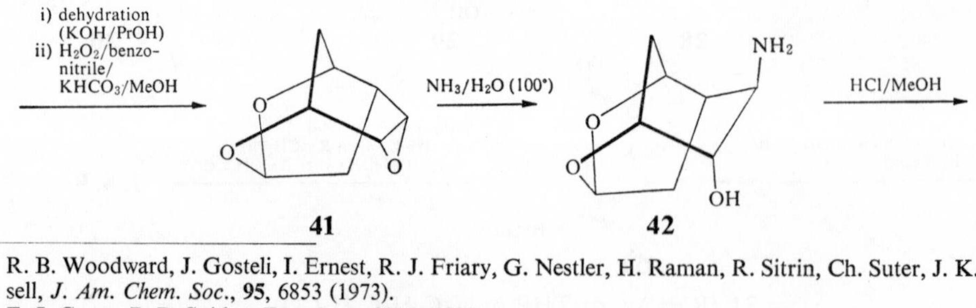

41   **42**

11) R. B. Woodward, J. Gosteli, I. Ernest, R. J. Friary, G. Nestler, H. Raman, R. Sitrin, Ch. Suter, J. K. Whitesell, *J. Am. Chem. Soc.*, **95**, 6853 (1973).
12) E. J. Corey, B. B. Snider, *Tetr. Lett.*, 3091 (1973).

43   NaNO₂/AcONa/aq. HOAc   **44 (R=Me)**   see Route I   PGF₂α **5**

The second route involves a novel approach to allylic functionalization, utilizing the reaction of *N*-phenyltriazolinedione with olefin (e.g., **45** ⟶ **46**).[12]

45   i) (*i*-Bu)₂AlH
                  ii) cyclohexanol/
                      BF₃/CH₂Cl₂
                  iii) N-phenyltri-
                      azolinedione   **46**   i) Me₂SO₄/THF
                                                                       ii) OsO₄/THF/py
                                                                       iii) Δ/KOH/aq. MeOH

47   H₂/Pt/MeOH   **48**

NaNO₂/aq. HOAc   **44 (R=cyclohexyl)**   see Route I $\begin{cases} \text{PGE}_2\ \mathbf{2} \\ \text{PGF}_{2\alpha}\ \mathbf{5} \end{cases}$

Route III[13]

PGE₁ was synthesized in an optically active form by a route involving the conjugate addition of a cuprate reagent to the cyclopentenone derivative **49**. Addition proceeded exclusively from the less hindered side of the cyclopentenone ring.

49   (R₁=THP; R₂=–CHMeOEt or   **50**
            R₁=R₂=Si(Me)₂–*t*Bu)

49: (+)-**49** (R_1 = Si(Me)$_2$-tBu) was prepared from readily available 2-(6-carbomethoxyhexyl) cyclopentan-1,3,4-trione by a route including microbial reduction.

50: The cuprate reagent **50** (R_2 = Si(Me)$_2$-tBu) was prepared from the corresponding vinyl iodide, lithium powder, and tri-n-butylphosphine-copper iodide complex.

An alternative synthesis of PGE$_1$ by conjugated addition of vinylcopper reagent to a substituted cyclopentenone, giving rise to **52**, has also been reported.[15] **52** was converted to PGE$_1$ in several steps.

Route IV[16]

Cross-coupling of a cuprate reagent **57** with an allylic electrophile **56** was successfully utilized in the stereospecific synthesis of PGA$_2$ **60**. In this synthesis, the attack of the reagent was stereochemically controlled by the bulky substituent at C-9.

13) C. J. Sih, J. B. Heather, G. P. Peruzzotti, P. Price, R. Sood, L-F. HsuLee, *J. Am. Chem. Soc.*, **95**, 1676 (1973); C. J. Sih, P. Price, R. Sood, R. G. Salmon, G. P. Peruzzotti, M. Casey, *ibid.*, **94**, 3643 (1972).
14) See also A. F. Kluge, K. G. Untch, J. H. Fried, *J. Am. Chem. Soc.*, **94**, 7827 (1972).
15) F. S. Alvarez, D. Wren, A. Prince, *J. Am. Chem. Soc.*, **94**, 7823 (1972).
16) E. J. Corey, J. Mann, *J. Am. Chem. Soc.*, **95**, 6832 (1973).

56+57 ⟶ 58: The cross-coupling gave the product formed only by SN2-type reaction. The absence of SN2′ product was interpreted in terms of steric shielding by the dimethyl-*t*-butylsilyloxy group and preference for *cis* stereochemistry in the SN2′ process.

60 ⟶ 2: PGA$_2$ can be converted into PGE$_2$ by stereocontrolled epoxidation followed by reduction.[17]

Route V[18,19]

PGF$_{2\alpha}$ was synthesized by a route involving regiospecific opening of the epoxide ring in **63** with (S)-3-*t*-butyloxy-1-octynylmethoxymethylalane **64**, prepared from methoxymethylchloro-alane, (S)-3-*t*-butyloxy-1-octyne and *n*-BuLi, yielding exclusively the alkynylated product **65**. PGF$_{3\alpha}$ was also synthesized by an analogous sequence.

17) E. J. Corey, H. E. Ensley, *J. Org. Chem.*, **38**, 3187 (1973); cf. G. L. Bundy, W. P. Schneider, F. H. Lincoln J. E. Pike, *J. Am. Chem. Soc.*, **94**, 2123 (1972).
18) J. Fried, J. C. Sih, C. H. Lin, P. Dalven, *J. Am. Chem. Soc.*, **94**, 4343 (1972); J. Fried, C. H. Lin, J. C. Sih, P. Dalven, G. F. Cooper, *ibid.*, **94**, 4342 (1972).
19) J. Fried, J. C. Sih, *Tetr. Lett.*, 3899 (1973).

64: When alkynylation reagent prepared from dimethylchloroalane was used, regiospecificity of the reaction was not observed.[19]

65 ⟶ 69; 67 ⟶ 70: Direct conversion of **65** and **67** into the lactones **69** and **70**, respectively, was achieved by catalytic oxidation.[19]

Route VI[20]

A practical and stereochemically controlled synthesis of PGE_1 has been developed. In this synthesis, the compound **76** was obtained in 21–26% yield from **73**.

---

20) M. Miyano, M. A. Stealey, *Chem. Commun.*, 180 (1973); M. Miyano, C. R. Dorn, R. A. Mueller, *J. Org. Chem.*, **37**, 1810 (1972).

73 was optically resolved via $(R)(-)$-α-methoxyphenylacetic ester and the 3R isomer was converted into $(-)$-PGE$_1$.

Route VII[21]

A new synthetic approach based on a biogenetic-type ring closure[22] of an acyclic triene intermediate such as **81** has been published. The lactone **59**, a useful intermediate for prostaglandin synthesis, was prepared by BF$_3$-catalyzed cyclization of the epoxide **81** in 15% yield.

80 ⟶ 81: The *cis, cis, trans* triene moiety in **81** was constructed in a geometrically specific manner by hydrogenation of the diyne-ene intermediate **80**.

81 ⟶ 59: The substituents at C-12/C-8/C-9 in **59** was oriented in the *trans/cis* arrangement. Equal amounts of the two C-15 epimers were formed.

Route VIII[23]

21) E. J. Corey, G. W. J. Fleet, M. Kato, *Tetr. Lett.*, 3963 (1973).
22) See also J. Martel, E. Toromanoff, J. Mathieu, G. Nomine, *Tetr. Lett.*, 1491 (1972).
23) E. J. Corey, N. H. Andersen, R. M. Carlson, J. Paust, E. Vedejs, I. Vlattas, R. E. K. Winter, *J. Am. Chem Soc.*, **90**, 3245 (1968).

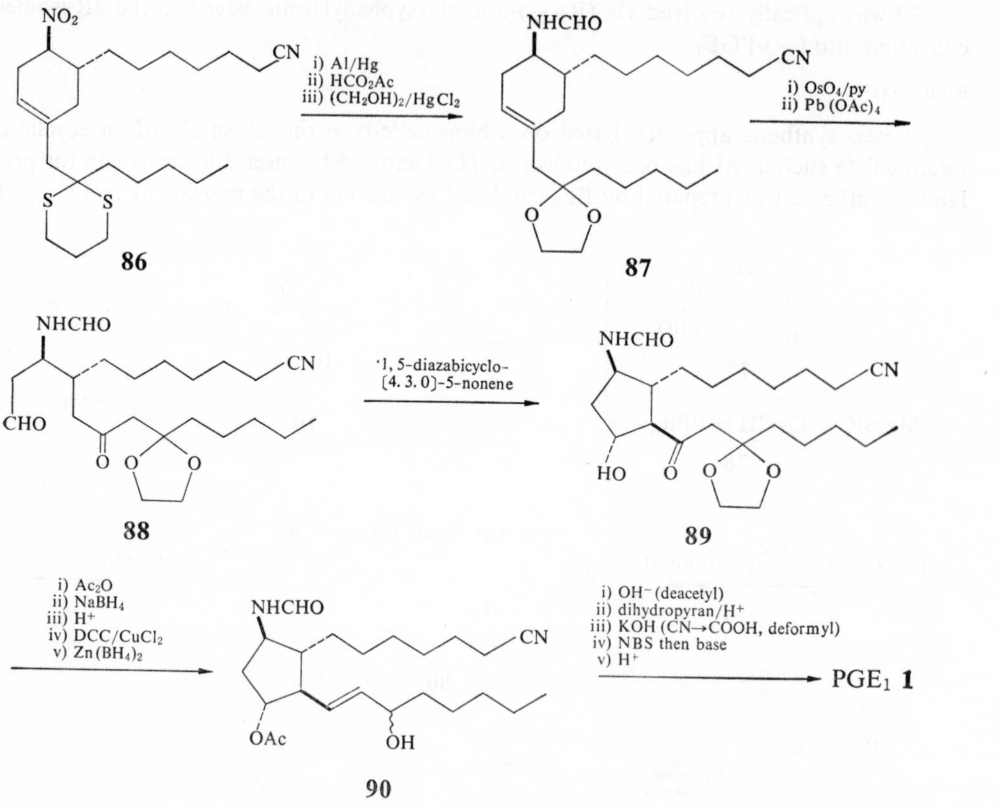

86

i) Al/Hg
ii) HCO₂Ac
iii) (CH₂OH)₂/HgCl₂

87

i) OsO₄/py
ii) Pb(OAc)₄

88

'1,5-diazabicyclo-[4.3.0]-5-nonene

89

i) Ac₂O
ii) NaBH₄
iii) H⁺
iv) DCC/CuCl₂
v) Zn(BH₄)₂

90

i) OH⁻ (deacetyl)
ii) dihydropyran/H⁺
iii) KOH (CN→COOH, deformyl)
iv) NBS then base
v) H⁺

⟶ PGE₁ **1**

84,85 ⟶ 86: Diels-Alder addition of **84** to **85** proceeded readily and in high yield to give **86** as the major product.

89 ⟶ 90: DCC/CuCl₂ is a useful reagent for the dehydration of β-hydroxy carbonyl compounds.

90 ⟶ 1: Transformation of the animo group to ketone was performed by bromination (to the *N*-bromo derivative) and dehydrobromination followed by hydrolysis of the resulting imine.

Route IX[24,25]

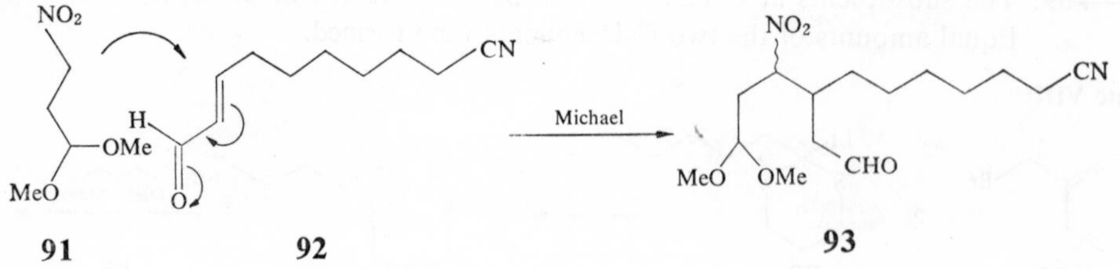

91   **92**   Michael ⟶   **93**

24) E. J. Corey, I. Vlattas, N. H. Andersen, K. Harding, *J. Am. Chem. Soc.*, **90**, 3247 (1968); *ibid.*, **90**, 5947 (1968).

25) E. J. Corey, I. Vlattas, K. Harding, *J. Am. Chem. Soc.*, **91**, 535 (1969).

i) (MeO)$_2$ $\overset{O}{\overset{\|}{P}}CH\overset{O}{\overset{\|}{C}}$-C$_5H_{11}$
ii) (CH$_2$OH)$_2$/H$^+$

94

i) Al/Hg
ii) HCO$_2$Ac
iii) TsOH/acetone

95 (mixture of 4 isomers)

i) Zn(BH$_4$)$_2$
ii) dihydropyran
iii) KOH

(for 9α, 11α -isomer)

96   PGE$_1$ **1**
(+15-epi-PGE$_1$)

94 — SnCl$_4$/acetone →

97

i) Zn(BH$_4$)$_2$
ii) base

98 (mixture of 15α-and 15β-OH derivatives)

Al/Hg
(15α-OH)

99

see Route VII → PGE$_1$ **1**

94 ⟶ 95 ⟶ 1: Treatment of the formylamino bis-ketal derived from **94** with TsOH gave a mixture of four stereoisomers of structure **95** which can be seperated by chromatography. **95** with 9α and 11α substituents was converted into PGE$_1$ and 15-epi-PGE$_1$ by the reactions used in Route VIII.[24]

94 ⟶ 97: The nitro ketal **94** was cyclized by SnCl$_4$ into **97**, which is essentially free from the undesired 11β-isomer.[25]

97 ⟶ 98: The nitro group was oriented in the more stable β-position in **98** after base treatment.[25]

99 ⟶ 1: The amine **99** was resolved via the (−)-α-bromocamphor-π-sulfonic acid salt. The result *levo* amine was then converted to natural (−)-PGE$_1$.[25]

Other routes[30]

PGF$_{1\alpha}$ and PGE$_1$ were synthesized, albeit in low yield, by non-stereospecific routes involving the acid-catalyzed rearrangement of the epoxide **103** or solvolytic rearrangement of the diol dimesylate **104** which has a bicyclo[3,1,0]hexane skeleton, as the key step.[26] Methyl esters of PGE$_2$[27] and PGE$_3$[28] were also synthesized by analogous sequences.

A stereospecific synthesis of PGE$_1$ starting with **105** through intermediate **106** and **107** was reported.[29]

26) G. Just, C. Simonovitch, F. H. Lincoln, W. P. Schneider, U. Axen, G. B. Spero, J. E. Pike, *J. Am. Chem. Soc.*, **91**, 5364 (1969); W. P. Schneider, U. Axen, F. H. Lincoln, J. E. Pike, J. L. Thompson, *ibid.*, **91**, 5372 (1969).

27) E. S. Ferdinandi, G. Just, *Can. J. Chem.*, **49**, 1071 (1971); W. P. Schneider, *Chem. Commun.*, 304 (1969).

28) U. Axen, J. L. Thompson, J. E. Pike, *Chem. Commun.*, 602 (1970).

29) D. Taub, R. D. Hoffsommer, C. H. Kuo, H. L. Slates, Z. S. Zelawski, N. L. Wendler, *Tetr.*, **29**, 1447 (1973).

30) For the reviews see J. S. Bindra, R. Bindra, *Progr. Drug Res.*, **17**, 410 (1973).

7.21 STRUCTURE OF LACTOBACILLIC ACID

$$CH_3 (CH_2)_5 \underset{\underset{H}{|12}}{\overset{S}{C}} \overset{CH_2}{\underset{}{\diagup \diagdown}} \underset{\underset{H}{|11}}{\overset{R}{C}} (CH_2)_9\text{-COOH}$$

lactobacillic acid $1^{1-3)}$

mp: 29.5–30° α_D: +0.25°
ir (μ): 9.8, 1.63, 2.20,
ms: 310 (M$^+$ of 1 methyl ester, $C_{20}H_{38}O_2$)

$$1 \xrightarrow{H_2/Pt} Me\,(CH_2)_5 \overset{Me}{\overset{|}{C}}HCH_2\,(CH_2)_9\,COOH \quad + \quad Me\,(CH_2)_5\,CH_2\overset{Me}{\overset{|}{C}}H\,(CH_2)_9\,COOH$$

$$\mathbf{2} \qquad\qquad\qquad\qquad \mathbf{3}$$

$$+ \; Me\,(CH_2)_{17}\,COOH$$

i) CH$_2$N$_2$
ii) CrO$_3$ **4**

$$Me\,(CH_2)_5 \overset{10}{-\!\!\triangle\!\!-} \underset{O}{\overset{||}{C}}\text{-}(CH_2)_8\text{-COOMe} \quad + \quad Me\,(CH_2)_4\text{-}\underset{O}{\overset{||}{\overset{13}{C}}}\overset{}{-\!\!\triangle\!\!-}(CH_2)_9\text{-COOMe}$$

$$\mathbf{5} \quad M_D: -55° \qquad\qquad\qquad \mathbf{6} \quad M_D: +75°$$
$$\text{uv: } 288\ (\Delta\varepsilon + 0.51)$$

1) Hydrogenation of **1** yielded the methyl-branched fatty acids **2** and **3**, and the normal C$_{19}$ acid **4**.[1]

2) Mild chromic oxidation of **1** methyl ester yielded the α-cyclopropyl ketones **5** and **6**. Comparison of the cd spectrum of **6** with those of ketones having known configurations established the configurations at C-11 and C-12 in **1** as R and S, respectively.[2]

Remarks

Lactobacillic acid **1** was isolated from *Lactobacillus arabinosus*,[1] *Brucella abortus*, *B. melitensis*[3] and other bacteria. Cyclopropane fatty acids containing 17, 19 and 21 carbon atoms were also found in various Gram-positive and Gram-negative bacteria.[3]

1) For earlier studies, see, K. Hoffmann *et al.*, *J. Am. Chem. Soc.*, **72**, 4328 (1950); *ibid.*, **80**, 5717 (1958); *ibid.*, **76**, 1799 (1954); *ibid.*, **79**; 3608 (1957).
2) J. F. Tocanne, *Tetr.* **28**, 363 (1972); cf. J. F. Tocanne, R. G. Bergmann, *ibid.*, **28**, 373 (1972).
3) O. W. Thiele, C. Lacave, J. Asselineau, *European J. Biochem.*, **7**, 393 (1969).

7.22 SYNTHESIS OF STERCULIC ACID

$$CH_3\text{-}(CH_2)_7\text{-}C \underset{\overset{\diagup \diagdown}{CH_2}}{=\!=} C\text{-}(CH_2)_7\text{-}COOH$$

0.8

sterculic acid **1**[1]

ir: 1872, 1008 (cyclopropene)

Methyl sterculate was synthesized from methyl stearolate **2** via the cyclopropene carboxylic acid ester **3**. Decarbonylation of the ester **3** was achieved via the corresponding acid chloride **4**.[2] Direct decarbonylation of **3** was performed with fluorosulfonic acid, yielding methyl sterculate in 60–65% overall yield.[3]

Me $(CH_2)_7\text{-}C \equiv C\text{-}(CH_2)_7 COOMe$ $\xrightarrow{N_2CHCOOEt/Cu}$ Me $(CH_2)_7\text{-}C \underset{\overset{\diagup \diagdown}{\overset{C-H}{\mid}{COOEt}}}{=\!=} C\text{-}(CH_2)_7 COOMe$

2   **3**

$\Big\downarrow$ i) FSO_3H
ii) $NaBH_4$

i) KOH
ii) $(COCl)_2$ $\xrightarrow{\hspace{2cm}}$ Me $(CH_2)_7\text{-}C \underset{\overset{\diagup \diagdown}{\overset{CH}{\mid}{COCl}}}{=\!=} C\text{-}(CH_2)_7 COCl$ $\xrightarrow[\text{iii) } NaBH_4]{\begin{array}{l}\text{i) } ZnCl_2\\ \text{ii) MeOH}\end{array}}$ methyl sterculate

4

Remarks

Sterculic acid **1** was isolated from the seed oil of *Sterculia foetida*. **1** and malvalic acid, a C-18 homolog of **1**, have been found widely in various species of Malvales. Sterculinic acid **5**, the first fatty acid to have both an acetylene and a cyclopropane ring, was found in *Sterculia alata*.[4] The feeding of *Sterculia* oil, **1** itself or malvalic acids to hens caused discoloration of eggs, cessation of egg production and suppression of the hatchability of the eggs.

$$HC \equiv C(CH_2)_7\text{-}C \underset{\overset{\diagup \diagdown}{CH_2}}{=\!=} C\text{-}(CH_2)_6 COOH$$

sterculinic acid **5**

1) For a review, see F. L. Carter, V. L. Frampton, *Chem. Rev.*, **64**, 497 (1964).
2) W. J. Gensler, M. B. Floyd, R. Yanase, K. W. Pober, *J. Am. Chem. Soc.*, **92**, 2472 (1970); cf. W. J. Gensler, K. W. Pober, D. M. Solomon, M. B. Floyd., *J. Org. Chem.*, **35**, 2301 (1970).
3) J. L. Williams, D. S. Sgoutas, *J. Org. Chem.*, **36**, 3064 (1971).
4) A. W. Jevans, C. Y. Hopkins, *Tetr. Lett.*, 2168 (1968).

7.23 STRUCTURES OF MYCOLIC ACIDS

$\left.\begin{matrix}0.6\\-0.3\end{matrix}\right\}$ cyclopropane H

$$CH_3\text{-}(CH_2)_{19}\text{-}CH\text{——}CH\text{-}(CH_2)_{14}\text{-}CH\text{——}CH\text{-}(CH_2)_{13}\text{-}\underset{\underset{C_{24}H_{49}}{|}}{\overset{\overset{OH}{|}}{CH}}\text{-}CH\text{-}COOMe$$

methyl hominomycolate-I **1**[1]

$\left.\begin{matrix}0.6\\-0.3\end{matrix}\right\}$ cyclopropane H

$$CH_3\text{-}(CH_2)_{17}\text{-}CH\text{——}CH\text{-}(CH_2)_{12}\text{-}\underset{\underset{Me}{|}}{CH}\text{-}CH\text{——}CH\text{-}(CH_2)_{17}\text{-}\underset{\underset{C_{22}H_{45}}{|}}{\overset{\overset{OH}{|}}{CH}}\text{-}CH\text{-}COOMe$$

methyl avimycolate-I **2**[1]

1) Pyrolysis of the methyl mycolates **1** and **2** yielded the meromycolals **3** and **4**, respectively.

2) On treatment with BF_3 in MeOH, the cyclopropane rings in meromycolane were cleaved, giving the dimethoxy derivative.

3) Analysis of the mass fragmentations of mycolates, meromycolals, meromycolanes and the dimethoxy derivatives was important in their structural elucidation.

$$\mathbf{1} \xrightarrow{\text{pyrolysis}} \quad \overset{\xrightarrow{487\ (C_{34}H_{63}O)}}{Me\text{—}(CH_2)_{19}\text{—}CH\text{—}CH\text{—}(CH_2)_{14}\text{—}CH\text{—}CH\text{—}(CH_2)_{13}\text{—}CHO}$$

hominomeromycolal-I **3**

$$\mathbf{2} \xrightarrow{\text{pyrolysis}} \quad Me\text{—}(CH_2)_{17}\text{—}CH\text{—}CH\text{—}(CH_2)_{12}\text{—}\underset{\underset{Me}{:}}{CH}\text{—}CH\text{—}CH\text{—}(CH_2)_{17}\text{—}CHO$$

543 $(C_{38}H_{71}O)$

307 $(C_{21}H_{39}O)$

avimeromycolal-I **4**
ms: 796 $(M^+, C_{56}H_{108}O)$

$$\mathbf{3\ (or\ 4)} \xrightarrow[\text{iii) LAH}]{\overset{\text{i) LAH}}{\text{ii) MsCl/py}}} \text{meromycolane} \xrightarrow{BF_3/MeOH} \text{dimethoxy derivative}$$

---

1) D. E. Minnikin, N. Polgar, *Chem. Commun.*, 916 (1967); *Tetr. Lett.*, 2643 (1966); and references therein.
2) See also, A. H. Etémadi, F. Pinte, J. Markovits, *Bull. Soc. Chim. Fr.*, 195 (1967); C. Asselineau, G. Tocanne, J. F. Tocanne, *ibid.*, 1455 (1970); and references therein.

Remarks

Methyl hominomycolate-I **1** is a major homologue of mycolic acid-I, isolated from the human strain of tubercle bacilli, *Mycobacterium tuberculosis* var. *hominis*. **2** was isolated from the avian strain *M. avium* as a major homologue.

Other mycolic acids

The mycolic acids fraction from human tubercle bacilli was found to contain, in addition to mycolic acid-I, mycolic acid-II and -III. Methyl mycolate-II was further separated into IIa and IIb; methyl mycolate-III was also separated, into IIIa and IIIb.[3]

$$\text{II} \begin{cases} \underset{\substack{\text{Me OMe} \\ | \quad | }}{\text{Me (CH}_2)_{17}\text{ CHCH-(CH}_2)_{16}} \triangle \text{(CH}_2)_{17}\underset{\substack{\text{OH} \\ | }}{\text{-CH-}}\underset{\substack{| \\ R }}{\text{CH-COOMe}} \\ \qquad\qquad \textbf{7} \ (\textit{cis } \text{IIa}) \\ \\ \underset{\substack{\text{Me OMe} \\ | \quad |}}{\text{Me-(CH}_2)_{17}\text{ CH-CH-(CH}_2)_{16}\text{-}}\underset{\substack{\text{Me} \\ |}}{\text{CH}} \triangle \text{(CH}_2)_{16}\underset{\substack{\text{OH} \\ |}}{\text{-CH-}}\underset{\substack{| \\ R}}{\text{CH-COOMe}} \\ \qquad\qquad \textbf{8} \ (\textit{trans}, \text{ IIb}) \end{cases}$$

$R = C_{24}H_{49}$ for the most abundant homologues

$$\text{III} \begin{cases} \underset{\substack{\text{Me O} \\ | \quad \|}}{\text{Me (CH}_2)_{17}\text{ CH-C-(CH}_2)_{16}} \triangle \text{(CH}_2)_{19}\underset{\substack{\text{OH} \\ |}}{\text{-CH-}}\underset{\substack{| \\ R}}{\text{CH-COOMe}} \\ \qquad\qquad \textbf{9} \ (\textit{cis}, \text{ IIIa}) \\ \\ \underset{\substack{\text{Me O} \\ | \quad \|}}{\text{Me (CH}_2)_{17}\text{-CH-C-(CH}_2)_{16}\text{-}}\underset{\substack{\text{Me} \\ |}}{\text{CH}} \triangle \text{(CH}_2)_{18}\underset{\substack{\text{OH} \\ |}}{\text{-CH-}}\underset{\substack{| \\ R}}{\text{CH-COOMe}} \\ \qquad\qquad \textbf{10} \ (\textit{trans}, \text{ IIIb}) \end{cases}$$

The mycolic acids fraction from *Mycobacterium phlei* contain the keto mycolic acids **11**, **12** and **13**.[4]

$$\underset{\substack{\text{Me O} \\ | \quad \|}}{\text{Me (CH}_2)_{17}\text{-CH-C-(CH}_2)_{31}\text{-}}\underset{\substack{\text{OH} \\ |}}{\text{CH-}}\underset{\substack{| \\ R}}{\text{CH-COOH}} \qquad \textbf{11}$$

$$\underset{\substack{\text{Me O} \\ | \quad \|}}{\text{Me (CH}_2)_{17}\text{-CH-C-(CH}_2)_{14}\text{-}}\underset{\substack{ \\ \text{cis}}}{\text{CH}} = \text{CH-(CH}_2)_{17}\underset{\substack{\text{OH} \\ |}}{\text{-CH-}}\underset{\substack{| \\ R}}{\text{CH-COOH}} \qquad \textbf{12}$$

$$\underset{\substack{\text{Me O} \\ | \quad \|}}{\text{Me (CH}_2)_{17}\text{-CH-C-(CH}_2)_{14}\text{-}}\underset{\substack{\text{Me} \\ |}}{\text{CH-}}\underset{\substack{ \\ \text{trans}}}{\text{CH}} = \text{CH-(CH}_2)_{16}\underset{\substack{\text{OH} \\ |}}{\text{-CH-}}\underset{\substack{| \\ R}}{\text{CH-COOH}} \qquad \textbf{13}$$

$R = C_{22}H_{45}$

3) D. E. Minnikin, N. Polgar, *Chem. Commun.*, 1172 (1967).
4) K. Kusamran, N. Polgar, D. E. Minnikin, *Chem. Commun.*, 111 (1972).

7.24 BIOSYNTHESIS OF ACETYLENIC COMPOUNDS

A number of acetylenic natural products with 9 to 18 carbon atoms, derived biogenetically through the acetate-malonate pathway, have been isolated from plants and fungi.[1]

Based on evidence obtained mainly by precursor-incorporation experiments, a possible biosynthetic pathway from stearic acid has been proposed, as outlined below.[2]

$$Me-(CH_2)_{16}-COOH \longrightarrow Me-(CH_2)_7-CH=CH-(CH_2)_7-COOH \longrightarrow$$
stearic acid **1**   oleic acid **2**

$$Me-(CH_2)_4-CH=CH-CH_2-CH=CH-(CH_2)_7-COOH$$
linoleic acid **3** $\downarrow$

$$Me-(CH_2)_4-C\equiv C-CH_2-CH=CH-(CH_2)_7-COOH$$
crepenynic acid **4**

$\downarrow$

$$Me-CH_2-CH_2-CH=CH-C\equiv C-CH_2-CH=CH-(CH_2)_7-COOH$$
dehydrocrepenynic acid **5**

$\downarrow$

$$Me-CH=CH-C\equiv C-C\equiv C-CH_2-CH=CH-(CH_2)_7-COOH$$
6

$\downarrow$

$$Me-C\equiv C-C\equiv C-C\equiv C-CH_2-CH=CH-(CH_2)_7-COOH \rightarrow \rightarrow C_{18}-acetylenes$$
7

$\downarrow$

$$Me-C\equiv C-C\equiv C-C\equiv C-CH_2-CH=CH-(CH_2)_3-COOH \rightarrow \rightarrow C_{14}-acetylenes$$
8

$\downarrow$

$$Me-C\equiv C-C\equiv C-C\equiv C-CH_2-CH=CHCH_2-COOH \rightarrow \rightarrow C_{12}-acetylenes$$
9

$\downarrow$

$$Me-C\equiv C-C\equiv C-C\equiv C-CH=CH-COOH \rightarrow \rightarrow C_{10}-acetylenes$$
10

The above pathway involves both dehydrogenation processes from less unsaturated to more unsaturated acids and chain-shortening steps starting from C_{18} acids (**6** or **7**) leading to $C_{10}-C_{14}$ acetylenes. The chain-shortening steps may proceed via β-oxidation processes. Direct oxidative fission at the C_{18} stage may also lead to C_{10} acids.

Acetylenic compounds having odd numbers of carbon atoms, e.g. C_{17}, C_{13}, C_{11} and C_9, may arise through reactions involving α-oxidation as well as ω-oxidation processes.

The biosynthetic intermediates produced in reactions of the above type are often subject to further modification, e.g. oxidation, reduction, decarboxylation, etc., leading to a wide variety of naturally occurring acetylenic compounds.[3] However, acetylenic metabolites which are not derived from fatty acid intermediates are also known, including terpenoids, carotenoids and others.[4]

1) For reviews, see F. Bohlmann, *Fortschr. Chem. Org. Naturst.*, **25**, 1 (1967); E. R. H. Jones, *Chem. Brit.*, **2**, 6, (1966); J. D. Bu'Lock, *Progr. in Org. Chem.*, **6**, p 86, 1964.
2) J. D. Bu'Lock, G. N. Smith, *J. Chem. Soc.*, 332 (1967).
3) F. Bohlmann, T. Burkhardt, C. Zdero, *Naturally Occuring Acetylenes*, Academic Press, 1973.
4) R. M. Ross, *Phytochemistry*, **11**, 221 (1972).

7.25 POLYACETYLENIC HYDROCARBON

$H_3C-C\equiv C-C\equiv C-C\equiv C-C\equiv C-C\equiv C-CH=CH_2$

tridecapentayne-ene **1**[1]

uv: 411, 378.5, 352, 329.5, 309, 286.5, 271.5, 265
ir: 2220 ($C\equiv C$),
        1870, 962, 933 (double bond)

1) Upon hydrogenation, **1** gave tridecane absorbing 11 moles of hydrogen.
2) Uv absorption maxima of **1** are situated in an intermediate position between those of dialkyl-hexayne and divinyltetrayne.
3) **1** contains one double bond, as indicated by ir absorption.

Remarks

Tridecapentayne-ene **1** is an extremely unstable compound and was first isolated from *Helipterum* species. It has been found that this hydrocarbon is widely distributed in various members of inulea and others, albeit in small quantity. Usually, more stable poly-thiophene derivatives of **1** can be obtained.[2,3]

C_{13}-hydrocarbons with tetrayne-diene **2, 3** triyne-triene **4, 5** and diyne-tetraene **6** systems are also known.

$$Me-(C\equiv C)_4-CH=CH-CH=CH_2$$
2[3]

$$Me-(C\equiv C)_3-CH=CH-CH=CH-CH=CH_2$$
4[5]

$$Me-CH=CH-(C\equiv C)_4-CH=CH_2$$
3[4]

$$Me-CH=CH-(C\equiv C)_3-CH=CH-CH=CH_2$$
5[6]

$$Me-CH=CH-(C\equiv C)_2-CH=CH-CH=CH-CH=CH_2$$
6[7]

1) J. S. Sörensen, D. Holme, E. T. Borlaug, N. A. Sörensen, *Acta. Chem. Scand.*, **8**, 1769 (1954).
2) J. S. Sörensen, J. T. Mortensen, N. A. Sörensen, *Acta Chem. Scand.*, **18**, 2182 (1964); see also F. Bohlmann, C. Zdero, *Chem. Ber.*, **105**, 2604 (1972) and references therein.
3) F. Bohlmann, K. M. Kleine, C. Arndt, *Chem. Ber.*, **97**, 2125 (1964).
4) S. Liaaen-Jensen, N. A. Sörensen, *Acta Chem. Scand.*, **15**, 1885 (1961).
5) F. Bohlmann, S. Postulka, J. Ruhnke, *Chem. Ber.*, **91**, 1642 (1958); F. Bohlmann, C. Arndt, H. Bolnowski, H. Jastrow, K. M. Kleine, *ibid.*, **95**, 1320 (1962).
6) J. S. Sörensen, N. A. Sörensen, *Acta Chem. Scand.*, **12**, 756 (1958).
7) J. S. Sörensen, N. A. Sörensen, *Acta Chem. Scand.*, **8**, 1741 (1954).

7.26 THIOPHENE DERIVATIVES OF POLYACETYLENES

7.10 (d. 3.8)
7.20

2.08

5.4–6.0

$CH_3-C \equiv C$ ——(thiophene)—— $(C \equiv C)_2\text{-}CH = CH_2$

1[1]
uv: 357.5, 338, 274, 264, 258

7.15 (m, 2H) } thiophene H
6.69 (m, 3H)

2.05

$CH_3-C \equiv C$ —(thiophene)— $C \equiv C$ —(thiophene)

2[2]
uv: 359, 353, 341.5, 334.5, 329, 320, 255

7.16

7.27

5.64
5.80

$CH_3\text{-}(C \equiv C)_2$ —(thiophene)— $C \equiv C\text{-}CH = CH_2$

2.02

3[3]

6.14

uv: 357, 332, 253

6.52 (dq, 2H) } thiophene H
6.75–7.0 (m, 3H)

5.3–6.0

CH_3 —(thiophene)—(thiophene)— $C \equiv C\text{-}CH = CH_2$

2.41

4[4]
uv: 349, 250

Many thiophene derivatives, e.g. **1–4**, which are formally adducts of H_2S with conjugated bisacetylene units, have been found in certain species of plants. Co-occurrence of these compounds with tridecapentayne-ene **7** (in minor quantities) is usual.

A chemical analogy to the formation of the thiophene ring from a polyacetylene chain has been reported (**5 → 6**).[5]

$Me\text{-}(C \equiv C)_4\text{-}Me$ $\xrightarrow{Na_2S}$ Me —(thiophene)— $(C \equiv C)_2\text{-}Me$

5 **6**

Remarks

The biogenesis of **1–4**, **8** and **9** from the precursor hydrocarbon **7** was proposed, and was partially confirmed by biosynthetic experiments.[2,6]

$Me\text{-}(C \equiv C)_5\text{-}CH = CH_2$

7

1 → 2

3 → 4

[O]

—(thiophene)—(thiophene)— $C \equiv C\text{-}CH = CH_2$ **8**

—(thiophene)—(thiophene)—(thiophene)

9

1) F. Bohlmann, K-M, Kleine, C. Arndt, *Chem. Ber.*, **97**, 2125 (1964).
2) F. Bohlmann, C. Zdero, *Chem. Ber.*, **105**, 1245 (1972).
3) F. Bohlmann, C. Arndt, K-M. Kleine, H. Bornowski, *Chem. Ber.*, **98**, 155 (1965).
4) F. Bohlmann, E. Berger, *Chem. Ber.*, **98**, 883 (1965); F. Bohlmann, C. Zdero, *Chem. Ber.*, **104**, 958 (1971).
5) K. E. Schulte, J. Reisch, L. Horner, *Chem. Ber.*, **95**, 1943 (1962).
6) F. Bohlmann, M. Wotschokowsky, U. Hinz, W. Lucas, *Chem. Ber.*, **99**, 984 (1966).

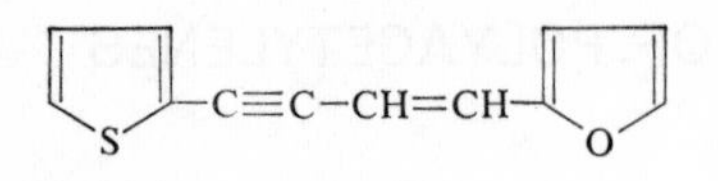

10[7)] 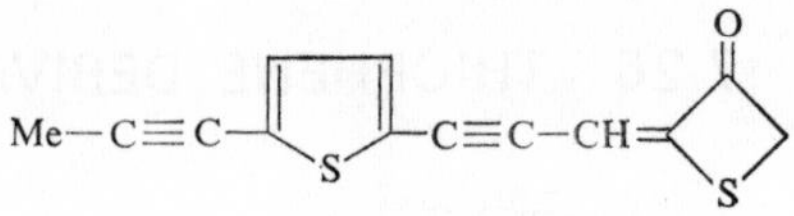   **11**[8)]

The thiophene derivatives **10** and **11**, containing furan and thietanone moieties, respectively, have also been isolated.

Synthesis

Acetylenic thiophene derivatives can be synthesized by a route involving the condensation of an iodothiophene with a copper salt of an acetylene. Two examples of this type of synthesis are shown below.

Synthesis of 3[9)]

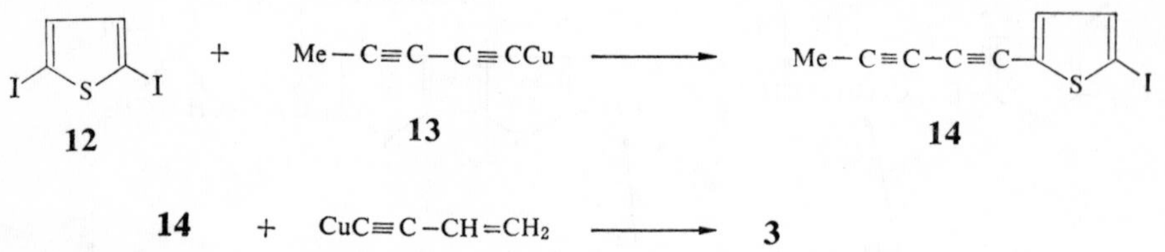

12   + $Me-C\equiv C-C\equiv CCu$ $\longrightarrow$ $Me-C\equiv C-C\equiv C-$ **14**

    13

14   + $CuC\equiv C-CH=CH_2$ $\longrightarrow$ **3**

Synthesis of 2[10)]

12   + $CuC\equiv C-CH_2-OTHP$ $\longrightarrow$ $-C\equiv C-CH_2-OTHP$

   15   **16**

16 + $CuC\equiv C-$ **17** $\xrightarrow{H^+}$ $HOCH_2C\equiv C-$ **18** $-C\equiv C-$ $\xrightarrow[\text{ii) LAH}]{\text{i) TsCl/py}}$ **2**

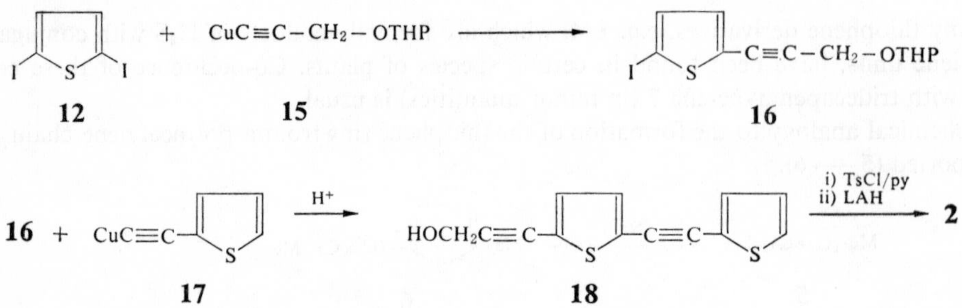

7) F. Bohlmann, C. Zdero, *Chem. Ber.*, **106**, 845 (1973); and references therein.
8) F. Bohlmann, W. Skuballa, *Chem. Ber.*, **106**, 497 (1973).
9) F. Bohlmann, P. Blaskiewicz, E. Bresinsky, *Chem. Ber.*, **101**, 4163 (1968).
10) F. Bohlmann, C. Zdero, H. Kapteyn, *Chem. Ber.*, **106**, 2755 (1973).

7.27   NATURALLY OCCURING CUMULENES

7.06 }
7.03 }

1.86 (dd, 6.6, 1.1)

CH₃

H   H

O

1[1]

uv: 380 (ε 24200), 264 (ε 5000)
ir: 1795, 1760, 1100, 1065, 880, 2098, 2045,
      1960 (–CH=C=C=C)
      1630, 730 (*cis* double bond)

1.00 (t, 7)   7.32   6.18

2.28

H

1.55

5.92 (t, 7)

2[2]

ms: 162.068 (M⁺ $C_{10}H_{10}O_2$)
uv: 337

1 $\xrightarrow[\text{ii) } h\nu]{\text{i) } H_2/\text{Lindlar}}$ Me–CH=CH–CH=CH–CH= O ⟵ Me–CH=CH–C≡C–CH= O

3   **4**

1) Catalytic hydrogenation of **1** afforded γ-hexylbutyrolactone, absorbing 5 moles of hydrogen.
2) Partial hydrogenation over Lindlar catalyst followed by uv irradiation yielded all-*trans* **3**. The same compound was obtained from the known matricarialactone **4**.
3) The *cis* configuration of the double bond in **1** was indicated by the ir spectrum and the nmr coupling constant.

Remarks

Compound **1**, isolated from *Conyza bonariensis*, was the first naturally occurring cumulene to be isolated. **1** was also found in several species of *Astereae*, which also produce matricaria-ester and matricarialactone **4**. The cumulene **2** was isolated from *Erigeron canadensis*.

The biogenesis of **1** was proposed to be as follows.

H⁺   H

Me–CH=CH–C≡C–C≡C–C   H
                            C
                     O=C
                        ‖
                        O

5   ⟶   **1**

Other examples of natural cumulenes containing a thioether linkage, e.g. **6** and **7**, were obtained from *Anthemis austriaca*. Biosynthetic studies of the cumulene **7** have been reported.[5]

H H

Me–C=C=C=CH–C≡C–C=C–COOMe
   |
   SMe

6[3]

Me–C=C=C=CH–
   |
   SMe   O   O

7[4]

1) F. Bohlmann, H. Bornowski, C. Arndt, *Chem. Ber.*, **98**, 2236 (1965).
2) F. Bohlmann, C. Zdero, *Tetr. Lett.*, 2465 (1970).
3) F. Bohlmann, C. Zdero, *Chem. Ber.*, **104**, 1329 (1971).
4) F. Bohlmann, K. M. Klein, C. Arndt, *Ann. Chem.*, **694**, 149 (1966).
5) F. Bohlmann, P-D. Hopf, *Chem. Ber.*, **106**, 3772 (1973).

7.28   POLYACETYLENES WITH CYCLIC ENOL ETHER SYSTEMS

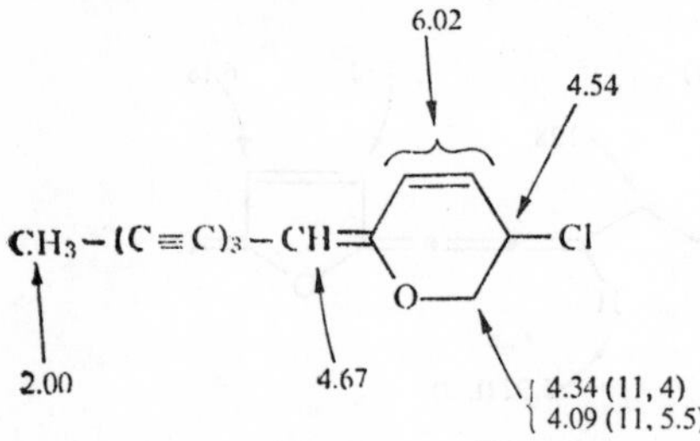

1[1]

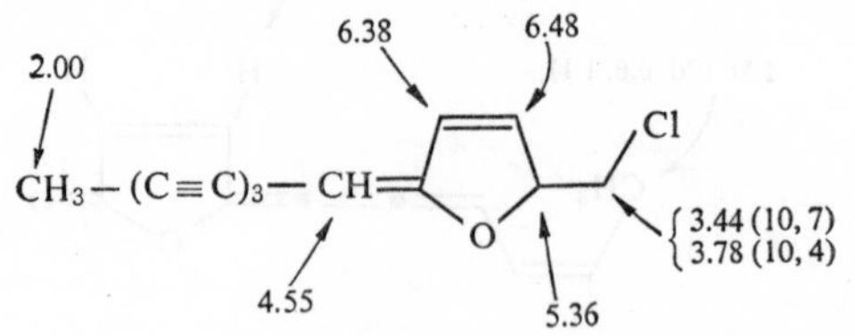

2[1]

Chlorine-containing polyacetylenic compounds **1** and **2** were synthesized via the common intermediate **5** by the following route.[2]

Synthesis[2]

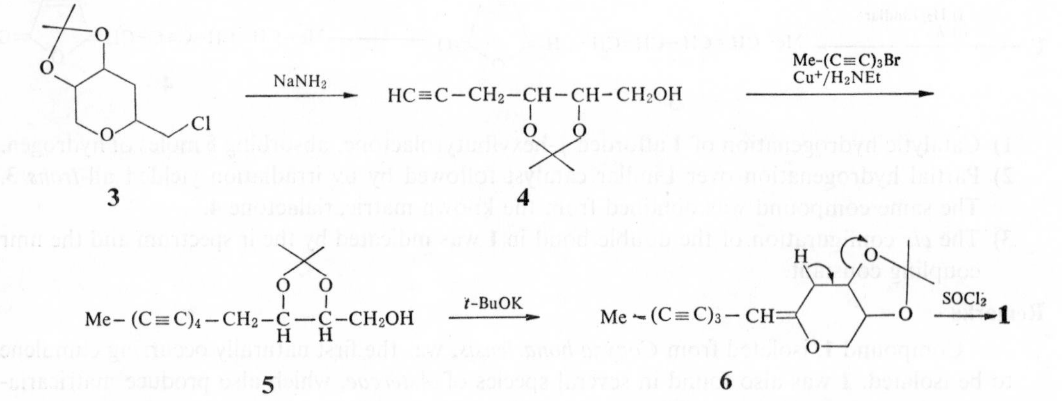

Remarks

Compounds **1** and **2** were isolated from various members of *Anaphalis* and *Gnaphalium*.[1] Polyacetylenic compounds containing oxygen heterocyclic rings such as γ- and δ-lactones, furan or tetrahydropyranyl are also known.

1) F. Bohlmann, C. Arndt, C. Zdero, *Chem. Ber.*, **99**, 1648 (1966); F. Bohlmann, C. Arndt, *ibid.*, **98**, 1416 (1965).
2) F. Bohlmann, R. Weber, *Chem. Ber.*, **105**, 3036 (1972).

7.29 POLYACETYLENES WITH SPIROKETAL LINKAGES

uv (*cis*-1): 317 (ε 18200), 263 (ε 4400), 248 (ε 4700), 235.5 (ε 10200)
ir: 2190, 2150 (triple bond), 1640, 1590 (enol ether)

uv (*cis*-2): 317, 263, 248, 235, 225, 217
uv (*trans*-2): 319, 310, 268, 253, 237, 217

1) **1** and **2** both exist in *cis* and *trans* forms with respect to the enol ether linkages. Irradiation of these compounds causes *cis-trans* isomerization of the enol double bond.

Synthesis

Compound **1** has been synthesized via the keto alcohol **9**, which was considered to be a biosynthetic intermediate. **2** was synthesized by an analogous method (Route II) via **10**.[2] Biosynthetic pathway along this sequence was partly verified by feeding experiments.[3]

Route I

Route II

1) F. Bohlmann, P. Herbst, C. Arndt, H. Schönowsky, H. Gleinig, *Chem. Ber.*, **94**. 3193 (1961); F. Bohlmann, P. Herbst, I. Dohrmann, *ibid.*, **96**, 226 (1963).
2) F. Bohlmann, G. Florentz, *Chem. Ber.*, **99**, 990 (1966).
3) F. Bohlmann, H. Schulz, *Tetr. Lett.*, 1801, 4795 (1968); F. Bohlmann, R. Jente, W. Lucas, J. Laser, H. Schulz, *Chem. Ber.*, **100**, 3183 (1967).

7.30 BIOSYNTHESIS OF PHENYL ACETYLENES

A group of acetylenic compounds containing aromatic rings, e.g. frutescin **1**, capillarin **2** and their analogues, are known. The biosynthetic pathway to these substances has been studied by precursor-incorporation experiments.

frutescin **1**[1)]   capillarin **2**[2)]

The distribution of radioactive carbons in the phenyl moiety when [1-^{14}C]- and [2-^{14}C]-acetate were used as substrates suggested the intermediacy of straight-chain fatty acids in the biosynthesis of such compounds.[3)]

The mechanism of cyclization of the polyacetylenic chain was proposed to be as follows, and this was supported by feeding experiments with labeled substrates.[4)]

1) Incorporation of the labeled substrate **4** in **1, 2,** and **3** was shown by feeding experiments using *Chrysanthemum frutescens.*

2) When labeled **6** was fed to *Coreopsis lanceolata*, it was shown to be incorporated into **8.**

Remarks

Frutescin **1** was isolated from *C. frutescens* and capillarin **2** from *Artemisia capillaris, A. dracunculus* and others.

1) F. Bohlmann, K-M, Kleine, *Chem. Ber.,* **95**, 602 (1962).
2) F. Bohlmann, K-M, Kleine, *Chem. Ber.,* **95**, 39 (1962).
3) J. R. F. Fairbrother, E. R. H. Jones, V. Thaller, *J. Chem. Soc.,* 1035 (1967).
4) F. Bohlmann, R. Jente, W. Lucas, J. Laser, H. Schulz, *Chem. Ber.,* **100**, 3183 (1967).

7.31   SYNTHESIS OF EXALTONE AND MUSCONE

exaltone **1**

muscone **2**

The macrocyclic odoriferous substances exaltone **1** and muscone **2** were prepared by acyloin-type condensation of long-chain aliphatic esters,[1] until the important synthetic intermediate cyclododecanone **3** became available recently in large amounts and at low cost. Cyclododecanone **3** became available via cyclododecatriene, which was obtained from butadiene by cyclotrimerization.[2]

Route I[3,4]

The first route involves the addition of a three-carbon unit to a 12-membered ring ketone by Stobbe condensation followed by ring closure.[3] The bond cleavage of the resulting bicyclo-[10,3,0]pentadecane system was achieved either by ozonolysis (**6 → 7**) or fragmentation of the 1,3-diol monotosylate **10**.[4]

1) M. Stoll, J. Hulstkamp, *Helv. Chim. Acta*, **30**, 1815 (1947); M. Stoll, A. Rouve, *ibid.*, **30**, 1822 (1947); M. Stoll, *ibid.*, **30**, 1837 (1947); V. Prelog, L. Frenkiel, M. Kobelt, P. Barman, *ibid.*, **30**, 1741 (1947).
2) G. Wilke, *Angew. Chem.*, **69**, 397 (1957); *ibid.*, **75**, 10 (1963).
3) K. Biemann, G. Büchi, B. H. Walker, *J. Am. Chem. Soc.*, **79**, 5558 (1957).
4) G. Ohloff, J. Becker, K. H. Schulte-Elte, *Helv. Chim. Acta*, **50**, 705 (1967).

Route II[5,6)]

1 and 2 were also synthesized from the intermediates 5 and 16, respectively, by a route involving Eschenmoser fragmentation as the key step.

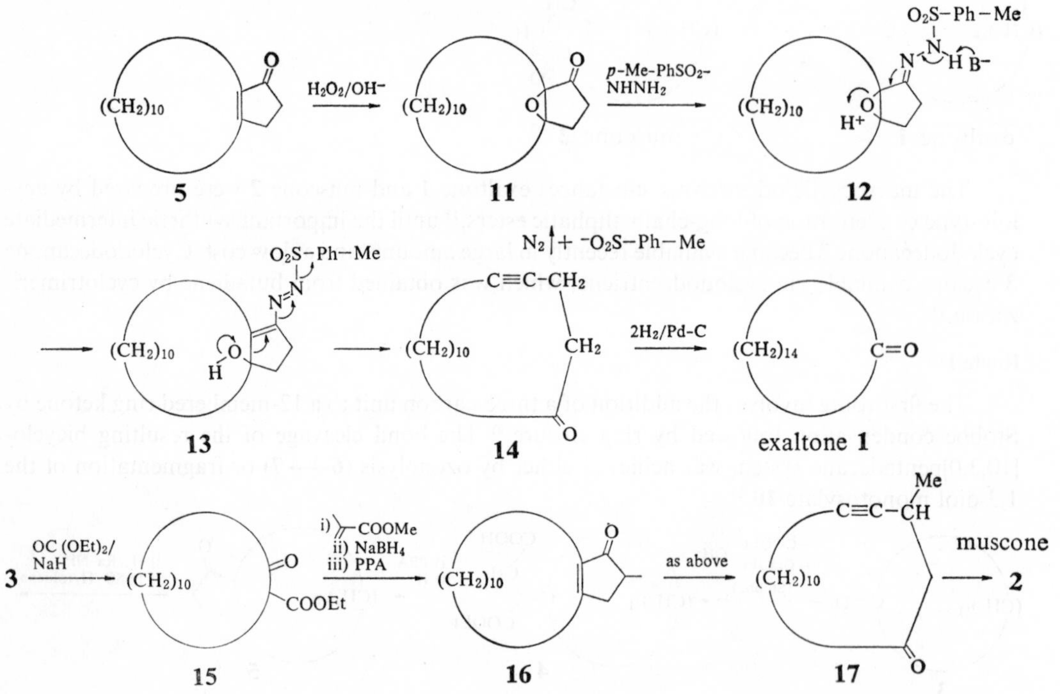

11 ⟶ 14: The tosylhydrazone of the epoxyketone 11 was converted into the acetylenic ketone 14 by a novel type of fragmentation via 12 and 13. The generality and wide utility of this type of reaction have now been recognized.

Route III[7)]

Photolysis of the bicyclic ketones 18 and 22 afforded the macrocyclic esters 20 and 24, respectively, through the ketenic intermediates 19 and 23. 1 and 2 were derived from 20 and 24, respectively, in several steps.

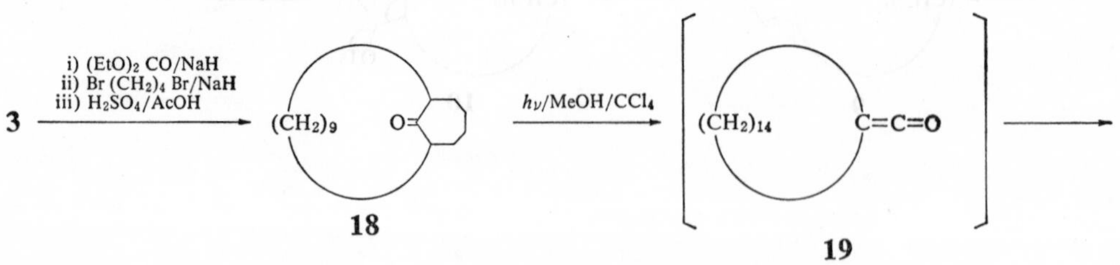

5) A. Eschenmoser, D. Felix, G. Ohloff, *Helv. Chim. Acta*, **50**, 708 (1967).
6) D. Felix, J. Schreiber, G. Ohloff, A. Eschenmoser, *ibid.*, **54**, 2896 (1971).
7) H. Nozaki, T. Mori, R. Noyori, *Tetr. Lett.*, 779 (1967); H. Nozaki, H. Yamamoto, T. Mori, *Can. J. Chem.*, **47**, 1107 (1969).

$$\longrightarrow \quad (CH_2)_{14} \quad CH-COOMe \quad \xrightarrow{\substack{\text{i) OH}^- \\ \text{ii) Hg(OAc)}_2/\text{benzene}/\text{I}_2 \\ \text{iii) Hg(OAc)}_2/\text{HOAc then OH}^- \\ \text{iv) Na}_2\text{Cr}_2\text{O}_7}} \quad \text{exaltone } \mathbf{1}$$

20

19 ⟶ 1: Oxidation of the ketene **19** generated by photolysis of **18** in hexane solution with oxygen afforded **1** directly.

$$(CH_2)_{12} \quad C=O \quad \xrightarrow[\text{ii) H}_2/\text{Pd}]{\text{i) } \substack{Cl \\ Cl} =/\text{NaH}} \quad (CH_2)_{10} \quad O= \quad \xrightarrow{h\nu/\text{MeOH}/\text{CCl}_4} \quad \left[(CH_2)_{12} \quad \substack{Me \\ C \\ \| \\ C=O} \right]$$

21   **22**   **23**

$$\longrightarrow \quad (CH_2)_{12} \quad \substack{Me \\ CH \\ | \\ CH \\ COOMe} \quad \xrightarrow{\text{as above}} \quad \text{muscone } \mathbf{2}$$

24

Route IV[8]

The dodecatrienylnickel complex **25** was converted to the bis-π-allylnickel complex **26**, absorbing one mole equivalent of allene. **26** was reacted with carbon monoxide, giving rise to the cyclic hydrocarbons **27** and **28** accompanied by the macrocyclic ketone **29** (4–5%). Muscone **2** was derived from **29** on hydrogenation.

25   **26**   **27**   **28**   **29**

$$\mathbf{29} \xrightarrow{\text{H}_2/\text{Pd-C}} \text{muscone } \mathbf{2}$$

Remarks

Muscone **2** is an odoriferous principle of natural musk from the musk deer, *Moschus moschiferus*. The analogous compound, civetone (9-cycloheptadecen-1-one) was obtained from the secretion of the civet cat. Exaltone **1** is a synthetic material possessing an intense musk odor.

8) R. Baker, B. N. Blackett, R. C. Cookson, *Chem. Commun.*, 802 (1972).

7.32 STRUCTURE OF RUBRATOXINS

rubratoxin-A **1**[1]

1) Rubratoxin-A was shown to be a dihydro derivative of the naturally occuring bisanhydride rubratoxin-B **2**. Oxidation of **1** affords rubratoxin-B **2** and 20-oxo-rubratoxin-B. The latter was also obtained from **2** by oxidation.[1,2] The structure of **2** was established by x-ray crystallographic analysis.[2]

mp: 204–206° α_D: 87°
ms: 458.193 ($[M-COO-H_2O]^+$ for $C_{25}H_{30}O_8$)
uv: 204 (ε 31900), 225 (infl.), 252 (ε 4430)
ir: 1852, 1815, 1770, 1728, 1712, 1700

Remarks

1 and **2** are toxic metabolites isolated from *Penicillium rubrum*,[1-3] which is responsible for fatal poisoning of livestock and poultry fed infected cereals. The name "nonadrides" has been proposed for compounds such as compounds **1-7**, containing cyclononadiene fused with bis-anhydride moieties.[4] Other nonadrides so far known are glauconic acid **3**[5] and glaucanic acid **4**[5] isolated from *Penicillium glaucum*, byssochlamic acid **5** from *Byssochlamys fulva*[6] and scytalidin **6** from *Scytalidium* species.[7] Heveadride 7, an isomer of byssochlamic acid **5** has been isolated from *Helminthosporium heveae*.[8]

glauconic acid **3** (R=OH) byssochlamic acid **5**   scytalidin **6**   heveadride **7**
glaucanic acid **4** (R=H)

1) M. O. Moss, F. V. Robinson, A. B. Wood, *J. Chem. Soc.*, C, 619 (1971); and references therein.
2) G. Büchi, K. M. Snader, J. D. White, J. Z. Gougoutas, S. Singh, *J. Am. Chem. Soc.*, **92**, 6638 (1970).
3) S. Natori, S. Sakaki, H. Kurata, S. Udagawa, M. Ichinoe, M. Saito, M. Umeda, K. Ohtsuka, *Appl. Microbiol.*, **19**, 613 (1970).
4) D. H. R. Barton, J. K. Sutherland, *J. Chem. Soc.*, 1769 (1965).
5) D. H. R. Barton, L. M. Jackman, L. R. Hahn, J. K. Sutherland, *J. Chem. Soc.*, 1772 (1965); L. D. S. Godinho, J. K. Sutherland, *ibid*, 1779 (1965).
6) J. E. Baldwin, D. H. R. Barton, J. K. Sutherland, *J. Chem. Soc.*, 1787 (1965).
7) G. M. Strunz, M. Kakushima, M. A. Stillwell, *J. Chem. Soc.* P-1, 2280, (1972).
8) R. I. Crane, P. Hedden, J. MacMillan, W. B. Turner, *J. Chem. Soc.* C, 194 (1973).

7.33 REARRANGEMENT OF GLAUCONIC ACID

Glauconic acid **1**, a cyclononadiene with bis-anhydride functions, undergoes a variety of skeletal rearrangements under pyrolytic conditions.

Rearrangement of glauconic acid[1]

Cope rearrangement, followed by two successive fragmentations, yields **5** and **6**.

Rearrangement of 1-acetate[2]

The acetate and tetrahydropyranyl derivative of **1** rearrange to **8** and **11**, respectively, under pyrolytic conditions.

1) D. H. R. Barton, J. K. Sutherland, *J. Chem. Soc.*, 1769 (1965).
2) D. H. R. Barton, L. M. Jackman, L. R-Hahn, J. K. Sutherland, *J. Chem. Soc.*, 1772 (1965).

7.34 SYNTHESIS OF BYSSOCHLAMIC ACID

The first total synthesis of nonadride has been accomplished by the route shown below. Unreactive aromatic rings which serve as latent anhydride moieties are used in this synthesis.[1)]

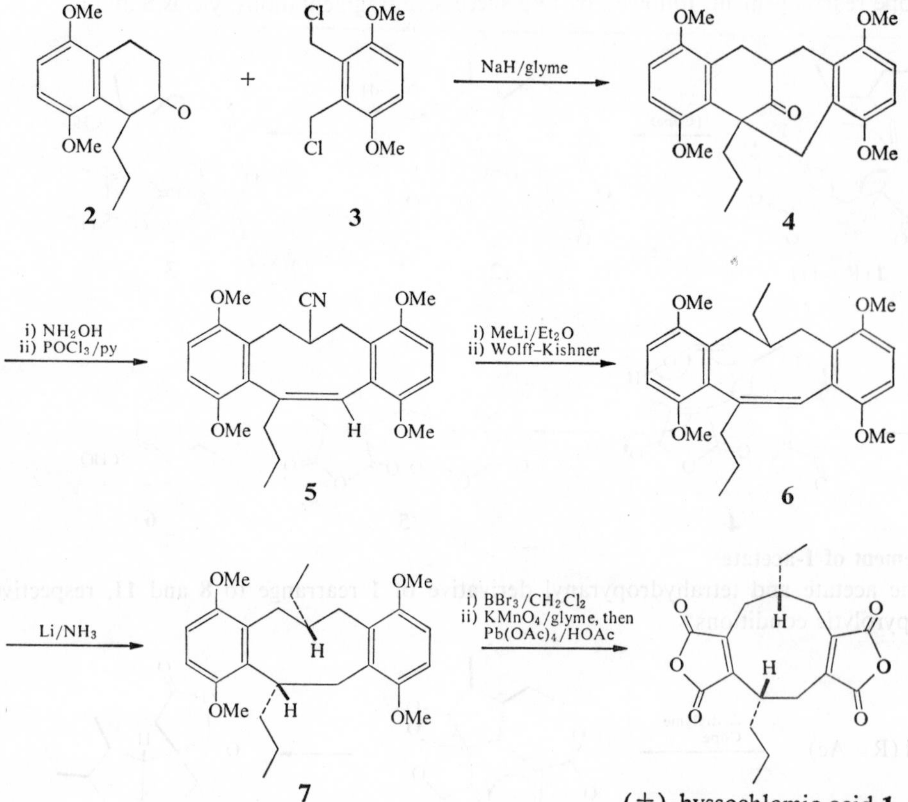

(±)-byssochlamic acid **1**

2,3⟶4: Cycloalkylation of **2** and **3** affords the bicyclo[4,3,1]decane system in one step.

6⟶7: Reduction of **6** yields **7** stereospecifically, with *cis* oriented alkyl substituents in a nine-membered ring. This can be explained in terms of the favorable conformation of **6** and its transition state.

1) G. Stork, J. M. Tabak, J. F. Blount, *J. Am. Chem. Soc.*, **94**, 4735 (1972).

7.35 BIOSYNTHESIS OF GLAUCONIC ACID

Glauconic acid **1**, a member of the nonadrides (cyclononadienes with bis-anhydride functions) is considered to be formed by the coupling of two C_9 units with identical carbon skeletons. This view is supported by the experimental evidence described below.

glauconic acid **1** (R=OH)
glaucanic acid **2** (R=H)

1) The ^3H-labeled compound **3** was shown to be incorporated into glauconic acid **1** by feeding experiments using *Penicillium purpurogenum*. Equal distribution of the label at C-4 and C-13 in **1** was shown by degradative work.[1]

2) The ^{14}C-labeled substrate **4** was incorporated into **1** more efficiently (51.5%). The labeled atoms were shown to be located at C-7 and C-16.[3]

3) Labeled glaucanic acid **2** was incorporated into **1**, indicating that **2** is a precursor of **1**.

4) Treatment of **4** with triethylamine (or NaH) in DMF afforded **5**, a C-12 epimer of glaucanic acid **2**, in 4% yield. This provides a laboratory analogy of the biosynthesis of **2**. For this dimerization, an attractive explanation involving [6+4]cycloaddition of the anion **4′** with the cisoid anhydride **4** has been proposed.[2,3]

Remarks

Byssochlamic acid **7** would arise from a similar coupling of a C_9 unit as shown.

1) J. L. Bloomer, C. E. Moppett, J. K. Sutherland, *Chem. Commun.*, 619 (1965); C. E. Moppett, J. K. Sutherland, *ibid.*, 772 (1966); J. L. Bloomer, C. E. Moppett, J. K. Sutherland, *J. Chem. Soc.*, C, 588 (1968)
2) R. K. Huff, C. E. Moppett, J. K. Sutherland, *Chem. Commun.*, 1192 (1968).
3) R. K. Huff, C. E. Moppett, J. K. Sutherland, *J. Chem. Soc*, P-1, 2584 (1972).

7.36   STRUCTURE AND SYNTHESIS OF PYRENOPHORIN

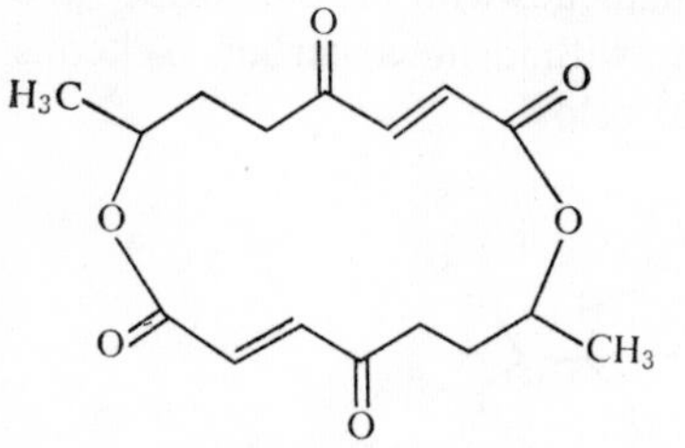

pyrenophorin **1**[1]

mp: 175° α_D: −50°
ms: 308 (M+ $C_{16}H_{20}O_6$)
uv: 220 (ε 23200)

1) The presence of a γ-keto-α,β-unsaturated lactone system is indicated by the ir, uv and nmr spectra.

2) The nmr coupling constant ($J = 15.5$ Hz) indicates that the double bond is *trans*.

3) Hydrolysis followed by oxidation yielded 4,7-dioxooctanoic acid **4**.[2]

4) Since the molecular weight is 308, **1** must be the dilactone of a C_8 hydroxy acid.

5) Compound **7**, obtained from *Stemphyllium radicinum*, might be an artifact derived from **1** by intramolecular Michael reaction.[3]

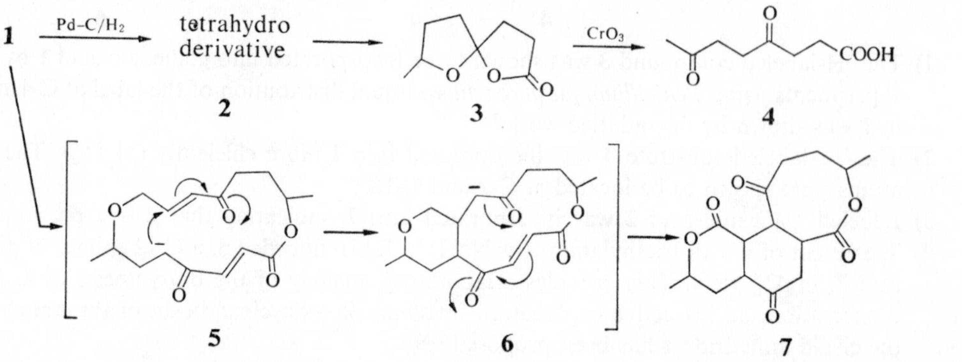

Synthesis[4]

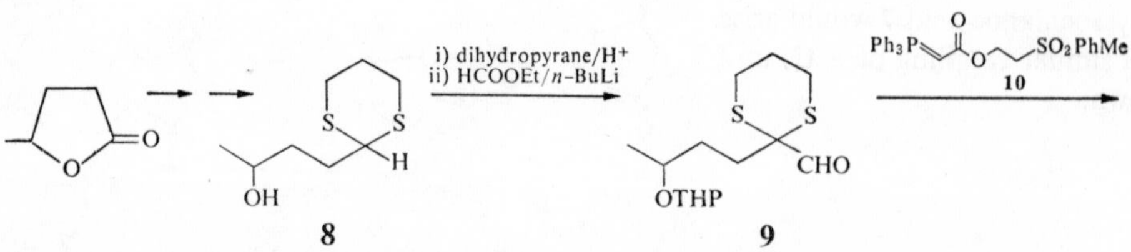

1) S. Nozoe, K. Hirai, K. Tsuda, K. Ishibashi, M. Shirasaka, J. F. Grove, *Tetr. Lett.*, 4675 (1965).
2) D. C. Aldridge, J. F. Grove, *J. Chem. Soc.*, 3239 (1964).
3) J. F. Grove, *J. Chem. Soc.*, C, 2261 (1971).
4) E. W. Colvin, T. A. Purcell, R. A. Raphael, *Chem. Commun.*, 1031 (1972).

i) H+
ii) BrAcX
iii) PPh₃/NaOH

11 → **12**

9 →

13

i) H+
ii) 1,5-diazabicyclo [4,3,0]non-5-ene
iii) bisimidazol-1-yl ketone
iv) 1,5-diazabicyclo [4,3,0]non-5-ene
v) NCS/AgNO₃

→ **1**

9,10 ⟶ 11; 9,12 ⟶ 13: These Wittig condensations gave only the *trans* isomers **11** and **13**, respectively.

13 ⟶ 1: The *p*-toluenesulfonylethyl protecting group can be selectively removed under mild conditions, and the imidazole derived from the hydroxy acid is smoothly cyclized to a 16-membered dilactone by a catalytic amount of base. The product is a 1 : 1 mixture of the diastereoisomers, the less polar of which was identical with naturally occurring **1**.[4]

Remarks

Pyrenophorin is an antifungal metabolite isolated from the plant pathogenic fungi *Pyrenophora avenae*[1] and *Stemphyllium radicinum*.[2,3] Pyrenophorol, the tetrahydro derivative of **1**, was obtained from *Byssochlamys nivea*.[5]

The macrocyclic dilactone colletodiol **14**, which is structurally and biogenetically related to pyrenophorin, has been isolated from culture filtrates of *Colletotrichum capsici*.[6-8] Colletoketol (10-deoxy analogue of **14**), colletol (11-deoxycolletodiol), and colletallol (10-deoxycolletodiol) have also been obtained from the same source.[9] The absolute configuration of **1** was assigned as (2R, 8R, 10R, 11S).[9] Vermiculine, a 16-membered dilactone was obtained from *Penicillium vermiculatum*.[10]

colletodiol **14**

5) Z. Kis, P. Furger, H. P. Sigg, *Experientia*, **25**, 123 (1969).
6) J. MacMillan, R. J. Pryce, *Tetr. Lett.*, 5497 (1968).
7) J. W. Powell, W. B. Whalley, *J. Chem. Soc.* C, 911 (1969).
8) J. F. Grove, R. N. Speake, G. Ward, *J. Chem. Soc.* C, 230 (1966).
9) J. MacMillan, T. J. Simpson, *J. Chem. Soc.* P-1, 1487 (1973).
10) R. K. Boeckman Jr., J. Fayos, J. Clardy, *J. Am. Chem. Soc.*, **96**, 5954 (1974).

7.37   STRUCTURE OF TETRANACTIN

tetranactin 1[1]

mp: 105–106° α_D(CHCl$_3$): 0°
ms: 792 (M$^+$ C$_{44}$H$_{72}$O$_{12}$); peaks at 199, 383, 397, 595
nmr in CDCl$_3$

1) Alkaline hydrolysis of **1** afforded homononactic acid **2** as the sole product, indicating that **1** consists of four units of homononactic acid **2** linked to form a cyclic polyester.

2) X-ray studies indicated that the tetranactin molecule is either a racemic or meso form.

homononactic acid **2**

Remarks

Tetranactin **1** is a miticidal antibiotic isolated from *Streptomyces aureus*. Analogous compounds, nonactin,[2] monactin, dinactin, trinactin,[3] and higher analogs[4] were also obtained from *Streptomyces* sp. All these compounds, designated as macrotetrolides, are macrocyclic tetralactones consisting of any combination of four of nonactic acid, homononactic acid and bishomononactic acid.

1) K. Ando., Y. Murakami, Y. Nawata, *J. Antibiotics*, **24**, 418 (1971).
2) R. Corbaz, L. Ettlinger, E. Gäumann, W. Keller-Schierlein, F. Kradolfer, L. Neipp, V. Prelog, H. Zähner, *Helv. Chim. Acta*, **38**, 1445 (1955); J. Dominguez, J. D. Dunitz, H. Gerlach, V. Prelog, *ibid.*, **45**, 129 (1962).
3) H. Beck, H. Gerlach, V. Prelog, W. Voser, *Helv. Chim. Acta*, **45**, 620 (1962).
4) H. Gerlach, R. Hütter, W. Keller-Schierlein, J. Seibl, H. Zähner, *Helv. Chim. Acta*, **50**, 1782 (1967).

7.38 STRUCTURES OF THE POLYETHER ANTIBIOTICS

A number of antibiotics containing several cyclic ether linkages in the molecule have recently been isolated from *Streptomyces* species. Their structures were elucidated by x-ray crystallographic analysis of heavy metal salts of the compounds.

These compounds are unusual in that their alkali salts are soluble in organic solvents, but not in water. Crystallographic studies revealed that these compounds have a cyclic conformation such that the metal ion is "wrapped up" by the polyetherial oxygen atoms.[1]

monensic acid **1**[2]

grisorixin **2** (R=H)[3]
nigericin **3** (R=OH)[4]

antibiotic X–206 **4**[5]

antibiotic X–537A **5**[6]

1) For example, see L. K. Steinrauf, E. W. Czerwinski, M. Pinkerton, *Biochem. Biophys. Res. Commun.*, **45**, 1279 (1971); cf. W. K. Lutz, H. K. Wipf, W. Simon, *Helv. Chim. Acta*, **17**, 1741 (1970).
2) A. Agtarap, J. W. Chamberlin, M. Pinkerton, L. K. Steinrauf, *J. Am. Chem. Soc.*, **89**, 5737 (1967).
3) P. Gachon, A. Kergomard, H. Veschambre, C. Esteve, T. Staron, *Chem. Commun.*, 1421 (1970); M. Alleaume, D. Hickel, *ibid.*, 1422 (1970).
4) T. Kubota, S. Matsutani, *J. Chem. Soc.*, C, 695 (1970); L. K. Steinrauf, M. Pinkerton, J. W. Chamberlin, *Biochem. Biophys. Res. Commun.*, **33**, 29 (1968).
5) J. F. Blount, J. W. Westley, *Chem. Commun.*, 927 (1971).
6) J. W. Westley, R. H. Evans Jr., T. Williams, A. Stempel, *Chem. Commun.*, 71 (1970); S. M. Johnson, S. J. Liu, J. Henrick, I. C. Paul, *ibid.*, 72 (1970); P. G. Schmidt, A. H.-J. Wang, I. C. Paul, *J. Am. Chem. Soc.*, **96**, 6189 (1974).

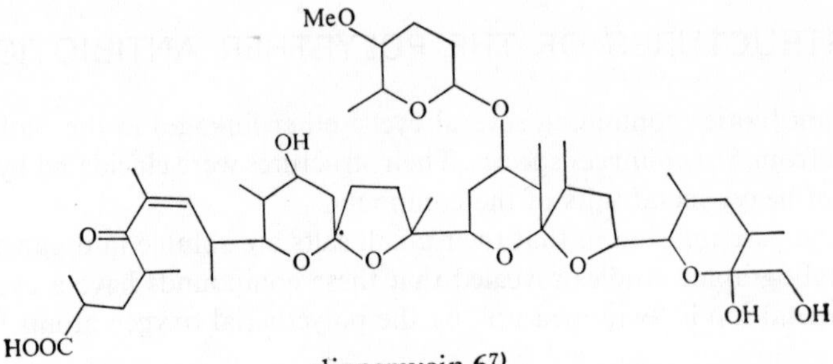

dianemycin 6[7]

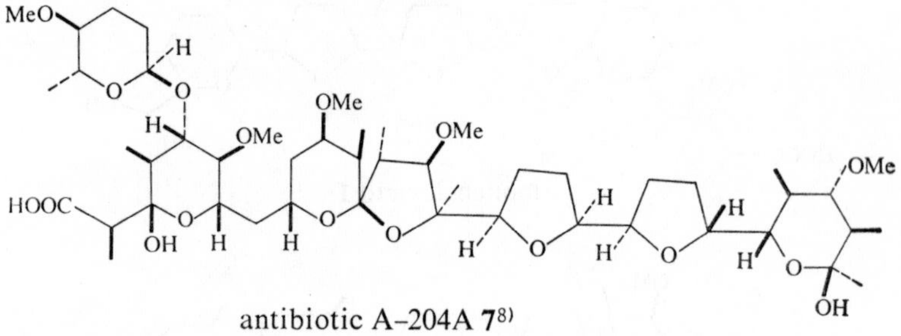

antibiotic A–204A 7[8]

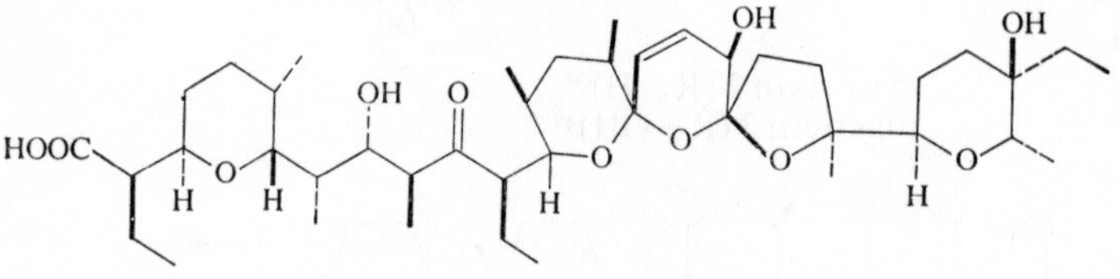

salinomycin 8[9]

Remarks

The polyether antibiotics show broad-spectrum antimicrobial activity against Gram-positive bacteria, mycobacteria and fungi, and also show anticoccidial activity. An incorporation study[11] using ^{14}C-labeled acetate, propionate and butyrate as well as ^{13}C-nmr analysis[10] of the [1-^{13}C]butyrate-derived compound indicated that the carbon skeleton of the antibiotic X-537A arises from five acetate units (⟋), four propionate units (⌄), and three butyrate units (⋀), as illustrated below.

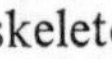

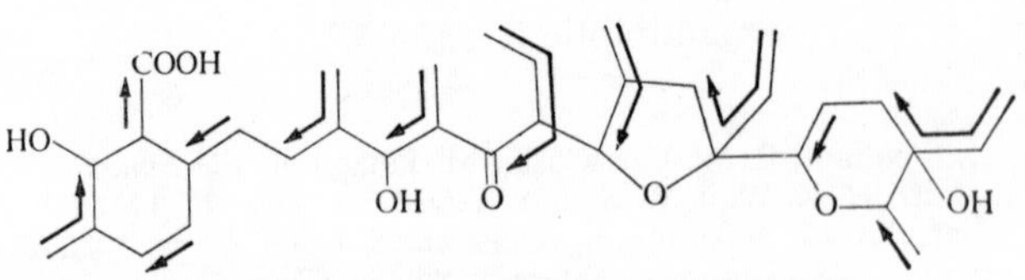

7) E. W. Czerwinski, L. K. Steinrauf, *Biochem. Biophys, Res. Commun.*, **45**, 1284 (1971).
8) N. D. Jones, M. O. Chaney, J. W. Chamberlin, R. L. Hamill, S. Chen, *J. Am. Chem. Soc.*, **95**, 3399 (1973), cf. T. J. Petcher, H. P. Weber, *Chem. Commun.*, 697 (1974).
9) H. Kinashi, N. Otake, H. Yonehara, S. Sato, Y. Saito, *Tetr. Lett.*, 4955 (1973).
10) J. W. Westley, D. L. Pruess, R. G. Pitcher, *Chem. Commun.*, 161 (1972).
11) J. W. Westley, R. H. Evans Jr., D. L. Pruess, A. Stempel, *Chem. Commun.*, 1467 (1970).

7.39 BIOSYNTHESIS OF THE MACROLIDES

In considering the structures of the macrolide antibiotics, it is immediately obvious that these compounds differ from common fatty acid derivatives in containing many branching methyl groups in the aglycone moiety.

Tracer experiments using labeled acetate or propionate have revealed that the building blocks of these compounds are actually derived from acetate and/or propionate units combined in head-to-tail manner.[1-4] The interpolation of propionate units into the chains is characteristic of all of macrolide antibiotics. However, in polyene macrolides with 26- to 38-membered lactones, the ratio of acetate units to propionate units tends to be far larger than in the 12- to 16-membered macrolides.

A few examples are shown below:

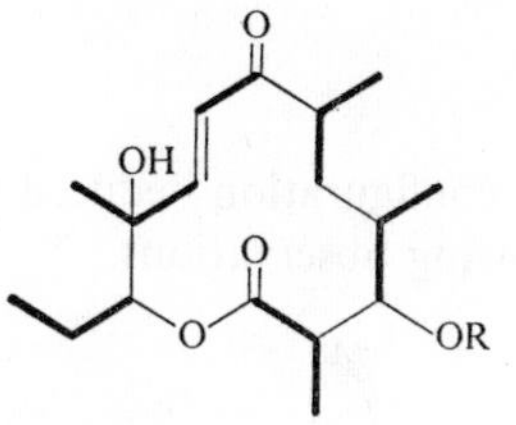

methymycin **1** 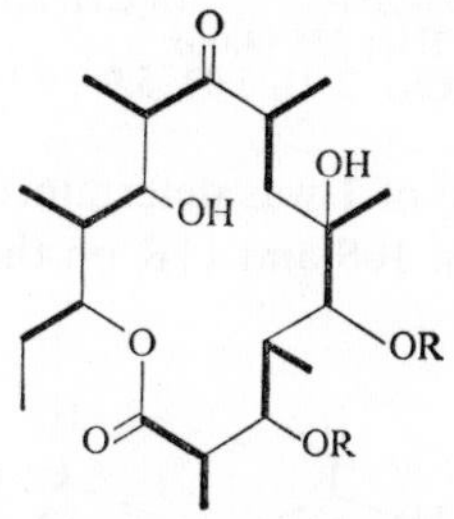   erythromycin B **2**

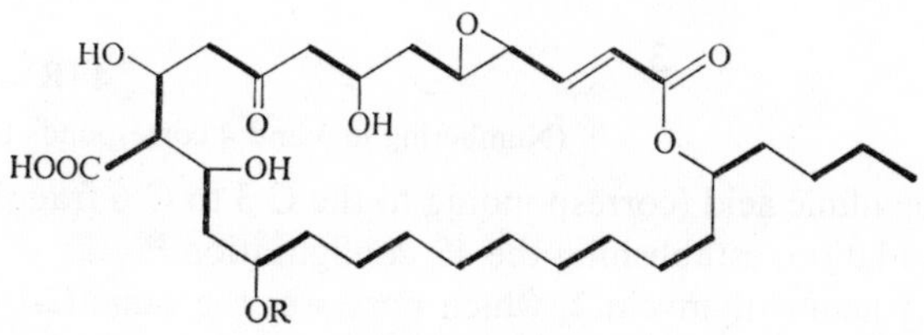

lucensomycin **3**

Methymycin contains five propionyl units and one acetyl unit,[1] erythromycin has seven propionyl units,[2] and lucensomycin has two propionyl units and twelve acetyl units.[3]

1) A. J. Birch, C. Djerassi, J. D. Dutcher, J. Majer, D. Perlman, E. Pride, W. Rickards, P. J. Thomson, *J. Chem. Soc.*, 5274 (1964).
2) T. Kaneda, J. C. Butte, S. B. Taubman, J. W. Corcoran, *J. Biol. Chem.*, **237**, 322 (1962).
3) D. G. Manwaring, R. W. Rickards, G. Gaudiano, V. Nicolella, *J. Antibiotics*, **22**, 545 (1969).
4) For a review, see J. W. Corcoran, M. Chick, *Biochemistry of the Macrolide Antibiotics* (ed. J. F. Snell), p. 159, Academic Press, 1966.

7.40   STRUCTURE OF METHYMYCIN, A 12-MEMBERED MACROLIDE

methymycin **1**

mp: 203–205° α_D(CHCl$_3$): +79°
uv (EtOH): 225 (4.06)
ir (CHCl$_3$): 2.93, 5.82, 5.95, 6.14

The structure of **1** was determined and the absolute configuration deduced as 2R, 3S, 4S, 6R, C-8/C-9 *trans*, 10S and 11R on the basis of the following observations.[1–3]

(Numbering in **3** and **4** corresponds to that of the parent skeleton.)

1) (−)-α-methyllevulinic acid (corresponding to the C-3 to C-6 fragment of **1**) was obtained by oxidative degradation, establishing the 4S configuration.[2]

2) Degradation of neomethymycin **2**, which possesses the same C-1 to C-7 stereochemistry as **1**, afforded *meso-α,α'*-dimethylglutaric acid indicating that 4-Me and 6-Me are *cis*.[2]

3) 3-H and 4-H were shown to be in an antiperiplanar configuration in the lactonic acid **3**, which was obtained from **1** by oxidation, on the basis of the coupling constant ($J_{2,3}=10$ Hz).[1]

4) The relative stereochemistry of 2-H and 3-H in the acetonide **4** was deduced to be *cis* from analysis of the nmr spectrum of **4**.[1]

5) The oxidation of **1** yielded (+)-2R,3R-*erythro*-2,3-dihydroxy-2-methylpentanoic acid, indicating the 10S and 11R configurations of **1**.[1]

6) The *trans* geometry of the double bond was deduced from its coupling constant ($J_{8,9}=16$ Hz).[1]

Remarks

Methymycin **1** is an antibiotic obtained from *Streptomyces venezuelae*. The closely related compound neomethymycin was isolated from the same source.[3]

1) D. G. Manwaring, R. W. Rickards, R. M. Smith, *Tetr. Lett.*, 1029 (1970); R. W. Richards, R. M. Smith, *ibid.*, 1025 (1970).
2) C. Djerassi, O. Halpern, D. I. Wilkinson, E. J. Eisenbraun, *Tetr.* **4**, 369 (1958).
3) C. Djerassi, O. Halpern, *Tetr.* **3**, 255 (1958).

7.41 STRUCTURE OF ERYTHROMYCIN, A 14-MEMBERED MACROLIDE

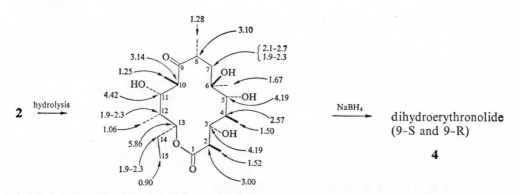

erythromycin-A **1** (R=OH)
erythromycin-B **2** (R=H)

data for **2**:
mp: 191–193°
uv: 285 (ε 31)
ir: 5.84, 5.94
nmr in py-d₅

erythronolide-B **3**

2 →(hydrolysis)

2 →(NaBH₄) dihydroerythronolide (9-S and 9-R)

4

Coupling constants (Hz) for 3[3)]

$J_{2,3}$	10.5	a, a	$J_{11,12}$	10.0	e, e
$J_{3,4}$	1.0	a. e	$J_{12,13}$	1.0	e, a
$J_{4,5}$	2.5	e, a	$J_{13,14}$	9.0	
$J_{7,7}$	14.8	gem.			
$J_{10,11}$	1.0	a, e	$J_{13,14'}$	5.0	

1) For structure elucidation, see K. Gerson *et al.*, *J. Am. Chem. Soc.*, **78**, 6396 (1956); *ibid.*, **79**, 6062, 6070 (1957).
2) D. R. Harris, S. G. McGeachin, H. H. Mills, *Tetr. Lett.*, 679 (1965).
3) T. J. Perun, R. S. Egan, *Tetr. Lett.*, 387 (1969).

Conformation of erythronolide-B

The conformations in solution of the 14-membered lactone ring compounds erythronolide-B **3** and 9(S)- and 9(R)-dihydroerythronolide-B **4** have been determined, based on nmr studies and applying the Karplus equation for the vicinal coupling constants[3] and aromatic solvent-induced shifts.[4-6] In addition to nmr studies, chemical evidence, ir and cd spectra[5] have yielded information regarding the conformation of such compounds in solution.[14]

The preferred conformation of erythronolide-B in solution can be represented by **3a**.[3,5,7] The conformation depicted is essentially identical with that found in the crystalline state.[2]

3a

Remarks

A number of macrolide antibiotics with 14-membered rings are known in addition to the erythromycin series macrolides. These include narbomycin,[8] picromycin,[9] oleandomycin,[10] lankamycin,[11] megalomycin-A,[12] kujimycin-A, albocycline, etc.[13]

4) T. J. Perun, R. S. Egan, J. R. Martin, *Tetr. Lett.*, 4501 (1969).
5) L. A. Mitscher, B. J. Slater, T. J. Perun, P. H. Jones, J. R. Martin, *Tetr. Lett.*, 4505 (1969); R. S. Egan, T. J. Perun, J. R. Martin, L. A. Mitscher, *Tetr*, **29**, 2525 (1973).
6) P. V. Demarco, *Tetr. Lett.*, 383 (1969); *J. Antibiotics*, **22**, 327 (1969).
7) W. D. Celmer, *Antimicrobial Agents and Chemotherapy*, 144 (1965).
8) V. Prelog, A. M. Gold, G. Talbot, A. Zamojski, *Helv. Chim. Acta*, **45**, 4 (1962).
9) R. W. Rickards, R. M. Smith, J. Majer, *Chem. Commun.*, 1049 (1968); H. Muxfeldt, S. Shrader, P. Hansemn, H. Brockmann, *J. Am. Chem. Soc.*, **90**, 4748 (1968); R. E. Hughes, H. Muxfeldt, C. Tsai, J. J. Stezowski, *ibid.*, **92**, 5267 (1970).
10) W. D. Celmer, *J. Am. Chem. Soc.*, **87**, 1797 (1965).
11) R. S. Egan, J. R. Martin, *J. Am. Chem. Soc.*, **92**, 4129 (1970).
12) A. K. Mallans, R. S. Jaret, H. Reimann, *J. Am. Chem. Soc.*, **91**, 7506 (1969); R. S. Jaret, A. K. Mallams, *J. Chem. Soc.* P-1, 1374 (1973); R. S. Jaret, A. K. Mallams, H. F. Vernay, *J. Chem. Soc.*, C, 1389 (1971).
13) H. Umezawa, *Index of Antibiotics from Actinomyces*, University of Tokyo Press, 1967; W. D. Celmer, *J. Am. Chem. Soc.*, **87**, 1799, 1801 (1965); *ibid.*, **88**, 5028 (1966); and references therein.
14) See also R. S. Egan, J. R. Martin, T. J. Perun, L. A. Mitscher, *J. Am. Chem. Soc.*, **97**, 4578 (1975); J. G. Nourse, J. D. Roberts, *ibid.*, **97**, 4584 (1975).

7.42 STRUCTURE OF LEUCOMYCIN, A 16-MEMBERED MACROLIDE

leucomycin-A$_3$ **1**

mp: 120–121°
uv: 231.5 (ε 29100)
ir: 1728, 1746 (C=O), 1230 (Ac)
pK_a: 6.70

   The structures and absolute configurations of the leucomycins were elucidated by chemical studies,[1-3] including the interrelation of **1** and magnamycin-B, and by x-ray studies.[4] The conformation in solution has been determined.[1]

1) MnO$_2$ oxidation of **1** gave magnamycin-B, possessing an $\alpha,\beta,\gamma,\delta$-unsaturated carbonyl group (uv: 279.5 nm).

2) The coupling constants of the vicinal protons were as follows: $J_{8,9}=4.2$, $J_{9,10}=8.9$, $J_{10,11}=15.4$, $J_{11,12}=10.0$ and $J_{12,13}=15.2$ Hz. The large coupling constants between 10-H/11-H and between 12-H/13-H indicate a *trans-trans* configuration of the double bonds.

3) The absolute configuration at C-9 of **1** was determined by applying the benzoate and Mill's rules.

4) Acidic rearrangement of **1** yielded demycarosyl-isoleucomycin-A$_3$, an allylic rearrangement product whose structure was determined by x-ray analysis.

5) From cd, ir and nmr data, and x-ray studies, the preferred conformation of the aglycone of **1** in solution is thought to be as shown in **1a**.[1]

1) S. Omura, A. Nakagawa, N. Yagisawa, Y. Suzuki, T. Hata, *Tetr.*, **28**, 2839 (1972).
2) S. Omura, H. Ogura, T. Hata, *Tetr. Lett.*, 1267 (1967).
3) S. Omura, A. Nakagawa, M. Katagiri, T. Hata, M. Hiramatsu, T. Kimura, K. Naya, *Chem. Pharm. Bull.*, **18**, 1501 (1970).
4) M. Hiramatsu, A. Furusaki, T. Noda, K. Naya, Y. Tomie, I. Nitta, T. Watanabe, T. Tabe, J. Abe, S. Omura, T. Hata, *Bull. Chem. Soc., Japan*, **43**, 1966 (1970).

1a

Remarks

Eight analogues of leucomycins were isolated from *Streptomyces kitasatoensis*. A number of 16-membered macrolides are known, such as spiramycin,[5] carbomycins (magnamycins),[5,6] cirramycin,[7] chalcomycin,[8] tylosin,[9] YL-704,[10] rosamycin,[11] B-58941,[12] etc.[13]

5) S. Omura, A. Nakagawa, M. Ohtani, T. Hata, H. Ogura, K. Furukata, *J. Am. Chem. Soc.*, **91**, 3401 (1969).
6) R. B. Woodward, L. S. Weiler, P. C. Dutta, *J. Am. Chem. Soc.*, **87**, 4662 (1965); W. D. Celmer, *ibid.*, **88**, 5028 (1966).
7) H. Tsukiura, M. Konishi, M. Sasa, T. Naito, H. Kawaguchi, *J. Antibiotics*, **22**, 89 (1969).
8) P. W. K. Woo, H. W. Dion, Q. R. Bartz, *J. Am. Chem. Soc.*, **86**, 2724 (1964).
9) R. B. Morin, M. Gorman, R. L. Hamill, P. V. Demarco, *Tetr. Lett.*, 4737 (1970).
10) M. Suzuki, I. Takamori, A. Kinumaki, Y. Sugasawa, T. Okuda, *Tetr. Lett.*, 435 (1971).
11) H. Reimann, R. S. Jaret *Chem. Commun.*, 1270 (1972).
12) T. Suzuki N. Sugita, M. Asai, *Chem. Lett.*, 789 (1973); T. Suzuki, E. Mizuta, N. Sugita, *ibid.*, **793 (1973)**; T. Suzuki, *ibid.*, 799 (1973).
13) H. Umezawa, *Index of Antibiotics from Actinomyces*, University of Tokyo Press, 1967; W. D. Celmer, *J. Am. Chem. Soc.*, **87**, 1799, 1801 (1965); *ibid.*, **88**, 5028 (1966); and references therein.

7.43 STRUCTURE OF VENTURICIDINE,
A 20-MEMBERED MACROLIDE

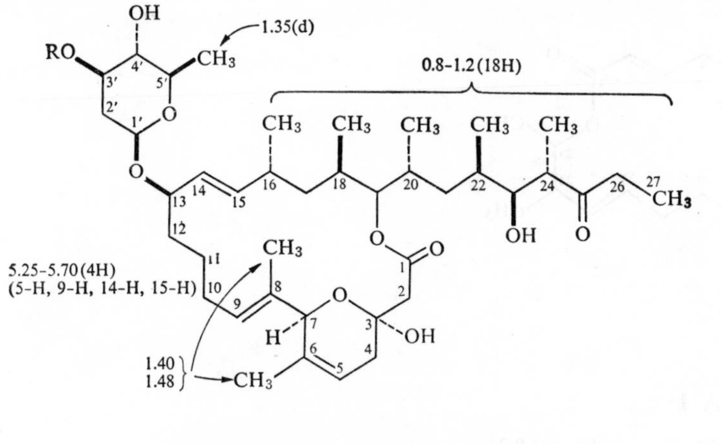

venturicidine A (R=NH$_2$CO) **1**[1]
venturicidine B (R=H)   **2**

mp: 145–147°
ms: 540 (M, C$_{41}$H$_{67}$NO$_{11}$ – 191(C$_7$H$_{13}$NO$_5$)
uv: 280–300 (unconjugated C=O)
nmr in CDCl$_3$

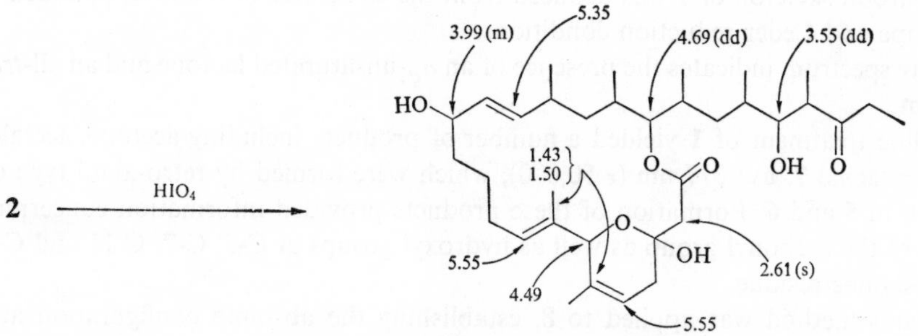

aglycone **3**
ms: 576 (C$_{34}$H$_{56}$O$_7$)

The complete structure of venturicidine was established by x-ray crystallographic analysis of the *p*-iodobenzenesulfonate ester. Chemical degradation and spectroscopic studies of **1** and **2** are fully consistent with the structures shown.

Remarks

Venturicidine A and B, antifungal antibiotics with a 20-membered ring lactone, were isolated from *Streptomyces aureofaciens*.

1) M. Brufani, L. Cellai, G. Musu, W. Keller-Schierlein, *Helv. Chim. Acta*, **55**, 2329 (1972).

7.44 STRUCTURE OF TETRIN,
A 26-MEMBERED MACROLIDE

tetrin-A **1**[1]

mp: > 350° α_D (py): +8.3°
ms (pentaacetyldecahydro-**1**): 883 (M−18, $C_{44}H_{71}NO_{18}$)
uv: 210 (ε 14500), 278 (ε 30100), 303 (ε 78300), 290 (ε 55300), 318 (ε 75500)
ir: 3500, 3300 (NH, OH), 2700 (COOH), 1710 (C=O), 1625 (C=C)

1) The molecular formula $C_{34}H_{51}NO_{13}$ was deduced for **1** from mass spectral data of some derivatives.

2) The carbon skeleton of **1** was deduced from the structures of **3** and **4**, obtained respectively by Cope and Ceder reduction conditions.

3) The uv spectrum indicates the presence of an $\alpha\beta$-unsaturated lactone and an all-*trans* tetraene system.

4) Alkaline treatment of **1** yielded a number of products including acetone, acetaldehyde and the pentaenal **7**, uv: 377 nm (ε 50,000), which were formed by retro-aldol type cleavages as shown in **5** and **6**. Formation of these products provided information concerning the positions of the carbonyl group as well as hydroxyl groups at C-5, C-7, C-11 and C-13, and the mycosamine residue.

5) Horeau's method was applied to **8**, establishing the absolute configuration at C-25 of **1**.

6) Additional structural evidence was provided by the nmr spectrum of *N*-acetyl tetrin-A **2**.

7) The C-9 ketone and 13-OH in **1** and the related antibiotics, pimaricin and leucensomycin, form cyclic hemiketal linkages in solution.[2]

Remarks

The tetraene antifungal macrolide tetrin-A **1** and the closely related compound tetrin-B[3] were isolated from *Streptomyces* species. The compounds pimaricin[4] and leucensomycin[5] are known to have tetraene macrolide structures similar to that of **1**.

1) R. C. Pandey, V. F. German, Y. Nishikawa, K. L. Rinehart Jr., *J. Am. Chem. Soc.*, **93**, 3738 (1971); and references therein.
2) C. N. Chong, R. W. Rickard, *Tetr. Lett.*, 5053 (1972).
3) K. L. Rinehart Jr., W. P. Tucker, R. C. Pandey, *J. Am. Chem. Soc.*, **93**, 3747 (1971).
4) B. T. Golding, R. W. Richards, W. E. Meyer, J. B. Patrick, M. Baker, *Tetr. Lett.*, 3551 (1966).
5) G. Gaudiano, P. Bravo, A. Quilico, B. T. Golding, R. W. Rickards, *Tetr. Lett.*, 3567 (1966).

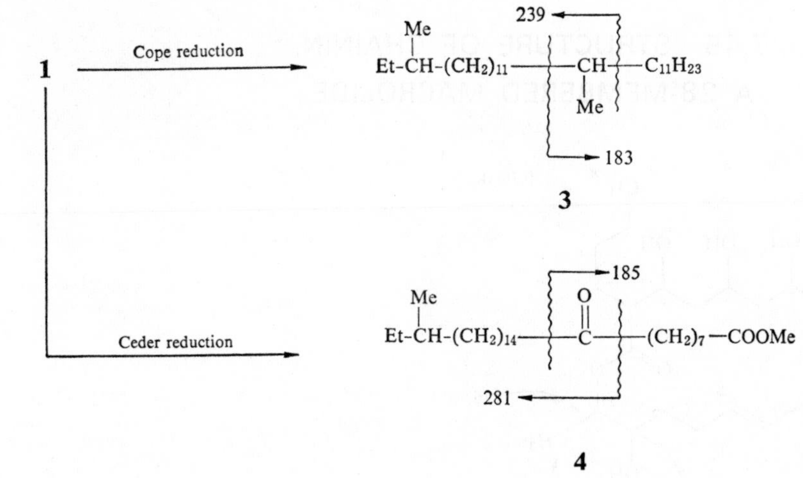

3

4

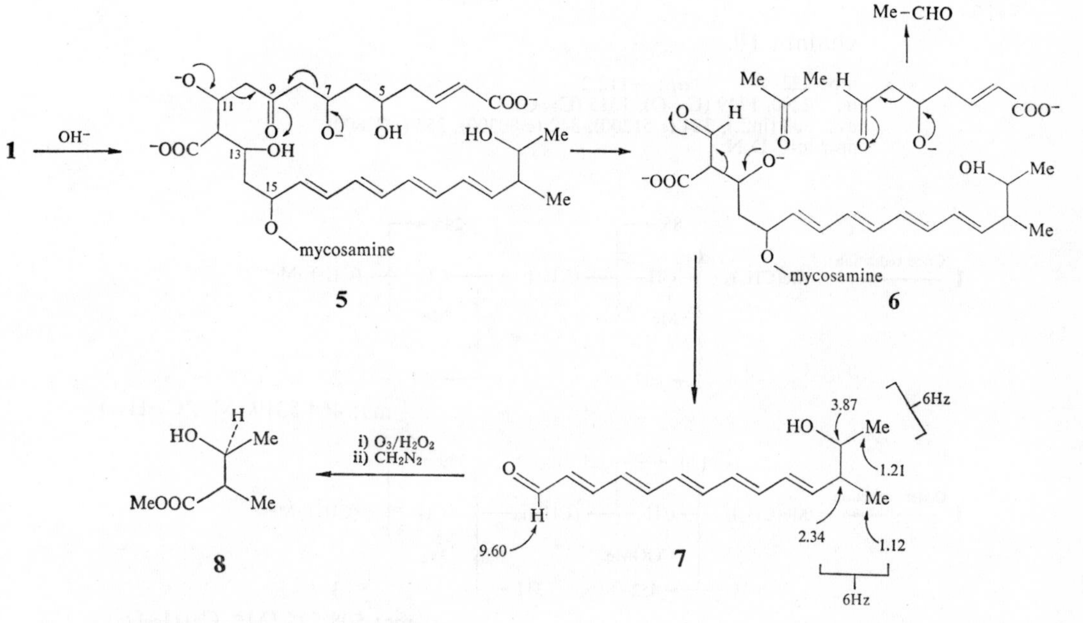

5

6

7

8

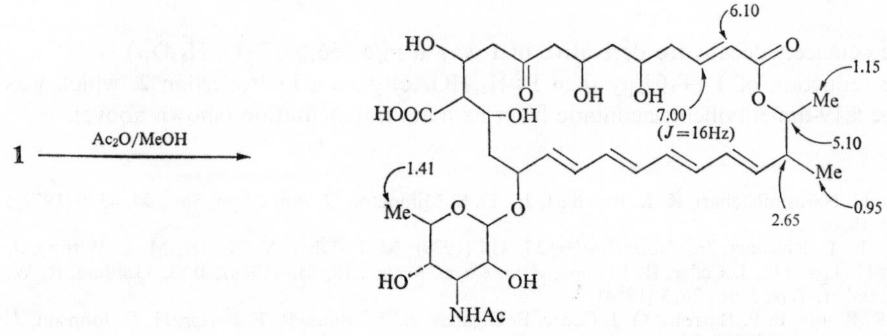

N-acetyl tetrin-A 2

7.45 STRUCTURE OF CHAININ,
A 28-MEMBERED MACROLIDE

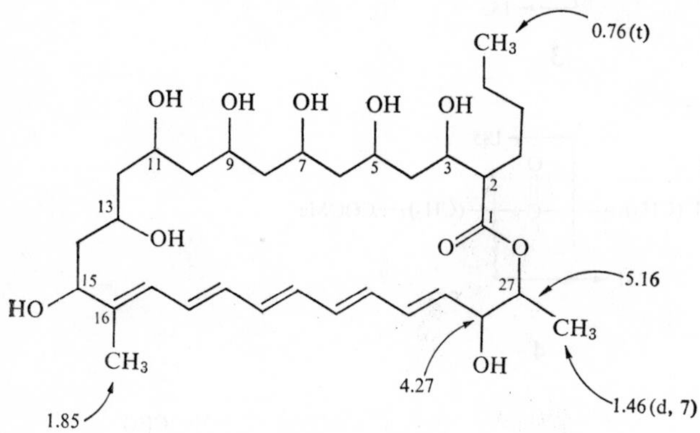

chainin **1**[1]

mp: 222–4° α_D: −112.2°
ir: 2290, 1749 (C=O), 1615 (C=C)
uv: 308 (infl.), 324 (ε 51200), 342 (ε 80300), 358 (ε 77800)
nmr in C_5D_5N

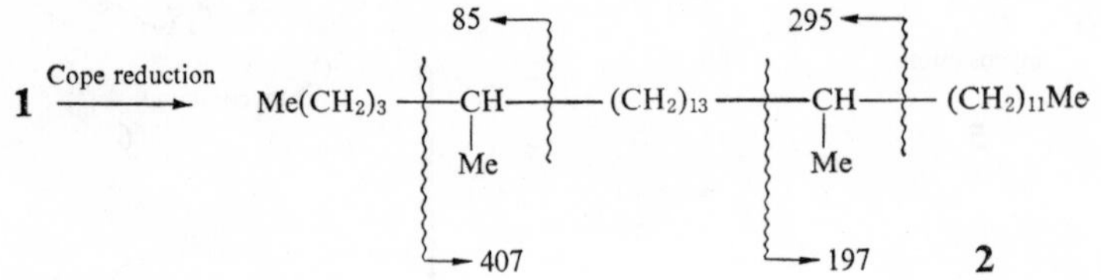

1 $\xrightarrow{\text{Cope reduction}}$ $Me(CH_2)_3$—CH—$(CH_2)_{13}$—CH—$(CH_2)_{11}Me$ **2**

Me   Me

ms: 464.5319 (M^+, $C_{33}H_{68}$)

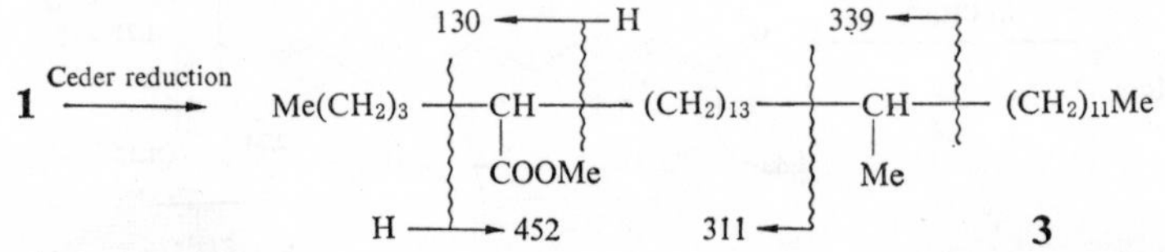

1 $\xrightarrow{\text{Ceder reduction}}$ $Me(CH_2)_3$—CH—$(CH_2)_{13}$—CH—$(CH_2)_{11}Me$ **3**

COOMe   Me

ms: 508.524 (M^+, $C_{34}H_{68}O_2$)

1) M^+ for the octaacetyldecahydro derivative of **1** was at m/e 956.5317 ($C_{49}H_{80}O_{18}$).
2) Exhaustive reduction of **1** (TsCl/py then $Pt/H_2/HOAc$) gave a hydrocarbon **2**, which was found to be 5,19-dimethylhentriacontane from its mass fragmentation (shown above).

1) R. C. Pandey, N. Narasimhachari, K. L. Rinehart, Jr., D. S. Millington, *J. Am. Chem. Soc.*, **94**, 4306 (1972); and references therein.
2) R. C. Pandey, K. L. Rinehart, Jr., *J. Antibiotics*, **23**, 414 (1970); M. L. Dhar, V. Thaller, M. C. Witting, *J. Chem. Soc.*, 842 (1964); O. J. Ceder, R. Ryhage, *Acta. Chem. Scand.*, **18**, 558 (1964); B. T. Golding, R. W. Rickards, M. Baker, *Tetr. Lett.*, 2615 (1964).
3) A. C. Cope, R. K. Bly, E. P. Burrous, O. J. Ceder, E. Ciganek, B. T. Gillus, R. F. Porter, H. E. Johnson, *J. Am. Chem. Soc.*, **84**, 2170 (1962).

3) Mass fragmentation of the ester **3** obtained by Ceder reduction of **1** revealed the position of the carboxyl group.

4) Base treatment of **1** afforded hexanal and acetoaldehyde, indicating the presence of an unsubstituted *n*-butyl group at C-2.

5) The uv spectrum reveals the presence of an all-*trans* monomethyl-substituted pentaene system in **1**.

6) **1** did not consume periodate, indicating the absence of 1,2-glycol. Thus the remaining eight hydroxyl groups are located at 1,3-positions relative to each other.

Remarks

Chainin **1** is a new antibiotic isolated from *Chainia* sp. The closely related antibiotics filipins and lagosin (fungichromin) are also known.[2,3]

7.46 STRUCTURE OF FLAVOFUNGIN, A 32-MEMBERED MACROLIDE

flavofungin **1**[1]

mp: 210° α_D: −85 to −90°
uv (MeOH): 363, 254
ir (KBr): 3400, 2945, 1710, 1680, 1620, 1232, 1122, 1020, 840

1) In the mass spectrum of the octaacetyl derivative **5**, successive loss of eight acetic acid units from the molecular ion (m/e 986, $C_{52}H_{74}O_{18}$) was observed.

2) The carbon skeleton of **1** was established by reduction to the parent hydrocarbon **3**.

3) That the position of lactonization is at C-31 was shown by mass fragmentation of the deoxylactone **4** as shown.

4) LAH reduction of **1** gave a product containing the pentaene chromophore (uv : 349, 331, 318), indicating that the carbonyl is conjugated with the polyene system.

5) The position of the isolated double bond was shown to be at C-28 by the formation of **6**.

6) **1** did not consume periodate, indicating the absence of 1,2-glycol. Treatment of the octatosylate of **2** with LiBr in DMF gave a product containing the heptaene chromophore, suggesting 1,3-relationships of the original hydroxyl groups.

7) That 27-OH is allylic, was shown by hydrogenolysis of **5**, yielding the heptaacetylperhydro derivative.

1) R. Bognár, S. Makleit, K. Zsupán, B. O. Brown, W. J. S. Lockley, T. P. Toube, B. C. L. Weedon, *J. Chem. Soc.*, **P-1**, 1848 (1972).

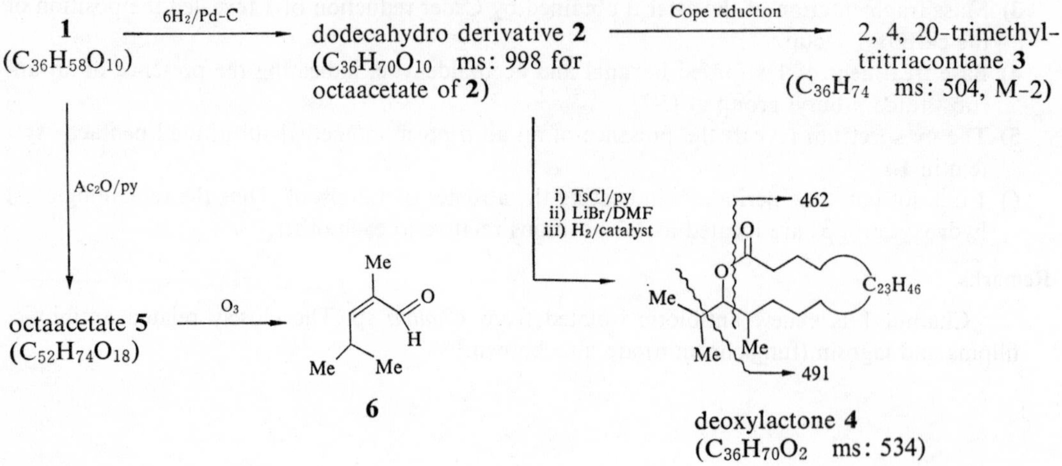
1
(C36H58O10)
6H2/Pd-C
dodecahydro derivative 2
(C36H70O10 ms: 998 for
octaacetate of 2)
Cope reduction
2, 4, 20-trimethyl-
tritriacontane 3
(C36H74 ms: 504, M-2)
Ac2O/py
i) TsCl/py
ii) LiBr/DMF
iii) H2/catalyst
octaacetate 5
(C52H74O18)
O3
Me
O
H
Me
Me
6
462
O
O
C23H46
Me
Me
Me
491
deoxylactone 4
(C36H70O2 ms: 534)

Remarks

Flavofungin **1**, a macrolide antifungal antibiotic was isolated from *Streptomyces flavofungini*. Mycoticin, which is assumed to have the same gross structure, was obtained from *S. ruber*.[2)]

---

2) H. H. Wasserman, J. E. Van Verth, D. J. McCaustland, I. J. Borowitz, B. Kamber, *J. Am. Chem. Soc.*, **89** 1535 (1967).

7.47 STRUCTURE OF AXENOMYCIN,
A 34-MEMBERED MACROLIDE

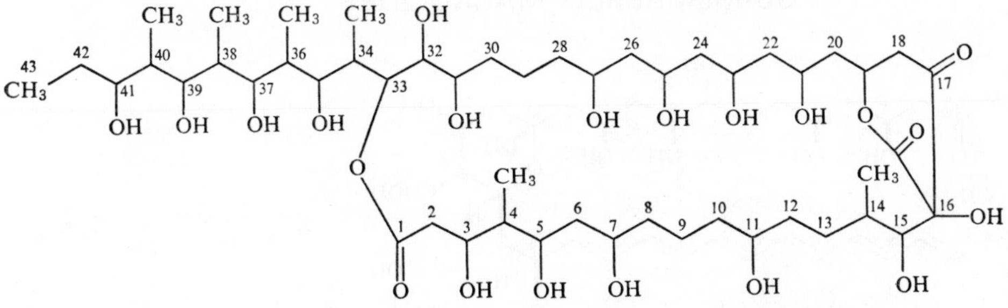

aglycone of axenomycin B **1**[1]

$C_{78}H_{126}O_{30}$
uv (MeOH): 250, 256, 267
ir: 1665

1) Methanolysis of **1** gave the compound **2**, a sugar moiety containing the naphthoquinone chromophore and the aglycone, axenolide **3**.

2) Periodate oxidation of **3** yielded three fragments **4**, **5** and **6** which contain all the carbon atoms of axenolide except the C-16 carboxyl carbon. The structure of **3** was deduced from the degradation products (numbering in **4**, **5**, and **6** corresponds to that of the parent skeleton).

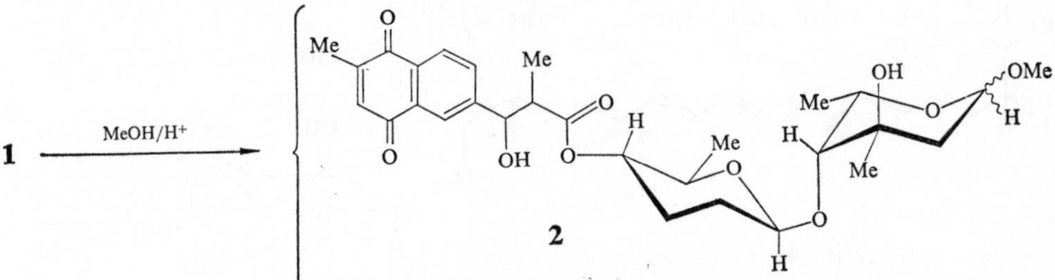

+ axenolide **3** (aglycone of **1**)

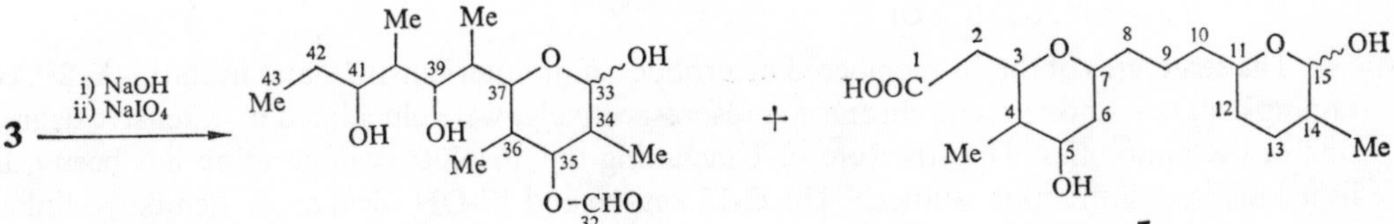

Remarks

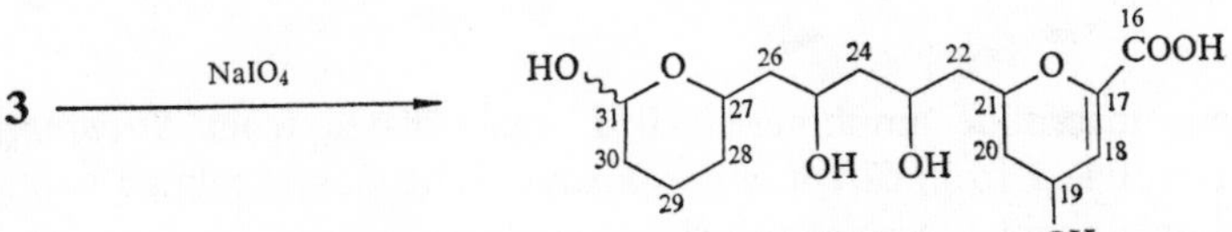

Axenomycins are macrolide antibiotics with a 34-membered ring produced by *Streptomyces lysandri*. They have activity against plateworms and yeasts.

1) F. Arcamone, G. Franceschi, B. Gioia, S. Penco, A. Vigevani, *J. Am. Chem. Soc.*, **95**, 2009 (1973).
2) F. Arcamone, W. Barbieri, G. Franceschi, S. Penco, A. Vigevani, *J. Am. Chem. Soc.*, **95**, 2008 (1973).

7.48 STRUCTURES OF AMPHOTERICIN AND NYSTATIN, 38-MEMBERED MACROLIDES

amphotericin-B **1**[1,3)]

mp: 170° α_D (DMF): +238°
uv (MeOH): 406, 382, 363 (heptaene)
ir: 5.83, 6.34 (C=O)

nystatin-A$_1$ **2**[2)]

mp: 160°
uv: 318.5, 304, 291, 230 (tetraene)
ir: 5.87, 6.37 (C=O)

The structures of the 38-membered macrolides, amphotericin-B **1**[1)] and nystatin-A$_1$ **2**[2)], containing heptaene and tetraene chromophores; respectively, were elucidated by extensive degradation of the antibiotics. The structure of **1** including the absolute configuration has been established by x-ray diffraction studies.[3)] The C-13 ketone and 17-OH form cyclic hemiketal linkages in **1** and **2**.[4)]

Remarks

The antifungal heptaene macrolide amphotericin-B **1** was isolated from *Streptomyces nodosus* and the tetraene macrolide **2** from *Streptomyces noursei*. The closely related heptaene macrolide candidin was obtained from *S. viridoflavus*.[5)]

1) E. Borowski, J. Zielinski, T. Ziminski, L. Falkowski, P. Kolodziejczyk, J. Golik, E. Jereczek, H. Aldercreutz, *Tetr. Lett.*, 3909 (1970); and references therein.
2) E. Borowski, J. Zielinskin, L. Falkowski, T. Ziminski, J. Golik, P. Kolodziejczyk, E. Jereczek, M. Gdulewicz, Y. Shenin, T. Kotienko, *Tetr. Lett.*, 685 (1971); and references therein.
3) W. Mechlinski, C. P. Schaffner, P. Ganis, G. Avitabile, *Tetr. Lett.*, 3873 (1970).
4) C. N. Chong, R. W. Rickards, *Tetr. Lett.*, 5053 (1972).
5) E. Borowski, *et al. Tetr. Lett.*, 1987 (1971).

CHAPTER 8

Sugars (carbohydrates)

8.1 INTRODUCTION

Sugars (carbohydrates) form one of the most abundant groups of natural products and include many physiologically active substances. Typical sugars have the composition represented by the general formula $C_nH_{2n}O_n$ and are classified into mono-, di-, ..., oligo-, and poly-saccharides. Recently, many sugar derivatives such as aminosugars, nucleosides, etc. have attracted increased attention, especially in the field of antibiotics.

In this chapter, owing to the space limitation, emphasis is given to sugar derivatives having unusual structures. Polysaccharides and polynucleotides are omitted and only two typical sugars of historical importance are included. The order of arrangement of topics is as follows: sugars, glycosides, aminosugars, nucleosides.

8.2   STRUCTURE OF D-GLUCOSE

α-D-glucose **1a**[1]

mp:146° $\alpha_D(H_2O):+111.2°$

β-D-glucose **1b**[1]

mp:148-150° $\alpha_D(H_2O):+17.5°$

Structural formula

Fittig and Baeyer (1868–70) proposed an aldehyde-alcohol formula **2** which was supported by a number of observations. These included the molecular formula (Tollens and Mayer, 1888); formation of a pentaacetate; reduction to sorbitol **3**; oxidation with bromine or nitric acid to yield gluconic acid **4**, which in turn yields hexanoic acid on reduction with hydriodic acid; and the formation of a cyanohydrin **5**, which gives rise to heptanoic acid on hydrolysis and reduction with hydriodic acid.

$$CH_2OH-CHOH-CHOH-CHOH-CHOH-CHO$$
2

$$CH_2OH-CHOH-CHOH-CHOH-CHOH-CH_2OH$$
3

$$CH_2OH-CHOH-CHOH-CHOH-CHOH-COOH$$
4

$$CH_2OH-CHOH-CHOH-CHOH-CHOH-CH(OH)CN$$
5

Configuration

E. Fischer (1891) applied the Le Bel-van't Hoff theory to the sugar series and established the configurations of glucose **1**, as well as other sugars, as follows (modern version of the Fischer proof by Hudson[2]). The OH at C-5 is on the right in the Fischer projection of **1** by convention for the D series; D-arabinosuccharinic acid **6** is optically active, and hence the OH at C-2 must be on the left; since D-glucose and D-mannose **7** can be prepared from D-arabinose **8** via the Kiliani synthesis, the OH at C-3 in **1** must be on the left; since D-glucose and D-mannose **7** are C-2 epimers, D-glucose **1** must have structure **A** or **B** (without establishment of the configuration at C-4); since both D-glucaric acid **9** and D-mannaric acid **10** are optically active, there can be no plane of symmetry at the center of the molecule, and hence the OH at C-4 in both **A** and **B** must be on the right; finally, since D-glucaric acid **9** can be obtained from the two aldohexoses, D-glucose and L-gulose **11**, D-glucose must have the structure **A** (**1**).

---

1) W. Pigman The Carbohydrates, p. 7–42, Academic Press, 1957.
2) C. S. Hudson, *Advances in Carbohydrate Chem.*, **3**, 1 (1948).

```
         COOH              CHO               CHO               COOH
   HO——2—H          HO——2—H             CHOH           H——2—OH
                                     HO——3—H
        CHOH              CHOH              CHOH              CHOH
   H——4—OH           H——4—OH          H——5—OH          H——4—OH
        COOH              CH₂OH             CH₂OH             COOH

          6                 8              1 and 7             12
   (optically active)                                  (optically inactive)
```

```
         COOH              CHO               CHO               COOH
   H——2—OH          H——2—OH          HO——2—H          HO——2—H
   HO——H             HO——H             HO——H             HO——H
   H——4—OH          H——4—OH          H——4—OH          H——4—OH
   H——OH             H——OH             H——OH             H——OH
        COOH              CH₂OH             CH₂OH             COOH

          9               A≡1               B≡7                10
```

```
        CH₂OH                             CH₂OH
   H——OH                           HO——H
   HO——H                           HO——H
   H——OH                           H——OH
   H——OH                           H——OH
        CHO                               CHO

   L-gulose 11                            B≡7
```

Ring structure

However, the open-chain formula **A** cannot account for the following observations: glucose gives a negative Schiff test; mutarotation of glucose solutions; the formation of two isomers of the pentaacetate, the methyl glycoside and even glucose itself (α and β). Colley (1870) and Tollens (1883) proposed respectively the 1,2-(**1c**) and 1,4-oxide (**1d**) structure for glucose to account the above facts. Conclusive proof of the ring structure as the 1,5-oxide (pyranose) (**1e**) was achieved by methylation and oxidation (Hirst, 1926), by bromine oxidation, giving gluconic δ-lactone (Isbell, 1932), and by periodate oxidation (Hudson, 1937).

1c   1d   1e

Configuration of the anomeric carbon atom

The relative configuration of the anomeric carbon atom was determined by measuring the conductivity of glucose freshly dissolved in boric acid solution. Boric acid forms a complex with vicinal *cis* hydroxyl groups. The conductivity of α-glucose in boric acid solution decreases with time whereas in the case of β-glucose **1b** the conductivity increases with time. α- and β-Glucopyranoses, **1a** and **1b** have, therefore, a hydroxyl group at C-1 *cis* and *trans*, respectively, to the hydroxyl group at C-2.

Absolute configuration

Absolute configuration was established by correlation with D-tartaric acid, whose absolute configuration was determined by the X-ray diffraction method (Bijvoet, 1954).

Conformation

The C-1 conformations, **1a** and **1b**, are preferable in solution since bulky groups tend to assume equatorial positions.

Remarks

D-Glucose, free or combined, is the most widely distributed of the sugars. It occurs free in fruits, honey, blood, lymph, cerebrospinal fluid, and urine and is a major component of many oligosaccharides such as sucrose, polysaccharides such as starch and cellulose, and glucosides.

8.3   STRUCTURE AND SYNTHESIS OF SUCROSE

sucrose **1**

1) **1** consumes three moles of HIO_4 (yielding one mole of HCOOH),[1] shows no mutarotation and gives a negative Fehling test and no osazone formation.

2) The action of acid or invertase on **1** produces glucose and fructose.

3) Maltase hydrolyzes **1** but emulsin does not, indicating an α-glucose unit.

4) Takainvertase is specific for β-fructofuranosides and it hydrolyzes **1**,[1] indicating a β-D-fructofuranose moiety.[2]

5) Hudson's rules support the structure **1**.[3]

6) The structure was confirmed by x-ray analysis.[4]

Synthesis[6]

The formation of a 1,6-anhydroglucose intermediate by back side attack of the 6-acetate on the epoxide facilitates the formation of the α-glucoside linkage.

In 1928, Pictet and Vogel[7] claimed to have achieved the synthesis of **1** by coupling tetra-O-acetyl-D-fructofuranose with tetra-O-acetyl-D-glucopyranose, but it was shown later that the product was actually isosucrose octaacetate.[8]

1) P. Fleury, J. Courtois, *Bull. soc. chim. France*, [5] **10**, 245 (1943); *Compt. rend.*, **216**, 65 (1943).
2) H. H. Schlubach, G. Rauchalles, *Ber.*, **58**, 1842 (1925).
3) F. Klages, R. Niemann, *Ann*, **529**, 185 (1937); M. L. Wolfrom, F. Shafizadeh, *J. Org. Chem.*, **21**, 88 (1956).
4) C. A. Beevers, W. Cochrane, *Proc. Roy. Soc.*, A. **190**, 257 (1947).
5) J. Avery, W. N. Haworth, E. L. Hirst, *J. Chem. Soc.*, 2308 (1927); W. N. Haworth, E. L. Hirst, A. Learner, *J. Chem. Soc.*, 2432 (1927).
6) R. U. Lemieux, G. Huber, *J. Am. Chem. Soc.*, **75**, 4118 (1953); *ibid.*, **78**, 4117 (1956).
7) A Pictet, H. Vogel, *Helv. Chim. Acta*, **11**, 436 (1928).
8) J. C. Irvine, J. W. H. Oldham, A. F. Skinner, *J. Am. Chem. Soc.*, **51**, 1279 (1929); J. C. Irvine, J. W. H. Oldham, *J. Am. Chem. Soc.*, **51**, 3609 (1929).

7   **8**   **9**

Remarks

Sucrose occurs almost universally throughout the plant kingdom. The commercial sources are sugar cane, sugar beets and maple trees.

8.4   STRUCTURE OF L-ASCORBIC ACID (VITAMIN C)

L-ascorbic acid
1

1) **1** is an optically active strong acid showing a positive $FeCl_3$ test, and forms a 2,4-dinitro-phenylhydrazone and a tetraacetate.

2) **1** gives furfural on acid treatment and L-idonic acid on catalytic hydrogenation.[1]

3) The methoxyl groups in **2** are stable toward alkali and hence there are two enolic hydroxyls in **1**.[2,3]

4) Oxidation of **1** with Cl_2 or Br_2 affords dehydroascorbic acid, which is not acidic, and which can be reduced to **1** with HI or H_2S.[4]

5) **2** gives an Na salt with NaOH (1 mole), indicating the presence of a lactone ring.

6) **7** gives a positive Weerman test, which indicates the presence of an α-hydroxyamide, and hence the structure is **7** and not **8**.[5]

1) F. Micheel, K. Kraft, *Z. Physiol. Chem.*, **218**, 280 (1933).
2) P. Karrer, H. Salomon, K. Schopp, R. Morf, *Helv. Chim. Acta*, **16**, 181 (1933); *Biochem. Z.*, **258**, 4 (1933).
3) P. Karrer, G. Schwarzonbach, K. Schopp, *Helv. Chim. Acta*, **16**, 302 (1933).
4) R. W. Herbert, E. L. Hirst, E. G. V. Percival, R. J. W. Reynolds, F. Smith, *J. Chem. Soc.*, 1270 (1933).
5) S. F. Dyke, The Carbohydrates, p. 16, Interscience, 1960.

Remarks

L-Ascorbic acid **1** is widely distributed in the plant and animal kingdoms. The normal daily requirement in man is 75–100 mg per day. **1** was first isolated from orange juice by Szent-Györgyi.[7]

---

6) F. Micheel, K. Kraft, *Z. Physiol. Chem.*, **215**, 215 (1933).

7) J. L. Svirbely, A. Szent-Györgyi, *Biochem. J.*, **27**, 279 (1933); A. Szent-Györgyi, *Biochem. J.*, **22**, 1387 (1928).

8.5   SYNTHESIS OF ASCORBIC ACID (VITAMIN C)

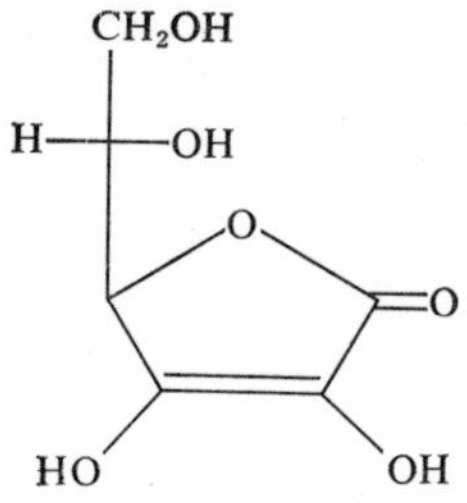

ascorbic acid **1**

Routes I and II, employing the Kiliani synthesis, involve the β-ketoacid intermediate **8**, whereas the α-ketoacid **15** is involved in Routes III and IV, which are more suitable for large-scale production. Either of these ketoacids can be isomerized to the corresponding enediol lactone.

Route I[1]

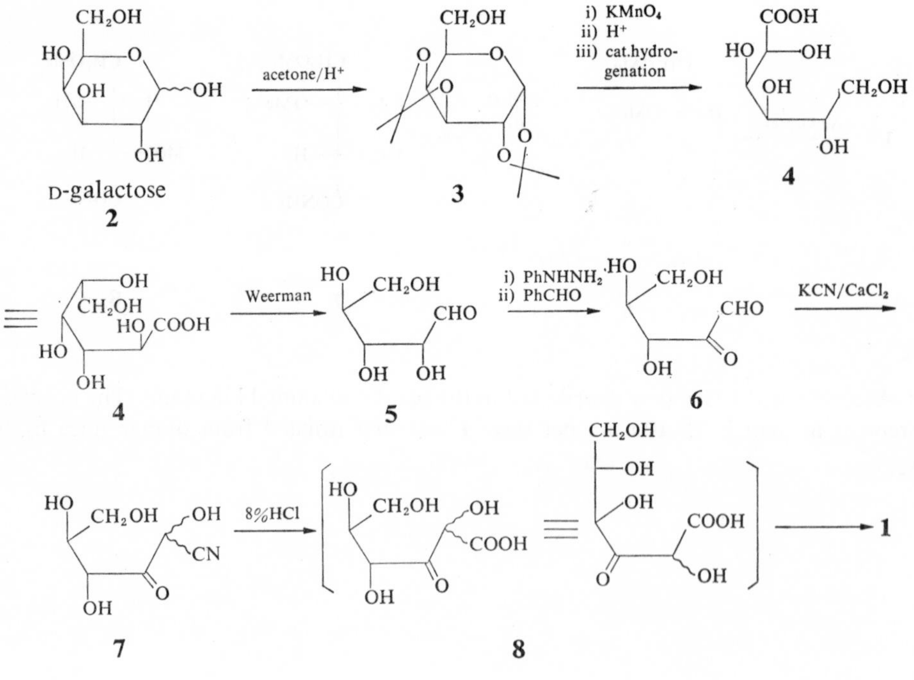

1) R. G. Ault, D. K. Baird, H. C. Carrington, W. N. Haworth, R. Herbert, E. L. Hirst, E. G. V. Percival, F. Smith, M. Stacey, *J. Chem. Soc.*, 1419 (1933); D. K. Baird, W. N. Haworth, R. W. Herbert, E. L. Hirst, F. Smith, M. Stacey, *ibid.*, 62 (1934); M. N. Haworth, E. L. Hirst, *J. Soc. Chem. Ind.*, **52**, 645 (1933); W. N. Haworth, E. L. Hirst, J. K. N. Jones, F. Smith, *J. Chem. Soc.*, 1192 (1934).

Route II[2]

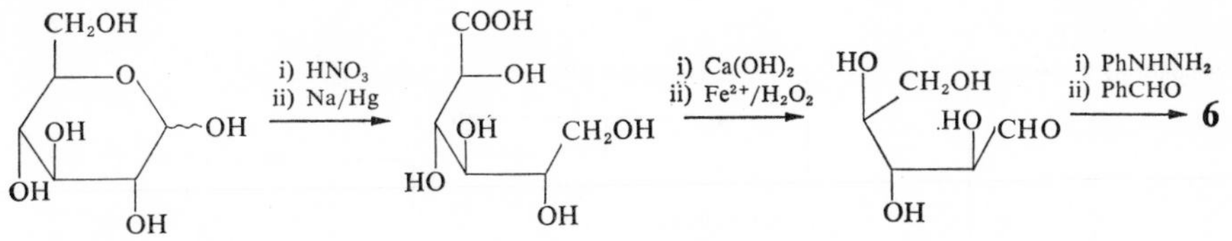

D-glucose **9** → **10** → **11** → **6**

i) HNO_3 / ii) Na/Hg

i) $Ca(OH)_2$ / ii) Fe^{2+}/H_2O_2

i) $PhNHNH_2$ / ii) $PhCHO$

Route III[3]

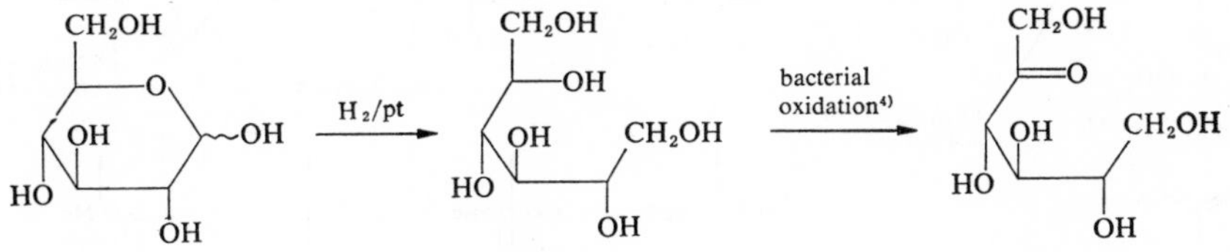

9 → sorbitol **12** → **13**

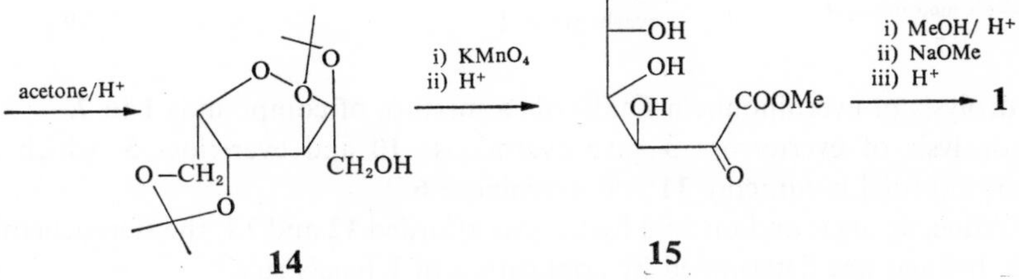

H_2/pt

bacterial oxidation[4]

acetone/H^+ → **14** → **15** → **1**

i) $KMnO_4$ / ii) H^+

i) $MeOH/H^+$ / ii) $NaOMe$ / iii) H^+

Route IV[5]

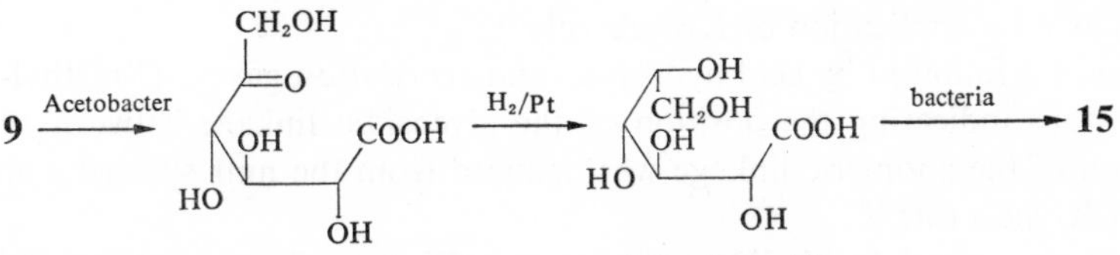

9 → **16** → L-idonic acid **17** → **15**

Acetobacter

H_2/Pt

bacteria

17 ⟶ 15: Some species of *Pseudomonas*, *Acetobacter* or *Aerobacter* can be used for this purpose.

2) T. Reichstein, A. Grussner, R. Oppenauer, *Helv. Chim. Acta*, **16**, 1019 (1933); *ibid.*, **21**, 561 (1938).
3) F. Smith, *The Vitamins: Chemistry, Physiology and Pathology* (ed. W. H. Sebrell, R. S. Harris), vol. 1. p. 188, Academic Press, 1954.
4) P. A. Wells, J. J. Stubbs, L. B. Lockwood, E. T. Roe, *Ind. Eng. Chem.*, **29**, 1385 (1937).
5) I. Hori, T. Nakatani, *J. Ferment. Technol.*, **31**, 72 (1953).

8.6 STRUCTURE OF EVERHEPTOSE

1) Acid hydrolysis of everninomicin D affords a mixture of compounds **1** to **7**.

2) Acid hydrolysis of evertetrose **3** gave evermicose **10** and evertriose **5**, which on further hydrolysis afforded D-curacose **11** and everninose **6**.[1]

3) **6** is non-reducing sugar and on acid hydrolysis afforded **12** and **13**; the stereochemistry of the anomeric linkage was determined by application of Klyne's rule.[2]

4) Methylation of **5** followed by hydrolysis afforded 2,3,6-tri-*O*-methyl-D-mannose, thus indicating the position of glycosidic linkage between **11** and **12**; the anomeric configuration was determined as β by application of Klyne's rule.[3]

5) Methylation of **3** followed by acid hydrolysis and acetylation gave 2-*O*-methyl-1,3-*O*-acetyl-D-curacose, thus indicating the position of the glycosidic linkage between **10** and **11**; the configuration of the anomeric linkage was deduced from the nmr spectrum and by the application of Klyne's rule.[1]

6) Treatment of compound **4** with diazomethane in THF gave **15** and **10**; the position of attachment of **15** to **10** was deduced from the nmr spectrum.[4]

7) **1** was treated with diazomethane in methanol to give **3** and compound **14**; the latter was converted with silica gel into the δ-lactone **15**, acetylation of which was accompanied by dehydration to give the α,β-unsaturated lactone **16**, indicating the glycosidic linkage to be at C-4 in the lactone.[4]

1) A. K. Ganguly, O. Z. Sarre, S. Szmulewicz, *Chem. Commun.*, 746 (1971).
2) A. K. Ganguly, O. Z. Sarre, J. Morton, *Chem. Commun.*, 1488 (1969).
3) A. K. Ganguly, O. Z. Sarre, *Chem. Commun.*, 911 (1970).
4) A. K. Ganguly, O. Z. Sarre, D. Greeves, J. Morton, *J. Am. Chem. Soc.*, **95**, 942 (1973).
5) H. Reimann, R. S. Jaret, O. Z. Sarre, *J. Antibiotics*, **22**, 131 (1969).
6) A. K. Ganguly, O. Z. Sarre, *Chem. Commun.*, 1149 (1969).
7) A. K. Ganguly, O. Z. Sarre, H. Reimann, *J. Am. Chem. Soc.*, **90**, 7129 (1968).

8) The negative Cotton effect of **16** indicates the R configuration at C-5 in **16**.[4]

9) Acid hydrolysis of **2** afforded evernitrose **8** and everninocin **7**; the stereochemistry of the anomeric position of **8** was assigned from application of Klyne's rule.[4]

10) **7**, which is identical with curacin from curamycin[8] and avilamycin,[9] was methylated with MeI/Ag$_2$O/DMF followed by hydrolysis with KOH/aq. EtOH and then H$^+$/H$_2$O to give D-oleandrose (2,6-dideoxy-3-O-methylglucose), indicating the position of attachment of the orsellic acid derivative.[5]

11) The anomeric configuration of **9** in **4** was determined from the nmr spectrum.[4]

Remarks[10]

Everninomicins, which are produced by *Micromonospora carbonacea*, are oligosaccharide antibiotics of high molecular weight, one of which, everninomicin D, yields on acid hydrolysis a mixture of products from which everheptose **1** was isolated. This antibiotic contains, besides various other unusual sugars, evernitrose **8**, which is the first naturally occurring nitro sugar to be isolated. The mechanism of the cleavage of **1** and **4** in methanol solution using diazomethane is not clear.[4] It appears that the presence of the free phenolic hydroxyl group contributes toward the stability of the molecule. Once the phenolic hydroxyl group is methylated, cleavage can occur simply by ester interchange with methanol.

8) O. L. Galmarini and V. Deulofeu, *Tetr.*, **15**, 76 (1962).

9) F. Buzzetti, F. Eisenberg, H. N. Grant, W. Keller-Schierlein, W. Voser, H. Zähner, *Experientia*, **24**, 320 (1968).

10) G. H. Wagmann, G. M. Luedemann, M. J. Weinstein, *Antimicrob. Ag. Chemother.*, **24**, 33 (1964).

8.7   STRUCTURE OF GANGLIOSIDE G$_1$

ganglio-N-tetraose **3**

ganglio-N-biose I **2**

(galNAc)

(gal)

lactose

sphingosine

HO CH$_2$OH   HO CH$_2$OH   CH$_2$OH   CH$_2$OH   CH$_2$(CH$_2$)$_{11}$CH$_3$

HO   OH   AcHN   OH   HO   OH   HC—OH

HOOC   (gal)   (glu)   HC—NH—C—(CH$_2$)$_{16}$CH$_3$

AcHN   HC—OH   CH$_2$ O

HC—OH   stearic acid

CH$_2$OH

N-acetylneuraminic acid (NANA) **4**   ganglioside G$_1$ **1**[1]

1) Hydrolysis of **1** with N/100 H$_2$SO$_4$ yielded NANA **4** and a new compound which was further hydrolyzed by N/10 H$_2$SO$_4$ to sphingosine, lactose and ganglio-N-biose I **2**.

2) Reduction of **2** with NaBH$_4$ followed by hydrolysis afforded galactose, indicating gal $\longrightarrow$ galNAc. Since periodate oxidation of **2** yielded galNAc, the point of attachment to galNAc cannot be C-5 or C-6. A positive Morgan-Elson reaction excluded the linkage at C-4, showing that gal is attached to galNAc at C-3.

3) **2** was hydrolyzed by β-galactosidase, indicating that the configuration at gal $\longrightarrow$ galNAc is β.

4) Acetolysis of **1** followed by hydrolysis with N/10 H$_2$SO$_4$ yielded ganglio-N-tetraose **3**.

5) Periodate oxidation followed by KBH$_4$ reduction of **3** yielded erythritol and threitol, indicating that C-4 of gal in lactose is attached to **2**, since only C-4-substituted galactose could produce threitol (erythritol from glu).

6) The β-configuration at galNAc $\longrightarrow$ gal was deduced from the α_D value (calculated for α, +93.6°; β, +22.9°: found, +13.9°).

7) NANA-ganglio-N-tetraose **5** was obtained by acetolysis of **1**.

8) Periodate oxidation of **5** yielded galactose and galNAc, indicating that NANA is attached to gal at C-3 or C-2. The latter can be excluded for steric reasons.

Remarks

Gangliosides are important components of ganglions, and have been isolated from beef brain and other ganglions. Since separation and crystallization of these complex lipids is ex-

1) R. Kuhn, H. Wiegandt, *Chem. Ber.*, **96**, 866 (1963); see also R. Kuhn, *Angew. Chem.*, **72**, 805 (1960); R. Kuhn, H. Wiegandt, H. Egge, *ibid.*, **73**, 580 (1961); R. Kuhn, H. Egge, *Chem. Ber.*, **96**, 3338 (1963); H. Wiegandt, G. Baschang , *Z. Naturforsh.*, **20b**, 164 (1965).

tremely difficult, it is noteworthy that the structure of such a complex ganglioside has been completely elucidated. Ganglioside G_1 (60 g), G_2 (45 g), G_3 and G_4 were obtained from 350 kg of beef brain by chromatographic separation.[1] Treatment of G_2, G_3 and G_4 with sialdase yielded G_1 and NANA, suggesting that these are composed of G_1 and NANA (in the ratios 1:1, 1:1 and 1:2, respectively). It was proposed that NANA in G_1 is not hydrolyzed by sialdase because of steric hindrance between the NANA $\longrightarrow$ gal bond and the galNAc $\longrightarrow$ gal bond, since they are in a vicinal *cis* configuration (see **8**) on the facing page.

8.8 STRUCTURE OF PIPTOSIDE

piptoside **1**[1]
ir: 3420, 3230, 1808, 1785
nmr in D_2O

1) The pentamethyl derivative **2** was formed from **1** by reaction with MeI/Ag_2O. It has no OH groups.

2) Alkali consumption (2 eq) and ir bands at 1808 and 1785 cm^{-1} indicate the presence of two γ-lactones.

3) Reaction of **1** with CH_2N_2 yielded a monomethyl derivative, indicating the presence of a hemiketal.

4) The anomeric signal indicates a β-glucosidic linkage.

3

uv (EtOH) : 270 (4.16)
224 (3.42)

$(-)$-2, 3-dimethyl-
succinic acid **4**
(abs. config. known)

piptosidin **5**

5) Piptosidin **5** forms a diacetate **6**: the nmr measurements are for the diacetate **6** in $CHCl_3$.

1) N. V. Riggs, J. D. Stevens, *Tetr. Lett.*, 1615 (1963); *Aust. J. Chem.*, **19**, 683 (1966).
2) N. V. Riggs, J. D. Stevens, *Aust. J. Chem.*, **15**, 305 (1962).

The relative stereochemistry at C-6 and C-7 was deduced from the J values, but the absolute stereochemistry of C-4 to C-7 was not determined.

Remarks

Piptoside **1** was isolated from New South Wales bitter vine, *Piptocalyx moorei*, but is itself tasteless.[2]

Biogenetically, **1** is considered to be a product of Michael addition of a 3-dehydrohexonic acid to angelic acid, followed by the formation of the glucoside.

8.9   STRUCTURE OF SENEGIN-IV

mp: 249–250°
α_D(MeOH): −20.2°
uv(EtOH): 315 (4.30)   senegin–IV **1**[1]

1) Acid hydrolysis of **1** gave one mole each of presenegenin **2**, *p*-methoxycinnamic acid **6**, glucose, galactose, fucose and xylose, and 3 moles of rhamnose.

2) The position of attachment of **6** on **5** was deduced from the structure of the fucoside **13** by comparison with the fucoside **12**.

3) Alkaline hydrolysis (1N KOH) of **7** gave the 23-Me ester of **3**, whose structure was deduced by comparison with an authentic sample,[2,3] indicating the position of attachment of the hexasaccharide **4**.

4) The position of attachment of rha-2 on rha-1 is deduced from the structure of the rhamnoside **11** by comparison with 4-*O*-Me-**7** obtained from senegin-III (des-(rha-2)-**1**).

5) The structure of **14** was deduced by methanolysis to tri-**16** and tetrasaccharides which were further degraded by Hakomori methylation followed by methanolysis (MeOH/HCl) to mixtures of known methylated monosaccharides.

6) The configuration of D-fucose was deduced as β from the nmr of the anomeric proton ($J=10$ Hz) of fucose in **19**.

7) The anomeric configuration of rha-1 was deduced as α from the $[M]_D$ difference between **19** and **20**.

8) The anomeric configurations of xyl and gal were deduced as β from the nmr (J value) of the anomeric protons of **14**.

9) The anomeric configurations of rha-2 and rha-3 were deduced as α from the $[M]_D$ difference between **8** and senegin-II Me ester, which lacks rha-2 and rha-3 in **7**.

1) J. Shoji, Y. Tsukitani, *Chem. Pharm. Bull.*, **20**, 424 (1972); Y. Tsukitani, J. Shoji, *ibid.*, **21**, 1564 (1973).
2) J. Shoji, S. Kawanishi, Y. Tsukitani, *Chem. Pharm. Bull.*, **19**, 1740 (1971); Y. Tsukitani, S. Kawanishi, J Shoji, *ibid.*, **21**, 791 (1973).
3) S. W. Pelletier, S. Nakamura, R. Soman, *Tetr.*, **27**, 4417 (1971).

Remarks

Senegae radix, a Chinese herb, is actually the dried roots of *Polygala senega* or *P. senega* var. *latifolia* (Polygalaceae) from which four kinds of saponin, i.e., senegin-I, -II, -III, and -IV, were isolated.[1,4] The structures of senegin-II and -III were deduced as **21**[2] and **22**[1] respectively. Presenegenin was identified as the genuine sapogenin of the senegins[5] and its structure was determined.[6]

glu⟶presenegenin 2⟵—fuc⟶ *p*-MeO-cinnamic acid **6**   glu⟵2⟵—fuc⟵6

      gal⟶xyl⟶rha-1   gal⟶xy⟶rha-1   rha-3

          senegin-II **21**   senegin-III **22**

4) J. Shoji, S. Kawanishi, Y. Tsukitani, *Yakugaku Zasshi* (Japan), **91**, 198 (1971).
5) I. Yoshioka, M. Fujio, M. Osamura, I. Kitagawa, *Tetr. Lett.*, 6303 (1966).
6) J. J. Dugan, P. de Mayo, *Tetr. Lett.*, 2567 (1964); *Can J. Chem.*, **42**, 491 (1964); *ibid.*, **43**, 2033 (1965); S. W. Pelletier, N. Adityachaudhury, *Tetr. Lett.*, 3065 (1964); S. W. Pelletier, N. Adityachaudhury, M. Tomaz, J. J. Raynalds, *J. Org. Chem.*, **30**, 4234 (1965); Y. Shimizu, S. W. Pelletier, *J. Am. Chem. Soc.*, **88**, 1544 (1966).

8.10 STRUCTURE OF NOVOBIOCIN

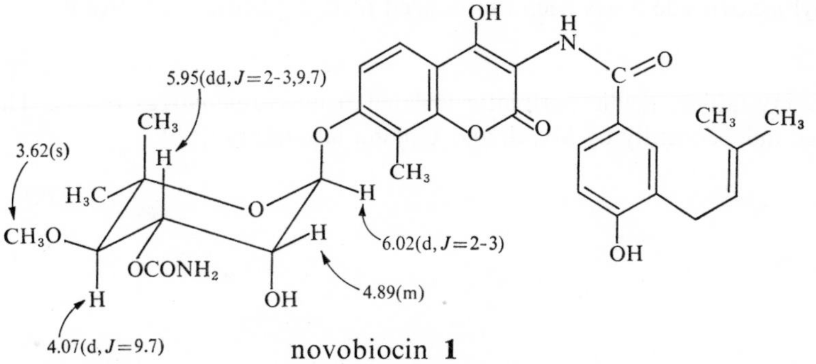

5.95(dd, J=2-3,9.7)

3.62(s)

4.07(d, J=9.7)

6.02(d, J=2-3)

4.89(m)

novobiocin **1**

$pK_{a'}$ (H$_2$O):4.3 and 9.1

1) Ir indicates the presence of –NHCO–, –OH, –O–, –C=C–(conj.), and –OCONH$_2$.

2) Hudson's isorotation rules suggest α-configuration of the glycosidic linkage.[1]

3) Nmr indicates that the pyranose ring exists in the **1** conformation.[2]

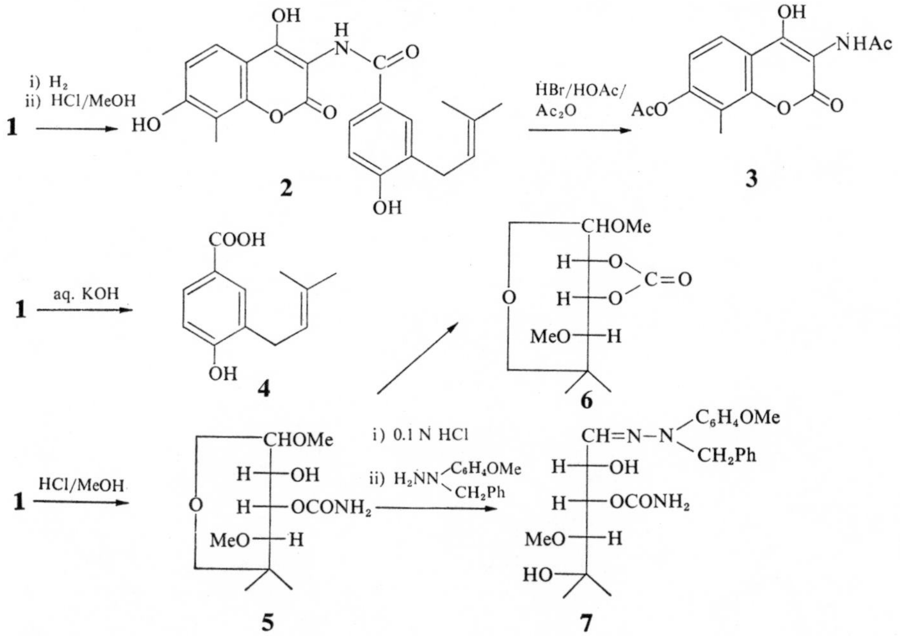

4) Structure **4** was confirmed by synthesis[3] and structure **3** was deduced by comparison with the synthetic desmethyl compound of **3**.[4]

1) E. Walton, J. O. Rodin, F. W. Holly, J. W. Richter, C. H. Shunk, K. Folkers, *J. Am. Chem. Soc.*, **82**, 1489 (1960).

2) B. T. Golding, R. W. Richards, *Chem. Ind.*, 1081 (1963).

3) E. A. Kaczka, C. H. Shunk, J. W. Richter, F. J. Wolf, M. M. Gasser, K. Folkers, *J. Am. Chem. Soc.*, **78**, 4125 (1956).

4) C. H. Shunk, C. H. Stammer, E. A. Kaczka, E. Walton, C. F. Spencer, A. N. Wilson, J. W. Richter, F. W. Holly, K. Folkers, *J. Am. Chem. Soc.*, **78**, 1770 (1956).

5) Formation of the cyclic carbonate **6** indicates *cis* configuration between 2-OH and 3-OH.[5]

6) The negative rotation of the hydrazone **7** allows assignment of the 2-OH to the right in the Fischer projection formula.[5]

7) Methyl 3-*O*-carbamyl-α-novioside **5** has been synthesized from a glucose derivative.[6]

Remarks

Novobiocin **1** is a crystalline, acidic antibiotic isolated from *Streptomyces niveus*. The structure was established independently by Merck and Upjohn laboratories.[3,4,7]

5) E. Walton, J. O. Rodin, C. H. Stammer, F. W. Holly, K. Folkers, *J. Am. Chem. Soc.*, **78**, 5454 (1956); *ibid.*, **80**, 5168 (1958).

6) B. P. Vaterlaus, J. Kiss, H. Spiegelberg, *Helv. Chim. Acta.*, **47**, 381 (1964).

7) H. Hoeksema, E. L. Caron, J. W. Hinman, *J. Am. Chem. Soc.*, **78**, 2019 (1956); *ibid.*, **79**, 3789 (1957).

8.11 STRUCTURE OF STREPTOMYCIN

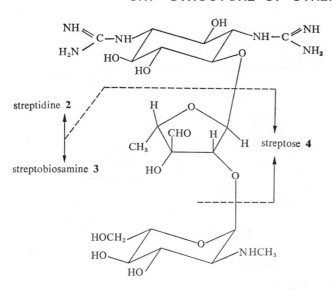

streptidine **2**

streptobiosamine **3**

streptose **4**

streptomycin **1**

1) Acid hydrolysis of **1** yields **2** and **3**[1-4] whereas 1 N NaOH gives maltol.[5]
2) The configuration of the glycoside linkages was suggested by calculations of optical rotations,[6] but has not been confirmed.[7]
3) The tertiary OH is free in **1**, since it yields a tetraacetate **11**.[8]
4) The structure and absolute configuration were confirmed by x-ray analysis of **1** oxime selenate.[18]

Structure of streptidine **2**[1-4]

1) **2** is optically inactive and hence a meso-form.
2) **2** gives a positive Sakaguchi test and two moles of guanidine on $KMnO_4$ oxidation, indicating the presence of two guanidino groups.
3) **5** consumes six moles of HIO_4, but yields no HCHO, whereas di-*N*-benzoyl-**5** consumes only two moles of HIO_4, indicating that **5** is 1,3-diamino-2,4,5,6-tetrahydroxycyclohexane. This was confirmed by the formation of **6**.
4) The formation of **6** (*dl*-form) shows the configuration in the 4,5,6 region to be *trans-trans*.[9]
5) The structures of **5** and **2** were confirmed by synthesis.

1) N. G. Brink, F. A. Kuehl Jr., K. Folkers, *Science*, **102**, 506 (1945).
2) R. L. Peck, R. P. Graber, A. Walti, E. W. Peel, C. E. Hoffhine Jr., K. Folkers, *J. Am. Chem. Soc.*, **68**, 29 (1946).
3) H. E. Carter, R. K. Clark Jr., S. R. Dickman, Y. H. Loo, J. S. Meek, P. S. Skell, W. A. Strong, J. T. Alberi, Q. R. Barty, S. B. Binkley, H. M. Crooks Jr., I. R. Hooper, M. C. Rebstock, *Science*, **103**, 53 (1946): H. E. Carter, R. K. Clark, S. R. Dickman, Y. H. Loo, P. S. Skell, W. A. Strong, *ibid.*, **103**, 540 (1946).
4) J. Fried, G. A. Boyack, O. Wintersteiner, *J. Biol. Chem.*, **162**, 391 (1946).
5) J. R. Schenck, M. A. Spielman, *J. Am. Chem. Soc.*, **67**, 2276 (1945).
6) M. L. Wolfrom, M. J. Cron, C. W. deWalt, R. M. Husband, *J. Am. Chem. Soc.*, **76**, 3675 (1954).
7) R. U. Lemieux, C. W. deWalt, M. L. Wolfrom, *J. Am. Chem. Soc.*, **69**, 1838 (1947).
8) N. G. Brink, F. A. Kuehl Jr., E. H. Flynn, K. Folkers, *J. Am. Chem. Soc.*, **68**, 2405 (1946).
9) O. Wintersteiner, A. Klingsberg, *J. Am. Chem. Soc.*, **70**, 885 (1948).

6) *N*-Dibenzoylstreptamine consumes two moles of HIO_4, whereas **9** consumes 1 mole, indicating that **2** is attached to **1** at the 4- or 6-position.[10]

$$2 \xrightarrow{Ba(OH)_2} \text{[structure]} \quad + \quad 4 \, NH_3 \; + \; 2 \, CO_2$$

streptamine **5**

i) acetyl.
ii) Δ

i) *N*-acetyl.
ii) Me_2SO_4/OH^-
iii) HCl

i) $KMnO_4$
ii) $MeOH/H^+$
iii) $NH_3/MeOH$

6

7

$$\begin{array}{cc}
CONH_2 & CONH_2 \\
H\text{—}OMe & MeO\text{—}H \\
MeO\text{—}H \quad + & H\text{—}OMe \\
CONH_2 & CONH_2
\end{array}$$

dl-dimethoxysuccinic
diamide **8**

i) $PhCOCl/py$
ii) hydrol.
iii) $MeSO_2Cl$
iv) NaI
v) Ra-Ni
vi) $Ba\text{-}(OH)_2/MeOH$

1

9

HIO_4
(1 mole)

10

Structure of streptobiosamine 3

1) A derivative of **3** yields methylamine on treatment with 20% NaOH.[11]
2) The structure of **13** was proved by synthesis of its optical antipode from D-glucosamine.[11]
3) The *cis* relationship of the hydroxyl groups in **12** is suggested by complex formation with H_3BO_3.[8]
4) The formation of **14** indicates that the two Me groups are in a vicinal position.[8]
5) Formation of MeCHO from **15** shows which Me group in **12** originated from the aldehyde group in **4**.[11]
6) The structure of **16**, which was confirmed by synthesis, shows that **4** belongs to the L-series.[12]
7) The formation of **19** from **17** indicates the configuration of the OH group at C-2 in **4**.[13]

10) F. A. Kuehl Jr., R. L. Peck, C. E. Hoffhine Jr., E. W. Peel, K. Folkers, *J. Am. Chem. Soc.*, **69**, 1234 (1947).
11) F. A. Kuehl Jr., E. H. Flynn, N. G. Brink, K. Folkers, *J. Am. Chem. Soc.*, **68**, 2679 (1966).
12) J. Fried, D. E. Walz, O. Wintersteiner, *J. Am. Chem. Soc.*, **68**, 2746 (1946).
13) H. L. Wolfrom, C. W. deWalt, *J. Am. Chem. Soc.*, **70**, 3148 (1948); F. A. Kuehl Jr., M. N. Bishop, E. H. Flynn, K. Folkers, *J. Am. Chem. Soc.*, **70**, 2613 (1948).

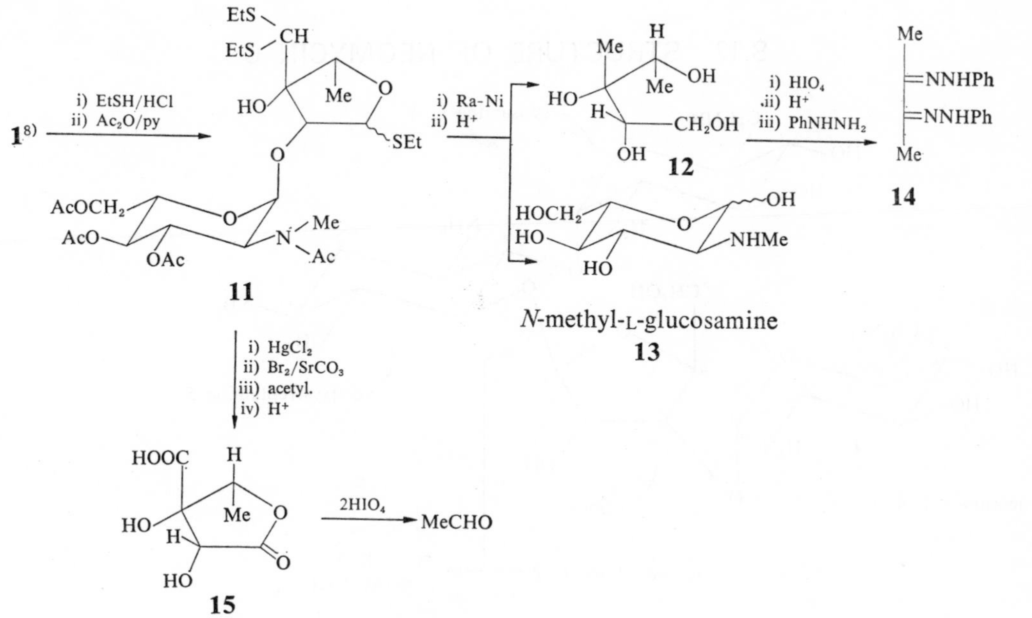

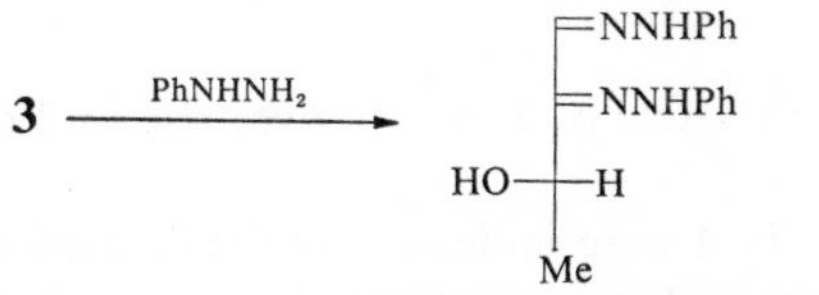

11

12

N-methyl-L-glucosamine

13

14

15

2HIO₄ → MeCHO

3 → PhNHNH₂ →

4-deoxy-L-erythrose
phenylosazone **16**

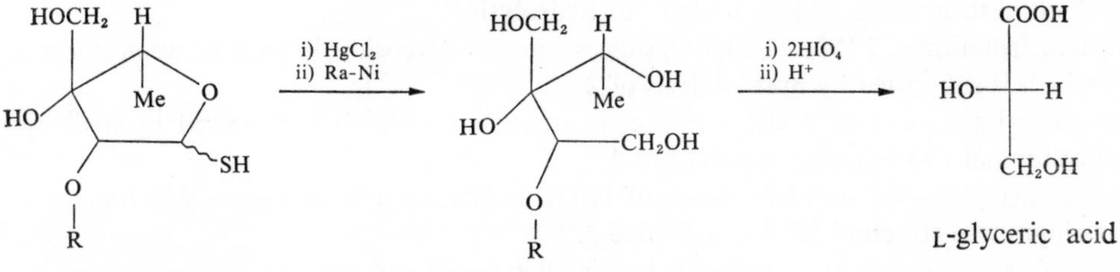

17 (dihydro-**11**)
R = *N*-acetylglucosamine

18

L-glyceric acid

19

Remarks

Streptomycin **1** was first isolated from *Streptomyces griseus*.[14] It is active against gram-negative bacteria and is used as an antimicrobial and antituberculous agent. **1** is usually accompanied by a minor component, mannosidostreptomycin (streptomycin B).[15] Catalytic hydrogenation of **1** gives dihydrostreptomycin (CHO in **1**⟶CH₂OH), which is as active as **1**.[16] Hydroxystreptomycin (Me in **1**⟶CH₂OH) is produced by *Streptomyces reticuli*.[17]

14) A. Schatz, Bugie, S. A. Waksman, *Proc. Soc. Exp. Biol. Med.*, **55**, 66 (1944).
15) J. Fried, E. O. Titus, *J. Biol. Chem*, **168**, 391 (1947).
16) J. Fried, O. Wintersteiner, *J. Am. Chem. Soc.*, **69**, 79 (1947).
17) S. Hosoya *et al.*, *Japan J. Exptl. Med.*, **20**, 327 (1949).
18) S. Neidle, D. Rogers, M. B. Hursthouse, *Tetr. Lett.*, 4725 (1968).

8.12   STRUCTURE OF NEOMYCIN C

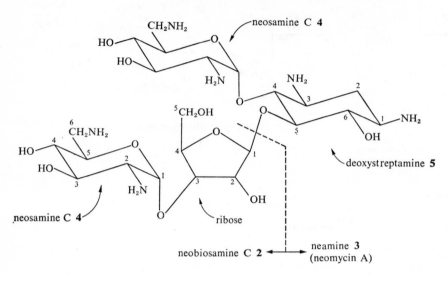

neomycin C **1**

1) Methanolysis of **1** yields **3** and the methyl glycoside of **2**, which gives rise to **2** on acid hydrolysis.[1]

2) The configurations in the 1 and 5 positions in **4** were indicated by the formation of **7** and **8**,[2] and the specific rotation of **4** suggests a D-glucose configuration at the 3 and 4 positions.[3] The structure of **4** was confirmed by synthesis.[4]

3) Reduction of **2** by $NaBH_4$ followed by N-benzoylation yields N,N'-dibenzoylneobiosaminol C, which in turn yields ribitol on vigorous hydrolysis.[2]

4) The large rotation of **2** (M_D: +33,700) suggests an α-D-glycosidic linkage between ribose and **4**,[3] which is supported by nmr analysis of **2**.

5) The methyl glycoside of **2** yields ribose on treatment with HIO_4 followed by hydrolysis, indicating that C-3 in ribose is linked to **4**.[5]

6) Tetra-N-acetyl-3 consumes two moles of HIO_4, indicating a 4- or 6-glycoside linkage of **5** and a pyranose structure for **4** in neamine **3**.[6]

7) Hudson's rules suggest an α-glycoside linkage between **4** and **5**.[6]

8) Acid hydrolysis of poly-O-methyl-hexa-N-acetylneomycin C, obtained from hexa-N-acetylneomycin C by treatment with MeI/BaO/DMF followed by Ag₂O/MeI, yields 2,5-dimethyl-D-ribose and N,N'-diacetyl-6-O-methyldeoxystreptamine, thus confirming the furanose structure of ribose and the position of attachment of ribose to **5**.[7]

1) K. L. Rinehart Jr., P. W. K. Woo, A. D. Argoudelis, A. M. Giesbrecht, *J. Am. Chem. Soc.*, **79**, 4567 (1957).
2) K. L. Rinehart Jr., P. W. K. Woo, *J. Am. Chem. Soc.*, **80**, 6463 (1958).
3) K. L. Rinehart Jr., P. W. K. Woo, A. D. Argoudelis, *J. Am. Chem. Soc.*, **80**, 6461 (1958).
4) K. L. Rinehart Jr., M. Hichens, K. Striegler, K. R. Rover, T. P. Culbertson, S. Tatsuoka, S. Horii, T. Yamaguchi, H. Hitomi, A. Miyake, *J. Am. Chem. Soc.*, **83**, 2964 (1961).
5) K. L. Rinehart Jr., A. D. Argoudelis, T. P. Culbertson, W. S. Chilton, K. Streigler, *J. Am. Chem. Soc.*, **82**, 2970 (1960).
6) H. E. Carter, J. R. Dyer, P. D. Shaw, K. L. Rinehart Jr., M. Hichens, *J. Am. Chem. Soc.*, **83**, 3723 (1961).
7) K. L. Rinehart Jr., M. Hichens, A. D. Argoudelis, W. C. Chilton, H. E. Carter, M. P. Ceorgiadis, C. P. Shaffner, R. F. Schillings, *J. Am. Chem. Soc.*, **84**, 3218 (1962).

9) The absolute configuration of **5** was determined by the cuprammonium method[8] using *N,N'*-diacetyl-6-*O*-methyldeoxystreptamine.[9]

10) The β-ribofuranoside linkage is assumed on the basis of nmr analysis.[10]

$$N,N'\text{-dibenzoyl-}\mathbf{4} \xrightarrow{\text{NaBH}_4}$$

```
        CH_2OH
    H——NHCOPh
   HO——H
    H——OH
    H——OH
       CH_2NHCOPh
           6
```

$$\xrightarrow{\text{HIO}_4}$$

```
   ^1CH_2OH
H—^2—NHCOPh
   ^3CHO
        7
```

methyl glycoside of
N,N-dibenzoyl-**2**

$$\xrightarrow[\text{iii) hydrol.}]{\substack{\text{i) HIO}_4 \\ \text{ii) Br}_2}}$$

```
   ^4COOH
H—_5—OH
   ^6CH_2NH_2
```

isoserine
8

Remarks

Streptomyces fradiae produces neomycin complex[11] which can be separated into neomycin A (neamine), neomycin B and neomycin C. Neomycin B has neosamine B attached to ribose in place of neosamine C.

The paromycins, kanamycins and zygomycins are antibiotics similar to the neomycins

neomycin B **9**

8) R. E. Reeves, *Advan. Carbohydrate Chem.*, **6**, 107 (1951).
9) M. Hichens, K. L. Rinehart Jr., *J. Am. Chem. Soc.*, **85** 1547 (1963).
10) K. L. Rinehart Jr., W. S. Chilton, M. Hichens, W. von Phillipsborn, *J. Am. Chem. Soc.*, **84**, 3216 (1962).
11) S. A. Waksman, H. A. Lechevalier, *Science*, **109**, 305 (1949).

SUGARS (CARBOHYDRATES)

110

8.13 SYNTHESIS OF KASUGAMYCIN

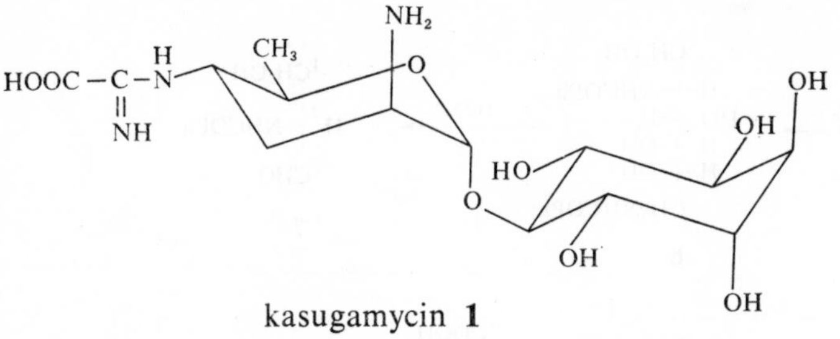

kasugamycin **1**

Kasugamycin, found in 1965,[2] is an antibiotic produced by *Streptomyces kasugaensis*, and exhibits a strong preventive action against rice blast. The structure was established in 1966 by chemical[3] and x-ray crystallographic[4] studies.

Route I[1]

This was the first total synthesis of kasugamycin, but the step **8 ⟶ 1** was originally developed by Suhara *et al.*[3] Compound **2** was synthesized from D-glucose in five steps, and hence the final product **1** has the correct absolute configuration.

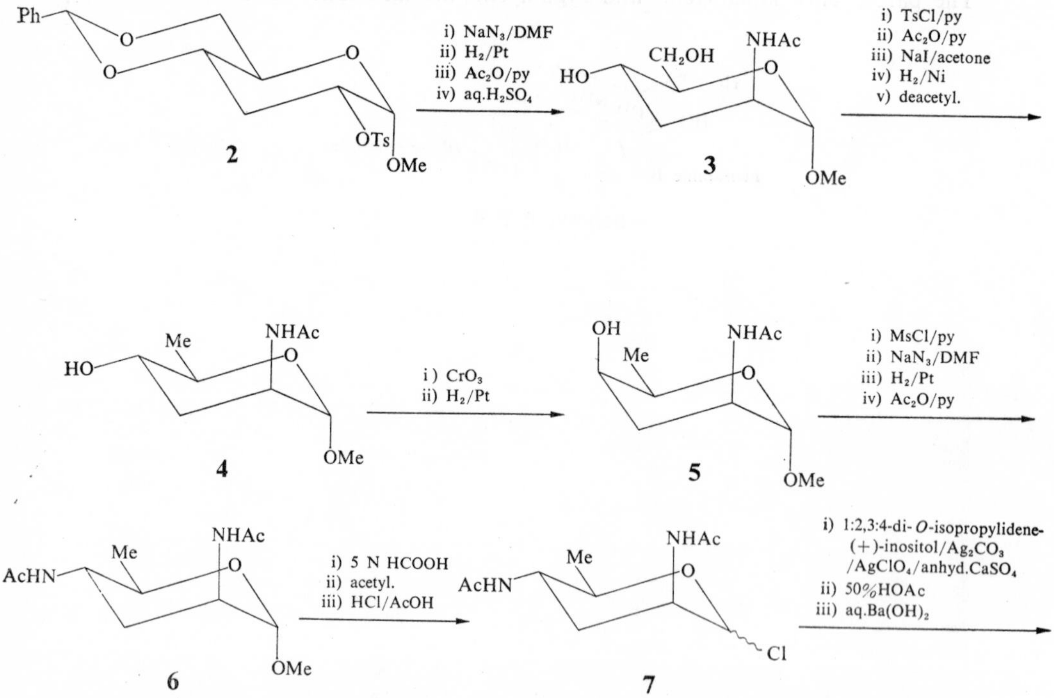

1) M. Nakajima, H. Shibata, K. Kitahara, S. Takahashi, A. Hasegawa, *Tetr. Lett.*, 2271 (1968).
2) H. Umezawa, Y. Okami, T. Hashimoto, Y. Suhara, M. Maeda, T. Takeuchi, *J. Antibiotics*, **18A**, 101 (1965).
3) Y. Suhara, K. Maeda, H. Umezawa, M. Ohno, *Tetr. Lett.*, 1239 (1966).
4) T. Ikekawa, H. Umezawa, Y. Iitaka, *J. Antibiotics*, **19A**, 49 (1966).

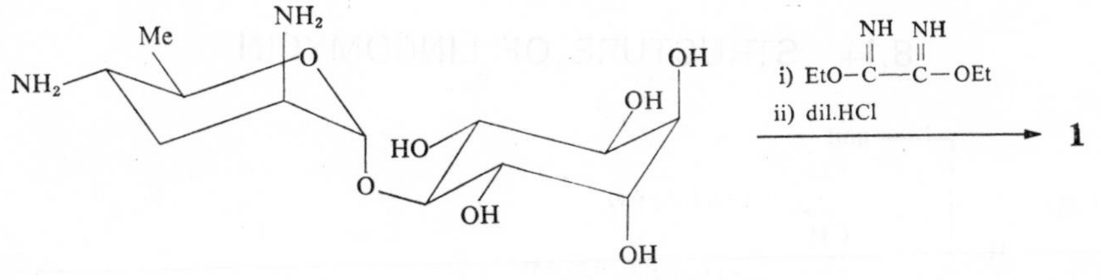

8

Route II[5]

The key step in this synthesis is the introduction of amino groups by stereoselective addition of NOCl to the double bond, followed by reduction. Another significant step is **15**⟶**16**, which involves the resolution of the originally racemic material without using any resolving reagent by the introduction of optically active inositol to form the diastereomers **16** and **17**.

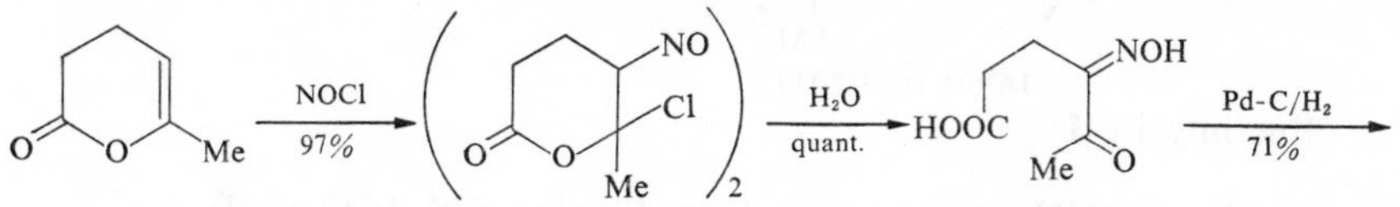

9   **10**   **11**

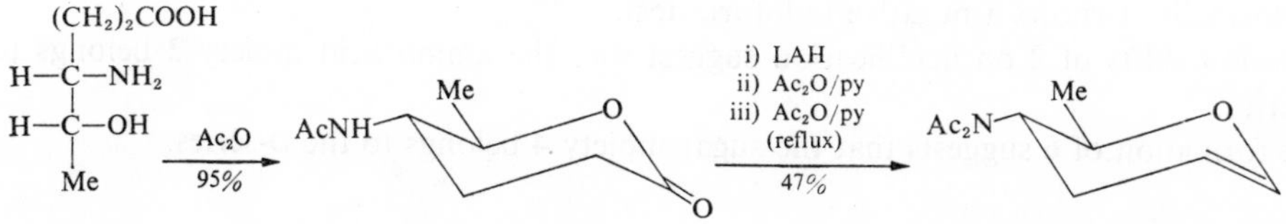

12   **13**   **14**

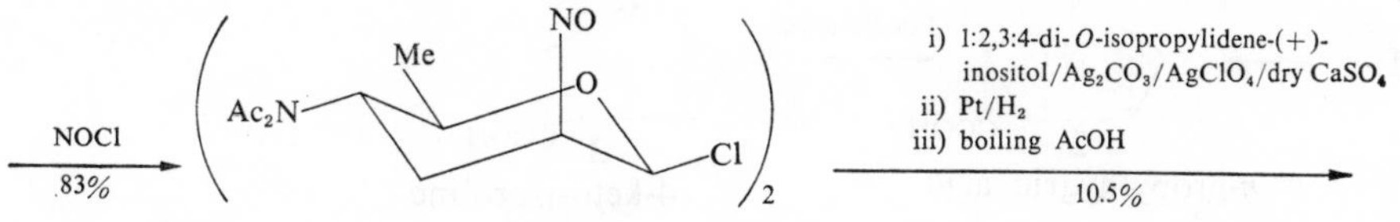

15

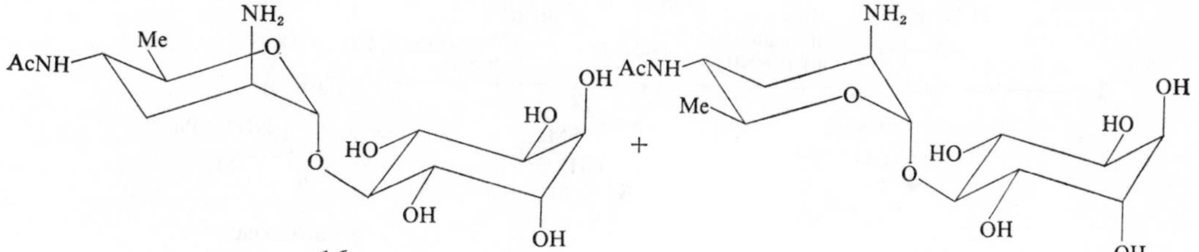

16   **17**

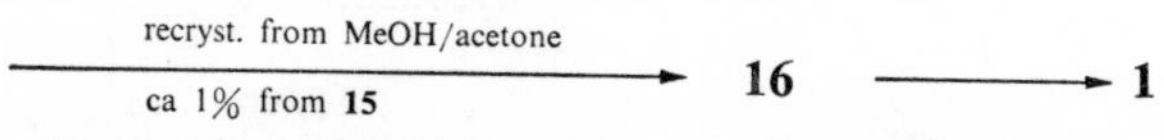

recryst. from MeOH/acetone
⟶ **16** ⟶ **1**
ca 1% from **15**

5) Y. Suhara, F. Sasaki, K. Maeda, H. Umezawa, M. Ohno, *J. Am. Chem. Soc.*, **90**, 6560 (1968); Y. Suhara, F. Sasaki, G. Koyama, K. Maeda, H. Umezawa, M. Ohno, *ibid.*, **94**, 6501 (1972).

8.14   STRUCTURE OF LINCOMYCIN

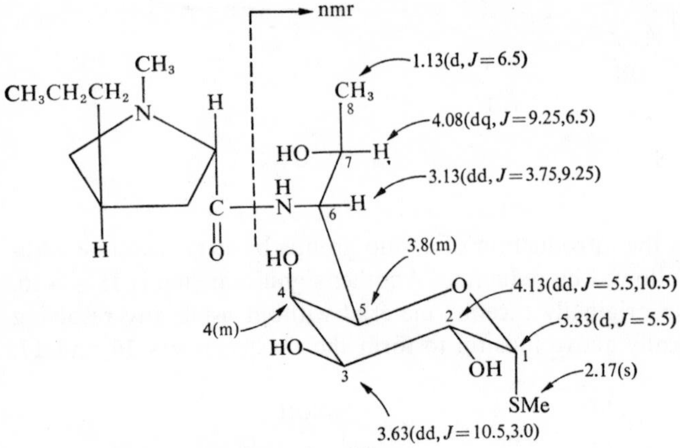

lincomycin **1**[1]

1) The presence of a –CONH– group was indicated by ir (1530, 1640 cm^{-1}).

2) The nmr spectrum was measured with the sugar moiety **4** in D_2O,[2] and indicated the presence of axial –SMe group.

3) Abnormally, **1** shows a negative iodoform test.

4) Rotation shifts of **2** on acidification suggest that the amino acid moiety **2** belongs to the L-series.

5) The formation of **6** suggests that the sugar moiety **4** belongs to the D-series.

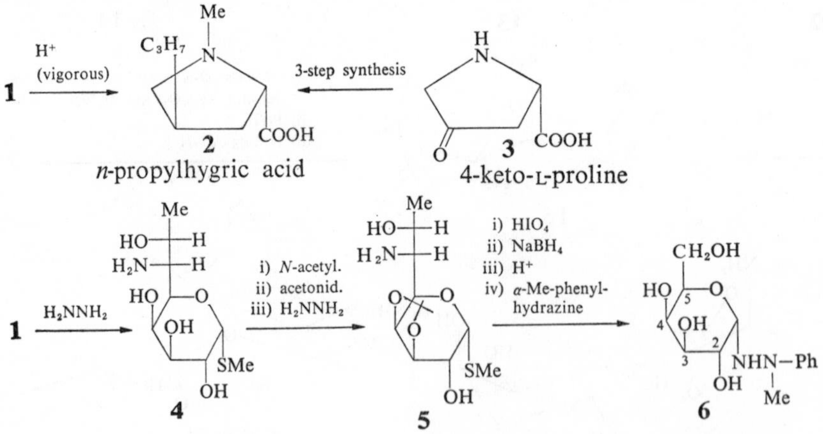

D-galactose
α-methylphenyl-
hydrazone

1) H. Hieksema, B. Bannister, R. D. Birkenmeyer, F. Kagan, B. S. Magerlein, F. A. MacKeller, W. Schroeder, G. Slomp, R. R. Herr, *J. Am. Chem. Soc.*, **86**, 4223 (1964); *ibid.*, **89**, 2444, 2448, 2454, 2459 (1967).
2) G. Slomp, F. A. MacKellar *J. Am. Chem. Soc.*, **89**, 2454 (1967).

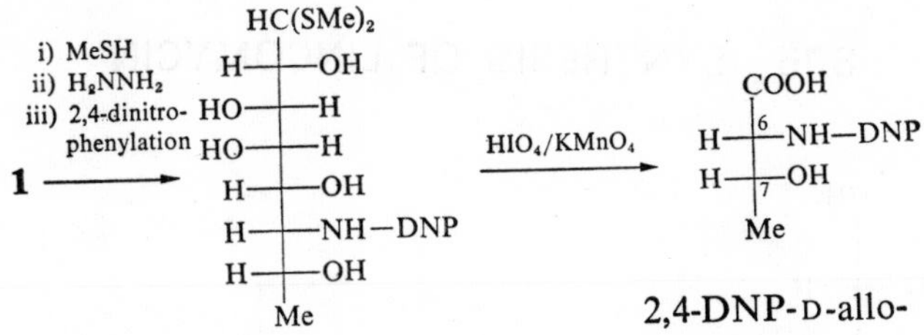

i) MeSH
ii) H₂NNH₂
iii) 2,4-dinitro-
phenylation

1 →

$HIO_4/KMnO_4$

7

2,4-DNP-D-allo-
threonine **8**

Remarks

Lincomycin **1** is an antibiotic with activity against gram-positive organisms. It was isolated from *Streptomyces lincolnēnsis* var. *lincolnensis*.

Celesticetin **9** is a congener of lincomycin.

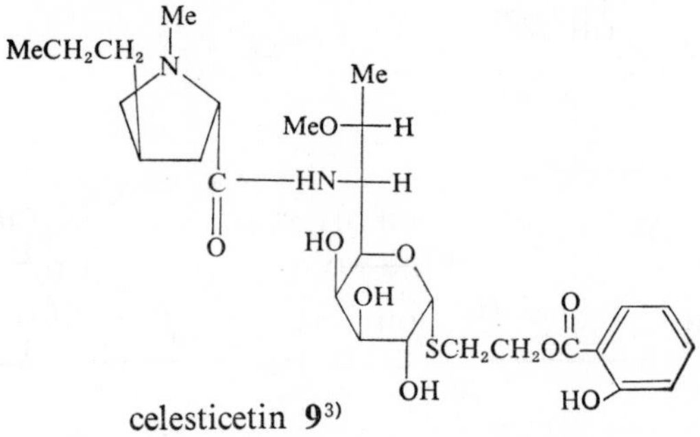

celesticetin **9**[3)]

---

3) H. Hoeksema, *J. Am. Chem. Soc.*, **86**, 4224 (1964); *ibid.* **90**, 755 (1968).

8.15   SYNTHESIS OF LINCOMYCIN

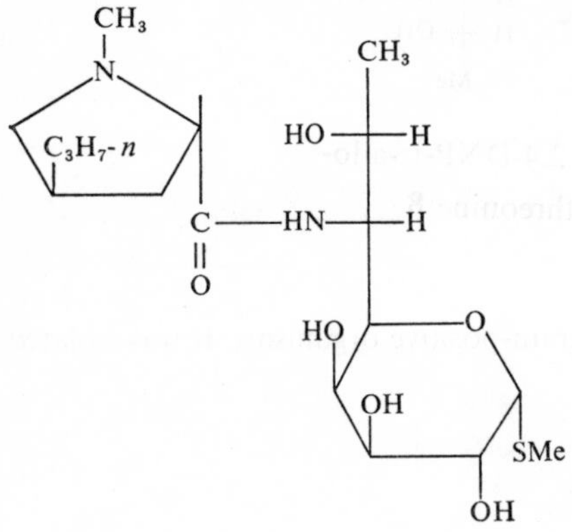

lincomycin **1**

Route I[1)]

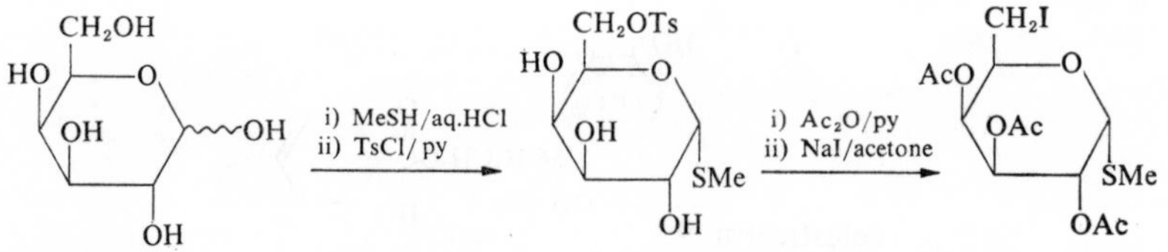

i) MeSH/aq.HCl
ii) TsCl/py

2

i) Ac$_2$O/py
ii) NaI/acetone

3

NaNO$_2$/DMF
20%

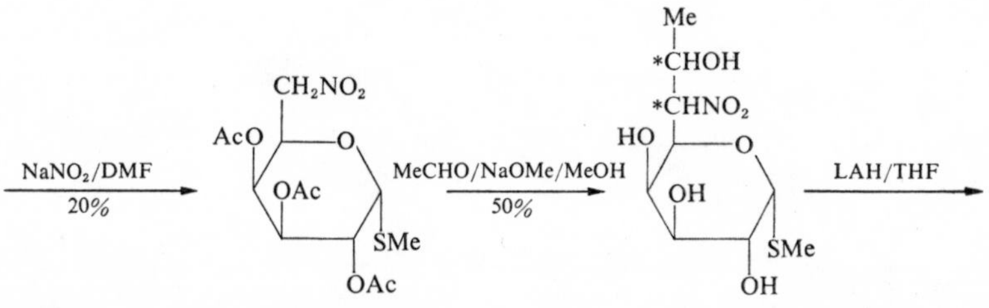

4

MeCHO/NaOMe/MeOH
50%

5 (mixture of isomers)

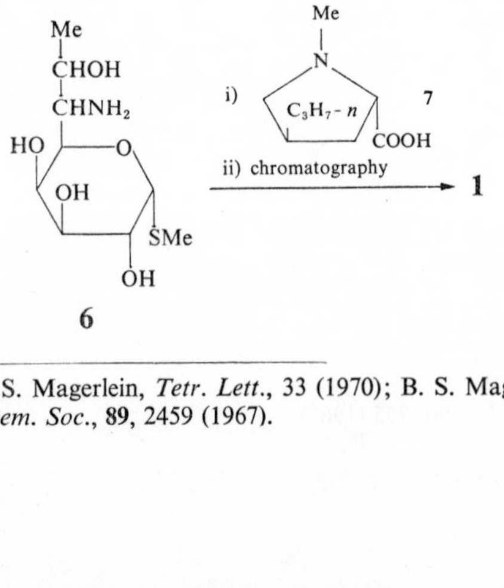

LAH/THF

i)

ii) chromatography

6

7

→ **1**

D-galactose

---

1) B. S. Magerlein, *Tetr. Lett.*, 33 (1970); B. S. Magerlein, R. D. Birkenmeyer, R. R. Herr, F. Kagan, *J. Am. Chem. Soc.*, **89**, 2459 (1967).

A significant point of this procedure is the introduction of the *S*-methyl group in the first stage of the synthesis.

Route II[2]

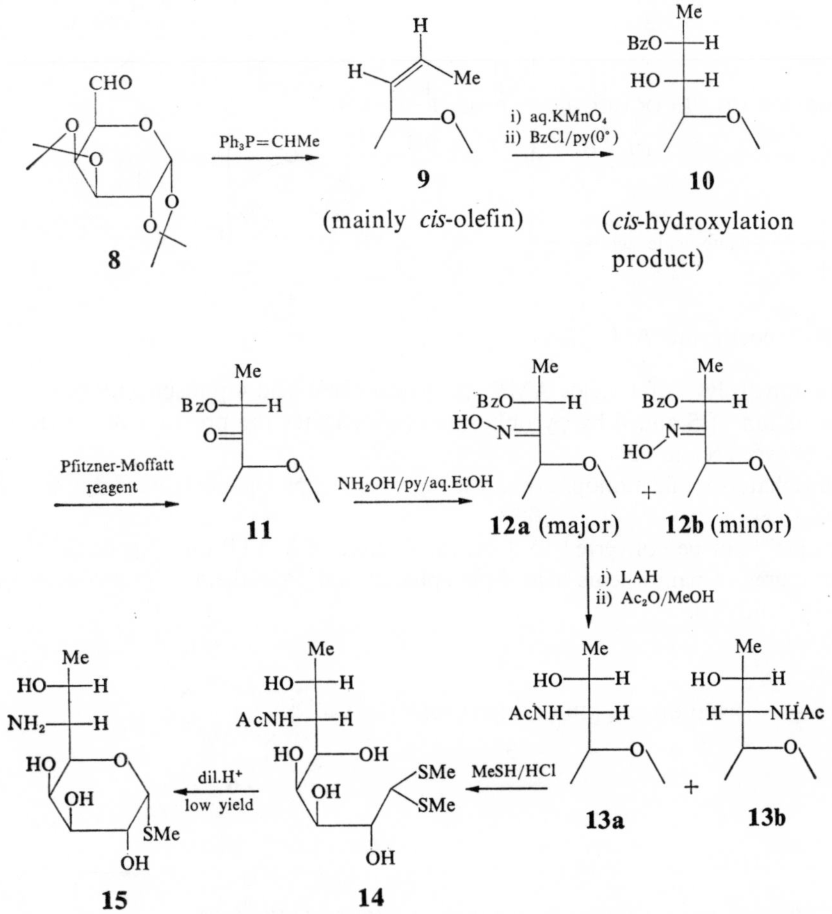

9
(mainly *cis*-olefin)

10
(*cis*-hydroxylation product)

11

12a (major)   12b (minor)

13a   +   13b

15   14

2) G. S. Howarth, W. A. Szarek, J. K. N. Jones, *J. Chem. Soc.* C, 2218 (1970).

8.16 STRUCTURE AND SYNTHESIS OF COENZYME A

pantetheine
3

$$CH_3 \quad O \quad O$$

$$HSCH_2CH_2NHCOCH_2CH_2NHCOCHCCH_2O-P-O-P-O-CH_2$$

$$HO \ CH_3 \quad OH \ OH$$

pantothenic acid
2

coenzyme A 1

1) Mild acid hydrolysis of 1 yields AMP and pantothenic acid 4-phosphate.[1]
2) The formation of 5 and 6 by pyrophosphatase indicates the presence of a pyrophosphate linkage between 5 and 6.[2]
3) A 3'-phosphate-specific monophosphatase removes one phosphate from C-3 of the adenosine moiety.
4) Pantetheine[5] can be converted to 1 on incubation with ATP and pigeon liver extracts.[6]
5) The structures of pantothenic acid 4-phosphate[3] and pantetheine-4-phosphate[4] were confirmed by synthesis.

$$1^{7)} \xrightarrow{H^+/H_2O} HSCH_2CH_2NH_2 + HOOCCH_2CH_2NH_2 + H_3PO_4 +$$

$$\qquad\qquad\qquad\quad 4 \qquad\qquad\qquad 5$$

6

$$1 \xrightarrow{pyrophosphatase} \begin{array}{c} Me \quad O \quad O \\ HSCH_2CH_2NHCOCH_2CH_2NHCOCHCCH_2OPOH + HOPOCH_2 \\ HO \ Me \quad OH \quad HO \end{array}$$

pantetheine-4-phosphate 7

adenosine-3',5'-diphosphate
8

---

1) J. Baddiley, E. M. Thain, J. Chem. Soc., 2253, 3421 (1951); 3783 (1952).
2) G. D. Novelli, N. O. Kaplan, F. Lipmann, Federation Proc., 9, 209 (1950); T. P. Wang, L. Shuster, N. O. Kaplan, J. Am. Chem. Soc., 74, 3204 (1952); T. P. Wang, N. O. Kaplan, J. Biol. Chem., 206, 311 (1954).
3) J. Baddiley, E. M. Thain, J. Chem. Soc., 246, 3421 (1951); T. E. King, E. M. Strong, J. Biol. Chem., 189, 315 (1951).
4) J. Baddiley, E. M. Thain, J. Chem. Soc., 1610 (1953).
5) G. M. Brown, E. E. Snell, J. Am. Chem. Soc., 75, 1691 (1953); J. Biol. Chem., 198, 375 (1952).
6) R. A. McRorie, W. L. Williams, J. Bacteriol., 61, 737 (1951).

Remarks

Coenzyme A was first discovered by Lipmann in 1945 during studies of the enzymatic acetylation of sulphanilamide.[8] A rich source of **1** is provided by *Streptomyces fradiae*.[7]

1 participates in many biochemical reactions, e.g. those involved in the Krebs cycle. Pantothenic acid belongs to the vitamin B group of compounds.

Synthesis

Route I[9]

14 ⟶ 1: The key step in this total synthesis is the application of a newly developed synthesis for asymmetric pyrophosphates. Hydrolysis of the cyclic phosphate gives two prod-

7) W. H. deVries, W. M. Govier, J. S. Evans, J. D. Gregory, G. D. Novelli, M. Soodak, F. Lipmann, *J. Am. Chem. Soc.*, **72**, 4838 (1950); J. Baddiley, E. M. Thain, *J. Chem. Soc.*, 2253 (1951); J. D. Gregory, G. D. Novelli, F. Lipmann, *J. Am. Chem. Soc.*, **74**, 854 (1952).

8) F. Lipmann, *J. Biol. Chem.*, **160**, 173 (1945); F. Lipmann, N. O. Kaplan, *J. Biol. Chem.*, **162**, 743 (1946).

9) J. G. Moffatt, H. G. Khorana, *J. Am. Chem. Soc.*, **81**, 1265 (1959); *ibid.*, **83**, 663 (1961).

ucts, **1** and **15**, separation of which was achieved by chromatography using ECTEOLA cellulose powder.

Route II[10]

$$\mathbf{10} \xrightarrow[\text{ii) } H_2O_2]{\text{i) Na/liq.NH}_3} (-SCH_2CH_2NHCOCH_2CH_2NHCOCHCCH_2O-\overset{\overset{\displaystyle O}{\|}}{\underset{\displaystyle OH}{P}}-CH)_2$$

with Me and HO Me substituents

16

i) condensation
ii) RNase T$_2$
iii) reduction ⟶ **1**

17

(PhO)$_2$POCl/Bu$_3$N ⟶

18

18 ⟶ 1: Takadiastase ribonuclease T$_2$ (RNase T$_2$) is known to split adenosine 2′,3′-cyclic phosphate to the 3′-phosphate.[11] Applying this enzyme, only CoA is produced, in 63% overall yield, whereas Khorana's method gives only 15% yield after tedious separation of iso-CoA by chromatography.

Route III[12]

$$NCCH_2CH_2NHCOCHCCH_2OH \xrightarrow[\text{ii) } H_2/Pd]{\text{i) (PhCH}_2O)_2POCl} NCCH_2CH_2NHCOCHCCH_2O-\overset{\overset{\displaystyle O}{\|}}{\underset{\displaystyle OH}{P}}-OH \xrightarrow[\text{ii) RNase T}_2]{\text{i) } 14/py}$$

with HO Me substituents

19   **20**

21

NH$_2$CH$_2$CH$_2$SH ⟶

22

$$\xrightarrow{H^+} \mathbf{1}$$

10) A. M. Michelson, *Biochim. Biophys. Acta*, **50**, 605 (1961); *ibid.*, **93**, 71 (1964).

11) M. Naoi-Tada, K. Sato-Asano, F. Egami, *J. Biochem.* (Tokyo), **46**, 757 (1959).

12) M. Shimizu, O. Nagase, S. Okada, Y. Hosokawa, H. Tagawa, Y. Abiko, T. Suzuki, *Chem. Pharm. Bull.*, **15**, 655 (1967); O. Nagase, *ibid.*, **15**, 648 (1967); M. Shimizu, G. Ohta, O. Nagase, S. Okada, Y. Hosokawa, *ibid.*, **13**, 180 (1965).

21 ⟶ 22 ⟶ 1: The key step of this method is the introduction of thiazoline which is subsequently hydrolyzed.

Route IV[13)]

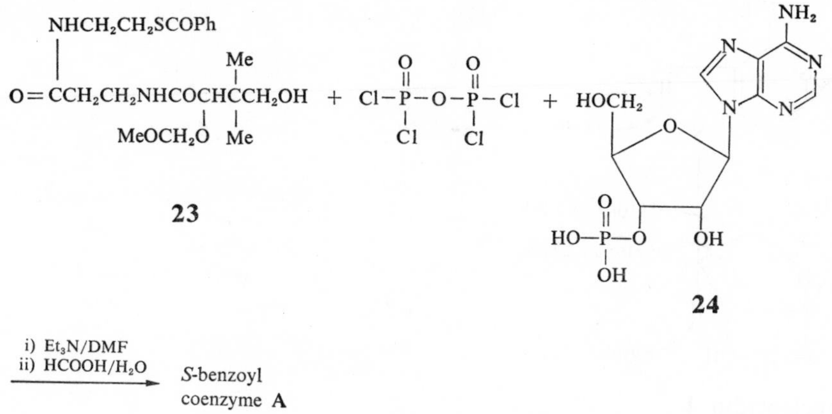

23   **24**

i) Et₃N/DMF
ii) HCOOH/H₂O ⟶ *S*-benzoyl
                      coenzyme **A**

25

13) W. Gruber, F. Lynen, *Ann. Chem.*, **659**, 139 (1962).

8.17   STRUCTURE OF NUCLEOCIDIN

nucleocidin **1**

pmr in py-d_5
ms: 652.2158 (M^+ of tetra-TMS of **1**; calc. 652.2182)
α_D(EtOH/0.1 N HCl): $-33.3°$

uv(0.1 N HCl): 225–227 ($E_{1cm}^{1\%}$ 392)

uv(0.1 N NaOH): 259 ($E_{1cm}^{1\%}$ 406)

^{19}F nmr in py-d_5 (56.4 MHz, ref. $CFCl_3$): 119.6 ppm (d, t, $J=9.0, 18.0$)

1) Acid hydrolysis of **1** yields adenine and a reducing sugar.

2) $BaSO_4$ was detected from **1** after treatment with $Ba(NO_2)_2$ and dil. acid at room temperature or $BaCl_2$ and 2N HCl at 100°, but not after treatment with $BaCl_2$ and dil. acid at room temperature.

3) Sulfamic acid was detected by paper chromatography after hydrolysis of **1** with Dowex 50.

4) The position of the SO_2NH_2 group was deduced from the mass spectrum fragmentation pattern of the tetra-TMS derivative of **1** (presence of fragment *a* and absence of fragment *b*: fragment *b* is characteristic of the mass spectra of 5′-TMS derivatives of furanose nucleosides) and from a consideration of the chemical shift of the –OCH_2– group (0.6 ppm downfield compared with that of adenosine).

5) The location of the fluorine atom was determined from nmr data.

Remarks

Nucleocidin is an antitrypanosomal antibiotic, and was isolated from *Streptomyces calvus* in 1957.[1,2] The major difficulty in the determination of its structure was to find out the presence of the fluorine atom. Nucleocidin was the first naturally occurring fluoro-sugar derivative to be isolated, and the first example of a naturally occurring *N*-unsubstituted sulfamic acid derivative.

1) S. O. Thomas, V. L. Singleton, J. A. Lowery, R. W. Sharpe, L. M. Pruess, J. N. Porter, J. H. Mowat, N. Bohonos, *Antibiotics Ann.*, 716 (1956–7).
2) R. I. Hewitt, A. R. Gumble, L. H. Taylor, W. S. Wallace, *Antibiotics Ann.*, 722 (1956–7).
3) C. W. Waller, J. B. Patrick, W. Fulmor, W. E. Meyer, *J. Am. Chem. Soc.*, **79**, 1011 (1957).
4) J. B. Patrick, W. E. Meyer, *Abstracts, 156th National Meeting of the Am. Chem. Soc.*, Atlantic City, N. J., Sept., 1968.
5) G. O. Morton, J. E. Lancaster, G. E. Van Lear, W. Fulmor, W. E. Meyer, *J. Am. Chem. Soc.*, **91**, 1535 (1969).

8.18   STRUCTURE AND SYNTHESIS OF ARISTEROMYCIN

aristeromycin **1**[1]

$\alpha_D: -52.5°$

uv(H_2O):262(4.167)

1) The uv spectrum of **1** suggests an adenosine derivative.

2) **1** contains a *cis* diol moiety and yields a penta-acetate.

3) **1** is not an *N*-glycoside, since it is not hydrolysed by H_2SO_4.

4) The nmr spectrum indicates the presence of two HOCH<, N–CH<, –CH<, –CH$_2$– and –CH$_2$OH groups.

5) The structure and stereochemistry including the absolute configuration were determined by x-ray analysis of the hydrobromide.[1]

Remarks

Aristeromycin **1** was isolated from *Streptomyces citricolor* in 1967.[2] It is a new antibiotic which inhibits the growth of *Piricularia oryzae* and *Xanthomonas oryzae*, which are phyto-pathogenic organisms of rice and other plants. It is an unusual carbocyclic analogue of adenosine.

Compound **1** had been synthesized in 1966[3] before the isolation of aristeromycin in the hope that the replacement of the oxygen atom in the furanose ring of adenosine with a methylene group might lead to analogues with interesting biochemical and therapeutic properties.

Synthesis[3]

---

1) T. Kishi, M. Muroi, T. Kusaka, M. Nishikawa, K. Kamiya, K. Mizuno, *Chem. Commun.*, 852 (1967).
2) T. Kusaka, H. Yamamoto, M. Shibata, M. Muroi, T. Kishi, K. Mizuno, *J. Antibiotics* (Japan), **21**, 225 (1968).
3) Y. F. Shealy, J. D. Clayton, *J. Am. Chem. Soc.*, **88**, 3885 (1966).

8.19 STRUCTURE OF POLYOXIN A

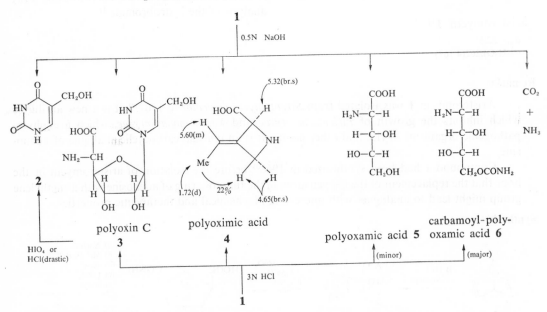

polyoxin A **1**

1) The furanosyl substituent in **3** is placed at N-1 of the uracil from the uv spectrum.
2) The nmr of *N*-acetyl-**3** shows that the amino group in **3** is at C-5′ (quartet at 4.69 (5′-H) becomes a doublet ($J=5.5$) on spin-decoupling of the NH_2 protons or addition of D_2O).
3) Positive Cotton effect and cd maximum of **3** indicate a β-configuration at the anomeric position.
4) The absolute configuration of **3** at C-5′ was determined from the ord of the *N*-dithiocarbamate derivative.
5) The configuration of the sugar was determined by a reaction sequence leading to β-D-allose (HNO_2; Pt/H_2; Rh/H_2; HCl/MeOH; $NaBH_4$; H^+).
6) Ozonolysis of **4** followed by hydrogenation (Pd/H_2) yielded $MeOOCCH_2NHCH_2COOH$.
7) The absolute configuration of **4** was determined from the ord of the *N*-dithiocarbamate derivative.
8) The structure of **5** was determined by the transformation of **5** into the known compound 2-acetamido-2-deoxy-β-L-xylose by acetylation, then Na/Hg reduction.

9) The formation from **6** of a five-membered unsaturated lactone containing the carbamyl group (2-acetamido-4-carbamyloxymethyl-4-hydroxybuten-2-oic acid lactone) indicated that the carbamyl group in **6** is at C-5.

10) The sequence of the three amino acids was established by a deamination process.

Remarks

Polyoxin complex, an antifungal antibiotics mixture, is in practical use as an agricultural fungicide in Japan. Twelve components, polyoxin A, B, C, D, E, F, G, H, I, J, K, and L, were isolated and the structures of all twelve components were determined.[2]

These compounds were isolated from *Streptomyces cacaoi* var. *asoensis*.[1]

1) K. Isono, J. Nagatsu, Y. Kawashima, S. Suzuki, *Agr. Biol. Chem.*, **29**, 848 (1965).
2) K. Isono, S. Suzuki, *Tetr. Lett.*, 1133 (1968); *ibid.*, 203 (1968); *Agr. Biol. Chem.*, **30**, 813, 815 (1966).

8.20 SYNTHESIS OF POLYOXIN J

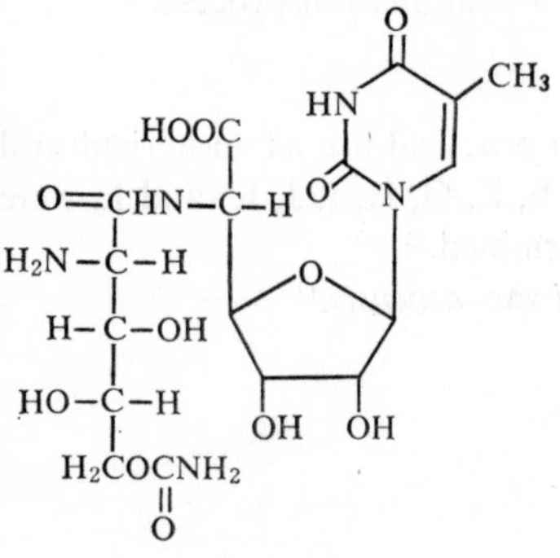

polyoxin J **1**[1]

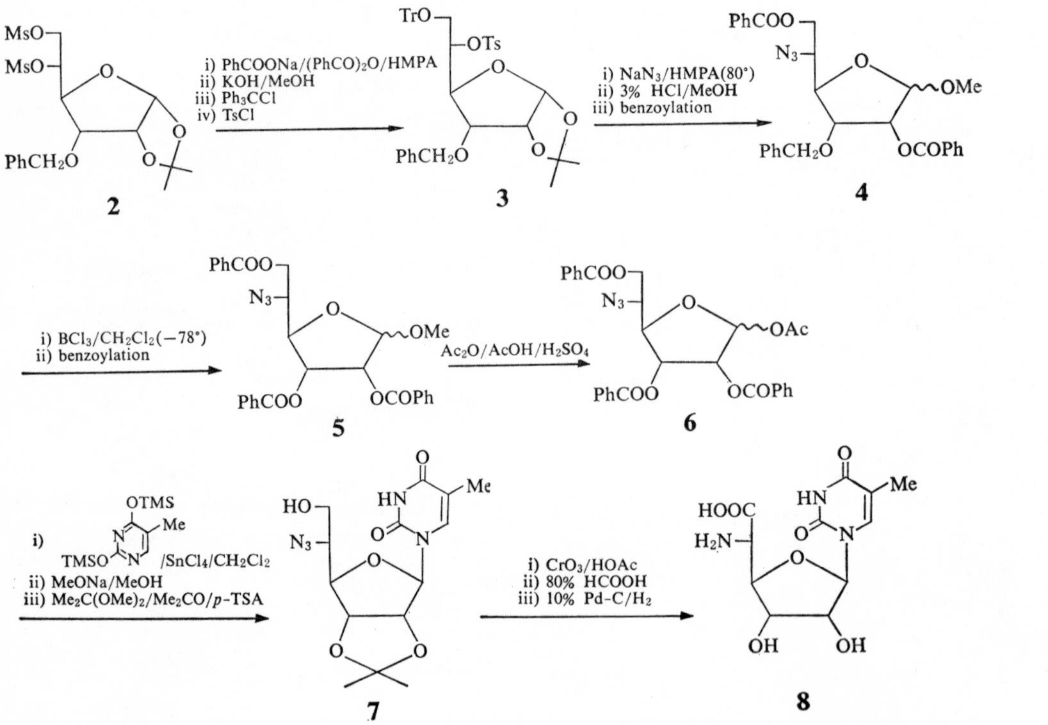

(Tr = Trityl)

1) H. Ohrui, H. Kuzuhara, S. Emoto, *Tetr. Lett.*, 4267 (1971); H. Kuzuhara, S. Emoto, *ibid.*, 5051 (1973); H. Kuzuhara, H. Ohrui, S. Emoto, *ibid.*, 5055 (1973).

4 ⟶ 5: The removal of the benzyl protecting group without reduction of the azide group is one of the key steps in this synthesis.

6 ⟶ 7: In reaction i) only the β-isomer is formed (almost quantitatively); using less of the catalyst led to the formation of the α-isomer.

Remarks

The polyoxin complex is an excellent agricultural fungicide isolated from the culture broth of *Streptomyces cacaoi* var. *asoensis*.[2] Polyoxin J has the simplest structure among the components of polyoxin complex.

2) S. Suzuki, K. Isono, J. Nagatsu, T. Mizutani, K. Kawashima, T. Mizuno, *J. Antibiotics* (Japan), **A18**, 131 (1965).

8.21 STRUCTURE OF AMICETIN AND
SYNTHESIS OF PLICACETIN

amicetin **1**[1)]

1) Brief acid hydrolysis of **1** yields amicetamine **2** and cytimidine **8**, which is further hydrolysed by 6 N HCl to cytosine, *p*-aminobenzoic acid and α-methylserine **5**.

2) Alkaline hydrolysis of **1** yields cytosamine **3**.

3) Methanolysis of **1** yields cytimidine **8**, the methyl glycoside of amicetose **7**, and the methyl glycoside of amosamine **6**. The two methyl glycosides are readily hydrolysed to **7** and **6**, respectively.

4) **7** gives positive iodoform and reducing tests, though its methyl glycoside does not.

5) Periodate oxidation of **7** yields acetaldehyde and succinic dialdehyde.

6) **6** gives positive iodoform and reducing tests, and yields dimethylamine on heating with alkali.

7) HIO_4 consumption, pK_a measurements and other studies suggest that **6** is 6,X-dideoxy–X-dimethylamino–aldohexose (X=unknown).

8) The structure and stereochemistry of **6** were established by synthesis from D-galactose.[2)]

9) The stereochemistry of the glycoside linkages was deduced by analysis of the nmr spectra of the anomeric protons in **3**.[3)]

10) Methylserine **5** consumes one mole of HIO_4, yielding formaldehyde and pyruvic acid.[4)]

11) The structure of **5** was confirmed by synthesis.[4)]

Synthesis of plicacetin 4

1) C. L. Stevens, K. Nagarajan, T. H. Haskell, *J. Org. Chem.*, **27**, 2991 (1962).
2) C. L. Stevens, P. Blumbergs, F. A. Daniher, *J. Am. Chem. Soc.*, **85**, 1552 (1963).
3) S. Hanessian, T. H. Haskell, *Tetr. Lett.*, 2451 (1964).
4) E. H. Flynn, T. W. Hinman, E. L. Caron, D. O. Woolf, *J. Am. Chem. Soc.*, **75**, 5867 (1953).

12 + 18 ⟶ 3: The key step of this synthesis is the stereospecific formation of the α-disaccharide linkage.

Remarks

Compound **1** has been isolated from *Streptomyces vinaceus-drappus* and *Str. plicatus.*[9]

Amicetin **1**, bamicetin, and plicacetin **3** are three structurally similar amino sugar nucleoside antibiotics. They are potent inhibitors of *in vitro* protein synthesis, inhibit the KB strain of human epidermoid carcinoma cells and increase the survival time of mice with leukemia-82. They also inhibit gram-positive and gram-negative bacteria as well as mycobacteria broth both *in vitro* and *in vivo*.

5) C. L. Stevens, J. Nemec, G. H. Rausford, *J. Am. Chem. Soc.*, **94**, 3280 (1972).
6) C. L. Stevens, N. A. Nielsen, P. Blumbergs, *J. Am. Chem. Soc.*, **86**, 1894 (1964).
7) C. L. Stevens, N. A. Nielsen, P. Blumbergs, K. G. Taylor, *J. Am. Chem. Soc.*, **86**, 5695 (1964).
8) T. H. Haskell, *J. Am. Chem. Soc.*, **80**, 747 (1958).
9) J. W. Hinman, E. L. Caron, C. DeBoer, *J. Am. Chem. Soc.*, **75**, 5864 (1953).

8.22 STRUCTURE OF BLASTICIDIN S

blastidic acid **2** ⟷ cytosinine **3**

2.93(s)

6.15(d, $J=7.5$)

7.73(d, $J=7.5$)

*6.45–6.00(br,2H)

blasticidin S 1[1)]

mp: 253–255° $\alpha_D(H_2O)$: +108°
uv(0.1 N HCl): 247 (ε 13,400)
uv(0.1 N NaOH): 266 (ε 8,850)
$pK_a(H_2O)$: 2.4, 4.6, 8.0, > 12

1) Controlled acid hydrolysis of **1** yields blastidic acid **2** and cytosinine **3**.

2) Assignement of the L-form of blastidic acid is based on ord comparisons with authentic β-amino acids.

3) Hydrogenation of cytosinine **3** yields cytosine and α-carboxy-β-aminotetrahydropyran **6**.

4) The position of linkage of **2** and **3** was determined from the following reation sequence and from a consideration of the pK_a of **1**.

1 → i) H₂ ii) OH⁻ → **4** → H⁺ → **5** + **6**

5) C-1′, C-4′, and C-5′ were determined as follows.

ozonolysis after appropriate protection

3

7

8
(erythro-D)

HNO₂

meso-tartaric acid **9**

gougerotin **10**

1) N. Otake, S. Takeuchi, T. Endo, H. Yonehara, *Tetr. Lett.*, 1405, 1411 (1965); *Agr. Biol. Chem.*, **30**, 126 (1966); J. J. Fox, K. A. Watanabe, *Tetr. Lett.*, 897 (1966); H. Yonehara, N. Otake, *ibid.*, 3785 (1966).

6) **7** is identical with that from cytidine, indicating the β-configuration at C-1′.

7) Formation of **8** and **9** established the 4′(S),5′(S)-configuration.

Remarks

Blasticidin S **1**, which was isolated from *Streptomyces griothromogeus*, is a new antibiotic which inhibits the virulent fungus *Piricularia oryzae*, a serious cause of rice blast disease in Japan. The structure of **1** was confirmed by x-ray analysis.[2]

Cytosinine **3** was synthesized from D-galactose.[3] Blasticidin S **1** was synthesized by coupling of **2** and **3**.[4]

Streptomyces gougerotii produces[5] a similar antibiotic, gougerotin, whose structure was determined as **10**.[6] A total synthesis of **10** was achieved by Fox *et al.*[7]

2) S. Onuma, Y. Nawata, Y. Saito, *Bull. Chem. Soc. Japan*, **39**, 1091 (1966).

3) T. Kondo, H. Nakai, T. Goto, *Tetr. Lett.*, 1881 (1972); *Tetr.*, **29**, 1801 (1973).

4) H. Yonehara, N. Otake, *Antimicrobial Agents and Chemotherapy*, 855 (1966).

5) T. Kanzaki, E. Higashide, H. Yamamoto, M. Shibata, K. Nakazawa, H. Iwasaki, T. Takewaka, A. Miyake, *J. Antibiot.*, Ser. A., **15**, 93 (1962).

6) J. J. Fox, Y. Kuwada, K. A. Watanabe, *Tetr. Lett.*, 6029 (1968); K. A. Watanabe, M. P. Kotick, J. J. Fox, *Chem. Pharm. Bull.*, **17**, 416 (1969).

7) K. A. Watanabe, E. A. Falco, J. J. Fox, *J. Am. Chem. Soc.*, **94**, 3272 (1972).

CHAPTER 9

Carboaromatic and Related Compounds

9.1 INTRODUCTION

This chapter deals with a group of natural products generally called 'phenolic compounds' or simply 'phenolics', though some may exhibit neutral or acidic character due to the absence of a free phenolic group or the presence of an acidic group. The compounds containing carbo-aromatic rings, such as benzene, naphthalene and anthracene, substituted generally by one or more hydroxyl groups or the biogenetic equivalents are the first representatives of this group (section nos. 2 through 17). If these compounds are oxidized to form 1,2- or 1,4-diketones, the products are deeply colored and are called quinones.[1] They form the second set of compounds (section nos. 18 through 26).

According to formal chemical classification, oxygen heterocyclic compounds such as furan, pyran and coumarin derivatives are treated separately from carboaromatic compounds. However the chemical properties, especially the phenolic properties, of naturally occurring O-heterocyclic compounds[2] are quite similar to those of the carboaromatic compounds. The biogenetic origins of the two groups are also inseparable. Thus the O-heterocyclic compounds are included in this chapter and account for the bulk of this group (section nos. 27 through 59).

Naturally occurring phenolic compounds were studied very extensively even in the early stages of natural product chemistry. Natural dyes such as alizarin and plant pigments such as flavonoids are typical cases. Most of the these compounds have structures with aromatic properties and, after the establishment of the structures of aromatic rings by Kekúle, final conclusions on the structures of some important members of this group were drawn as early as the last century. Thus the chemistry of phenolic compounds is sometimes assumed to be rather old-fashioned. However, new attractive topics in this field are always appearing as a result of fresh discoveries in nature, and we can say that the chemistry of phenolic compounds is both an old and a new field in natural product chemistry.

As a result, this chapter covers a quite large area of accumulated data and the selection of a limited number of topics has been a difficult task. Thus, many have been selected for historical or pedagogical reasons.

Although the chemical characteristics of all the compounds in the group are rather similar, the biosynthetic origins of phenolic compounds are very diverse. The main pathways can be summarized as follows:

The acetate-malonate pathway (see Chapter 1) is quite widespread in mold metabolism and varieties of carboaromatic and O-heterocyclic mold metabolites are biosynthesized from one acetyl and a number of malonyl units (see section nos. 5–7, 11, 12, 22–24, 28, 31-37, 40, 41, 43 of this chapter). The biosynthesis of such phenolic compounds has been extensively studied[3] and some examples are shown below (section nos. 5, 6, 12, 28, 37, 41 and 43 of this chapter). In higher plants some phenolic compounds are also assumed to be formed by this pathway (section nos. 2, 3, 8 and 9 of this chapter).

The isoprenoids formed by the mevalonic acid pathway (see Chapter 1) are sometimes modified by oxidative reaction to aromatic compounds and appear as natural products. The simplest cases are p-cymene 1 and thymol 2. Since these compounds are strictly monoterpenes, sesquiterpenes and so on, no topics of this type have been selected in this chapter. Examples of aromatic sesquiterpenes and diterpenes are included in Volume I.

---

1) R. H. Thomson, *Naturally Occurring Quinones*, Academic Press, 1971.
2) F. M. Dean, *Naturally Occurring Oxygen Ring Compounds*, Butterworths, 1963.

Me

1 *p*-cymene (R=H)
2 thymol (R=OH)

Me Me

The last but most important pathway for the biosynthesis of phenolic compounds in higher plants is the shikimic acid pathway (see Chapter 1). Cinnamic acid derivatives (**5, 6, 7**) are directly formed from aromatic amino acids such as **4**. Simple C_6–C_3 compounds such as anethole **10**, safrole **11** and eugenol **12**, components of some kinds of essential oils, and *p*-coumaryl **13**, coniferyl **14** and sinapy alcohols **15**, monomeric precursors of lignins (section no. 16), are also derived from the aromatic amino acids. Simple C_6–C_1 compounds such as *p*-hydroxybenzoic acid and gallic acid are formed directly from shikimic acid **3** or from the C_6–C_3 compounds by the loss of C_2 units. Some *O*-heterocyclic compounds such as coumarins (e. g. **8** and **9**) (section no. 38) and benzofurans (section no. 27) have the same biosynthetic origin. By the condensation of two or more C_6–C_3 units, lignans (section nos. 13, 14) and other polyphenolic compounds (section nos. 16, 17, 30) are derived.

COOH

sugars ⟶ ⟶

HO   OH
   OH
shikimic acid **3**

COOH

NH₂
phenylalanine **4**

−NH₃

COOH

cinnamic acid **5**

⟶

COOH

HO   R
p-coumaric acid (R=H) **6**

⟶ (R=OH) ⟶ (R=O−glu) ⟶
2,4-dihydroxy-
cinnamic acid **7** 2-glycoside of **7**

R'O   O   O

umberiferone **8** (R′=H)
helniarin   **9** (R′=Me)

MeO—⟨ ⟩—CH=CH—Me

10

O
   ⟨ ⟩—CH₂−CH=CH₂
O

11

MeO
HO—⟨ ⟩—CH₂−CH=CH₂

12

HO—⟨ ⟩—CH=CH−CH₂OH

13

HO—⟨ ⟩—CH=CH−CH₂OH
   OMe

14

OMe
HO—⟨ ⟩—CH=CH−CH₂OH
   OMe

15

There exist many compounds formed by the combination of these pathways. The condensation of three acetate units with the C_6–C_3 unit leads to the formation of stilbenes **18**, chalcones **19** and all flavonoids (section nos. 44–59 of this chapter), the largest group of *O*-heterocyclic compounds.

The participation of prenyl unit(s) is frequently observed for shikimate-derived phenolic compounds, as in the case of furocoumarins (section no. 39), for acetate-malonate derived phenolics as shown in section nos. 8–12 and 31 and for flavonoids as shown in section nos. 52, 55 and 58.

Oxidative modification of molecular frameworks is frequently observed with most phenolic compounds and the formation of quinonoids (section nos. 18–26) is an example of such modification. In the quinones the compounds shown in section nos. 18, 20, 21 and 26 of this chapter are shikimate-derived and those in 19, 22, 25 are acetate-malonate derived. Intermolecular and intramolecular oxidative coupling are important in the biosynthesis of phenolic compounds and examples of compounds thus formed are shown in section nos. 16, 22, 25, 28, 32, 43 and 49.

The phenolic compounds are distributed quite widely from microorganisms to higher plants and in animals (section nos. 19 and 25). Although rich sources for those derived from acetate-malonate are the lower fungi, compounds from the same precursors also appear in higher plants. On the other hand, shikimate-derived phenolic compounds have rarely been isolated from lower fungi, in contrast to their frequent occurrence in the Basidiomycetes and higher plants. In the latter, shikimate-derived phenolics are quite abundant; the absence or presence of lignin forms the distinction between herbaceous plants and woody plants. Some of the shikimate-derived phenolic compounds, especially flavonoids, are assumed to be good markets for chemical phyllogeny or systematics.[4]

Most of the phenolic compounds have been assumed to be 'waste products' and biological activities of the compounds have not attracted much attention compared with those of alkaloids and steroids. However, some antibiotics such as tetracyclines (section no. 7) and anthracyclinones (e.g., 20) and purgative principles of aloe, senna and rhubarb (e.g., 21) are well-known examples of the medicinal uses of phenolic compounds. Some naturally occurring quinones such as phylloquinone (section no. 20) and ubiquinones (section no. 18) play important roles in the oxidation-reduction systems of all kinds of living things. Even simple phenol-carboxylic acids, isoflavones and quinones such as juglone (22) are now known to show allelopathic effects. It appears that the biological activities of phenolic compounds are much more diverse than previously expected.

3) J. H. Richards, J. B. Hendrickson, *The Biosynthesis of Steroids, Terpenes, and Acetogenins*, Benjamin, 1964; W. B. Turner, *Fungal Metabolites*, Academic Press, 1971.

4) T. Swain, *Chemistry in Botanical Classification* (ed. G. Bendz, J. Santesson), p. 81, Nobel Foundation and Academic Press, 1974; E. C. Bath-Smith, *ibid.*, p. 93; J. B. Harborne, *ibid.*, p. 103; and references cited therein.

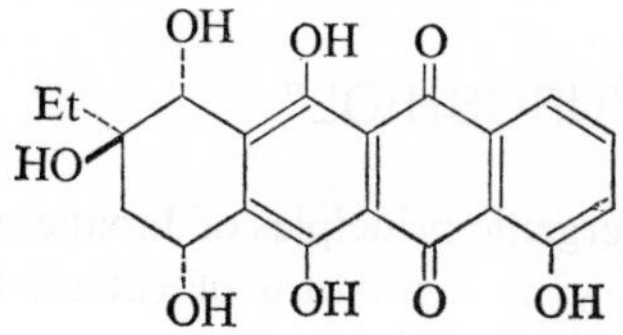

α-rhodomycinone **20**

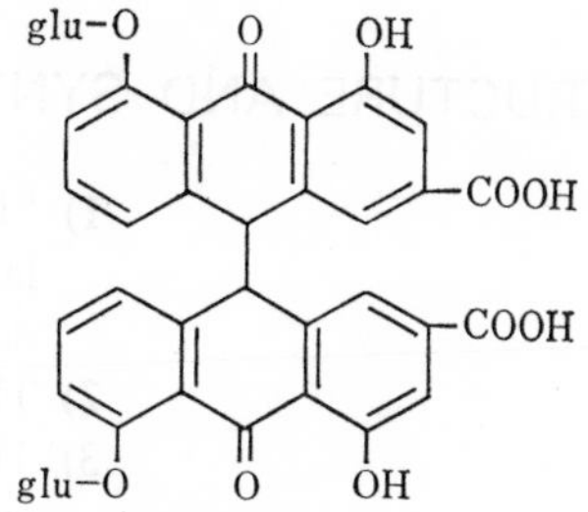

sennoside A and B **21**

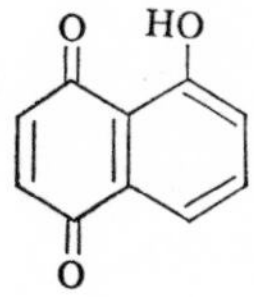

22

Since the chemistry of phenolic compounds have been extensively studied for a long time, there have been several comprehensive treatises coverring the field[1,2,5-7] There are also many excellent reviews dealing with particular topics, some being listed in the text.

5) T. A. Geissman (ed.), *The Chemistry of Flavonoid Compounds*, Pergamon Press, 1962.
6) T. A. Geissman, D. H. G. Crout, *Organic Chemistry of Secondary Metabolism*, Freeman, Cooper & Co., 1969.
7) W. D. Ollis (ed.), *Recent Developments in the Chemistry of Natural Phenolic Compounds*, Pergamon Press, 1961.

9.2   STRUCTURE AND SYNTHESIS OF "URUSHIOL"

OH
—OH
—R

urushiol

1) "Urushiol", the allergenic principles of Japanese lac[1] and poison ivy, is a mixture of catechol derivatives with C_{15} side chains (1–5).
2) The structure 4 was elucidated by ozonolysis.[2]
3) The double bonds were assigned as *cis* from the ir spectra.[2]

1 (R=–($C_{15}H_{31}$)
2 (R=–(CH_2)$_7$CH=CHC$_6$H$_{13}$)
3 (R=–(CH_2)$_7$CH=CHCH$_2$CH=CHC$_3$H$_7$)
4 (R=–(CH_2)$_7$CH=CHCH$_2$(CH=CH)$_2$Me
5 (R=–(CH_2)$_7$CH=CHCH$_2$CH=CHCH$_2$CH=CH$_2$

$$4 \longrightarrow \text{dimethyl ether} \xrightarrow{O_3} \begin{cases} \text{MeCHO} \\ \text{OHCCH}_2\text{CHO} \end{cases}$$

maleic anhydride adduct **6** $\xrightarrow[\text{ii) KMnO}_4]{\text{i) O}_3}$

OMe
—OMe
—(CH_2)$_7$COOH

7

Synthesis of 2[3]

$$C_6H_{13}C\equiv CH + I(CH_2)_6Cl \xrightarrow{\text{NaNH}_2/\text{liq. NH}_3} \quad C_6H_{13}C\equiv C(CH_2)_6Cl$$
   8   **9**   **10**

H_2/Ni

OH
—OH
—CHO

11

OC$_7$H$_7$
—OC$_7$H$_7$
—CHO

12

$+ \quad C_6H_{13}CH=CH(CH_2)_6Cl$

13

$\xrightarrow[\text{ii) H}_2\text{O}]{\text{i) Mg/Et}_2\text{O}}$

OC$_7$H$_7$
—OC$_7$H$_7$
—CH(CH_2)$_6$CH=CHC$_6$H$_{13}$
OH

14

$\xrightarrow{\text{KHSO}_4/\Delta}$

OC$_7$H$_7$
—OC$_7$H$_7$
—CH=CH(CH$_2$)$_5$CH=CHC$_6$H$_{13}$

15

$\xrightarrow{\text{Na/BuOH}}$ **2**

Remarks

"Urushiol", isolated from Japanese lac, *Rhus verniciflua* (Japanese name: urushi, Anacardiaceae),[1] is now known to be a mixture of compounds (1–4).[2] A similar mixture (1–3 and 5) was isolated from ivy, *Rhus radicans*.[3–5]

1) R. Majima, *Chem. Ber.*, **42**, 1418 (1909); *ibid.*, **55B**, 172 (1922).
2) S. V. Sunthankar, C. R. Dawson, *J. Am. Chem. Soc.*, **76**, 5070 (1954).
3) B. Loev, C. R. Dawson, *J. Org. Chem.*, **24**, 980 (1959); *J. Am. Chem. Soc.*, **78**, 6095 (1956).
4) W. F. Symes, C. R. Dawson, *J. Am. Chem. Soc.*, **76**, 2959 (1954).
5) K. H. Markiewitz, C. R. Dawson, *J. Org. Chem.*, **30**, 1610 (1965).

9.3 PHLOROGLUCIDES IN MALE FERN

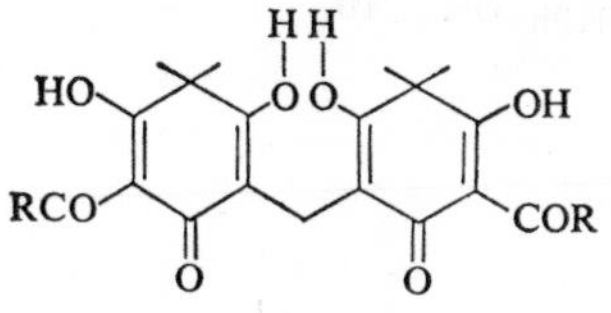

AA (R=Me)
PP (R=Et)
BB (R=Pr)

albaspidin **1**

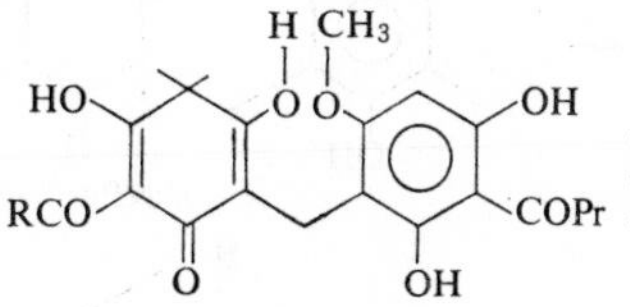

AB(R=Me)
PB (R=Et)
BB (R=Pr)

desaspidin **2**

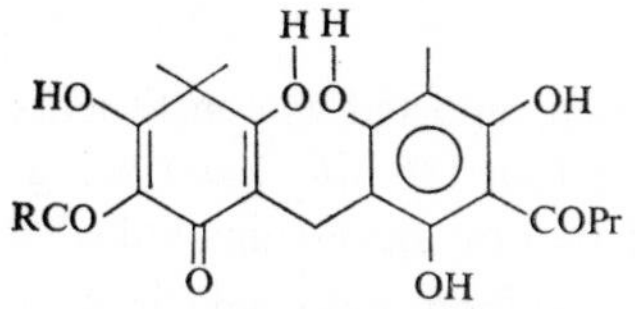

AB(R=Me)
PB (R=Et)
BB (R=Pr)

flavaspidic acid **3**

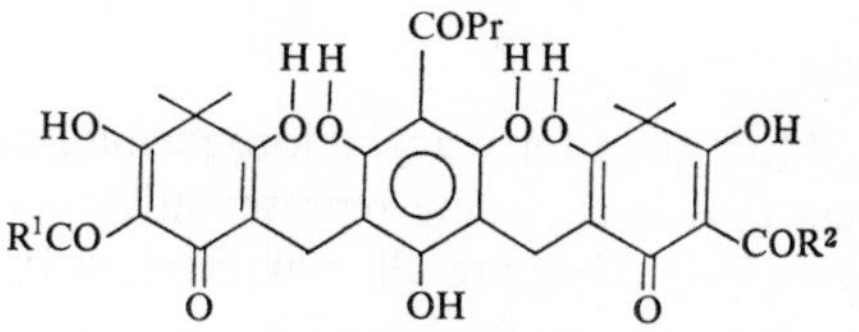

filixic acid **4**

BBB (R^1=R^2=Pr)
PBP (R^1=R^2=Et)
PBB (R^1=Et, R^2=Pr)
ABB (R^1=Me, R^2=Pr)
ABP (R^1=Me, R^1=Et)
ABA (R^1=R^2=Me)

1) 'Reductive alkaline cleavage' (**5**⟶**7, 8**)[1] was used most extensively for elucidation of the structures. The easy cleavage of the C–C bond can be visualized as a reverse Michael reaction. However, the reaction is complicated by the reversibility of the cleavage step: the reaction between **6** and **7** forms a *C*-methylated derivative of **5**, thus leading to phloroglucinols carrying methyl groups on ring positions that were free in the starting material.

2) Another reaction, the coupling of a diazo compound to phloroglucide (**5**⟶**7, 10**),[2] has the advantage over reductive alkaline cleavage of being more straightforward.

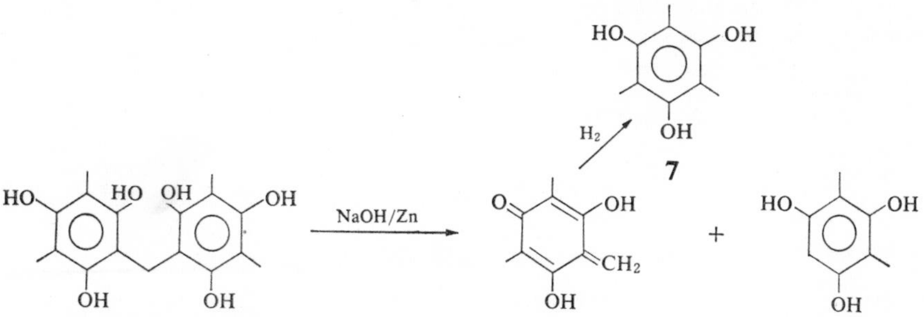

7

H$_2$

5   NaOH/Zn ⟶   **6**   +   **8**

1) R. Boehm, *Ann. Chem.*, **302**, 171 (1898).
2) R. Boehm, *Ann. Chem.*, **318**, 253 (1901).

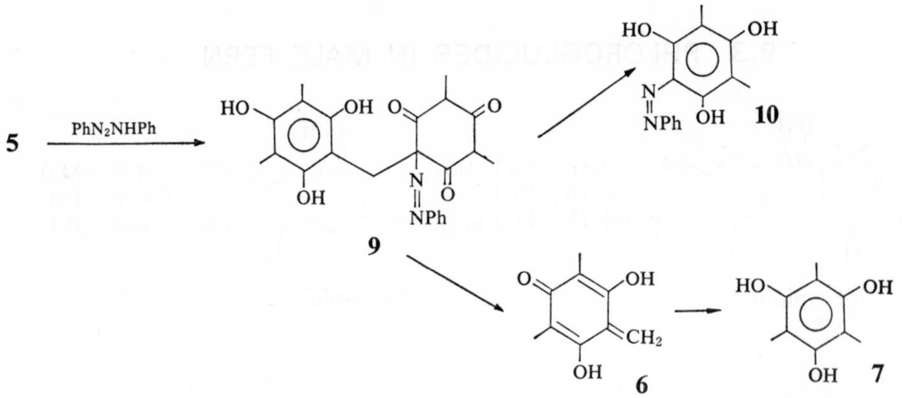

Remarks

More than forty acylphloroglucinol derivatives have been reported as constituents of the genus *Dryopteris*, some representative constituents of male fern (*D. filix-mas*) being shown above (**1–4**). They are all characterized by the presence of two or more rings linked through methylene groups. The rings fall into three main types, derived from phloroglucinol (or its *O*- and *C*-methyl derivatives), 3,5-dihydroxy-4,4-dimethyl-2,5-cyclohexadiene (filicinic acid **11**), or 6-propyl–2*H*-pyran–2,4(3*H*)-dione **12**.[3]

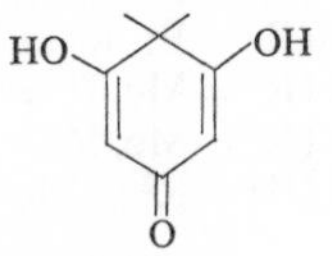

filicinic acid **11**

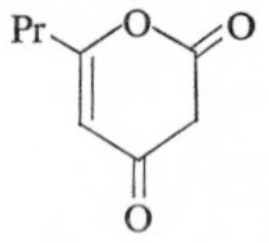

6-propyl–2H–pyran–2,4(3H)–
dione **12**

3) G. Berti, F. Bottari, *Progress in Phytochemistry* (ed. L. Reinhold, Y. Liwschitz), vol. 1, p. 589, Wiley, 1968.

9.4 BIOGENETIC-TYPE SYNTHESIS OF POLYKETIDES

The polyketide route, one of the most important routes in the biosynthesis of naturally occurring phenolic compounds, is now supported by a large number of tracer experiments. In addition, the synthesis of highly reactive poly-β-keto acids has been carried out to determine their reactivity in cyclization *in vitro*.

Synthesis of triketone[1]

This represents the earliest work on the synthesis of poly-β-ketone, and was reinvestigated following the detailed proposal of the "acetate theory".[2]

1 $\longrightarrow$ 2: Decarboxylation and recyclization of dehydroacetic acid **1** afforded the pyrone **2**, which is regarded as the masked form of triketone **3**.

3 $\longrightarrow$ 4: **4** is obtained by the intramolecular aldol cyclization of **3**.

3 $\longrightarrow$ 5,6,7: These products are formed by the intermolecular cyclization of two molecules of **3**, and their structures have been confirmed by Birch.[3]

Synthesis of dihydropinosylbin[3]

This was the first report demonstrating the cyclization-aromatization processes considered to be involved in polyketide biosynthesis. However, the direct cyclization of **9** to dihydropinosylbin **12** was unsuccessful.

1) J. N. Collie, *J. Chem. Soc,*, **91**, 1806 (1907).
2) A. J. Birch, F. W. Donovan, *Aust. J. Chem.*, **36**, 360 (1953).
3) A. J. Birch, D. W. Cameron, R. W. Richards, *J. Chem. Soc.*, 4395 (1960).

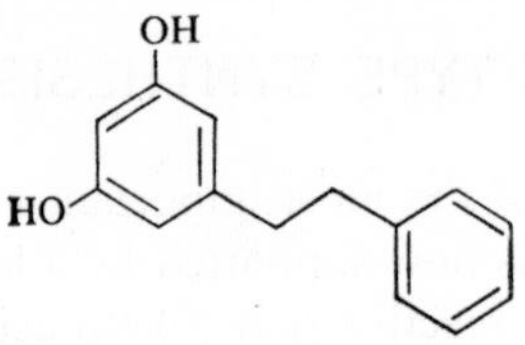

12

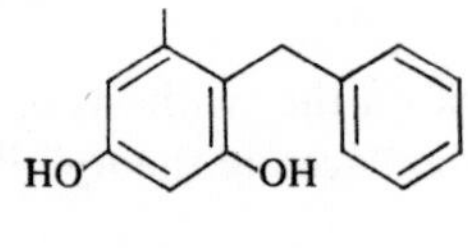

13

10 ⟶ 12,13: Two different modes of aldol-type cyclization (**a** and **b**) afforded dihydropino-sylbin **12** together with **13**.

Synthesis of dipyrones as masked poly-β-keto acids[4,5]

The effectiveness of pyrones as masked poly-β-keto acids was recognized in the synthesis of dihydropinosylbin.[3] Dipyrone **15**, a potential source of tetraketide, has been synthesized from triacetic acid pyrone **14**.

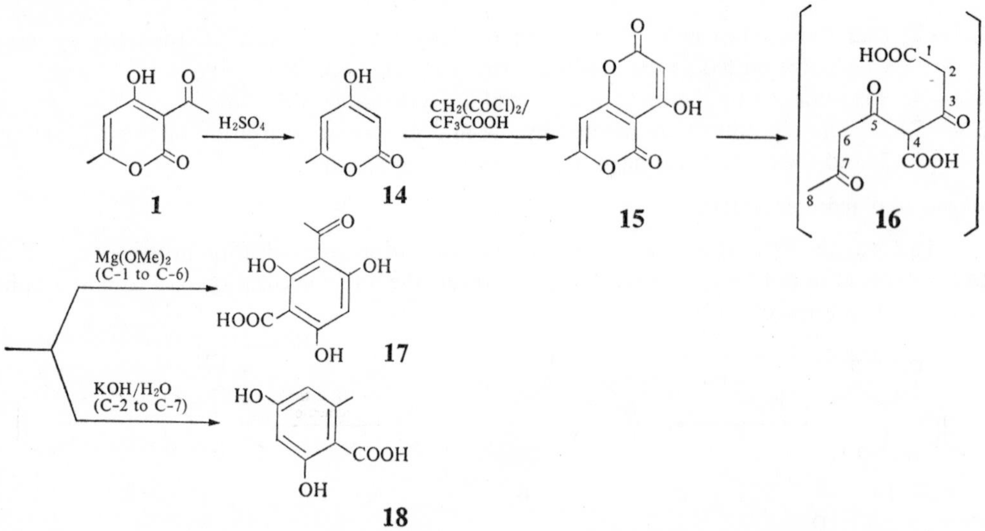

14 ⟶ 15: Malonyl chloride is an effective reagent for building up a condensed dipyone system which is a masked form of **16**.

15 ⟶ 17,18: Mg(OMe)$_2$ was found to catalyze Claisen condensation to phloroacetophenone **17**. Standard basic conditions, such as KOH, yielded orsellinic acid **18**.

4) T. Money, F. W. Comer, G. R. B. Webster, I. G. G. Wright, A. I. Scott, *Tetr.*, **23**, 3435 (1967).
5) J. L. Douglas, T. Money, *Tetr.*, **23**, 3545 (1967); J. L. Douglas, T. Money, *Can. J. Chem.*, **45**, 1990 (1967).

Synthesis of tetraacetic acid and its masked form[6,7)]

15 → 19: Selective hydrolysis was achieved by short treatment with base.

21 → 22: Compound **21**, obtained by the catalytic hydrogenation of **20**, yielded 6-methyl salicylic acid **22** on treatment with KOH, as had been expected on the basis of biosynthetic studies on **22**.

 3 → 23: The anion of **3**, generated by treatment with Li–diisopropyl amide, was carboxylated with CO_2.

24 → 25,26: Aldol condensation of **24** gave **25**, whereas Claisen condensation of **24** gave **26**.

Remarks

Tetraacetic acid lactone **20** was first obtained from cultures of *Penicillium stipitatum*[8)] in the presence of ethionine. It was later synthesized from dipyrone **15**. Ethionine acts to block C_1 introduction into enzyme-bound tetraacetic acid, and also blocks the cyclization process. As a result, accumulated tetraacetic acid is released into the culture medium in the form of the lactone **20**. This was the first demonstration of the presence of a poly-β-keto acid in polyketide biosynthesis. Unmasked tetraacetic acid was found to show the expected cyclization reactivity.

6) T. T. Howanth, G. P. Murphy, T. M. Harris, *J. Am. Chem. Soc.*, **91**, 517 (1969).
7) H. Guilford, A. I. Scott, K. Skingle, M. Yalpani, *Chem. Commun.*, 1127 (1968).
8) R. Bentley, P. M. Zwitkowits, *J. Am. Chem. Soc.*, **89**, 676 (1967).

9.5   BIOSYNTHESIS OF PATULIN

Major pathway for patulin biosynthesis[1]

6-methylsalicylic acid **1**

toluquinol **3**   *m*-cresol **2**

m-hydroxybenzyl alcohol **4**   **5**   **6**

gentisyl alcohol **7**   gentisyl aldehyde **8**   gentisic acid **9**

patulin **11**   pre-patulin **10**

1) P. I. Forrester, G. M. Gaucher, *Biochemistry*, **11**, 1102 (1972); *ibid.*, **11**, 1108 (1972).
2) S. W. Tannenbaum, E. W. Bassett, *J. Biol. Chem.*, **234**, 1861 (1959); *Biochem. Biophys. Acta.*, **40**, 535 (1960).
3) J. D. Bu'Lock, D. Hamilton, M. A. Hulme, A. J. Powell, H. M. Smalley, D. Shephard, G. N. Smith, *Can. J. Microbiol.*, **11**, 765 (1965); J. D. Bu'Lock, D. Shephard, D. J. Winstanley, *ibid.*, **15**, 279 (1969); A. I. Scott, M. Yalpani, *Chem. Commun.*, 945 (1967).

Although the conversion of 6-methylsalicylic acid into patulin was demonstrated by means of a tracer technique using radioisotopes,[2] and many metabolites such as *m*-cresol, *m*-hydroxy-benzyl alcohol, *m*-hydroxybenzaldehyde or toluquinol were subsequently identified as the most likely intermediates after 6-methylsalicylic acid in *Penicillium urticae*,[3] the details of the route of patulin biosynthesis remain obscure.

The major biosynthetic route has recently been postulated from results obtained by a kinetic pulse-labeling study in which labeled acetate and other pertinent metabolites were fed to cultures and the kinetics of incorporation of the radioactivity into subsequent metabolites including patulin were examined.[1]

9.6   BIOSYNTHESIS OF FUNGAL TROPOLONES

The biosynthesis of fungal tropolones from acetate, malonate and methionine is outlined below.

stipitatonic acid **1**   stipitatic acid **2**

sepedonin **5**   puberulonic acid **3**   puberulic acid **4**

The labeling pattern was confirmed by degradation reactions,[1-7] inhibition experiments by the addition of ethionine,[8,9] enzymatic studies,[10] and ^{13}C-nmr.[11] The degradation reactions employed for **1** and **2** are as follows:

Remarks

The seven-membered, aromatic tropolone ring system occurs naturally in mold metabolites (**1–5**), in essential oils of *Cupressaceae* (hinokitiol (β-thujaplicin)), and in the alkaloid colchicine and related compounds. The biogenetic origins of the tropolone rings in each of the three groups are entirely different, and fungal tropolones have been proved to be closely related to the acetogenin orsellinic acid, which is distributed widely in molds. Stipitatonic **1** and stipitatic acids **2** have been isolated from *Penicillium stipitatum*, puberulonic **3** and puberulic acids **4** from *P. aurantio-virens*, *P. puberulum*, *P. johannioli*, and *P. cyclopium-viridicatum* series, and sepedonin **5** from *Sepedonium chrysospermum*.[12]

1) R. Bentley, *Biochim. Biophys, Acta*, **29**, 666 (1958).
2) L. D. Ferretti, J. H. Richards, *Proc. Nat. Acad. Sci., U. S.*, **46**, 1438; *Biochem. Biophys. Res. Comm.*, **2**, 107 (1960).
3) R. Bentley, *J. Biol. Chem.*, **238**, 1889, 1895 (1963).
4) I. G. Andrew, W. Segal, *J. Chem. Soc.*, 607 (1964).
5) A. I. Scott, H. Guilford, E. Lee, *J. Am. Chem. Soc.*, **93**, 3534 (1971).
6) S. Takenaka, S. Seto, *Agric. Biol. Chem.*, **35**, 862 (1971).
7) A. I. Scott, E. Lee, *J. Chem. Soc. Chem. Commun.*, 655 (1972); A. I. Scott, K. J. Wiesner, *ibid.*, 1075 (1972).
8) R. Bentley, J. G. Graphery, J. G. Keil, *Arch. Biochem. Biophys.*, **111**, 80 (1965).
9) R. Bentley, P. M. Zwitkowits, *J. Am. Chem. Soc.*, **89**, 676, 681 (1967).
10) R. Bentley, C. P. Thiessen, *J. Biol. Chem.*, **238**, 3811 (1963).
11) A. G. McInnes, D. G. Smith, L. C. Vining, L. Johnson, *Chem. Commun.*, 325 (1971).
12) W. B. Turner, *Fungal Metabolites*, Academic Press, 1971.

9.7 SYNTHESIS OF OXYTETRACYCLINE

This interesting synthesis of terramycin (oxytetracycline) **1**[1)] was accomplished by Muxfeldt and his collaborators.

1

2,3 ⟶ 4: Diels-Alder reaction ensures a *cis* ring junction. The confguration of the acetoxyl group is also determined by the endo approach during cycloaddition.

4 ⟶ 6: The participation of the acetoxyl group in the Grignard reaction determines the configuration of the newly introduced methyl group. The reaction proceeds through **5**.

9 ⟶ 10: Ozonolysis followed by treatment with sodium carbonate gave a mixture of isomers containing the formyl group.

10 ⟶ 12: The configuration of the formyl group is converted to a stable form, as in **12**. The protective group for the phenol is changed to the more suitable methoxymethyl group, which can be removed easily with dilute acids.

12 ⟶ 13: Thioazlactone,[2] more reactive than azlactone, was utilized in this synthesis.

13 ⟶ 15: The keto-amide reagent was synthesized from dimethyl 3-oxoglutarate by treatment with ammonia, followed by acid hydrolysis of the enamine. The reaction proceeds through **14** and gives the more stable product **15**.

16 ⟶ 1: The thiobenzoyl group was removed by hydrolysis of the *S*-methylimine which was obtained by the action of methyl iodide on the thiobenzoyl compound. **1** showed 50% of the antibiotic activity of natural tetracycline.

Remarks

A number of tetracyclines including terramycin have been isolated from *Streptomyces* sp. and their medical value as broad-spectrum antibiotics has been established. The structure of tetracycline was established by chemical methods[3] and was later confirmed by x-ray analysis.[4] Another synthesis of a tetracycline analogue has been reported.[5]

1) H. Maxfeldt, G. Hardtmann, F. Kathawala, E. Vedejs, J. B. Moobery, *J. Am. Chem. Soc.*, **90**, 6534 (1968); and references cited therein.
2) H. Maxfeldt, J. Behling, G. Grethe, W. Roglski, *J. Am. Chem. Soc.*, **89**, 4991 (1967).
3) F. A. Hochstein, C. R. Stpehns, L. G. Conover, P. P. Ragna, R. Pasternack, P. N. Gordon, F. J. Pilgrim, K. J. Brunings, R. B. Woodward, *J. Am. Chem. Soc.*, **75**, 5455 (1953).
4) Y. Takeuchi, M. J. Buerger, *Proc. Nat. Acad. Sci. U.S.*, **46**, 1366 (1960); H. Cid-Dresdner, *Z. Kristallogr.*, **121**, 170 (1965).
5) J. J. Korst, J. D. Johnston, K. Butler, E. J. Bianco, J. H. Conover, R. B. Woodward, *J. Am. Chem. Soc.*, **90**, 439 (1968); H. Maxfeldt, W. Rogalski, *ibid.*, **87**, 933 (1965); A. I. Gurevich, M. G. Karapetyan, M. N. Kolosov, V. G. Korobko, V. V. Onoprienko, S. A. Popravko, M. M. Shemiakin, *Tetr. Lett.*, 131 (1967).

9.8   STRUCTURES OF HOP CONSTITUENTS

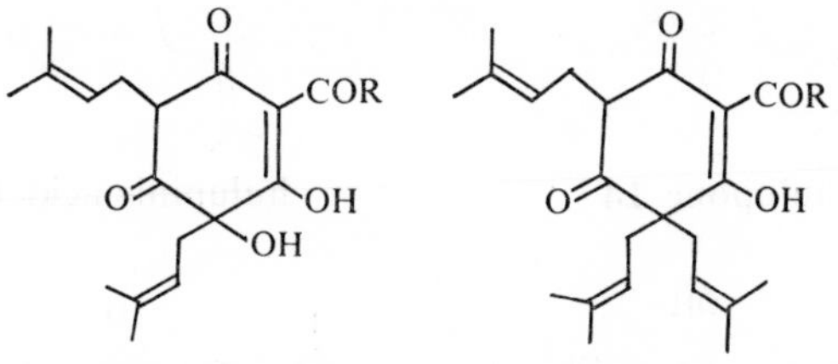

humulone   **1** (R= -CH$_2$CHMe$_2$)   lupulone   **4** (R= -CH$_2$CHMe$_2$)

cohumulone **2** (R= -CHMe$_2$)   colupulone **5** (R= -CHMe$_2$)

Me   Me

adhumulone **3** (R= -CHCH$_2$Me)   adlupulone **6** (R= -CHCH$_2$Me)

1) The structures were determined by various degradation reactions.[1,2]

Hydrogenation accompanied by the loss of a prenyl unit, and isomerization under alkaline conditions are shown below.[3-6]

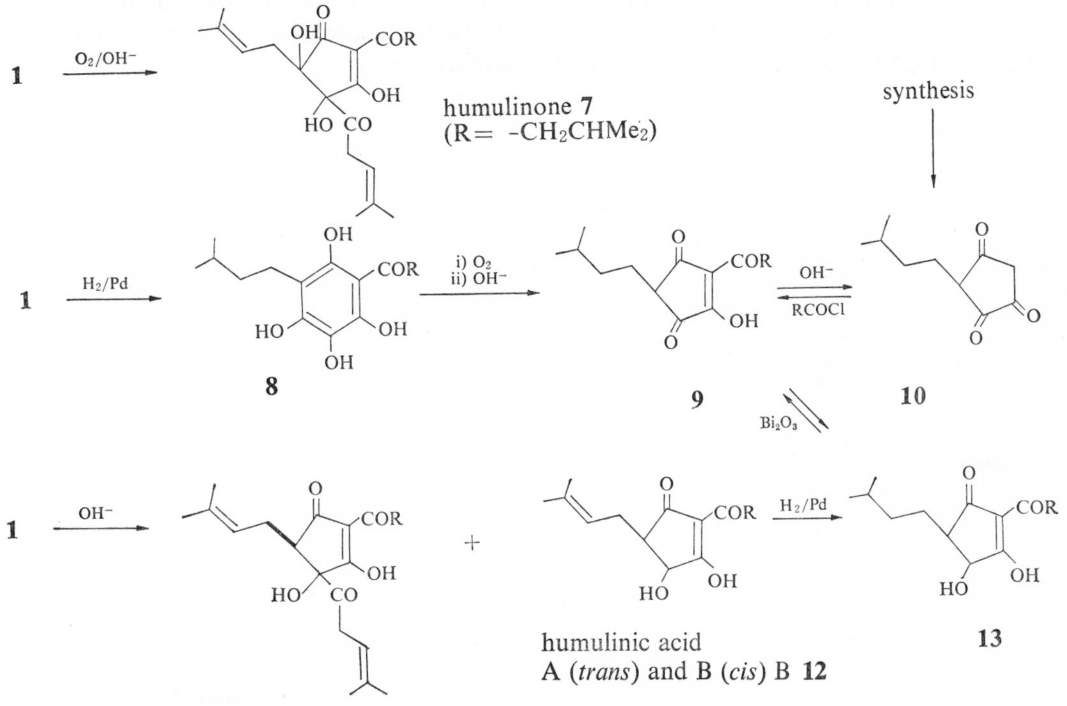

isohumulone A (α-OH) and B (β-OH) **11**

humulinic acid A (*trans*) and B (*cis*) B **12**

1) R. Stevens, *Chem. Rev.*, 19 (1967).
2) P. R. Ashurst, *Fortschr. Chem. Org. Naturstoffe*, **25**, 63 (1967).
3) W. Wöllmer, *Ber.*, **49**, 104 (1916).

4 $\xrightarrow{O_2/Na_2SO_3}$ hulupone **14**

$\xrightarrow{O_2}$ hulupinic acid **15**

4 $\xrightarrow{H_2/Pd}$ **16**

$\xrightarrow{O_2/Pb^{2+}}$ tetrahydrohumulone **17**

Remarks

Hop resins are obtained from the cones of the female plant of *Humulus lupulus* (Moraceae) and are used for the preparation of beer. The phloroglucinol derivatives, humulone, lupulone and their derivatives, are the bitter principles of the resin and also of beer. The migration products, **7, 11, 14,** and **15** are contained in beer as artifacts.[1,2]

4) A. H. Cook, G. Harris, *J. Chem. Soc.*, 1873 (1950); A. H. Cook, G. A. Howard, C. A. Slater, *J. Inst. Brewing*, **62**, 220 (1956).
5) R. Stevens, D. Wright, *J. Inst. Brewing*, **67**, 496 (1961).
6) J. S. Burton, R. Stevens, J. A. Elvidge, *J. Chem. Soc.* 952 (1964).

9.9   STRUCTURE OF TETRAHYDROCANNABINOL

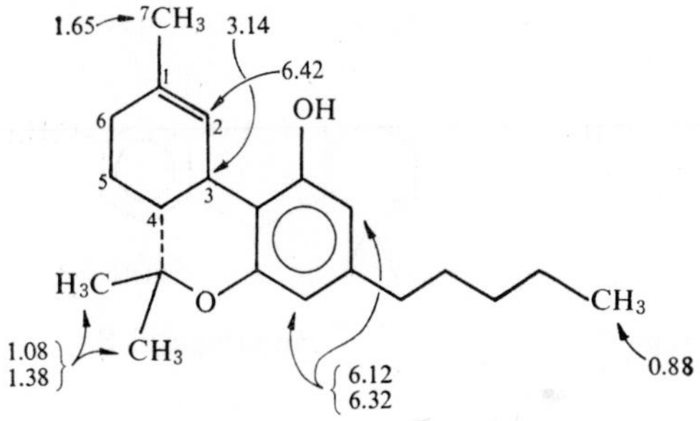

Δ^1-tetrahydrocannabinol **1**[1]

liquid
bp (0.05 mm): 155–157° α_D (CHCl$_3$): −150°
uv (EtOH): 278 (ϵ 2,040), 282 (ϵ 2,075)

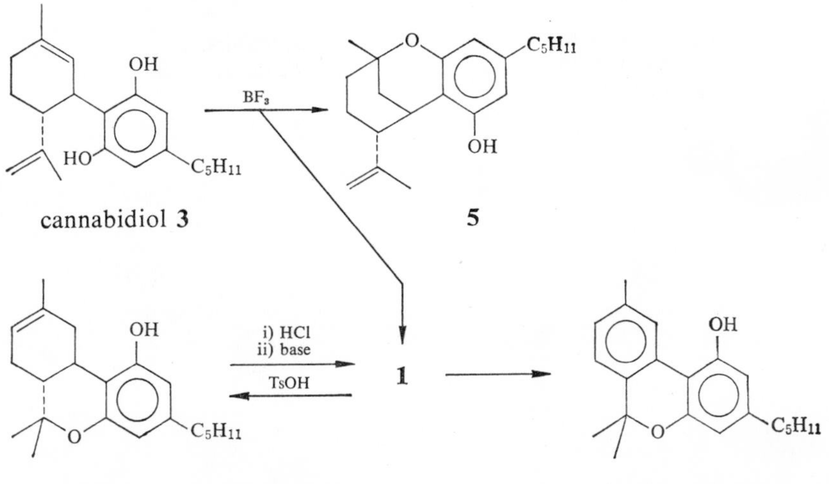

cannabidiol **3**

5

2

cannabinol **4**

1) Dehydrogenation of **1** gave cannabinol **4**.
2) Treatment of **1** with TsOH yields $\Delta^{1(6)}$-tetrahydrocannabinol[2].
3) **2** was converted to **1** by hydrochlorination followed by dehydrochlorination.
4) Reaction of cannabidiol **3** with BF$_3$-etherate afforded a mixture of **1** (60%) and Δ^8-iso-tetrahydrocannabinol **5**.

Remarks

The resin of the female plant of *Cannabis sativa* has been used as a medicine and a psychotomimetic drug, and it has been known as hashish or marihuana.[2] Δ^1-Tetrahydrocannabinol

1) Y. Gaoni, R. Mechoulam, *J. Am. Chem. Soc.*, **93**, 217 (1971).
2) For a review, see R. Mechoulam, Y. Gaoni, *Fortschr. Chem. Org. Naturstoffes.*, **25**, 174 (1967); R. Mechoulam. *Science*, **168**, 1159 (1970).

1 is an active constituent isolated from the resin. The crude fraction extracted from *Cannabis* was found to contain **1, 2, 3, 4**, cannabicyclol **6**, cannabichromene **7**, and cannabigerol **8**.[1] Recently the presence of the acids **9** and **10** in the living plant and the formation of the related compounds by light and heat was confirmed.[3]

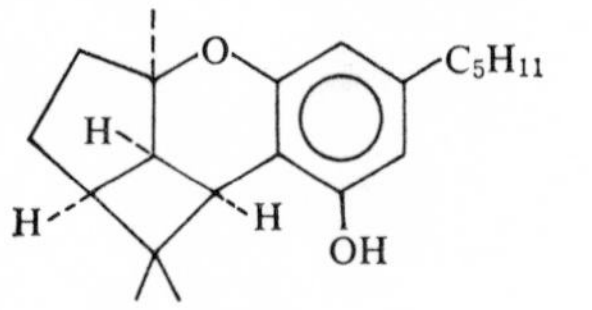

cannabicyclol **6** 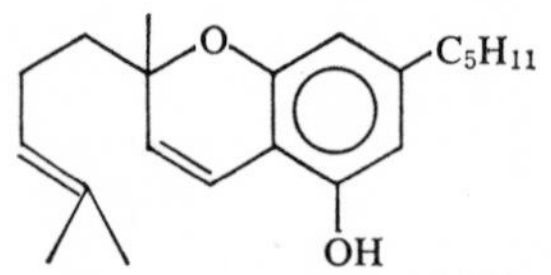   cannabichromene **7** 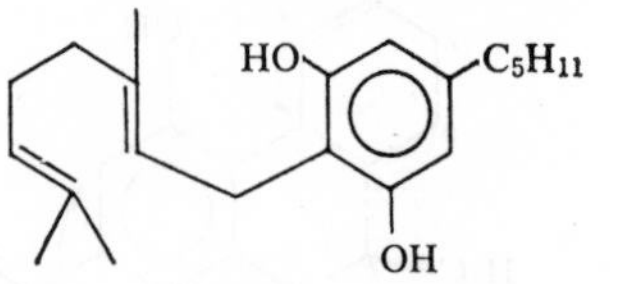   cannabigerol **8**

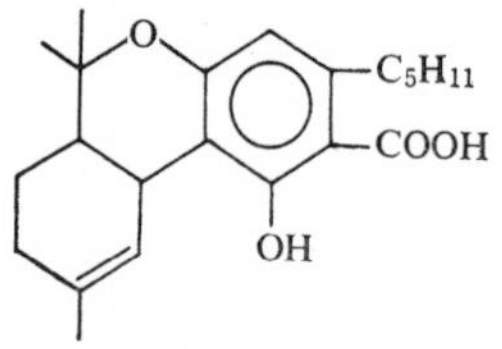

tetrahydro-
cannabinolic
acid **9**

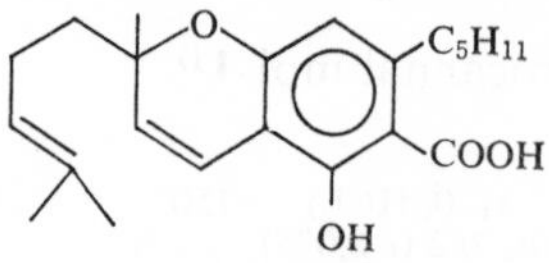

cannabichro-
menic acid **10**

3) Y. Shoyama, T. Fujita, T. Yamauchi, I. Nishioka, *Chem. Pharm. Bull.* (Tokyo),**16**, 1157 (1968); Y. Shoyama, T. Yamauchi, I. Nishioka, *ibid.*, **18**, 1327 (1970); Y. Shoyama, R. Oku, T. Yamauchi, I. Nishioka, *ibid.*, **20**, 1927 (1972); Y. Shoyama, K. Kuboe, I. Nishioka, T. Yamauchi, *ibid.*. **20**, 2072 (1972).

9.10   SYNTHESIS OF TETRAHYDROCANNABINOL

Condensation of citral **3** with the lithium salt of olivetol dimethyl ether **4** followed by reaction with p-TsCl afforded cannabidiol dimethyl ether **5** in 7% yield. This was converted into Δ^1-tetrahydrocannabinol (THC) **1** by demethylation and successive acid treatment. Reaction of **3** and olivetol with 1% BF$_3$ in CH$_2$Cl$_2$ gave **1** in 20% yield.[2]

Route I[1]

i) stand at room
temperature
ii) TsCl/py

i) MeMgI
ii) HCl

3   4   5   1

Route II[3]

On treatment of a mixtures of (−)-*cis* or (−)-*trans*-verbenol **6** and olivetol with p-TsOH, (−)-**7** was obtained as a major product (60% yield). This was converted to (−)-$\Delta^{1(6)}$-THC **2** by treatment with BF$_3$. $\Delta^{1(6)}$-THC **2** could be converted to Δ^1-THC **1** by hydrochlorination and dehydrochlorination.

p-TsOH

BF$_3$

(−)-**6**   olivetol   (−)-**7**

i) HCl
ii) t-AmOK[4]   (−)-**1**

(−)-**2**

Route III[4]

(+)-*trans*-p-menthadien-1-ol **8** (or its (+)-*cis* isomer) was converted to (−)-cannabidiol **9** by the action of DMF neopentyl acetal, or weak acid, in the presence of olivetol. Treatment

1) R. Mechoulam, P. Braun, Y. Gaoni, *J. Am. Chem. Soc.*, **94**, 6159 (1972); R. Mechoulam, Y. Gaoni, *ibid.*, **87**, 3273 (1965).
2) See also E. C. Taylor, K. Lenard, Y. Shvo, *J. Am. Chem. Soc.*, **88**, 367 (1966).
3) R. Mechoulam. P. Braun, Y. Gaoni, *J. Am. Chem. Soc.*, **89**, 4552 (1967).
4) T. Petrzilka, W. Haefliger, C. Sikemeier, G. Ohloff, A. Eschenmoser, *Helv. Chim. Acta*, **50**, 719 (1967); T. Petrzilka, W. Haefliger, C. Sikemeier, *ibid.*, **52**, 1102 (1969).

of a mixture of **8** and olivetol with *p*-TsOH afforded $\Delta^{1(6)}$-THC **2** via the thermodynamically more unstable Δ^1-isomer **1**.

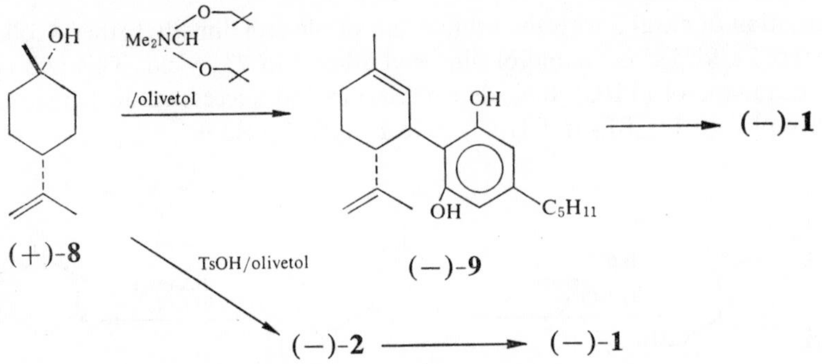

(+)-**8**   TsOH/olivetol   (−)-**9**

(−)-**2** ⟶ (−)-**1**

Route IV[5]

One-step stereospecific synthesis of (−)-**1** of high optical purity was achieved by the reaction of (+)-*trans*-2-carene oxide **10** and olivetol in the presence of BF₃.

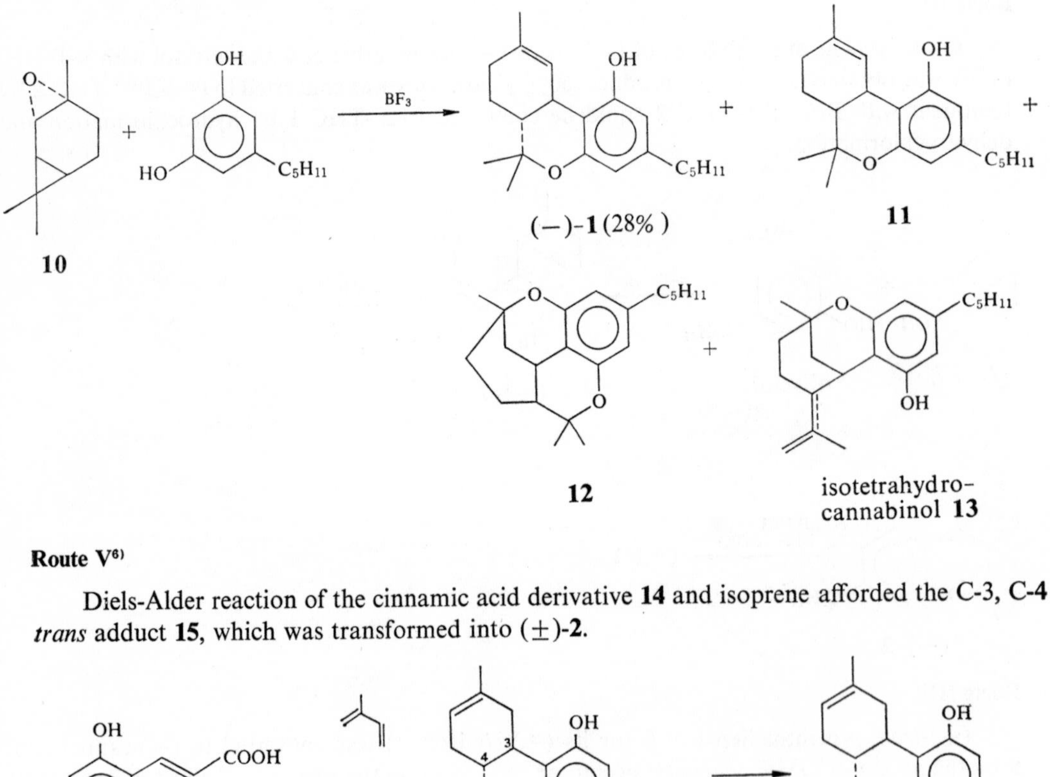

10   (−)-**1**(28%)   **11**

12   isotetrahydro-cannabinol **13**

Route V[6]

Diels-Alder reaction of the cinnamic acid derivative **14** and isoprene afforded the C-3, C-4 *trans* adduct **15**, which was transformed into (±)-**2**.

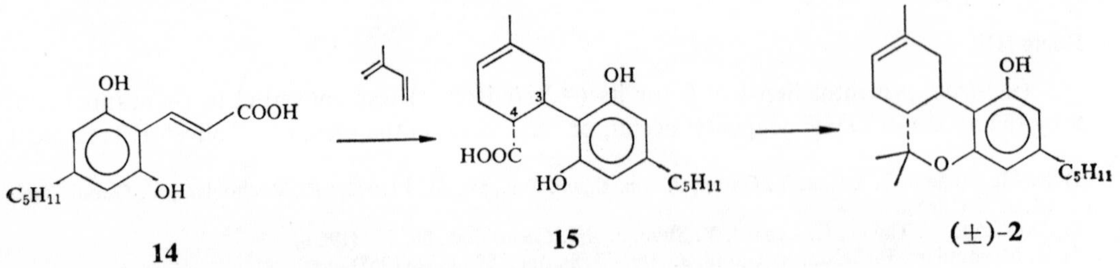

14   **15**   (±)-**2**

5) R. K. Razdan, G. R. Handrick, *J. Am. Chem. Soc.*, **92**, 6061 (1970).
6) T. Y. Jen, G. A. Hughes, H. Smith, *J. Am. Chem. Soc.*, **89**, 4551 (1967).

Route VI[7]

($\pm$)-**1** was synthesized by a route involving von Pechmann condensation (**16**+olivetol$\longrightarrow$ **17**) followed by base catalized cyclization of the resulted coumarin **17**.

7) K. E. Fahrenholtz, M. Lurie, R. W. Kierstead, *J. Am. Chem. Soc.*, **89**, 5934 (1967).

9.11 TRIPRENYL PHENOLS AND TRIPRENYL QUINONES

A group of natural products biogenetically derived via coupling reactions of farnesyl pyrophosphate (or partially cyclized forms thereof) with aromatic substrates are known. Examples are shown below.

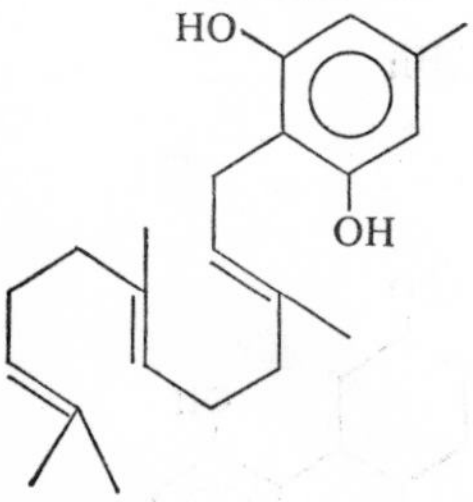

grifolin 1

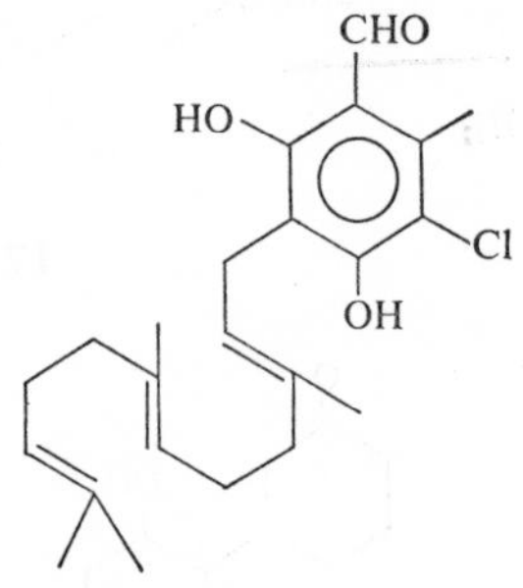

LL-Z 1272-α 2

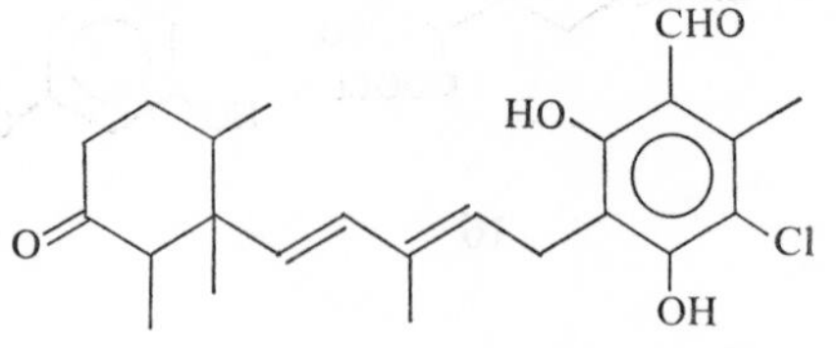

ascochlorin (=LL-Z 1272-γ) 3

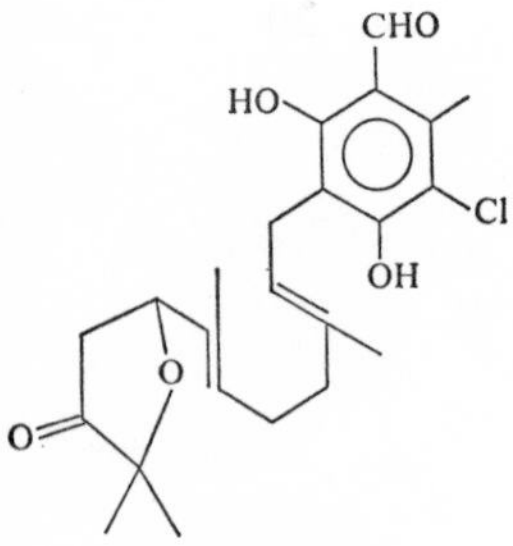

ascofuranone 4

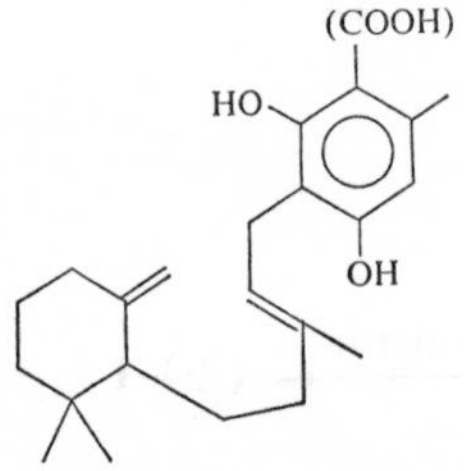

presiccanochromene-A 5 (and corresponding acid)

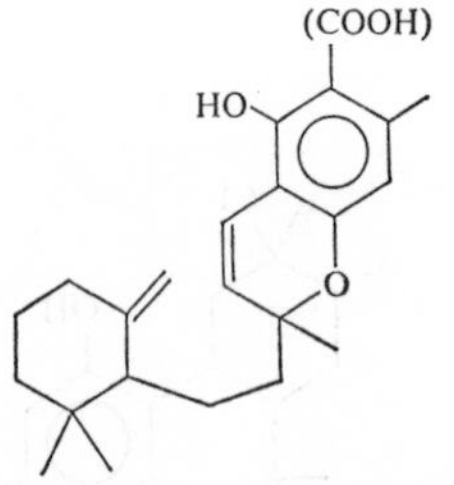

siccanochromene-A 6 (and corresponding acid)

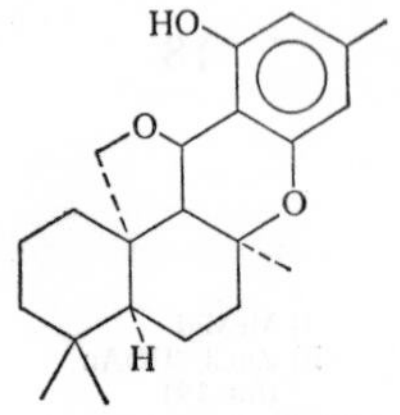

siccanin 7

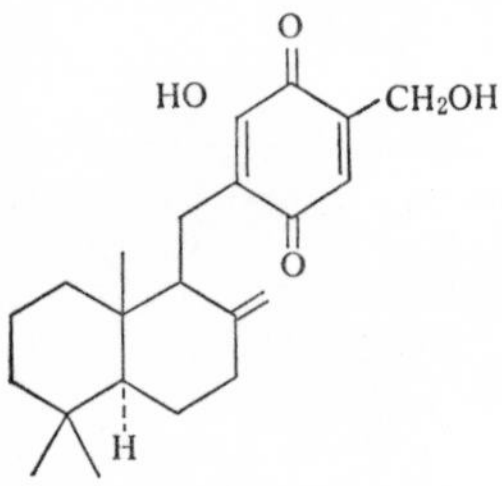

trauranin 8

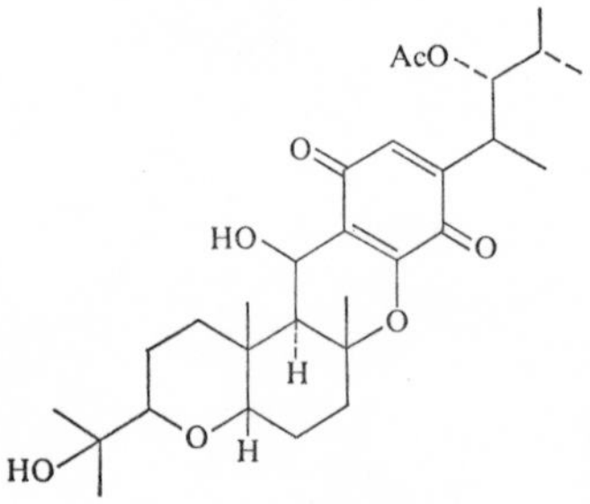

cochlioquinone-A 9

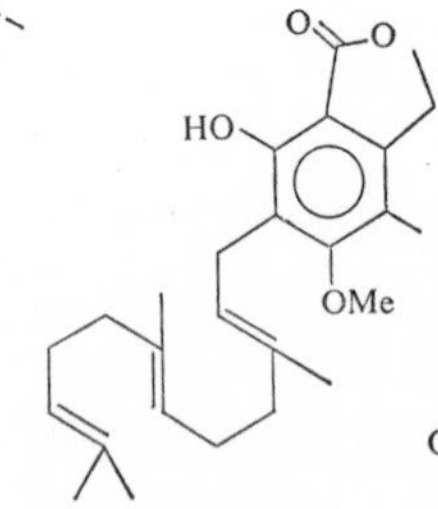

10

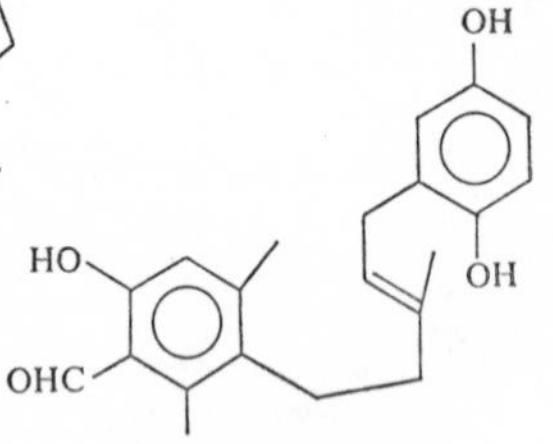

panicein-B$_3$ 11

Remarks

1 was isolated from *Grifola confluence.*[1] The chlorine-containing antibiotics 2, 3, and 4 and many other related substances were isolated from *Asccochyta viciae,*[3,4] *Cylindrocladium*

1) T. Goto, H. Kakisawa, Y. Hirata, *Tetr.* **19**, 2079 (1963).
2) G. A. Ellestad, R. H. Evans Jr., M. P. Kunstmann, *Tetr.*, **25** 1323 (1969).
3) Y. Nawata, Y. Iitaka, *Bull. Chem. Soc. Japan*, **44**, 2652 (1971).
4) H. Sasaki, T. Okutomi, T. Hosokawa, Y. Nawata, K. Ando, *Tetr. Lett.*, 2541 (1972).

ilicicola,[6)] *Nectria coccinea*[5)] and unidentified *Fusarium* sp.[2)] **5, 6**, and **7** were obtained from *Helminthosporium siccans*,[7)] **8** was from *Neurospora sitophila*,[8)] and **9** was isolated from *Cochliobolus miyabeanus*.[9)] **10** was found in *Penicillium brevi-compactum* as a biosynthetic precursor of mycophenolic acid.[10)]

A number of prenyl-quinols or quinones (*e.g.* **11**) have been isolated from the marine sponge, *Halichondria panicea*[11)].

9.12   BIOSYNTHESIS OF SICCANIN

siccanin **1**[1,2)]

mp: 138° α_D (CHCl$_3$): $-150°$
ms: 342 (M$^+$ C$_{22}$H$_{30}$O$_3$)
uv (EtOH): 278 (sh.), 285 (ε 1800)
ir (CHCl$_3$): 3500, 1633, 1575
nmr in CDCl$_3$

The biogenesis of the antibiotic siccanin **1** was postulated as shown overleaf on the basis of tracer experiments. Siccanin has an unusual *cis-syn-cis* set of ring junctions.

9-11
5) D. C. Aldridge, A. Barrow, R. G. Foster, M. S. Large, H. Spencer, W. B. Turner, *J. Chem. Soc.*, Perbin I, 2136 (1972).
6) S. Hayakawa, H. Minato, K. Katagiri, *J. Antibiotics*, **24**, 653 (1971); A. Kato, K. Ando, G. Tamura, K. Arima, *ibid.*, **23**, 168 (1970).
7) K. Hirai, K. T. Suzuki, S. Nozoe, *Tetr.* **27**, 6057 (1971); S. Nozoe, K. T. Suzuki, *ibid.*, **27**, 6063 (1971).
8) K. Kawashima, K. Nakanishi, M. Tada, H. Nishikawa, *Tetr. Lett.*, 1227 (1964); K. Kawashima, K. Nakanishi, H. Nishikawa, *Chem. Pharm. Bull.*, **12**, 796 (1964).
9) J. R. Carruthers, S. Corrini, W. Fedeli, C. G. Casinori, C. Galeffi, A. N. Vaccaro, A. Scala, *Chem. Commun.*, 164 (1971).
10) L. Canonica, W. Kroszczynoki, B. H. Ranzi, B. Rinedonc, C. Scolastico, *Chem. Commun.*, 257 (1971).
11) G. Cimino, S. De Stefano, L. Minale, *Tetr.*, **29**, 2565 (1973).
9-12
1) K. Hirai, S. Nozoe, K. Tsuda, Y. Iitaka, K. Ishibashi, M. Shirasaka, *Tetr. Lett.*, 2177 (1967).
2) K. Hirai, K. T. Suzuki, S. Nozoe, *Tetr.*, **27**, 6057 (1971).

farnesyl pyrophosphate **2**

γ-monocyclo-
farnesyl pyro-
phosphate **3**

presiccanochromenic
acid **4**

siccanochromenic acid **5**

siccanochromene-A **6**

siccanochromene-B **7**

⟶ **1**

1) The formation of γ-monocyclofarnesol (**3**, H instead of PP) from mevalonic acid or farnesyl pyrophosphate was demonstrated using a cell-free system prepared from *Helminthosporium siccans*.[3]

2) Siccanochromene-A **6** was formed under the same conditions in the presence of orsellinic acid.[4]

3) Presiccanochromenic acid **4** and siccanochromenic acid **5** were isolated from the extracts as minor constituents,[5] both of which were shown to be converted into **6** by cell-free preparation.[4]

4) Siccanochromenes **6** and **7** were found in the extracts. Tritium-labeled **6** and **7** were shown to be incorporated into siccanin **1** by incubation with a growing cell system of the fungi.[6]

Remarks

Siccanin **1** was isolated from *Helminthosporium siccans*, a plant-pathogenic fungi, and exhibits marked activity against *Trichophyton* sp.

3) K. T. Suzuki, N. Suzuki, S. Nozoe, *Chem. Commun.*, 527 (1971).
4) K. T. Suzuki, S. Nozoe, *Chem. Commun.*, 1166 (1972).
5) S. Nozoe, K. T. Suzuki, *Tetr.* **27**, 6063 (1971); and references therein.
6) K. T. Suzuki, S. Nozoe, *Chem. Commun.*, 527 (1971).

9.13   STRUCTURE OF PODOPHYLLOTOXIN

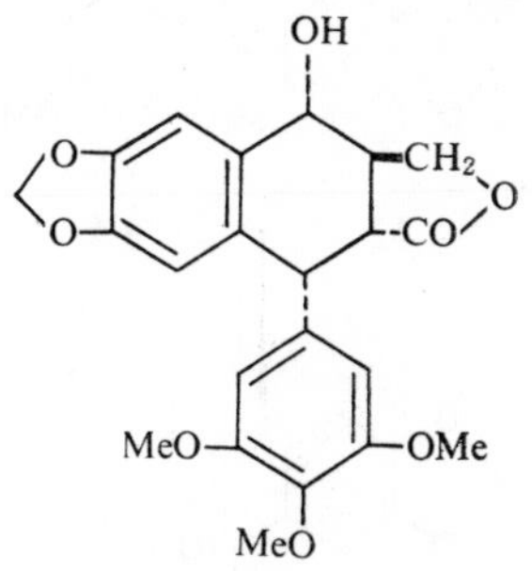

podophyllotoxin **1**

mp: 183–184° α_D (EtOH): −94.8°
ir (CHCl$_3$): 1785
nmr of β-peltatin methyl ether **5** (CDCl$_3$): 3.70
  (3×OMe), 4.1 (OMe), 5.82 (–O–CH$_2$–O–),
  6.21 (arom. H), 6.31 (2×arom. H)

1) The structure **1** was chiefly determined on the basis of oxidation reactions.[1,2]
2) The absolute configuration of **1** was established by epimerization reactions[1,2] and by the x-ray study.[3]

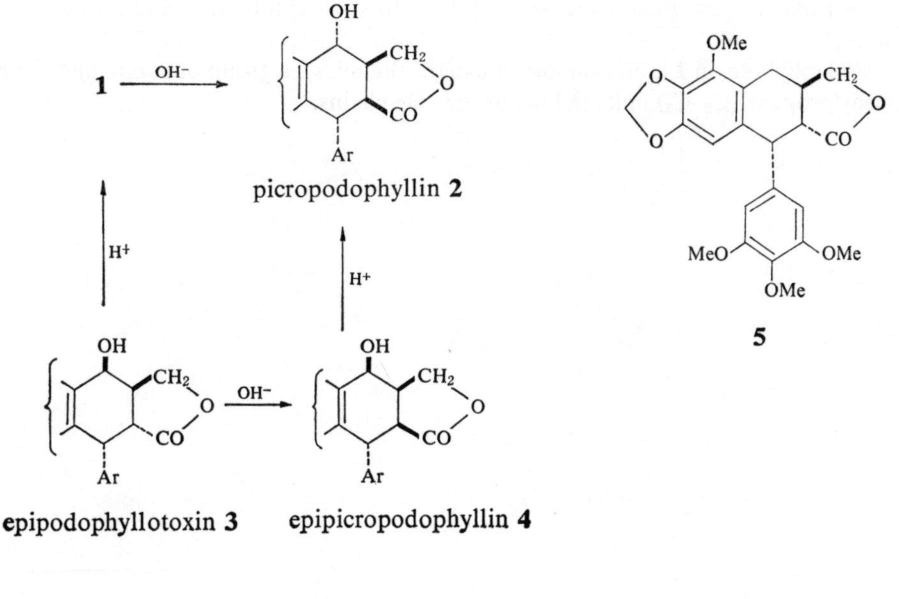

picropodophyllin **2**

epipodophyllotoxin **3**   epipicropodophyllin **4**

5

1) J. L. Hartwell, W. E. Detty, *J. Am. Chem. Soc.*, **72**, 246 (1950).
2) J. L. Hartwell, A. W. Schrecker, *Fortschr. Chem. Org. Naturstoffe.*, **15**, 83 (1958).
3) T. J. Petcher, H. P. Webber, M. Kuhn, A. von Wartbury, *J. Chem. Soc.*, Perkin II, 288 (1973).

1 ⟶

OH
CH₂OH
COOH

KMnO₄ (room temp.)

MeO   OMe
OMe

KMnO₄ (100°)

KMnO₄ (50°)

KMnO₄ (60°)

6

Zn/KOH/ MeOH

MeO   OMe
OMe

7

O

MeO   OMe
OMe

12

+

COCl

MeO   OMe
OMe

13

COOH

OMe
OMe

10

COOH
COOH

9

COOH

O

MeO   OMe
OMe

8

Cu/quinoline

O

MeO   OMe
OMe

11

Remarks

Podophyllotoxin **1**, its isomer, picropodophyllin **2**, and derivatives of **1** and **2** were first isolated from *Podophyllum peltatum* (Berberidaceae)[2] but have been shown to occur rather widely in other families. The structure **1** was firmly established by total synthesis.[4] Podophyllotoxin **1** and its derivatives have been reported to show cytotoxicity and tumor-inhibitory activity.[5–7]

Podophyllotoxin **1** is an example of lignans (lignoids), a group of compounds composed of two phenylpropane (C₆–C₃) units linking at the side chains.

4) W. J. Gensler, C. D. Gatsonis, *J. Am. Chem. Soc.*, **84**, 1748 (1962).
5) J. L. Hartwell, M. J. Schaer, *Cancer Res.*, **7**, 716 (1947).
6) S. M. Kupchan, J. C. Hemingway, J. R. Knox, *J. Pharm. Sci.*, **54**, 659 (1965).
7) H. Stähelin, *Planta Medica*, **22**, 336 (1972); H. Limburg, *ibid.*, **22**, 348 (1972).

9.14 STRUCTURE AND SYNTHESIS OF FUTOENONE

futoenone **1**

mp: 197° α_D (CHCl$_3$): $-58°$
ms: 340 (M$^+$), 177, 163 (base peak), 147, 135
uv (EtOH): 257 (ε 12,900), 285 (ε 8800)
ir (Nujol): 1655, 1622, 1522

1) The presence of the 3,4-methylenedioxyphenyl moiety was suggested by the following observations: the uv spectrum of the mixture of epimeric alcohols obtained by LAH reduction

of **1** $\searrow C \diagdown \begin{smallmatrix} H \\ OH \end{smallmatrix}$ in place of C=O in **1**); m/e 163 and 135 in the mass spectrum; the nmr data.

2) The C-7 to C-11 sequence was determined by detailed examination of the nmr, including
partial decoupling and observation of nuclear Overhauser effects. The relative stereochemistry
was also proposed on the basis of the coupling constants ($J_{7\alpha,8\alpha}$=ca. 0.0, $J_{8\alpha,9\alpha}$=ca. 0.5,
$J_{9\alpha,10\beta}$=11.5, $J_{10\beta,11\alpha}$=11.2).[1]

3) The spiro-dienone structure suggested by uv and nmr was confirmed by the reaction sequence
1 $\longrightarrow$ 2 $\longrightarrow$ 6 $\longrightarrow$ 1, which includes a dienone-phenol type rearrangement and recyclization.[1,2]

2 (R$_1$=R$_2$=Ac)

3 (R$_1$=R$_2$=H)

4 (R$_1$=H, R$_2$=CH$_2$Ph)

6 (R$_1$=Ts, R$_2$=H) $\xleftarrow{\text{H}_2/\text{Pd-C}}$ **5** (R$_1$=Ts ,R$_2$=CH$_2$Ph)

1) A. Ogiso, M. Kurabayashi, H. Mishima, M. C. Woods, *Tetr. Lett.*, 2003 (1968); M. C. Woods, I. Miura, A.
Ogiso, M. Kurabayashi, H. Mishima, *ibid.*, 2009 (1968).

2) A. Ogiso, M. Kurabayashi, S. Takahashi, H. Mishima, M. C. Woods, *Chem. Pharm. Bull.*, **18**, 105 (1970).

1 → i) Lemieux oxidation ii) methylation → **7** + **8**

Synthesis[2]

9 → i) Br₂ ii) NaOH → **10** (R=COOH) → **11** (R=CH₂OH) → **12** (R=CHO)

+ CH₂COMe → **13** →

14 → NaBH₄ → **4**

Remarks

Futoenone[1,2] **1** was isolated from the leaves and stems of *Piper futokadzura* (Piperaceae) together with futoamide **15**,[3] futoxide **16**[4] (crotepoxide),[5] and futoquinol **17**.[6]

15   **16**   **17**

3) S. Takahashi, M. Kurabayashi, A. Ogiso, H. Mishima, *Chem. Pharm. Bull.*, **17**, 1225 (1969).
4) S. Takahashi, *Phytochemistry*, **8**, 321 (1969).
5) S. M. Kupchan, R. J. Hemingway, P. Coggan, A. T. McPhail, G. A. Sim, *J. Am. Chem. Soc.*, **90**, 2982 (1968).
6) S. Takahashi, A. Ogiso, *Chem. Pharm. Bull.*, **18**, 100 (1970).

9.15   STRUCTURE OF HOPEAPHENOL

hopeaphenol **1** (R=H)
dibromo derivative **2** (R=Br)

mp: 351°
α_D (EtOH): −407°
uv: 281 (ε 14,500)

1) The structure **1** was determined by x-ray analysis of the decamethyl ether of the dibromo derivative **2**.[1,2]

2) The structure **1** is made up of four 3,5,4′-trihydroxystilbene units.

Remarks

Hopeaphenol **1** has been isolated from *Hopea odorata*[3] and related species[4,5] (Dipterocarpaceae) and provides an example of a condensed polyphenol of high molecular weight with an established structure.

1) P. Coggan, T. J. King, S. C. Wallwork, *Chem. Commun.*, 439 (1966).
2) P. Coggan, A. T. McPhail, S. C. Wallwork, *J. Chem. Soc.* B, 884 (1970).
3) P. Coggan, N. F. James, F. E. King, T. J. King, R. J. Molyneux, J. W. W. Morgan, K. Sellars, *J. Chem. Soc.* 406 (1965).
4) R. Madhav, T. R. Seshadri, G. B. V. Subramanian, *Tetr. Lett.*, 2713 (1965).
5) R. Madhav, T. R. Seshadri, G. B. V. Subramanian, *Phytochemistry*, **6**, 1155 (1967).

9.16   STRUCTURE AND BIOGENESIS OF LIGNIN

A Model Structure of Spruce Lignin[1]

The structure below was proposed[1] on the basis of degradation reactions, analyses of functional groups,[2,3] and nmr analyses.[4]

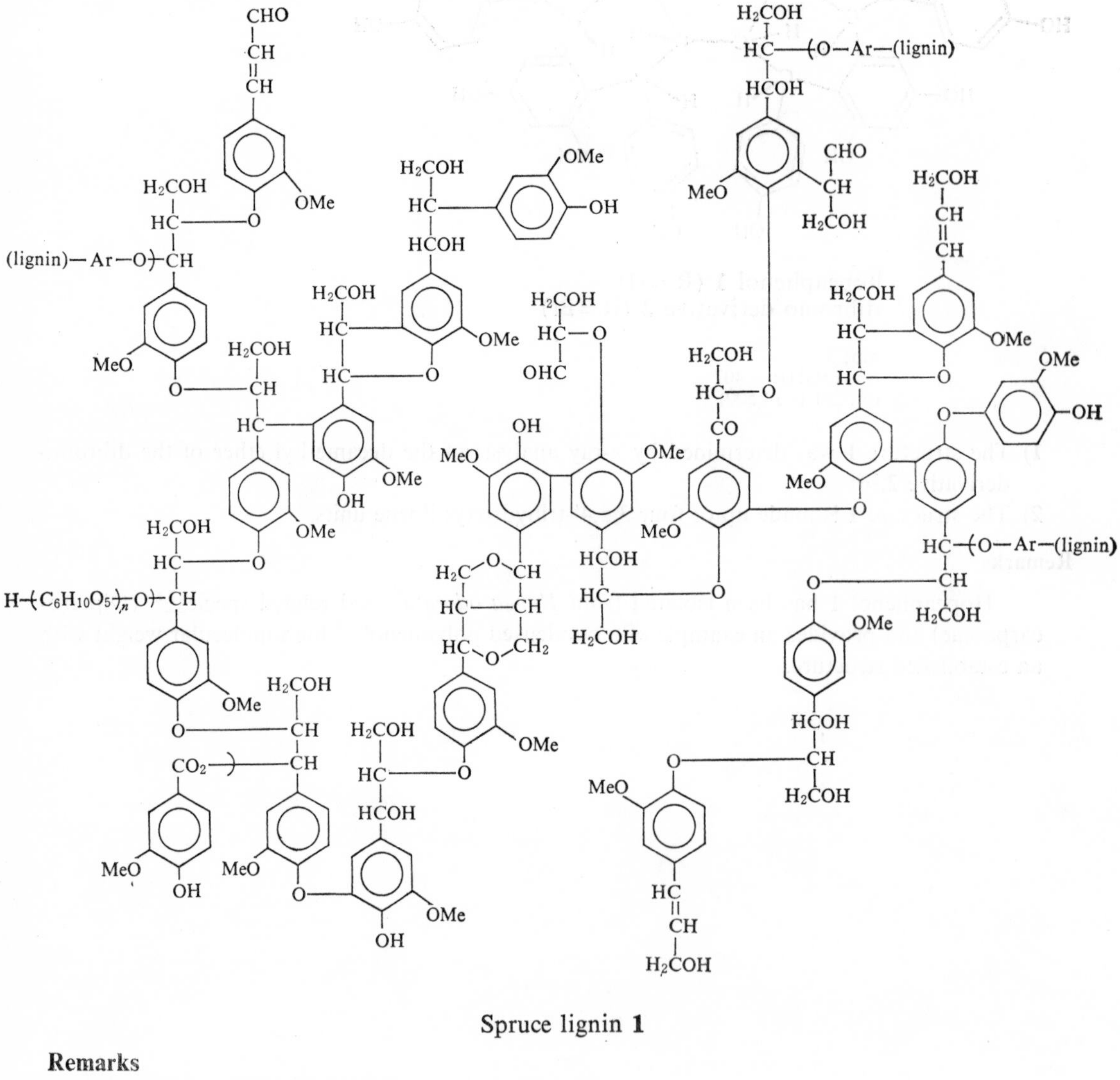

Spruce lignin **1**

Remarks

Lignin, the most abundant carbon compound existing in nature except for cellulose, consists of very closely related polymeric substances, rather than a single, unique material. Thus,

1) E. Adler, S. Larsson, K. Lundquist, G. E. Mikshe, *Abst. Intern. Wood Chem. Symp.*, 1969.
2) E. Adler, H. D. Becker, T. Ishihara, A. Stamvik, *Holzforsch,,* **20**, 3 (1966).
3) E. Adler, K. Lundquist, *Acta Chem. Scand.*, **15**, 223 (1961).
4) C. H. Ludwig, B. J. Nist, J. N. McCarthy, *J. Am. Chem. Soc.*, **86**, 1186, 1196 (1964).

isolation in a pure state, free from carbohydrate but without any denaturation, is practically impossible. Experimental proof of the role of the alcohols **10**, **11** and **12** as the natural precursors of lignin, and studies on artificial lignin prepared therefrom[5-9] have contributed much to the clarification of the structure. The biogenesis of lignin is outlined below, partly on the basis of such studies.

Lignin is a complex polymer composed mainly of C_6-C_3 units: *p*-coumaryl **10**, coniferyl **11**, and sinapyl alcohols **12**. The initial stage of the coupling of the precursors is assumed to proceed as shown below.[5] However, the lack of optical activity of lignin led to the suggestion that further polymerization proceeds without enzymatic control.

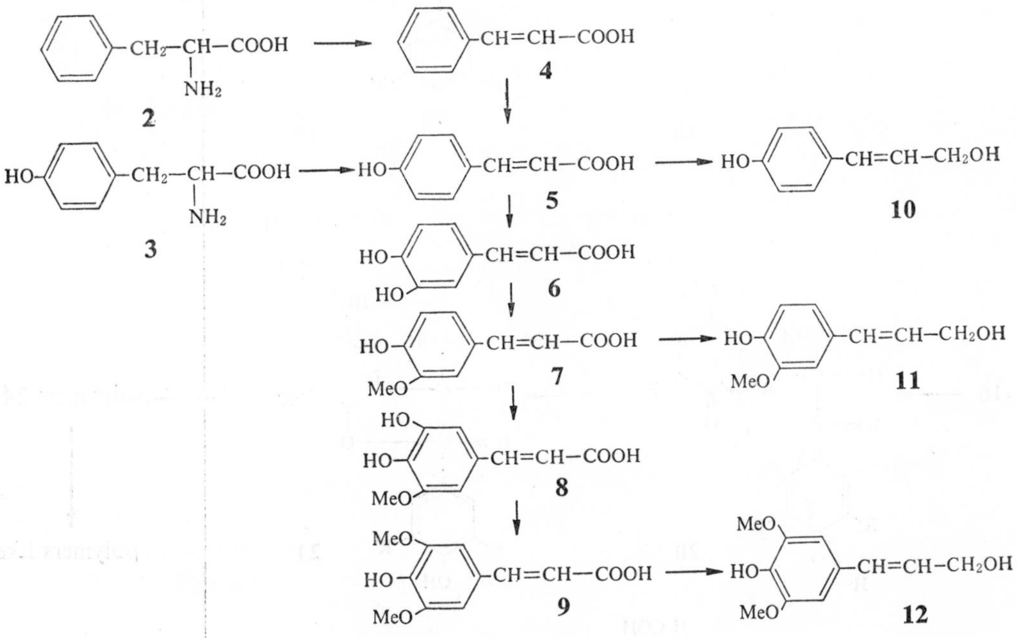

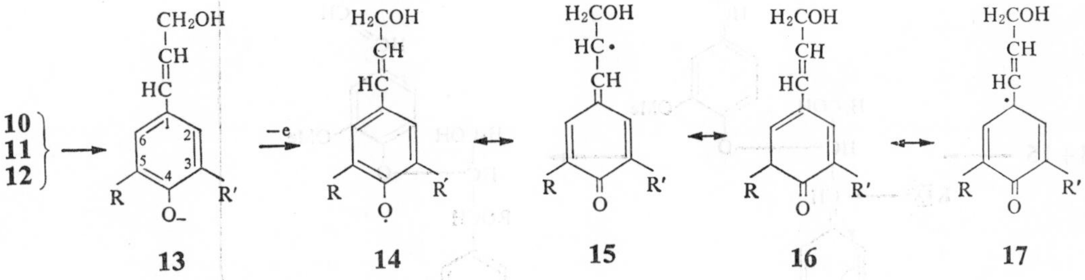

5) J. M. Harkin, *Recent Advances in Phytochemistry* (ed. M. K. Seikel, V. C. Runeckles), vol. 2, p. 35, North-Holland, 1969.
6) K. Freudenberg, *Fortschr. Chem. Org. Naturstoffe.*, **20**, 41 (1962).
7) K. Freudenberg, *Holzforsch.*, **18** (3), 166 (1964).
8) K. Freudenberg, *Science*, **148**, 595 (1965).
9) J. M. Harkin, *Oxidative Phenol Coupling* (ed. W. I. Taylor, A. R. Battersby), p. 243, Marcel Dekker, 1967.

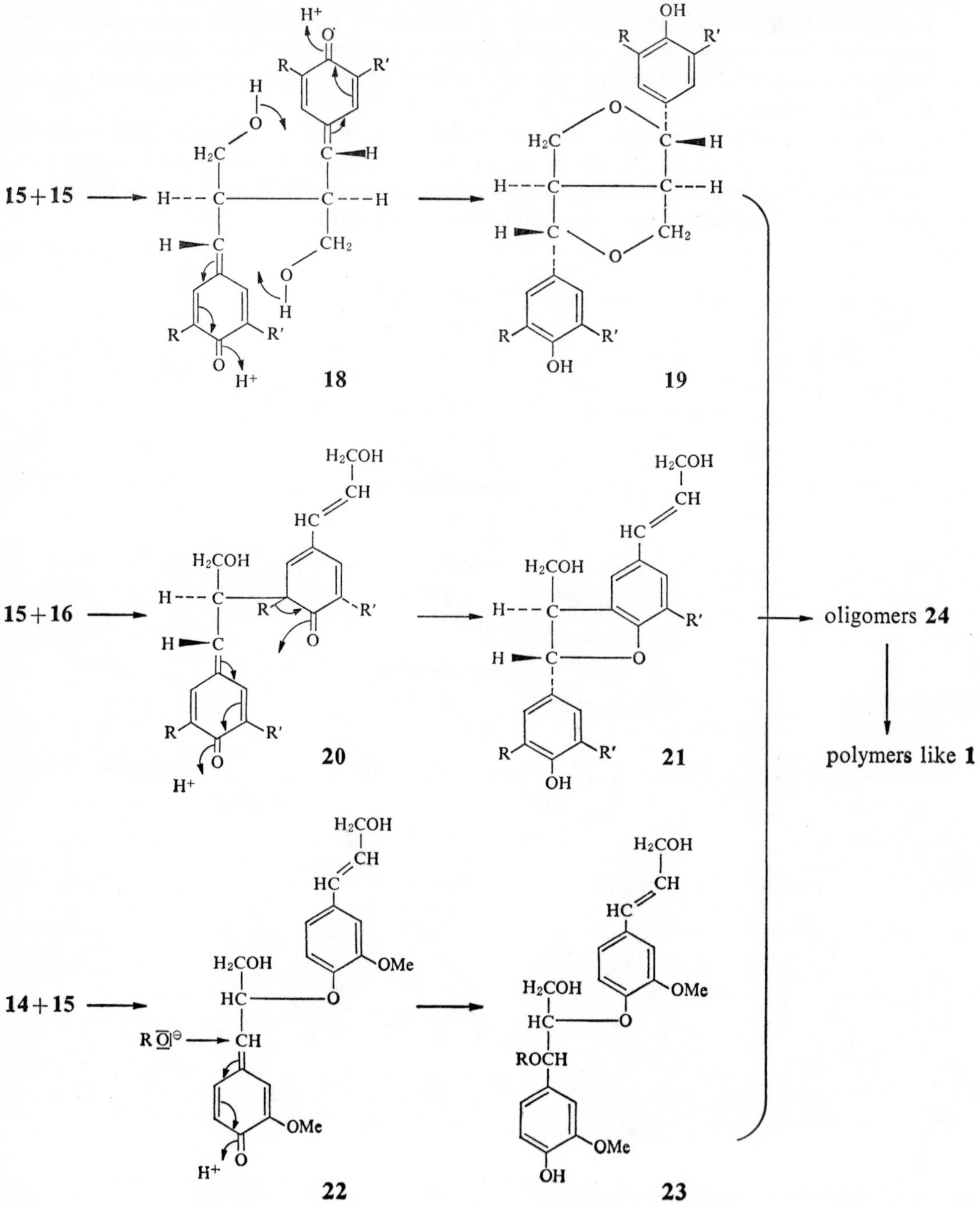
15+15
18
19
15+16
20
21
14+15
22
23
oligomers 24
polymers like 1

9.17 STRUCTURES OF CHINESE GALLOTANNINS

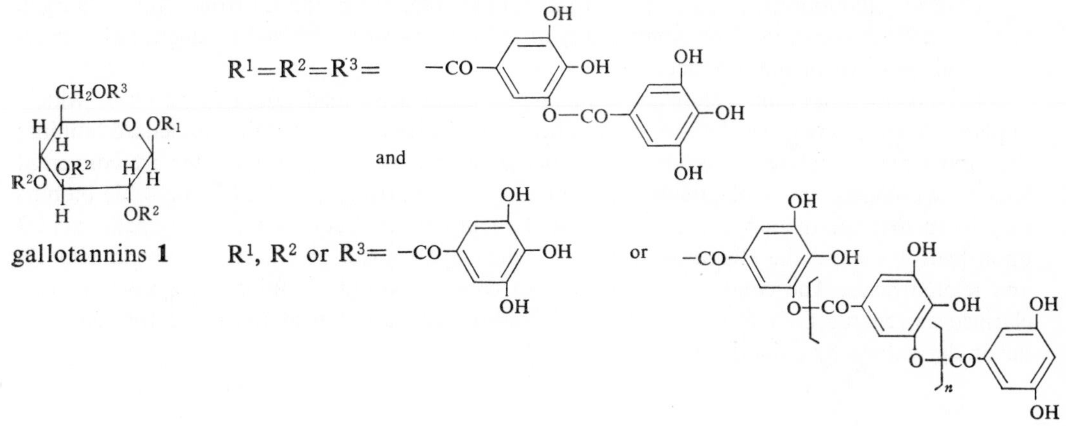

$R^1 = R^2 = R^3 =$

and

gallotannins **1** R^1, R^2 or $R^3 = -CO-$ or

Some of the degradative reactions of gallotannins are illustrated below.

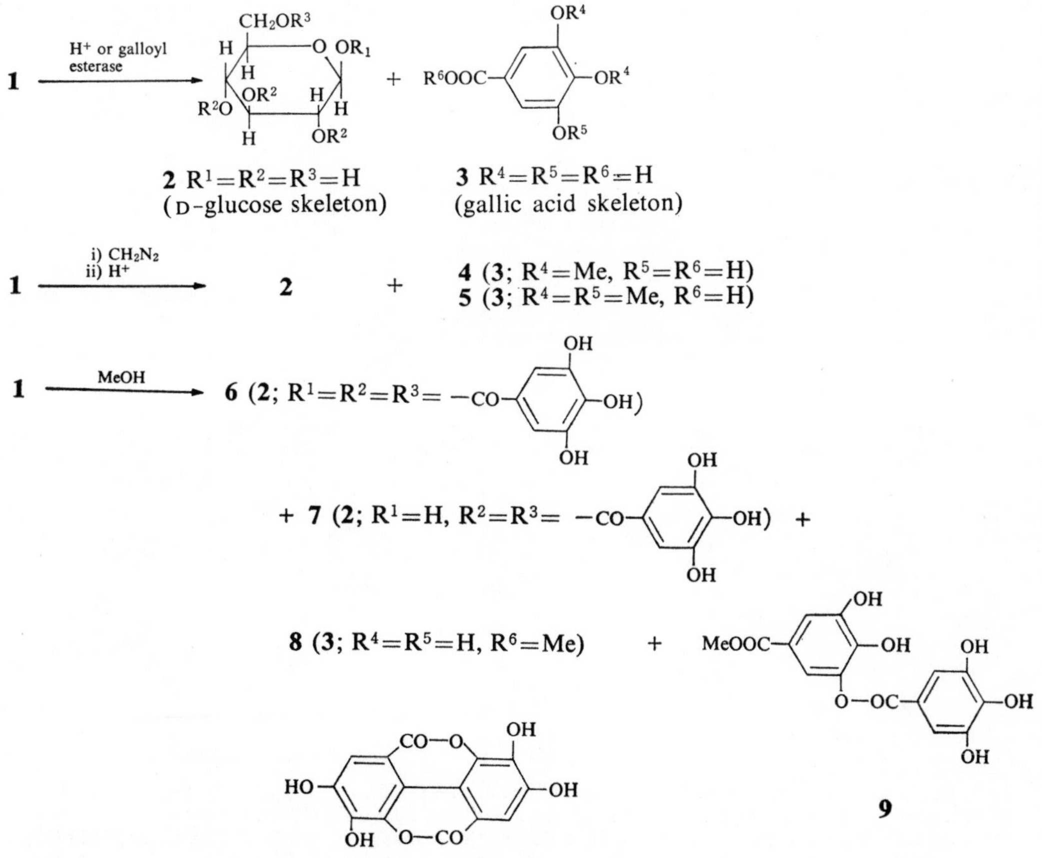

1 $\xrightarrow{\text{H}^+ \text{ or galloyl esterase}}$ $+$ R^6OOC-

2 $R^1 = R^2 = R^3 = H$ **3** $R^4 = R^5 = R^6 = H$
(D-glucose skeleton)   (gallic acid skeleton)

1 $\xrightarrow[\text{ii) H}^+]{\text{i) CH}_2\text{N}_2}$ **2** $+$ **4** (**3**; $R^4 = $Me, $R^5 = R^6 = $H)
5 (**3**; $R^4 = R^5 = $Me, $R^6 = $H)

1 $\xrightarrow{\text{MeOH}}$ **6** (**2**; $R^1 = R^2 = R^3 = -CO-$)

$+$ **7** (**2**; $R^1 = $H, $R^2 = R^3 = -CO-$) $+$

8 (**3**; $R^4 = R^5 = $H, $R^6 = $Me) $+$ MeOOC-

9

ellagic acid **10**

1) The ratio of **6** and **7** is 10:1.

Remarks

   Chinese gallotannins, often referred to as tannic acid, are produced from Chinese nutgall formed on *Rhus semialata*. The components of the tannins were clarified by degradation reactions and synthesis of the products.[1-4]

   Polyphenol compounds showing tanning properties are called tannins and occur widely in plants. Tannins are generally divided into (a) the hydrolysable and (b) the condensed tannins; the former are hydrolysed by acids or enzymes to simple units, while the latter are converted into more complex, insoluble, colored products called phlobaphens. The hydrolysable tannins may be further subdivided into two groups: the gallotannins such as **1** forming gallic acid **3** upon hydrolysis and the ellagitannins which, when hydrolysed, give a sugar, ellagic acid **10**, and related acids. The condensed tannins, also called flavonoid tannins, have not yet been clarified as regards their essential structural features and are considered to be formed from flavan derivatives by oxidative coupling.[4]

1) R. D. Haworth, K. Jones, H. J. Rogers, *Proc. Chem. Soc.*, 8 (1958).
2) E. Haslam, R. D. Haworth, K. Jones, H. J. Rogers, *J. Chem. Soc.*, 1829 (1961).
3) E. Haslam, R. D. Haworth, S. D. Mills, H. J. Rogers, R. Armitage, T. Searle, *J. Chem. Soc.*, 1836 (1961).
4) E. Haslam, R. D. Haworth, *Progress in Organic Chemistry* (ed. J. Cook, W. Carruthers), vol. 6, p. 1, Butterworth, 1964.

9.18   STRUCTURES OF UBIQUINONES (COENZYME Q)

ubiquinones n **1**

(*n*=1 to 12)
uv (EtOH): 275, 405
ir: 1653, 1608

1) The redox properties and uv absorption indicated a benzoquinone structure.[1,2]
2) Ir, the molecular weight, the hydrogen uptake, perbenzoic acid titration, and the nmr suggested the nature of the polyisoprenoid chain.[1,2]
3) The arrangement of the substituents was established by oxidative degradations.[1,2]
4) Under basic or photochemical conditions **1** isomerizes to chromenols **7**.[1,2]

ubichromenols *n* **7**

Remarks

Ubiquinones **1** exist very widely in bacteria, fungi, algae, higher plants, vertebrates and invertebrates.[3] They are localized mainly in mitochondria and play a major role in respiratory electron transport.[3] The synthesis[4] and biosynthesis[3] of these compounds have also been extensively studied.

1) R. A. Morton, U. Gloor, O. Schindler, G. M. Wilson, L. H. Chopard-dit-Jean, F. W. Hemming, O. Isler, W. M. F. Leat, J. F. Pennock, R. Rüegg, U. Schwieter, O. Wiss, *Helv. Chim. Acta*, **41**, 2343 (1958).
2) R. L. Lester, F. L. Crane, Y. Hatefi, *J. Am. Chem. Soc.*, **80**, 4751 (1958); D. E. Wolf, C. H. Hoffman, N. R. Trenner, B. G. Arison, C. H. Shunk, B. O. Linn, J. F. McPherson, K. Folker, *ibid.*, **80**, 4752 (1958).
3) *Biochemistry of Quinones* (ed. R. A. Morton), Academic Press, 1965.
4) O. Isler, H. Mayer, R. Rüegg, J. Würsch, *Vitamins and Hormones*, **24**, 331 (1966).

9.19   SPINOCHROMES, NAPHTHOQUINONES FROM SEA URCHINS

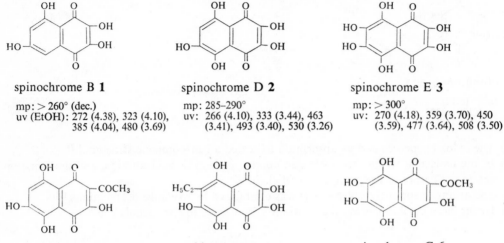

spinochrome B **1**

mp: > 260° (dec.)
uv (EtOH): 272 (4.38), 323 (4.10),
            385 (4.04), 480 (3.69)

spinochrome D **2**

mp: 285–290°
uv: 266 (4.10), 333 (3.44), 463
      (3.41), 493 (3.40), 530 (3.26)

spinochrome E **3**

mp: > 300°
uv: 270 (4.18), 359 (3.70), 450
      (3.59), 477 (3.64), 508 (3.50)

spinochrome A **4**

mp: 192–193°
uv: 251 (3.98), 270 (3.95), 317
      (3.94), 490 (3.49), 520 (3.52)

echinochrome A **5**

mp: 222–223°
uv: 260 (4.32), 343 (3.99), 470
      (3.88), 490 (3.87), 530 (3.66)

spinochrome C **6**

mp: 246–248°
uv: 240 (4.12), 285 (4.21), 330
      (3.95), 463 (3.77), 504 (3.66)

Remarks

Nearly thirty naphthaquinones have been isolated from sea urchins (*Echinodermata*), chiefly from their spines; they were therefore named spinochromes. The considerable confusion regarding their identities and structures has now been eliminated by recent studies.[1-7] The six naphthaquinones **1–6** are representative and are the most widely distributed among *Echinodermata*.[8] The accumulated spectral data (uv,[9] nmr,[10] mass[11]) for these derivatives are good guides in naphthoquinone chemistry.

1) J. Gough, M. D. Sutherland, *Tetr. Lett.*, 269 (1964).
2) C. W. J. Chang, R. E. Moore, P. J. Scheuer, *Tetr. Lett.*, 3557 (1964).
3) R. E. Moore, H. Singh, P. J. Scheuer, *J. Org. Chem.* **31**, 3645 (1966).
4) H. A. Anderson, J. Smith, R. H. Thomson, *J. Chem. Soc.*, 2141 (1965).
5) C. W. J. Chang, R. E. Moore, P. J. Scheuer, *J. Am. Chem. Soc.*, **86**, 2959 (1964).
6) I. Singh, R. E. Moore, C. W. J. Chang, P. J. Scheuer, *J. Am. Chem. Soc.*, **87**, 4023 (1965).
7) I. Singh, R. E. Moore, C. W. J. Chang, R. T. Ogata, P. J. Scheuer, *Tetr.*, **24**, 2969 (1968).
8) P. J. Scheuer, *Chemistry of Marine Natural Products*, p. 82, Academic Pres, 1973.
9) I. Singh, R. T. Ogata, R. E. Moore, C. W. J. Chang, P. J. Scheuer, *Tetr.*, **24**, 6053 (1968).
10) R. E. Moore, P. J. Scheuer, *J. Org. Chem.*, **31**, 3272 (1966).
11) D. Becher, C. Djerassi, R. E. Moore, H. Singh, P. J. Scheuer, *J. Org. Chem.*, **31**, 3650 (1966).

9.20 SYNTHESIS OF PHYLLOQUINONE (K) (VITAMIN K₁)

The synthesis of phylloquinone (K)[1] (vitamin K_1) **5** by the condensation of **1**[2,3] or 2-methyl-1,4-naphthoquinone[4] with phytol **3**[3,4] or phytyl bromide[2] was reported shortly after it was first isolated. A synthesis from **1** and isophytol has also been reported.[5]

1   **2** (R=Ac or PhCO)   **3**

4   phylloquinone **5**

2,3——→4: Partial esterification of **1** to **2**,[6] together with the use of BF₃ in place of anhydrous oxalic acid[3] to effect the condensation, greatly improved the yield of the reaction.

Remarks

Phylloquinone (K) (vitamin K₁) **5** was first isolated from alfalfa and its function is now known to involve the maintainance of adequate levels of prothrombin and other clotting factors. It is generally associated with the green parts of higher plants and has also been detected in various chlorophyll-free tissues of higher plants and algae.

The 2′,3′-*trans* configuration of natural phylloquinone **5** was confirmed by comparison of the nmr spectra of the synthetic *trans* and *cis* isomers.[7]

6 (R=Me)
8 (R=H)

7

From bacteria a number of related compounds, the homologs of menaquinones (MK-n)[1] **6**, II-dihydromenaquinones (MK-n (II-H₂))[1] **7**, and demethylmenaquinones (DMK-n)[1] **8**, have been isolated and shown to be involved in electron transport.[8]

1) Nomenclature and abbreviations proposed by IUPAC-IUB commission.
2) S. B. Binkley, L. C. Cheney, W. F. Holcomb, R. W. McKee, S. A. Thaver, D. W. MacCorquodale, E. A. Doisy, *J. Am. Chem. Soc.*, **61**, 2558 (1939).
3) L. F. Fieser, *J. Am. Chem. Soc.*, **61**, 2559, 3467 (1939).
4) H. J. Almquist, A. A. Klose, *J. Am. Chem. Soc.*, **61**, 2557 (1939).
5) O. Isler, K. Doebel, *Helv. Chim. Acta*, **37**, 225 (1954).
6) R. Hirschmann, R. Miller, N. L. Wendler, *J. Am. Chem. Soc.*, **76**, 4592 (1954).
7) L. M. Jackman, R. Rüegg, G. Ryser, C. von Planta, U. Gloor, H. Meyer, P. S. Schudel, M. Kofler, O. Isler, *Helv. Chim. Acta*, **48**, 1332 (1965).
8) *Biochemistry of Quinones* (ed. R. A. Morton), Academic Press, 1965; J. C. Wallwork, F. L. Crane, *Progress in Phytochemistry*, vol. 2, p. 267, Wiley, 1970.

9.21   BIOSYNTHESIS OF SHIKIMATE-DERIVED NAPHTHOQUINONES

The biosynthesis of naphthoquinones is believed to occur via the pathways shown below.[1]

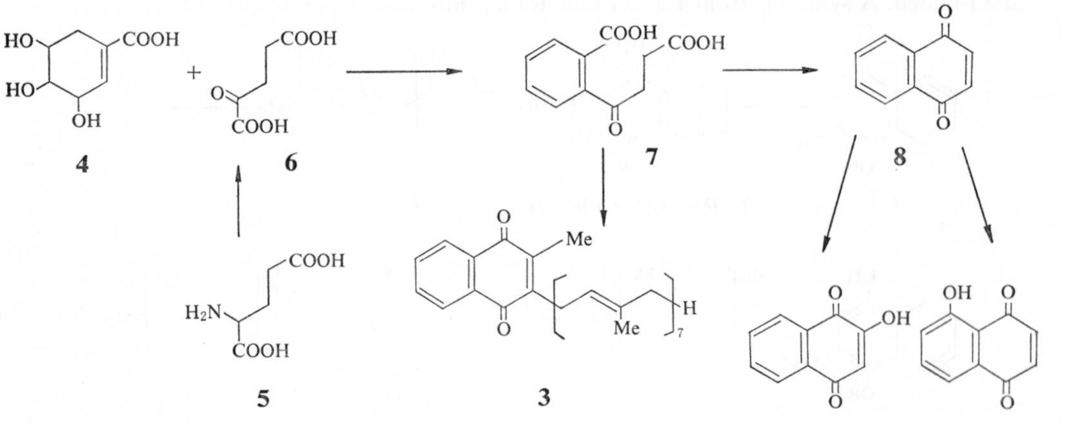

Tracer studies

The incorporation of shikimate **4** was observed into lawsone **1** in *Impatiens balsamina*,[2] into juglane **2** in *Juglans regina*,[3] as well as into vitamin K_2 in *Bacillus megaterium*.[4] In all cases, the ring atoms and carboxyl carbon were introduced into the aromatic nucleus and carbonyl carbon, respectively, of the quinone. An equal distribution of ^{14}C between the two carbonyls of **1** and **2** indicate the presence of a symmetrical intermediate.

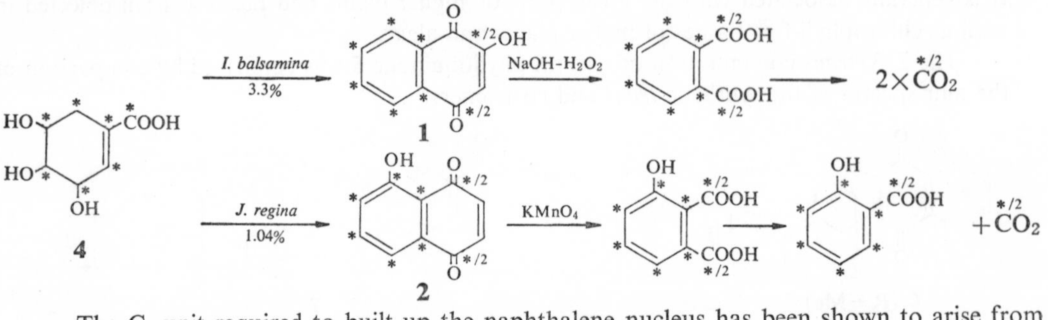

The C_3 unit required to built up the naphthalene nucleus has been shown to arise from glutamate by the specific incorporation of [2-^{14}C]-glutamate into **1**.[5] A similar result was obtained with vitamin K_2.[6]

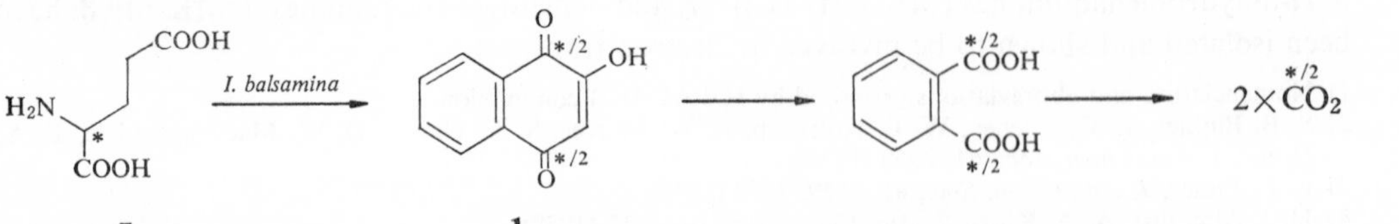

1) M. H. Zenk, *Pharmacognosy and Phytochemistry* (ed. H. Wagner, L. Hörhammer), p. 314–346, Springer Verlag, 1971.
2) E. Leistner, M. H. Zenk, *Tetr. Lett.* 475 (1967); Z. Naturforsch., **22b**, 865 (1967).
3) E. Leistner, M. H. Zenk, *Z. Naturforsch.*, **23b**, 259 (1968)
4) E. Leistner, H. H. Schmidt, M. H. Zenk, *Biochem. Biophys. Res. Commun.*, 845 (1967).

Experiments with asymmetrically labeled shikimate 4[7] showed that the pro-S hydrogen at C-2 and the hydrogen at C-6 were lost, and the pro-R hydrogen at C-2 was retained during the biosynthesis. Hydroxylation of the aromatic ring in the biosynthesis of juglone is not accompanied by NIH shift.

An expected intermediate formed by the condensation of 4 and 6 is o-succinylbenzoic acid 7, which showed a high and specific incorporation into 1[8] and 3.[9] In the latter case [arom. carboxyl-^{14}C]o-succinylbenzoic acid labeled only one of the cabonyls, and the participation of symmetrical intermediate such as naphthoquinone is excluded.

Naphthoquinone (8), a symmetrical compound, showed a high and specific incorporation into 1 and 2.[3]

5) J. M. Campbell, *Tetr. Lett.*, 4777 (1969).
6) D. J. Robins, J. M. Campbell, R. Bentley, *Biochem. Biophys. Res. Commun.*, 1081 (1970).
7) K. H. Sharf, M. H. Zenk, D. K. Onderka, M. Carroll, H. G. Floss, *Chem. Commun.*, 576 (1971).
8) P. Dansette, R. Azerad, *Biochem. Biophys. Res. Commun.*, 1090 (1970).
9) I. M. Campbell, D. J. Robins, M. Kelsey, R. Bentley, *Biochemistry*, **16**, 3069 (1971).
10) S. Gatenbeck, R. Bentley, *Biochem. J.*, **94**, 478 (1968).
11) K. H. Bolkart, M. H. Zenk, *Z. Pflanzenphysiol.*, **61**, 356 (1969).

Remarks

Other than above route, there are two further pathways known for the biosynthesis of naphthoquinones. Javanicin is derived via the polyketide pathway[10] and cimaphylin from homogentisate plus mevalonic acid.[11]

9.22 ANTHRAQUINONES OF *Penicillium islandicum*

chrysophanol **1** (R₁=R₂=H)

skyrin **5**

chrysophanol **1** ($R_1=R_2=H$)
emodin **2** ($R_1=OH$, $R_2=H$)
islandicin **3** ($R_1=H$, $R_2=OH$)
catenarin **4** ($R_1=R_2=OH$)
nmr measured for peracetates[1]

Dimeric anthraquinones

Dimeric anthraquinones containing almost all possible combinations of monomeric anthraquinones have been isolated, all being linked through the 5–5′ positions. The names of some of the dimers are given below. For instance, **2–3** indicates a dimer composed of emodin and islandicin linked through the 5–5′ positions.

dianhydrorugulorsin **1–1**   skyrin **2–2**
auroskyrin **1–2**   rhodoislandin B **2–3**
roseoskyrin **1–3**   aurantioskyrin **2–4**
rhodoislandin A **1–4**   punicoskyrin **3–4**

All the dimeric anthraquinones showed a positive cd curve, indicating restricted rotation. The dimers can be reductively cleaved to the corresponding monomers by means of sodium dithionite.

Remarks

Dimeric anthraquinones such as skyrin have been isolated from various fungi[2] and lichens.[3]

1) Y. Ogihara, N. Kobayashi, S. Shibata, *Tetr. Lett.*, 1881 (1968).
2) S. Shibata, S. Natori, S. Udagawa, *List of Fungal Products*, p. 82–83, University of Tokyo Press, 1964.
3) C. F. Culberson, *Chemical and Botanical Guide to Lichen Products*, p. 188–189, University of North Carolina Press,

9.23 STRUCTURES OF RUGULOSIN, LUTEOSKYRIN AND RUBROSKYRIN

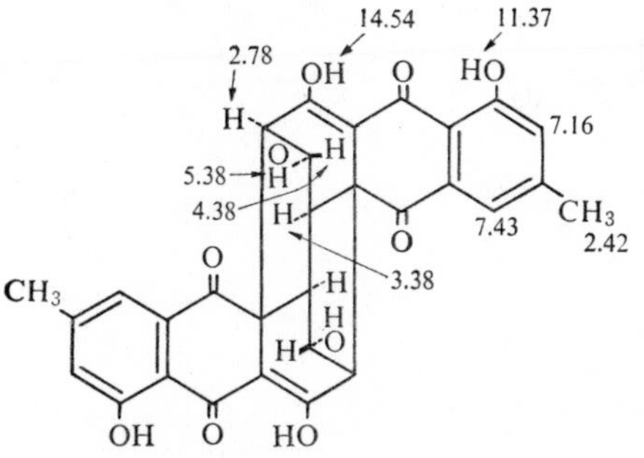

(+)-rugulosin **1**

mp: 290° (dec.)
α_D (dioxane): +492°
ir (Nujol): 3360 (OH), 1690 (C=O), 1620 (C=O chelated)
nmr in DMSO-d_6

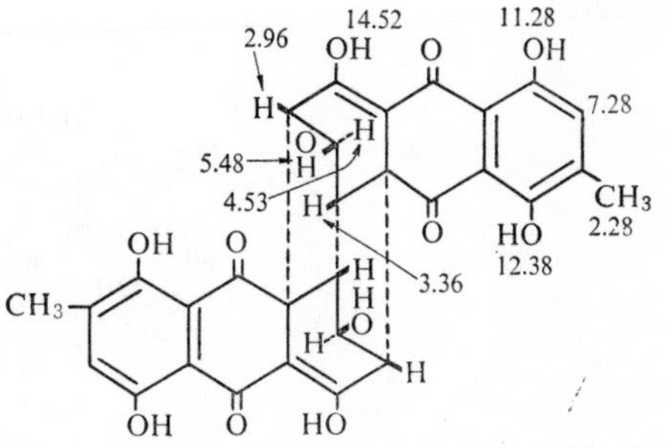

(−)-luteoskyrin **2**

mp: 273° (dec.)
α_D (acetone): −880°
ir (Nujol): 3378 (OH), 1623 (C=O)
nmr in DMSO-d_6

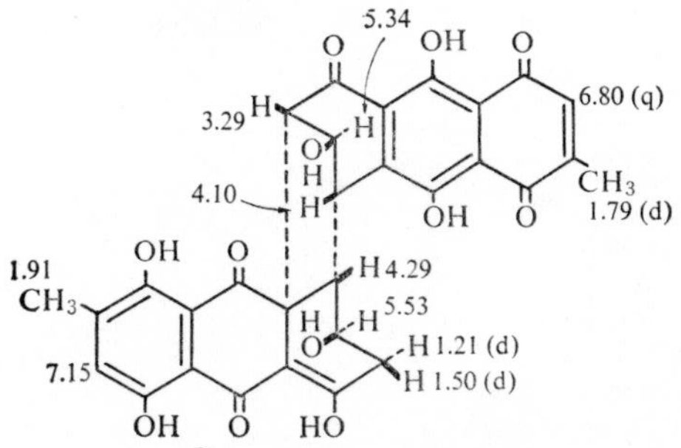

(−)-rubroskyrin **3**

uv (CHCl$_3$): 274, 415, 435, 530, 540
ir (Nujol): 3559 (OH), 3341 (OH), 1706 (C=O), 1623 (C=O)
nmr for the triacetate

1) Mild dehydration of these compounds yields optically active bisanthraquinones.[1,2]
2) Pyrolysis of **1** yielded a 1:1 mixture of chrysophanol and emodin. **2** and **3** gave a similar mixture of islandicin and catenarin. The same fragmentation was observed in the ms, where the monomeric anthraquinones appear as base peaks.

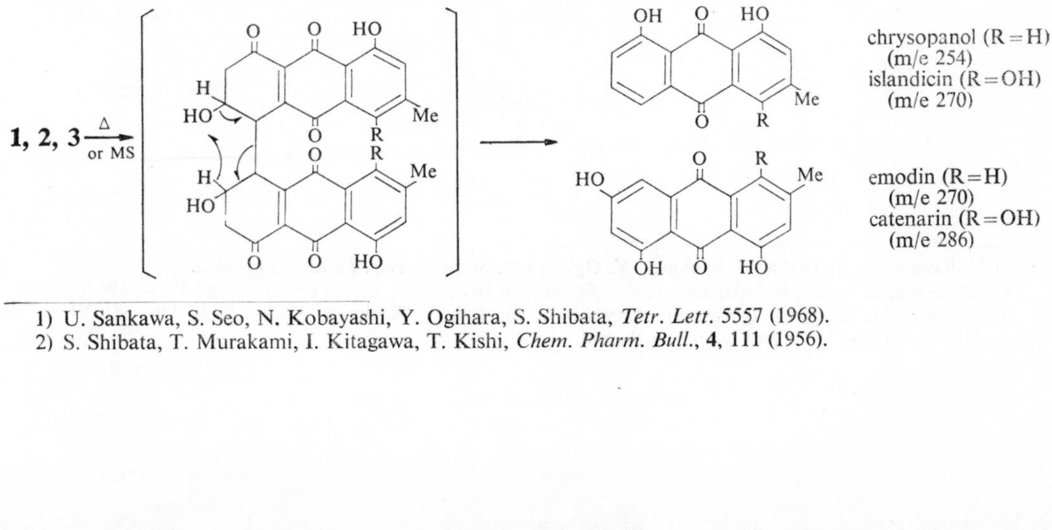

1, 2, 3 $\xrightarrow{\Delta \text{ or MS}}$

chrysopanol (R=H) (m/e 254)
islandicin (R=OH) (m/e 270)

emodin (R=H) (m/e 270)
catenarin (R=OH) (m/e 286)

1) U. Sankawa, S. Seo, N. Kobayashi, Y. Ogihara, S. Shibata, *Tetr. Lett.* 5557 (1968).
2) S. Shibata, T. Murakami, I. Kitagawa, T. Kishi, *Chem. Pharm. Bull.*, **4**, 111 (1956).

3) On treatment with pyridine, **3** afforded **2** by intramolecular Michael reaction.

4) The structures deduced on the basis of chemical and spectroscopic evidence were finally confirmed by x-ray analysis of **4**, a bromination product of tetrahydroregulosin.[3]

4

Remarks

1 has been isolated from various *Penicillia*, along with skyrin. It is classed as a mycotoxin. **2** was first isolated from *Penicillium islandcum* NRRL 1036.[4] It was identified as one of the toxic factors in rice infected with this organism, and causes damage to the livers of experimental animals.[5]

3) N. Kobayashi, Y. Iitaka, U. Sankawa, Y. Ogihara, S. Shibata, *Tetr. Lett.*, 6135 (1968).

4) S. Shibata, S. Natori, S. Udagawa, *List of Fungal Products*, p. 85, University of Tokyo Press, 1964.

5) M. Saito, M. Enomoto, T. Tatsuno, *Microbial Toxins* (ed. A. Ciegler, S. Kadis, S. J. Ali), vol. VI, p. 299, Academic Press, 1971; and references therein.

9.24   STRUCTURES OF FLAVOSKYRIN AND RELATED
MODEL COMPOUNDS

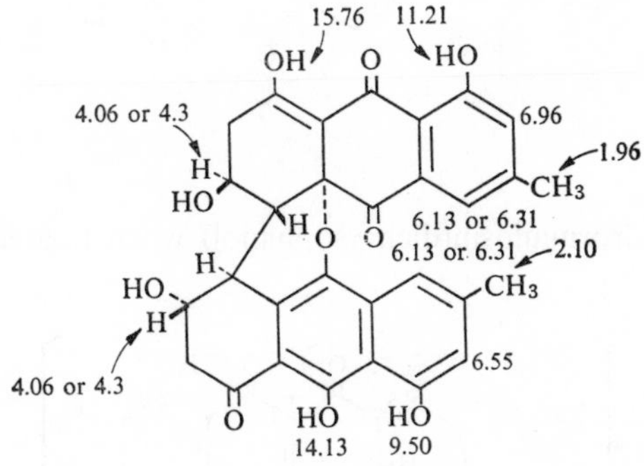

flavoskyrin **1**[1-3]

mp: 215° (dec.)
α_D (dioxane): −295°
uv (dioxane): 267, 303, 312, 328, 368, 414, 433 (sh.)
ir (KBr): 1715, 1623, 1590

1) Model compounds **2a**, **2b** and **2c** were synthesized from the corresponding anthraquinones via the dihydroketoanthraquinones **5**, which were converted into **2** by treatment with a catalytic quantity of pyridine.[3]

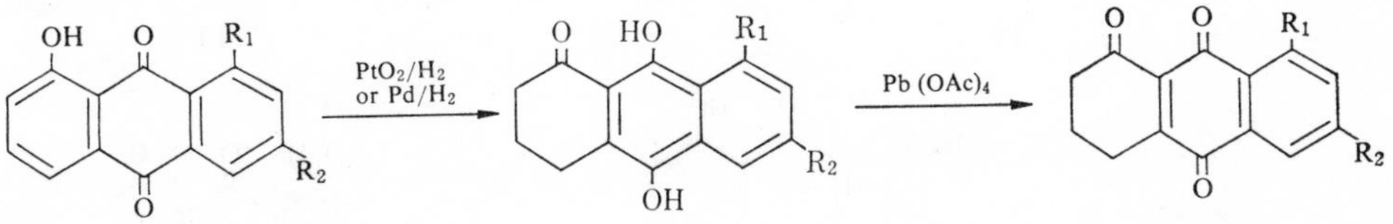

3a ($R_1=R_2=H$)
 b ($R_1=OH$, $R_2=H$)
 c ($R_1=OH$, $R_2=Me$)

2) The pyrolysis of **1** affords chrysophanol **7** and water with quantitative yield.[1] The mass spectrum also shows m/e 254, corresponding to **7** as a base peak. The fission of **1** is caused by retro-Diels-Alder reaction in both cases.

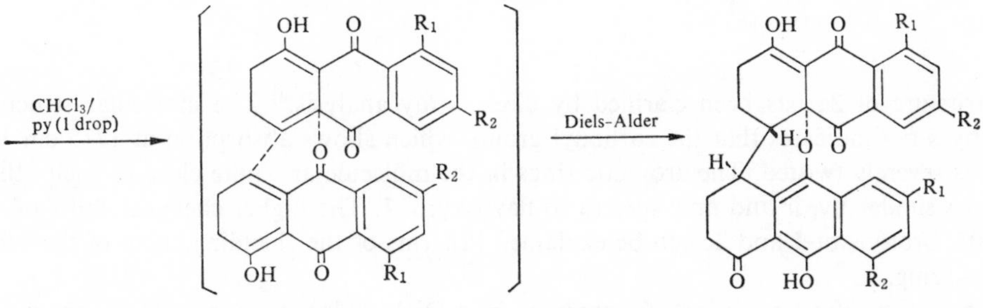

1) B. H. Howard, H. Raistrick, *Biochem. J.*, **65**, 56 (1954).
2) S. Seo, Y. Ogihara, U. Sankawa, S. Shibata, *Tetr. Lett.*, 735 (1972).
3) S. Shibata, T. Murakami, I. Kitagawa, T. Kishi, *Chem. Pharm. Bull.*, **4**, 111 (1956); S. Seo, U. Sankawa, S. Shibata, *Tetr. Lett.*, 731 (1972).

3) 1 afforded (−)rugulosin **12** along with dianhydrorugulosin(dichrysophanol) when treated with pyridine.[2)]

4) The structure of **2a** has been clarified by direct x-ray analysis.[4)] The molecular structure given by x-ray indicates that the carbonyl group, which shows absorption at 1715 cm^{-1} in the ir, is severely twisted. The aromatic rings in the molecule are quite close to each other.

5) 3c shows similar uv, ir and nmr spectra to flavoskyrin **1**. The higher chemical shifts of the aromatic protons of **1** and **2c** can be explained in terms of the shielding effect of the other aromatic ring.

6) The mechanism of dimerization of **5** seems to be a Diels-Alder reaction with *exo* approach. This will also valid in the biosynthesis of **1**, where dihydroemodin **13**, participates as a precursor.

Remarks

   1 was first isolated from *Penicillium islandicum* NRRL 1175.[1)] A monomeric structure was proposed for **1** and **2**,[5,6)] but this was later revised to **1** following x-ray analysis of **2a**.

---

4) U. Sankawa, Y. Iitaka, unpublished.
5) T. Ikekawa, T. Kishi, *Chem. Pharm. Bull.*, **8**, 889 (1960).
6) K. Zahn, H. Koch, *Ber. Deutsch. Chem. Ges.*, **71**, 172 (1938).

9.25 STRUCTURES OF APHIDIDAE COLORING MATTERS

In each species of insect there are four aphins to be considered: a protoaphin **a** present in the living insect, a xanthoaphin **b** into which protoaphin is converted by enzymatic action on the death of the insect; a chrysaphin **c**, produced when xanthoaphin is kept in solution; and an erythroaphin **d**, which is formed by a similar transformation of chrysoaphin and which represents the moderately stable end product of the series. The conversions **b**⟶**c**⟶**d** occur spontaneously in crude extracts of the insects and are accelerated by heat or by acids or alkalis.[1]

protoaphin-fb **1**

uv(CHCl$_3$): 267, 421, 446, 485, 520, 560, 586
ir (Nujol): 3430, 2930, 1673, 1645, 1613, 1590

xanthoaphin-fb **2**

uv (CHCl$_3$): 258, 282, 358, 379, 406, 430, 459

chrysoaphin-fb **3**

uv (CHCl$_3$): 268, 380, 402, 457, 486
ir: 3300, 2967, 1643, 1633, 1590

erythroaphin-fb **4**

uv(CHCl$_3$): 421, 447, 485, 521, 563, 589
ir: 3300, 2700 (br), 1625, 1575

1) The structure of protoaphin-fb **1**, isolated from *Aphis faboe*, was established by reductive cleavage to form **5** and **6** and by further oxidation reactions.[2,3]

2) The structure of erythroaphin-fb **4** was proposed on the basis of physical data, the formation of perylene derivatives by Zn–dust fusion, the formation of mellitic acid on oxidation with nitric acid, and comparison of the reactivity with that of 4,9-dihydroxyperylene-3,10-quinone.[4-6]

1) J. P. E. Human, A. W. Johnson, S. F. MacDonald, A. R. Todd, *J. Chem. Soc.*, 477 (1950).
2) D. W. Cameron, R. I. J. Cromartie, D. G. I. Kingston, Lord Todd, *J. Chem. Soc.*, 51 (1964).
3) D. W. Cameron, D. G. I. Kingston, N. Sheppard, Lord Todd, *J. Chem. Soc.*, 98 (1964).
4) B. R. Brown, A. W. Johnson, J. R. Quayle, A. R. Todd, *J. Chem. Soc.*, 107 (1954).
5) B. R. Brown, A. Calderbank, A. W. Johnson, B. S. Joshi, J. R. Quayle, A. R. Todd, *J. Chem. Soc.*, 959 (1955).
6) D. W. Cameron, R. I. T. Cromartie, Y. K. Hamied, P. M. Scott, Lord Todd, *J. Chem. Soc.*, 62 (1964).

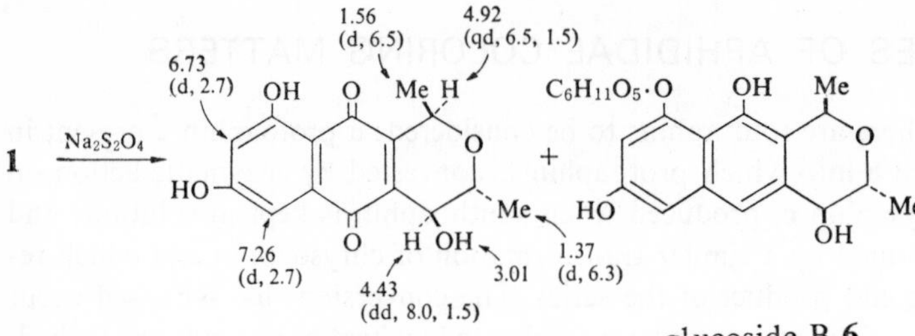

quinone A **5**   glucoside B **6**

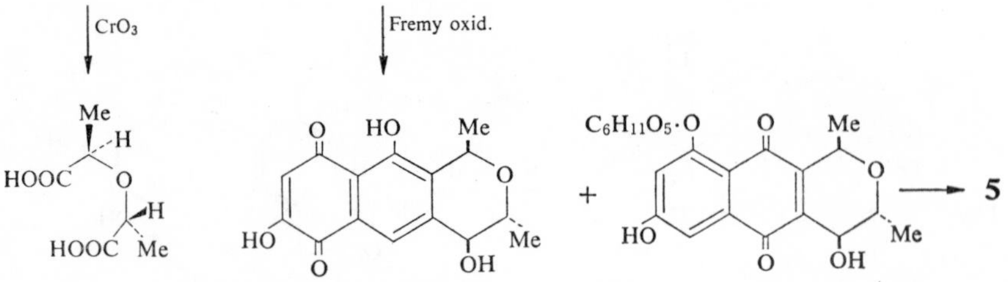

7   **8**   **9**

Remarks

The fluorescent coloring matters occurring in the hemolymph of various species of Aphidi-
diae[7] or materials derived therefrom during the course of separation are known as aphins.[1]
Aphin-fb (from *Aphis faboe*) series is the best characterized, as shown above. Protoaphin-sl
(from *Tuherolachnus salignus*) is an isomer of **1**.[2,8] Another series, protodactynaphin-jc, isolated
from *Dactynotus jaceae*, affords an equilibrium mixture of rhodo- and xanthodactynaphin, **10**
and **11**, respectively, by enzymatic hydrolysis.[9]

For details of the chemistry of aphins, see ref. 10.

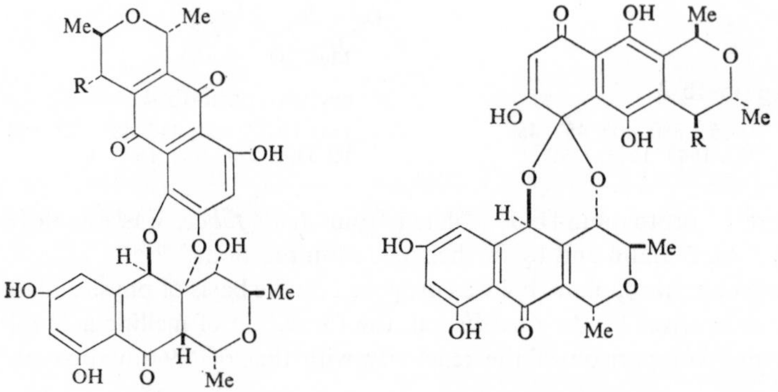

10   **11**

rhododactynaphin-jc-1 (R=OH)   xanthodactynaphin-jc-1 (R=OH)
rhododactynaphin-jc-2 (R=H)   xanthodactynaphin-jc-2 (R=H)

7) H. Duewell, J. P. E. Human, A. W. Johnson, S. F. MacDonald, A. R. Todd, *J. Chem. Soc.*, 3304 (1950).
8) B. R. Brown, A. Calderbank, A. W. Johnson, S. F. MacDonald, J. R. Quayle, A. R. Todd, *J. Chem. Soc.*,
   954 (1955).
9) J. H. Bowie, D. W. Cameron, *J. Chem. Soc.* C, 720 (1967).
10) R. H. Thomson, *Naturally Occurring Quinones*, p. 597, Academic Press, 1971.

9.26 BIOSYNTHESIS OF RUBIACEOUS ANTHRAQUINONES

The overall biosynthetic pathway to these compounds is believed to follow the scheme shown below.[1]

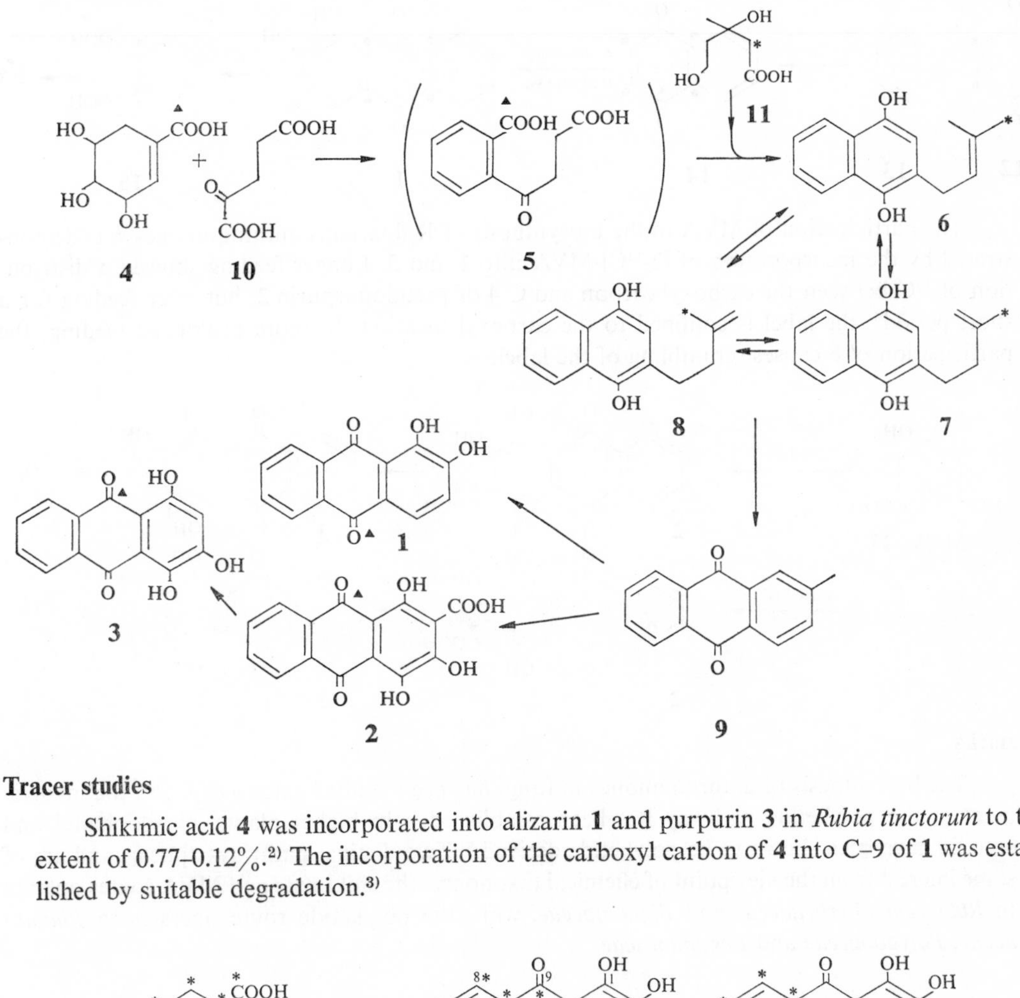

Tracer studies

Shikimic acid **4** was incorporated into alizarin **1** and purpurin **3** in *Rubia tinctorum* to the extent of 0.77–0.12%.[2] The incorporation of the carboxyl carbon of **4** into C–9 of **1** was established by suitable degradation.[3]

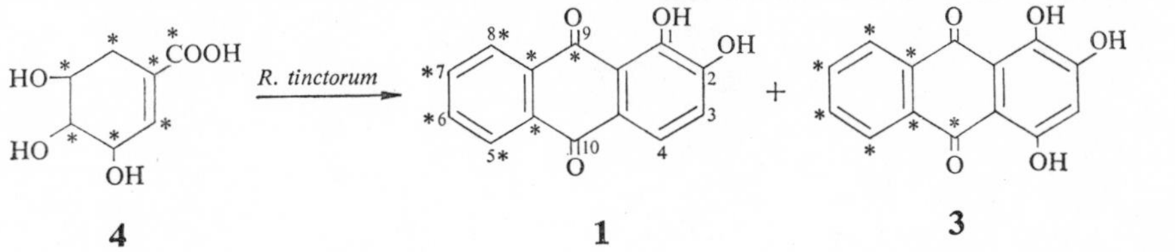

1) R. H. Thomson, *Naturally Occurring Quinones*, p. 10, Academic Press, 1971.
2) E. Leistner, M. H. Zenk, *Tetr. Lett.*, 475 (1967); *Z. Naturforsch.* **22b**, 865 (1967).
3) M. H. Zenk, *Pharmacognosy and Phytochemistry* (ed. H. Wagner, L. Hörhammer), p. 314–346, Springer Verlag, 1971.

Naphthoquinone **14** and *o*-succinylbenzoate were incorporated into **1** with reasonable effi-
ciency.[4] However, Naphthoquinone or its hydroquinone itself was not assumed to be the normal
intermediate of the biosynthesis since radioactive **4** and **5** labeled only one of the two anthraquinone carbonyls.

12   **13**   **14**   **1**   **15**

The participation of MVA in the biosynthesis of Rubiaceous anthraquinones was demonstrated by the incorporation of [2-^{14}C]-MVA into **1** and **3**. Longer feeding showed a distribution of ^{14}C between the carboxyl carbon and C-4 of pseudopurpurin **2**, but after feeding for a
short period[6] the label is confined to the carboxyl location. In more prolonged feeding, the
participation of **8** causes scrambling of the label.

MVA **11**   **2**   **3**

2   **3**

Remarks

The biosynthesis of anthraquinones in fungi has been studied extensively, and the involvement of the polyketide pathway has been established.[7] In higher plants, chrysophanol and
emodin have been shown to be polyketide derived.[8] Two distinct pathways appear, and are of
some interest from the viewpoint of chemical taxonomy: the shikimate and MVA routes operate
in *Rubiaceae*, *Verbenaceae* and *Bignoniaceae*, while the polyketide route operates in *Rhamn-
aceae*, *Polygonaceae* and *Leguminoseae*.

4) E. Leistner, M. H. Zenk, *Tetr. Lett.*, 861 (1968); E. Leistner, *Phytochem.*, **12** 337 (1973).
5) A. R. Burnett, R. H. Thomson, *J. Chem. Soc.* (C), 2457 (1968).
6) E. Leistner, M. H. Zenk, *Tetr. Lett.* 1395 (1968); *Abstracts IUPAC th International Symposium Chemistry Natural Products* (London), p. 113, 1968.
7) See W. B. Turner, *Fungal Metabolites* p. 156–168, Academic Press, 1971.
8) A. Meynaud, A. Ville, H. Pacneco, *Compt. Rend.*, **266D**, 1783 (1968); E. Leistner, M. N. Zenk, *Chem. Commun.*, 210 (1969).

9.27 BIOSYNTHESIS OF BENZOFURANS IN *Stereum subpileatum*

The overall biosynthetic pathway of benzofurans in this organism is summarized below.[1]

Tracer studies

No direct incorporation of MVA has been demonstrated, but the origin of the furan ring seems to involve the prenylation of *p*-hydroxybenzaldehyde or its equivalent, since the MVA origin of the furan ring has been established in higher plants.[2]

The incorporation of phenylalanine into **4**, together with its labeling pattern suggest that **2** is a key intermediate in benzofuran biosynthesis.

The other C_2 unit in **4**, **5** and **6** can be derived by an acetoin condensation with pyruvate. **1** can be obtained from **2** by a Darkin reaction followed by methylation.

Remarks

Compound **1** was isolated from *Stereum subpileatum* (Basidiomycetes), which was found growing on beer barrels. It is responsible for the characteristic odor of the fungus. The other benzofurans were isolated during the course of biosynthetic studies.

1) J. D. Bu'Lock, A. T. Hudson, B. Kaye, *Chem. Commun.*, 814 (1967).
2) H. G. Floss, U. Mothes, *Phytochemistry*, **5**, 169 (1966).
3) J. H. Birkinshaw, P. Chaplen, W. P. K. Findlay, *Biochem. J.*, **66**, 188 (1957).

9.28 BIOSYNTHESIS OF USNIC ACID

The biosynthetic route to usnic acid is suggested by the synthetic route, involving phenolic oxidative coupling.

Biogenetic-type synthesis[1]

The involvement of two molecules of **2** in the biosynthesis of usnic acid **1** was first suggested by Schöpf and Ross.[2] The synthesis of **1** was performed by phenolic oxidative coupling of **2** and is the earliest demonstration of phenolic oxidative coupling reactions. It provided strong evidence for the participation of **2** and the radicals **3** in the biosynthesis of **1**.

4 ⟶ 5: An ether linkage is formed spontaneously under these reaction conditions and **5** was obtained as a reaction product. Oxidation of **2** with horse radish peroxidase also yielded **5**.[3,4]

[14]C Tracer experiments[4]

The distribution patterns of ^{14}C from various labeled compounds have been determined by suitable degradation reactions. The participation of **2** was confirmed by feeding experiments with synthetic ^{14}C labeled methylphloroacetophenone **2**. However, **7** was not incorporated, indicating that the C-1 methyl of **2** is introduced prior to the formation of the aromatic ring i.e., at the polyketomethylene stage.

1) D. H. R. Barton, A. M. Deflorin, D. E. Edward, *J. Chem. Soc.*, 530 (1965).
2) D. Schöpf, F. Ross, *Ann. Chem.*, **546**, 1 (1941).
3) A. Penttila, H. M. Fales, *Chem. Commun.*, 656 (1966).
4) H. Taguchi, U. Sankawa, S. Shibata, *Chem. Pharm. Bull.*, **17**, 2054 (1969).

The radical coupling reaction of **2** must be stereospecific since naturally occurring **1** is mainly found in the optically active form. Contrary to the case *in vitro*, the formation of the ether linkage *in vivo* must be controlled enzymatically, since some lichens produce only isousnic acid **8**.

Remarks

Usnic acid **1** is widely distributed among lichens, and occurs as *l*-, *d*- and quasi-racemic forms. **1** is also obtained from cultures of mycobionts of *Ramalina* sp.

9.29 MASS SPECTRA OF LICHEN SUBSTANCES

Depsides[1]

A characteristic feature of the mass spectra (ms) of depsides is the lack of high intensity peaks between the molecular ion and the monomer ions. The ms of atranolin **1** shows high intensity peaks corresponding to the acid **2** and phenol parts **3, 4**. It is sometimes difficult to detect the molecular ions due to their low intensity.

As is the case in the ms of evernic acid methyl ester **5**, the phenol peaks due to **7** and **8** are sometimes too small to be detectable, but they appear in the anion ms as higher intensity peaks. Effective use of anion ms as well as cation ms provides complete information.

2 m/e: 179 (42%)

atranolin **1**
ms: M+ 374 (1.6%)

3 m/e: 196 (54%)   **4** m/e: 164 (100%)

—MeOH

evernic acid Me-ester **5**
ms: M+ 346 (+, 65%; −, 20%)

6 m/e: 165 (+, 100%; −, <10%)

7
m/e: 182 (+, <10%; −, 100%) **8** m/e: 150 (+, 10%; −, 30%)

Depsidones[1]

Contrary to the case with depsides, depsidones show relatively strong molecular peaks, accompanied by fragment ions arising from the loss of functional groups (**10** and **14**). Fragmentation involving an ether linkage gives characteristic information about the aromatic rings contained in depsidones (**11** and **12**).

Depsidones with longer side chains, e.g. **13**, show characteristic loss of side chains.

---

1) S. Huneck, C. Djerassi, D. Becher, M. Barber, N. Von Arcenne, K. Steinfelder, R. Tümmler, *Tetr.*, **24**, 2707 (1967).

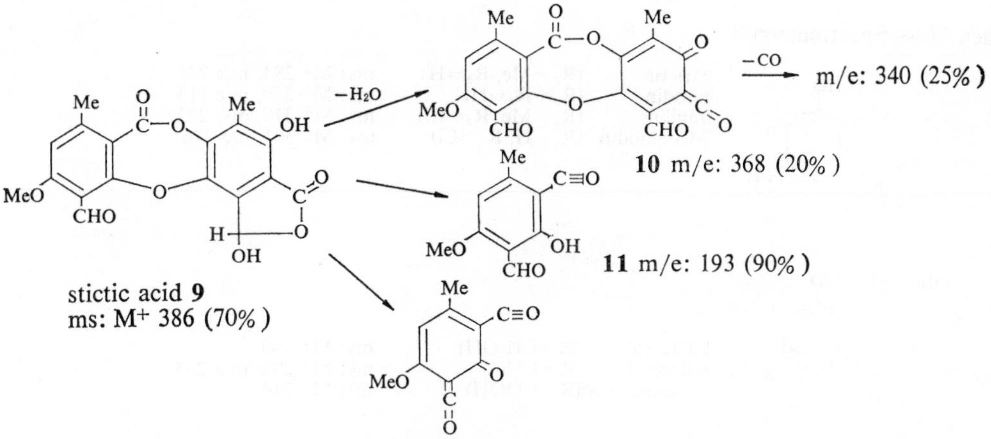

stictic acid **9**
ms: M⁺ 386 (70%)

10 m/e: 368 (20%)

11 m/e: 193 (90%)

12 m/e: 191 (55%)

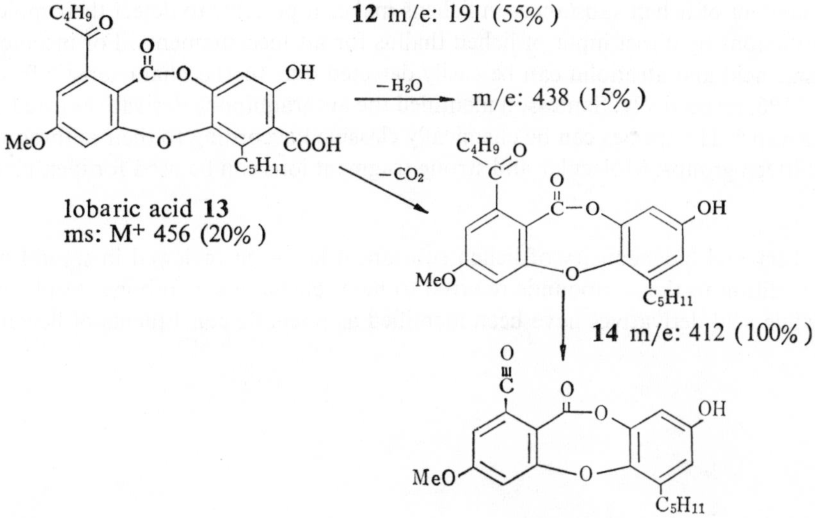

$-CO$ m/e: 340 (25%)

lobaric acid **13**
ms: M⁺ 456 (20%)

$-H_2O$ m/e: 438 (15%)

14 m/e: 412 (100%)

15 m/e : 355 (20%)

Usnic acid[1]

Fission of the B ring gives two characteristic fragments, **17** and **18**.

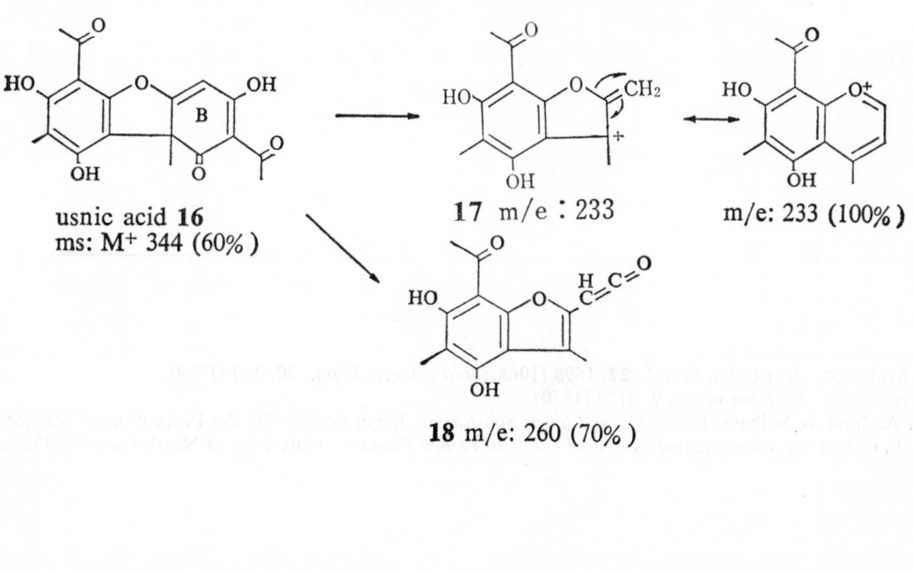

usnic acid **16**
ms: M⁺ 344 (60%)

17 m/e : 233

m/e: 233 (100%)

18 m/e: 260 (70%)

Lichen Mass Spectrometry[2]

parietin	$(R_1=Me, R_2=H)$	ms: M^+ 284, m/e 241
emodin	$(R_1=R_2=H)$	ms: M^+ 270, m/e 213
fragilin	$(R_1=Me, R_2=Cl)$	ms: M^+ 318, m/e 275
7-Cl-emodin	$(R_1=H, R_2=Cl)$	ms: M^+ 304 m/e 276

fallacinol	$(R=CH_2OH)$	ms: M^+ 300
fallacinal	$(R=CHO)$	ms: M^+ 298, m/e 297
parietinic acid	$(R=COOH)$	ms: M^+ 314

The high content of lichen substances in lichens makes it possible to detect their molecular and characteristic ions by direct input of lichen thallus for ms measurements. For instance, the presence of usnic acid and atranolin can be easily detected due to the characteristic fragment ions at 233 and 196, respectively. Santesson identified the anthraquinone derivatives listed above in 230 *Caloplaca* sp.[3] The species can be chemically classified according to their anthraquinone content into thirteen groups. Molecular and strong fragment ions can be used for identification.

Remarks

The chemistry and biochemistry of lichen substances has been reviewed in several monographs.[4,5] In addition to the compounds referred to here, chromone, xanthone, terphenylquinone and pulvinic acid derivatives have been identified as phenolic constituents of lichens.

2) J. Santesson, *Acta Chim. Scand.*, **22**, 1698 (1968); *ibid.*, *Arkiv. Kemi.*, **30**, 363 (1969).
3) J. Santesson, *Phytochemistry*, **9**, 2149 (1970).
4) Y. Asahina, S. Shibata, *Chemistry of Lichen Substances*, Japan Society for the Promotion of Science, 1954.
5) C. F. Culberson, *Chemical and Botanical Guide to Lichen Products*, University of North Carolina Press, 1969.

9.30 STRUCTURE OF VARIEGATIC ACID

variegatic acid **1**

softened at 235° (dec.)
uv (10% EtOH): 262 (4.28), 386 (3.97)
ir (KBr): 3410, 2500–2650, 1754, 1678, 1612, 1590

1) Clusters of red needles of **1** are obtained on recrystallization of the gummy extract from water or HCOOH.

2) **1** is insoluble in light petroleum, but is soluble in all polar solvents.

3) **1** yields a tetraacetate **2**, $C_{26}H_{18}O_{12}$, on treatment with acetic anhydride and H_2SO_4. Treatment of **2** with hot MeOH yields the methyl ester **3**, and teratment with CH_2N_2 yields the methyl ester **4**.

2 (R=OAc)
5 (R=H)

3 (R^1=H, R^2=Me)
4 (R^1=R^2=Me)

4) The formation of the acid **3** (R^2=H) and the ester **3** by the action of aqueous or methanolic alkali on the acetate **2** is similar to the reaction of pulvinic acid anhydride **5**, which supports the assignments of the structures **2** and **3**.

5) Oxidation of **2** with CrO_3 yields 3,4-diacetoxybenzoic acid **6**.

Remarks

Compound **1** has been isolated from all four species of *Boletus*, i.e. *B. variegatus*, *B. bovinus*, *B. erythrupus*, and *B. appendiculatus*.[1] The phenomenon of the blueing of the flesh of boleti on bruising has long been of interest to chemists and biologists. The phenolic compound, boletol, responsible for the reaction was isolated by Kögl *et al.*[2] and identified as an anthraquinone derivative. The physical properties of **1** are comparable with those reported for boletol, but the chemical properties described for boletol are incompatible with the tetronic acid structure of

1) P. C. Beaumont, R. L. Edwards, G. C. Elsworthy, *J. Chem. Soc.* C, 2968 (1968).

1.[1,3] Further, no trace of boletol has been found in these species although repeated examinations were carried out on fungus extracts. It may therefore be concluded that the main metabolites responsible for the blueing of boleti are hydroxytetronic acids such as **1**. A solution of **1** in 10% ethanol produces a blue color with dilute alkali or potato extract.

Several related compounds have recently been isolated from higher fungi and also from *Aspergillus terreus.*[4]

1 and related compounds have been synthesized by demethylation of the condensation products from methoxybenzyl cyanide and diethyl oxalate in the presence of NaOEt or NaH.[1]

The biosynthesis of **1** occurs via terphenylquinones **7**, involving the simple dimerisation of two C_6–C_3 units as shown below.[5]

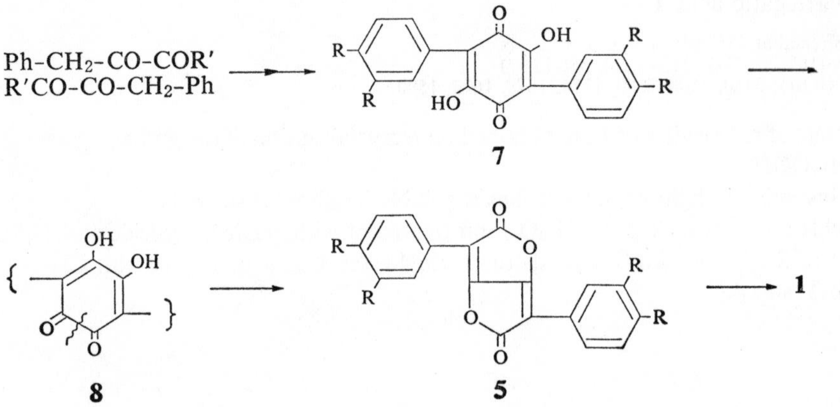

Ph-CH$_2$-CO-COR′
R′CO-CO-CH$_2$-Ph   ⟶

7

8   **5**   ⟶ **1**

---

2) F. Kögl, W. B. Deijs, *Ann. Chem.*, **515**, 10, 23 (1934); G. Bertrand, *Compt. Rend.*, **133**, 1233 (1901); *ibid.*, **134**, 124 (1902).
3) W. Steglich, W. Further, A. Prox, *Z. Naturforsch.*, **B23**, 1044 (1968).
4) e.g. A. Bresinsky, H. Besl, W. Steglich, *Phytochemistry*, **13**, 271 (1974); N. Ojima, S. Takenaka, S. Seto, *ibid.*, **14**, 573 (1975).
5) G. Read, L. C. Vining, *Chem. Ind.*, 1547 (1959); K. Mosbach, *Acta Chem. Scand.*, **21**, 2331 (1967).

9.31 STRUCTURE OF MYCOPHENOLIC ACID

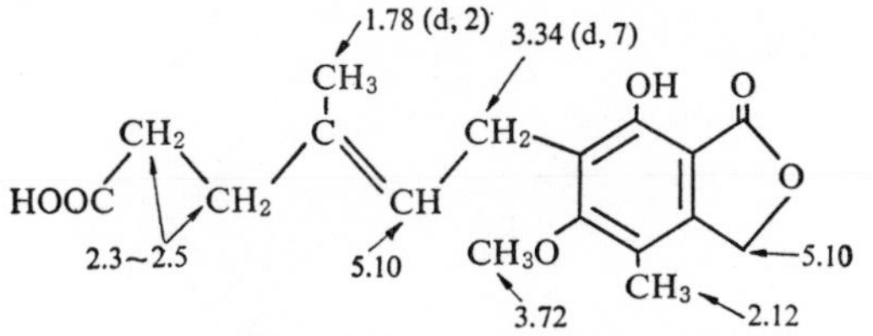
1.78 (d, 2)
3.34 (d, 7)
2.3~2.5
5.10
3.72
5.10
2.12

mycophenolic acid **1**

colorless needles mp: 141°
uv (EtOH): 305 (3.63)
ir (KBr): 3420, 1750, 1710, 1625

1) The structure was elucidated by the following reactions.[1-3]

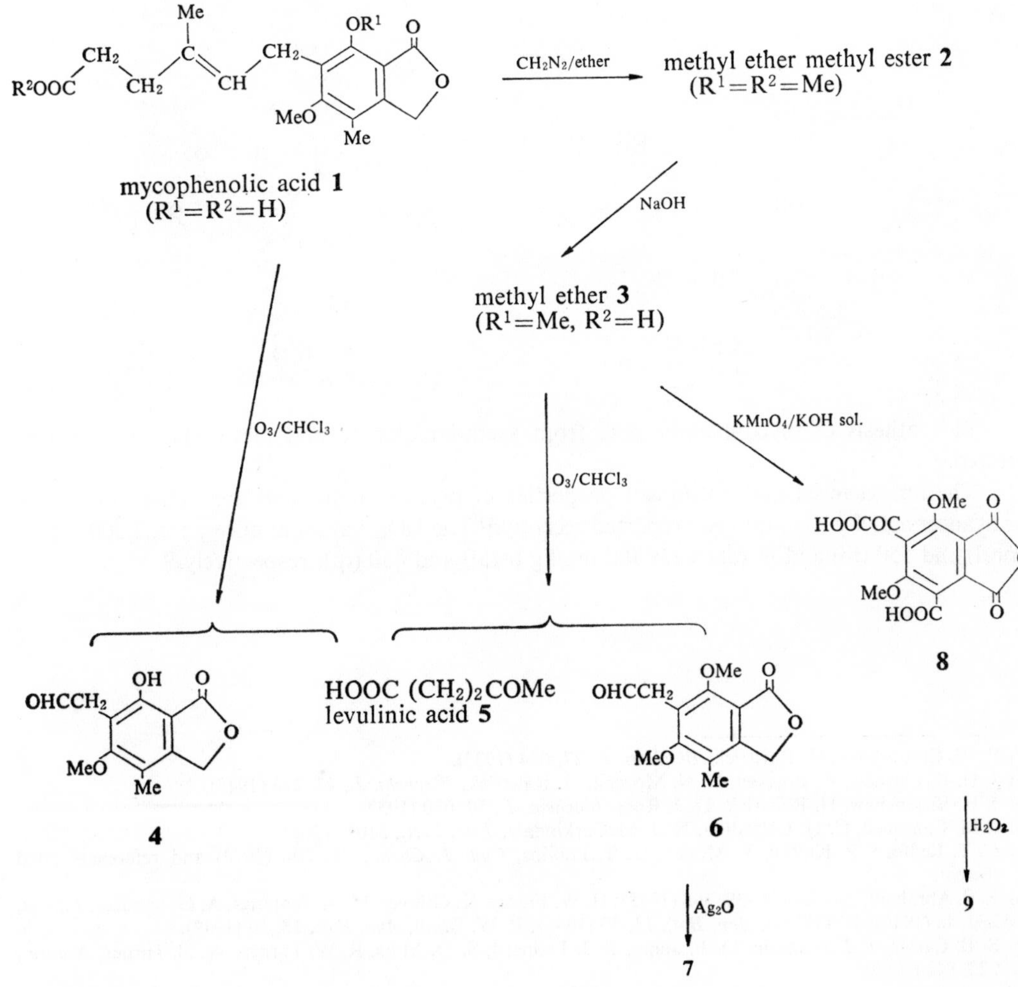
CH2N2/ether
methyl ether methyl ester 2
(R1=R2=Me)
mycophenolic acid 1
(R1=R2=H)
NaOH
methyl ether 3
(R1=Me, R2=H)
O3/CHCl3
O3/CHCl3
KMnO4/KOH sol.
HOOC (CH2)2COMe
levulinic acid 5
4
6
8
Ag2O
7
H2O2
9

7   **9**

Remarks

Mycophenolic acid was first named and isolated by Alsberg and Black in 1913 from a strain of *Penicillium stoloniferum* (which may be regarded as a synonym of *P. brevi-compactum*) found on moldy Italian maize.

Three metabolites closely related to mycophenolic acid were isolated from *P. brevi-compactum* together with this acid.[4] These were identified as mycophenolic acid ethyl ester, mp: 88–90°, a threo isomer **10**, $C_{17}H_{20}O_7$, mp: 218–220° and a carboxylic acid **11**, $C_{17}H_{18}O_6$, mp: 163–165°.

10

$C_{17}H_2O_7$
mp: 218-220°

11

$C_{17}H_{18}O_6$
mp: 163-165°

Biosynthesis of mycophenolic acid from acetate-malonate and mevalonate was demonstrated.[5]

The antibacterial and antifungal properties of mycophenolic acid are well-known[6] and anticancer activity has also been reported recently.[7] The LD$_{50}$ values in mice were, 2,500 mg/kg (oral) and 550 (ip) and in rats were 700 mg/kg (oral) and 450 (ip), respectively.[7]

1) P. W. Clutterbuck, H. Raistrick, *Biochem. J.*, **27**, 654 (1933).
2) J. H. Birkinshaw, A. Bracken, E. N. Morgan, H. Raistrick, *Biochem. J.*, **43**, 216 (1948).
3) J. H. Birkinshaw, H. Raistrick, D. J. Ross, *Biochem. J.*, **50**, 630 (1952).
4) I. M. Campbell, C. H. Calzadilla, N. J. McCorkindale, *Tetr. Lett.*, 5107 (1966).
5) C. T. Bedford, P. Knittel, T. Money, G. T. Phillips, *Can. J. Chem.*, **51**, 694 (1973) and references cited therein.
6) E. P. Abraham, *Biochem. J.*, **39**, 398 (1945); H. W. Florey, K. Gilliver, M. A. Jennings, A. G. Sandres, *Lancet*, 250, 46 (1946); K. Gillivei, *Ann. Bot.*, **13**, 59 (1946); P. W. Brian, *Ann. Bot.*, **13**, 59 (1949).
7) S. B. Carter, T. J. Franklin, D. F. Jones, B. J. Leonard, S. D. Mills, R. W. Turner, W. B. Turner, *Nautre.*, 223, 848 (1969).

9.32 SYNTHESIS OF GRISEOFULVIN

Two synthetic routes to griseofulvin are illustrated below.

Route I

The key step of this synthesis is the oxidative coupling of **9**, which yields dehydrogriseofulvin **10**. The yields of the reactions involved are very high.[1]

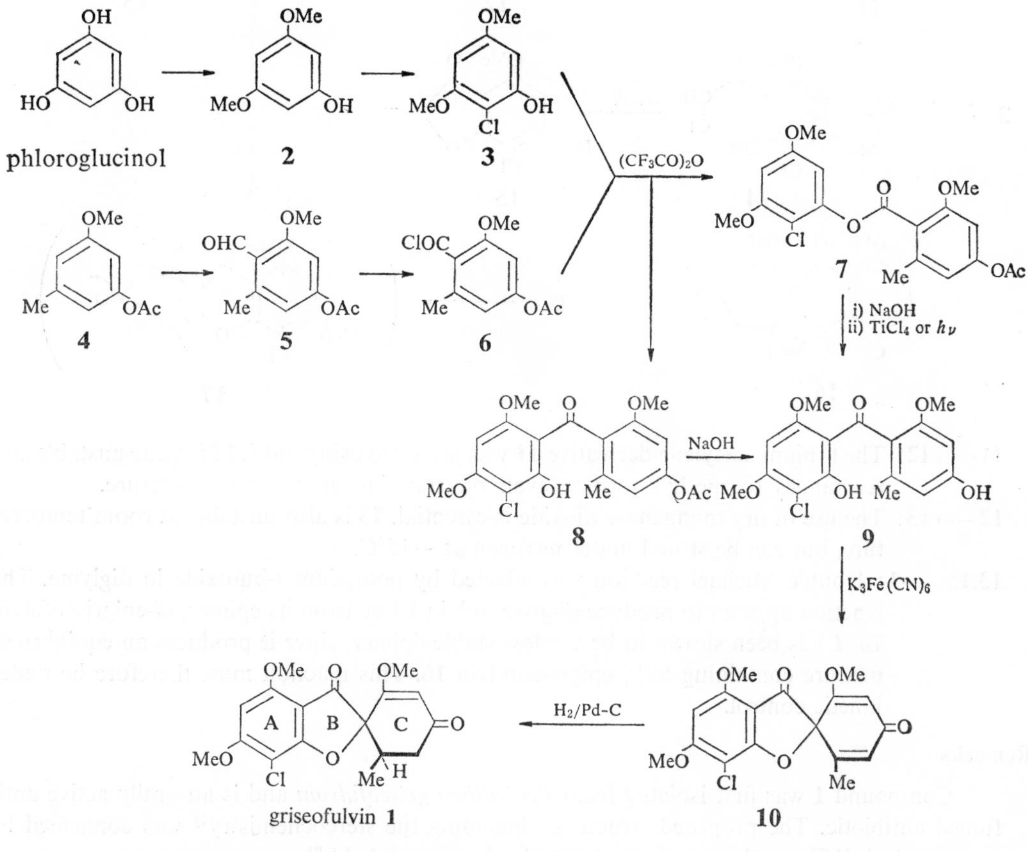

phloroglucinol   **2**   **3** $(CF_3CO)_2O$

4   **5**   **6**

7

i) NaOH
ii) TiCl$_4$ or $h\nu$

8   NaOH $\longrightarrow$ **9**

K$_3$Fe(CN)$_6$

griseofulvin **1** $\xleftarrow{\text{H}_2/\text{Pd-C}}$ **10**

3,6 ⟶ 7,8: In high concentrations, nuclear acylation occurred to the extent of 30–35%. **7** and **8** were readily separated by recrystallization.

7 ⟶ 9: The acetyl group was removed by mild hydrolysis and the resulting phenol ester was subjected to Fries rearrangement. TiCl$_4$ in nitrobenzene or photochemical treatment gave satisfactory results.

9 ⟶ 10: Potassium ferricyanide (on addition of **9** to it) gave nearly quantitative yield. Heterogeneous reaction conditions (MnO$_2$ in ether/acetone or Pb(OAc)$_4$ in the same solvent) gave 95–100% yield.

10 ⟶ 1: The use of a high catalyst ratio and of a relatively nonpolar solvent resulted in high yields of **1** (85–90%).

1) C. H. Kuo, R. D. Hoffsommer, H. L. Slates, D. Taub, N. L. Wendler, *Chem. Ind.*, 1627 (1960); *Tetr.* **19**, 1 (1963).

Route II

This synthesis is based on the double Michael addition of ethynyl vinyl ketone to a highly active methylene compound. The spiran ring system and the carbon ring were built up in a single reaction step.[2]

11 ⟶ 12: The lithium acetylene derivative 11 was prepared using PhLi. 12 is quite unstable and rearranges to methyl sorbate even on standing at room temperature.

12 ⟶ 13: The use of dry manganase dioxide is essential. 13 is also unstable at room temperature, but can be stored under nitrogen at −15°C.

13,15 ⟶ 1: Double Michael reaction was effected by potassium t-butoxide in diglyme. The reaction appears to produce dl-griseofulvin 1 free from its epimer, dl-epigriseofulvin 16. 1 has been shown to be the less stable epimer, since it produces an equilibrium mixture containing 60% epigriseofulvin 16. This reaction must therefore be under kinetic control.

Remarks

Compound 1 was first isolated from *Penicillium griseofulvum* and is an orally active antifungal antibiotic. The proposed structure[3] including the stereochemistry[4] was confirmed by x-ray analysis.[5] Two other syntheses have also been reported.[6,7]

2) G. Stork, M. Tomasz, *J. Am. Chem. Soc.*, **84**, 310 (1962); *ibid.*, **86**, 471 (1964).
3) J. F. Grove, J. McMillan, J. P. C. Mulholland, M. A. T. Rogers, *J. Chem. Soc.*, 3977 (1952).
4) J. McMillan, *J. Chem. Soc.*, 1823 (1959).
5) W. A. Brown, G. A. Sim, *J. Chem. Soc.*, 1050 (1963).
6) A. Brossi, M. Baumann, M. Getecke, E. Kyburz, *Helv. Chim. Acta*, **43**, 1444, 2071 (1960).
7) A. C. Day, J. Nabney, A. I. Scott., *J. Chem. Soc.*, 4067 (1961).

9.33 STRUCTURE OF SCLEROTIORIN

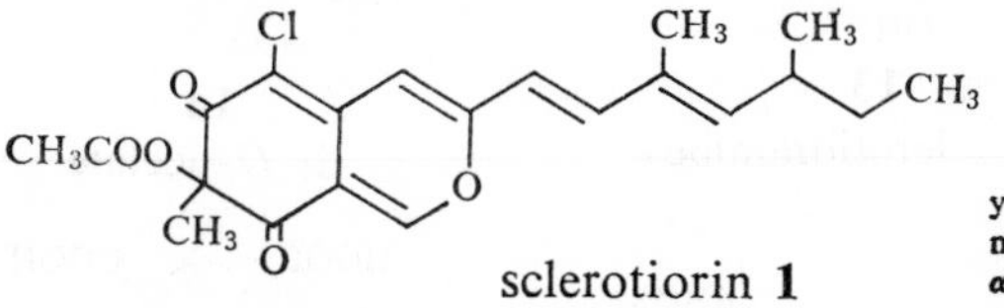

sclerotiorin **1**

yellow crystals
mp: 205–206° (under sublimation)
α_D: (CHCl$_3$): +500°

1) Degradation of **1** with aq. NaHCO$_3$ or dil. NaOH yields hydrochloric, formic and acetic acids, together with 4,6-dimethylocta-2,4-dienoic acid **2** and 2,4-dimethylhex-2-enal **3**.

2) Tetrahydrosclerotiorin **5**, mp: 142–144°, α_D: +213°, is a mixture of two diastereoisomers, α; mp: 118°, α_D: +206°, and β; mp: 159°, α_D: +230°.

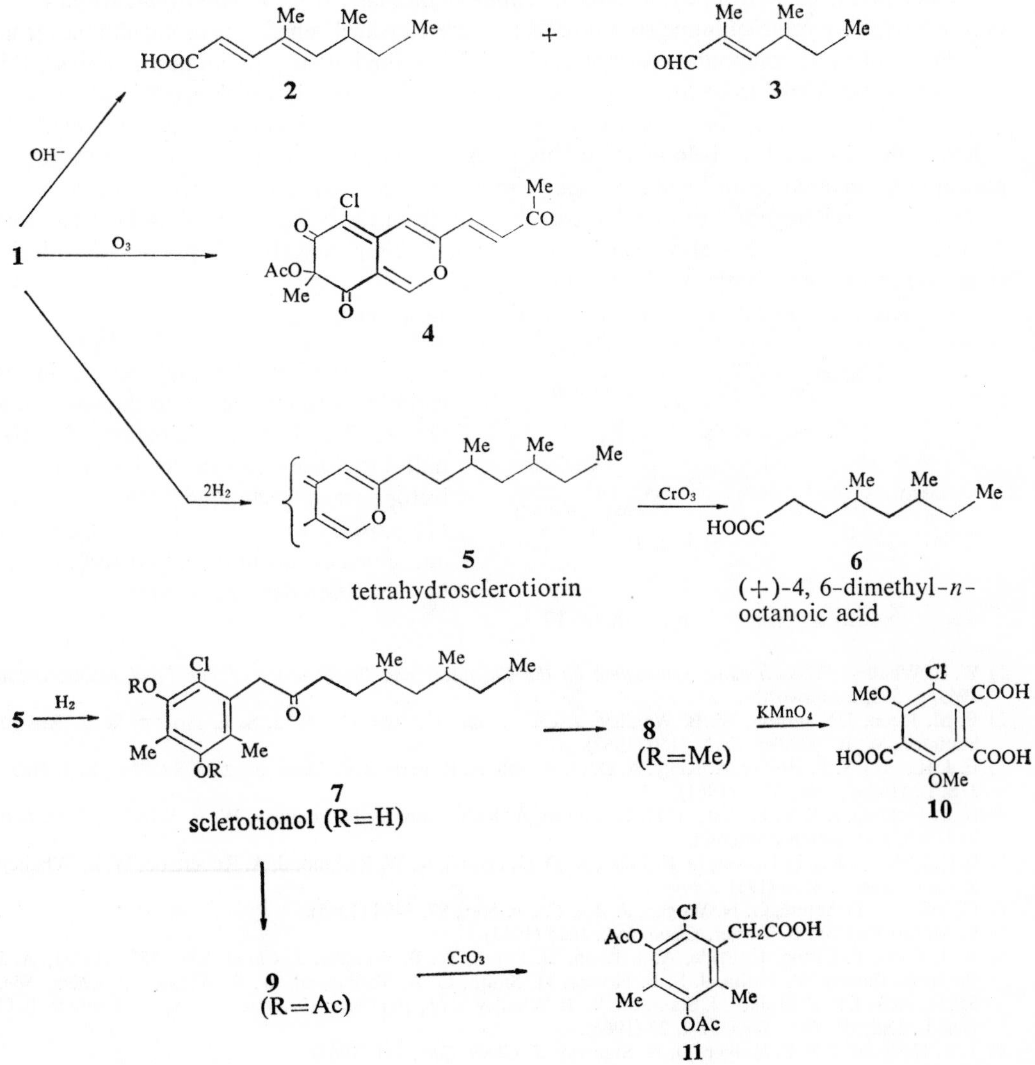

2

3

4

5
tetrahydrosclerotiorin

6
(+)-4, 6-dimethyl-n-octanoic acid

7
sclerotionol (R=H)

8
(R=Me)

9
(R=Ac)

10

11

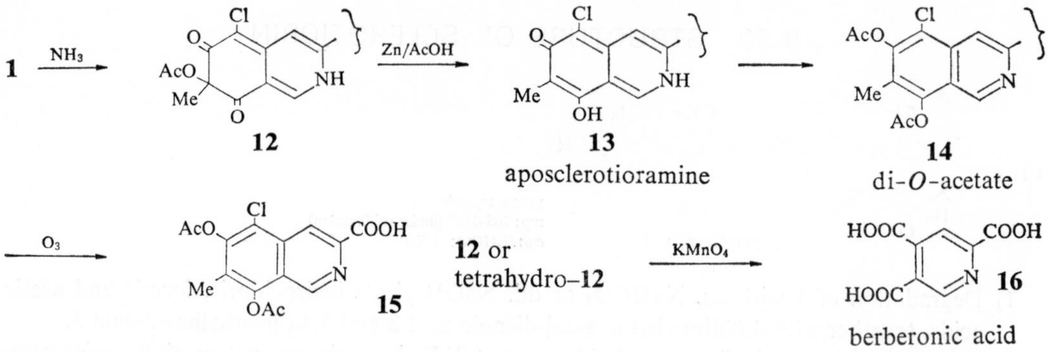

12

13
aposclerotioramine

14
di-*O*-acetate

15   12 or
tetrahydro–12 $\xrightarrow{KMnO_4}$

16
berberonic acid

Remarks

1 was first isolated from *Penicillium sclerotiorum* and later from *P. multicolor*. Mold metabolites having a pyrano-quinone structure like **1** were classified generically as azaphilones from the affinity of these compounds for ammonia, yielding vinylogous γ-pyridones. However, this term is now considered to be too restrictive. For instance, monascin which is obviously a member of this group, cannot be classed as an azaphilone since it is inert towards ammonia.[1] As regards other metabolites belonging to this group, rotiorin[2] from the same fungus, rubropunctatin[3], monascorubrin[4] and monascin[5] from *Monascus purpureus* and *M. rubiginosus* are all known, as well as mitorubrin and mitrorubrinol[6] from *Penicillium rubrum*. A laevorotatory sclerotiorin, $\alpha_D : -482°$ has also been isolated from *P. hirayamae*. It is 7-epi-(+)-sclerotiorin, since (+)-dienoic acid is obtained upon hydrolysis.[7]

For reference, nmr data for mitorubrin **17** are shown below.[6]

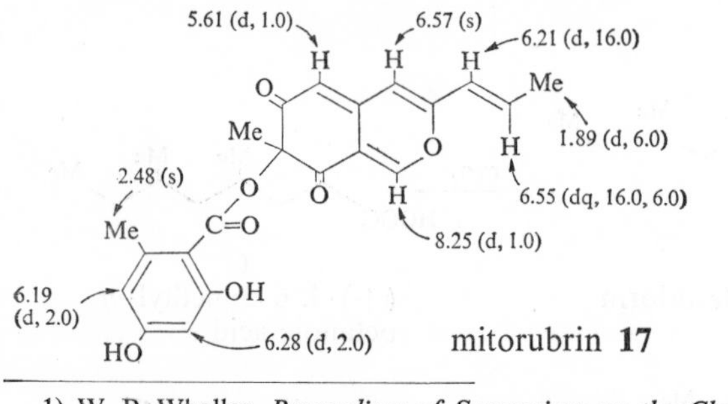

mitorubrin **17**

The biosynthesis of sclerotiorin and rotiorin through the "acetate–malonate pathway" and the participation of a C_1– unit have been shown by means of the isotope tracer technique.[8] The formation of a β-oxo-lactone system in rubropunctatin and monascorubrin from two acetate units has also been demonstrated.[9]

1) W. B. Whalley, *Proceedings of Symposium on the Chemistry and Biochemistry of Fungi and Yeasts, Dublin 1963*, p. 565, Butterworths.
2) F. M. Dean, J. Staunton, W. B. Whalley, *J. Chem. Soc.* C, 3004 (1959); J. S. E. Holker, W. J. Ross, J. Staunton, W. B. Whalley, *ibid.*, 4150 (1962).
3) E. J. Haws, J. S. E. Holker, A. Kelly, A. D. G. Powell, A. Robertson, *J. Chem. Soc.*, 3598 (1959); E. J. Haws, J. S. E. Holker, *ibid.*, 3820 (1961).
4) B. C. Fielding, J. S. E. Holker, A. D. G. Powell, A. Robertson, D. N. Stanway, W. B. Whalley, *Tetr. Lett.* 24 (1960) and references therein.
5) B. C. Fielding, J. S. E. Holker, D. F. Jones, A. D. G. Powell, K. W. Richmond, A. Robertson, W. B. Whalley, *J. Chem. Soc.* C, 4579 (1961).
6) G. Büchi, J. D. White, G. N. Wogan, *J. Am. Chem. Soc.*, **87**, 3484 (1965).
7) E. M. Gregory, W. B. Turner, *Chem. Ind.*, 1625 (1963).
8) A. J. Birch, P. Fitton, E. Pride, A. J. Ryan, H. Smith, W. B. Whalley, *J. Chem. Soc.*, 4576 (1958); A. J. Birch, A. Cassera, P. Fitton, J. S. E. Holker, H. Smith, G. A. Thompson, W. B. Whalley, *J. Chem. Soc.*, 3583 (1962); J. S. E. Holker, J. Staunton, W. B. Whalley, *ibid.*, 16 (1964); G. Jaureguiberry, M. Lenfant, B. C. Das, E. Lederer, *Tetr. Suppl.*, 8-I, 27 (1966).
9) J. R. Hadfield, J. S. E. Holker, D. N. Stanway, *J. Chem. Soc.*, 751 (1967).

9.34   STRUCTURE OF CITREOVIRIDIN

citreoviridin **1**

yellow crystals   mp: 107–111° (from MeOH)
uv (EtOH): 388 (ϵ 48,000), 294 (ϵ 27,100), 286 (sh,
       ϵ 24,600), 234 (ϵ 10,200), 204 (ϵ 17,000)
ir (KBr): 3500, 1702, 1689, 1654, 1626, 1562, 1531,
         1452, 1405, 1249, 1069, 999, 821, 811

1) The structure was elucidated from the physical data of **1** and its degradation products.

Remarks

Citreoviridin was first isolated by Hirata[1] from a strain of *Penicillium toxicarium* (*P. citreo-viride*), and the structure was determined as described above. The fungus *P. toxicarium* was isolated as so-called "yellowed rice fungus" from Formosan rice, and the moldy rice was found to be highly toxic to animals.[3] Uraguchi[4] investigated the toxicity of the fungus and found that the symptoms of acute poisoning in animals were quite similar to the reported clinical manifestations of acute cardiac beriberi (shoshin-kakke) which was prevalent in Japan in the past. Recently the compound has also been isolated from moldy rice infested by *P. citreo-viride*.[2] The LD_{50} of pure toxin was about 20 mg/kg for mice (sc).[5]

Citreoviridin has also been isolated as the toxin component of *Penicillium pulvillorum* grown on moistened maize meal in South Africa.[6]

1) Y. Hirata, *Nippon Kagaku Zasshi*, **68**, 63, 74, 104 (1947).
2) N. Sakabe, T. Goto, Y. Hirata, *Tetr. Lett.*, 1825 (1964).
3) I. Miyake, *Nisshin Igaku*, **34**, 161 (1947).
4) K. Uraguchi, *Microbial Toxins* (ed. A. Ciegler, S. Kadis, S. J. Ajl), vol. **IV**, p. 367, Academic Press, 1971.
5) Y. Ueno, *Mycotoxins in Human Health* (ed. I. F. H. Purchase), p. 115, MacMillan, 1971.
6) D. W. Nagel, P. S. Steyn, D. B. Scott, *Phytochemistry*, **11**, 627 (1972).

9.35 STRUCTURE OF OOSPONOL

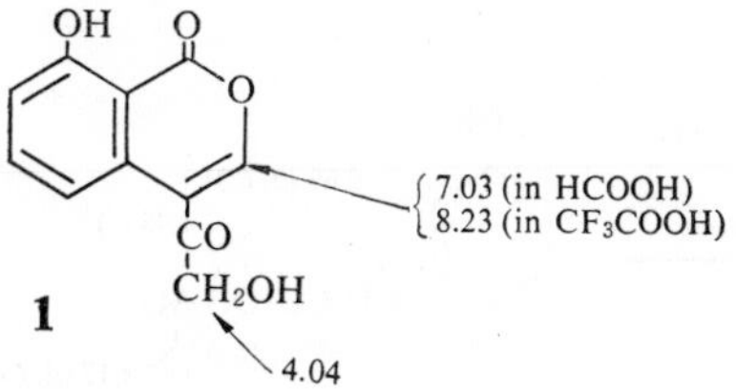

OH
O
O
CO
CH2OH
7.03 (in HCOOH)
8.23 (in CF3COOH)
4.04

1

colorless needles mp: 176°
uv (EtOH): 233 (4.10), 255 (3.90), 335 (3.77)
ir (Nujol): 3400, 3070, 1710, 1670, 1470, 870

1) The structure was elucidated from the physical data and the degradations shown below.[1,2]

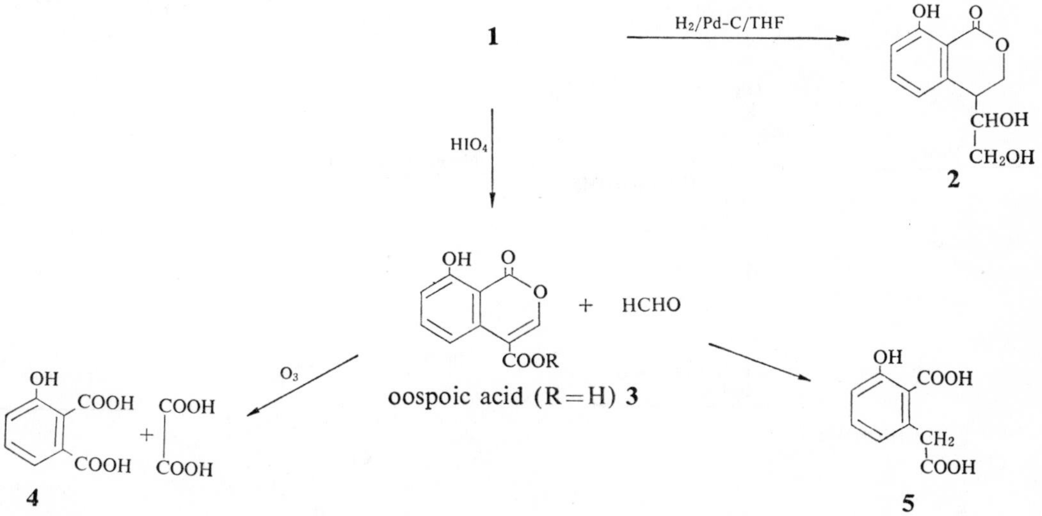

1
H2/Pd-C/THF
CHOH
CH2OH
2
HIO4
+ HCHO
COOR
oospoic acid (R=H) 3
O3
COOH
COOH
COOH
COOH
4
COOH
CH2
COOH
5

Remarks

Oosponol was obtained together with oospolactone and oospoglycol from the mycelium of *Oospora astringenes* which was isolated from the air in the rooms of asthmatic patients.[3] Later it was also isolated from *Gloeophyllum striatum* as a dopamine β-hydroxylase inhibitor.[4] This compound was shown to be toxic to mice and rats (LD_{50} 40 mg/kg ip, 280 mg/kg per *os* to mice, 92 mg/kg ip, 250 mg/kg per *os* to rats). It caused severe skin rash and bronchitis to subjects after several contacts, although no irritation was caused at the first contact.[4]

Synthesis of oosponol was reported by several authors.[2,5,6]

Biosynthetic studies of oosponol using ^{14}C–labeled tracer were also carried out and the participation of the "acetate–malonate pathway" in the formation of the skeleton and of a C_1-unit in the formation of the alcohol group was reported.[7]

Studies of the constrictive effect of this compound on tracheal muscle have been made.[8]

1) I. Yamamoto, K. Nitta, Y. Yamamoto, *Agr. Biol. Chem.*, **26**, 486 (1962).
2) K. Nitta, J. Imai, I. Yamamoto, Y. Yamatomo, *Agr. Biol. Chem.*, **27**, 817 (1963).
3) I. Yamamoto, *Agr. Biol. Chem.*, **25**, 400 (1961).
4) H. Umezawa, H. Iinuma, M. Ito, M. Matsuzaki, T. Takeuchi, *J. Antibiotics*, **25**, 239 (1972).
5) M. Shiozaki, K. Mori, M. Matsui, *Agr. Biol. Chem.*, **32**, 42 (1968).
6) M. Uemura, T. Sakan, *Chem. Commun.*, 921 (1971).
7) K. Nitta, Y. Yamamoto, T. Inoue, T. Hyodo, *Chem. Pharm. Bull.*, **14**, 363 (1966).
8) Y. Kobayashi, S. Ohashi, M. Yamaguchi, I. Yamamoto, *Nisshin Igaku*, **49**, 660 (1962).

9.36 SYNTHESIS OF OCHRATOXIN A

Route I[1]

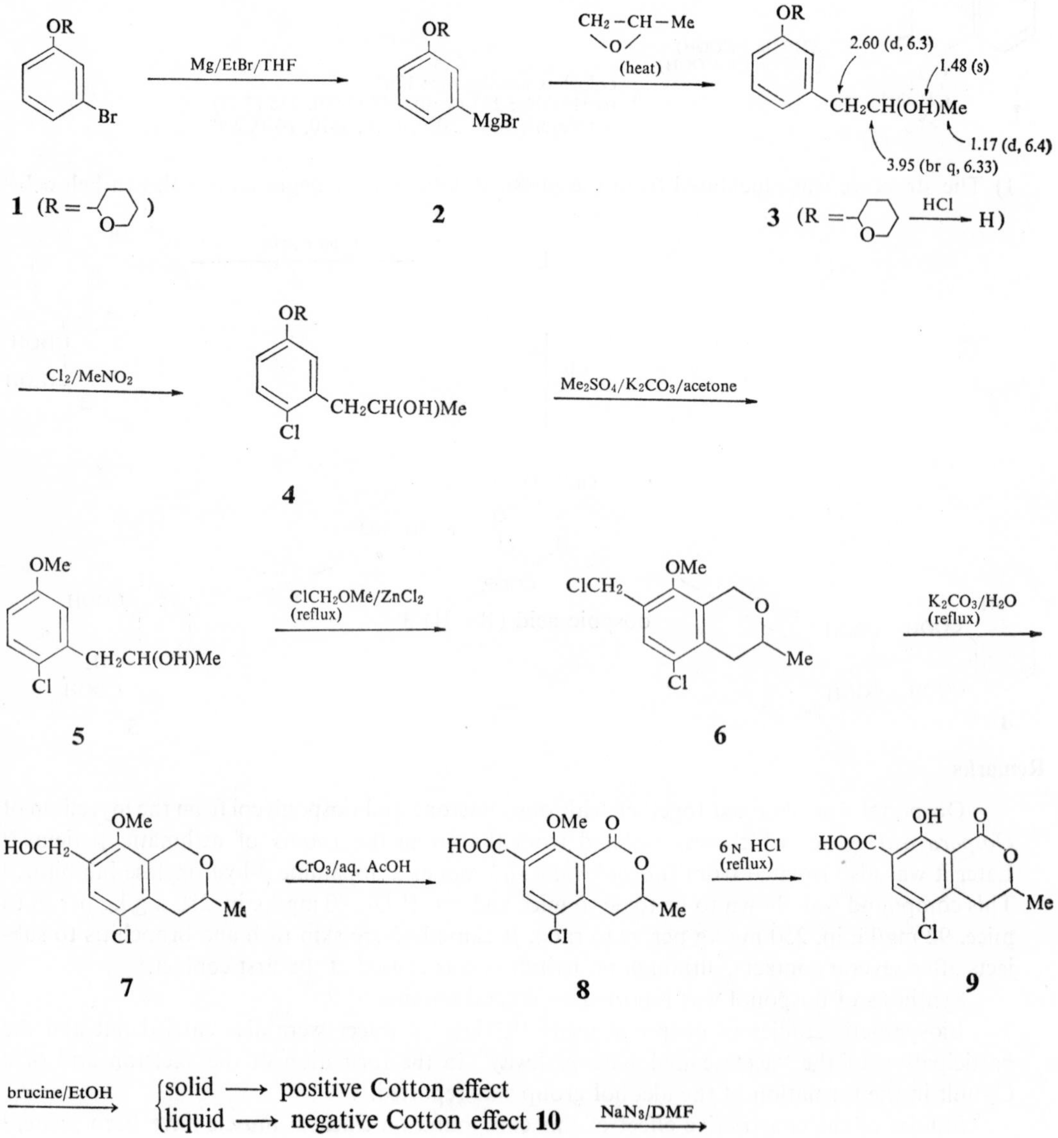
OR
Br
1 (R = THP)
Mg/EtBr/THF
MgBr
2
CH2-CH-Me
O
(heat)
OR
2.60 (d, 6.3)
1.48 (s)
CH2CH(OH)Me
1.17 (d, 6.4)
3.95 (br q, 6.33)
3 (R = THP HCl → H)
Cl2/MeNO2
CH2CH(OH)Me
Cl
4
Me2SO4/K2CO3/acetone
OMe
5
ClCH2OMe/ZnCl2
(reflux)
ClCH2
Me
6
K2CO3/H2O
(reflux)
HOCH2
7
CrO3/aq. AcOH
HOOC
8
6N HCl
(reflux)
OH
9
brucine/EtOH
solid → positive Cotton effect
liquid → negative Cotton effect 10
NaN3/DMF

1) P. S. Steyn, C. W. Holzapfel, *Tetr.*, **23**, 4449 (1967).

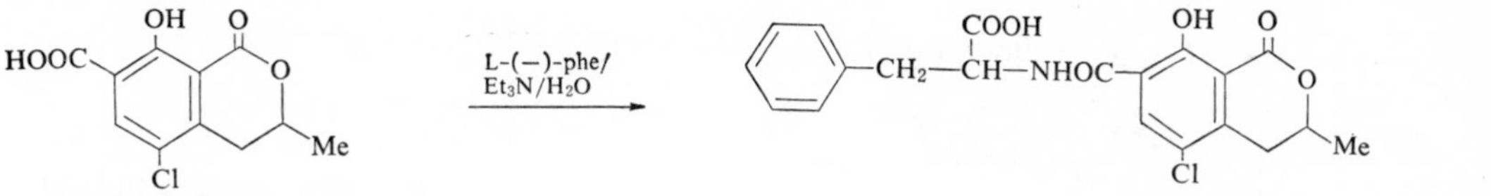

11

(−)-ochratoxin A **12**
mp: 90–93° (from benzene)
uv: 213 (ε 37,650), 332 (ε 6,500)

3 → 4: Chlorination of **3** gave a mixture of chloroisomers and **4** was separated on formamide-impregnated cellulose powder and silica chromato-plates.

Route II[2]

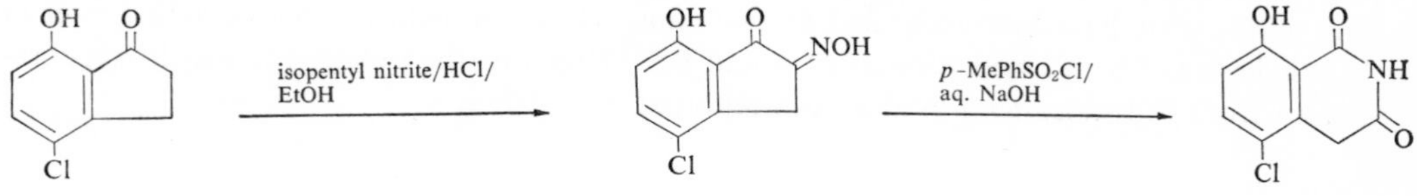

13   **14**   **15**

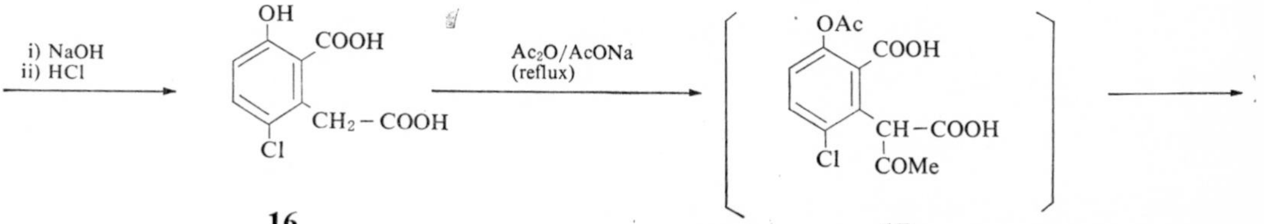

16   **17**

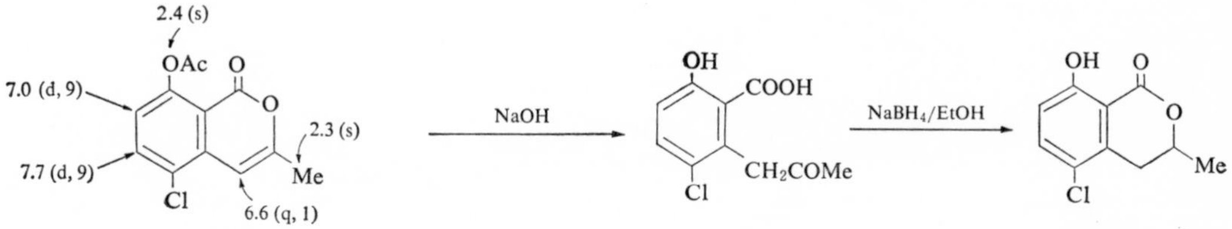

18   **19**   **20**

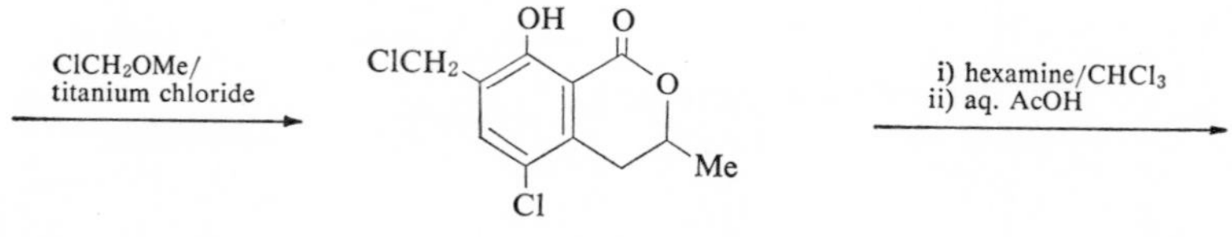

21

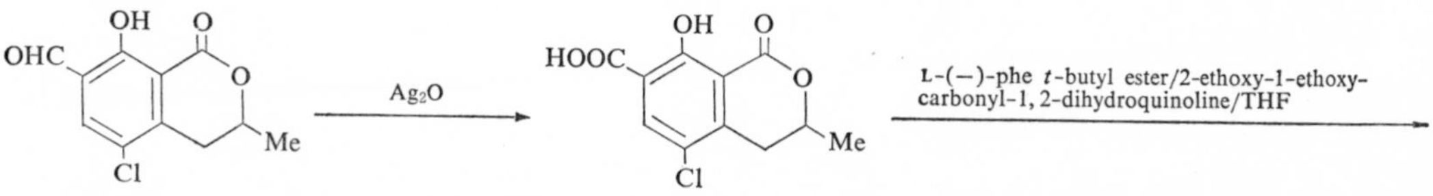

22   **23**

2) J. C. Roberts, P. Woollven, *J. Chem. Soc.* C, 278 (1970).

$\xrightarrow{\text{HBr/AcOH}}$ $(-, \pm)$-ochratoxin A $\xrightarrow{\text{prep. TLC}}$ $(-, -)$-ochratoxin A

24 mp: 71–75°
α_D^{22}: +38°

25 mp: 82–89°
α_D^{22}: −66°

(ca. 90% isomeric purity)

1 ⟶ 23: Overall yield was about 0.62%.

Remarks

Ochratoxin A, a highly toxic metabolite from some strains of *Aspergillus ochraceus* has been shown to be a 7-carboxy-5-chloro-8-hydroxy-3,4-dihydroamide.[3] *Penicillium viridicatum* has also been reported as a new source of ochratoxin A.[4] Besides ochratoxin A, some derivatives of this toxin have been identified as co-toxins. These are ochratoxin B, C (ethyl ester of A), ethyl ester of B and methyl esters of A and B.[3,5] Recently, the isolation of 4-hydroxyochratoxin A from *P. viridicatum* has also been reported.[6]

3) K. J. van der Merwe, P. S. Steyn, L. Fourie, *J. Chem. Soc.* C, 7083 (1965).
4) W. van Walbeek, P. M. Scott, J. Harwig, J. W. Lawrence, *Can. J. Microbiol.*, **15**, 1281 (1965).
5) P. S. Steyn, C. W. Holzapfel, *J. S. Afr. Chem. Inst.*, **20**, 186 (1967).
6) R. D. Hutchison, P. S. Steyn, *Tetr. Lett.*, 4033 (1971).

9.37 BIOSYNTHESIS OF SCLERIN

Possible biogenetic route

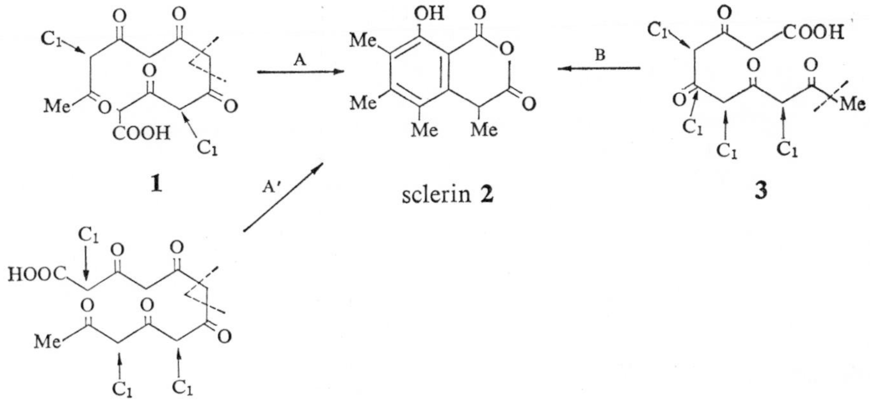

1   sclerin 2   3

4

Degradation of radioactive sclerin[1)]

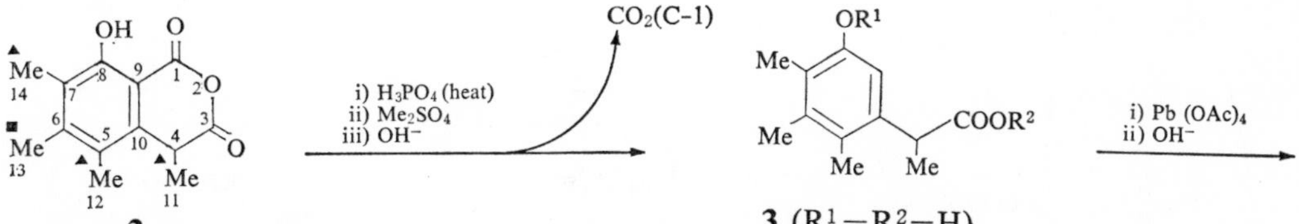

2

i) H₃PO₄ (heat)
ii) Me₂SO₄
iii) OH⁻

CO₂(C-1)

3 (R¹=R²=H)
4 (R¹=R²=Me)
5 (R¹=Me, R²=H)

i) Pb (OAc)₄
ii) OH⁻

▲ ■ ● = labeled with ¹⁴C

from H COONa
     Me COONa
     Me COONa

6 (R=Ac)
7 (R=H)

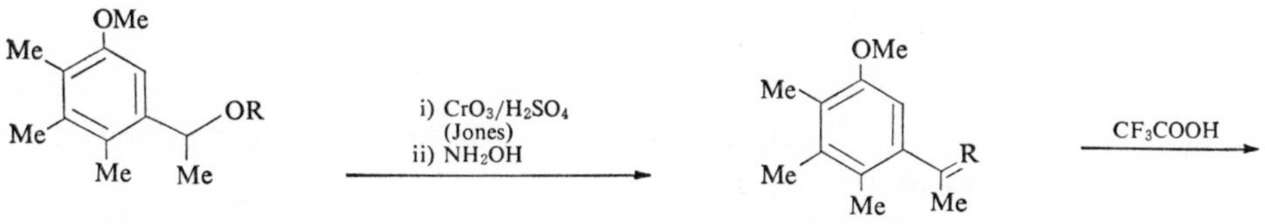

i) CrO₃/H₂SO₄
   (Jones)
ii) NH₂OH

8 (R=O)
9 (R=NOH)

CF₃COOH

1) T. Kubota, T. Tokoroyama, S. Oi, Y. Satomura, *Tetr. Lett.*, 631 (1969); T. Tokoroyama, T. Kubota, *J. Chem. Soc.* C, 2703 (1971).

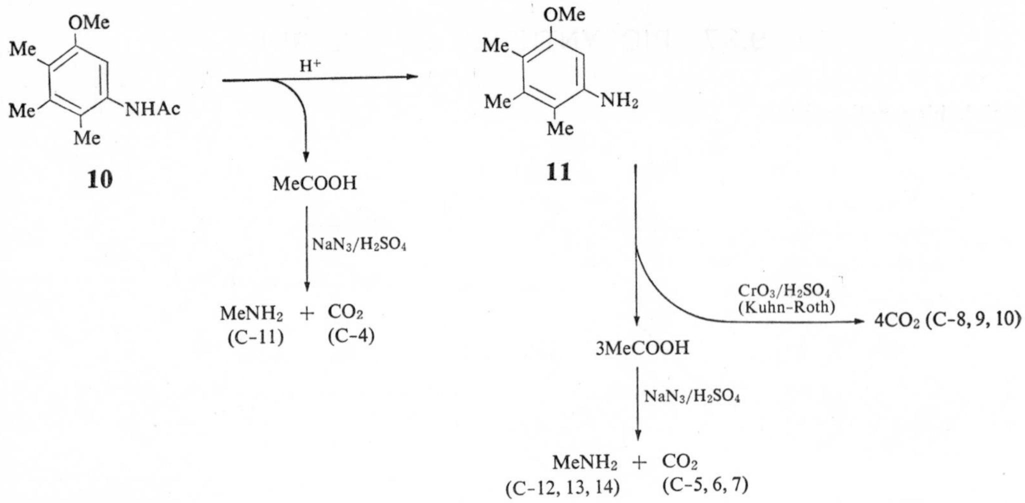

Biosynthetic pathway of *Sclerotinia* metabolite[1]

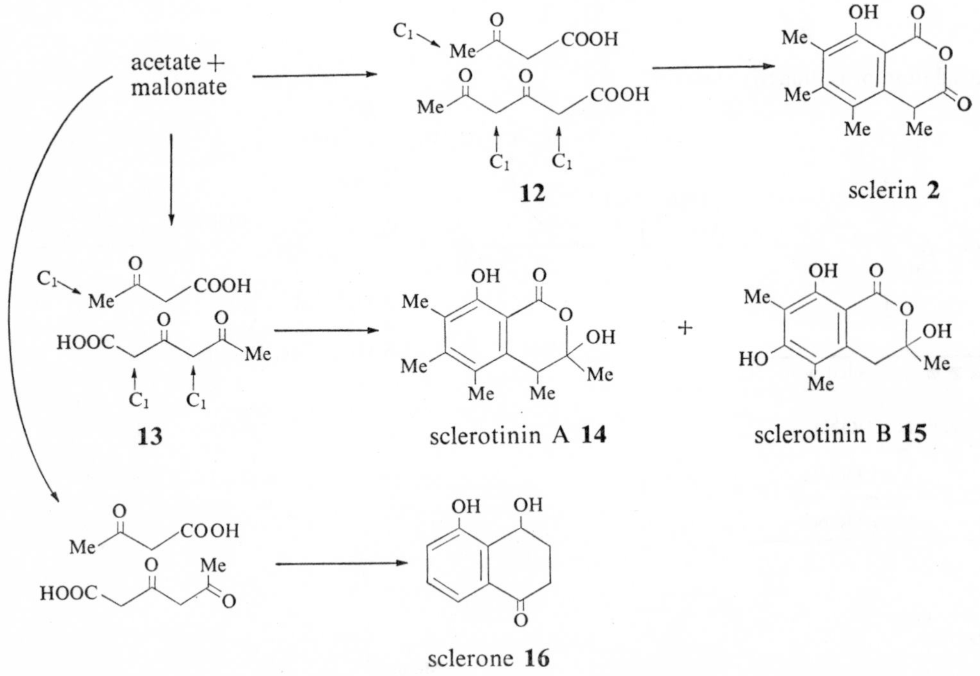

Sclerin has been isolated as a metabolite of a *Sclerotinia* fungus[2], which exhibits a hormone-like activity in some plants.[3]

Very recently an experiment using ^{13}C-nmr has been carried out and results supporting the route A' shown above has been obtained.[4]

2) T. Tokoroyama, T. Kamikawa, T. Kubota, *Tetr.*, **24**, 2345 (1968).
3) Y. Satomura, A. Sato, *Agr. Biol. Chem.*, **29**, 337 (1965).
4) M. Yamazaki, Y. Maebayashi, F. Katoh, Y. Koyama, T. Tokoroyama, 18*th Symposium on the Chemistry of Natural Products, Symposium Papers*, p. 218, 1974.

9.38 COUMARINS

Naturally occurring coumarins can be classified as follows:
a) Coumarins substituted with one or more hydroxyl and/or methoxyl groups in the benzene ring (e.g. umbelliferone **1**).
b) Coumarins substituted with isoprenoid residues (e.g. aurapten **2**, suberosin **3**, xanthyletin **4**, samidin **5**).
c) Furocoumarins (e.g. angelicin **6**, see 9.39 Furocoumarins).
d) 3-Phenylcoumarins (e.g. pachyrrhizin **7**).
e) 4-Substituted coumarins:
    4-Alkylcoumarins (e.g. mammein **8**).
    4-Hydroxycoumarins (e.g. dicoumarol **9**).
    4-Phenylcoumarins (e.g. dalbergin **10**; for more details, see p. 252).
f) 3-Phenyl–4-hydroxycoumarins (e.g. scandenin **11**).
g) 3,4-Benzocoumarins (e.g. ellagic acid **12**).

Spectral Properties

coumurrayin **13**[1]

uv (EtOH): 221 (4.19), 239 (3.77), 263 (3.99),
            329 (4.09)
ir: 1710
nmr in CDCl$_3$

auraptenol **14**[2]

uv (EtOH): 322 (4.17)
ir (CHCl$_3$): 1725

Remarks

Coumarins are a widely distributed and important class of natural compounds. Most of them are isolated from plants, especially Umbelliferae, Rutaceae and Leguminosae, and a few from animals or microorganisms. They absorb a wide range of ultraviolet light and generate intense fluorescence (usually blue). The occurrence of coumarin types a, b and c is very frequent while that of the others is rather rare.[3]

Biogenetically coumarins in general are derived from skikimic acid, but it should be noted that 3-phenylcoumarin derivatives (i.e. types d and f) belong to isoflavonoids, and 4-phenyl-coumarins to neoflavonoids (see p. 252). One may classify coumestans (see p. 248) as 3-phe-nylcoumarins at the same level.

Synthesis

Route I   (Pechmann reaction)

15   **16**   **17**

Route II   (Perkin reaction)

18   **19**

1) E. Ramstad, W. C. Lin, T. Lin, W. Koo, *Tetr. Lett.*, 811 (1968).
2) W. L. Stanley, A. C. Waiss, R. E. Lundin, S. H. Vannier, *Tetr.*, **21**, 89 (1965).
3) F. M. Dean, *Naturally Occurring Oxygen Ring Compounds*, p. 176, Butterworths, 1963.

9.39 FUROCOUMARINS

psoralene **1**   angelicin **2**   peucedanin **3**

1) The structures of furocoumarins were determined by classical degradative reactions such as oxidation, alkali fusion, and were confirmed by synthesis.[1]

2) The formation of the degradation product **5** shows that the rings in **1** and **3** are fused in a linear fashion, whereas the formation of **6** shows that ring fusion in **2** is angular.

oreoselone **7**

1) F. M. Dean, *Naturally Occurring Oxygen Ring Compounds*, p. 198, Butterworths, 1963; and references therein.

Synthesis

Route I

Route II

Route III

Route IV

Remarks

In general, nmr spectroscopy is a useful tool for the determination of furan ring structures. In the nmr spectrum of **25**, for example, long-range couplings have been observed as illustrated (i.e. $J_{3',8} = 1.0$, $J_{4,8} = 0.6$ Hz).[2,3] Benzene-induced shifts in the nmr spectra of coumarins have also been examined.[4] The extent of the shifts ($\Delta\delta(\text{ppm}) = \delta_{\text{CDCl}_3} - \delta_{\text{benzene}}$) depends on the position of the protons as shown in **26**. Results such as those mentioned above are useful in making comparisons with analogous compounds.

25

26

Experiments on the biosynthesis of a furocoumarin (sphondin **28**) showed that the furan ring is derived from mevalonic acid **27**.[5] Proposals for the biogenesis of benzofuran derivatives have been made (**30** $\longrightarrow$ **36**).[6,7]

27   28   29

30   31   32   33

34   35   36

2) E. V. Lassak, J. T. Pinhey, *J. Chem. Soc.* C, 2000 (1967).
3) J. A. Elvidge, R. G. Foster, *J. Chem. Soc.*, 590 (1963); *ibid.*, 981 (1964).
4) R. Grigg, J. A. Knight, P. Roffey, *Tetr.*, **22**, 3301 (1966).
5) H. G. Floss, U. Mothes, *Phytochemistry*, **5**, 161 (1966).
6) R. Aneju, S. K. Mukerjee, T. R. Seshadri, *Tetr.*, **4**, 256 (1958).
7) *Recent Developments in the Chemistry of Natural Phenolic Compounds* (ed. W. D. Ollis), p. 74, Pergamon Press, 1961.

9.40 STRUCTURE OF AFLATOXIN B₁

4.81 (dt, 2.5, 7)
5.53 (t, 2.5)
3.42 (t, 5.0)
2.61 (t, 5.0)
5.62 (t, 2.5)
6.89 (d, 7)
6.51 (s)
4.02 (s)
OCH₃

aflatoxin B₁ **1**

mp: 268–269°
α_D (CHCl₃): −558°
uv (EtOH): 220 (ε 25,600), 265 (ε 13,400),
              362 (ε 21,800)
ir (CHCl₃): 1760, 1684, 1632, 1598, 1562

1) The basal structure of the coumarin part of **1** was first suggested by uv and ir data and the bisfuran part was elucidated by comparison of the nmr spectra of **1** with that of sterigmatocystin.[1]

2) The position of the carbonyl in the five-membered ring was determined by comparison with synthetic model compounds, mainly by uv spectra.

3) The oxidation of **1** with 30% H₂O₂ in 1 N NaOH yielded succinic acid, whereas the oxidation of tetrahydrodesoxoaflatoxin B₁ **3** with KMnO₄ afforded glutaric acid together with succinic and malonic acids.[2]

1 $\xrightarrow{\text{H}_2/\text{Pd-C/}\ \text{EtOH or AcOH}}$

6.42 (d, 5.5)

$\xrightarrow{\text{H}_2/\text{Pd-C}}$

dihydroaflatoxin B₁ **2**[3]
(=aflatoxin B₂)
mp: 305° α_D(CHCl₃): −490°

OMe

6.30 (s)

tetrahydrodesoxoaflatoxin B₁ **3**

1) T. Asao, G. Büchi, M. M. Abdel Kader, S. B. Chang, E. L. Wick, G. N. Wogen, *J. Am. Chem. Soc.*, **85**, 1705 (1963); *ibid.*, **87**, 882 (1965).

2) D. A. van Dorp, A. S. van der Zijden, R. K. Beerthuis, S. Sparreboom, W. O. Ord, K. de Jong, R. Keuning, *Recueil Trav. Chim. des Pays-Bas*, **82**, 587 (1963).

3) K. J. van der Merwe, L. Fourie, de B. Scott, *Chem. Ind.*, 1660 (1963).

Remarks

Aflatoxin B₁ is a highly toxic and carcinogenic metabolite first isolated from a strain of *Aspergillus flavus*.[4] This exceedingly carcinogenic toxin has subsequently been isolated from a number of other *Aspergillus* and *Penicillium* species.[5] Related metabolites of the fungi in the *A. flavus* series, aflatoxin B₂, G₁, G₂,[1] M₁, M₂,[6] B₂ₐ, G₂ₐ,[7] B₃[8] (parasiticol),[9] GM₁[8], aspertoxin[10] and *O*-methylsterigmatocystin[11] have so far been isolated. The toxicity of these metabolites is, however, much less than that of aflatoxin B₁.

aflatoxin G₁ **4**

The x-ray analysis of aflatoxin G₁[12] is in complete agreement with the structure elucidated by Asao *et al.*,[1] and revealed a *cis* fusion of the two dihydrofuran rings.

4) R. Allcroft, R. B. A. Carnagham: *Chem. Ind.*, 50 1963).
5) For a review, see *Aflatoxin* (ed. L. A. Goldblatt), Academic Press, 1969.
6) C. W. Holzapfel, P. S. Steyn, I. F. H. Purchase, *Tetr. Lett.*, 2799 (1966).
7) M. F. Dutton, J. G. Heathcote, *Biochem. J.*, **101**, 21p (1966).
8) J. G. Heathcote, M. F. Dutton, *Tetr.*, **25**, 1497 (1969).
9) R. D. Stubblefield, O. L. Shotwell, G. M. Shannon, D. Weisleder, W. K. Rohwedder, *Agr. Food Chem.*, **18**, 391 (1970).
10) J. V. Rodricks, E. Lustig, A. D. Campbell, L. Stoloff, *Tetr. Lett.*, 2975 (1968); A. C. Waiss Jr., M. Wiley. D. R. Black, R. E. Lundin, *ibid.*, 3207 (1968).
11) H. J. Burkhardt, J. Forgacs, *Tetr.*, **24**, 717 (1968).
12) K. K. Cheung, G. A. Sim, *Nature*, **201**, 1185 (1964).

9.41 BIOSYNTHESIS OF STERIGMATOCYSTIN AND AFLATOXINS

Biogenetic hypothesis[1,2]

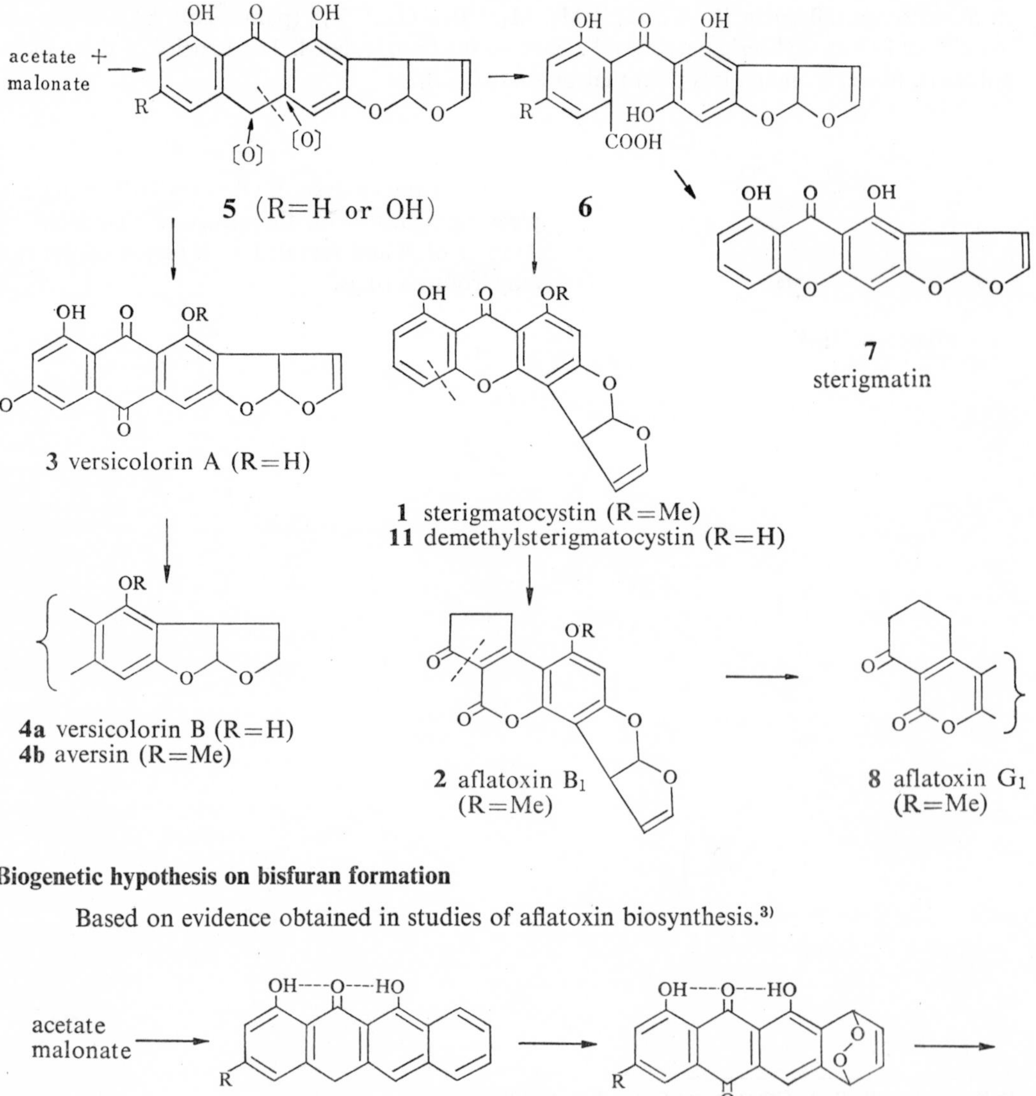

acetate +
malonate

5 (R=H or OH)

6

7
sterigmatin

3 versicolorin A (R=H)

1 sterigmatocystin (R=Me)
11 demethylsterigmatocystin (R=H)

4a versicolorin B (R=H)
4b aversin (R=Me)

2 aflatoxin B$_1$
(R=Me)

8 aflatoxin G$_1$
(R=Me)

Biogenetic hypothesis on bisfuran formation

Based on evidence obtained in studies of aflatoxin biosynthesis.[3]

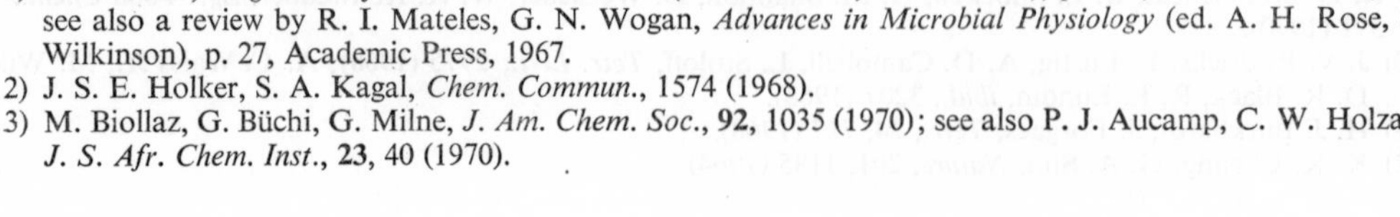

acetate
malonate

9 (R=H or OH)

10

1) R. Thomas, *Biogenesis of Antibiotic Substances* (ed. Z. Vaned, Z. Hostalek), p. 155, Academic Press, 1965; see also a review by R. I. Mateles, G. N. Wogan, *Advances in Microbial Physiology* (ed. A. H. Rose, J. F. Wilkinson), p. 27, Academic Press, 1967.
2) J. S. E. Holker, S. A. Kagal, *Chem. Commun.*, 1574 (1968).
3) M. Biollaz, G. Büchi, G. Milne, *J. Am. Chem. Soc.*, **92**, 1035 (1970); see also P. J. Aucamp, C. W. Holzapfel, *J. S. Afr. Chem. Inst.*, **23**, 40 (1970).

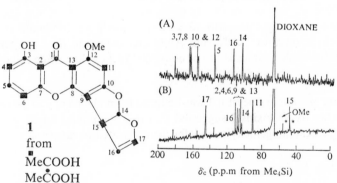

Experiments with ¹⁴C-tracer

The distribution of ^{14}C in the sterigmatocystin molecule is shown below.[4]

● have 10.9%, + have 9.9%, * have 9.3% of total activity. Unlabeled carbons each have 0.4% of total activity.

1

1 was isolated from a culture of *Aspergillus versicolor* containing Na acetate (1-¹⁴C). The distribution of radioactivity, which incorporated 0.5% from acetate, was determined by measuring the activities of the individual degradation products obtained[4]. Similar results of experiment of aflatoxin biosynthesis were obtained.[5]

Experiments with ¹³C-tracer[6]

1
from
■ MeCOOH
● MeCOOH

(A)

3,7,8 10 & 12   16 14
DIOXANE
5

(B)

2,4,6,9 & 13
17
16 14   11   15 OMe
* *

200   160   120   80   40   0
δ_c (p.p.m from Me₄Si)

The ¹³C- nmr spectra of sterigmatocystin in dioxane shown above are taken from Tanabe ref. 6. Spectrum A is from Me¹³COONa, 30 mg/ml with 63 scans of 5030 Hz at 200 sec/scan. Spectrum B is from ¹³MeCOONa, less than 27 mg/ml (saturated solution) with 1055 scans of 5030 Hz at 200 sec/scan. The labeling pattern in the structure of **1** indicates as [1-¹³C]-acetate precursor and as [2-¹³C]-acetate precursor.

* indicates C–C coupling between C-9 and C-15.

4) J. S. E. Holker, L. J. Mulheirn, *Chem. Commun.*, 1576 (1968).
5) M. Biollaz, G. Büchi, G. Milne, *J. Am. Chem. Soc.*, **90**, 5017, 5019 (1968); H. G. Raj, L. Viswamathan. H. S. R. Murthy, T. A. Venkitasubramian, *Experientia*, **25**, 1141 (1969).
6) M. Tanabe, T. Hamasaki, H. Seto, *Chem. Commun.*, 1539 (1970).

Conversion of averufin to aflatoxins

12 (0.293 Ci/mol)   **2** (0.200 Ci/mol)

Averufin **12** was isolated from a mutant of *Aspergillus parasiticus*,[7] and labeled with ^{14}C by addition of Na acetate [1-^{14}C] (0.5 mCi) to cultures of the mutant.[8]

Conversion of sterigmatocystins to aflatoxins

13 (X=OH)   * indicates position labeled with ^{14}C

aflatoxin B$_2$ **14**   aflatoxin G$_2$ **15**
(1.94% incorp.)   (0.41% incorp.)

5-Hydroxydihydrosterigmatocystin **13** [^{14}CH$_3$ labeled] was prepared from **12** by reduction, partial methylation with ^{14}CH$_3$I and Ag$_2$O and oxidation by Elb's persulfate method.[9]
1 [^{14}C labeled] was also confirmed to be converted to **2** by the resting cell of *A. parasiticus*.[10]

Remarks

The structure of sterigmatocystin **1** was deduced from chemical and x-ray studies.[11,12] The structural similarity to aflatoxins **2** suggested that this compound might be carcinogenic, and sarcoma and hepatocarcinogenicity were later demonstrated in rats.[13]

7) J. A. Donkersloot, R. I. Mateles, S. S. Yang, *Biochem. Biophys. Res. Commun.*, **47**, 1051 (1972).
8) M. T. Lin, D. P. H. Hsieh, R. C. Yao, J. M. McKeown, J. B. Robinson, L. J. Mulheirn, *Biochemistry*, **12.** 5167 (1963).
9) G. C. Elsworthy, J. S. E. Holker, *Chem. Commun.*, 1069 (1970).
10) D. P. H. Hsieh, M. T. Lin, R. C. Yao, *Biochem. Biophys. Res. Commun.*, **52**, 992 (1973).
11) Y. Hatsuda, S. Kuyama, *Nippon Nogei Kagaku Zasshi*, **28**, 989 (1954); E. Bullock, J. C. Roberts, J. G. Underwood, *J. Chem. Soc. C*, 4179 (1962).
12) N. Tanaka, Y. Katsube, Y. Hatsuda, T. Hamasaki, M. Ishida, *Bull. Chem. Soc. Japan*, 3635 (1970).
13) I. F. H. Purchase, *S. Afr. Med. J.*, **41**, 406 (1967); K. Terao, M. Yamazaki, K. Miyaki, *Shokuhin Eiseigaku Zasshi*, **14**, 272 (1973).

9.42 STRUCTURE OF TOVOXANTHONE

1.49 (s)
5.78 (d, 10.0)
7.95 (d, 10.0)
13.1
6.67 (dd, 1.0, 8.0)
7.45 (t, 8.0)
6.80 (s)
6.77 (dd, 1.0, 8.0)

tovoxanthone **1**

thick yellow needles (acetone) mp: 219°
uv (EtOH): 242 (ε 45,500), 265 (ε 36,750), 319 (ε 28,500)
uv (EtOH/NaOH): 248 (ε 45,250), 330 (ε 17,050)
uv (EtOH/AlCl$_3$): 243 (ε 46,500), 267 (ε 31,300), 334 (ε 27,150)
ir (KBr): 3330, 1640, 1605, 1572, 1453, 1298, 1235, 1154
ms: 310 (M 39%), 295 (100%), 282 (17%)
nmr in CDCl$_3$

1) The formula $C_{18}H_{12}O_3(OH)_2$ was elucidated by mass spectral and elemental analyses.[1]
2) The uv spectrum was compatible with a xanthone skeleton, also indicating the ortho (AlCl$_3$ shift) and para (NaOH shift) relationship of the OH groups and the carbonyl. This view was confirmed by ir: 1640 cm^{-1} ($>$C$=$O) and nmr: 13.1 ppm (OH).
3) The strong deshielding of the benzylic protons of dihydrotovoxanthone was also noteworthy.

1.86 (t)
3.48 (t)

Remarks

1 was isolated from the wood of *Tovomita choisyana* (Guttiferae) together with sitosterol, stigmasterol and betulinic acid by extraction with benzene followed by silica gel column chromatography.

The structure **1** was confirmed by synthesis of the 6-methoxyl derivative of **1** by direct reaction of 1,7-dihydroxy-6-methoxyxanthone with 2-methylbut-3-yn-2-ol and ZnCl$_2$.

1) S. J. Gabriel, O. R. Gottlieb, *Phytochemistry*, **11**, 3035 (1972).

9.43   BIOSYNTHESIS OF ERGOT PIGMENTS

A number of radioactively labeled acetate substrates, i.e. [1-^{14}C]-acetate, [2-^{14}C]-acetate and [2-^{3}H]-acetate, have been fed to *Claviceps* and shown to be well incorporated into the ergot pigments.

$$^{14}\text{MeCOONa} \xrightarrow{\textit{Claviceps purpurea}} \text{ergochrome AB} \quad \textbf{1}$$
$$\begin{pmatrix} 26\text{mg} \\ 1.41\ \text{mCi/mmole} \end{pmatrix} \qquad \begin{pmatrix} 200\text{mg} \\ 2495 \times 10^{-3}\ \text{mCi/mmol} \end{pmatrix}$$

Chemical degradation of ergochrome AB 1

Radioactivity of degradation products (spec. act., dpm/mmole)

Product	^{14}MeCOOH precursor	Me^{14}COOH precursor
Kuhn-Roth oxidation		
1	4.87×10^6	3.65×10^5
7	1.61×10^5	9.12×10^3
8	2.93×10^5	2.20×10^4
9	2.93×10^5	
10	$3.4\ \times 10^2$	2.20×10^4

Product	^{14}MeCOOH precursor	Me^{14}COOH precursor
HBr degradation		
1	9.6×10^5	3.65×10^5
4	4.9×10^3	1.28×10^4
5	2.62×10^5	4.99×10^4
6	3.68×10^4	1.5×10^2

[U-^{3}H]-Emodin prepared by Wilzbach's method and [U-^{14}C]-emodin produced biosynthetically by *Penicillium islandicum* were also administered to *Claviceps purpurea* and the incorporation of radioactivity into ergochromes was confirmed. The feeding of [U-^{14}C]-shikimic acid to the fungus, however, did not result in the incorporation of any radioactivity into ergochromes.[1,2]

Biogenetic scheme for ergochromes as "seco-anthraquinones"

acetate +
malonate $\longrightarrow$

11   **12**

endocrocin **13**   emodin **14**   **15**

16

The conditions required for Baeyer-Villiger oxidation for cleavage of the quinone ring were systematically studied, but all the anthraquinones studied were found to be inert. Therefore, reactions of this type seemed unlikely for the conversion of anthraquinone to xanthone *in vivo*.[3]

1) B. Franck, F. Hüper, D. Gröger, D. Erge, *Angew. Chem.*, **78**, 752 (1966); *Chem. Ber.*, **101**, 1954 (1968).
2) D. Gröger, D. Erge, B. Franck, U. Ohnsorge, H. Flash, F. Hüper, *Chem. Ber.*, **101**, 1970 (1968).
3) B. Franck, V. Radtke, U. Zeidler, *Angew. Chem.*, **79**, 935 (1967).

Hypothetical reactions of anthrones yielding benzophenones and anthraquinones may be represented as follows:

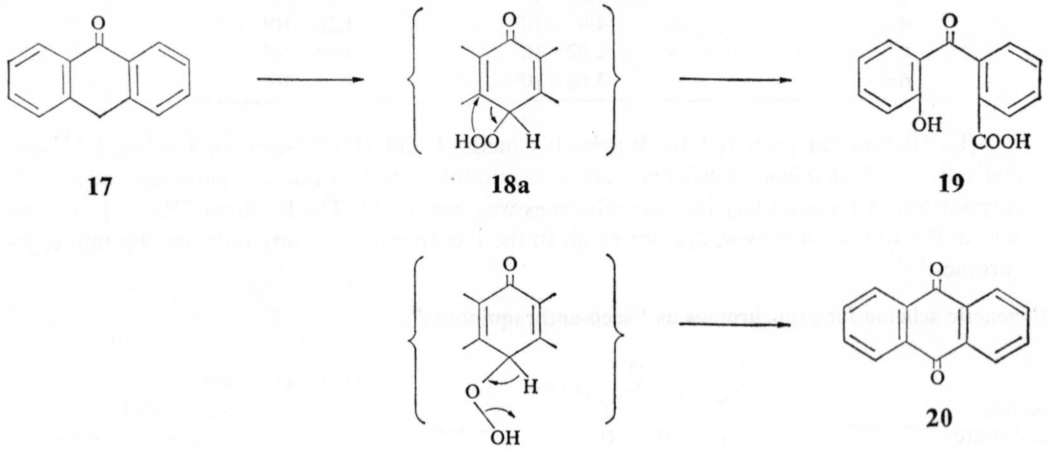

17   **18a**   **19**

18b

20

Model reactions for oxidative phenolic coupling

A model oxidative phenolic coupling reaction yielding a product which has the basic skeleton of half of an ergochrome molecule in shown below.[5]

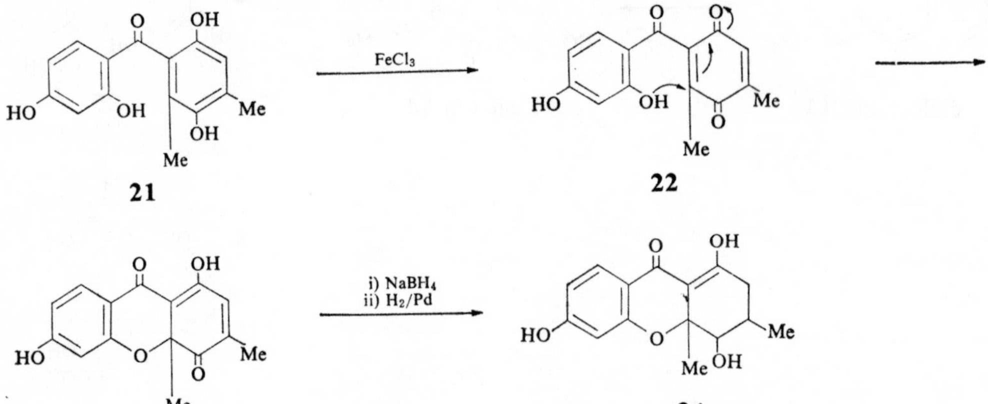

21   **22**

23   **24**

---

4) T. Money, *Nature*, **199**, 592 (1963).
5) B. Franck, *Angew. Chem. Intern. Ed.*, **8**, 251 (1969); B. Franck, H. Flasch, *Fortsh. Chem. org. Naturstoffe*, **30**, 151 (1973).

Structures of ergochromes obtained from ergot[5]

half-structures	combination (all linkages between biphenyl residues are 2,2′)[6]
A	A–A ergochrome AA (secalonic acid A)
	B–B ergochrome BB (secalonic acid B)
	C–C ergochrome CC (ergoflavin)
B	A–B ergochrome AB (secalonic acid C)
	A–C ergochrome AC (ergochrysin A)
C	B–C ergochrome BC (ergochrysin B)
	A–D ergochrome AD
	B–D ergochrome BD
	C–D ergochrome CD
D	D–D ergochrome DD

6) J. W. Hooper, W. Marlow, W. B. Whalley, A. D. Borthwick, R. Bowden, *Chem. Commun.*, 111 (1971); *J. Chem. Soc.* C, 3580 (1971).

9.44   CLASSES OF FLAVONOID COMPOUNDS

Flavonoid compounds are C_{15} compounds (exclusive of O-alkyl groups and secondary substitution) composed of two phenolic nuclei connected by a three-carbon unit. One of the largest classes of naturally occurring phenolic compounds, these materials are widely distributed in nature, mostly as their glycosides. Among the various types of flavonoids, chalcones, anthocyanidins and aurones are well-known as plant pigments.[1,2] The numberings of flavonoids are shown below, but it is to be emphasized that chalcones have unique numbering system. Natural chalcones always have a hydroxyl group at the 2′ position, except for echinatin (4,4′-dihydroxy-2-methoxychalcone),[3] licochalcones A and B,[4] isolated from the genus *Glycyrrhiza*.

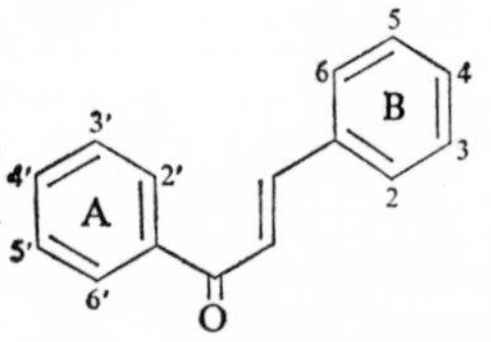

1
chalcones
(butin)

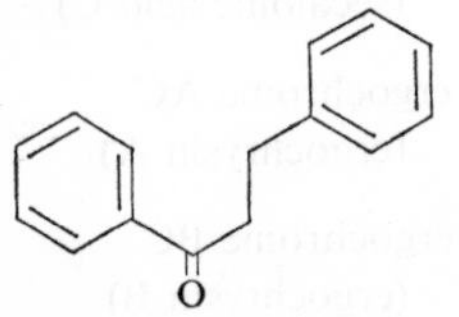

2
dihydrochalcones
(phlorizin)

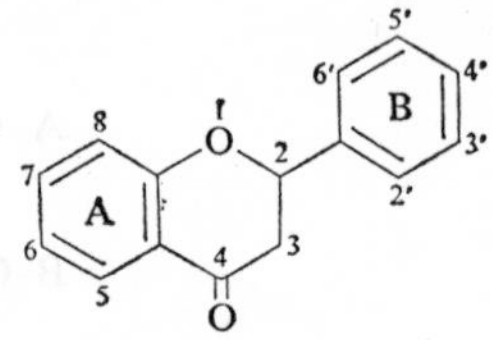

3
flavanones
(butein)

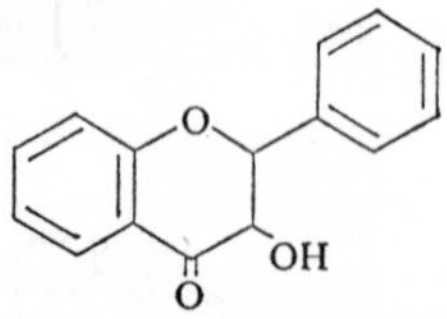

4
flavanonols
(fustin)

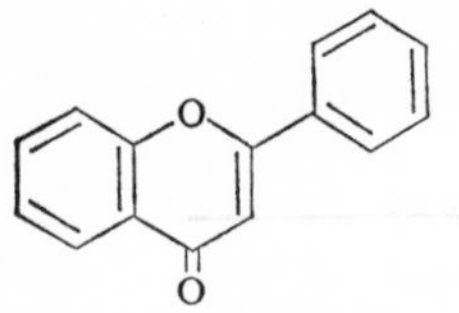

5
flavones
(chrysin)

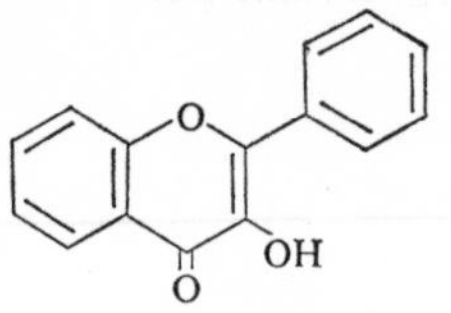

6
flavonols
(quercetin)

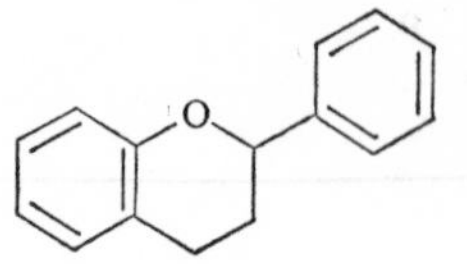

7
flavans

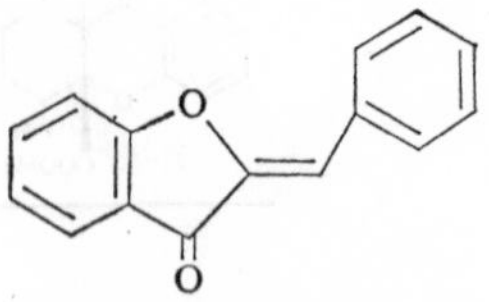

8
aurones
(sulphuretin)

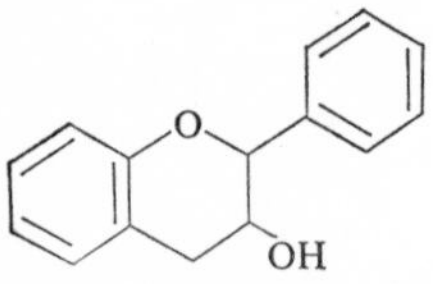

9
catechins
(catechin)

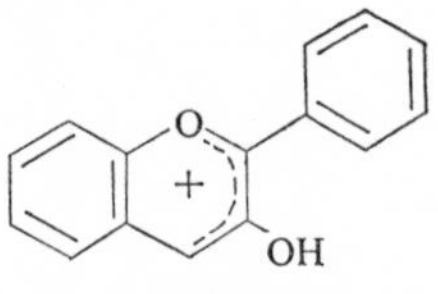

10
anthocyanidins
(cyanidin)
(glycosides: anthocyanins)

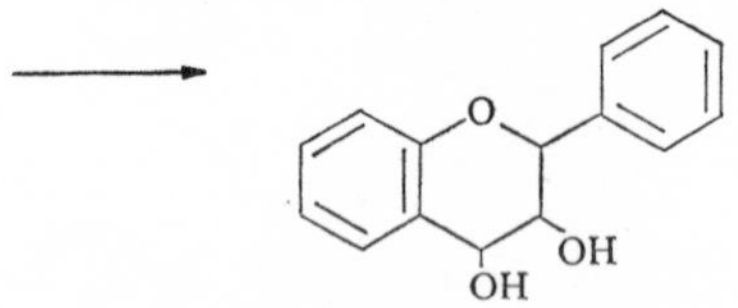

11
leucoanthocyanidins
(leucocyanidin)
(glycosides: leucoanthocyanins)

1) *The Chemistry of Flavonoid Compounds* (ed. T. A. Geissman), Pergamon Press, 1962.
2) F. M. Dean, *Naturally Occurring Oxygen Ring Compounds*, Butterworths, 1963.
3) T. Furuya, K. Matsumoto, M. Hikichi, *Tetr. Lett.*, 2567 (1971).
4) T. Saitoh, S. Shibata, *Tetr. Lett.*, 4461 (1975).

9.45   STRUCTURE OF PORIOLIDE

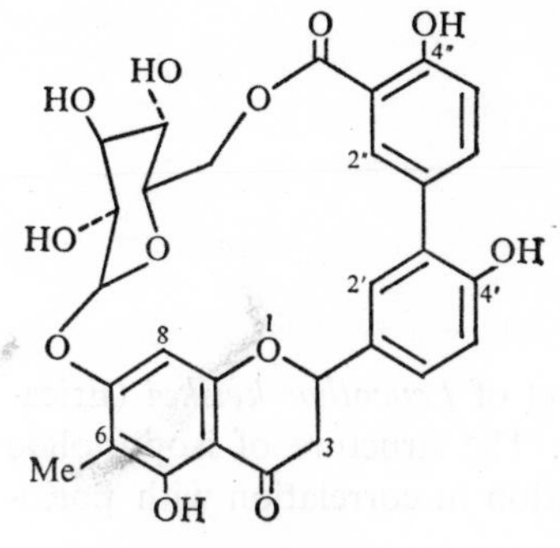

poriolide **1**

mp: 265° α_D (acetone): −334°
ms: 566 (M⁺ $C_{29}H_{26}O_{12}$)
uv (EtOH): 287 (ε 22,000)
ir (nujol): 3300, 1675, 1640, 1610, 1600, 1490

1) Treatment of **1** with NaOH and then HCl gives D-glucose.

2) The secondary methyl group of **2** at C-3 was confirmed by reduction of **2** with NaBH₄ followed by treatment with p-TsOH to give a styrene; a doublet at 0.94 ppm changed to a singlet at 2.06 ppm.

3) Degradative reaction of **2** with Ba(OH)₂ in boiling dioxane-water afforded two biphenyl-carboxylic acids **4** and **5** which were investigated after methylation.

4) Finally, the structure was established by x-ray analysis of **3**.

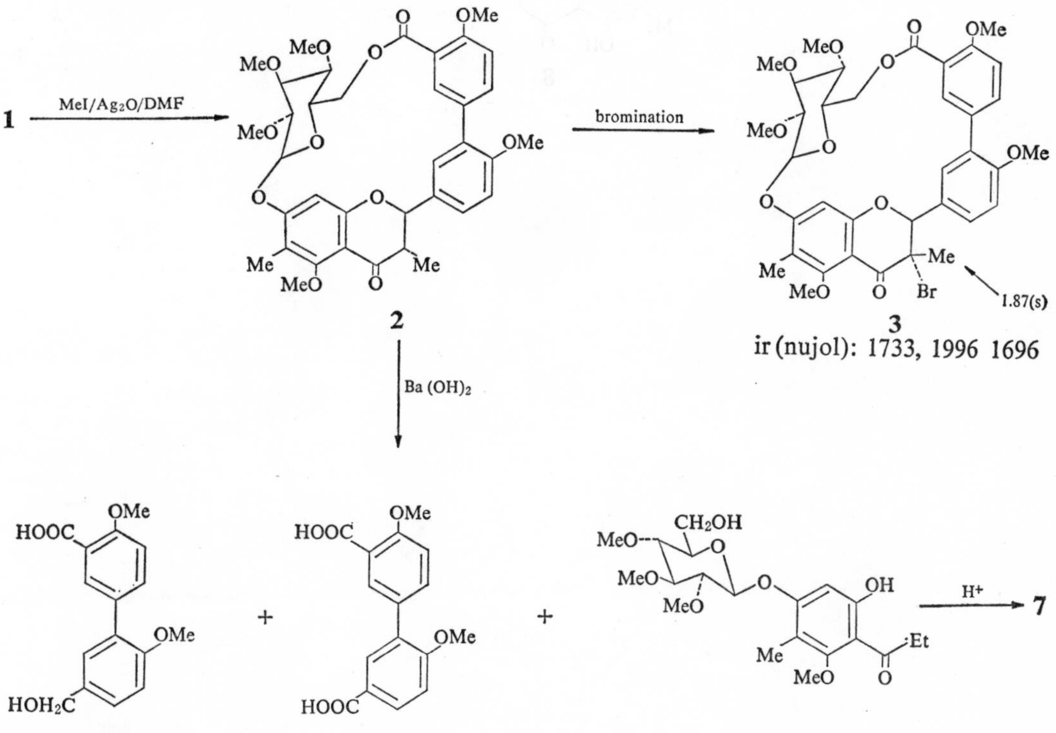

2

ir (nujol): 1733, 1996 1696

1.87(s)

3

Ba (OH)₂

4   +   **5**   +   **6** $\xrightarrow{H^+}$ **7**

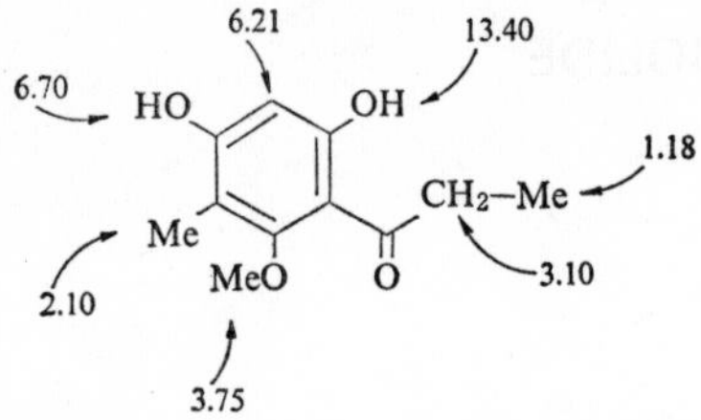

7

Remarks

Poriolide **1**[1] is a major constituent of the methanolic extract of *Leucothoe keiskei* (Erica-ceae) and shows toxicity (intravenous LD_{50} 1.0 mg/kg in mice). The structure of isoporiolide **8**, the second constituent, was determined by chemical degradation in correlation with porio-lide.

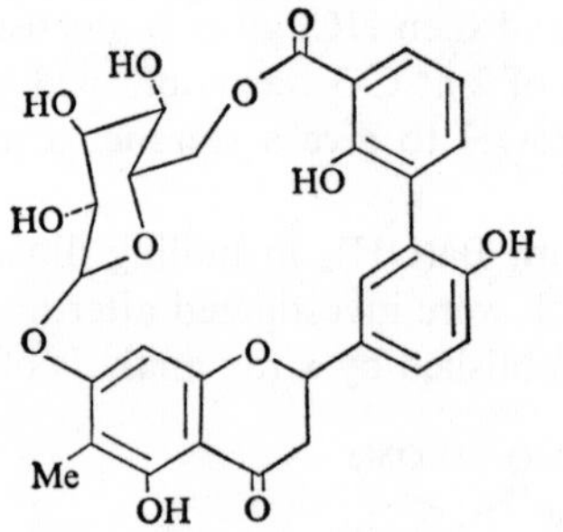

8

1) A. Ogiso, A. Sato, S. Sato, C. Tamura, *Tetr. Lett.*, 3071 (1972).

9.46 STRUCTURE OF FLAVONOLIGNANS FROM
Silybum marianum

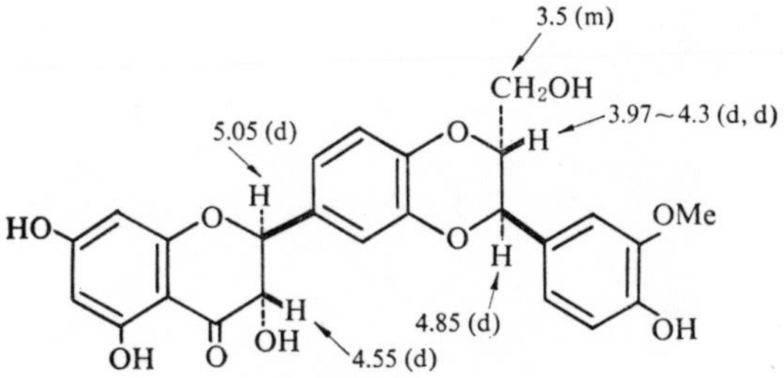

3.5 (m)

CH₂OH

5.05 (d)

3.97~4.3 (d, d)

OMe

HO

OH

H

4.85 (d)

4.55 (d)

OH O OH

silybin (silymarin) **1**

$C_{25}H_{22}O_{10}$
mp: 167° α_D^{20} (acetone): +11°
ir (KBr): 3440 (OH), 1638 (C=O), 1280 (OH),
        1160 (C–O–), 852, 830, 795 (C–H)
uv (MeOH): 288, 322

1) Dehydrogenation of **1** afforded the flavonol derivative **2**.[1,2] The degradation of **2** yielded quercetin **3**.[2]

2) The structure of **1** was finally established by the synthesis of **4**.[3,4]

3) Strong peaks were observed in the mass spectrum of **1** corresponding to retro-Diels-Alder fragmentation.

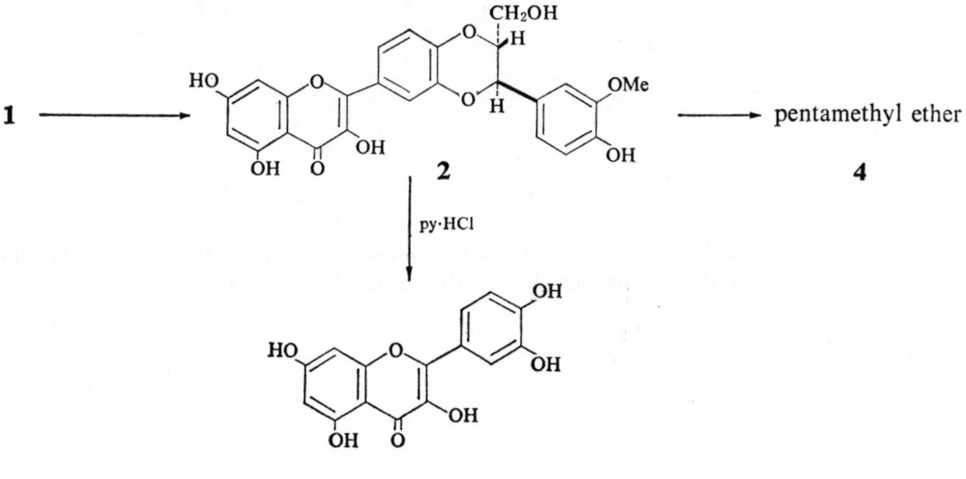

1 ⟶   (structure **2**)   CH₂OH OMe ⟶ pentamethyl ether

HO

OH O OH   **2**   OH   **4**

py·HCl

(structure **3**)

OH

HO

OH

OH O

3

---

1) B. Jniak, R. Hänsel, *Planta Medica*, **8**, 71 (1960).
2) H. Wagner, L. Hörhammer, R. Münster, *Arzneim. Forsch.*, **18**, 688 (1968).
3) A. Pelter, R. Hänsel, *Tetr. Lett.*, 2911 (1968).
4) R. Hänsel, J. Schultz, A. Pelter, *Chem. Commun.*, 195 (1972).

4) The structure of silydianin **5** was established by x-ray crystallographic analysis.[5] The mass spectrum of **5** also shows strong retro-Diels-Alder fragmentation peaks.

silydianin (compound E5) **5**[1]

$C_{25}H_{22}O_{10}$
mp: 191° α_D^{20} (EtOH): $+214°$
uv (MeOH): 205, 288, 322

Remarks

A Diels-Alder reaction betwen taxifolin orthoquinone **7** and coniferyl alcohol **8** has been suggested as a possible biosynthetic route to flavonolignans.[5] When the Diels-Alder reaction involves the dione of **7**, a silybin **1** type of structure is obtained whereas the involvement of diene affords the bridged ring system of silydianin **5**. From the viewpoint of their biogenesis, the name "flavonolignans" has been proposed for these classes of compounds.

Silybum marianum is known to have protective and curative effects against various hepatic toxa. Silybin **1** was shown to be the principle of this action.[6]

5) S. J. Abraham, S. Takagi, R. D. Rosenstein, R. Shiono, H. Wagner, L. Hörhammer, O. Telgmann, N. R. Farrsworth, *Tetr. Lett.*, 2675 (1970).
6) G. Hahn, H. D. Lehmann, M. Kürten, H. Uebel, G. Vogel, *Arzneim. Forsch.*, **18**, 698 (1968).

9.47 SYNTHESIS OF FLAVONES AND FLAVONOLS

There are several methods available for the synthesis of flavones or flavonols. The following three methods are very important and well-known.

The Baker-Venkataraman synthesis[1,2]

An *o*-benzoyloxyacetophenone **1** is isomerized to an *o*-hydroxy-β-diketone **2** by a base, by intramolecular ester condensation. By this method flavones may obtain directly, without the aid of acids, provided that weak bases are used for the isomerization step. The acetophenones are treated in benzene, toluene, or ether with sodium, sodium hydride, potassium hydroxide, sodamide or potassium carbonate. The method has been extended for the synthesis of flavonols by treatment of the intermediate β-diketone with performic acid, which hydroxylates the active methylene group, and the subsequent cyclization is spontaneous.

1

2

baicalein trimethyl
ether **3**

The Allan-Robinson (Kostanecki-Robinson) synthesis[3]

An *o*-hydroxyacetophenone **4** intended to provide ring A, is heated with a mixture of the sodium salt and the anhydride of an aromatic acid intended to provide ring B. Although acylation may occur twice, the unwanted 3-acyl substituent can be removed by very gentle alkaline hydrolysis (**9** ⟶ **6**). Better yields may be expected when using triethylamine as a catalyst instead of the salt of the acid. When the ketones have a methoxy group in the ω-position, this method gives 3-methoxyflavones. (**10** ⟶ **11**).

1) W. Baker, *J. Chem. Soc.*, 1381 (1933).
2) H. S. Mahal, K. Venkataraman, *J. Chem. Soc.*, 1767 (1934).
3) J. Allan, R. Robinson, *J. Chem. Soc.*, 2334 (1926).

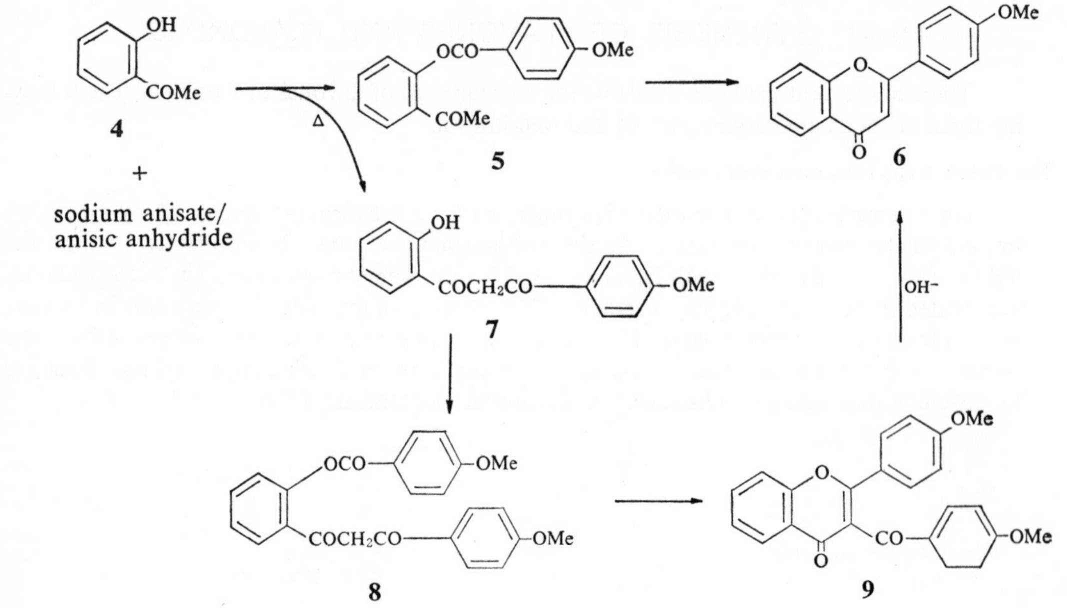

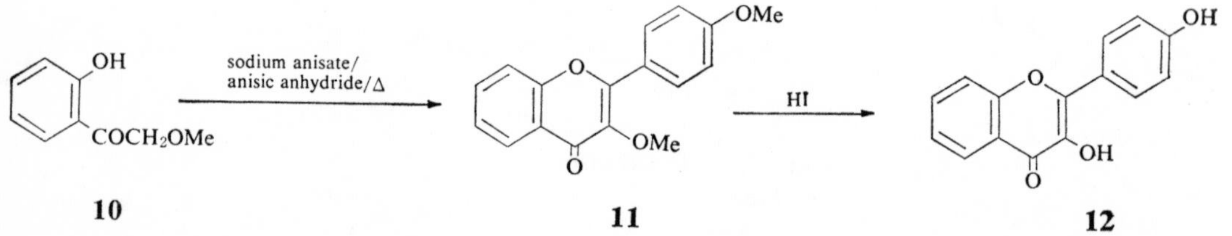

The Algar-Flynn-Oyamada synthesis[4,5]

A chalcone 13 can be converted to a flavonol 16 with hot alkaline hydrogen peroxide. The mechanism of the reaction has been explained as shown below. Dihydroflavonols (= flavanonols, e.g. 15) are intermidiates and could be isolated in some cases by carrying out the oxidation in the cold. If the chalcone contains a 6'-substituent, internal nucleophilic substitution appears to represent an alternative course, leading to an aurone (e.g. 20). The presence of a 4-hydroxy group in a 6'-methoxychalcone, however, encourages flavonol formation.

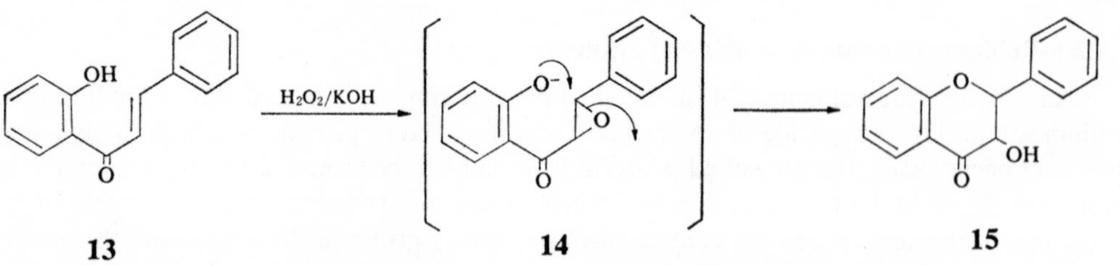

4) J. Algar, J. P. Flynn, *Proc. R. Irish Acad.*, **42B**, 1 (1934).
5) T. Oyamada, *Bull. Chem. Soc. Japan*, **10**, 182 (1935); cf. J. Gripenberg, *Flavones*, in *The Chemistry of Flavonoid Compounds* (ed. T. A. Geissman), p. 406, Pergamon Press, 1962; F. M. Dean, *Naturally Occurring Oxygen Ring Compounds*, p. 292, Butterworths, 1963.

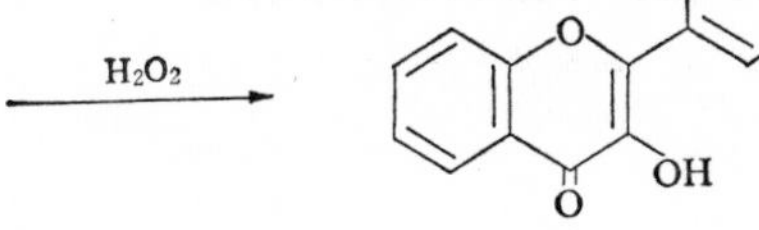

16

cf.

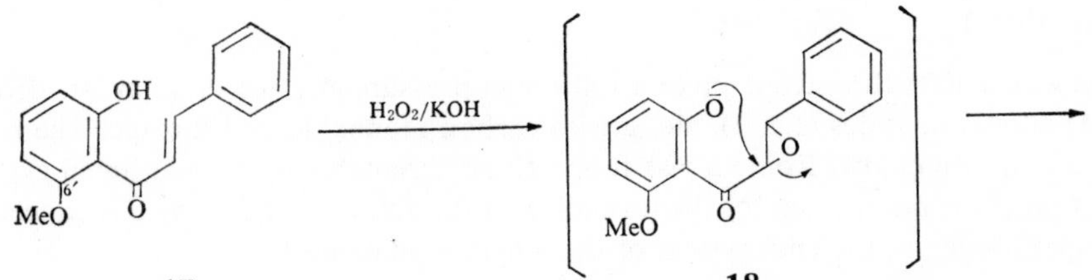

17   **18**

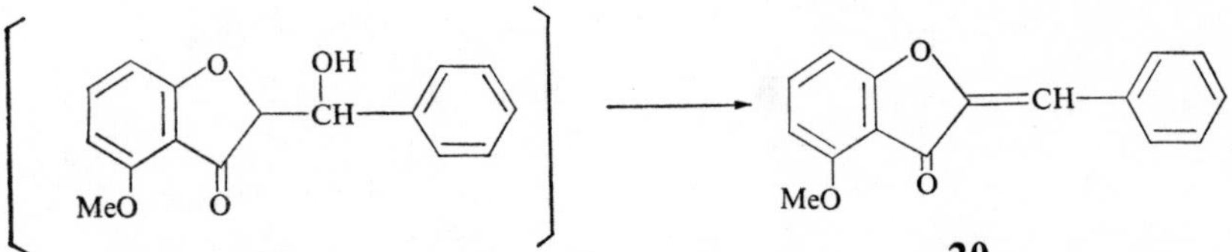

19   **20**

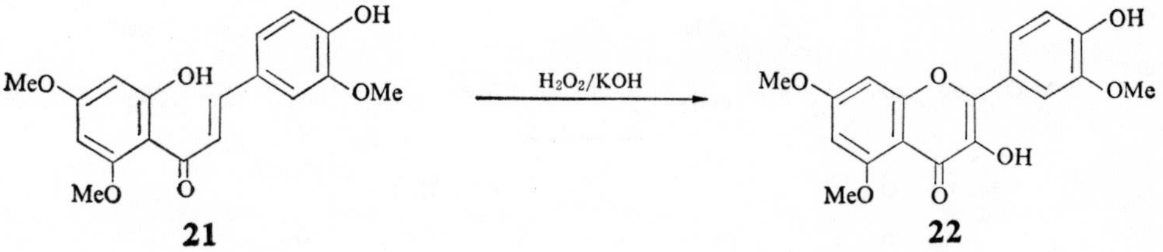

21   **22**

9.48 PHOTOCHEMICAL REACTION OF FLAVONOIDS

3-hydroxyflavones[1–3]

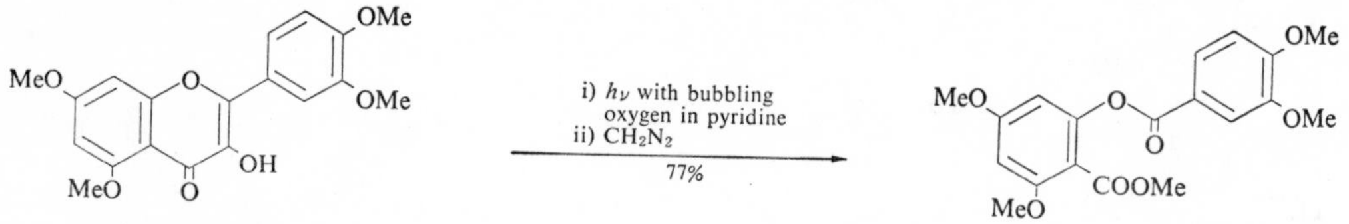

quercetin 5, 7, 3', 4'-
tetramethyl ether **1**

i) $h\nu$ with bubbling
oxygen in pyridine
ii) CH_2N_2

77%

2

Irradiation with a 300w tungsten lamp or a 100w high-pressure mercury lamp at 3-hydroxy-flavones (e.g. **1**) affords depsides (e.g. **2**), along with carbon monoxide and dioxide. The reaction mechanism is shown below. Route **a** was found to be acceptable by an additional experiment. The most probable mechanism for the formation of the ketohydroperoxide **3** is concerted addition of singlet oxygen to the enol system of the 3-hydroxyflavone **1**.

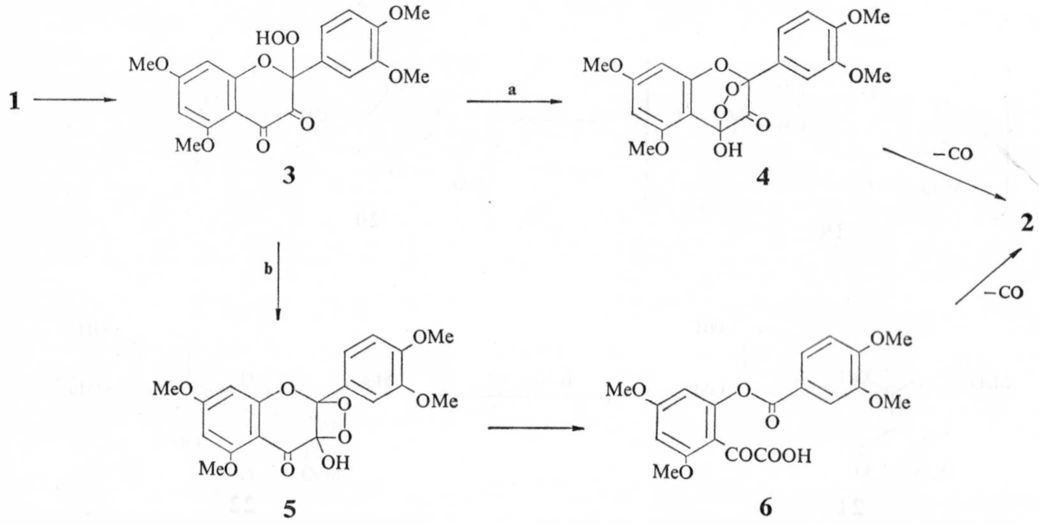

Since it has been shown that carbon monoxide is not oxidized to carbon dioxide under the conditions employed for the experiment, there are two mechanisms for generating carbon dioxide (**a** and **b**), of which mechanism **a** is thought at present to be more probable.[1,2]

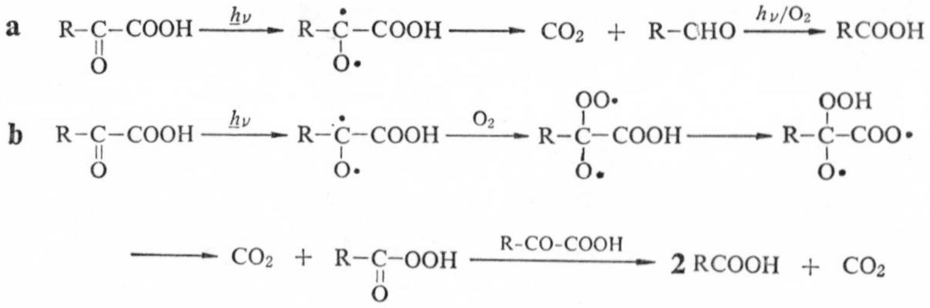

3-methoxyflavones[4-6)]

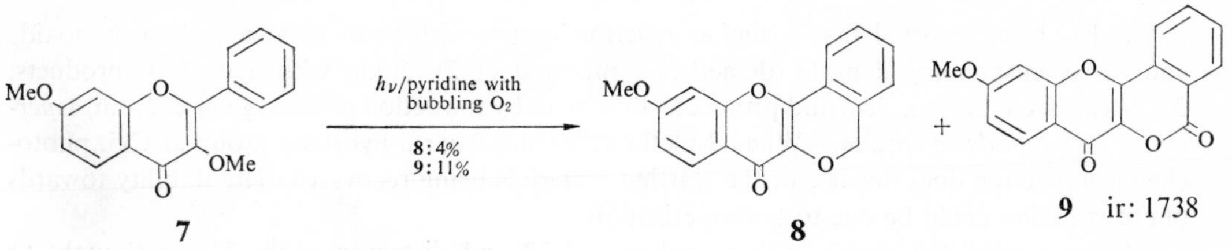

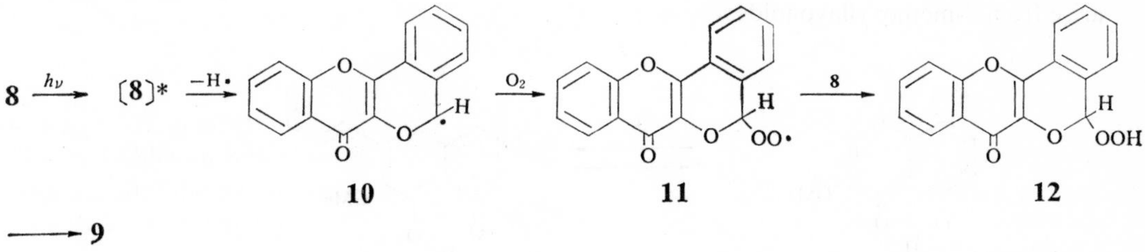

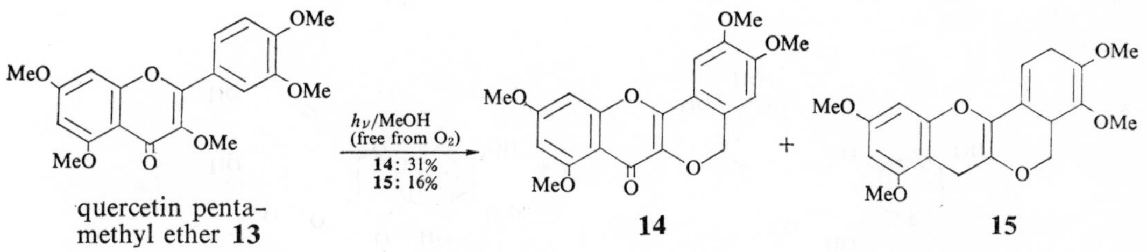

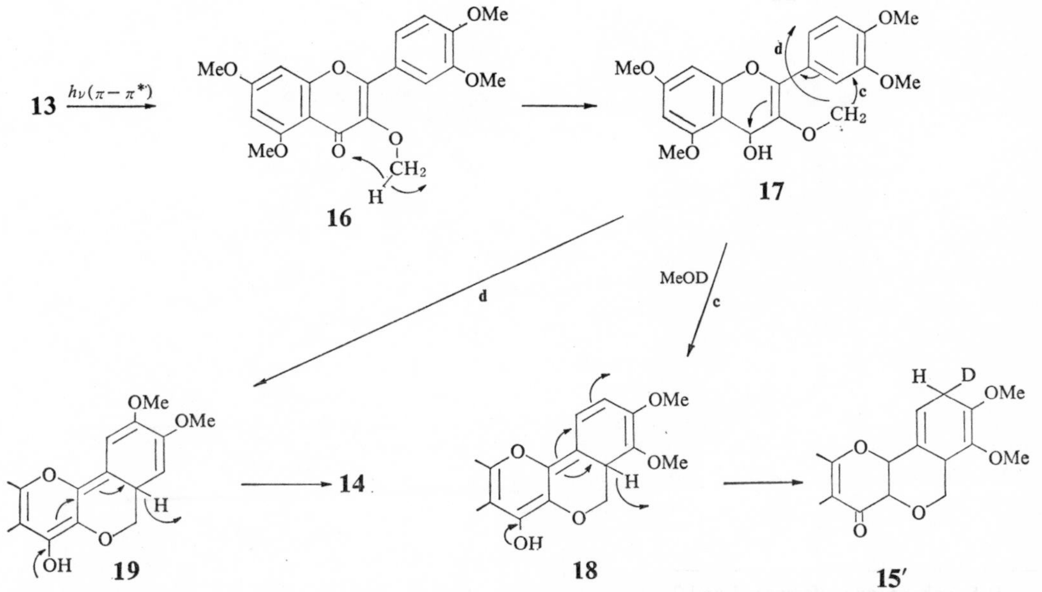

$7 \xrightarrow{h\nu/\text{pyridine with bubbling O}_2}$
8:4%
9:11%

7 **8** **9** ir: 1738

$8 \xrightarrow{h\nu} [8]^* \xrightarrow{-H\cdot} \quad \xrightarrow{O_2} \quad \xrightarrow{8}$

10 **11** **12**

$\longrightarrow$ **9**

quercetin penta-methyl ether **13** $\xrightarrow{h\nu/\text{MeOH (free from O}_2)}$ **14**: 31% **15**: 16%

14 **15**

$13 \xrightarrow{h\nu(\pi-\pi^*)}$

16 **17**

MeOD

19 $\longrightarrow$ **14** **18** $\longrightarrow$ **15'**

1) T. Matsuura, H. Matsushima, H. Sakamoto, *J. Am. Chem. Soc.*, **89**, 6370 (1967).
2) T. Matsuura, H. Matsushima, R. Nakashima, *Tetr.*, **26**, 435 (1970).

Remarks

It has been reported that a similar reaction occurs with rutin to give carbon monoxide and the corresponding depside (demethyl compound of **2**), along with hydrolysis products, 2,4,6-trihydroxybenzoic acid and protocatechuic acid, by the action of dioxygenases from *Asperigillus* and *Pullularia* species.[3] When 3-methoxyflavones bear a hydroxyl group at C-5, photochemical reaction does not occur, the starting materials being recovered. The stability towards photooxidation could be due to tautomerization.

Some natural flavonoids such as peltogynol **22** and distemonanthin **23** are thought to arise from 3-methoxyflavonoids.

3) See above, and references 1 and 2.
4) A. C. Waiss Jr., J. Corse, *J. Am. Chem. Soc.*, **87**, 2068 (1965).
5) A. C. Waiss Jr., R. E. Lundin, A. Lee, J. Corse, *J. Am. Chem. Soc.*, **89**, 6213 (1967).
6) T. Matsuura, H. Matsushima, *Tetr.*, **24**, 6615 (1968).

9.49   STRUCTURE OF ZEYHERIN

zeyherin **1**

1) The benzene-induced nmr spectrum of **4** showed very pronounced shifts ($\Delta\delta +0.51 \sim 0.61$) of all the phenolic methoxyl groups, indicating that the interflavonoid linkage is through the 8-position of the D ring.[1,2]

MeO: 3.60, 3.73 (2), 3.82 (2), 3.90
zeyherin heptamethyl ether **4**

mp: 200.5–201.5°
ms: 656 (M$^+$ C$_{37}$H$_{36}$O$_{11}$)
uv: 289, 333 (sh.)
ir: 1700

1) F. du R. Volsteedt, D. G. Roux, *Tetr. Lett.*, 1647 (1971).
2) A. Pelter, R. Warren, J. U. Usmani, M. Ilyas, M. Rahman, *Tetr. Lett.*, 4259 (1969).
3) W. Baker, W. D. Ollis, *Recent Developments in the Chemistry of Natural Phenolic Compounds* (ed. W. D. Ollis), p. 152, Pergamon Press, 1961; F. M. Dean, *Naturally Occuring Oxygen Ring Compounds*, p. 314 Butterworths, 1963; T. Kariyone, *Japan J. Pharmacog.*, **16**, 1 (1962).

Remarks

Zeyherin **1** was isolated together with **5**, **6** and **7** from the red heartwood of *Phyllogeiton zeyheri* (formerly *Rhamnus zeyheri*). There are many naturally occurring biflavonyls (bis-flavonoids) linked through various positions.[3]

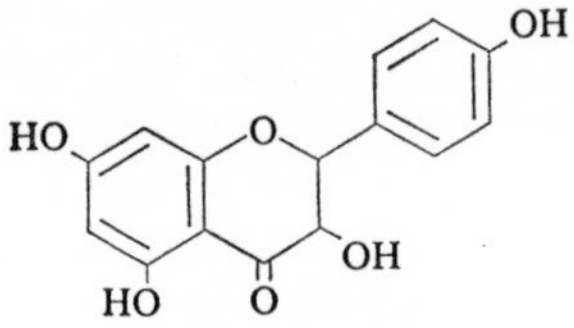

aromadendrin **5** 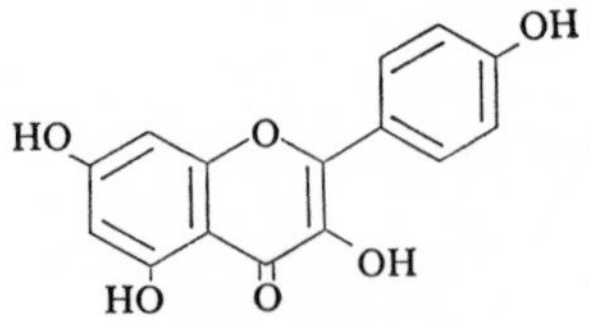   kaempferol **6**

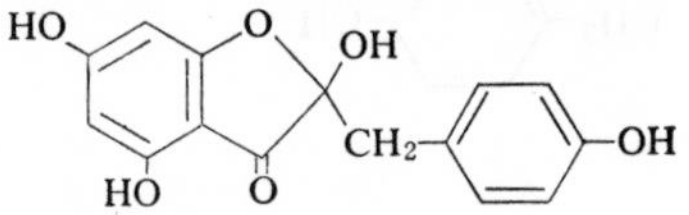

maesopsin **7**

9.50 ANTHOCYANINS AND PLANT COLOR

The anthocyanins are the principal red, violet and blue pigments of plants, especially in petals and fruits. They are 3- or 3,5-glycosides of anthocyanidins, the major types of which are shown below.

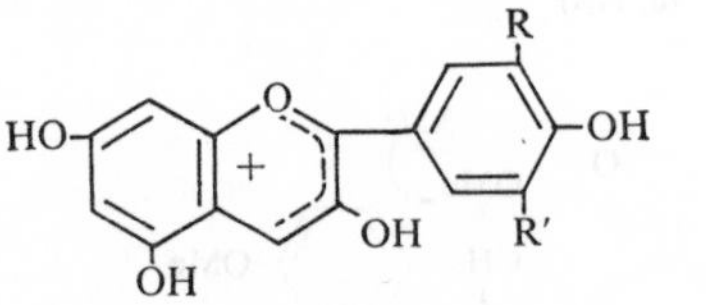

	R	R′
pelargonidin **1**	H	H
cyanidin **2**	H	OH
delphinidin **3**	OH	OH
peonidin **4**	H	OMe
malvidin **5**	OMe	OMe
petunidin **6**	OH	OMe

The strikingly brilliant colors of flowers, in the range deep red through purple to deep blue, are influenced by many factors. An increase in the number of hydroxyl groups deepens the visible color, as illustrated by the orange-red color of pelargonidin **1**, the deep red of cyanidin

2, and the bluish-red of delphinidin **3** derivatives. Methylation and glycosylation of the hydroxyl groups also influence the color. However the differences in absorption are not sufficient to account for the remarkable variation of flower color.

Absorption of anthocyanidins[1]

	uv (MeOH)	uv (HCl) of the glycoside (nm)
1	520	492
2	535	507
3	545	516
4	532	
5	542	
6	545	

Anthocyanins show amphoteric characteristics, as shown below, and the color of the solution changes according to the pH. However flower sap is invariably acidic irrespective of the color of the flower, the pH being around 5.5 in most cases.[2] Thus the contribution of pH to the flower color will be restricted.

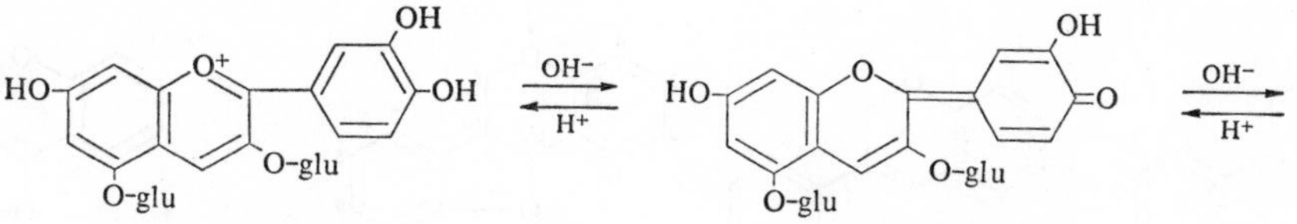

cyanin cation **7**
(pH<3, red)

cyanin base **8**
(pH 8.5, violet)

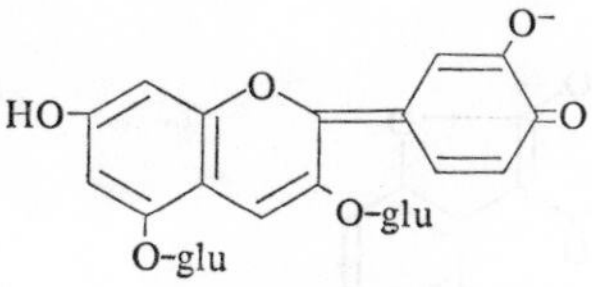

cyanin anion **9**
(pH>11, blue)

Recent studies have revealed that co-pigmentation with flavonoids, tannins and other plant pigments, together with metal-complexing are major factors affecting the variety of plant color. Complexes such as an organometallic complex containing Mg and of anthocyanins with flavonoids have been isolated and characterized.[3-5] A number of general references are available.[1,6-8]

1) J. B. Harborne, *Chemistry and Biochemistry of Plant Pigments* (ed. T. W. Goodwin), p. 247, Academic Press, 1965.
2) K. Shibata, K. Hayashi, T. Isaka, *Acta Phytochem.* (Japan), **15**, 17 (1949).
3) E. C. Bate-Smith, R. G. Westall, *Biochim. Biophys. Acta*, **4**, 427 (1950).
4) K. Hayashi, K. Takeda, *Proc. Japan Acad.*, **46**, 535 (1970).
5) S. Asen, R. M. Horowitz, *Phytochem.*, **13**, 1219 (1974).
6) K. Hayashi, *The Chemistry of Flavonoid Compounds* (ed. T. Geissman), p. 248, Pergamon Press, 1962.
7) J. B. Harborne, *Comparative Biochemistry of the Flavonoids*, Academic Press, 1967.
8) F. M. Dean, *Naturally Occurring Oxygen Ring Compounds*, p. 388, Butterworths, 1963.

9.51 CLASSES OF ISOFLAVONOID COMPOUNDS

Isoflavonoids possess rearranged flavonoid skeltons, the B ring being attached to the 3-position of the hetero ring.[1-3] Angolensin 12 is the only natural α-methyldeoxybenzoin so far isolated. The isoflavonoids are common constituents of plants of the Leguminosae family. Although the numbering system of the isoflavonoids is shown below, it should be noted that rotenoids 4 and pterocarpans 7 have a unique numbering. So far there is the only natural isoflavene, neorauflavene, reported.[4]

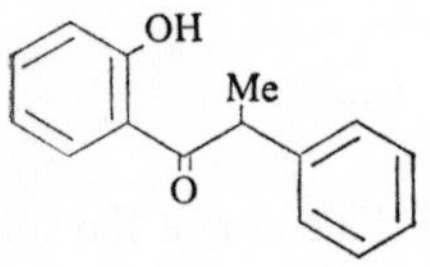

1
α-methyldeoxybenzoins
(angolensin)

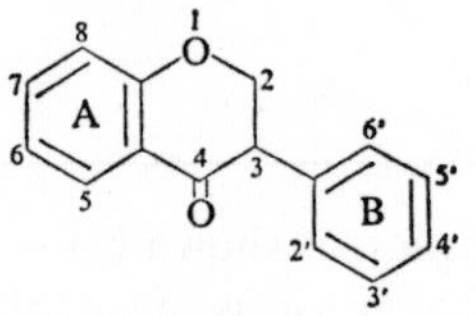

2
isoflavanones
(sophorol)

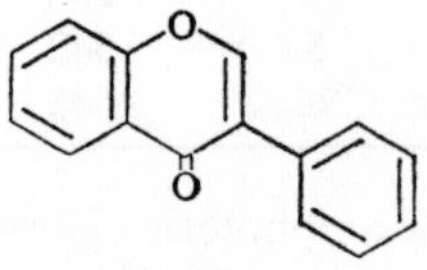

3
isoflavones
(formononetin)

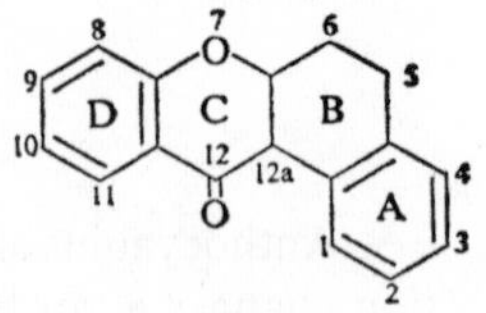

4
rotenoids
(rotenone)

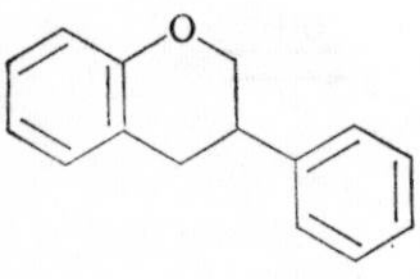

5
isoflavans
(vestitol)

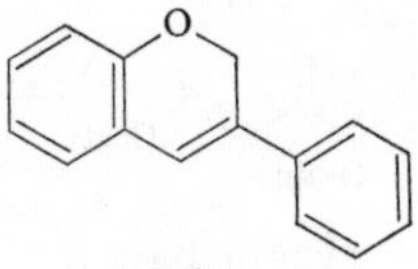

6
isoflavenes
(neorauflavene)

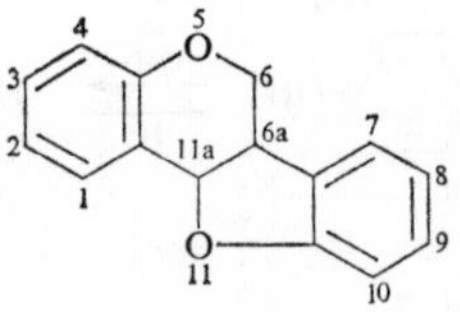

7
pterocarpans
(pterocarpin)

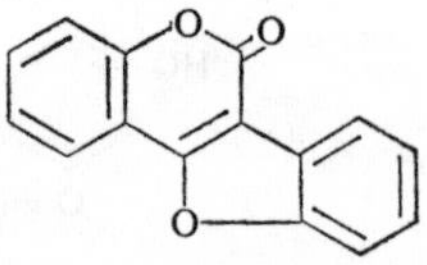

8
coumestans
(coumestrol)

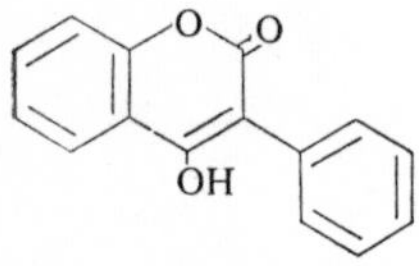

9
3-aryl-4-hydroxy-
coumarins
(scandenin)

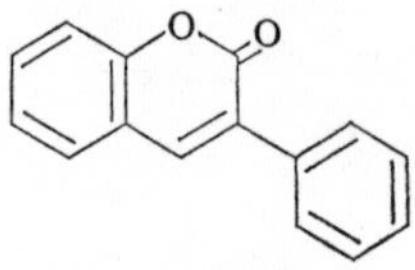

10
3-arylcoumarins
(pachyrrhizin)

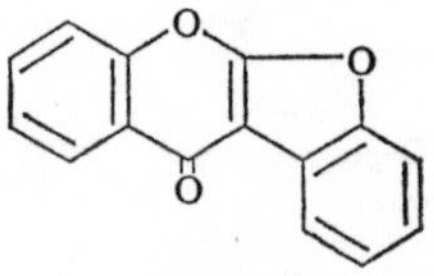

11
coumarano-
chromones
(lisetin)

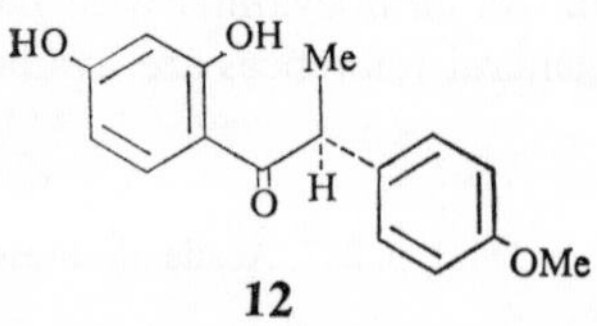

12

1) W. D. Ollis, *The Isoflavonoids*, in *The Chemistry of Flavonoid Compounds* (ed. T. A. Geissman), p. 353, Pergamon Press, 1962.
2) F. M. Dean, *Nautrally Occurring Oxygen Ring Compounds*, p. 366, Butterworths, 1963.
3) E. Wong, *Fortschr. Chem. Org. Naturst.*, **28**, 61 (1970).
4) A. J. Brink, G. J. H. Rau, J. P. Engelbrecht, *Tetr.*, **30**, 311 (1974).

9.52 STRUCTURE OF AURICULATIN

1.68   1.78
(d, 7)   (br. t, 7)
1.42   7.95
6.96 (d, 8)
6.42 (d.d, 8, 2)
H_β H_α OH   O
5.59 (d, 10)
6.69 (d, 10)   12.45   HO   OH   6.64
8.53   6.50 (d, 2)

auriculatin **1**

mp: 136–139° (monohydrate)
ms: 420 (M⁺ $C_{25}H_{24}O_6$)
uv (EtOH): 225 (4.46), 289 (4.63), 341 (inf.)
uv (EtOH/AlCl₃): 305 (4.8)
ir (CHCl₃): 3590, 3500–2500, 1651, 1621

1) The formation of a triacetate and a trimethyl ether (with Me_2SO_4) indicated the presence of three hydroxyl groups.

2) The ir, uv and nmr (singlet at 7.95, characteristic of 2-H) spectra, together with the reactions (e.g. **2**⇌**3**) indicated an isoflavone structure.

3) The presence of 5-OH was indicated firstly by the nmr peak at 12.45, secondly by the bathochromic shift (16 nm) of the uv maximum on addition of AlCl₃[1,2] and thirdly by the formation of a dimethyl ether on reaction with CH_2N_2.

4) The degradation products (**4–6**) indicate the locations of the hydroxyl groups and the two C_5 groupings of rings A and B, and these assignments were confirmed by analysis of the mass spectrum of **2**.

5% KOH
HCOOEt, Na

auriculatin tri-
methyl ether **2**

H_2O_2 (OH⁻)

3

30% KOH

i) 30% KOH
ii) HI

OMe
OMe
COOH
4

OMe
OMe
CH_2COOH
5

OH
dihydroisoosajinol **6**

1) L. Jurd, *Spectral Properties of Flavonoid Compounds*, in *The Chemistry of Flavonoid Compounds* (ed. T. A. Geissman), p. 107, Pergamon Press, 1962.
2) T. J. Mabry, K. R. Markham, N. B. Thomas, *The Systematic Identification of Flavonoids*, p. 171, Springer-Verlag, 1970.

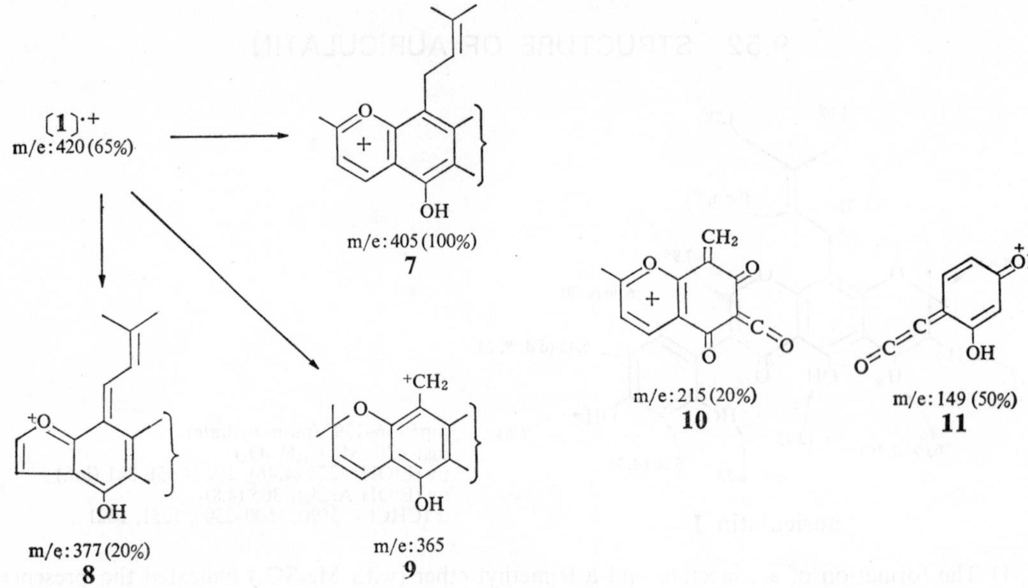

$[\mathbf{1}]^{\cdot +}$
m/e: 420 (65%)

m/e: 405 (100%)
7

m/e: 215 (20%)
10

m/e: 149 (50%)
11

m/e: 377 (20%)
8

m/e: 365
9

5) The orientation of the 2,2-dimethylchromene ring was established by analysis of the chemical shift differences of H_α on acetylation ($\Delta\delta + 0.26$).[3]

Remarks

The root of *Milletia auriculata* (Leguminosae) contains auriculatin along with sumatrol (rotenoid).[4] There are several isoflavones having very similar structures (**12 → 15**).

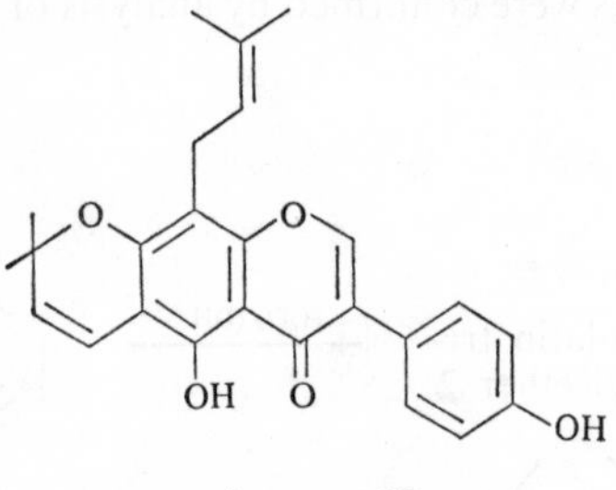

scandenone **12**

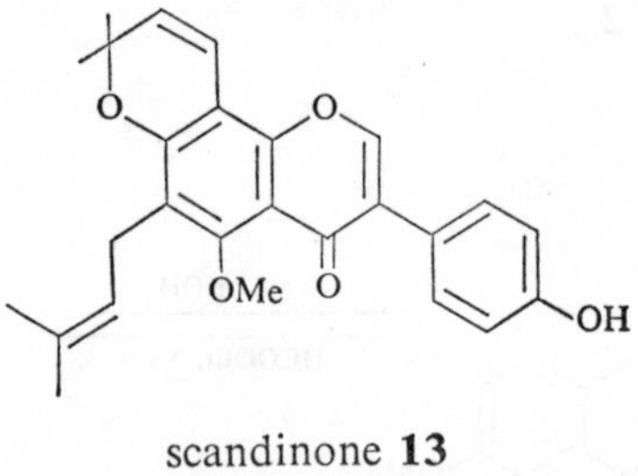

scandinone **13**

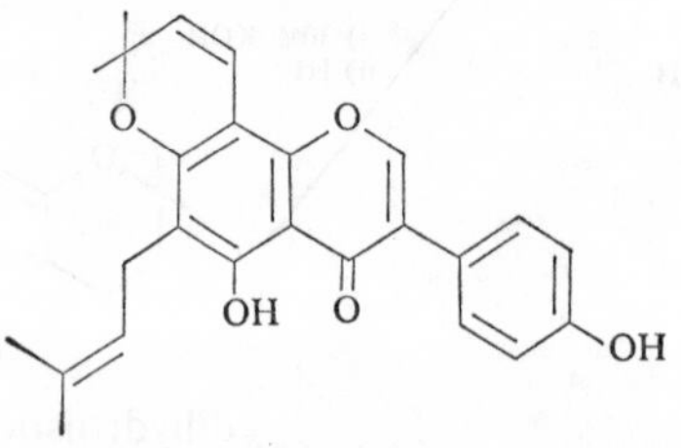

osajin **14**

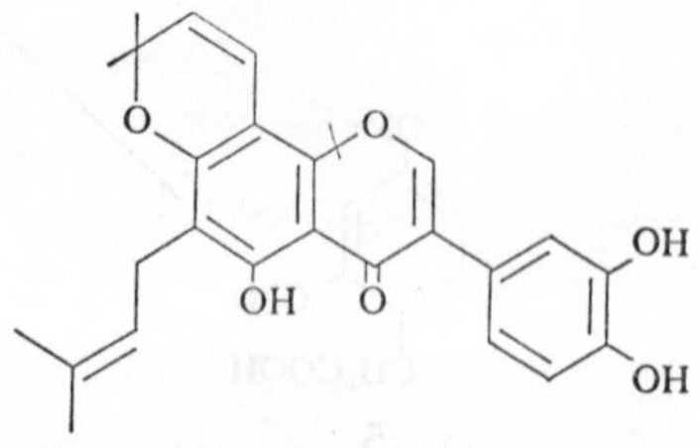

pomiferin **15**

3) A. Arnone, G. Cardillo, L. Merlini, R. Mondelli, *Tetr. Lett.*, 4201 (1967).
4) M. Shabbir, A. Zaman, L. Crombie, B. Tuck and D. A. Whiting, *J. Chem. Soc.*, C, 1899 (1968).

Leguminous plants are excellent sources of isoflavonoid compounds, and almost all of them which have isoprenoid residues have been isolated from Lotoidae (subfamily of Leguminosae).

The mechanism of biosynthesis of γ,γ-dimethylallylphenol and 2,2-dimethylchromene was proposed[5] to be as follows.

5) *Recent Developments in the Chemistry of Natural Phenolic Compounds* (ed. W. D. Ollis), Pergamon Press, 1961.

9.53 SYNTHESIS OF ISOFLAVONES

Ethyl formate method[1)]

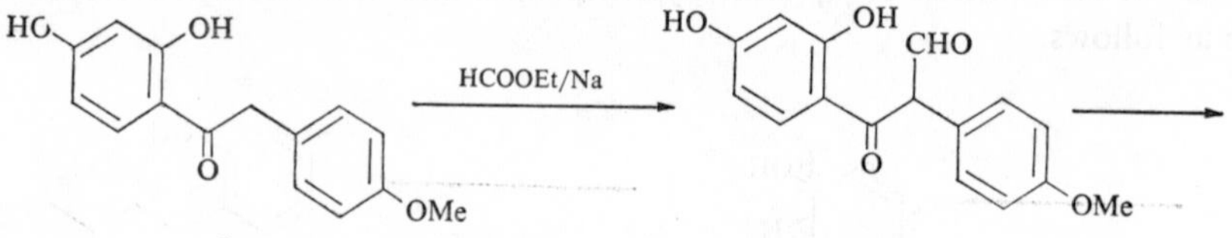

1 2

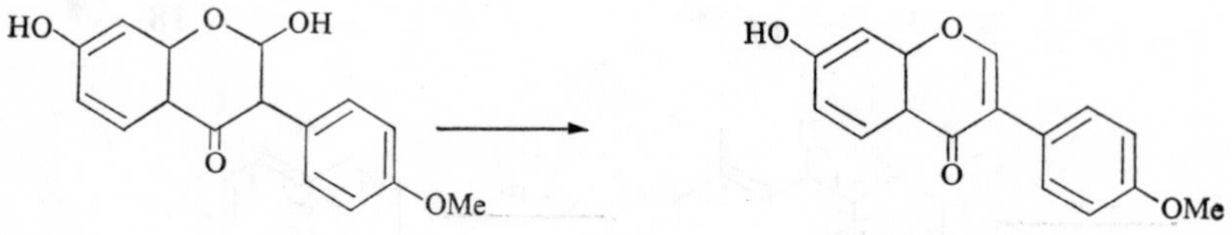

formononetin 4

3

Ethoxalyl chloride method[2)]

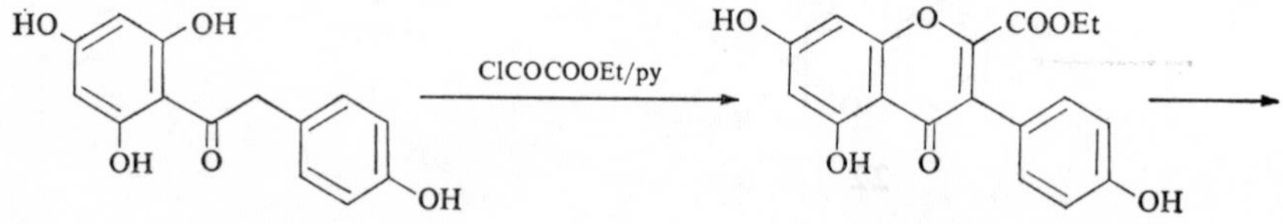

5 6

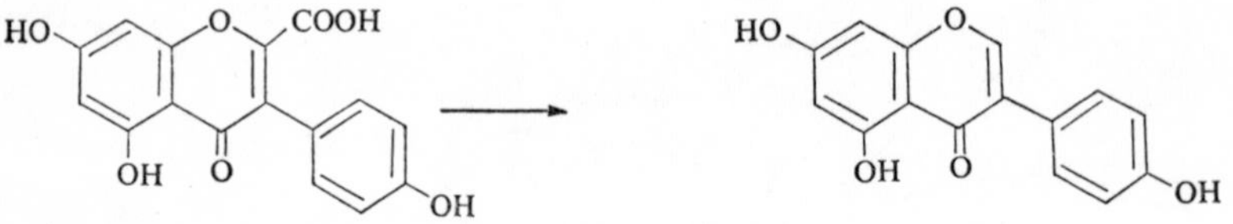

7 8

Ether formation[3)]

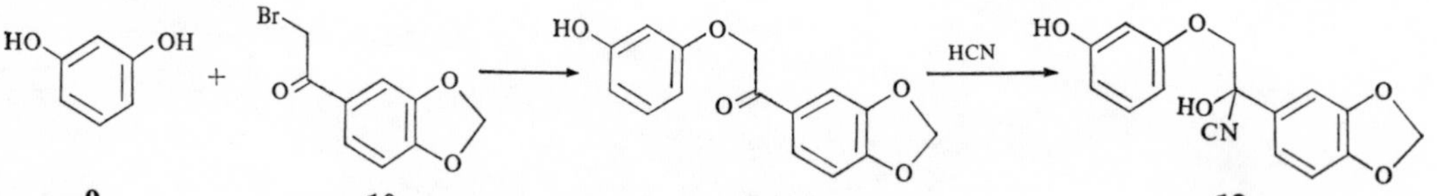

9 10 11 12

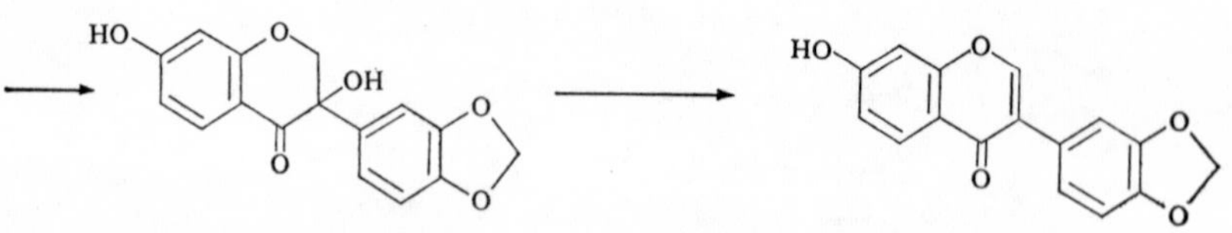

13 14

1) P. C. Joshi, K. Venkataraman, *J. Chem. Soc.*, 513 (1934).
2) W. Baker, J. B. Harborne, W. D. Ollis, *J. Chem. Soc.*, 1860 (1953).
3) E. Späth, E. Lederer, *Chem. Ber.*, **63B**, 743 (1930).

Benzil method[4]

15 →(BrCH₂COOEt)→ **16** →(OH⁻)→ **17**

→(NaOAc/Ac₂O)→ **18**

Epoxychalcone method[5]

19 + **20** →(OH⁻)→ **21** →

22 →(H₂O₂)→ **23** →(BF₃)→

24 →(HCl)→ afromosin **25**

Thallic acetate method[6]

26 →(Tl(OAc)₃/MeOH)→ **27**

4) W. B. Whalley, G. Lloid, *J. Chem. Soc.*, 3213 (1956).
5) A. C. Jain, P. D. Sarpal, T. R. Seshadri, *Indian J. Chem.*, **3**, 369 (1965).
6) W. D. Ollis, K. L. Ormand, I. O. Sutherland, *Chem. Commun.*, 1237 (1968).

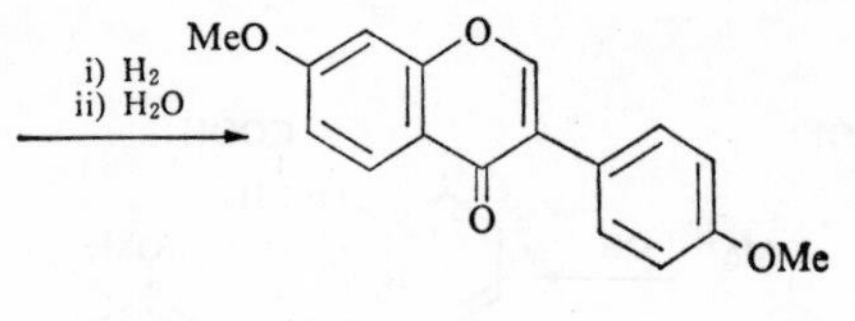

i) H₂
ii) H₂O
→

28

Enamine mehtod[7]

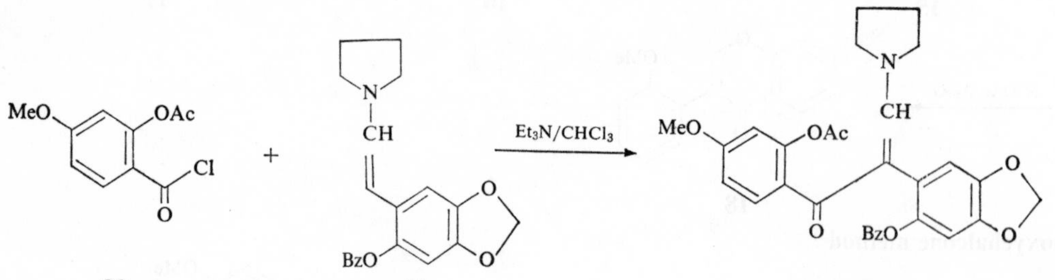

29 + **30** —Et₃N/CHCl₃→ **31**

—C₅H₁₁N/py→

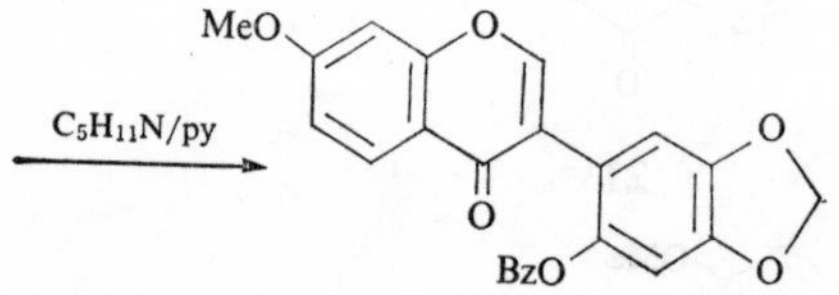

32

Wessely-Moser Rearrangement[8]

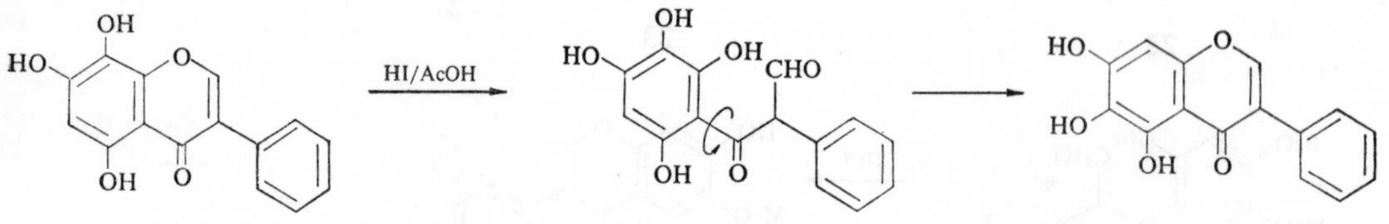

33 —HI/AcOH→ **34** → **35**

Remarks

In the ethyl formate and ethoxalyl chloride methods, deoxybenzoins (phenyl benzyl ke-tones) are starting materials. These can be synthesized as follows.[9]

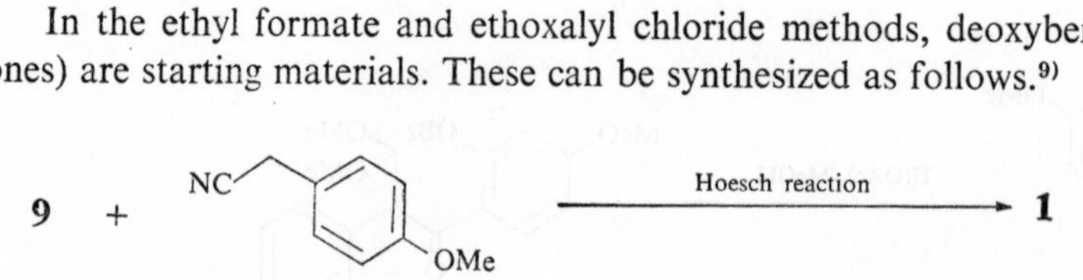

9 + **36** —Hoesch reaction→ **1**

7) M. Uchiyama, M. Matsui, *Agric. Biol. Chem.*, **31**, 1490 (1967).
8) W. Baker, I. Dunstan, J. B. Harborne, W. D. Ollis, R. Winter, *Chem. Ind.* 277 (1953).
9) W. D. Ollis, *The Isoflavonoids*, in *The Chemistry of Flavonoid Compounds* (ed. T. A. Geissman), **Pergamon** Press, 1962 and references therein.

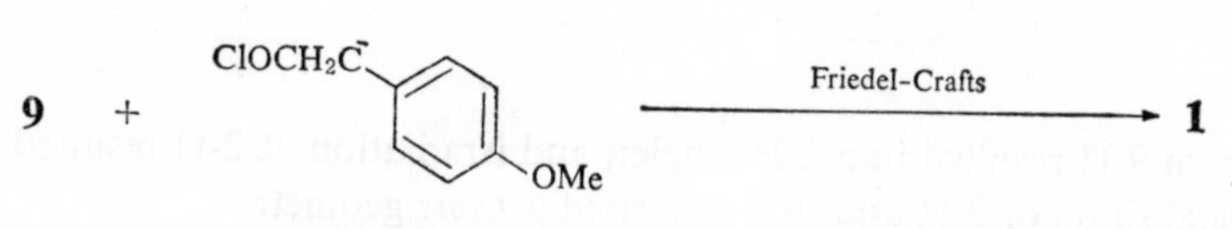

$$9 \;+\; \text{ClOCH}_2\text{C} \text{—} \text{OMe} \quad \xrightarrow{\text{Friedel–Crafts}} \quad 1$$

37

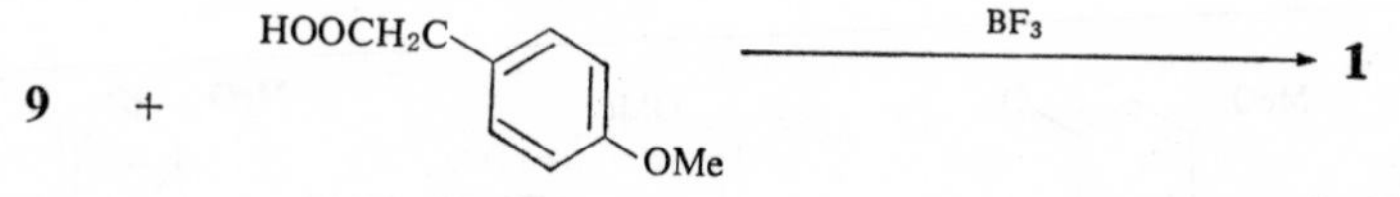

$$9 \;+\; \text{HOOCH}_2\text{C} \text{—} \text{OMe} \quad \xrightarrow{\text{BF}_3} \quad 1$$

38

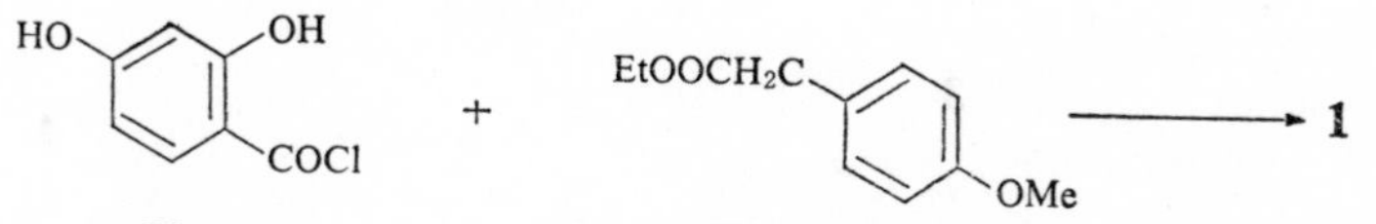

$$39 \;+\; 40 \quad \longrightarrow \quad 1$$

39   40

Protection of hydroxyl groups, e.g. by alkoxylation or benzylation, is necessary in some methods, otherwise degradation will occur during the reaction. The ethyl formate and ethoxalyl chloride methods have been widely used in many isoflavone syntheses. The epoxychalcone and thallic acetate methods provide interesting laboratory parallels to isoflavone biosynthesis.

9.54   STRUCTURES OF EUCOMIN AND EUCOMOL

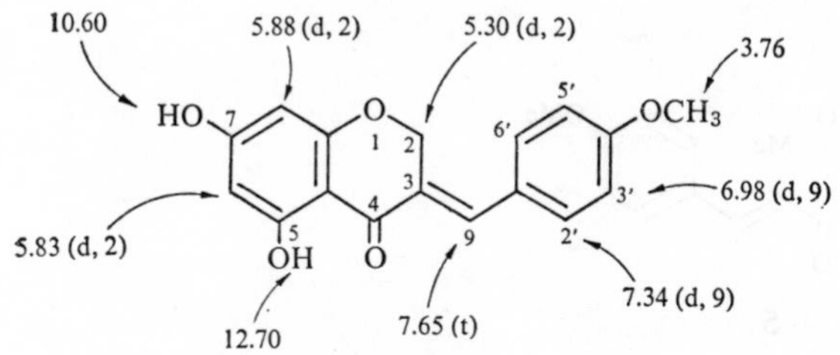

eucomin 1

yellow crystals mp: 194–196°
uv: 194 (3.38), 214 (3.31), 365 (3.33)
ir: 3320, 1645, 1630, 1600
nmr in DMSO-d_6

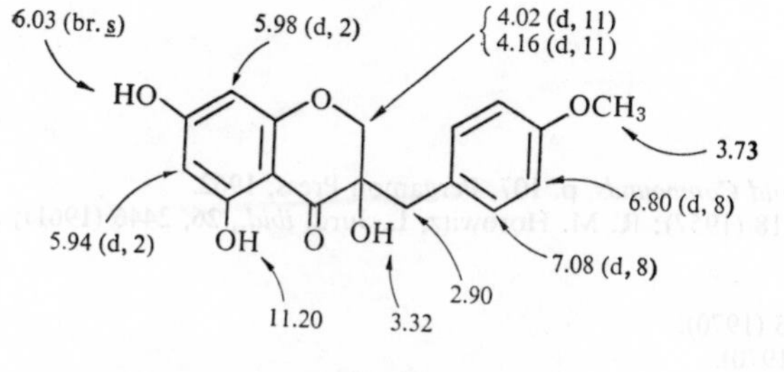

eucomol 2

colorless crystals
mp: 134.5–135° $\alpha_D{}^{25}$: $-32°$
uv (EtOH): 195 (3.84), 214 (3.49), 293 (3.29), 326
          (sh, 2.97)
ir (KBr): 3450, 3380–3305, 1635, 1610, 1585, 1510
nmr in CDCl$_3$

Eucomin

1) Irradiation at 9-H resulted in a 2-H singlet, and irradiation at 2-H resulted in a 9-H singlet. The chemical shifts of 2-H and 9-H indicated a *trans* geometry.
2) Photochemical reaction of eucomin dimethyl ether **3** yielded the *cis* isomer **4**.
3) The structure of **3** was confirmed by synthesis.

Eucomol

1) Addition of NaOAc results in a shift in the uv spectrum from 293 to 330 nm, and addition of $AlCl_3$ from 293 to 315, indicating 7-OH and 5-OH, respectively.[1,2]
2) An nmr singlet at 5.86 in DMSO-d_6 moved to 3.32 in $CDCl_3$, indicating *tert*-OH. The presence of *tert*-OH is confirmed by the formation of a diacetate and a triacetate.

Remarks

The homo-isoflavones are a new class of natural products isolated from the bulbs of *Eucomis bicolor* (Liliaceae).[3] So far, ten homo-isoflavones have been isolated from the genus *Eucomis* only.[4,5] Three of them are 3,9-dihydro derivatives. Punctatin (= 8-methoxyeucomin) is orange colored, while the other compounds are yellow to colorless. Biosynthetically, eucomin **1** is probably formed via the 2-methoxychalcone **5**.[6]

5

1) cf. T. A. Geissman, *The Chemistry of the Flavonoid Compounds*, p. 107, Pergamon Press, 1962.
2) L. Jurd, R. M. Horowitz, *J. Org. Chem.*, **22**, 1618 (1957): R. M. Horowitz, L. Jurd, *ibid.*, **26**, 2446 (1961); L. Jurd, *Phytochemistry*, **8**, 445 (1969).
3) P. Böhler, Ch. Tamm, *Tetr. Lett.*, 3479 (1967).
4) W. T. L. Sidwell and Ch. Tamm, *Tetr. Lett.*, 475 (1970).
5) R. E. Finckh, Ch. Tamm. *Experientia*, **26**, 472 (1970).
6) P. M. Dewick, *J. Chem. Soc., Chem. Comm.*, 438 (1973).

9.55   SYNTHESIS OF ROTENOIDS

Rotenone

Route I[1-3]

(±)-hydroxydi-
hydrotubanol **1**   **2**   (±)-derrisic acid **3**

4   dehydrorotenone **5**

rotenol **6**   rotenone **7**

1) Direct formation of **3** from **1**-tubanol has not been achieved, probably due to the hydro-
   chlorolytic lability of the allylaryl ethereal ring.
2) The key to this reaction is the application of DCC (dicyclohexylcarbodiimide).
3) The last two steps yield a mixture of stereoisomers of rotenone, from which the natural
   form was obtained by thermal isomerization.

Route II[4]

8   **9**   **10**   (−)-tubaic acid **11**

$\xrightarrow{\text{DCC,}}$ dehydrorotenone **5**

($\pm$)-Mundeserone

Route I[5]

resorcinol **12**   **2**   **13**

dehydromundeserone **14**   ($\pm$)-mundeserone **15**

Route II[7]

17

18   **19**

    The characteristic of this synthesis, following a possible path of biogenesis, is the application of a partial demethylation, first explored by the authors.[9]

---

1) M. Miyano, M. Matsui, *Agr. Biol. Chem.*, **22**, 128 (1958).
2) M. Miyano, A. Kobayashi, M. Matsui, *Agr. Biol. Chem.*, **25**, 673 (1961).
3) M. Miyano, *J. Am. Chem. Soc.*, **87**, 3962 (1965).
4) M. Miyano, *J. Am. Chem. Soc.*, **87**, 3958 (1965).
5) J. R. Herbert, W. D. Ollis, R. C. Russell, *Proc. Chem. Soc.*, 177 (1960).
6) S. Takei, S. Miyajima, M. Ohno, *Chem. Ber.*, **65**, 1041 (1932); A. Robertson, *J. Chem. Soc.*, 1163 (1933).
7) V. Chandrashekar, M. Krishnamurti, T. R. Seshadri, *Tetr.*, **23**, 2505 (1967).
8) H. Fukami, G. Sakata, M. Nakajima, *Agr. Biol. Chem.*, **29**, 82 (1965).
9) K. Aghoramurti, A. S. Kukla, T. R. Seshadri, *Curr. Sci.*, **30**, 218 (1961).
10) E. Wong, *Fortschr. Chem. Org. Naturst.*, **28**, 61 (1970); and references therein.
11) L. Crombie, M. B. Thomas, *J. Chem. Soc.*, C, 1796 (1967); L. Crombie, C. L. Green, D. A. Whiting, *J. Chem. Soc.*, C, 3029 (1968).

(±)-Elliptone

Route I[8)]

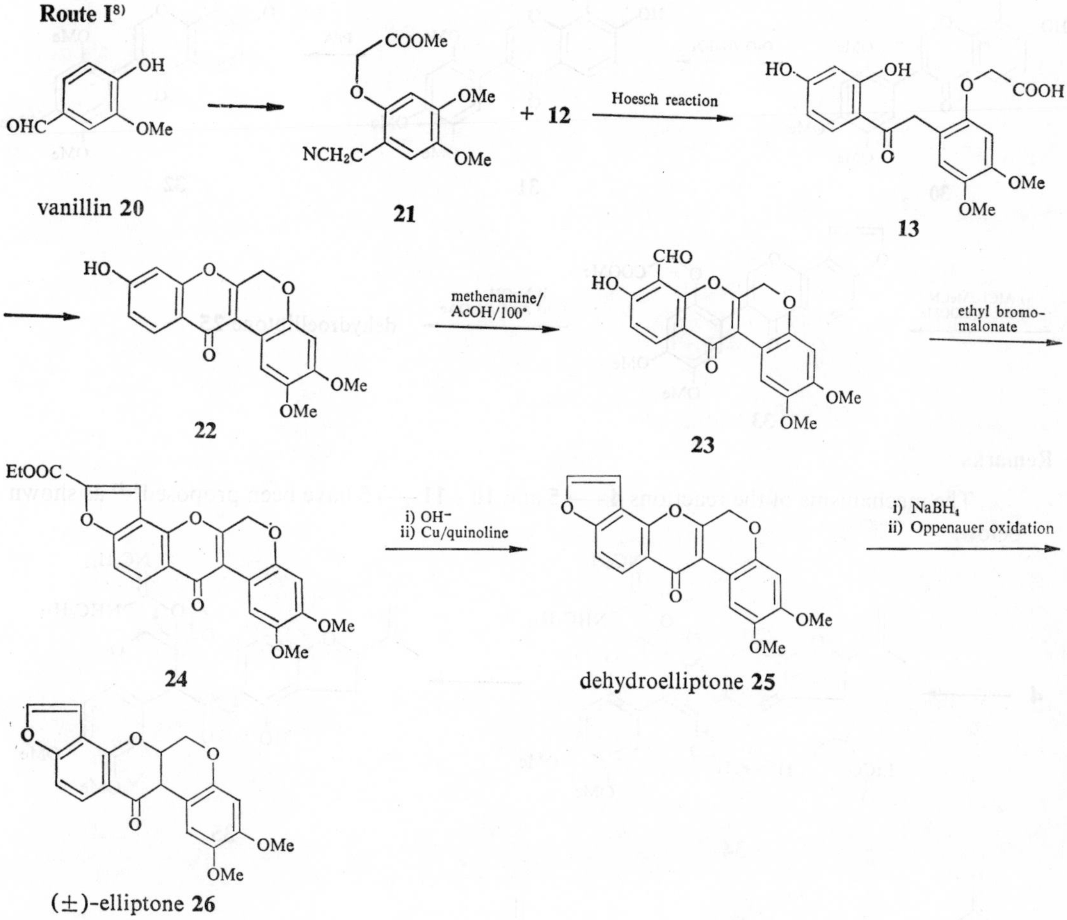
OH
OHC
OMe
vanillin 20
COOMe
O
OMe
NCH$_2$C
OMe
21
+ 12
Hoesch reaction
HO
OH
O
COOH
O
OMe
OMe
13
HO
O
O
O
OMe
OMe
22
methenamine/
AcOH/100°
CHO
HO
O
O
O
OMe
OMe
23
ethyl bromo-
malonate
EtOOC
O
O
O
O
OMe
OMe
24
i) OH$^-$
ii) Cu/quinoline
O
O
O
O
OMe
OMe
dehydroelliptone 25
i) NaBH$_4$
ii) Oppenauer oxidation
O
O
O
O
OMe
OMe
(±)-elliptone 26

Route II[7)]

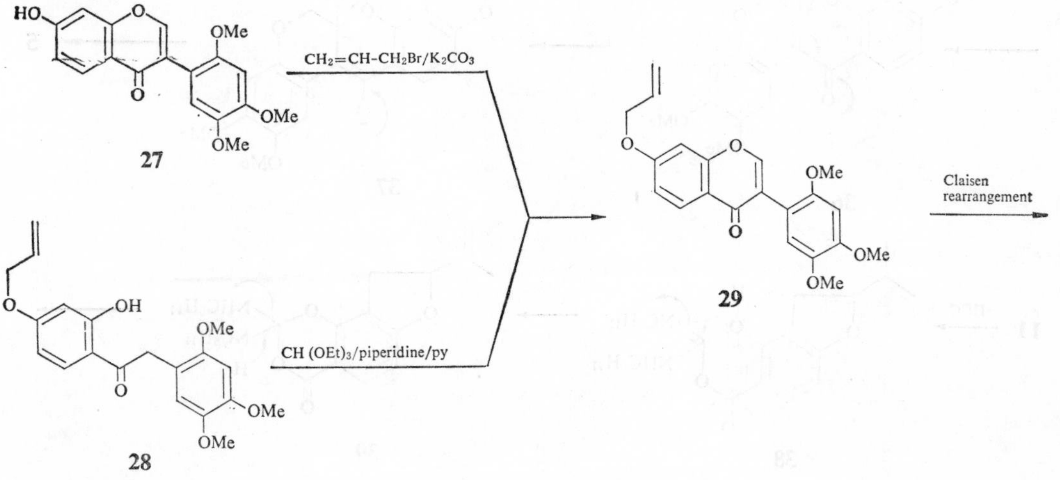
HO
O
OMe
O
OMe
OMe
27
CH$_2$=CH-CH$_2$Br/K$_2$CO$_3$
O
OH
OMe
O
OMe
OMe
28
CH (OEt)$_3$/piperidine/py
O
O
OMe
O
OMe
OMe
29
Claisen
rearrangement

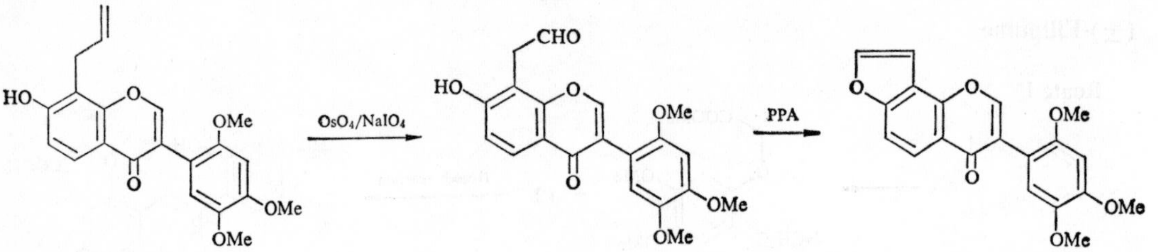

30          $OsO_4/NaIO_4$ 31   PPA   32

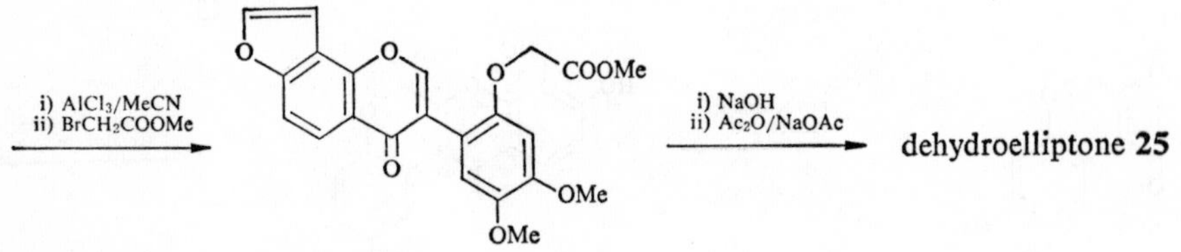

i) $AlCl_3/MeCN$
ii) $BrCH_2COOMe$

33

i) NaOH
ii) $Ac_2O/NaOAc$ dehydroelliptone 25

Remarks

The mechanisms of the reactions $4 \longrightarrow 5$ and $10+11 \longrightarrow 5$ have been proposed[3,4] as shown below.

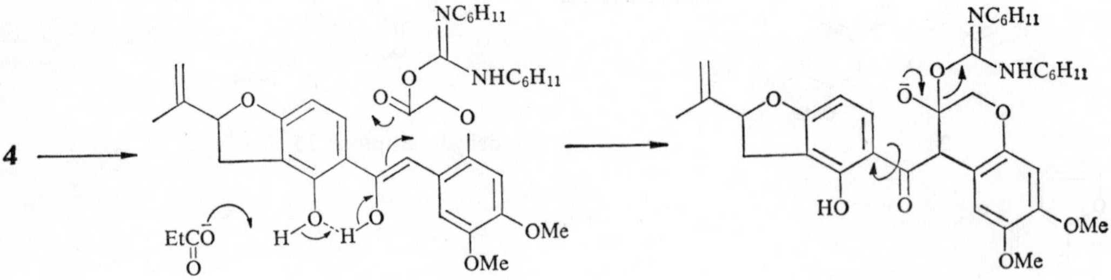

4 $\longrightarrow$

34   35

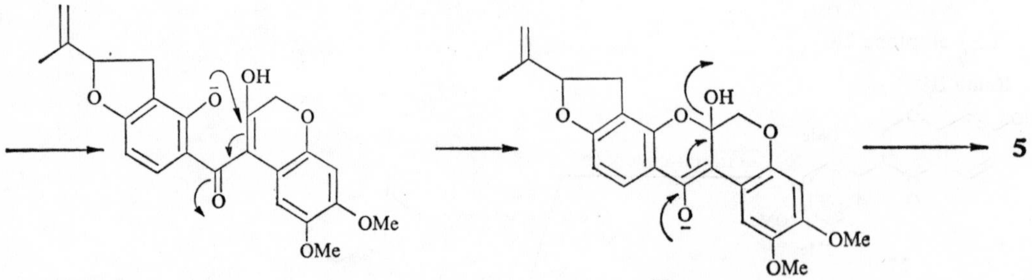

36   37 $\longrightarrow$ 5

11   DCC

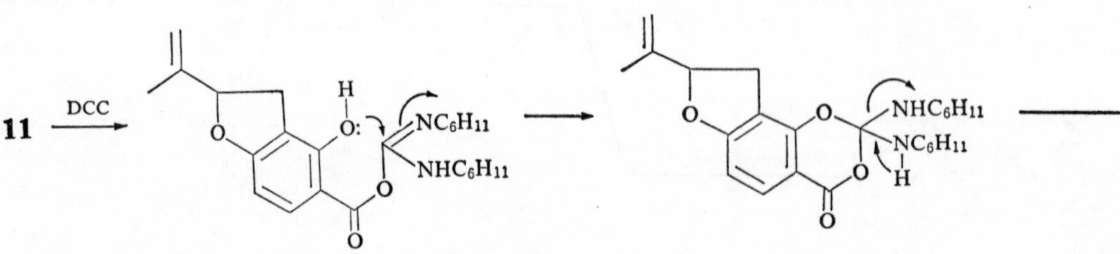

38   39 $\longrightarrow$

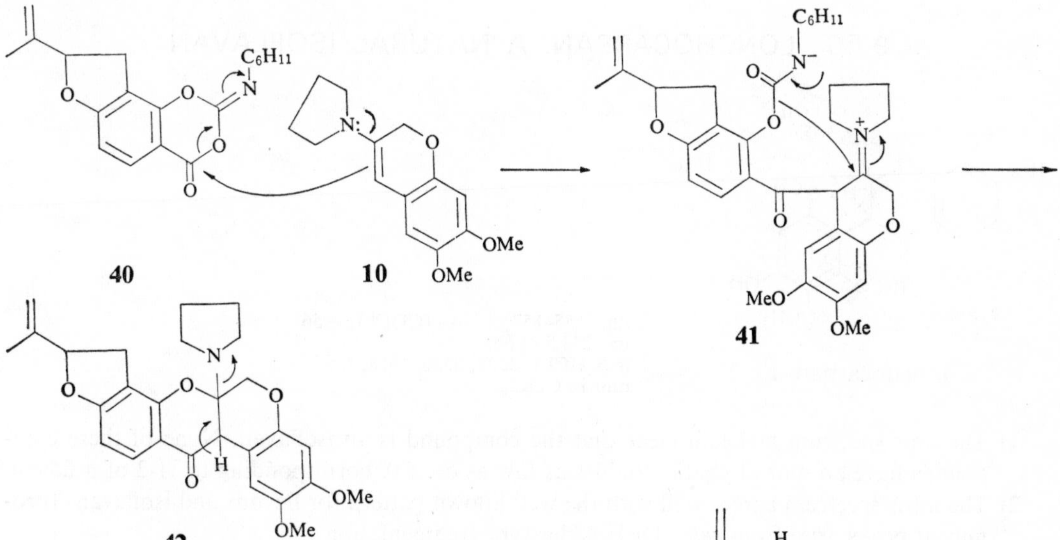

40   **10**

41

42

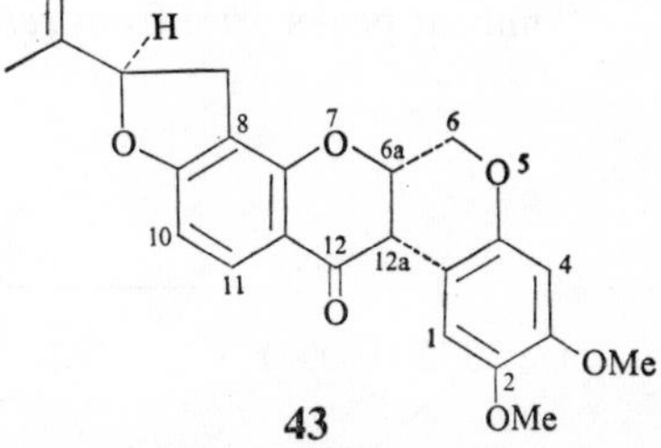

Several genera of the family Leguminosae contain in their roots a number of chromanochromanones having insecticidal activity. The most important compound is rotenone, which gives its name to the group. The numbering system and stereochemistry are shown in **43**. All rotenoids have the same configuration (S) at 6a and 12a.

43

Biogenetically, rotenoids are regarded as a class of isoflavonoids.[10] The feeding of ^{14}C-phenylalanine **44** labeled variously at the 1-, 2- or 3-positions to *Derris elliptica* resulted in the labeling of rotenone in the 12-, 12a- or 6a-positions.[11]

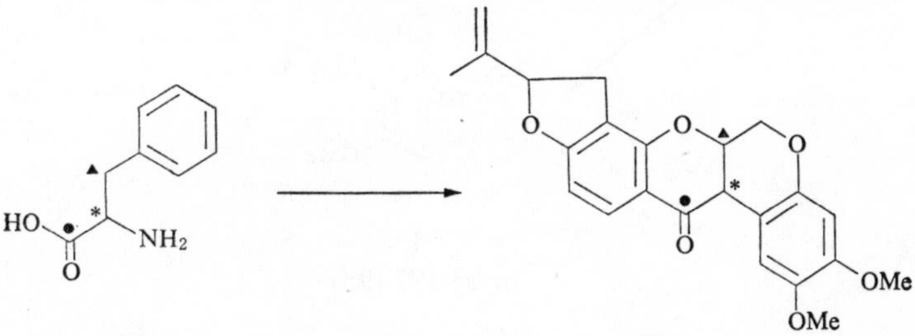

44   **45**

The probable biosynthetic sequence to rotenoids is shown below.[10] A 2'-methoxyl group of isoflavones is the key element for ring formation in rotenoids. [Me-^{14}C]-Methionine was incorporated into the 6-position of rotenone.[11]

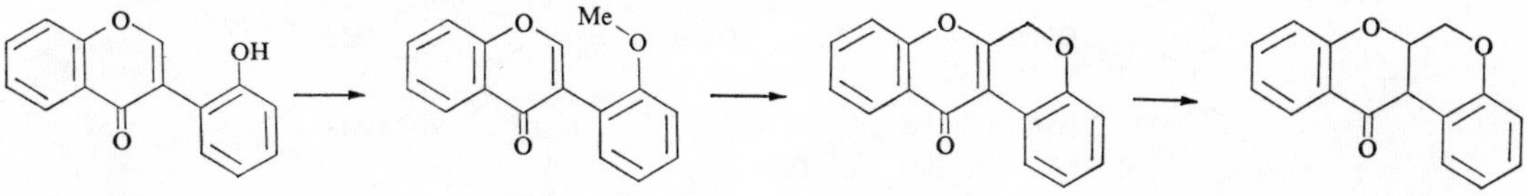

46   **47**   **48**   **49**

9.56   LONCHOCARPAN, A NATURAL ISOFLAVAN

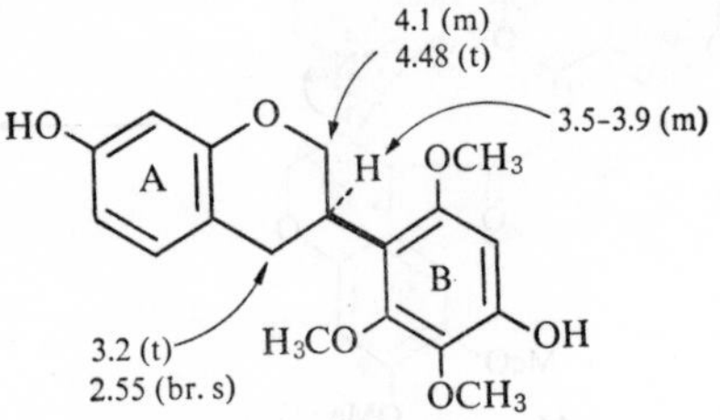

4.1 (m)
4.48 (t)

3.5–3.9 (m)

3.2 (t)
2.55 (br. s)

lonchocarpan **1**

mp: 155–157° α_D (CDCl$_3$): +56°
uv: 283.5 (3.66)
ir (CHCl$_3$): 3595, 3530, 1618, 1598
nmr in CDCl$_3$

1) The nmr spectrum makes it clear that the compound is an isoflavan. None of these compounds have an nmr aliphatic doublet as low as ca. 5.0, corresponding to H-2 of a flavan.

2) The mass spectrum agrees well with the well-known pattern for flavans and isoflavans (prominent peaks arise from *retro* Diels-Alder type fragmentation).

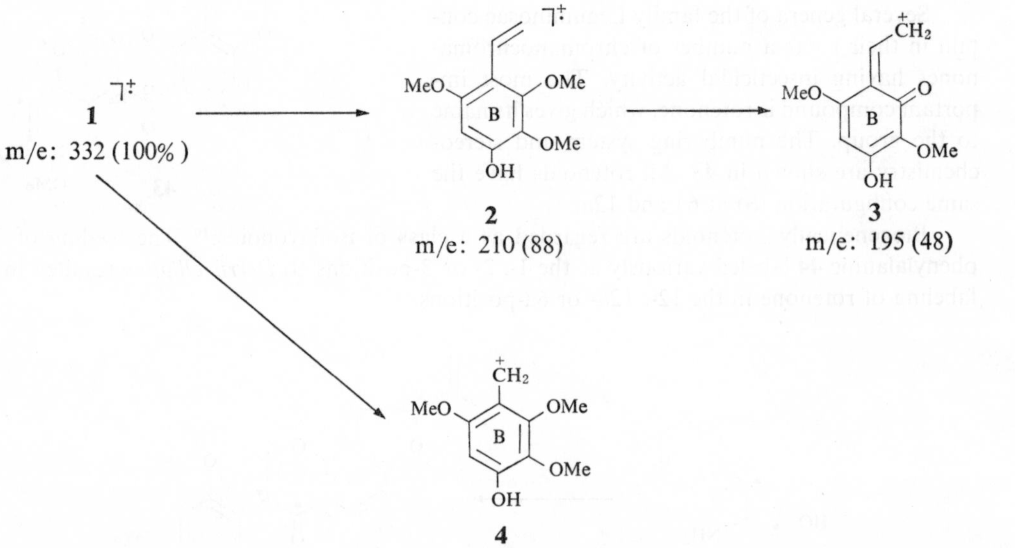

1
m/e: 332 (100%)

2
m/e: 210 (88)

3
m/e: 195 (48)

4
m/e: 197 (98)

3) As for the location of the *O*-function on the B ring, there were three possible arrangements (**5, 6, 7,**) for the dimethyl ether. The solvent-induced methoxy-proton shifts of lonchocarpan dimethyl ether were studied to solve the problem, by means of trideuteriomethylation.

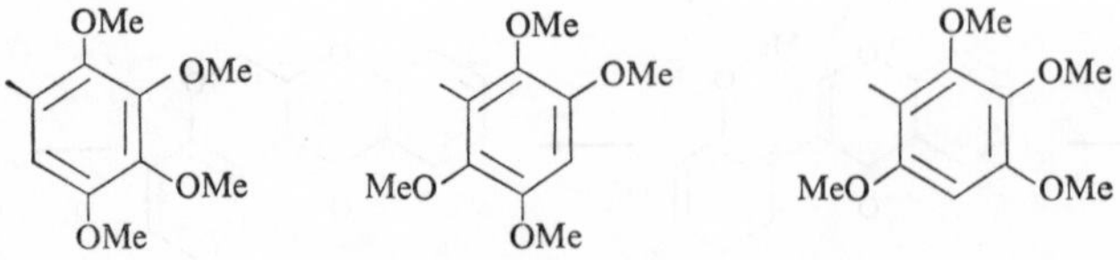

5   **6**   **7**

4) The structure of the dimethyl ether **12** was confirmed by synthesis.

5) To determine the position of the OH group on the B ring (two possible formula, **1** and **13**, still remained), the ir spectra of various phenolic substances were examined for hydrogen bonding.

6) The stereochemistry was established by ord.

8

9

uv: 232, 270, 280, 402
ir: 1790, 1660

10

uv: 228, 278, 313
ir: 3280, 1640

11

uv: 238, 247, 288, 303 (sh)
ir: 1640, 1615

(±)-lonchocarpan
dimethyl ether **12**

13

Remarks

The natural occurrence of isoflavans in plants was not reported until 1968. The only example of an isoflavan was equol **14**, isolated as an animal metabolite.[1] It is of interest to note, however, that three independent groups reported the isolation of isoflavan derivatives from plants at almost the same time.[2-4] (All their manuscripts were received in 1968). The sources of all these isoflavans are Legminosae plants. Lonchocarpan has been isolated from *Lonchocarpus laxiflorus*,[2] which is widely distributed throughout Africa.

equol **14**

1) G. F. Marrian, D. Beall, *Biochem. J.*, **29**, 1586 (1935).
2) A. Pelter, P. I. Amenechi, *J. Chem. Soc.* (C), 887 (1969).
3) S. Shibata, T. Saitoh, *Chem. Pharm. Bull.*, **16**, 1932 (1968).
4) K. Kurosawa, W. D. Ollis, B. T. Redman, I. O. Sutherland, *Chem. Commun.*, 1263 (1968).

9.57 STRUCTURE AND SYNTHESIS OF COUMESTEROL

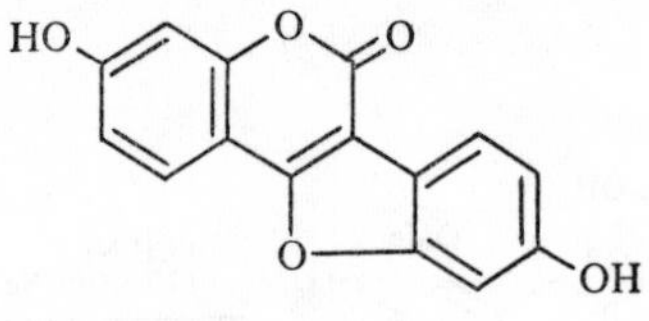

coumestrol $\mathbf{1}$[1–4]
mp: 385° (dec.)
uv: 244, 304, 343

1) Structure elucidation was carried out by degradation; methylative ring opening, hydrolysis, decarboxylation, and then oxidation.

Finally two fragments were obtained; the acid **5** from ring A, and the aldehyde **6** from ring B.

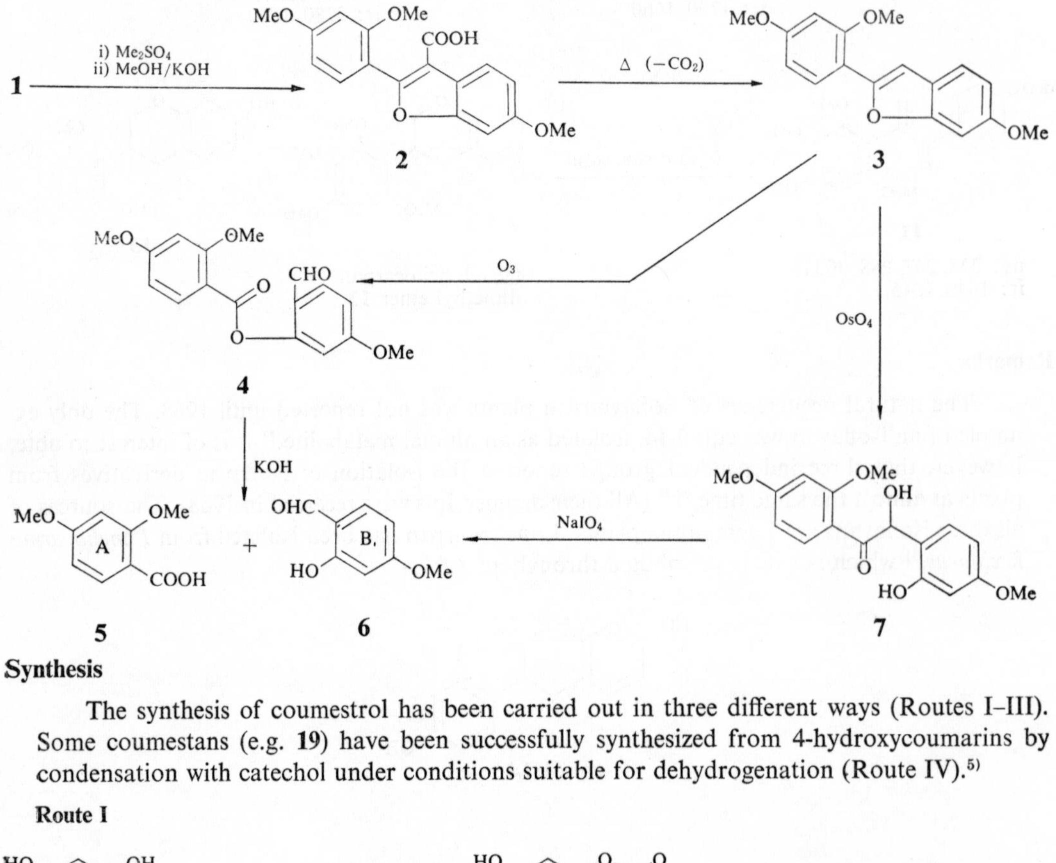

Synthesis

The synthesis of coumestrol has been carried out in three different ways (Routes I–III). Some coumestans (e.g. **19**) have been successfully synthesized from 4-hydroxycoumarins by condensation with catechol under conditions suitable for dehydrogenation (Route IV).[5]

Route I

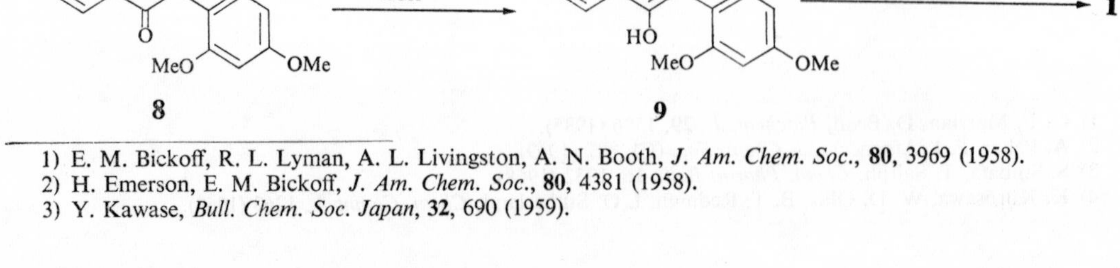

1) E. M. Bickoff, R. L. Lyman, A. L. Livingston, A. N. Booth, *J. Am. Chem. Soc.*, **80**, 3969 (1958).
2) H. Emerson, E. M. Bickoff, *J. Am. Chem. Soc.*, **80**, 4381 (1958).
3) Y. Kawase, *Bull. Chem. Soc. Japan*, **32**, 690 (1959).

Route II

Route III

Route IV

wedelolactone **19**

Remarks

The first known example of a natural coumestan, wedelolactone **19** was reported in 1957. Since then sixteen coumestans have been isolated, mostly from leguminous plants, though two were also isolated from a plant of the family Compositae. Coumestrol **1** from *Trifolium repens* and *Medicago sativa* is one of the earlier ones. As it has estrogenic activity, intensive studies have been carried out by many workers. From the point of view of biosynthesis, coumestans are related to isoflavonoid compounds.

4) L. Jurd, *J. Org. Chem.*, **29**, 3036 (1964).
5) H. W. Wanzlick, R. Gritzky, H. Heiderpriem, *Chem. Ber.*, **96**, 305 (1963).

9.58   STRUCTURE OF LISETIN

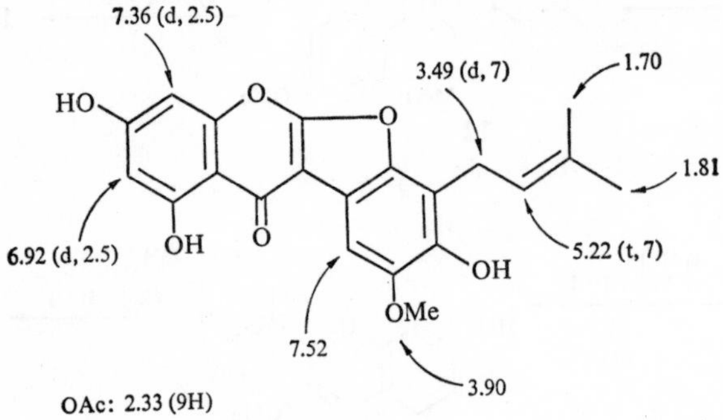

7.36 (d, 2.5)

HO

3.49 (d, 7)   1.70

1.81

6.92 (d, 2.5)   OH   O

5.22 (t, 7)

OH

OMe

7.52

OAc: 2.33 (9H)   3.90

lisetin **1**

mp: 283–286° (dec.)
uv (EtOH): 258 (ε 39,400), 284 (ε 23,200),
                  338 (ε 13,900)
ir (CHCl₃): 1653 (C=O)
nmr in CDCl₃ for **1**-acetate

1) Lisetin forms a triacetate and a trimethyl ether.

2) The nmr spectrum showed the presence of a γ,γ-dimethylallyl group.

3) An aromatic singlet at δ 7.52, appearing at lower field than normal, was assigned to 6′-H, which is specifically deshielded by the carbonyl function.

4) Comparison of the chemical shifts of a pair of doublets with $J=2.5$ Hz in lisetin triacetate δ 6.92 and 7.36) and lisetin trimethyl ether (δ 6.41 and 6.58) placed two of the hydroxyl groups at C-5 and C-7.

lisetin tri-
methyl ether **2**

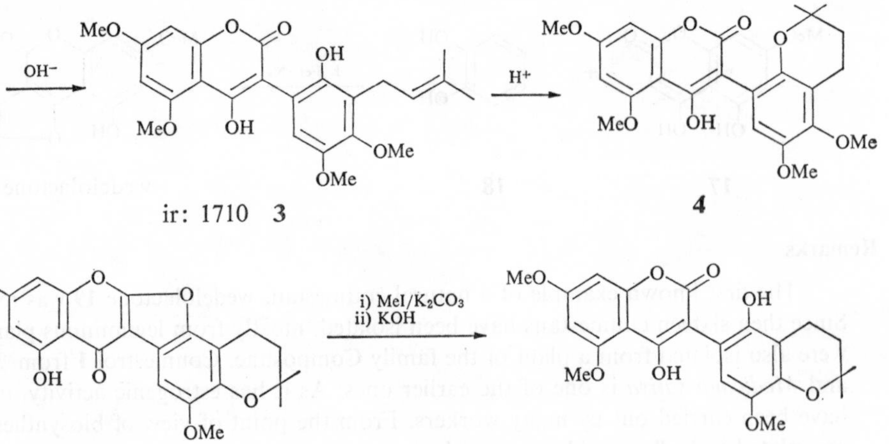

$\xrightarrow{OH^-}$

ir: 1710   **3**

$\xrightarrow{H^+}$

4

1 $\xrightarrow{H^+}$

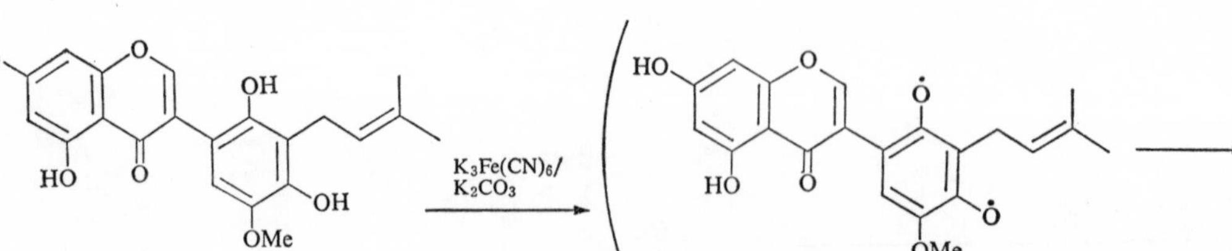

isolisetin **5**

$\xrightarrow[\text{ii) KOH}]{\text{i) MeI/K}_2\text{CO}_3}$

6

piscerythrone **7**

$\xrightarrow[\text{K}_2\text{CO}_3]{\text{K}_3\text{Fe(CN)}_6/}$

8

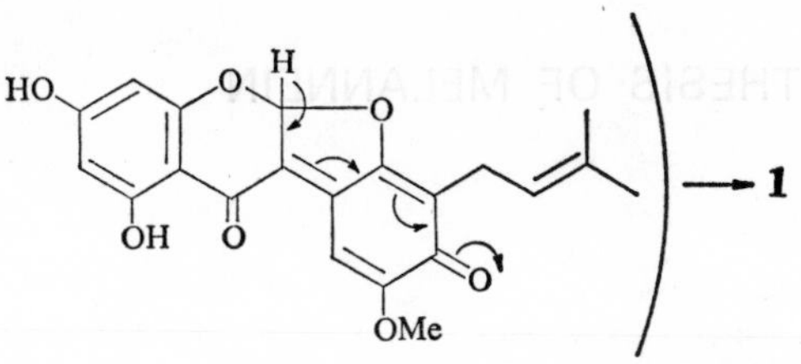

9

5) Since two isomers were obtained (**4** and **6**), the presence of a methoxyl group at C–5′ can be inferred.

6) The conversion of **7** to **1** by oxidative coupling is interesting for its possible biosynthetic implications.

Remarks

Lisetin was isolated from the root bark of Jamaican Dogwood, *Piscidia erythrina*, together with **7**.[1,2] It is the only example of a naturally occurring coumaronochromones so far, and should be characterized as an isoflavonoid.

1) C. P. Falshaw, W. D. Ollis, *Chem. Commun.*, 305 (1966).
2) C. P. Falshaw, W. D. Ollis, J. A. Moore K. Magnus, *Tetr. Suppl.*, **7**, 333 (1966).

9.59   STRUCTURE AND SYNTHESIS OF MELANNEIN

melannein **1**

yellow crystals mp: 221–223°
uv (EtOH): 234 (sh, 4.38), 256 (sh, 4.17),
          308 (sh, 4.00), 344 (4.09)
ir (KBr): 3289, 1664, 1616
nmr in CDCl$_3$ as diacetate

1) Oxidative degradation gives 2-hydroxy-3′,4,4′,5-tetramethoxybenzophenone **2** and veratric acid **3**.

2) The synthesis of melannein diethyl ether was carried out by Friedel-Crafts acylation and Perkin reaction.

3) The Friedel-Crafts reaction affects the demethylation of the *o*-methoxyl group.

$$\mathbf{1} \quad \xrightarrow[\text{ii) KMnO}_4\text{ (neutral)}]{\text{i) Me}_2\text{SO}_4/\text{K}_2\text{CO}_3}$$

2   +   veratric acid **3**

4   +   **5** $\xrightarrow{\text{Friedel-Crafts}}$ **6** $\xrightarrow{\text{NaOAc/Ac}_2\text{O}}$ **1**-diethyl ether **7**

Remarks

Melannein is a neoflavanoid isolated from the heartwood of *Dalbergia baroni*.[1] It gives a positive reaction in the Shinoda test[2] (Mg and HCl). This reductive reaction is well-known as a

1) B. J. Donnelly, D. M. X. Donnelly, A. M. O'Sullivan, *Tetr.*, **24**, 2617 (1968).
2) J. Shinoda, *J. Pharm. Soc. Japan*, **48**, 214 (1928).

specific color reaction for many classes of flavonoid compounds, but some xanthones are also positive.

A general term "neoflavanoids"[3] is used to describe a group of natural products which all have a 4-arylchroman type structure. This represents a class of naturally occurring C_{15}-phenolic compounds, cf. flavanoids (2-arylchroman derivatives), isoflavanoids (3-arylchroman derivatives).

Although dalbergiones **10** are not heterocyclic, they should also be called neoflavanoids, just as it is customary to regard chalcones as flavonoids and angolensin as an isoflavonoid.

The name "dalbergiones" arose since compounds of this type were first isolated from the *Dalbergia* genera (Leguminosae). Although 4-arylcoumarins **12** have been found in plants of Leguminosae and Guttiferae, the other types of neoflavonoids have been isolated only from leguminous plants.

There are three hypotheses on the biosynthesis of neoflavanoids: **a)** Just as the formation of the isoflavonoid skeleton involves a $2 \longrightarrow 3$ aryl migration, a second aryl migration $(3 \longrightarrow 4)$ occurs to form the neoflavanoids. **b)** Alkylation of a phenolic C_6-unit (or its polyketide equivalent) by cinnamyl pyrophosphate could lead to neoflavanoids as shown below. **c)** The key to this hypothesis is an intramolecular C-arylation of chalcones **13** with formation of the spiro intermediates **14** via an anionic process (single 1,3-aryl migration).[5]

Kunesch and Polonsky reported that $(\pm)$-[3-^{14}C]-phenylalanine was incorporated into calophyllolide **16** in *Calophyllum inophyllum*, and that 92% of the label was located at C-4[4] The results of this experiment excluded hypothesis **a)**, but are still compatible with **b)** and **c)**.

b)

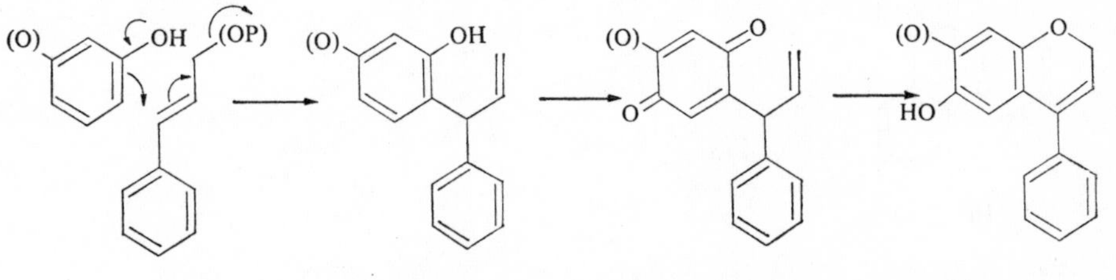

| 8 | 9 | dalbergiones **10** | neoflavenes **11** |

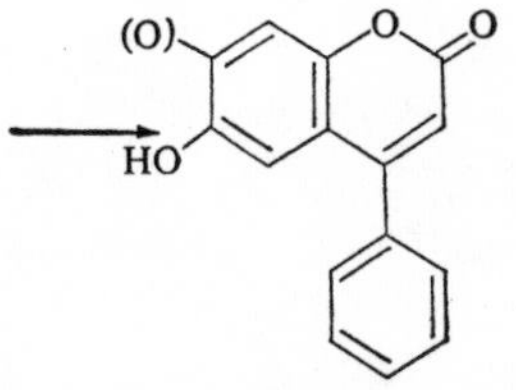

4-arylcoumarins **12**

3) W. D. Ollis, *Experientia*, **22**, 777 (1966).
4) G. Kunesch, J. Polonsky, *Chem. Commun.*, 317 (1967).
5) M. H. Benn, *Experientia*, **24**, 9 (1968).

c)

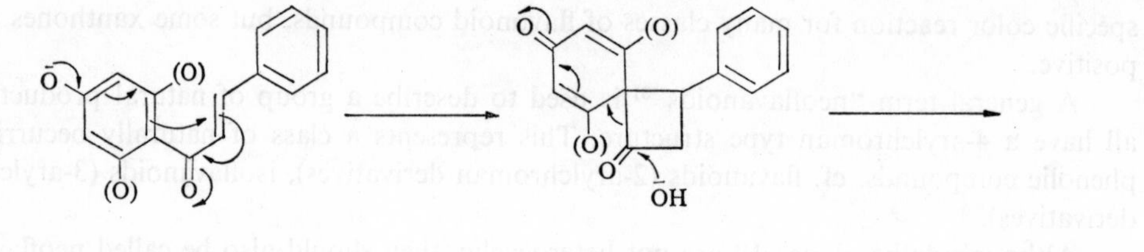

13

14

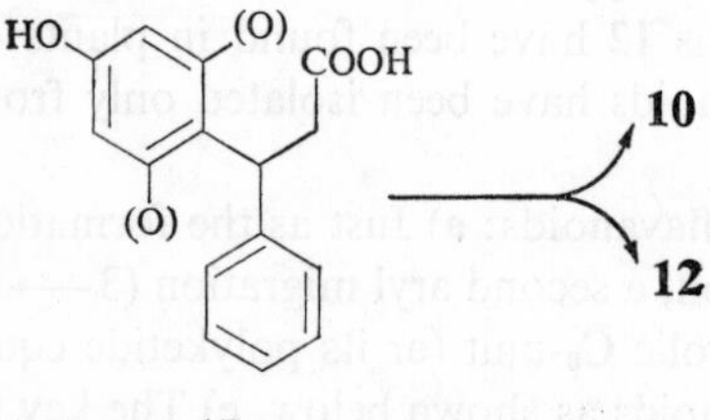

15

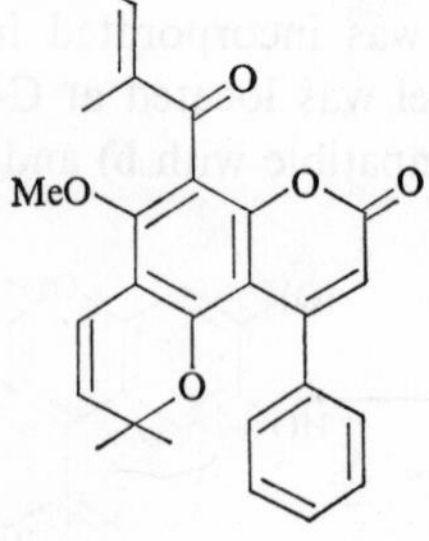

10

12

16

Alkaloids

10.1   INTRODUCTION

Nitrogenous bases found in plants are called alkaloids, and have been known for a long time as toxic principles. Their physiological activities attracted the attention of chemists from early times, and as a result, many alkaloids have been found in the last 100 years. Interestingly certain plant families do not appear to produce any compounds of this type. Elucidation of their structures was tedious due to their great complexity and many years were needed before the structures of some of these alkaloids were determined. A typical example is strychnine, the structure of which eluded organic chemists for more than 50 years until the correct structure was proposed by Robinson in 1948. Recent developments in separation techniques and spectroscopy as well as x-ray crystallography have facilitated research work and much effort has recently been directed toward total syntheses, based on a detailed understanding of the chemical reactions involved. Tracer techniques have opened up new facets of natural products chemistry, for instance, the elucidation of metabolic pathways, and the biosynthesis of alkaloids.

As the plant species in a single family or genus often produce bases of at least biogenetically similar structures, they have been conveniently classified by origin (e.g., amaryllidaceae alkaloids, veratrum alkaloids) as well as by structural types (e.g., benzylisoquinoline alkaloids, indole alkaloids). With the concept of biogenesis, another type of classification was developed. This is more useful in visualizing the relationship of groups already classified by structural types. A general classification of alkaloids is shown in the following schematic diagram.[1]

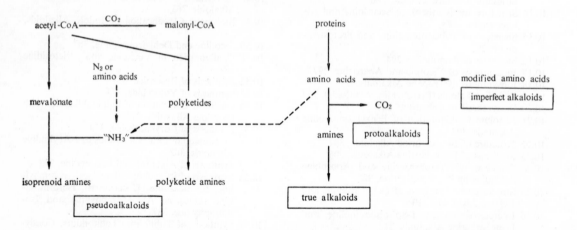

In this chapter, this type of classification is roughly followed: we discuss pseudoalkaloids, in which "ammonia" is incorporated into an isoprenoid or polyketide carbon skeleton, first, and then proceed to true alkaloids in which decarboxylated amino acids (protoalkaloids) are condensed with a non-nitrogenous structural moiety. Limitations of space make it impossible to cover all areas of alkaloid chemistry, particularly in relation to imperfect and protoalkaloids, and for further information in specific fields, readers should consult more comprehensive books[2] and reviews.[3]

---

1) M. Yamazaki, *Natural Products*, **68** (ed. Y. Kitahara, *et al.*) (in Japanese), Nankodo, 1968.
2) R. H. F. Manske, H. L. Holmes, *The Alkaloids, Chemistry and Physiology*, vols. 1–13, Academic Press; E. Leete, in *Biogenesis of Natural Products* (ed. P. Bernfeld), Pergamon Press, 1967.
3) For instance, Specialist Periodical Reports (*The Alkaloids*), The Chemical Society.

10.2   STRUCTURE OF DENDROBINE

dendrobine **1**

$C_{16}H_{25}O_2N$
mp: 135–136°
α_D: (EtOH abs.): −50.4°
pK'_a (H_2O): 7.80
ir(KBr): 1767

The structure of dendrobine **1** has been determined by three groups of chemists by chemical and spectroscopic methods.[1-3]

17

Remarks

Dendrobine **1** was first isolated from *Dendrobium nobile* Lindl.[4] Later, four related alkaloids, nobiline **2**,[1,2,5] dendroxine **3**,[6] dendramine **4**[7] and dendrine **5**[8] were isolated. These alkaloids are closely related to the picrotoxane sesquiterpenes[9] (see vol. 1). The biosynthesis of dendrobine and related compounds has been studied.[10]

1) T. Onaka, S. Kamata, T. Maeda, Y. Kawazoe, M. Natsume, T. Okamoto, F. Uchimaru, M. Shimizu, *Chem Pharm. Bull.*, **12**, 506 (1964).
2) S. Yamamura, Y. Hirata, *Tetr. Lett.*, 79 (1964); *Nippon Kagaku Zasshi*, **85**, 377 (1964).
3) Y. Inubushi, Y. Sasaki, Y. Tsuda, B. Yasui, T. Konida, J. Matsumoto, E. Katarao, J. Nakano, *Yakugaku Zasshi*, **83**, 1184 (1963); *Tetr.*, **20**, 2007 (1964); Y. Inubushi, Y. Sasaki, Y. Tsuda, J. Nakano, *Tetr. Lett.*, 1519 (1965).
4) H. Suzuki, I. Keimatsu, K. Ito, *Yakugaku Zasshi*, **52**, 1049 (1932); *ibid.*, **54**, 801 (1934).
5) T. Onaka, S. Kamata, T. Maeda, Y. Kawazoe, M. Natsume, T. Okamoto, F. Uchimura, M. Shimizu, *Chem. Pharm. Bull.*, **13**, 745 (1965).
6) T. Okamoto, M. Natsume, T. Onaka, F. Uchimura, M. Shimizu, *Chem. Pharm. Bull.*, **14**, 672 (1966).
7) Y. Inubushi, Y. Tsuda, E. Katarao, *Chem. Pharm. Bull.*, **14**, 668 (1966).
8) Y. Inubushi, J. Nakano, *Tetr. Lett.*, 2723 (1965); I. Granelli, K. Leander, *Acta. Chem. Scand.*, **24**, 1108 (1970).
9) L. A. Porter, *Chem. Rev.*, **67**, 441 (1967).
10) M. Bioliaz, D. Arigoni, *Chem. Commun.*, 633 (1969); A. Corbella, P. Gariboldi, G. Jommi, C. Scolastico, *ibid.*, 634 (1969); D. E. Edwards, J. L. Douglas, B. Mootoo, *Can. J. Chem.*, **48**, 2517 (1970); K. W. Turnbull, W. Acklin, D. Arigoni, *Chem. Commun.*, 598 (1972); A. Corbella, P. Gariboldi, G. Jommi, *ibid.*, 600 (1972); *ibid.*, 729 (1973).

10.3 SYNTHESIS OF DENDROBINE

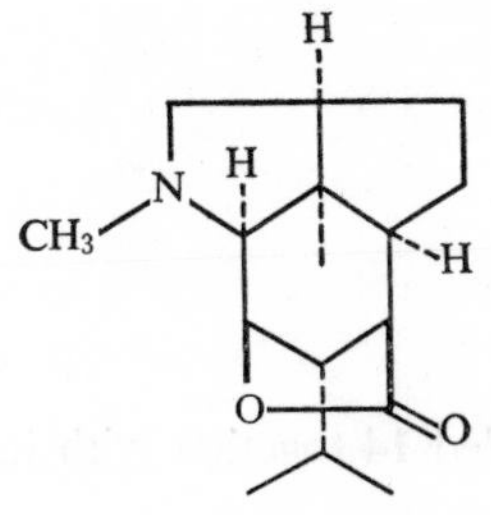

dendrobine **1**

Route I[1)]

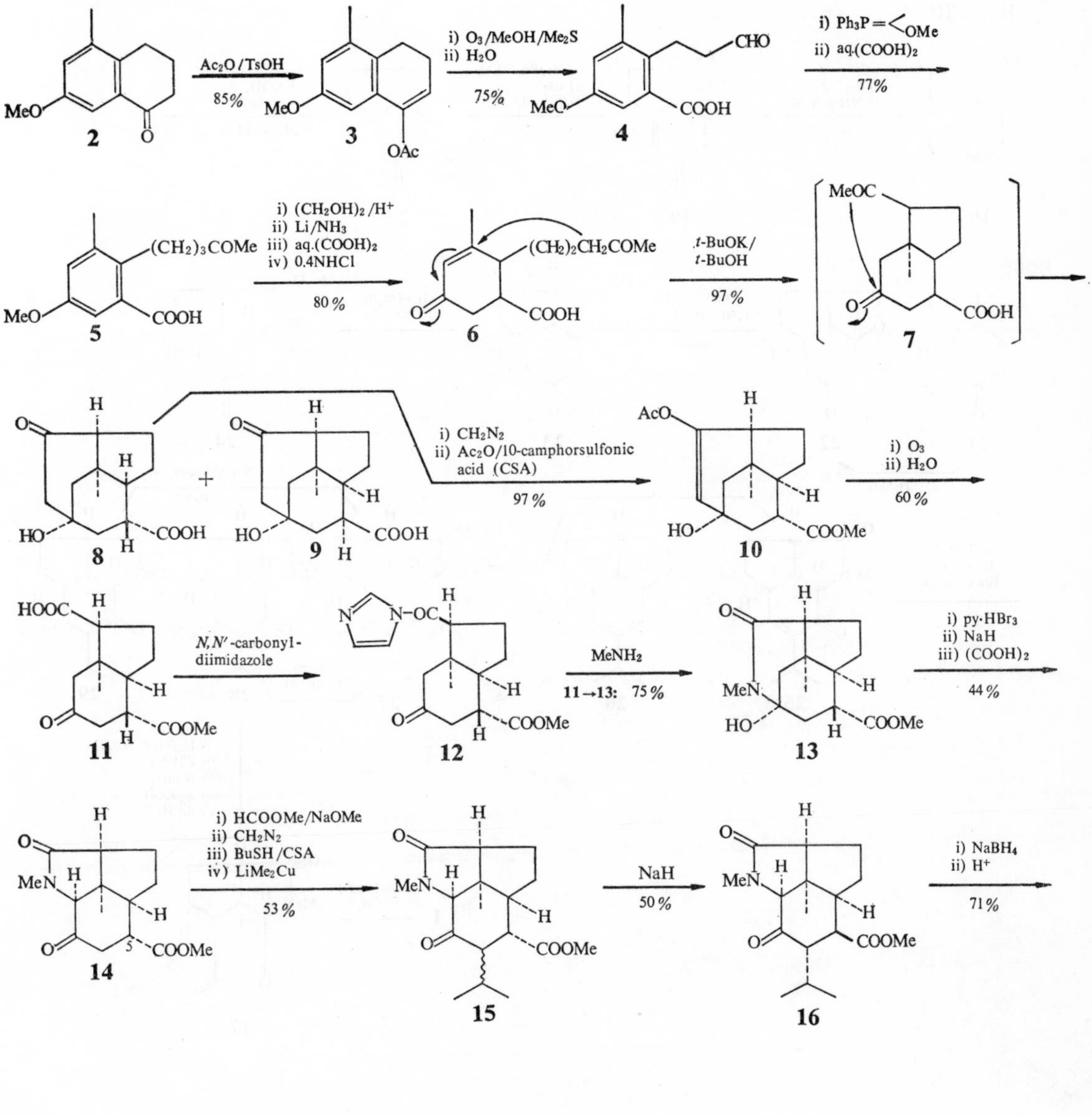

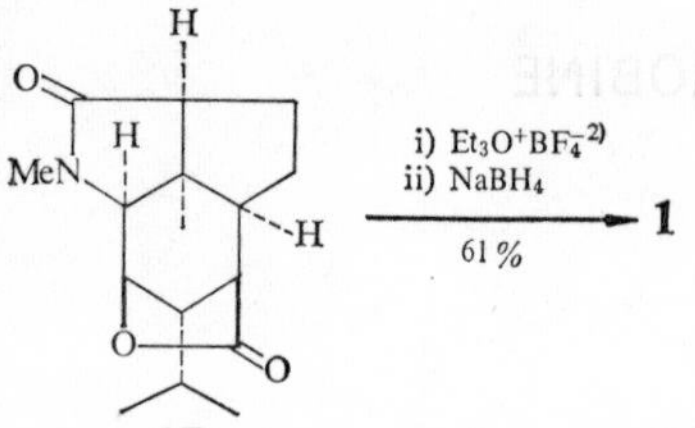

i) Et₃O⁺BF₄⁻²⁾ → written as: i) Et$_3$O$^+$BF$_4^{-2)}$
ii) NaBH$_4$

61%

17

6 ⟶ 8,9: Compounds **8** and **9** were obtained in a ratio of 8:1.

13 ⟶ 14: Anhydrous oxalic acid was used. Aqueous oxalic acid yielded **14** together with its C-5 epimer, the latter being predominant.

15 ⟶ 16: The product was a 1:1 mixture of **15** and **16**. Compound **16** and the product **17** obtained from it were identical with the products derived from natural **1**.[3]

Route II[4]

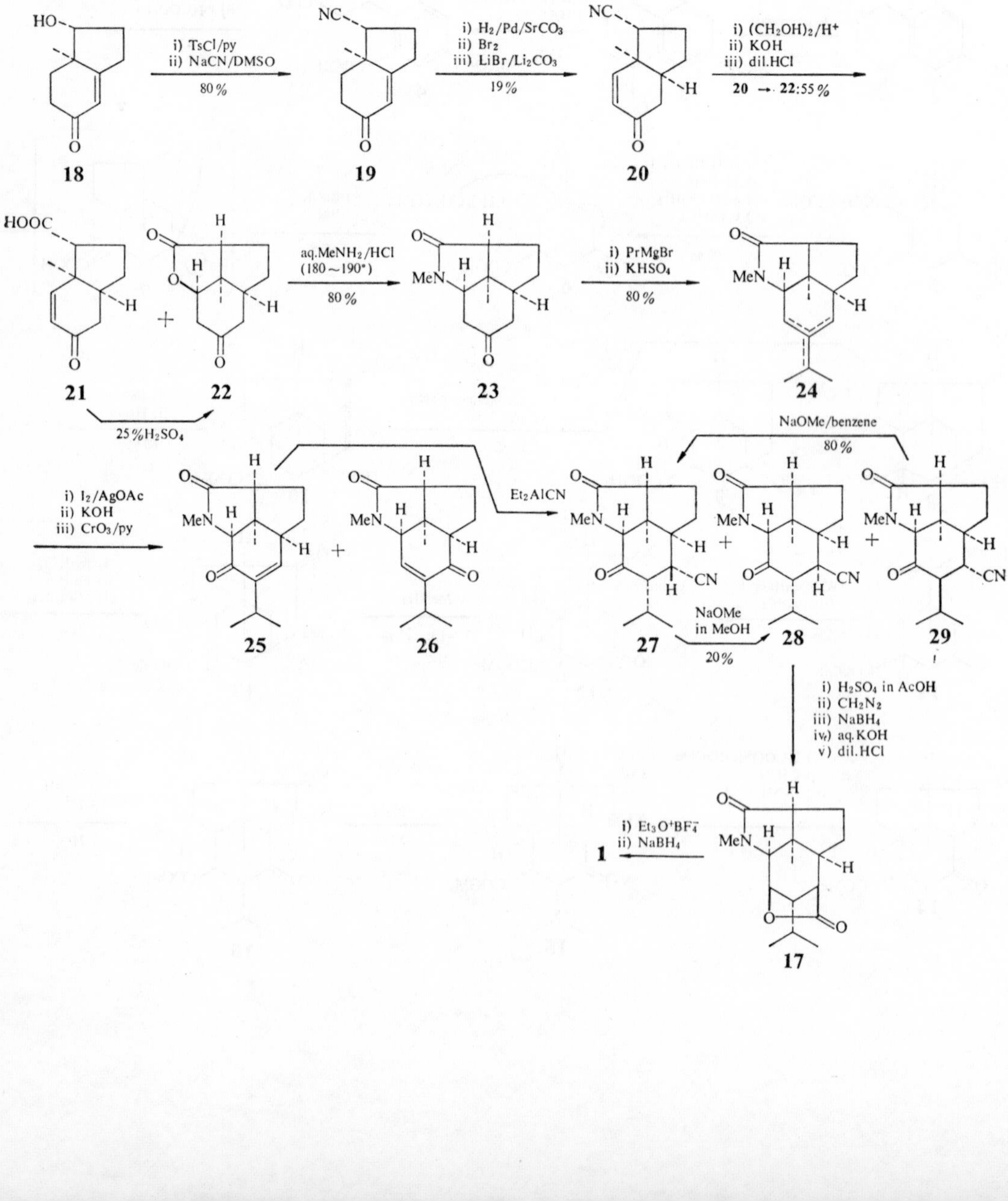

i) TsCl/py
ii) NaCN/DMSO
80%

18

i) H$_2$/Pd/SrCO$_3$
ii) Br$_2$
iii) LiBr/Li$_2$CO$_3$
19%

19

i) (CH$_2$OH)$_2$/H⁺
ii) KOH
iii) dil.HCl
20 → 22:55%

20

21 + **22**

25% H$_2$SO$_4$

aq.MeNH$_2$/HCl
(180∼190°)
80%

23

i) PrMgBr
ii) KHSO$_4$
80%

24

NaOMe/benzene
80%

i) I$_2$/AgOAc
ii) KOH
iii) CrO$_3$/py

25 + **26**

Et$_2$AlCN

27 + **28** + **29**

NaOMe
in MeOH
20%

i) H$_2$SO$_4$ in AcOH
ii) CH$_2$N$_2$
iii) NaBH$_4$
iv) aq.KOH
v) dil.HCl

i) Et$_3$O⁺BF$_4$⁻
ii) NaBH$_4$

1

17

18 ⟶ 19: A *cis* substitution reaction occurred.

19 ⟶ 20: The hydrogenation product was a mixture of *cis* and *trans* isomers in a ratio of 97.3:2.7. Compound **20** was also prepared by the following route.

23 ⟶ 24: The Grignard reaction was performed at −60 to −70°. At higher temperatures only the enolate corresponding to **23** precipitates out.

25 ⟶ 27,28,29: Nagata's hydrocyanation[5] afforded **27**, **28** and **29** in 18%, 29% and 20% yields, respectively.

1) K. Yamada, M. Suzuki, Y. Hayakawa, K. Aoki, H. Nakamura, H. Nagase, Y. Hirata, *J. Am. Chem. Soc.*, **94**, 8278 (1972).
2) R. F. Borch, *Tetr. Lett.*, 61 (1968).
3) S. Yamamura, Y. Hirata, *Tetr. Lett.*, **79** (1964); T. Onaka, S. Kamata, T. Maeda, Y. Kawazoe, M. Natsume, T. Okamoto, F. Uchimura, M. Shimizu, *Chem. Pharm. Bull.*, **12**, 506 (1964); Y. Inubushi, Y. Sasaki, Y. Tsuda, B. Yasui, T. Konita, J. Matsumoto, E. Katarao, J. Nakano, *Tetr.*, **20**, 2007 (1964).
4) Y. Inubushi, T. Kikuchi, T. Ibuka, K. Tanaka, I. Saji, K. Tokane, *Chem. Commun.*, 1252 (1972); *Chem. Pharm. Bull.*, **22**, 349 (1974).
5) W. Nagata, M. Yoshika, *Tetr. Lett.*, 1913 (1966).

10.4 STRUCTURE OF GARRYA ALKALOIDS

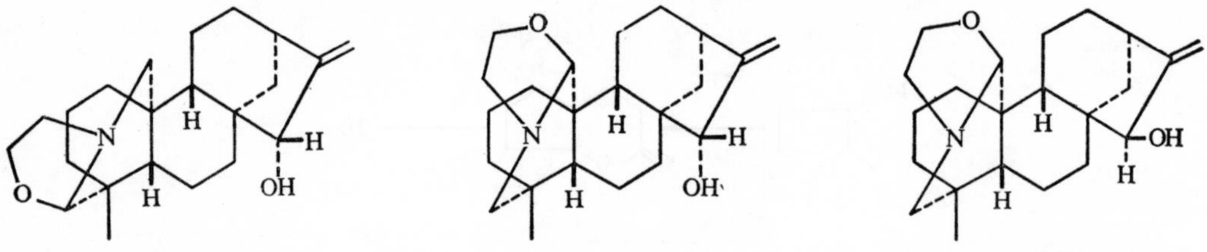

garryine **1**   veatchine **2**   garryfoline **3**

The structures of these alkaloids were established by extensive chemical investigations.[1,2]

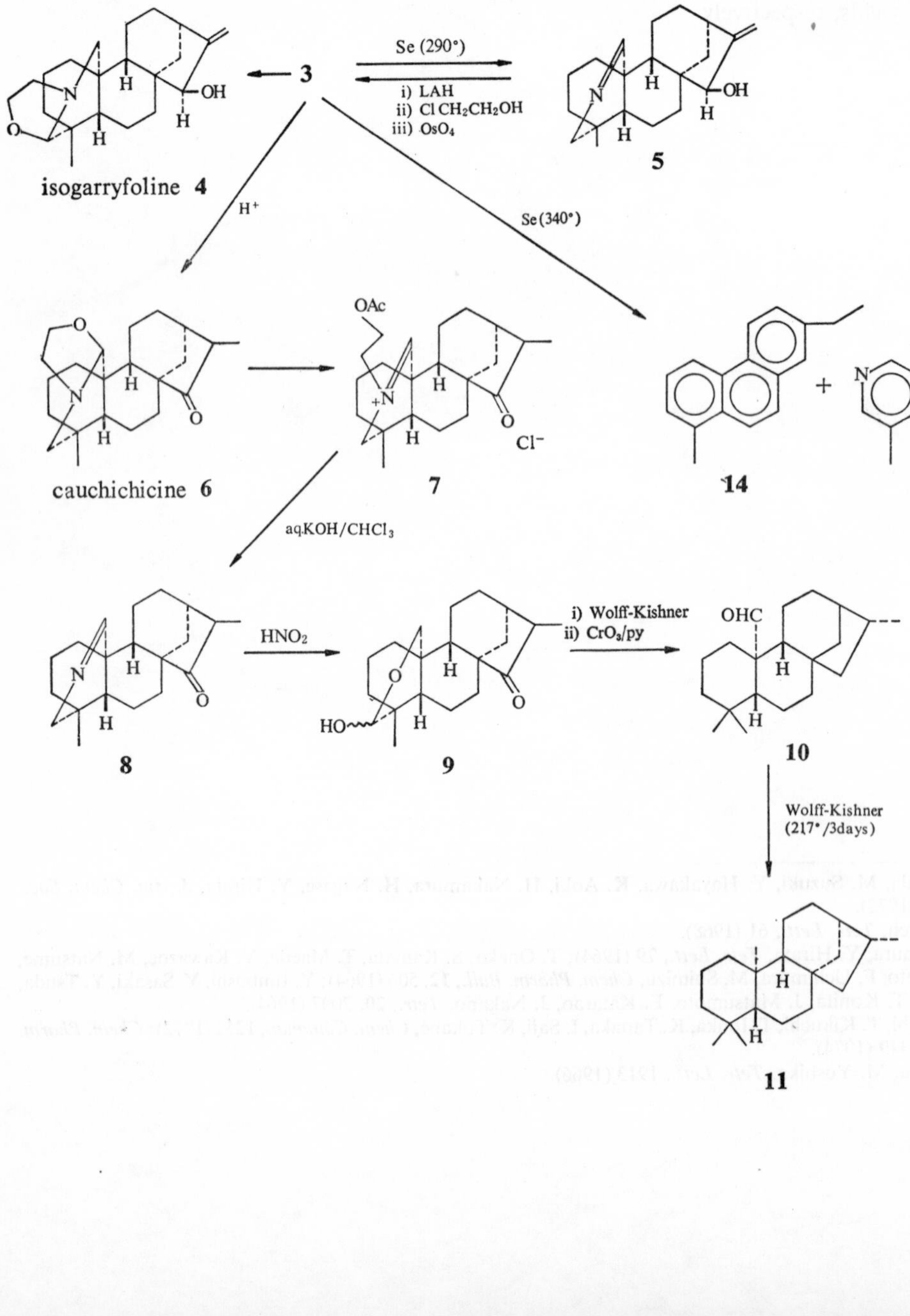

isogarryfoline **4**

Se (290°)

i) LAH
ii) ClCH₂CH₂OH
iii) OsO₄

5

H⁺

Se(340°)

cauchichicine **6**

7

14   +   **15**

aq.KOH/CHCl₃

8   HNO₂   **9**

i) Wolff-Kishner
ii) CrO₃/py

10

Wolff-Kishner
(217°/3days)

11

1) K. Wiesner, M. Götz, D. L. Simmons, R. A. Fahmy, K. S. Atwal, H. W. Lu, H. Tsai, T. Y. R. Tsai,
   H. J. Liu (1982).
2) S. W. Pelletier (1980).
3) S. Yunusov, S. Y. Yunusov, Nat. Prod., 29 (1980); Z. Djarmati, R. Fraser-Reid, K. Ishiguro, M. Sorm, K. Wiesner,
   T. Giessman, P. Perel, A. Coulson, Chem. Pharm. Bull., 73, 9527 (1984); Y. Yamamura, W. Sasaki, Y. Tsuda,
   S. Yamaura, Y. Watanabe, Y. Yamamura, T. Kaneko, J. Nat. Prod., 596, 29 (1980).
4) Y. Tsuda, T. Kaneko, Y. Sata, M. Tanaka, L. Sato, W. Watanabe (1981); K. Ishiguro, M. Sorm, Sato,
   W. Yamamura, T. Kaneko, S. Yamaura, S. Yamamura, J. Nat. Prod., 1911 (1980).

3 ⟶ 4: This conversion probably proceeds as shown below. A similar conversion of **2** (to **1**) also occurs.

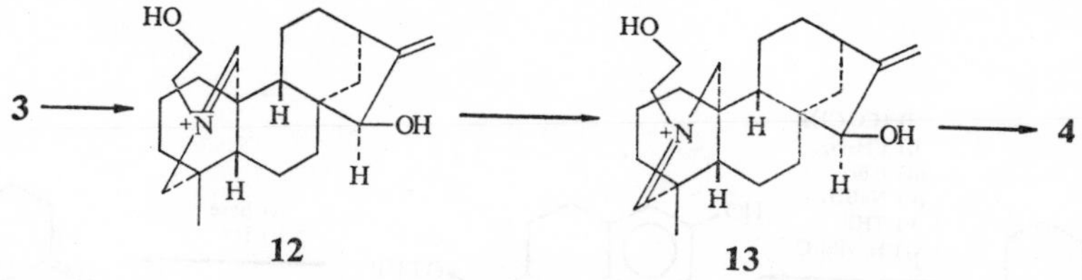

3 ⟶ 6: This type of conversion is widely observed in allyl alcohols such as this, e.g. **1** and **2**, but the rate of reaction depends on the configuration of the hydroxyl group. The reaction sometimes proceeds catalytically with palladium/carbon.

7 ⟶ 8: **8** is in equilibrium with its allylic azomethine.

10 ⟶ 11: **11** is identical with (−)-β-dihydrokaurene[3] (= stevane B[4]) obtained from (−)-kaurene of known absolute configuration. The same absolute configuration was proposed on the basis of optical rotatory dispersion studies.[5] Atisine was independently shown by chemical correlation with abietic acid[6] to have the antipodal configuration.

Remarks

The alkaloids **1**, **2** and **3** were isolated from *Garrya veatchii* and *Garrya laurifolia* (Coridaceae).

1) K. Wiesner, Z. Valenta, *Progress in the Chemistry of Organic Natural Products* (ed. L. Zechmeister), vol. 16, p. 26, Springer Verlag, (1958); S. W. Pelletier, L. H. Keith, *The Alkaloids, Chemistry and Physiology* (ed. R. H. F. Manske), vol. 12, p. 2, Academic Press, (1970).

2) H. Vorbrueggen, C. Djerassi, *J. Am. Chem. Soc.*, **84**, 2990 (1962).

3) L. H. Briggs, B. F. Cain, B. R. Davis, J. K. Wilmshurst, *Tetr. Lett.*, No. 8, 8 (1959).

4) F. Dolder, H. Licht, E. Mossetig, P. Quitt, *J. Am. Chem. Soc.*, **82**, 246 (1960).

5) C. Djerassi, P. Quitt, E. Mossetig, R. C. Cambie, R. S. Rutledge, L. H. Briggs, *J. Am. Chem. Soc.*, **83**, 3720 (1961).

6) L. H. Zalkow, N. N. Girotra, *J. Org. Chem.*, **28**, 2037 (1963); W. A. Ayer, C. E. McDonald, G. G. Iverach, *Tetr. Lett.*, 1095 (1963).

10.5   SYNTHESIS OF VEATCHINE AND GARRYINE

Route I[1,2]

4 → 5: This procedure had been developed previously.[3] Only one stereoisomer of 4 under-
went cyclization to 5.

5 → 6: 6 was also obtained by a similar carbomethoxylation of the benzoate of 5. Carbo-

1) S. Masamune, *J. Am. Chem. Soc.*, **86**, 288 (1964).
2) S. Masamune, *J. Am. Chem. Soc.*, **86**, 290 (1964).
3) S. Masamune, *J. Am. Chem. Soc.*, **83**, 1009 (1961).

methoxylation using dimethylcarbonate occurred at the other side of the carbonyl group. The stereochemistry of **6** was confirmed by chemical correlation.

9 ⟶ 10: The first two steps represent the Barton reaction.[4]

11 ⟶ 12: The second step represents allylic oxygenation by singlet oxygen.[5]

13 ⟶ 2 ⟶ 1: This followed a route previously established with compounds of natural origin.[6]

Route II[7,8]

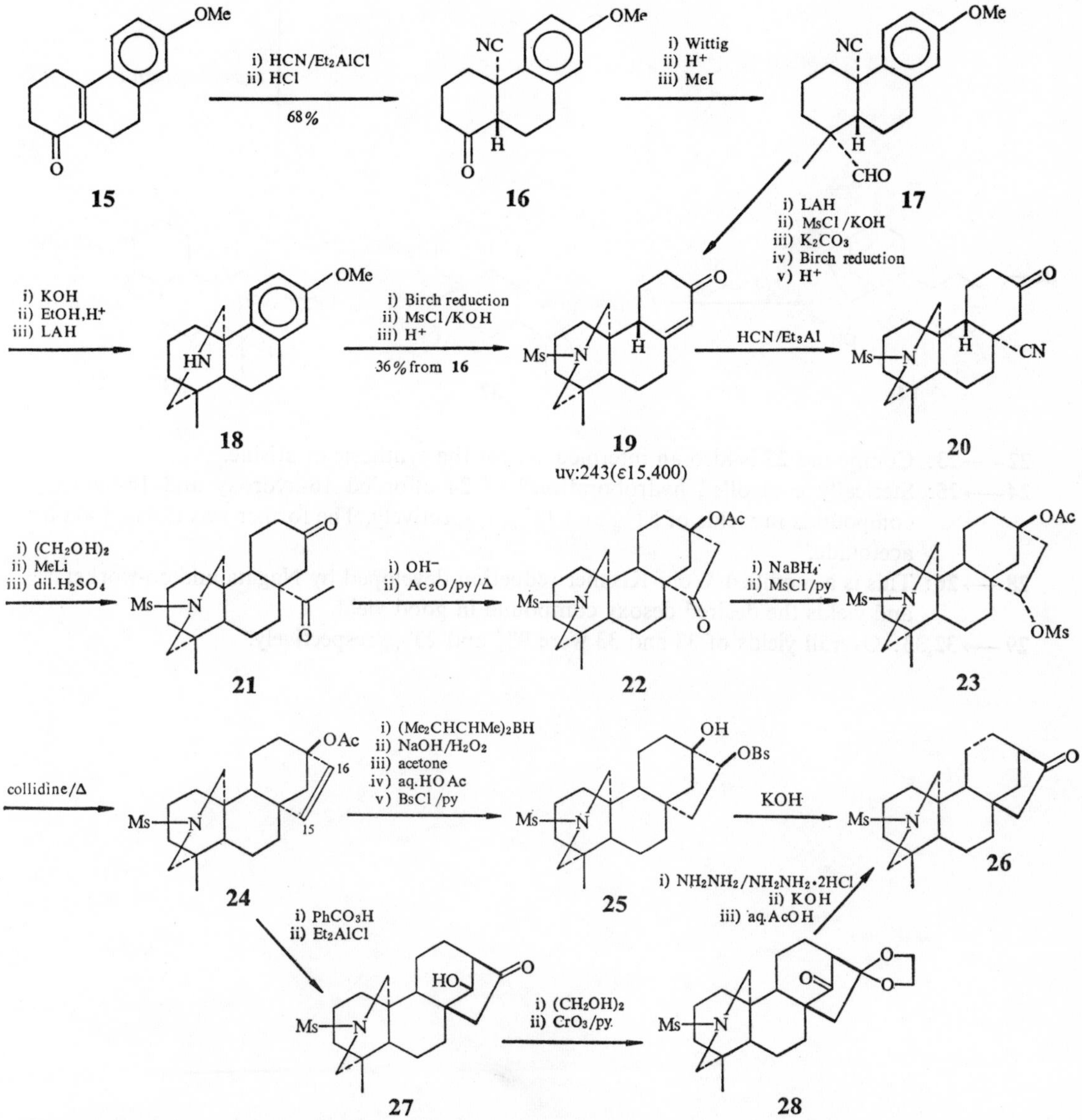

uv:243(ε15,400)

4) For reviews, see A. L. Nussbaum, C. H. Robinson, *Tetr.*, **17**, 35 (1961); M. Akhtar, *Advan. Photochem.*, **2**, 263 (1964).

5) For reviews, see C. S. Foote, *Accounts Chem. Res.*, **1**, 104 (1968); K. Gollnick, *Advan. Photochem.*, **6**, 1 (1968).

6) K. Wiesner, W. I. Taylor, S. K. Fidgor, M. F. Bartlett, J. R. Armstrong, E. A. Edward, *Chem. Ber.*, **86**, 800 (1953).

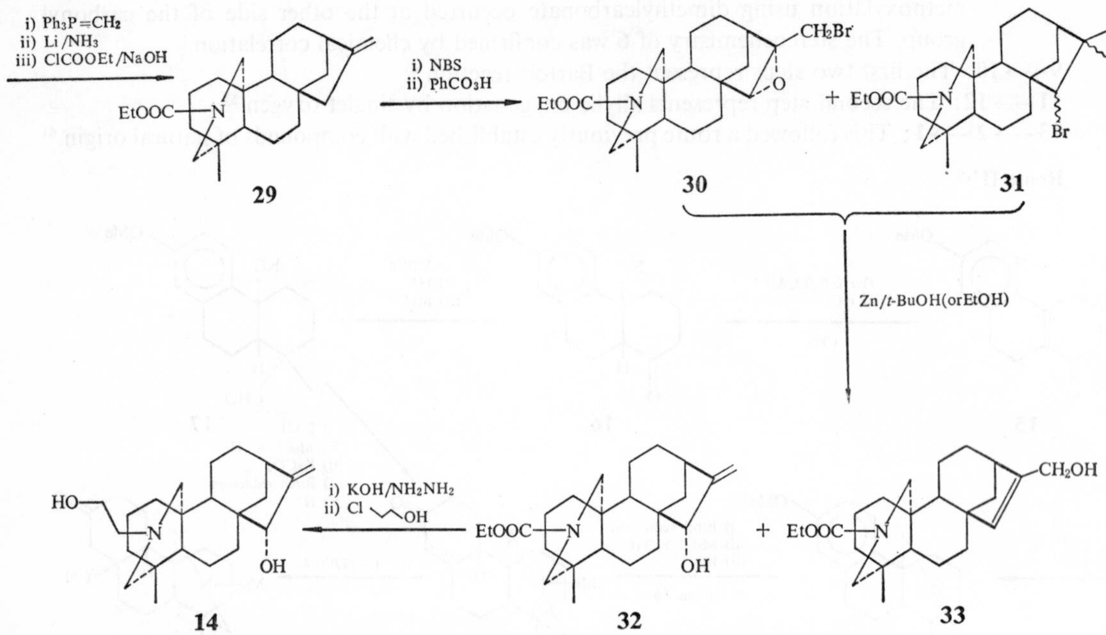

i) Ph$_3$P=CH$_2$
ii) Li/NH$_3$
iii) ClCOOEt/NaOH

29

i) NBS
ii) PhCO$_3$H

EtOOC—N **30** + EtOOC—N **31** (CH$_2$Br, O, Br)

Zn/t-BuOH(orEtOH)

HO **14** OH

i) KOH/NH$_2$NH$_2$
ii) Cl⌣OH

EtOOC—N **32** OH + EtOOC—N **33** CH$_2$OH

22 ⟶ 23: Compound **23** is also an intermediate for the synthesis of atisine.

24 ⟶ 25: Sterically controlled hydroboration[9] of **24** afforded 16-hydroxy and 15-hydroxy compounds in yields of 53% and 18%, respectively. The former was isolated via its acetonide.

28 ⟶ 26: This is a modified Wolff-Kishner reduction developed by Nagata and co-workers[10] and yields the desired desoxy compound in good yield.

29 ⟶ 32,33: Overall yields of **32** and **33** were 9% and 23%, respectively.

7) W. Nagata, T. Sugasawa, M. Narisada, T. Wakabayashi, Y. Hayase, *J. Am. Chem. Soc.*, **85**, 2341 (1963).

8) W. Nagata, M. Narisada, T. Wakabayashi, T. Sugasawa, *J. Am. Chem. Soc.*, **89**, 1499 (1967).

9) H. C. Brown, G. Zweifel, *J. Am. Chem. Soc.*, **82**, 3222 (1960); G. Zweifel, N. R. Aryangar, H. C. Brown, *ibid.*, **85**, 2072 (1963).

10) W. Nagata, H. Itazaki, *Chem. Ind.*, 1194 (1964).

10.6   CHEMICAL CORRELATION OF ATISINE AND
ISOATISINE WITH VEATCHINE

After determination of the structures of veatchine **3** and other Garrya alkaloids, the structures of atisine **1** and isoatisine **2** (aconitum alkaloids) were postulated. The correlation shown below proved the identity of the stereochemistry of most of the carbon atoms in the two groups of compounds.[1,2]

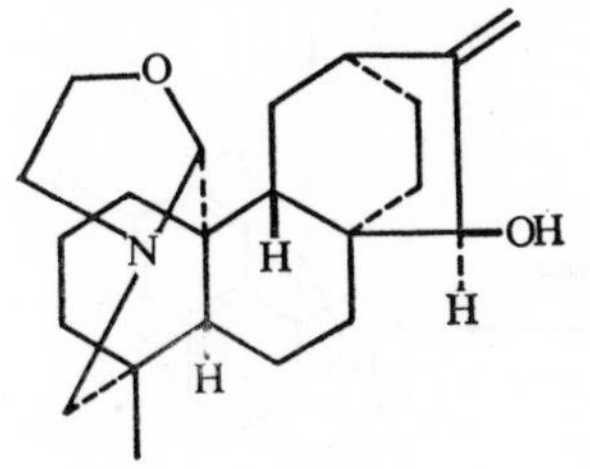

atisine **1** 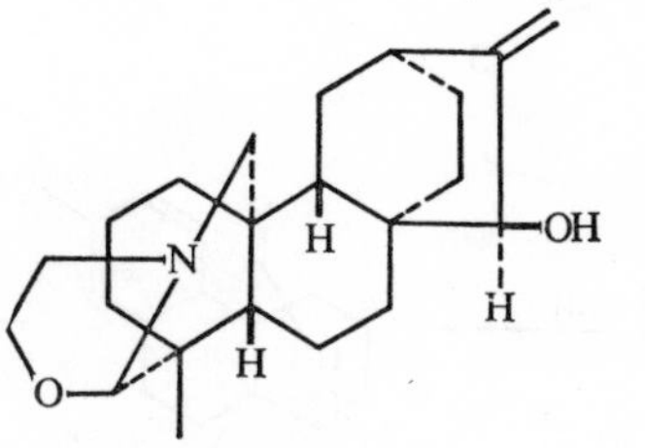   isoatisine **2** 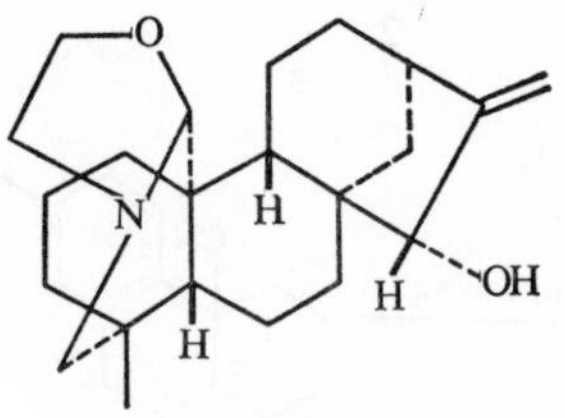   veatchine **3**

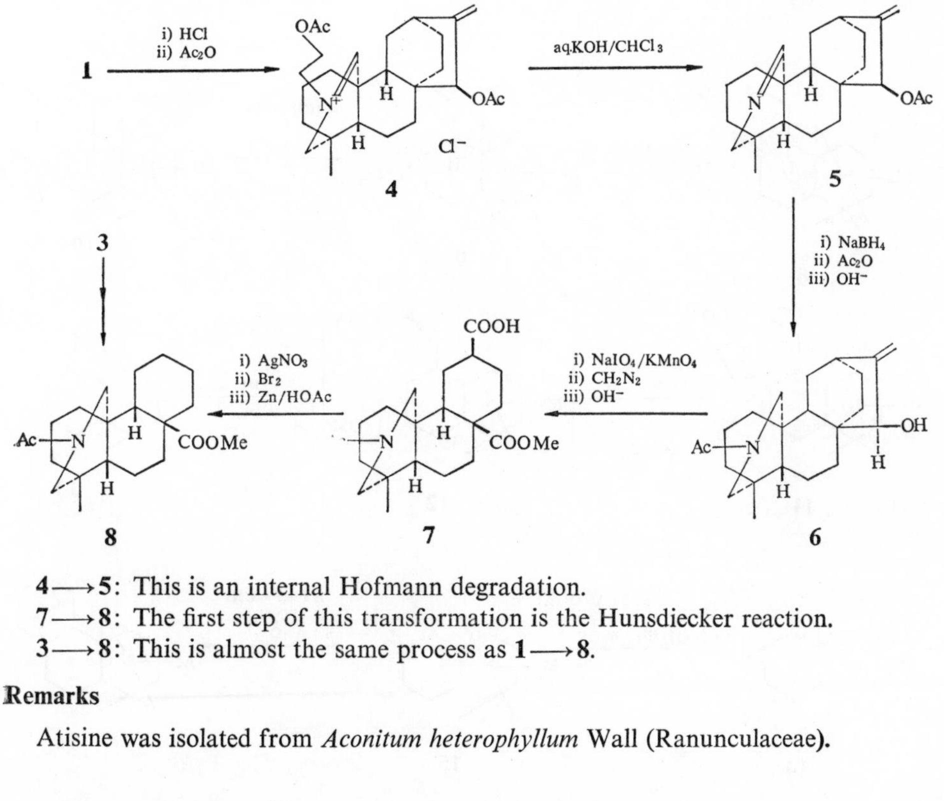

4 ⟶ 5: This is an internal Hofmann degradation.

7 ⟶ 8: The first step of this transformation is the Hunsdiecker reaction.

3 ⟶ 8: This is almost the same process as **1 ⟶ 8**.

Remarks

Atisine was isolated from *Aconitum heterophyllum* Wall (Ranunculaceae).

1) D. Dvornik, O. E. Edwards, *Can. J. Chem.*, **35**, 860 (1957).

2) H. Vorbrueggen, C. Djerassi, *J. Am. Chem. Soc.*, **84**, 2990 (1962).

10.7 SYNTHESIS OF ATISINE

Route I[1]

1) W. Nagata, T. Sugasawa, M. Narisada, T. Wakabayashi, Y. Hayase, *J. Am. Chem. Soc.*, **85**, 2342 (1963); *ibid.*, **89**, 1483 (1967).

2 ⟶ 3: The synthesis of **2** has already been described.[2] For this reaction the less basic die-
thylaluminium chloride was more effective than triethylaluminium, and the product
was a mixture of *cis* and *trans* isomers. After repeated recrystallization and treatment
with hydrochloric acid, **3** was obtained in 68% yield.

5 ⟶ 6 ⟶ 7: When aqueous alcohol was used as a solvent for **5**, the product was a 1:1 mixture
of **6** (R=H) and **6** (R=Me or Et). The *N*-methyl derivative of **7** has also been synthesized.[3]

8 ⟶ 9: The product was a 60:1 mixture of *trans* and *cis* compounds. The configuration of the
cyano group was assigned from a consideration of the ir absorption of –CN and the
dipole moment.

17 ⟶ 18: Epoxidation of the crude bromide **17** afforded a mixture of epimeric epoxides which
were separated after treatment with Zn/AcOH. The overall yields of **18** and **19** from
16 were 14% and 10%, respectively.

18 ⟶ 1: This conversion has been accomplished using the optically active compound.[4]

2) R. Robinson, E. Schlittler, *J. Chem. Soc.*, 1288 (1935); G. Stork, *J. Am. Chem. Soc.*, **69**, 2936 (1947); A. J.
Birch, H. Smith, R. E. Thornton, *J. Chem. Soc.*, 1339 (1957).

3) I. Iwai, A. Ogiso, B. Shimizu, *Chem. Ind.*, 1288 (1962); *Chem. Pharm. Bull.*, **11**, 770, 774 (1963); I. Iwai,
A. Ogiso, *Chem. Ind.*, 1084 (1963); *Chem. Pharm. Bull.*, **12**, 820 (1964).

4) S. W. Pelletier, W. A. Jacobs., *J. Am. Chem. Soc.*, **78**, 4144 (1956).

Route II[5,6]

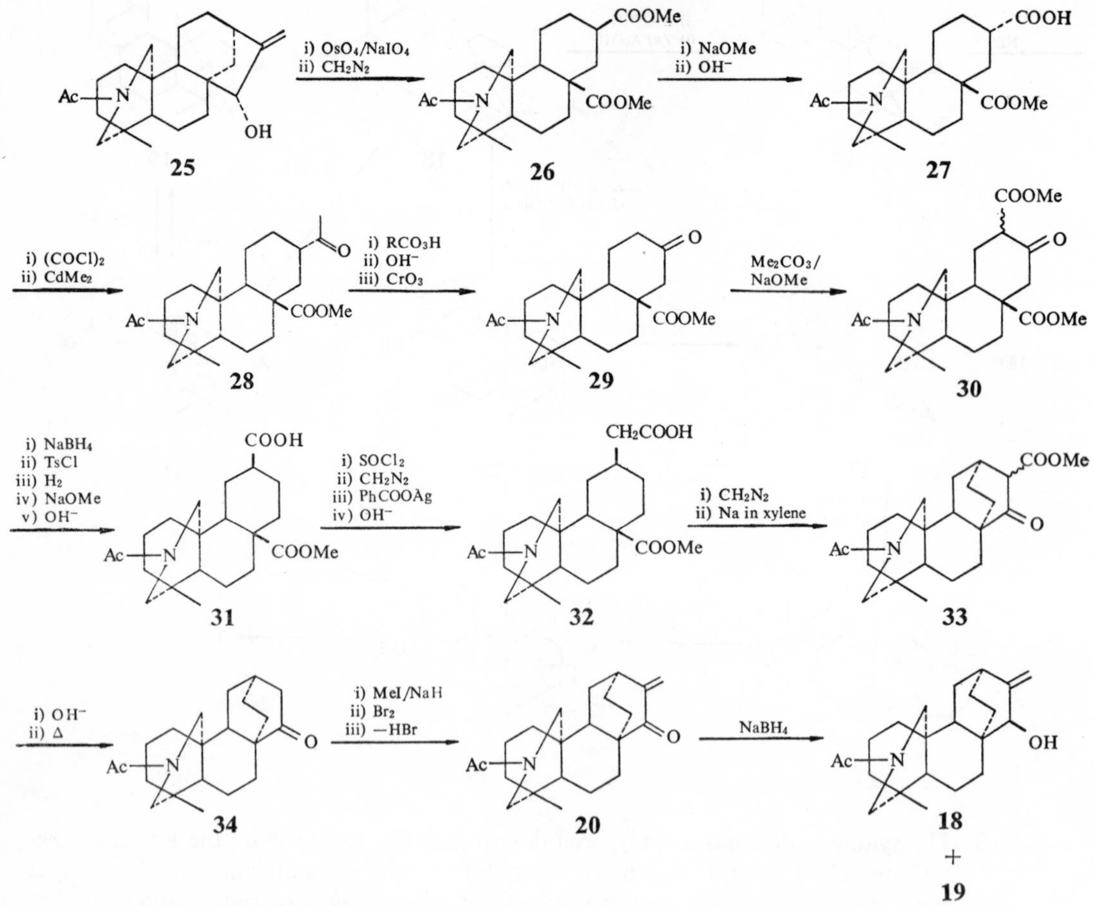

i) OsO₄/NaIO₄
ii) CH₂N₂

i) NaOMe
ii) OH⁻

25

26

27

i) (COCl)₂
ii) CdMe₂

i) RCO₃H
ii) OH⁻
iii) CrO₃

Me₂CO₃/
NaOMe

28

29

30

i) NaBH₄
ii) TsCl
iii) H₂
iv) NaOMe
v) OH⁻

i) SOCl₂
ii) CH₂N₂
iii) PhCOOAg
iv) OH⁻

i) CH₂N₂
ii) Na in xylene

31

32

33

i) OH⁻
ii) Δ

i) MeI/NaH
ii) Br₂
iii) —HBr

NaBH₄

34

20

18
+
19

Compound **25** was prepared from veatchine[7] and since veatchine itself has been synthesized, the present route constitutes a formal synthesis of atisine. Compound **31** was identical with an authentic sample derived from **1**.[8]

5) S. Masamune, *J. Am. Chem. Soc.*, **86**, 291 (1964).
6) S. W. Pelletier, P. C. Parthasarathy, *Tetr. Lett.*, 205 (1963).
7) H. Vorbrueggen, C. Djerassi, *J. Am. Chem. Soc.*, **84**, 2990 (1962); K. Wiesner, J. R. Armstrong, M. F. Wartlett, J. A. Edwards, *ibid.*, **76**, 6068 (1954).
8) S. W. Pelletier, *J. Am. Chem. Soc.*, **82**, 2398 (1960).

10.8 STRUCTURE OF ACONITINE

aconitine **1**

Aconitine **1** and its *N*-methyl analog mesaconitine are highly toxic alkaloids isolated from various *Aconitum* species. After efforts over a long period by many organic chemists,[1] their structures were determined both chemically[2] and by x-ray analysis.[3]

1——→4: Oxonitine **4** was also obtained from mesaconitine. The evidence for the presence of the *N*-formyl group is strong.[2,4]

14——→15: Since the two double bonds in **14** are facing each other, the reaction proceeds stereospecifically to yield **15**.

1) E. S. Stern, *The Alkaloids, Chemistry and Physiology* (ed. H. F. Manske), vol. 4, p. 274, Academic Press, (1954); *ibid.*, vol. 7, p. 473 (1960).
2) K. Wiesner, M. Götz, D. L. Simmons, L. R. Fowler, F. W. Batchelor, R. F. C. Brown, G. Buchi, *Tetr. Lett.*, 15 (1959); F. W. Batchelor, R. F. C. Brown, G. Buchi, *ibid.*, 1 (1960); K. Wiesner, M. Götz, D. L. Simmons, L. R. Fowler, *Coll. Czech. Chem. Commun.*, **28**, 2462 (1963).
3) M. Przybylska, L. Marion, *Can. J. Chem.*, **37**, 1116 ,1843 (1959).
4) R. B. Turner, J. P. Teschke, M. S. Gibson, *J. Am. Chem. Soc.*, **82**, 5182 (1960).

10.9   STRUCTURE OF JERVINE

jervine **1**

$C_{27}H_{39}O_3N$
mp: 238°
α_D(EtOH): −47°
uv: 252(ϵ 14,000) 360(ϵ 70)
mur of diacetate:

0.89(3H, d, J=7)	2.02(OAc)
1.03(3H, S)	2.10(NAc)
1.07(3H, d, J=6)	4.67(1H, br)
2.25(3H, S)	5.43(1H)

The structure of jervine was studied for a long time by many groups of chemists[1] and finally a C-nor-D-homo steroid structure **1** (except for the stereochemistry) was proposed[2] based on accumulated degradation data, as well as the structure of veratramine **11**, which had been correlated with **1**.[3] The stereochemistry of the carbocyclic ring system was established by chemical correlation with hecogenin **3**.[4] However, the configuration at the side chain remained unresolved[5] until the structure of veratrobasine **10**, which was identical with jervin-11β-ol, was determined by x-ray analysis[6,7] although a total synthesis had been accomplished.[8]

hecogenin **3**

1) For reviews, see L. F. Fieser, M. Fieser, *Steroids*, p. 870, Reinhold, 1959; S. M. Kupchan, A. W. By, *Alkaloids* (ed. R. H. F. Manske), vol. 10, p. 193, Academic Press, 1968.
2) J. Fried, O. Wintersteiner, M. Moore, B. M. Iselin, A. Klingsberg, *J. Am. Chem. Soc.*, **73**, 2970 (1951).
3) T. Masamune, Y. Mori, M. Takasugi, A. Murai. *Tetr. Lett.*, 913 (1964).
4) H. Mitsuhashi, Y. Shimizu, *Tetr.*, **19**, 1027 (1963); T. Masamune, M. Takasugi, Y. Mori, *Tetr. Lett.*, 489 (1965).

1 → i) MeI ii) fragmentation → **6** → Wolff-Kishner → **7** → **8** (antipodal to 7) ← D-(+)-citronellal **9**

veratrobasine **10** → HCl → veratramine **11** → i) Ac₂O ii) H₂/Pd iii) CrO₃ → **12**

(13 and 19 are considered to be epimeric at C-17.)

---

5) T. Masamune, M. Takasugi, A. Murai, K. Kobayashi, *J. Am. Chem. Soc.*, **89**, 4521 (1967); S. Okuda, K. Tsuda, H. Kataoka, *Chem. Ind.*, 512 (1961); J. W. Scott, L. J. Durham, H. A. P. de Jongh, U. Burckhardt, W. S. Johnson, *Tetr. Lett.*, 2381 (1967).

6) S. M. Kupchan, M. I. Suffness, *J. Am. Chem. Soc.*, **90**, 2730 (1968).

7) G. N. Reeke, Jr., R. L. Vincen, W. N. Lipscomb, *J. Am. Chem. Soc.*, **90**, 1663 (1968).

8) For the synthesis of jervine, see 10.10.

Remarks

Jervine **1** was isolated from *Veratrum album* together with various *Veratrum* alkaloids.[9] Jervine and veratramine belong to the Jerveratrum group, some members of which are illustrated below.

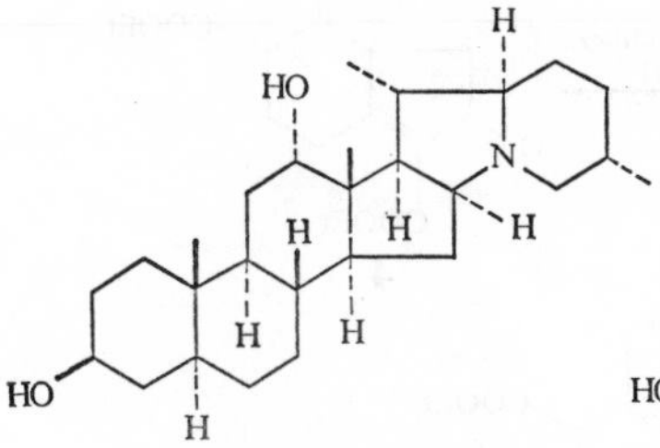

rubijervine **21**[10] 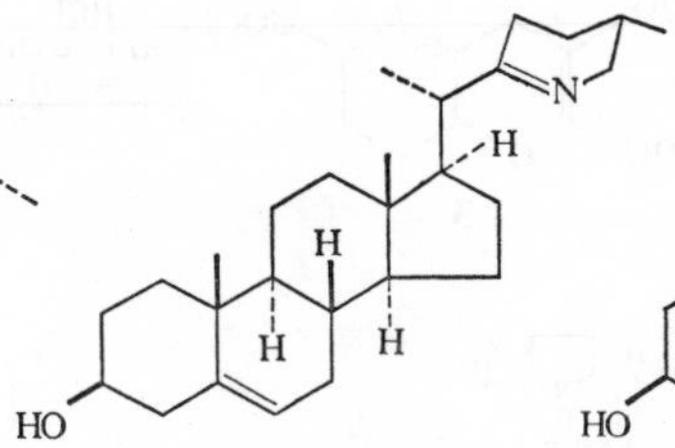   verazine **22**[11] 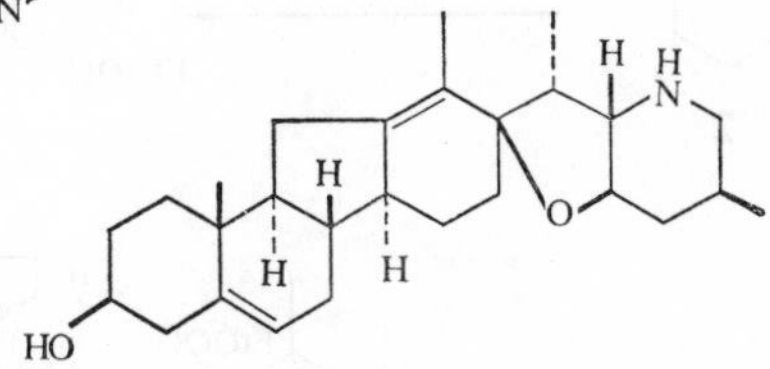   11-deoxojervine **23**[3]

9) E. Simon, *Pogg. Ann.*, **41**, 569 (1837).
10) E. Höhne, K. Schreiber, H. Ripperger, H. H. Worch, *Tetr.*, **22**, 673 (1966).
11) G. Adam, K. Schreiber, J. Tomko, A. Vassová, *Tetr.*, **23**, 167 (1967).

10.10 SYNTHESIS OF JERVINE

A representative jerveratrum alkaloid jervine **1** has recently been synthesized.[1,2)]

1) W. S. Johnson, J. M. Cox, D. W. Graham, H. W. Whitelock, Jr., *J. Am. Chem. Soc.*, **89**, 4524 (1967).
2) T. Masamune, M. Takasugi, A. Murai, K. Kobayashi, *J. Am. Chem. Soc.*, **89**, 4521 (1967).

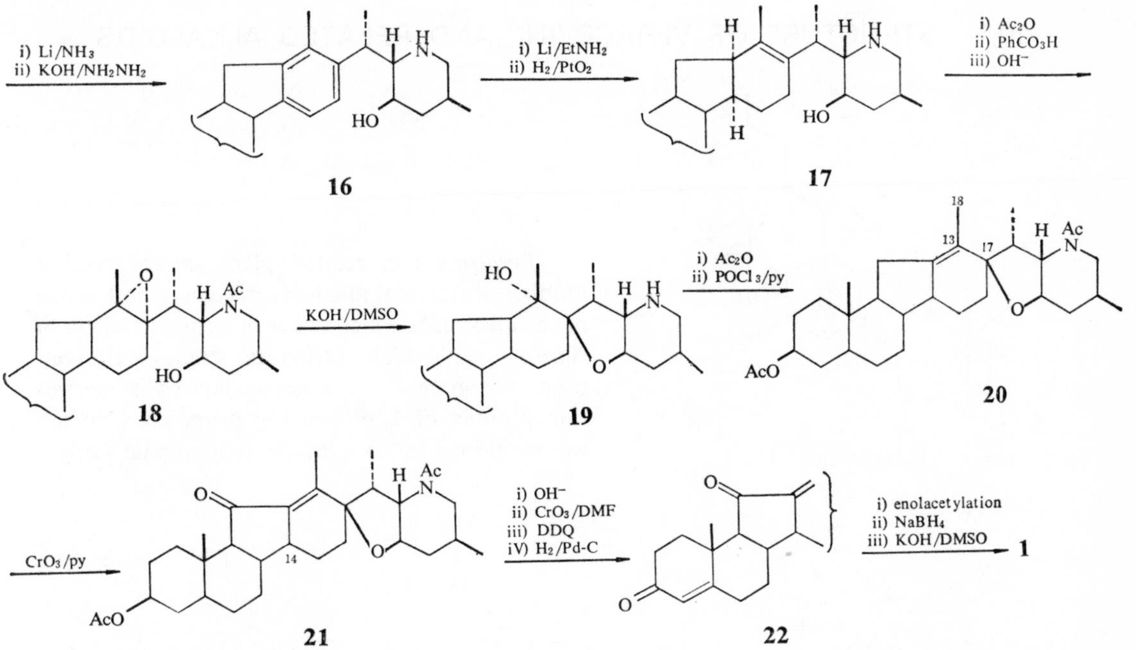

i) Li/NH₃
ii) KOH/NH₂NH₂

16

i) Li/EtNH₂
ii) H₂/PtO₂

17

i) Ac₂O
ii) PhCO₃H
iii) OH⁻

18

KOH/DMSO

19

i) Ac₂O
ii) POCl₃/py

20

CrO₃/py

21

i) OH⁻
ii) CrO₃/DMF
iii) DDQ
iV) H₂/Pd-C

22

i) enolacetylation
ii) NaBH₄
iii) KOH/DMSO

1

8 ⟶ 9: 9 was resolved as the 1-α-(1-naphthyl)ethylamine salt. The resulting optically active 9 was identical with the authentic compound derived from veratramine.[1,3]

11 ⟶ 12: 12 has also been obtained by the degradation of hecogenine[4] and veratramine.[5] It is a key intermediate in the synthesis of veratramine.[5]

13 ⟶ 14: 14 and its C-22 epimer were present in yields of 5% and 30%, respectively, in the diastereomeric mixture. However, the epimer can be equilibrated with NaOMe to give a 2:1 mixture with 14, so the overall yield of 14 was thus 15%.

The reagent for this reaction was obtained as follows:

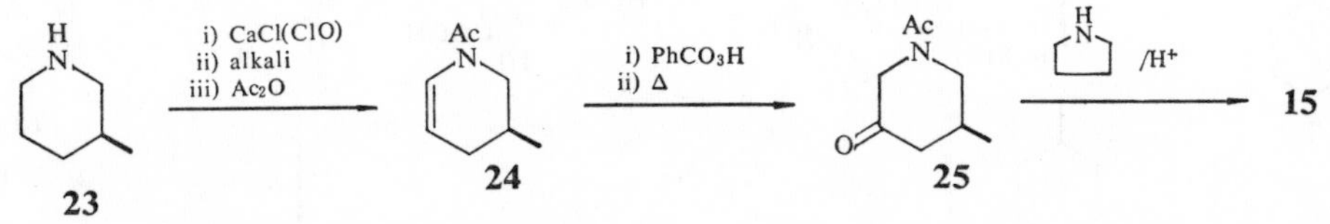

23

i) CaCl(ClO)
ii) alkali
iii) Ac₂O

24

i) PhCO₃H
ii) Δ

25

/H⁺

15

19 ⟶ 20: Dehydration afforded 20 (30%) with its $\Delta^{13(18)}$ isomer (40%). 20 is identical with the compound derived from 1.[6]

20 ⟶ 21: 21 was obtained in 1% yield. The major product was the 14-hydroxy compound.

3) R. W. Franck, G. P. Rizzi, W. S. Johnson, *Steroids*, 463 (1964).
4) H. Mitsuhashi, K. Shibata, *Tetr. Lett.*, 2281 (1964); W. F. Johns, I. Laos, *J. Org. Chem.*, 30, 4220 (1965).
5) W. S. Johnson, H. A. P. de Jongh, C. E. Coverdale, J. W. Scott, U. Burckhardt, *J. Am. Chem. Soc.*, 89, 4523 (1967).
6) T. Masamune, N. Sato, K. Kobayashi, I. Yamazaki, Y. Mori, *Tetr.*, 23, 1591 (1967).

10.11   STRUCTURE OF VERACEVINE AND RELATED ALKALOIDS

veracevine **1**

Veratrum and related plant species yield a number of steroidal alkaloids.[1] Veracevine **1** is one of the alkamines which occur in nature as esters of simple organic acids. Structural studies were concentrated on cevine **2**, a base-catalyzed isomerization product of **1**,[1,2] and the proposed structure was supported by an x-ray crystallographic study.[3]

1) V. Prelog, O. Jeger, (ed. R. H. F. Manske), *The Alkaloids*, vol. 3, p. 247, Academic Press (1953); *Idem., ibid.,* vol. 7, p. 363 (1960), S. M. Kupchan, W. A. Arnold, *ibid.,* vol. 10, p. 193 (1968).
2) D. H. R. Barton, O. Jeger, V. Prelog, R. B. Woodward, *Experientia,* **10**, 81 (1954); S. M. Kupchan, W. S. Johnson, S. Rajagopalan, *Tetr.,* **7**, 47 (1959).
3) W. T. Eeles, *Tetr. Lett.,* 24 (1960).

1) **2** consumes 2 moles of lead tetraacetate smoothly and 6.5 moles slowly. On the other hand, cevine triacetate consumes only one mole of the reagent. A model experiment[4] revealed that while β-dialkylaminoalcohols are resistant to oxidation, tertiary amino ketones consume more than one mole of the reagent at room temperature.

4) D. H. R. Barton, C. J. W. Brooks, J. S. Fawcett, *J. Chem. Soc.*, 2137 (1954).

2) The sequence **1** ⟶ **17** ⟶ **2** occurs during the hydrolysis of ester alkaloids.[5] Similar transformations are found with other closely related alkamines: germine ⟶ isogermine ⟶ pseudogermine;[6] zygadenine ⟶ isozygadenine ⟶ pseudozygadenine;[7] protoverine ⟶ isoprotoverine ⟶ pseudoprotoverine.[8,9]

3) The reaction **1** or **2** ⟶ **18** has been represented as follows:[10]

4) The conversion **20** ⟶ **21** provides strong evidence for the stereochemistry of the four tertiary hydroxyl groups.[11]

Ang=angeloyl

ir: 1730, 1715, 1695, 1639, 1597
uv(MeOH): 273, 325
uv(NaOH): 240, 282, 374

5) S. M. Kupchan, D. Lavie, C. V. DeLiwala, B. Y. A. Andoh, *J. Am. Chem. Soc.*, **75**, 5519 (1953); S. W. Pelletier, W. A. Jacobs, *ibid.*, **75**, 3248 (1953).
6) S. M. Kupchan, C. R. Narayanan, *J. Am. Chem. Soc.*, **81**, 1913 (1959).
7) S. M. Kupchan, C. V. DeLiwala, *J. Am. Chem. Soc.*, **75**, 1025 (1953).
8) H. Auterboff, F. Gunther *Arch. Pharm.*, **288**, 455 (1955).
9) S. M. Kupchan, C. I. Ayres, M. Neeman, R. H. Hensler, T. Masamune S. Rajagopalan, *J. Am. Chem. Soc.*, **82**, 2242 (1960).

5) 17 ⟶ 20: The rate of solvolysis of the acetate of **17** is 1000 times faster than that of **24**, where no participation is expected of either N or 20-OH. The participation of 20-OH was indicated by the solvolysis of **25**, which is 40 times faster than that of **24**. Therefore N in cevadine D-orthoacetate diacetate facilitates solvolysis by a factor of 25. This rate enhancement represents intramolecular general base catalysis of solvolysis presumably involving the structure **26** depicted below.[12]

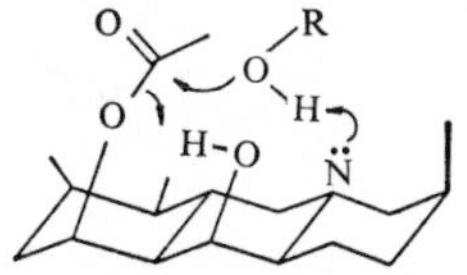

26

Remarks

The nmr spectra of various ceveratrum alkaloid derivatives have been studied.[13] A number of hypotensive ceveratrum alkaloids have been isolated and their structures determined.[1] These alkaloids are derivatives of several alkamines, including **1**. The structures of various relevant compounds are illustrated below.

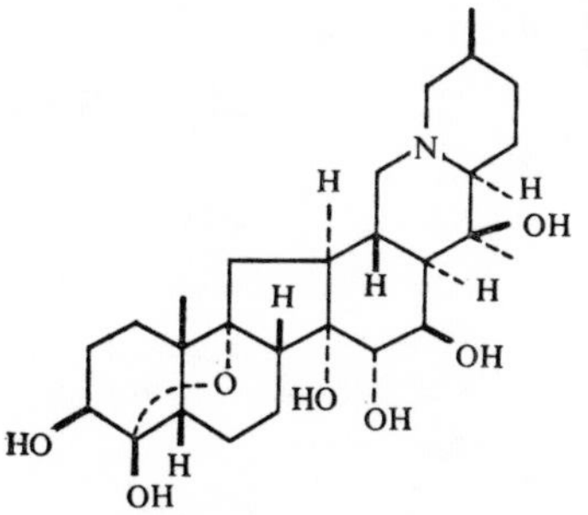

zygadenine **27**[14]

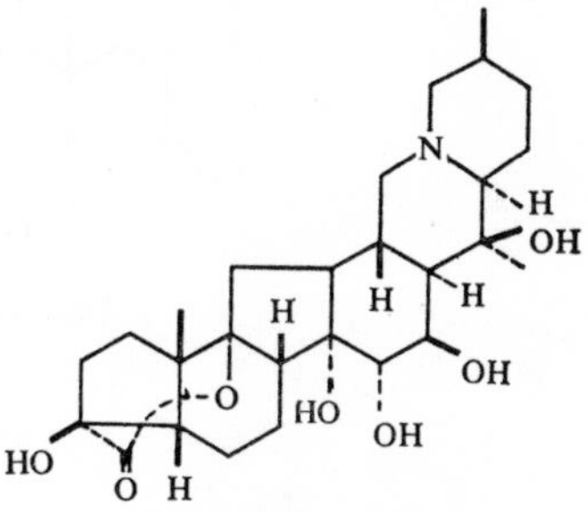

zygadenilic acid δ-lactone **28**[15]

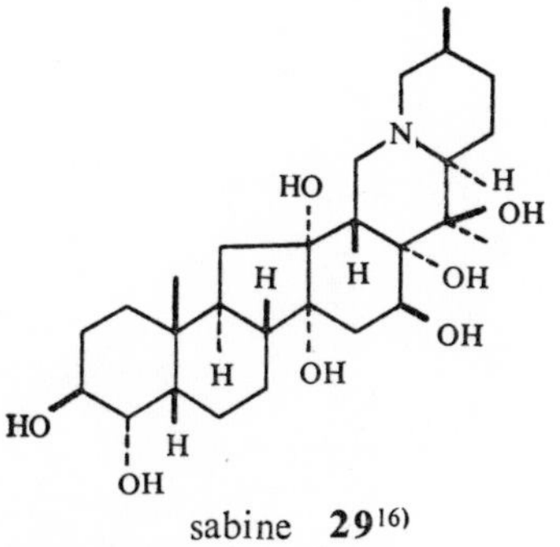

sabine **29**[16]

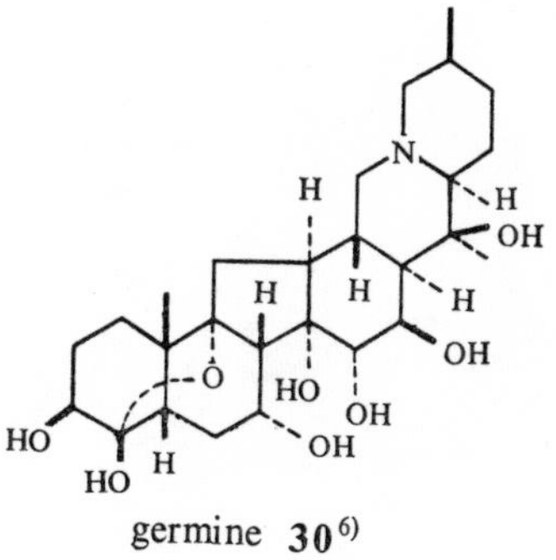

germine **30**[6]

10) S. M. Kupchan, D. Lavie, *J. Am. Chem. Soc.*, **77**, 683 (1955).
11) S. M. Kupchan, *J. Am. Chem. Soc.*, **77**, 686 (1955).
12) S. M. Kupchan, S. P. Eriksen, Y. T. Shen, *J. Am. Chem. Soc.*, **85**, 350 (1963); S. M. Kupchan, S. P. Eriksen, Y. T. S. Liang *ibid.*, **88**, 347 (1966).
13) S. Itô, J. B. Stothers, S. M. Kupchan, *Tetr.*, **20**, 913 (1964).
14) S. M. Kupchan, *J. Am. Chem. Soc.*, **81**, 1913 (1959).
15) B. Shimizu, *J. Pharm. Soc. Japan*, **78**, 444 (1958); **79**, 993 (1959).
16) S. M. Kupchan, N. Gruenfeld, N. Katsui, *J. Med. Pharm. Chem.*, **5**, 690 (1962).

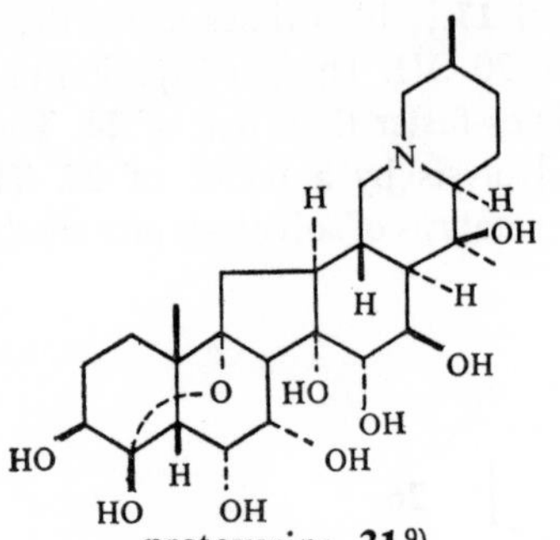

protoverine **31**[9)]

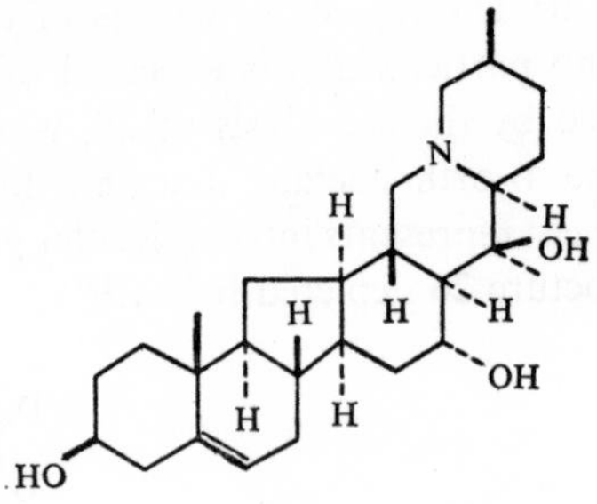

veramarine **32**[17)]

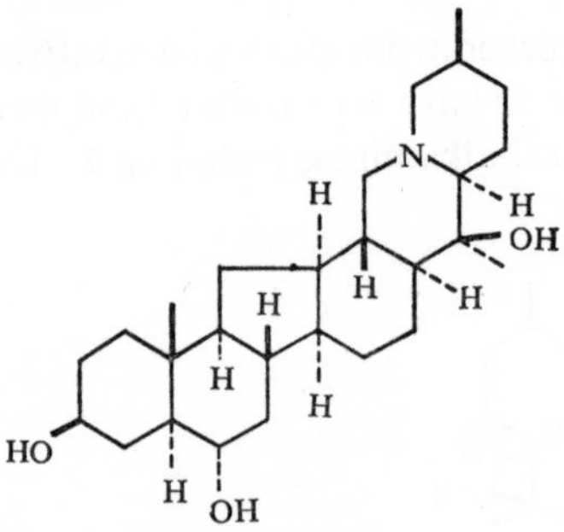

verticine **33**[18)]

17) S. Itô, T. Ogino, J. Tomko, *Coll. Czech. Chem. Comm.*, **33**, 4429 (1968).
18) S. Itô, M. Kato, K. Shibata, T. Nozoe, *Chem. Pharm. Bull. (Tokyo)*, **9**, 253 (1961); **11**, 1337 (1963); S. Itô, Y. Fukazawa, T. Okuda, Y. Iitaka, *Tetr. Lett.*, 5373 (1968).

10.12   STRUCTURE OF SERRATININE

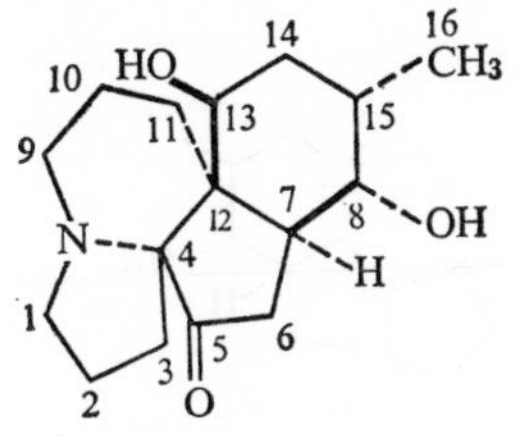

serratinine **1**[1,2]

mp: 244–245° α_D: −27.8°
ms: 279(M⁺), 251(M⁺-CO), 152($C_9H_{14}NO$)[1]
ir: 3472,   3436,   3185(br., OH),   1724(C=O),
     1427(CH)
pKa: 7.0

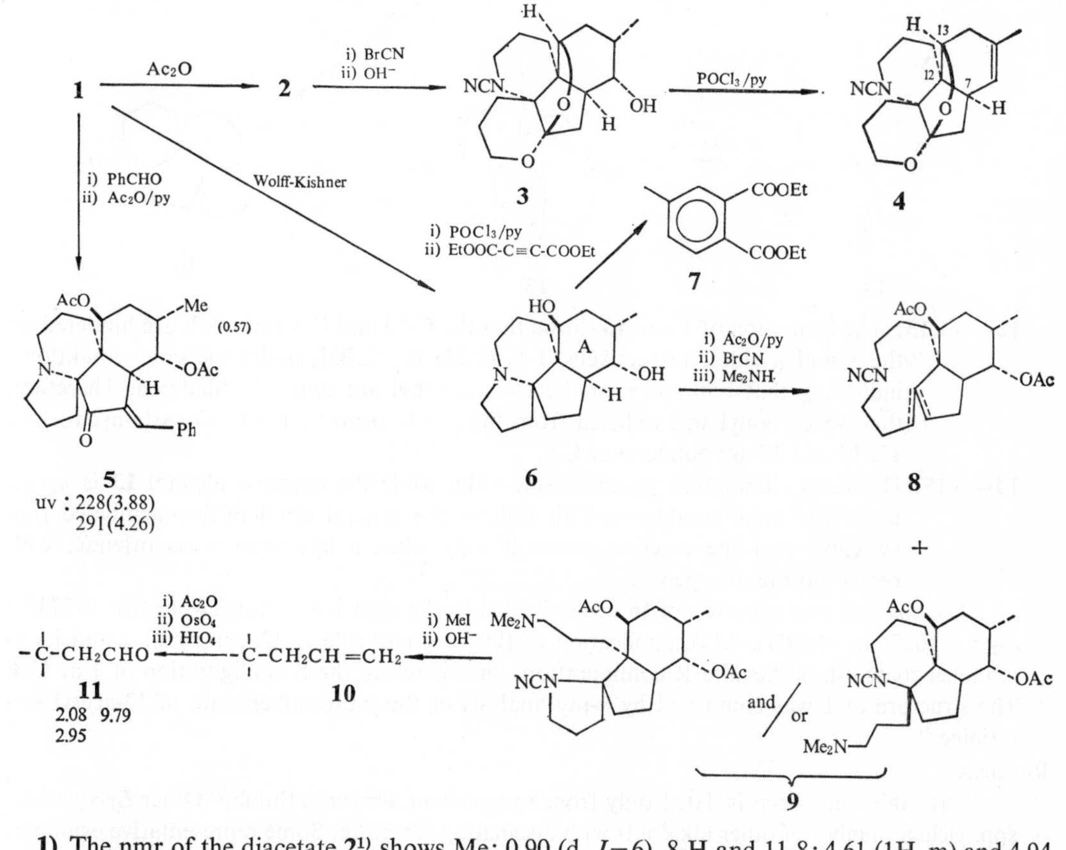

5
uv : 228(3.88)
291(4.26)

11
2.08 9.79
2.95

$$-\overset{|}{\underset{|}{C}}-CH_2CHO \longleftarrow -\overset{|}{\underset{|}{C}}-CH_2CH=CH_2$$

10

9

1) The nmr of the diacetate **2**[1] shows Me: 0.90 (d, J=6), 8-H and 11-8: 4.61 (1H, m) and 4.94 (1H, m).

6:   The α-aminoketone system in **1** was also suggested[1] by the large difference in pKa between **1** and deoxoserratinine **6** (pKa: 10.9).

6⟶7: Application of the Alder-Rickert rule indicated ring A to be six-membered.[3]

6⟶11: This established the presence of three methylene groups connecting the nitrogen and quaternary carbons.

1) Y. Inubushi, H. Ishii, B. Yasui, M. Hashimoto, T. Harayama, *Chem. Pharm. Bull.* **16**, 82 (1968).
2) Y. Inubushi, H. Ishii, B. Yasui, M. Hashimoto, T. Harayama, *Chem. Pharm. Bull.* **16**, 92 (1968).
3) K. Alder, H. F. Rickert, *Ann. Chem.*, **524**, 180 (1936).

2 ⟶ 3 ⟶ 4: This established the relative configurations at C-7, C-12 and C-13.[4]

1 ⟶ 5: The configuration of the methyl group was assigned on the basis of the upfield shift of its nmr signal going from **2** to **5**, probably due to the anisotropic effect of the benzene ring.[4]

12 ⟶ 1,13: The formation of **1** and **13** shows that the C-13 and C-5 carbonyls are hindered on the β and α sides, respectively. **1** is stable to NaBH$_4$ under the same conditions, indicating that both sides of the C-5 carbonyl are sterically hindered. Therefore, the C-5 carbonyl was reduced from the β side prior to the C-13 carbonyl to give **13**. **13** and **14** are epimeric at C-5.

14 ⟶ 15: Hofmann elimination proceeds smoothly, while the epimeric alcohol **13** is stable under the same conditions. This follows the general trend in five-membered ring systems[5] that the reaction proceeds only when a hydrogen is *cis* oriented with respect to the nitrogen.

The absolute configuration of **16** was indicated by the negative Cotton effect (ϕ_{308}: $+3735°$; ϕ_{273}: $-1023°$ in MeOH). Also, application of the benzoate rule to C$_8$-epimer of **1** and its 8-monobenzoate establishes the R configuration, and therefore the S configuration of **1** at C-8. The structure of **1** was confirmed by x-ray analysis of the *p*-bromobenzoate of 13-acetyl serratinine.[6]

Remarks

Serratinine has been isolated only from *Lycopodium serraturm* thunb.[7] Other *Lycopodium* spp. yield a number of other alkaloids with fascinating structure. Some representative examples are shown on the facing page.

4) Y. Inubushi, H. Ishii, B. Yasui, T. Harayama, *Chem. Pharm. Bull.* **16**, 101 (1968).
5) J. Sicher, J. Zavada, J. Krupicka, *Tetr. Lett.*, 1619 (1966); J. L. Coke, M. P. Cooke, Jr., *J. Am. Chem. Soc.*, **89**, 6701 (1967); *Tetr. Lett.* 2253 (1968).
6) K. Mishio, T. Fujiwara, K. Tomita, H. Ishii, Y. Inubushi, T. Harayama, *Tetr. Lett.*, 861 (1969).
7) Y. Inubushi, Y. Tsuda, H. Ishii, T. Sano, M. Hosokawa, T. Harayama, *Yakugaku Zasshi*, **84**, 1108 (1964); Y. Inubushi, H. Ishii, B. Yasui, T. Harayama, M. Hosokawa, R. Nishino, Y. Nakahara, *Yakugaku Zasshi*, **87**, 1397 (1967).
8) W. A. Harrison, M. Curcumelli-Rodostamo, D. F. Carson, L. R. C. Barclay, D. B. MacLean, *Can. J. Chem.*, **39**, 2086 (1961).
9) K. W. Wiesner, Z. Valenta, W. A. Ayer, L. R. Fowler, J. E. Francis, *Tetr.*, **4**, 87 (1958).
10) W. A. Ayer, G. G. Iverach, *Can. J. Chem.*, **38**, 1823 (1960); W. A. Ayer, J. A. Berezowsky, G. G. Iverach, *Tetr.*, **18**, 567 (1962).

lycopodine [8)]   annotinine [9)]   α-obscurine [10)]   serratinidine [11)]

cerunine [12)]   annopodine [13)]   alopecurine [14)]   lyconnotine [15)]

selagine [16)]   luciduline [17)]

A biogenetic scheme involving the condensation of two eight carbon polyketides (bold lines in the above formulae) has been proposed.[18)] In spite of a lack of experimental support, the hypothesis has been useful in the study of these alkaloids, and the numberings shown are based on it.[19)] Later, lysine was shown to be a biosynthetic precursor of lycopodine, and an alternative pathway involving the condensation of two pelletierine units formed from lysine and acetoacetate was proposed.[20)] There has been experimental support for the latter pathway in the biosynthesis of cerunine.[21)]

Mass spectral data of diagnostic value for determining the carbon skeleton are available.[15,22)]

11) H. Ishii, B. Yasui, R. Nishino, T. Harayama, Y. Inubushi, *Chem. Pharm. Bull.* **18**, 1880 (1970).
12) W. A. Ayer, J. K. Jenkins, K. Piers, S. Valverde-Lopez, *Can. J. Chem.*, **45**, 445 (1967).
13) W. A. Ayer, G. G. Iverach, J. K. Jekins, N. Masaki, *Tetr. Lett.*, 4597 (1968).
14) W. A. Ayer, B. Altenkirk, N. Masaki, S. Valverde-Lopez, *Can. J. Chem.*, **47**, 2449 (1969).
15) F. A. L. Anet, M. Z. Haq, N. H. Khan, W. A. Ayer, R. Hayatsu, S. Valverde-Lopez, P. Deslongchamps, W. Riess, M. Ternbah, Z. Valenta, K. Wiesner, *Tetr. Lett.*, 751 (1964); Z. Valenta, P. Deslongchamps, R. A. Ellison, K. Wiesner, *J. Am. Chem. Soc.*, **86**, 2533 (1964).
16) Z. Valenta, H. Yoshimura, E. F. Rogers, M. Ternbah, K. Wiesner, *Tetr. Lett.*, No. 10, 26 (1960); H. Yoshimura, Z. Valenta, K. Wiesner, *Tetr. Lett.*, No. 12, 14 (1960).
17) W. A. Ayer, N. Masaki, D. S. Nkunika, *Can. J. Chem.*, **46**, 3631 (1968).
18) H. Conroy, *Tetr. Lett.*, No. 10, 34 (1960).
19) K. Wiesner, *Fortsh. Chem. Org. Naturstoffe*, **20**, 271 (1962).
20) M. Castillo, R. N. Gupta, Y. K. Ho, D. B. MacLean, I. D. Spenser, *J. Am. Chem. Soc.*, **92**, 1074 (1970).
21) Y. K. Ho, R. N. Gupta, D. B. MacLean, I. D. Spenser, *Can. J. Chem.*, **49**, 3352 (1971).
22) D. B. MacLean, *Can. J. Chem.*, **41**, 2654 (1963); D. B. MacLean, M. Curcumelli-Rcdostamo, *Can. J. Chem.*, **44**, 611 (1966); W. A. Ayer, J. K. Jenkins, S. Valverde-Lopez, *Tetr. Lett.*, 220 (1964); Y. Inubushi, T. Ibuka, Y. Harayama, H. Ishii, *Tetr.*, **24**, 3541 (1968).

10.13 CHEMICAL CORRELATION OF SERRATININE WITH SERRATINIDINE AND FAWCETTIDINE

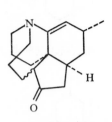

serratinine **1**

serratinidine **2**[1]

fawcettidine **3**[2]

mp: 232–234° α_D(EtOH): +224.2°
ms: 304(M$^+$ C$_{18}$H$_{28}$O$_2$N$_2$)
ir: 3310, 3110(OH and NH), 1659(amide I),
    1563(amide II)
pK_a: 6.4

C$_{16}$H$_{23}$ON (oil)
picrate mp: 222–223°
methiodide mp: 223–225°
ir: 1740(C=O)

The co-occurrence of various types of *Lycopodium* alkaloids implies the following biogenetic pathway for **1** and **2**, starting from the widely distributed lycodoline **4**.[3]

The common intermediate **5** (or its equivalent) has been prepared from **1**, and chemical correlation of **1**, **2** and **3** was achieved by the reaction sequence shown below.[2,3]

1) **12** was identical with a natural product called fawcettimine.[4] It exists as a mixture of valence-bond isomers (**12a** and **12b**). The existence of **12b** was proved by isolation of the derivative **13** of **12b**.

$$\mathbf{12a} \qquad \qquad \mathbf{12b} \qquad \qquad \mathbf{13}$$

Remarks

Fawcettidine **3** and fawcettimine **12** were isolated from *Lycopodium fawcettii*.[3] Co-occurrence of lycodoline **4**, serratinidine **2** and serratinine **1** in *L. serratum* was demonstrated.[5]

1) H. Ishii, B. Yasui, R. Nishino, T. Harayama, Y. Inubushi, *Chem. Pharm. Bull.*, **18**, 1880 (1970).
2) Y. Inubushi, H. Ishii, T. Harayama, R. H. Burnell, W. A. Ayer, B. Altenkirk, *Tetr. Lett.*, 1069 (1967).
3) W. A. Ayer, G. G. Iverach, *Can. J. Chem.*, **42**, 2514 (1964).
4) R. H. Burnell, *J. Chem. Soc.*, 3091 (1961); R. H. Burnell, B. S. Mootoo, *Can. J. Chem.*, **39**, 1090 (1961); R. H. Burnell, C. G. Chin, B. S. Mootoo, D. R. Taylor, *ibid.*, **41**, 3091 (1963).
5) Y. Inubushi, H. Ishii, B. Yasui, T. Harayama, M. Hosokawa, R. Nishino, Y. Nakahara, *Yakugaku Zasshi*, **87**, 1394 (1967).

10.14 STRUCTURE AND SYNTHESIS OF SECURININE
AND ALLOSECURININE

securinine **1**[1-4]

H 5.54(s)
H 6.42(q,J=9,1.5)
H 6.67(q.J=9,5)
3.86(t,J=5)

mp: 143–144°
α_D(EtOH): −1042.3°
uv(EtOH): 256(4.27),
330(3.24)
ir(CCl$_4$): 1840, 1760
(C=O), 1640(C=C)

allosecurinine **2**[3]

mp: 136–138°
α_D(EtOH): −1089°
uv(EtOH): 256(ϵ 18,200)
ir(CCl$_4$): 1792,1764(C=O)
pK_a': 7.17

i) KOH
ii) H$_2$/Ra-Ni

Δ/Pd-C

H$_2$/Pd

Δ/Zn

H$_2$/Pt

1   **3**   **4**

5
uv:215(4.33)
ir:1815,1770,1652

6
ir:1790

OH   OH   NH$_2$
Me

8   **9**   **10**

KMnO$_4$

COOH
COOH
7

1) The evidence[2-4] for the C-2 through C-5a arrangement and the α,β-unsaturated γ-lactone
system was obtained from uv, ir and nmr spectral data of **1** and **2**.

1) V. A. Muravéva, A. I. Ban'kovskii, *Doklady Akad. Nauk S.S.S.R.*, **110**, 998 (1956) (*Chem. Abstr.*, **51**, 8121a
   (1957)); A. D. Turov, Ya. A. Aleshkina, *Med. Prom. S.S.S.R.*, **11**, No. 1, 54 (*Chem. Abstr.*, **52**, 6724a (1958)).
2) S. Saito, K. Kotera, N. Sugimoto, Z. Horii, Y. Tamura, *Chem. Ind.*, 1652 (1962); S. Saito, N. Shigematsu,
   A. Ide, N. Sugimoto, Z. Horii, M. Hanaoka, Y. Yamawaki, Y. Tamura, *Tetr.*, **19**, 2085 (1963).
3) I. Satoda, M. Murayama, J. Tsuji, E. Yoshii, *Tetr. Lett.*, 1199 (1962).
4) T. Nakano, T. H. Yang, S. Terao, *Tetr.*, **19**, 609 (1963); *J. Org. Chem.*, **28**, 2619 (1963).

2) The reaction sequence **1** $\longrightarrow$ **3** $\longrightarrow$ **4** indicated the partial structure[2] shown by the bold line in **1**.

3) The formation of **7** from **6** and of **8**, **9** and **10** from **1** proved the bonding between C-10a and C-10b.[2]

4) The configurations at C-10a and C-10b in **1** were deduced[2] by examination of the ir spectra of **14** and **15**; **14** exhibits bands due to *trans*-quinolizidine (Bohlman bands) but lacks a band due to intramolecularly bonded OH. The reverse is true for **15**.

Absolute configuration[5]

5) Z. Horii, M. Ikeda, Y. Yamawaki, Y. Tamura, S. Saito, K. Kodera, *Tetr.*, **19**, 2101 (1963).

The formation of (+)-N-benzoyl pipecolic acid **18**[7] of known absolute configuration, as well as of **20** (and its salt) with positive ord curves, establishes the R configuration at C-10a.[6] S-(−)-Tetrahydropalmatine,[8] its salt and S-(−)-norcoralydine[9] all exhibit negative ord curves. In addition, **23** has a negative Cotton effect in its ord curve,[6] which proves C-10b to have S configuration, both from direct application of the octant rule and from comparison with Cotton effect of (−)-homocamphor **24**[10] An equatorial hydroxyl group adjacent to a carbonyl group is known to have no influence on the sign of a Cotton effect.[10] X-ray analysis of securinine hydro-bromide dihydrate confirmed the above conclusions.[11]

Remarks

All of the four diastereomeric isomers have been found in nature. While **1** was obtained from *Securinega suffruticosa* Rehd,[1] virosecurinine,[4] the antipode of **1**, was isolated from *S. virosa*, Pax. et Hoffm. Compound **2** was found as a minor component in leaves of *S. suffruticosa*, and viroallosecurinine,[12] the antipode of **2**, was obtained from *S. virosa*. Securinine **1** has a strychinine-like activity.

Synthesis of securinine[13]

---

7) J. W. Clark-Lewis, P. I. Mortimer, *J. Chem. Soc.*, 189 (1961).
8) G. G. Lyle, *J. Org. Chem.*, **25**, 1779 (1960).
9) A. Brossi, M. Baumann, F. Burkhardt, R. Richler, J. R. Frey, *Helv. Chim. Acta*, **45**, 2219 (1962).
10) W. Klyne, *Tetr.*, **13**, 29 (1961).
11) S. Imado, M. Shiro, Z. Horii, *Chem. Pharm. Bull.*, **13**, 643 (1965).
12) S. Saito, T. Iwamoto, T. Tanaka, C. Matsumura, N. Sugimoto, Z. Horii, Y. Tamura, *Chem. Ind.*, 1263 (1964); S. Saito, T. Tanaka, T. Iwamoto, C. Matsumura, N. Sugimoto, Z. Horii, M. Makita, M. Ikeda, Y. Tamura, *Yakugaku Zasshi*, **84**, 1126 (1964).
13) Z. Horii, M. Hanaoka, Y. Yamawaki, Y. Tamura, S. Saito, N. Shigematsu, K. Kotera, H. Yoshikawa, Y. Sato, H. Nakai N. Sugimoto, *Tetr.*, **23**, 1165 (1967).

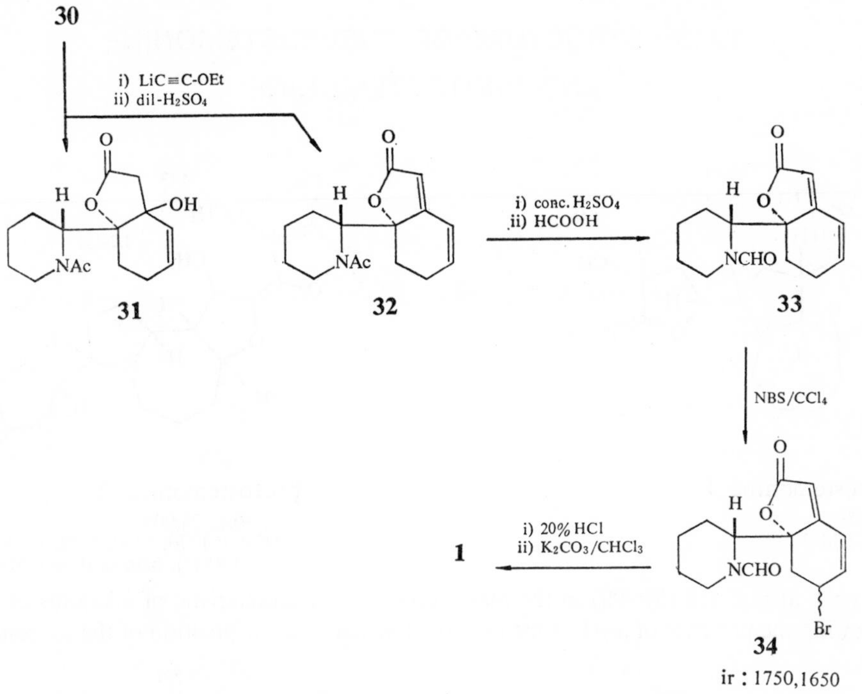

30

i) LiC≡C-OEt
ii) dil-H₂SO₄

31   **32**

i) conc. H₂SO₄
ii) HCOOH

33

NBS/CCl₄

i) 20% HCl
ii) K₂CO₃/CHCl₃

1

34
ir : 1750,1650

1) **25** was synthesized by the method of Jeager *et al.*[14]

2) The stereochemistry of **27** and **28** was confirmed by their identity with the degradation products of allosecurinine **2** and securinine **1**, respectively, which were obtained by the following reaction scheme.[15]

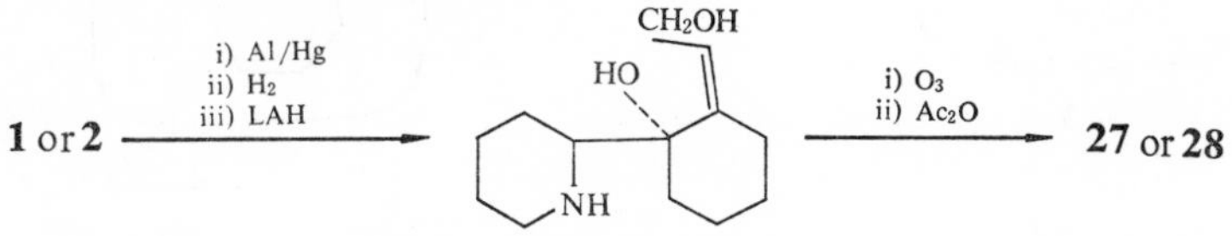

i) Al/Hg
ii) H₂
iii) LAH

CH₂OH

i) O₃
ii) Ac₂O

1 or 2 ———→ ———→ **27 or 28**

3) **29** was obtained as the sole product but the stereochemistry of the bromine atom was not established.

4) Elimination of hydrogen bromide from **29** with NaOH or EtONa gave a migrated product, probably through the following reaction sequence.

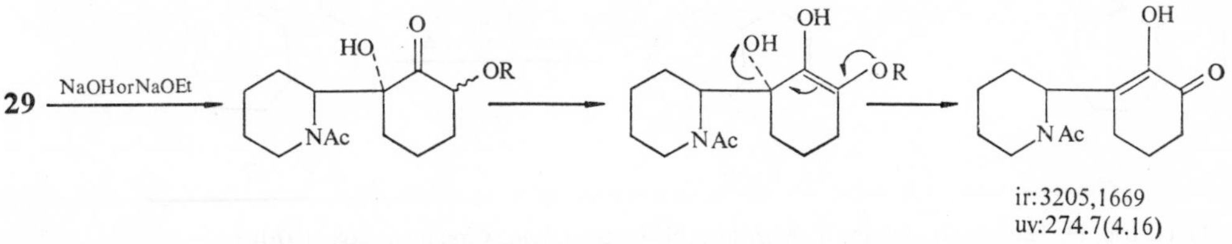

29 —NaOHorNaOEt→ ———→ ———→

ir:3205,1669
uv:274.7(4.16)

5) Condensation of **30** with LiC≡COEt afforded only the 1,2-addition product.

---

14) R. H. Jeager, H. Smith, *J. Chem. Soc.*, 160 (1955).
15) S. Saito, K. Kotera, N. Shigematsu, A. Ide, N. Sugimoto, Z. Horii, M. Hanaoka, Y. Yamawaki, Y. Tamura, *Tetr.*, **19**, 2085 (1963).

10.15   STRUCTURE OF TUBEROSTEMONINE
AND PROTOSTEMONINE

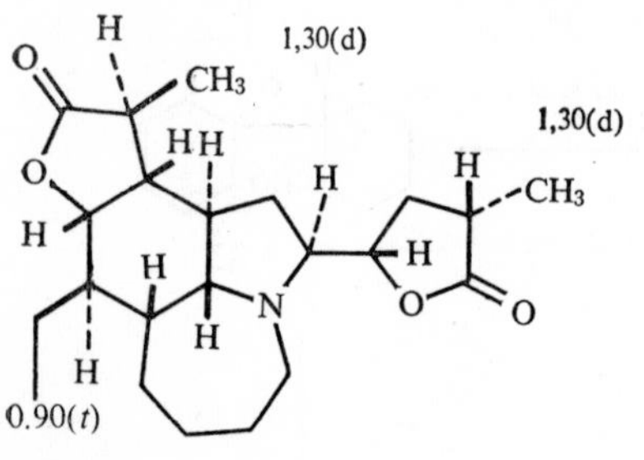

tuberostemonine **1**

ir: 1765
ms: 375(M$^+$)

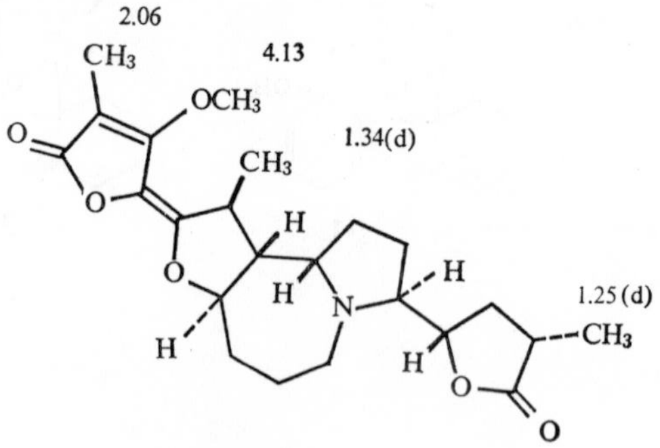

protostemonine **2**

uv: 305(ϵ 23,000)
ir(KBr): 1770, 1740, 1679, 1618
ms: 417(M$^+$), base peak m/e 318(M$^+$-99)

1) A peak at m/e 318 (M$^+$-99) in the mass spectrum is characteristic of alkaloids of this type, showing the presence of a γ-lactone moiety attached to the α-position of the pyrrolidine ring.

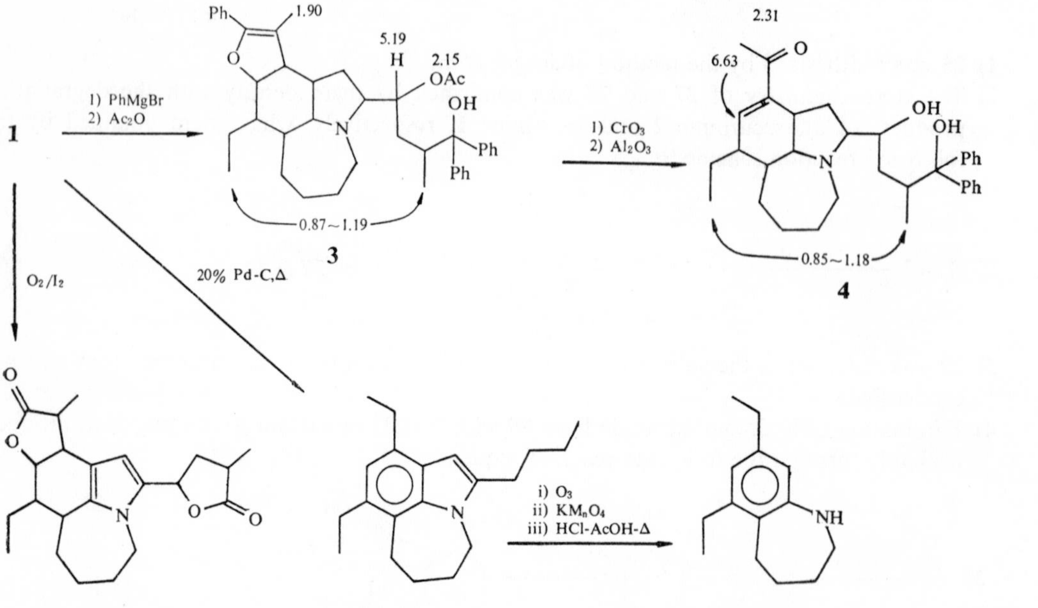

1) H. Irie, H. Harada, K. Ohno, T. Mizutani, S. Uyeo, *Chem. Commun.*, 268 (1970).
2) H. Koyama, K. Oda, *J. Chem. Soc.* B, 1330 (1970).
3) H. Schild, *Chem. Ber.*, **69**, 74 (1936).
4) K. Suzuki, *Yakugaku Zasshi*, **54**, 573 (1934).
5) M. Götz, T. Bögri, A. H. Gray, *Tetr. Lett.*, 707 (1961).
6) T. Shingu, Y. Tsuda, S. Uyeo, Y. Yamato, H. Harada, *Chem. Ind.*, 1191 (1962).
7) H. Harada, H. Irie, N. Masaki, K. Osaki, S. Uyeo, *Chem. Commun.*, 460 (1967).

i) HCl(gas)/benzene
2 ——————→
ii) H₂O

8

OH⊖ ————→

9

10

MnO₂

5.92(d,4)
6.24(d,4)

11

2 ——→ 8: The actual product is protostemonine
hydrate hydrochloride.[1]

9: Structure was elucidated by x-ray crystallographic analysis of its hydrobromide.[2]

Remarks

Tuberostemonine **1** was isolated by Schild[3] and Suzuki[4] from *Stemona tuberosa* (Stem-onaceae), which had been used as an insecticide for domestic animals in Japan. The structure of this alkaloid has been determined by Canadian[5] and Japanese[6] groups. The former group carried out a careful examination of the nmr spectra of the products obtained on Grignard reaction of the two γ-lactone moieties of this alkaloid while the latter group investigated the products obtained from its dehydrogenation. Final elucidation of the structure of the alkaloid was achieved by x-ray crystallographic analysis of its metho-bromide monohydrate.[7]

Investigation of the constituents of the same plant resulted in the isolation of stenine **12**,[8] oxotuberostemonine **13**[9] and iso-tuberostemonine.

Protostemonine **2** was isolated from *Stemona japonica* (stemonaceae),[10] and the structure determined by chemical[1,2] and x-ray analysis using symbolic addition method.[11] The related compound stemofoline **14** was isolated from leaves of the same plant, and the structure was elucidated by x-ray analysis of the hydrobromide.[12]

12

13

2.07
4.13
1.37(d)
4.25(m)
0.91(t)

14

8) S. Uyeo, H. Irie, H. Harada, *Chem. Pharm. Bull.*, **15**, 568 (1967).
9) C. P. Huber, S. R. Hall, E. N. Maslen, *Tetr. Lett.*, 4081 (1968).
10) H. Kondo, M. Satomi, *Yakugaku Zasshi*, **67**, 182 (1947).
11) H. Irie, K. Ohno, K. Osaki, T. Taga, S. Uyeo, *Chem. Pharm. Bull.*, **21**, 451 (1973).
12) H. Irie, N. Masaki, K. Osaki, T. Taga, S. Uyeo, *Chem. Commun.*, 1056 (1970).

10.16   SYNTHESIS OF ANNOTININE

annotinine **1**
mp:232°
$\alpha_D(CHCl_3)$:−27.5°

The structure of annotinine, a *lycopodium* alkaloid, was established after extensive chemical investigations[1] and was confirmed by x-ray crystallography.[2] A total synthesis of annotinine was reported recently,[3] following synthesis of the racemic form.[4]

2 → CH₂=CHCOOH/Δ → **3** → CH₂=C=CH₂/ $h\nu$ → **4**   ir : 1700,1640, 908 → (CH₂OH)₂/TsOH →

5 → H₂/Pd → **6** → i) TsOH / acetone ii) NaBH₄/aq-THF iii) MsCl/py → **7** → *t*-BuOK/DMSO →

8 → i) SeO₂/HOAc ii) KOH/MeOH iii) CrO₃/py → **9** → KCN/NH₄Cl/ DMF → **10** → H₂SO₄/MeOH →

1) Z. Valenta, F. W. Stonner, C. Bankiewicz, K. Wiesner, *J. Am. Chem. Soc.*, **78**, 2867 (1956); K. Wiesner, Z. Valenta, W. A. Ayer, L. R. Fowler, J. F. Francis, *Tetr.*, **4**, 87 (1958).
2) M. Przyolska, L. Marion, *Can. J. Chem.*, **35**, 1075 (1957).
3) K. Wiesner, L. Poon, I. Jirokovsky, M. Fishman, *Can. J. Chem.*, **47**, 433 (1969).
4) K. Wiesner, I. Jirokovsky, M. Fishman, C. A. Williams, *Tetr. Lett.*, 1523 (1967); K. Wiesner, I. Jirokovsky, *ibid.*, 2077 (1967); K. Wiesner, L. Poon, *ibid.*, 4937 (1967); Z. Koblicova, K. Wiesner, *ibid.*, 2563 (1967).
5) C. A. Grob, J. H. Wilkens, *Helv. Chim. Acta*, **48**, 808 (1965).
6) E. H. Bohme, Z. Valenta, K. Wiesner, *Tetr. Lett.*, 2441 (1965).

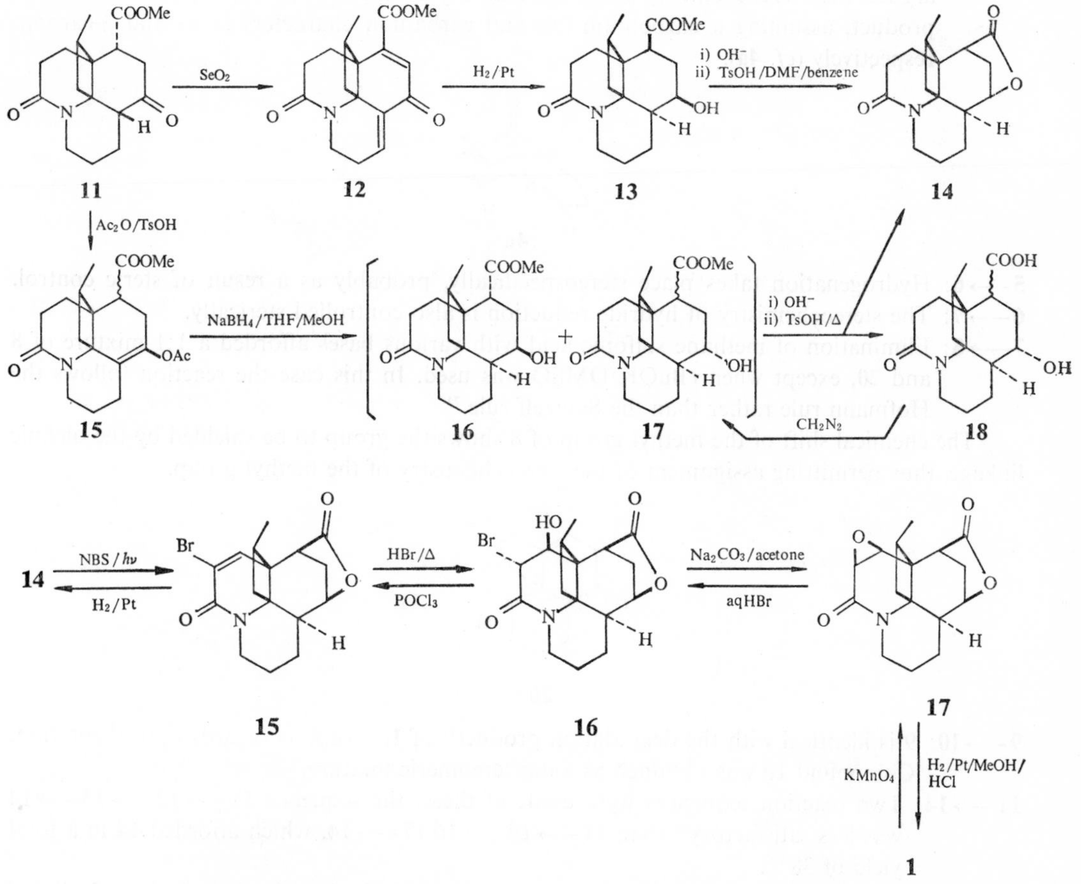

11 → SeO₂ → **12** → H₂/Pt → **13** → i) OH⁻ ii) TsOH/DMF/benzene → **14**

11 → Ac₂O/TsOH → **15**

15 → NaBH₄/THF/MeOH → [**16** + **17**] → i) OH⁻ ii) TsOH/Δ → **18**

18 → CH₂N₂ → **14**

14 → NBS/hν / H₂/Pt → **15**

15 → HBr/Δ / POCl₃ → **16**

16 → Na₂CO₃/acetone / aqHBr → **17**

17 → KMnO₄ / H₂/Pt/MeOH/HCl → **1**

2 → 3 → 4: **2** was prepared from acrylonitrile, methacrylate and ethyl acetoacetate by a previously known method.[5] **4** was obtained as the sole product from **3**. The regio- and stereochemistry of the photoaddition were investigated using model compounds,[6] and in addition, the structure of **4** was confirmed by the following reaction sequence.

If the allene had added in an isomeric fashion, the product corresponding to **18** would be **18a**, which after the sequence **18 → 19** would yield the corresponding hydroxyketolactam.

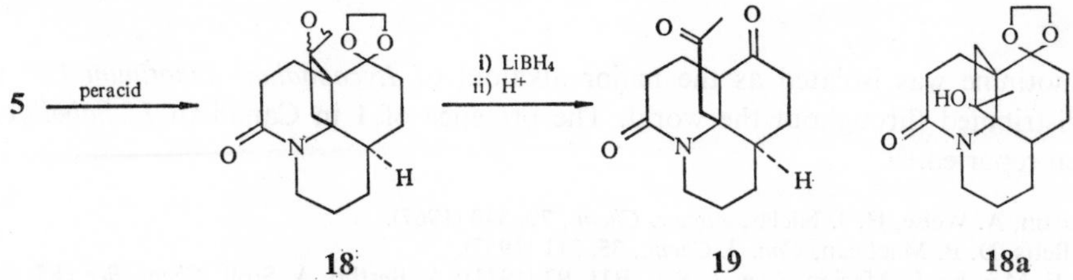

5 → peracid → **18** → i) LiBH₄ ii) H⁺ → **19**   **18a**

4: Configuration of the hydrogen indicated remains unsolved. However, this is not important in the synthesis since the hydrogen will be abstracted later (*cf.* **12**). The author argued the stereochemistry from the stability of excited state, but not from that of product, assuming a carbonium ion and carbanion characters at α- and β-carbon, respectively (*cf.* **4a**).

4a

5 ⟶ 6: Hydrogenation takes place stereospecifically, probably as a result of steric control.

6 ⟶ 7: The stereochemistry of hydride reduction is also controlled sterically.

7 ⟶ 8: Elimination of methane sulfonic acid with various bases afforded a 1:1 mixture of **8** and **20**, except when *t*BuOK/DMSO was used. In this case the reaction follows the Hofmann rule rather than the Saytzeff rule.[7]

The chemical shift of the methyl group of **8** shows the group to be shielded by the olefinic linkage, thus permitting assignment of the stereochemistry of the methyl group.

20

9 ⟶ 10: **9** is identical with the degradation product[1] of **1**, except as regards optical rotation. Compound **10** was obtained as a diastereomeric mixture.

11 ⟶ 14: Two reaction sequences were used: of these, the sequence **11 ⟶ 12 ⟶ 13 ⟶ 14** was less satisfactory[4] than **11 ⟶ 15 ⟶ 16,17 ⟶ 14**, which afforded **14** in a total yield of 38%.

Synthetic **11** was identical with the compound obtained by degradation of **1**,[1] and thus isomerization has occurred during the acid treatment of **10**. The free acid corresponding to **11** was optically resolved as the brucine salt. **14** was also identical with the specimen derived from **1**. It was correlated with **1** via a reversible reaction sequence.[8]

15 ⟶ 16,17: This process was rationalized in three stages: first, alcoholysis of **15** to the corresponding enolate ion; second, kinetically controlled protonation of the enolate to a ketoester isomeric with **11** at the tertiary carbon α to the keto group; third, non-stereospecific reduction of the keto group before isomerization can take place.

Remarks

Annotinine was isolated as the major alkaloid of *Lycopodium annotinum* L.[9] which is widely distributed throughout the world. The presence of **1** in Canadian L. *flabelliforme* has also been reported.[10]

7) D. Martin, A. Weise, H. J. Niclas, *Angew. Chem.*, **79**, 340 (1967).
8) E. E. Betts, D. B. MacLean, *Can. J. Chem.*, **35**, 211 (1957).
9) R. H. F. Manske, L. Marion, *Can. J. Res.*, **B21**, 92 (1943); A. Bertho, A. Stoll, *Chem. Ber.*, **85**, 663 (1952); R. H. F. Manske, L. Marion, *J. Am. Chem. Soc.*, **69**, 2126 (1947); O. Achmatowicz, W. Redewald, *Roczniki. Chem.*, **29**, 509 (1955); *ibid.*, **30**, 223 (1956).
10) S. N. Alam, K. A. H. Adams, D. B. MacLean, *Can. J. Chem.*, **42**, 2456 (1964).

10.17 BIOSYNTHESIS OF NICOTINE AND ANABASINE

＊ : labeled with ^{14}C

Pyrrolidine and piperidine rings in nicotine **12** and anabasine **11** are known to be derived from ornithine **7** and lysine **1** respectively,[1-5] and the pyridine ring originates from nicotinic acid[6] **6** in tobacco plants. γ-NH₂ of ornithine was proved to be utilized for the formation of the pyrrolidine ring.[7] However, the incorporation of α-N-methylornithine without loss of the methyl group was also shown.[3] In plants, nicotinic acid is known to be formed by the condensation of glycerol **4** and aspartic acid **5** whereas in animals and microorganisms it is derived from tryptophan.

1) A. A. Bothner-by, R. S. Schultz, R. F. Dawson, M. L. Solt, *J. Am. Chem. Soc.*, **84**, 52 (1962).
2) P. L. Wu, R. U. Byerrum, *Biochemistry*, **4**, 1628 (1965).
3) H. B. Schröter, D. Neumann, *Tetr. Lett.*, 279 (1966).
4) T. Kisaki, S. Mizusaki, E. Tamaki, *Arch. Biochem. Biophys.*, **117**, 677 (1966); S. Mizusaki, T. Kisaki, E. Tamaki, *Plant. Physiol.*, **43**, 93 (1968).
5) E. Leete, *J. Am. Chem. Soc.*, **89**, 7081 (1967).
6) M. L. Solt, R. F. Dawson, D. R. Christman, *Plant Physiol.*, **35**, 887 (1960); R. F. Dawson, D. R. Christman, M. L. Solt, A. P. Wolf, *Arch. Biochem. Biophys.*, **91**, 144 (1960).
7) E. Leete, E. G. Gros, T. J. Gilbertson, *Tetr. Lett.*, 587 (1964).

10.18 BIOSYNTHESIS OF TROPANE ALKALOIDS

ornithine **1**   **2**   hygrine **3**

Scopolia lurida   *Datura stramoniun D.metel*

cuscohygrine **5**   hyostiamine **4**

Remarks

The pyrrolidine ring in tropane alkaloids was shown by labeling experiments to be derived from ornithine **1**. Other carbons are derived from acetoacetate.[1-6]

1) **2** was incorporated into **4** and **5** in a good ratio but α-N-methylornithine was not.[4,6]

2) **3** was incorporated into **4** in 2.1% of ratio by 3 days feeding of *Datura stramonium*. **5** was labeled by administration of **3** to *Scopolia lurida* for 7 days.

1) E. Leete, *J. Am. Chem. Soc.*, **84**, 55 (1962); *Tetr. Lett.*, 1619 (1964).
2) H. W. Liebisch, W. Maier, H. R. Schütte, *Tetr. Lett.*, 4079 (1966).
3) J. Kaczkowski, H. R. Schütte, K. Mothes, *Biochim. Biophys. Acta*, **46**, 588 (1961).
4) F. E. Baralle, E. G. Gross, *Chem. Commun.*, 721 (1969); *Phytochem.*, **8**, 853 (1969).
5) D. G. O. Donovan, M. F. Keogh, *J. Chem. Soc.*, 223 (1969).
6) A. Ahmad, E. Leete, *Phytochem.*, **9**, 2345 (1970).

10.19   SENECIO ALKALOIDS (PYRROLIZIDINE ALKALOIDS)

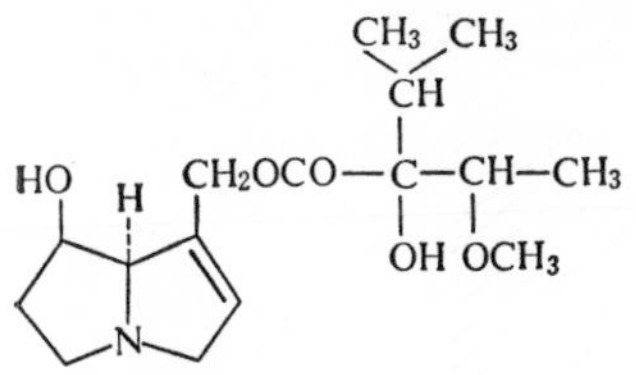

heliotrine **1**

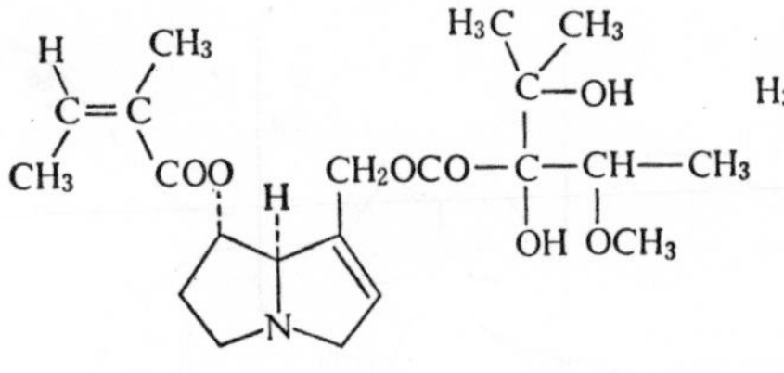

lasiocarpine **2**

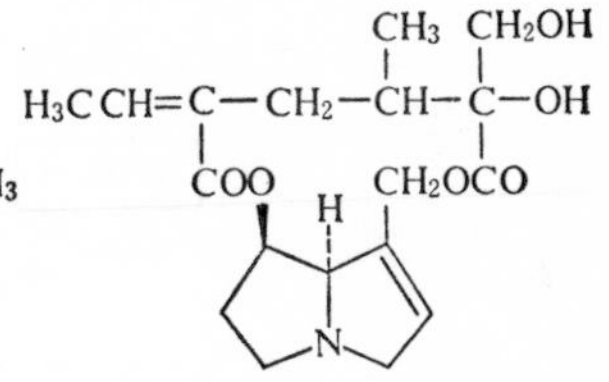

retrorsine **3**

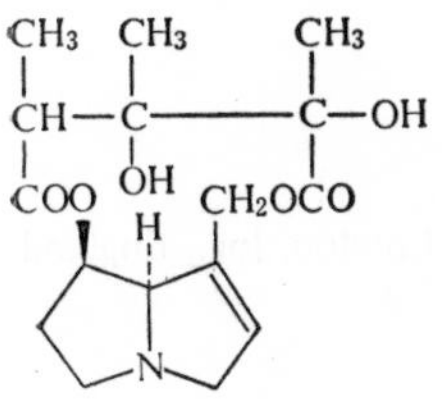

monocrotaline **4**

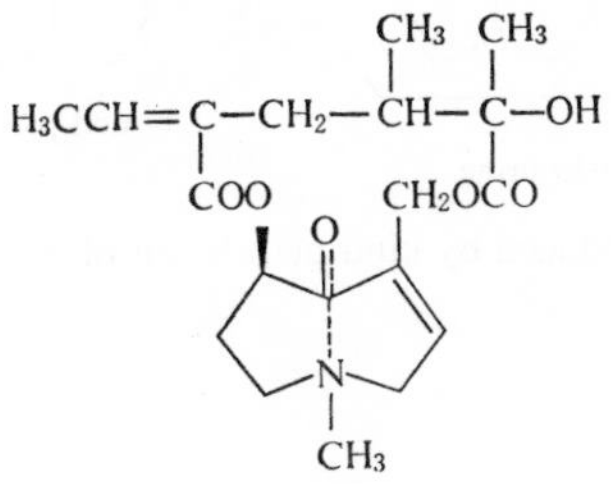

senkirkine **5**

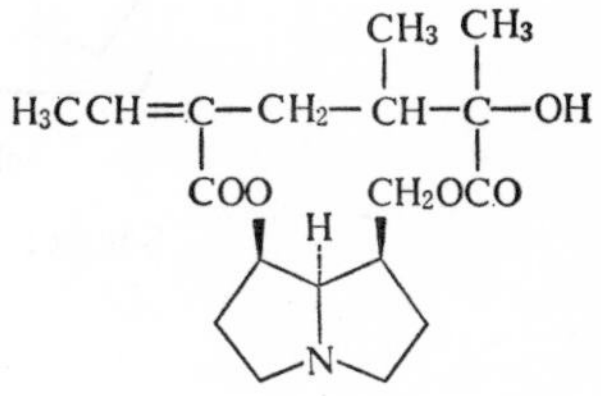

platyphylline **6**

This group of alkaloids, first isolated from *Senecio* spp. (Compositae), have been found in other genera of the Compositae, in many genera of the Boraginaceae, in some of the Leguminosae and some of the Gramineae. The alkaloids are esters of amino alcohols with a pyrrolizidine nucleus (necines) and aliphatic acids (necic acids). Most of the necines bear a hydroxymethyl group at C-1 and a hydroxyl group at C-7 with or without a double bond at C-1/C-2 on the pyrrolizidine ring. The necic acids are C_3–C_7 monocarboxylic acid and dibasic acid (succinic, glutaric and adipic acids) derivatives with one or more hydroxyl groups. Combinations of the necines and the necic acids amounting to over 100 compounds have so far been isolated and their chemistry is well-established.[1-6] Examples of monoesters (**1**), diesters (**2**), and cyclic diesters (**3–6**) are shown above.

Several plants containing alkaloids, such as *Senecio jacobaea* and *Crotalaria spectabilis*, are known to be poisonous and the poisoning of farm animals and occasionally humans in many countries has occurred. After the discovery of the pyrrolizidine alkaloids, their toxicology and pharmacology were extensively studied. In particular, some of the alkaloids and plant materials containing the alkaloids have been proved to be liver carcinogens in experimental animals.[7]

Recent studies on the metabolism of the alkaloids have shown that it is not the pyrrolizidine alkaloids themselves, but their pyrrole derivatives, formed as metabolites, which exhibit toxicity and a mechanism of action as alkylating agents was proposed, as shown overleaf.[8] This diagram

1) N. J. Leonard, *The Alkaloids* (ed. R. H. F. Manske, H. L. Holmes), vol. 1, p. 107, Academic Press, 1949.
2) F. L. Warren, *Fortschr. Chem. Org. Naturstoffe*, **XII**, p. 198, Springer-Verlag, 1955.
3) N. J. Leonard, *The Alkaloids* (ed. R. H. F. Manske, H. L. Holmes), vol. 6, p. 37, Academic Press, 1960.
4) F. L. Warren, *Fortschr. Chem. Org. Naturstoffe*, **XXIV**, p. 329, Springer-Verlag, 1966.
5) L. B. Bull, C. C. J. Culvenor, A. T. Dick, *The Pyrrolizidine Alkaloids*, North-Holland, 1968.
6) F. L. Warren, *The Alkaloids* (ed. R. H. F. Manske, H. L. Holmes), vol. 12, p. 245, Academic Press, 1970.
7) R. Schoental, *Cancer Res.*, **28**, 2237 (1968).
8) A. R. Mattocks, *Phytochemical Ecology* (ed. J. B. Harborne), p. 179, Academic Press, 1972.

represents the hypothetical fate of a reactive pyrrolic metabolite (dihydropyrrolizine ester) in the liver of a rat.

Source: ref. 8. Reproduced by kind permission of Academic Press, London, Inc., England.

10.20　BIOSYTHESIS OF EPHEDRINE

ephedrine **5**

1) $*$ and $\blacktriangle$ indicate positions labeled with ^{14}C.

2) Compound **1** with aromatic ^{3}H labeling was 0.0085% incorporated into **5** in September. **2** was 0.0017% incorporated into **5**.

3) **3** was 0.12% incorporated into **5** in May and 0.33% in June. The corresponding benzaldehyde was also incoporated (0.016%) into **5**.

4) The C_2–N unit for the transformation **3**⟶**4** originates from aspartate and formate.

Remarks

　　Phenylalanine **1** was assumed to be a direct precursor of ephedrine **5** on the basis of efficient incorporation of [3-^{14}C] phenylalanine.[1-3] However, recent investigations have shown[4] that **5** is formed from a C_6-C_1 unit, C_2-N unit and C_1 units, as shown. This is different from the biosynthetic pathways of hordenine **6** and mescaline **7**, which are derived from tyrosine by decarbonylation and subsequent methylation.

hordenine **6**　　　　　　　　mescaline **7**

1) S. Shibata, I. Imaseki, *Chem. Pharm. Bull.*, **4**, 277 (1956).
2) S. Shibata, I. Imaseki, M. Yamazaki, *Chem. Pharm. Bull.*, **5**, 71, 594 (1957); *ibid.*, **7**, 449 (1959).
3) I. Imaseki, S. Shibata, M. Yamazaki, *Chem. Ind.*, 1625 (1958).
4) K. Yamasaki, U. Sankawa, S. Shibata, *Tetr. Lett.*, 4099 (1969). K. Yamasaki, T. Tamaki, S. Uzawa, U. Sankawa, S. Shibata, *Phytochem.*, **12**, 2877 (1973).

10.21 ABSOLUTE CONFIGURATION OF BENZYLISOQUINOLINE ALKALOIDS

The absolute configuration of the only asymmetric center in these alkaloids was established by correlation with *d*-aspartic acid, as shown below.[1–3]

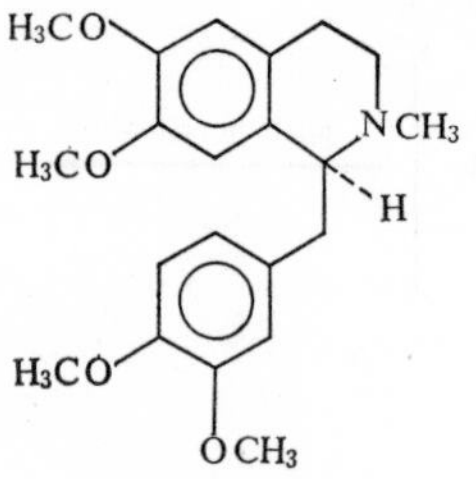

H₃CO
H₃CO
NCH₃
H
H₃CO
OCH₃

(+)-landanosine
(+)-1

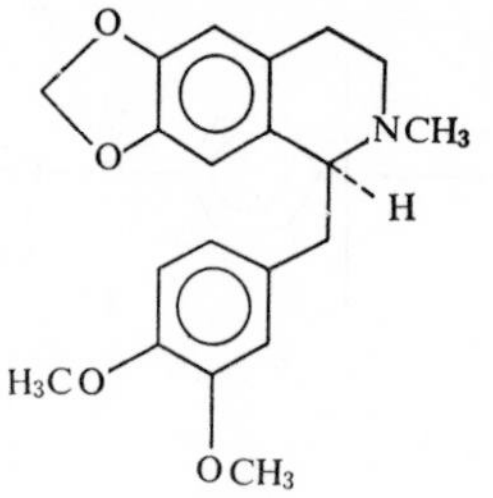

O
O
NCH₃
H
H₃CO
OCH₃

(+)-romneine
(+)-2

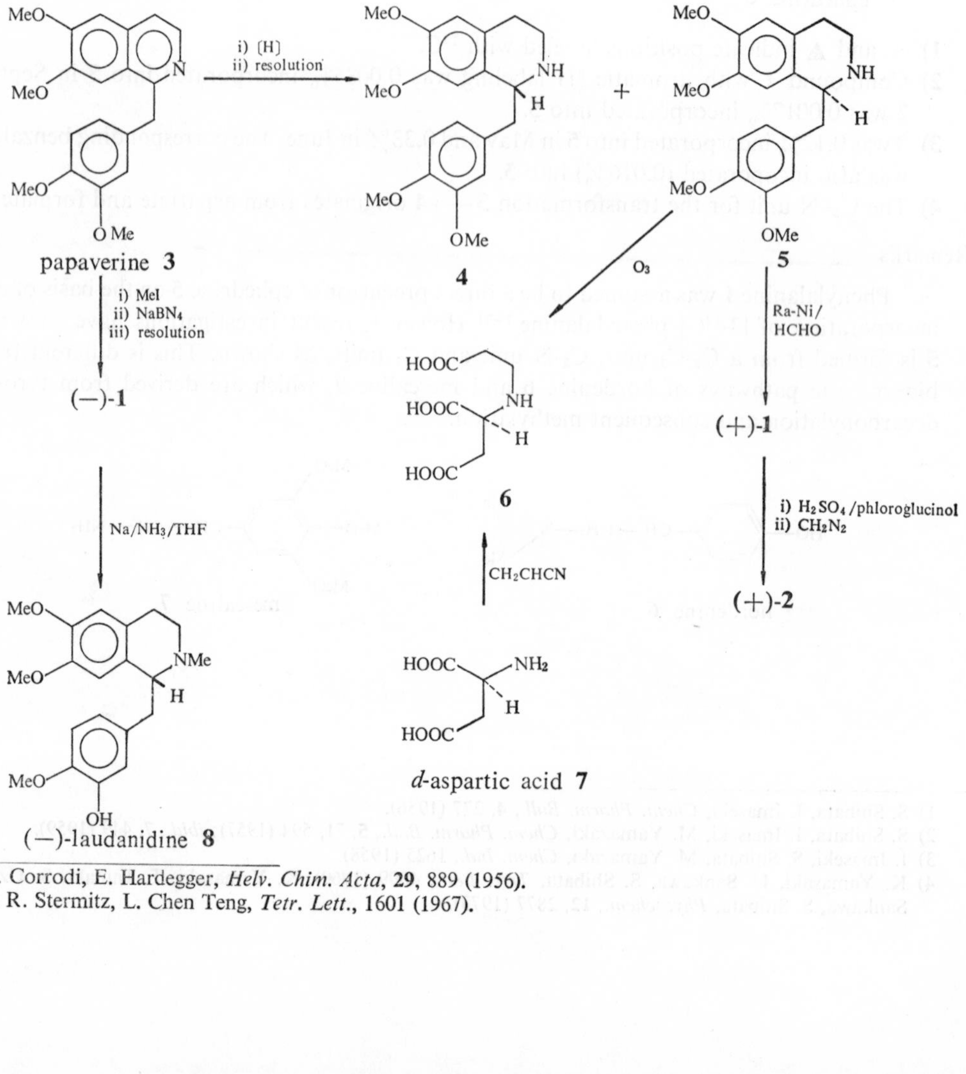

MeO
MeO
N
MeO
OMe
papaverine 3
i) [H]
ii) resolution
MeO
MeO
NH
H
MeO
OMe
4
+
MeO
MeO
NH
H
MeO
OMe
5
i) MeI
ii) NaBN₄
iii) resolution
(−)-1
O₃
HOOC
HOOC
NH
H
HOOC
6
Ra-Ni/
HCHO
(+)-1
Na/NH₃/THF
CH₂CHCN
i) H₂SO₄/phloroglucinol
ii) CH₂N₂
(+)-2
MeO
MeO
NMe
H
MeO
OH
HOOC
NH₂
H
HOOC
d-aspartic acid 7

(−)-laudanidine 8

1) H. Corrodi, E. Hardegger, *Helv. Chim. Acta*, **29**, 889 (1956).
2) F. R. Stermitz, L. Chen Teng, *Tetr. Lett.*, 1601 (1967).

8

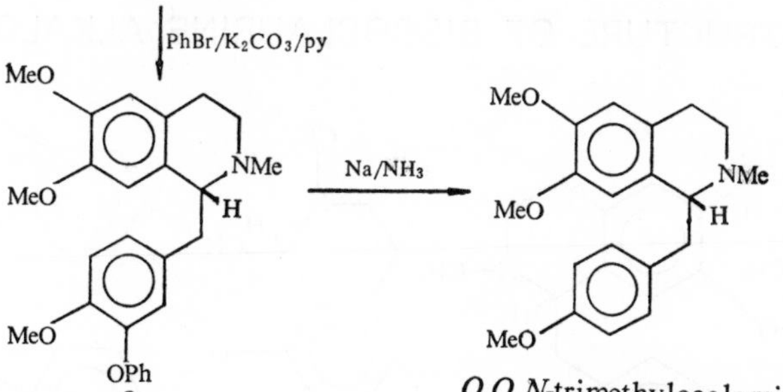

PhBr/K$_2$CO$_3$/py

MeO

MeO

NMe

H

MeO

OPh

9

Na/NH$_3$

MeO

MeO

NMe

H

MeO

O,O,N-trimethylcoclamine

10

3——4: The first step was an electrolytic reduction. For the second step, N-acetyl-l-leucine and di-p-toluyltartaric acid were used.

9——10: O-methoxydiphenyl ether is known to be cleaved by Na in liquid ammonia to yield anisole and phenol.[6-8]

Remarks

1 was obtained from *Papaver somniferum*[4] and **2** from *Romneya coulteri* var. *trichocalyx*.[5] The correlation described above also established the absolute stereochemistry of dimeric biscoclaurine alkaloids,[9] as well as those of other coclaurine-type bases. Some examples are shown below.

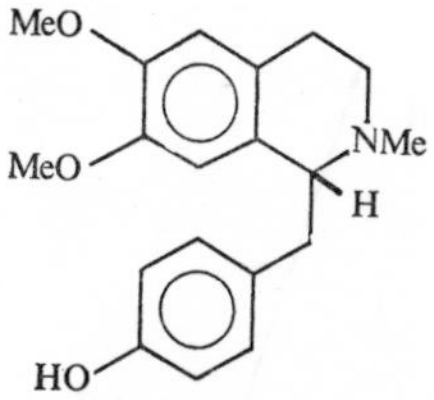

MeO

MeO

NMe

H

HO

l-armepavine **11**[10,11]

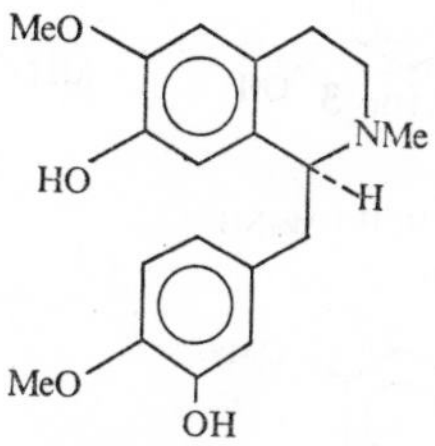

MeO

HO

NMe

H

MeO

OH

d-reticuline **12**[12]

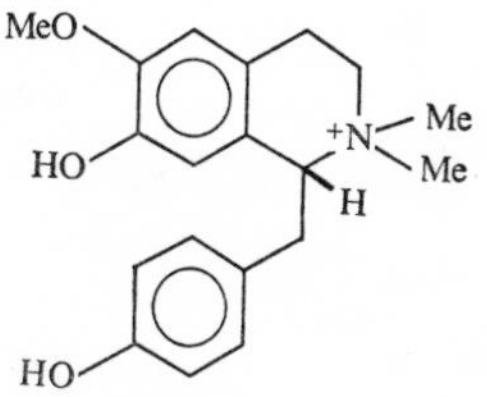

MeO

HO

+N—Me

Me

H

HO

l-magnocurarine **13**[10,13]

3) M. Tomita, J. Kunitomo, *Yakugaku Zasshi*, **82**, 734 (1962).
4) J. Kunitomo, E. Yuge, Y. Nagai, R. Fujitani, *Chem. Pharm. Bull.*, **16**, 364 (1968).
5) F. R. Stermitz, L. Chen, J. White, *Tetr.*, **22**, 1095 (1966).
6) P. A. Sartorette, F. J. Sowa, *J. Am. Chem. Soc.*, **59**, 603 (1937); A. L. Kranzfelder, J. J. Verbane, F. J. Sowa, *ibid.*, **59**, 1488 (1937); F. C. Weker, F. J. Sowa, *ibid.*, **60**, 94 (1938).
7) M. Tomita, Y. Inubushi, H. Niwa, *Yakugaku Zasshi*, **72**, 206 (1952).
8) Y. Inubushi, *Yakugaku Zasshi*, **72**, 220 (1952).
9) M. Tomita, *Fortschr. Chem. Org. Naturstoffe* (ed. L. Zechmeister), vol. 13, p. 175, Springer-Verlag, 1952; M. Curcumelli-Rodostamo, *The Alkaloids* (ed. R. H. F. Manske), vol. 13, p. 304, Academic Press, 1971.
10) M. Tomita, J. Kunitomo, *Yakugaku Zasshi*, **82**, 734 (1962).
11) R. Konovalova, S. Yunusoff, A. Orechoff, *Chem. Ber.*, **68**, 2158, 2277 (1935); *J. Gen. Chem.* (U.S.S.R.), **10**, 641 (1940).
12) A. R. Battersby, G. W. Evans, R. O. Martin, M. E. Warren, Jr., H. Rapoport, *Tetr. Lett.*, 1275 (1965).
13) M. Tomita, Y. Inubushi, M. Yamagata, *Yakugaku Zasshi*, **71**, 1069 (1951).

10.22   STRUCTURE OF BISCOCLAURINE ALKALOIDS

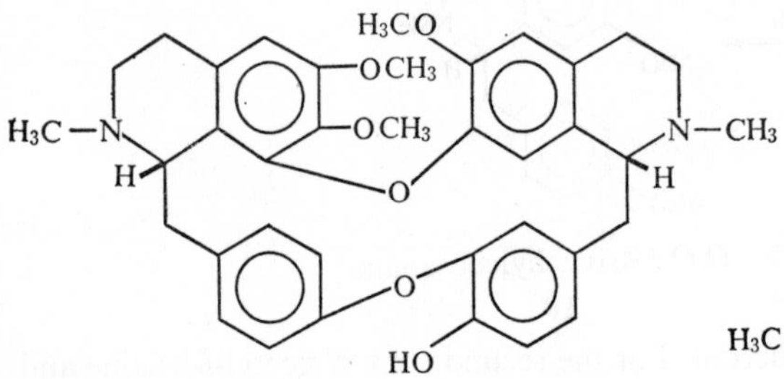

oxycanthine **1**[1]
mp:216—217°
α_D:+279°

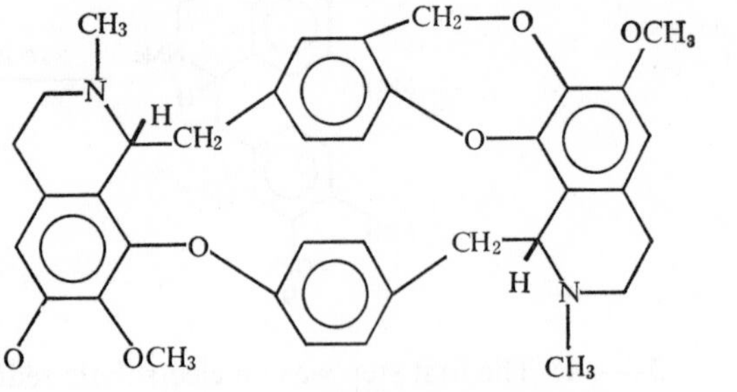

insularine **2**[2]
mp: 293°dec.(as the methiodide)
α_D: +27.9°

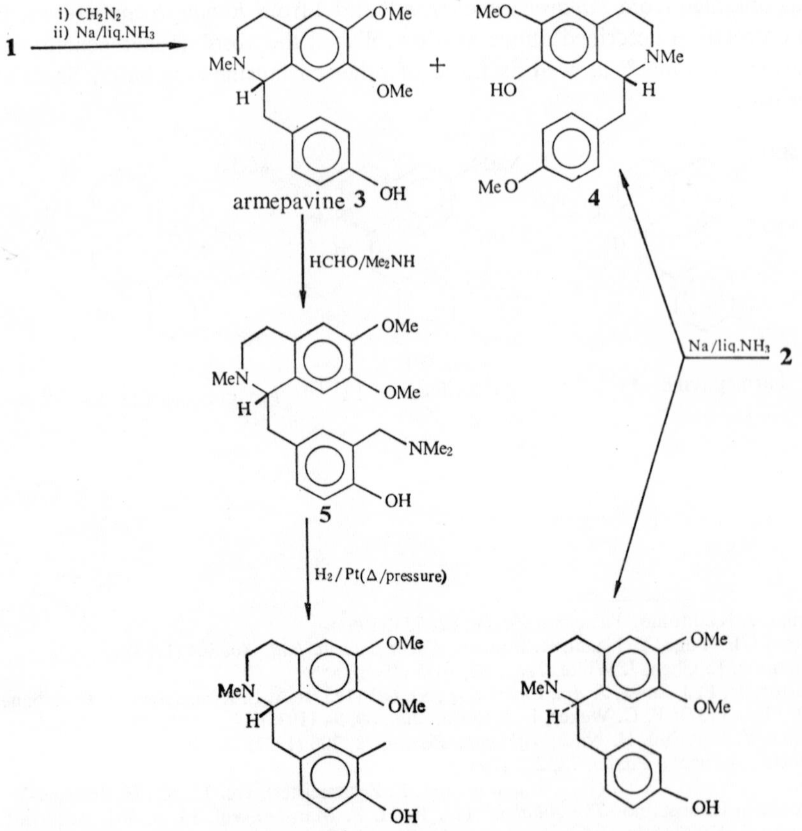

1 → i) CH_2N_2 ii) $Na/liq.NH_3$ →

armepavine **3** OH   +   **4**

HCHO/Me_2NH

5

$H_2/Pt(\Delta/pressure)$

$Na/liq.NH_3$ **2**

homoarmepavine **6**[7]

6b (antipode of **6**)

1) E. Fujita, *Yakugaku Zasshi*, **72**, 213 (1952).
2) J. Kunitomo, *Yakugaku Zasshi*, **82**, 1152 (1962).

1) Normal methods of molecular weight determination such as the Rast method or determination of freezing point depression usually give values corresponding to half the actual molecular weight.

2) The reactions **1 ⟶ 3,4** and **2 ⟶ 4,6b** follow basic studies[1,3] on the cleavage of o-methoxydiphenyl ether by Na and liquid ammonia. Cleavage takes place only at the C-O bond next to the methoxyl group. In the following case, however, the methylenedioxy group is cleaved to give a phenol.[4,5]

cepharanthine **7** → **8** + **4**

One of the ortho methoxy groups has been cleaved by replacing Na with Li.

9 → **10**

Remarks

Bisccolaurine alkaloids (bisbenzylisoquinoline alkaloids) occur in various families of plants.[6] d-Tubocurarine **11** and cepharanthine **7** have medical applications.

Biscoclaurine alkaloids have a variety of structural types, some of which are shown overleaf. However, all of these are considered to be biogenetically derived from benzylisoquinoline via phenolic oxidative coupling.

3) M. Tomita, E. Fujita, F. Murai, *Yakugaku Zasshi*, **71**, 226, 1035 (1951); *ibid.*, **71**, 301 (1951); E. Fujita, F. Murai, *ibid.*, **71**, 1039 (1951); M. Tomita, Y. Inubushi, H. Niwa, *ibid.*, **72**, 211 (1952); M. Tomita, E. Fujita, T. Abe, *ibid.*, **72**, 384 (1952); Y. Inubushi, K. Nomura, E. Nishimura, M. Yamamoto, *ibid.*, **78**, 1189 (1958); Y. Inubushi, K. Nomura, *ibid.*, **79**, 838 (1959); *ibid.*, **81**, 7 (1961); *ibid.*, **82**, 696, 1341 (1962).
4) M. Tomita, K. Kondo, *Yakugaku Zasshi*, **77**, 1019 (1957).
5) M. Tomita, Y. Sasaki, *Chem. Pharm. Bull.*, **1**, 105 (1953); *ibid.*, **2**, 89, 375 (1954); Y. Sasaki, H. Ohnishi, N. Sato, *ibid.*, **3**, 178 (1955); Y. Sasaki, *ibid.*, **3**, 250 (1955); J. Kunitomo, *Yakugaku Zasshi*, **82**, 981 (1962).
6) M. C. Rodostamo, *The Alkaloids* (ed. R. H. F. Manske), vol. 13, p. 303, Academic Press, 1971.
7) A. J. Everett, L. A. Lowe, S. Wilkinson, *Chem. Commun.*, 1020 (1970).
8) J. H. Mason, A. S. Howard, W. I. Taylor, M. J. Vernengo, I. R. C. Bick, P. S. Clezy, *J. Chem. Soc.*, 1948 (1967); I. R. C. Bick, J. H. Bowie, J. H. Mason, D. H. Williams, *ibid.*, 1951 (1967); K. Abe, J. H. Mason, *ibid.*, 1957 (1967).
9) M. Tomita, Y. Inubushi, *Chem. Pharm. Bull.*, **3**, 7 (1955); Y. Inubushi, K. Nomura, *Tetr. Lett.*, 1133 (1962).
10) M. Tomita, H. Furukawa, S.-T. Lu, S. M. Kupchan, *Tetr. Lett.*, 4309 (1965); *Chem. Pharm. Bull.*, **15**, 959 (1967).

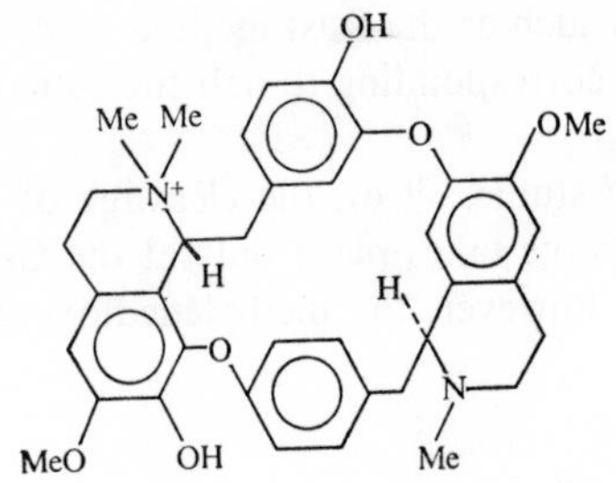

tubocurarine **11**[7)]

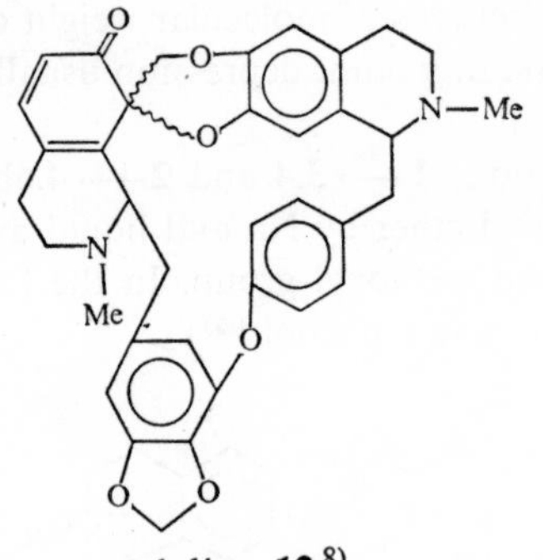

repanduline **12**[8)]

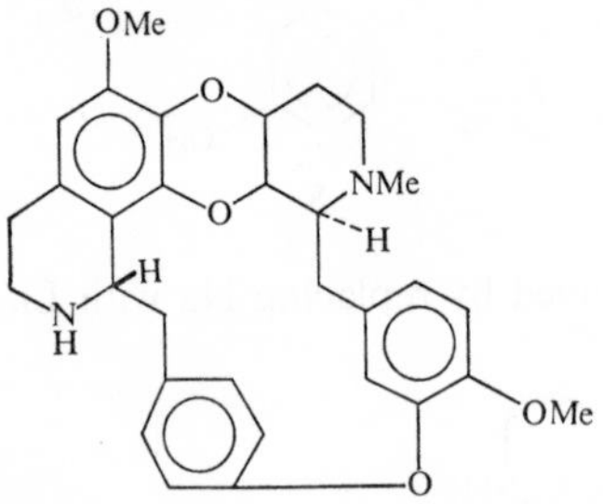

trilobine **13**[9)]

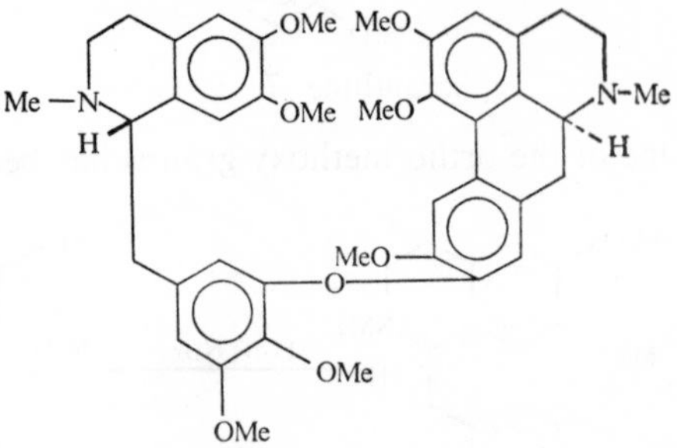

thalicarpine **14**[10)]

10.23 SYNTHESIS OF BISCOCLAURINE ALKALOIDS

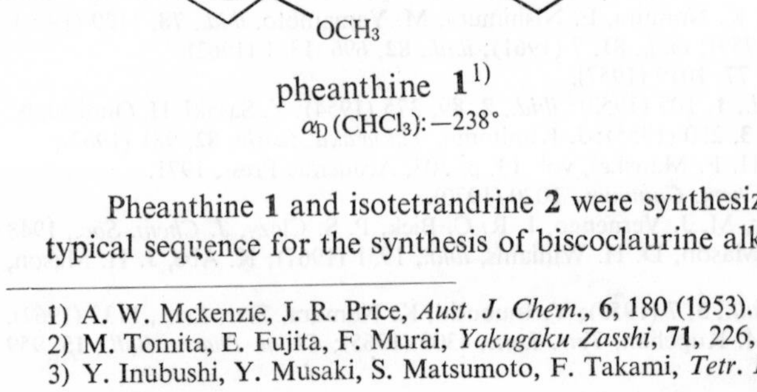

pheanthine **1**[1)]

α_D (CHCl$_3$): $-238°$

Pheanthine **1** and isotetrandrine **2** were synthesized[3)] by the following method, which is a typical sequence for the synthesis of biscoclaurine alkaloids.

---

1) A. W. Mckenzie, J. R. Price, *Aust. J. Chem.*, **6**, 180 (1953).
2) M. Tomita, E. Fujita, F. Murai, *Yakugaku Zasshi*, **71**, 226, 1035 (1951).
3) Y. Inubushi, Y. Musaki, S. Matsumoto, F. Takami, *Tetr. Lett.*, 3399 (1968); *J. Chem. Soc. C*, 1547 (1969).

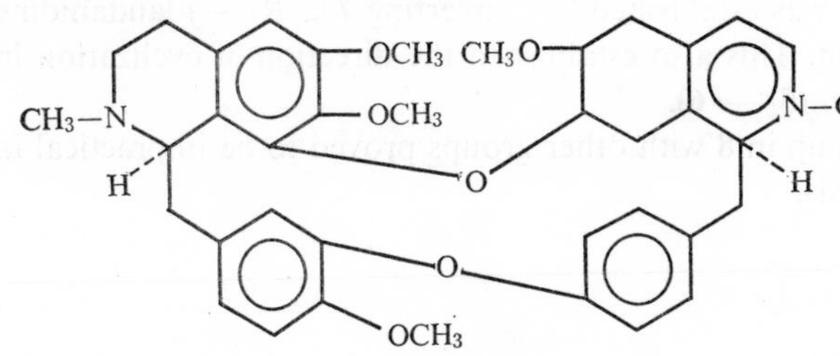
OCH3
CH3O
CH3–N
OCH3
N–CH3
H
O
H
O
OCH3

isotetrandrine **2**[2)]
mp: 182°
α_D(CHCl$_3$):146°

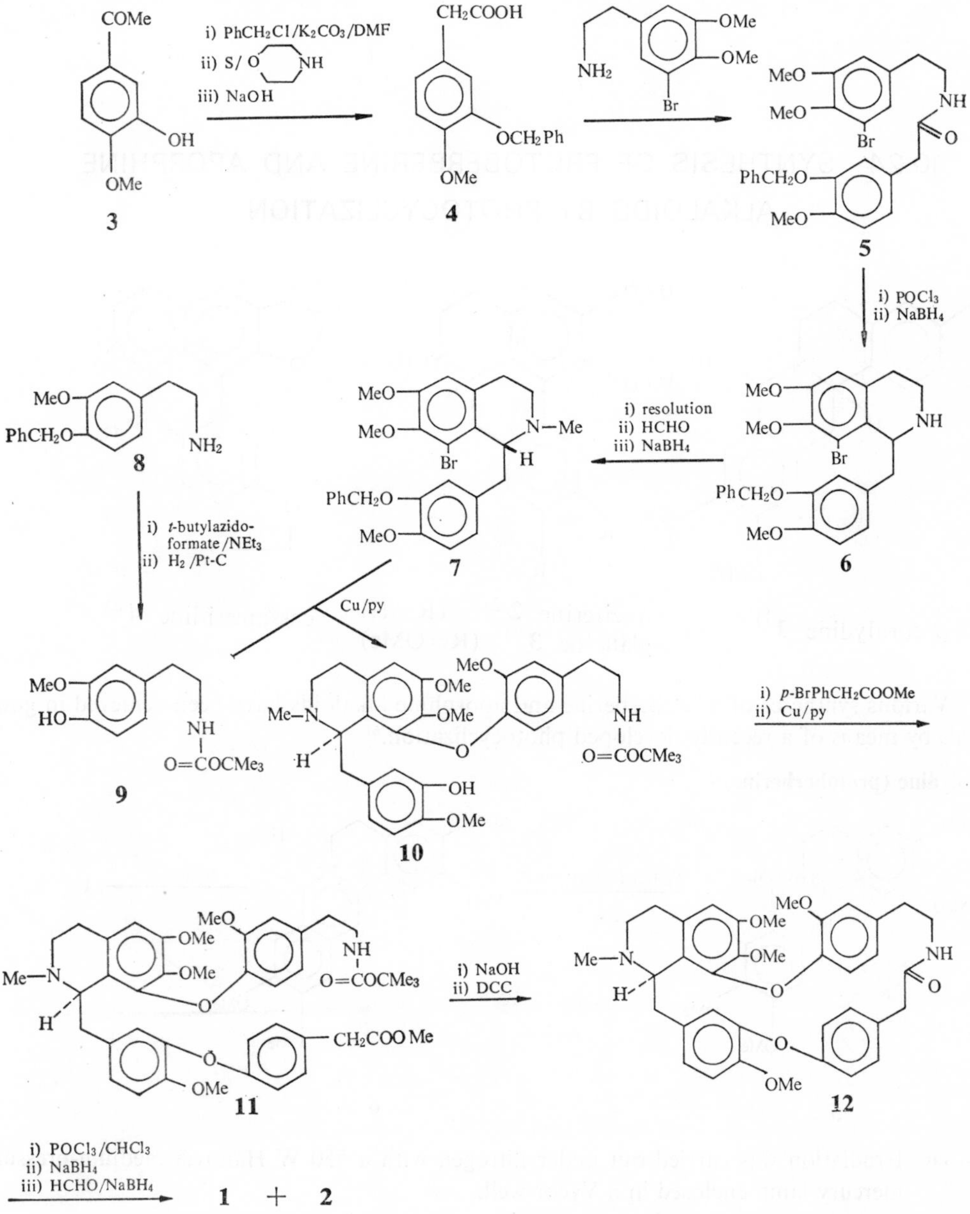
COMe
OH
OMe
3
i) PhCH2Cl/K2CO3/DMF
ii) S/ O NH
iii) NaOH
CH2COOH
OCH2Ph
OMe
4
NH2
OMe
OMe
Br
MeO
MeO
Br
NH
O
PhCH2O
MeO
5
i) POCl3
ii) NaBH4
MeO
MeO
Br
NH
PhCH2O
MeO
6
i) resolution
ii) HCHO
iii) NaBH4
MeO
MeO
N–Me
Br
H
PhCH2O
MeO
7
MeO
PhCH2O
NH2
8
i) t-butylazido-formate/NEt3
ii) H2/Pt-C
MeO
HO
NH
O=COCMe3
9
Cu/py
MeO
OMe
OMe
Me–N
H
O
NH
O=COCMe3
OH
OMe
10
i) p-BrPhCH2COOMe
ii) Cu/py
MeO
OMe
OMe
Me–N
H
O
NH
O=COCMe3
CH2COOMe
O
OMe
11
i) NaOH
ii) DCC
MeO
OMe
OMe
Me–N
H
O
NH
O
O
OMe
12
i) POCl3/CHCl3
ii) NaBH4
iii) HCHO/NaBH4
1 + 2

6⟶7: The absolute configuration of 7 was established by converting 7 to R-(−)-laudanidine of known absolute configuration. This also established the direction of cyclization in the Bischler-Napieralski reaction (5⟶6).

8⟶9: The protection of the amino group in 8 with other groups proved to be impractical in subsequent stages of the synthesis.

10.24   SYNTHESIS OF PROTOBERBERINE AND APORPHINE ALKALOIDS BY PHOTOCYCLIZATION

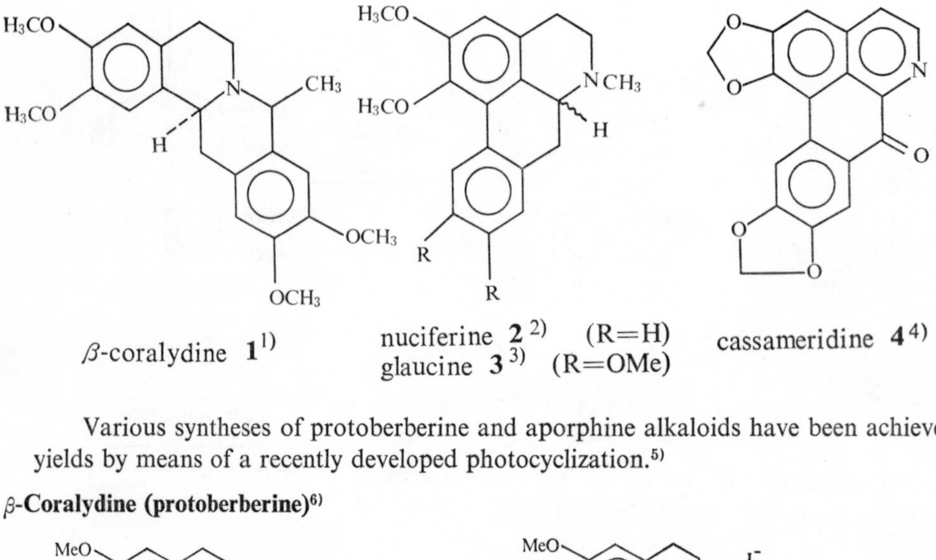

β-coralydine 1[1]   nuciferine 2[2]   (R=H)   cassameridine 4[4]
                      glaucine 3[3]   (R=OMe)

Various syntheses of protoberberine and aporphine alkaloids have been achieved in good yields by means of a recently developed photocyclization.[5]

β-Coralydine (protoberberine)[6]

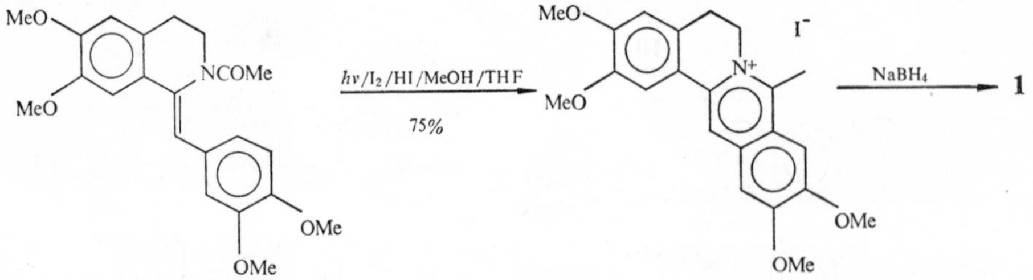

$hv/I_2/HI/MeOH/THF$

75%

$NaBH_4$ ⟶ 1

5   6

5⟶6: Irradiation was carried out under nitrogen with a 550 W Hanovia medium-pressure mercury lamp enclosed in a Vycor well.

Nuciferine,[7] glaucine[7,8] and cassameridine[9] (aporphines)

7 (X=Cl or Br,
R=H or OMe)

8

9   10   11

AcO$_2$H

4

$7 \longrightarrow 8$; $9 \longrightarrow 10$: These reactions involve a photochemical *cis-trans* isomerization, a conrotatory electrocyclization and subsequent elimination of HX.[8] CaCO$_3$ and
*t*BuOK were used to neutralize the hydrogen halides formed. The latter,
however, seemed to be superior, because it facilitated the elimination of HX.

7 or 9   12   13   −HX   8 or 10

$8 \longrightarrow 2$ or 3; $10 \longrightarrow 11$: LAH in ether afforded a mixture of the *N*-formyl and *N*-methyl
compounds.[8]

1) W. Awe, J. Thum, H. Wichman, *Arch. Pharm.*, **293**, 907 (1960).
2) J. Kunitomo, M. Kamimura, *Yakugaku Zasshi*, **84**, 1100 (1964).
3) T. Tomita, K. Furukawa, *Yakugaku Zasshi*, **83**, 293 (1963).
4) M. P. Cava, K. V. Rao, B. Douglas, J. A. Weisbach, *J. Org. Chem.*, **33**, 2443 (1968).
5) F. B. Mallory, C. S. Wood, J. T. Gordon, *J. Am. Chem. Soc.*, **86**, 3094 (1964); N. C. Yang, G. R. Lenz,
A. Shani, *Tetr. Lett.*, 2941 (1966); S. M. Kupchan, R. M. Kanojia, *ibid.*, 5353 (1966); N. C. Yang, A. Shani,
G. R. Lenz, *J. Am. Chem. Soc.*, **88**, 5369 (1966); M. P. Cava, S. C. Havlicek, *Tetr. Lett.*, 2625 (1967).
6) G. R. Lenz, N. C. Yang, *Chem. Commun.*, 1136 (1967).
7) M. P. Cava, S. C. Havlicek, A. Lindert, R. J. Spangler, *Tetr. Lett.*, 2937 (1966).
8) M. P. Cava, M. J. Mitchell, S. C. Havlicek, A. Lindert, R. J. Spangler, *J. Org. Chem.*, **35**, 175 (1970).
9) M. P. Cava, P. Stern, K. Wakisaka, *Tetr.*, **29**, 2245 (1973).

10.25   STRUCTURE AND SYNTHESIS OF OCHOTENSIMINE AND RELATED ALKALOIDS

ochotensimine **1**[1]
uv: 226(e25,700),287(e13,100)

ochotensine **2**

1) The signals of the two protons at C-8 were observed, after spin-decoupling experiments, as an AB-type quartet centered at 6.80 ($J = 18$ Hz).
2) The structure of **2** was determined by x-ray crystallographic analysis of its methiodide.[2,3]
3) The absolute configuration of ochotensine and related alkaloids has been determined by application of the aromatic chirality method.[4]

Remarks

Ochotensimine **1** and ochotensine **2** were isolated from *Corydalis ochotensis* Turcz. **2** was found to be 3-demethylochotensimine, and was converted to **1** by methylation with diazomethane.[1] Related alkaloids having a 1-spirobenzylisoquinoline skeleton have been isolated from *Fumaria officinalis*, and their structures have been elucidated spectroscopically.[5]

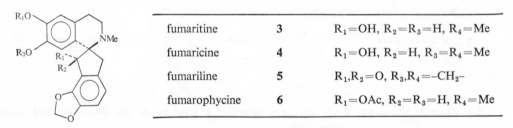

fumaritine	**3**	R_1=OH, R_2=R_3=H, R_4=Me
fumaricine	**4**	R_1=OH, R_2=H, R_3=R_4=Me
fumariline	**5**	R_1,R_2=O, R_3,R_4=–CH$_2$–
fumarophycine	**6**	R_1=OAc, R_2=R_3=H, R_4=Me

Nuclear Overhauser effects observed with fumariline[6]

The structure of fumariline **5** was elucidated by the observation of nuclear Overhauser effects, as shown on the facing page.

1) S. McLeon, M-S. Lin, *Tetr. Lett.*, 3819 (1964).
2) S. McLean, M-S. Lin, *Tetr. Lett.*, 185 (1966).
3) S. McLean, M-S. Lin, *Can. J. Chem.*, **44**, 2449 (1966).
4) M. Shamma, J. L. Moniot, R. H. F. Manske, W. K. Chan, K. Nakanishi, *Chem. Commun.*, 310 (1972).
5) J. K. Saunders, R. A. Bell, C-Y. Chen, D. B. MacLean, R. H. F. Manske, *Can. J. Chem.*, **46**, 2873 (1968).
6) J. K. Saunders, R. A. Bell, C-Y. Chen, D. B. MacLean, R. H. F. Manske, *Can. J. Chem.*, **46**, 2876 (1968).

Irradiated proton	Observed proton	% Area increase
9–H	1–H	22
5–H	4–H	25
N–Me	9–H	7
9–H	10–H	19

Synthesis of ochotensine 2[7,8] and the related alkaloids, fumariline 5, fumaritine 3 and fumaricine 4[5,9] has been performed by Pictet-Spengler condensation of the appropriate indanedione (e.g. 8) and 3-hydroxy-4-methoxyphenylethylamine 7.

12 $\longrightarrow$ 13: α-Aminoketone functionality of 12 was protected by formation of the 4-hydroxyimidazolidine-2-one derivative for the methylation of the catechol moiety in the next step.

7) S. McLeon, M. Lin, J. Whelen, *Tetr. Lett.*, 2425 (1968).
8) H. Irie, T. Kishimoto, S. Uyeo, *J. Chem. Soc*, C, 3051 (1968).
9) T. Kishimoto, S. Uyeo, *J. Chem. Soc*, C, 2600 (1968); *ibid.*, 1644 (1971).

10.26   INTERCONVERSION OF 1-SPIROISOQUINOLINE AND PROTOBERBERIN ALKALOIDS

The conversion of protoberberine into spirobenzylisoquinoline alkaloids has been achieved.[1,2] This type of rearrangement may be similar to that occurring in nature.[1]

6⟶7: 7, which shows no C-Me signal, was characterized as the diacetate.

1) M. Shamma, C. D. Jones, *J. Am. Chem. Soc.*, **91**, 4009 (1969); *ibid.*, **92**, 4943 (1970).
2) H. Irie, K. Akagi, S. Tani, K. Yabusaki, H. Yamane, *Chem. Pharm. Bull.*, **21**, 855 (1973).

10 ⟶ 13: Photolysis of **10** in EtOH in the presence of NaH (one eq.) yielded **13** in 45% yield via the photoinduced rearrangement shown.

A similar rearrangement in the reverse direction was achieved by photolysis of the N-nor compound **14**, yielding xylopinine **15** in 80% yield together with the oxo-compound **16** in 5% yield.[2]

10.27 PHTHALIDE ISOQUINOLINE ALKALOIDS, SYNTHESIS AND ABSOLUTE CONFIGURATION

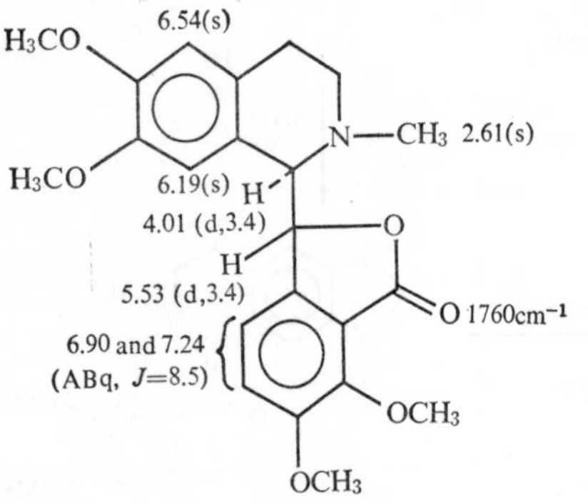

cordrastine I **1**

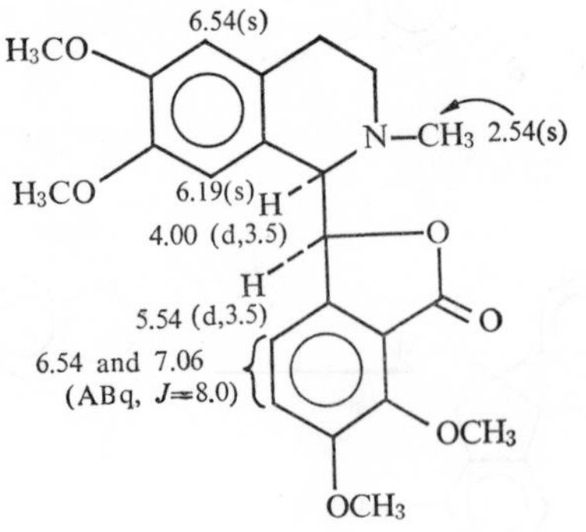

cordrastine II **2**

Conversion of 1-spiroisoquinolines to phthalide isoquinolines

This conversion has been achieved as part of a synthesis of cordrastine I and II.[1]

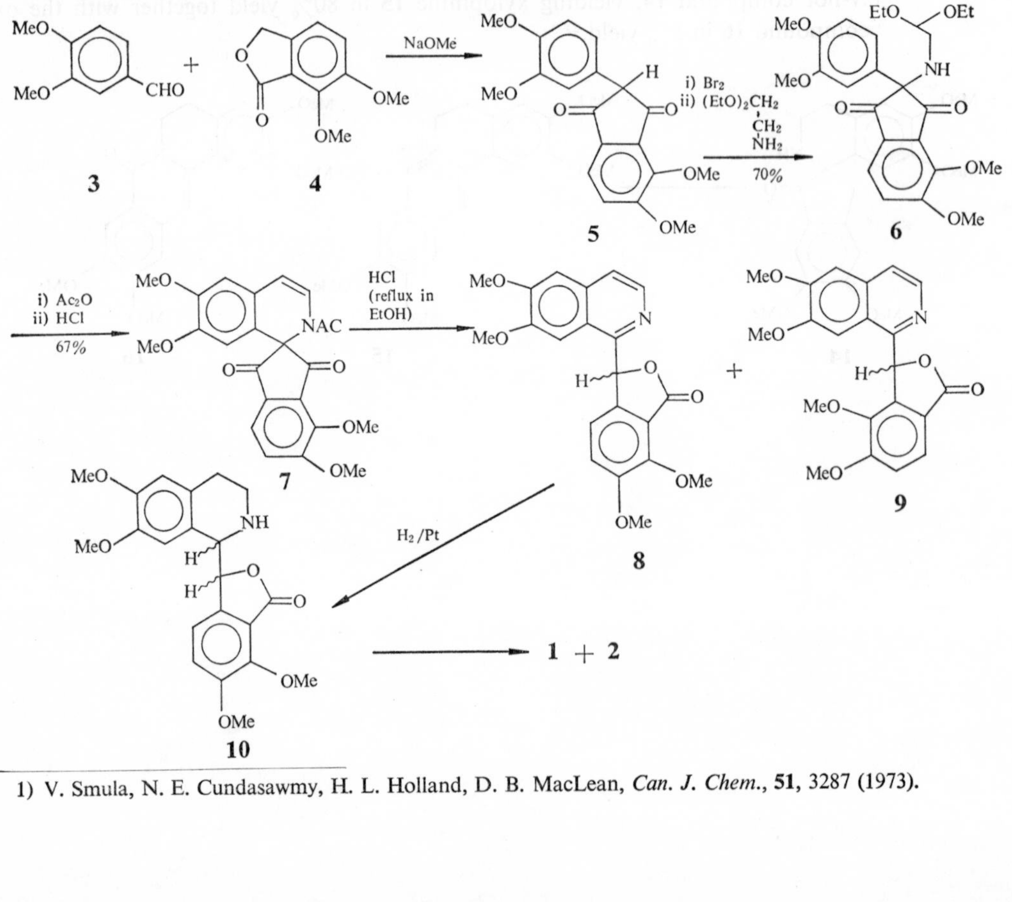

1) V. Smula, N. E. Cundasawmy, H. L. Holland, D. B. MacLean, *Can. J. Chem.*, **51**, 3287 (1973).

7 ⟶ 8,9: A mechanism for the formation of the phthalide isoquinoline alkaloid **8** from **7** has been proposed as shown below. A similar process with cleavage of the other C-C bond leads to **9**. The yields of **8** and **9** were 40 and 27%, respectively.

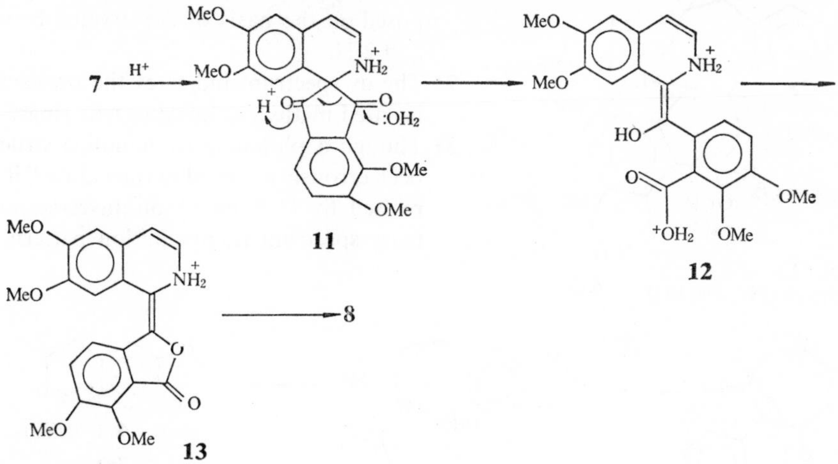

11

12

13

Absolute configuration

The absolute configuration of the phthalide isoquinolines has been established by correlation with tetrahydroberberine, as shown below.

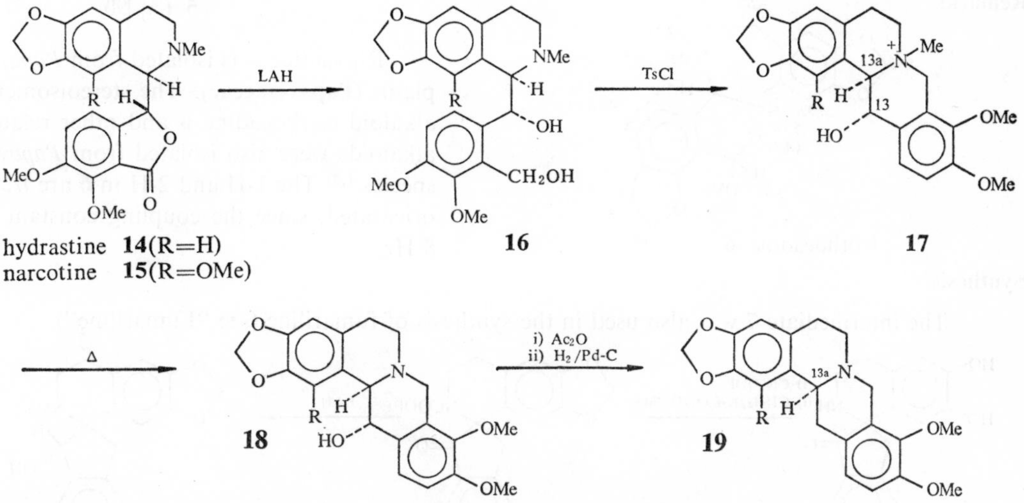

hydrastine **14(R=H)**
narcotine **15(R=OMe)**

16

17

18

i) Ac₂O
ii) H₂/Pd-C

19

1) The configuration of OH in **18** was determined by comparison of its ir and nmr spectra with those of its stereoisomers.[2] Intramolecular hydrogen bonding and Bohlman bands[3] due to *trans* quinolizidine are of particular diagnostic value.

2) The absolute configuration of tetrahydroberberines **19** at C-13a was previously determined.[4]

2) M. Ohta, H. Tani, S. Morozumi, *Tetr. Lett.*, 859 (1963); *Chem. Pharm. Bull.*, **12**, 1072 (1964); M. Ohta, H. Tani, S. Morozumi, S. Kodaira, *Tetr. Lett.*, 1857 (1963); *Chem. Pharm. Bull.*, **12**, 1080 (1964); A. R. Battersby, H. Spencer, *J. Chem. Soc.*, 1087 (1965).

3) F. Bohlmann, *Chem. Ber.*, **91**, 2157 (1958); *ibid.*, **92**, 1798 (1959); E. Wenkert, D. K. Roychardhuri, *J. Am. Soc.*, **78**, 6417 (1956).

4) H. Corrodi, E. Hardegger, *Helv. Chim. Acta*, **39**, 889 (1956).

10.28   STRUCTURE AND SYNTHESIS OF RHOEADINE

rheoadine **1**
uv: 205(4.91),240(3.96),292(3.94)

1) The *cis* junction of the B and D rings was proposed on the basis of the *J* value between 1-H and 2-H.

2) The uv spectrum indicates the presence of two isolated methylenedioxybenzene rings.

3) Though a phthalide isoquinoline structure had been errously assigned to rhoeadine,[1] it was later revised to **1**[2–4] based on investigation of its mass spectrum fragmentation pattern.

2   **3**

4 (m/e177)

5 (m/e 191)

Remarks

isorhoeadine **6**

Rhoeadine was isolated from *Papaver* plants (Papaveraceae). The stereoisomeric alkaloid isorhoeadine **6** and other related alkaloids were also isolated from *Papaver* species.[4,5] The 1-H and 2-H in **6** are *trans* orientated, since the coupling constant is 8 Hz.

Synthesis[6,7]

The intermediate **7** was also used in the synthesis of fumariline (see "Fumariline").

7   **8**   **9**

1) W. Awe, W. Winkler, *Naturwiss.*, **47**, 107 (1960).
2) F. Santavy, M. Maturova, A. Nemeckova, M. Horak, *Collect. Czech. Chem. Commun.*, **24**, 3493 (1959); *ibid.*, **25**, 1901 (1960).
3) F. Santavy, M. Maturova, A. Nemeckova, H. B. S. Schroter, H. Proestisilova, V. Preininger, *Planta Med.*, **8**, 167 (1960).
4) F. Santavy, J. L. Kaul, L. Hruban, L. Dolejs, V. Hanus, K. Blaha, A. D. Cross, *Collect. Czech. Chem. Commun.*, **30**, 335, 3479 (1965).

10   **11**   **12**

8 ⟶ 9: LAH reduction yields **9** as the sole product.

9 ⟶ 10,11: This Wagner-Meerwein rearrangement produces **10** and **11** in a 1:1 ratio.

11 ⟶ 1: This conversion has been carried out using naturally occurring **11**.[8]

Narcotine **13** was converted to the rhoeadine-type base **20** by following the sequence which was proposed to occur in nature.[9]

narcotine **13**   **14**   **15**

16   **17**

18   **19**   **20**

4) The structure of **15** was confirmed by an x-ray analysis of its rubidium salt dihydrate.

5) LiBH$_4$ reduction of **17** gave **18** stereoselectively.

5) F. Santavy, *The Alkaloids* (ed. R. H. F. Manske), vol. 12, p. 398–416, Academic Press, 1970.
6) H. Irie, T. Kishimoto, S. Uyeo, *J. Chem. Soc.* C, 3051 (1968).
7) H. Irie, S. Tani, H. Yamane, *Chem. Commun.*, 1713 (1970).
8) F. Santavy, J. L. Kaul, L. Hruben, L. Dolejs, V. Hanus, K. Blaha, A. D. Cross, *Coll. Czech. Chem. Commun.* 30, 3479 (1965).
9) W. Klötzer, S. Teitel, J. F. Blount, A. Brossi, *J. Am. Chem. Soc.*, 93, 4321 (1971).

10.29   CHEMICAL CORRELATION OF BENZYLISOQUINOLINE AND APORPHINE ALKALOIDS

(+)-Orientaline and isothebaine[1]

This shows a typical correlation of a benzylisoquinoline with an aporphine alkaloid.

(+)-orientaline **1**
α_D(HCl/H$_2$O): +53.5°

(−)-orientalinone[2] **2**
α_D: −62.2°

orientalinol **3**

(+)-isothebaine[3] **4**

1) The absolute configuration of **1** was determined by correlation with (+)-laudanosine.[4]
2) **2** exhibits a typical ABX system in the nmr spectrum due to the dienone protons.[5]
3) A tracer experiment with (+)-³H-orientaline using *P. orientale* established that it was indeed the precursor of **4**.[1]

(−)-Laudanosine and (−)-glaucine[6]

(−)-laudanosine **5**

6

(−)-glaucine **7**[7]

1) A. R. Battersby, T. H. Brown, J. H. Clements, *J. Chem. Soc.*, 4550 (1965).
2) M. Tomita, H. Furukawa, *Yakugaku Zasshi*, **82**, 1458 (1962).
3) K. W. Bentley, S. F. Dyke, *J. Org. Chem.*, **22**, 429 (1957).
4) H. Corrodi, E. Hardigger, *Helv. Chim. Acta*, **39**, 889 (1956).
5) R. Barner, A. Boller, J. Borgulya, E. G. Herzog, W. von Philipsborn, C. von Planta, A. Fürst, H. Schmid, *Helv. Chim. Acta.*, **48**, 94 (1965); K. Bernauer, *ibid.*, **46**, 1783 (1963); *ibid.*, **47**, 2122 (1964), L. J. Haynes, K. L. Stuart, D. H. R. Barton, G. W. Kirby, *Proc. Chem. Soc.*, 280 (1963), *ibid.*, 261 (1964).
6) F. Faltis, E. Alder, *Arch. Pharm.*, **284**, 281 (1951).
7) J. Go, *Yakugaku Zasshi*, **50**, 122, 937, 940 (1930).

10.30 STRUCTURE AND SYNTHESIS OF HASBANANE ALKALOIDS

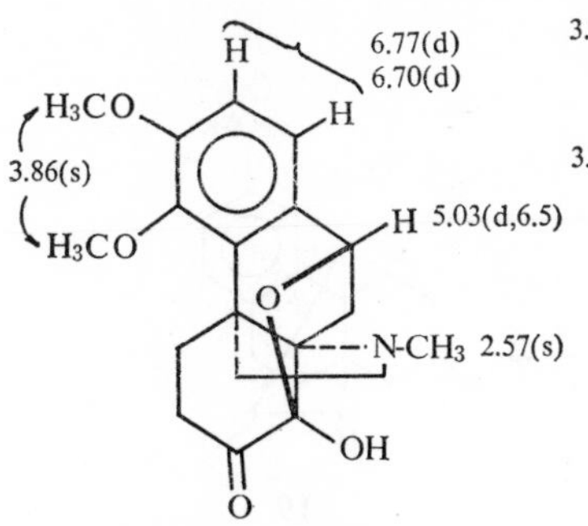

metaphanine **1**
mp: 232(acetone/CHCl₃)
ms: M⁺ 345
ir: 1730
pKa': 6.03

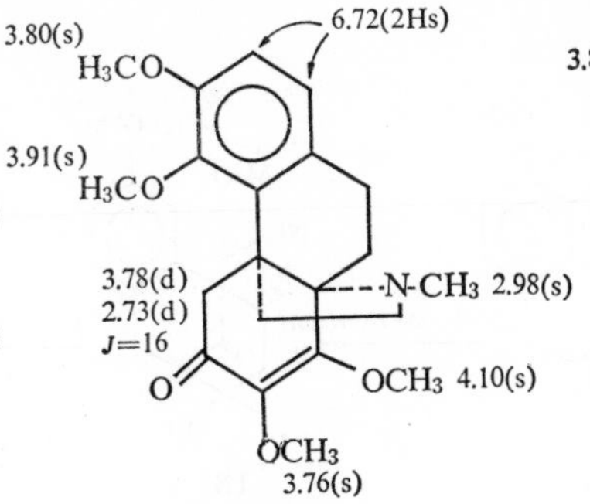

6.77(d)
6.70(d)
3.86(s)
5.03(d,6.5)
N-CH₃ 2.57(s)

hasbanonine **2**
mp: 116°
ir: 1664

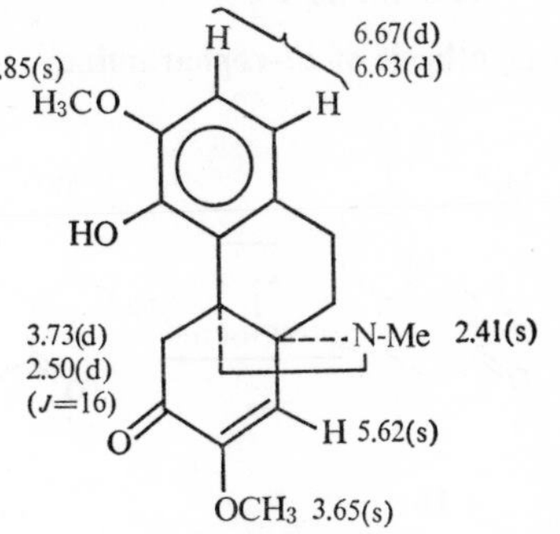

3.80(s)
3.91(s)
6.72(2Hs)
3.78(d)
2.73(d)
$J=16$
N-CH₃ 2.98(s)
OCH₃ 4.10(s)
OCH₃
3.76(s)

cepharamine **3**
mp: 186–187°
ms: M⁺ 329
α_D: −248°(CHCl₃)
uv: 259nm
ir: 3470, 1685, 1630

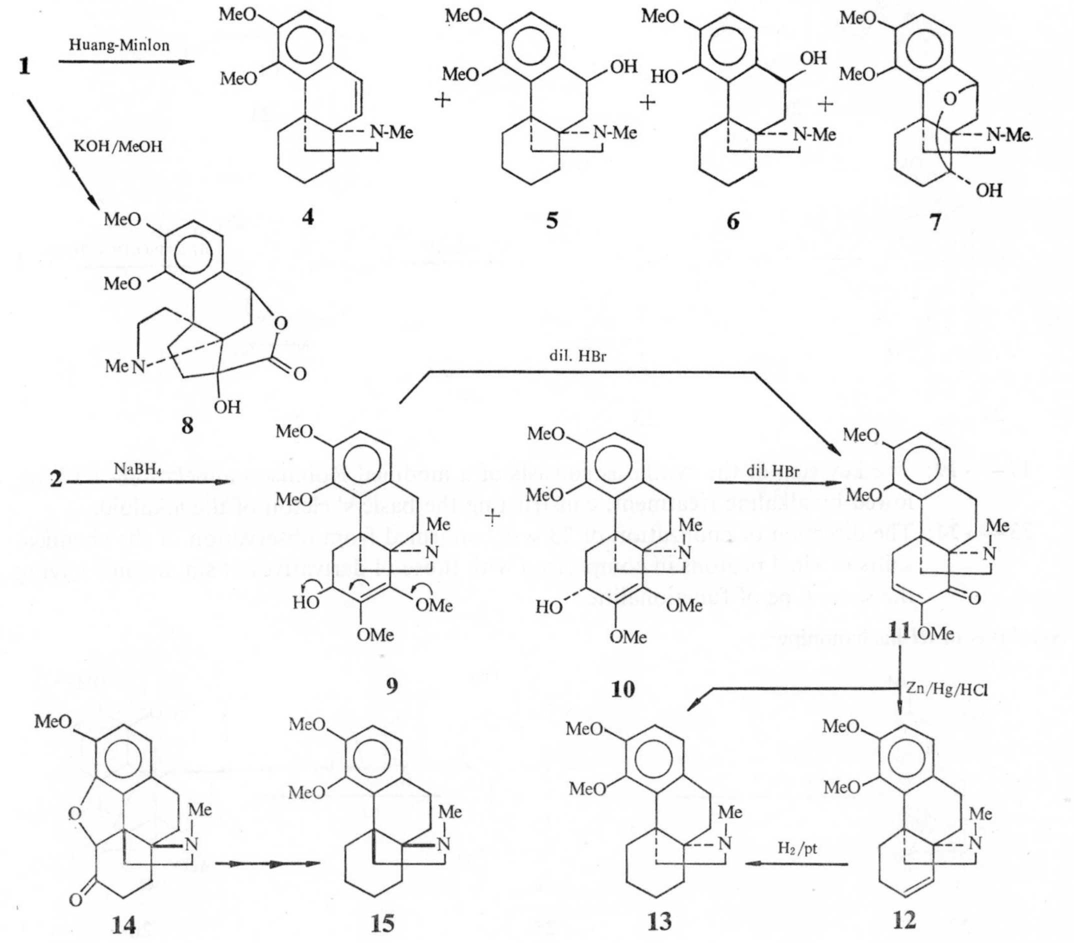

3.85(s)
6.67(d)
6.63(d)
3.73(d)
2.50(d)
($J=16$)
N-Me 2.41(s)
H 5.62(s)
OCH₃ 3.65(s)

1

Huang-Minlon → **4**

+ **5** + **6** + **7**

KOH/MeOH → **8**

dil. HBr

2 NaBH₄ → **9** + **10** dil. HBr → **11** OMe

Zn/Hg/HCl

14 → → **15**   **13** ← H₂/pt **12**

13 is the enantiomer of **15** which was derived from codeine through dihydroindolino-codeinone **14**.

Synthesis of *dl*-cepharamine[1]

17 ⟶ 19: The key step of the synthesis consists of a modified Robinson annelation of **17** followed by alkaline treatment, constructing the basic skeleton of the alkaloid.

23 ⟶ 24: The direction of enolization of **23** was confirmed from observation of the chemical shifts of vinyl protons in comparison with those of derivatives of sinomenine having the same type of functionalities.

Synthesis of *dl*-hasbanonine[2]

Synthesis of *dl*-metaphanine[3]

1) Y. Inubushi, T. Ibuka, M. Kitano, *Tetr. Lett.* 1611 (1969); Y. Inubushi, T. Ibuka, M. Kitano, *Chem. Pharm. Bull.*, **19**, 1820 (1971).

2) T. Ibuka, K. Tanaka, Y. Inubushi, *Tetr. Lett.*, 4811 (1970).

3) T. Ibuka, K. Tanaka, Y. Inubushi, *Tetr. Lett.*, 1393 (1972).

36   **37**

34 ⟶ 35: Hydride reduction of **34** in the usual manner provided the undesired epimeric alcohol which could not be used to construct the hemiacetal grouping of metaphanine.

Remarks

Metaphanine **1**[4) was isolated from *Stephania japonica* (Menispermaceae) along with hasbanonine **2**.[5) Cepharamine **3**[6) was isolated from *Stephania cepharantha* Hayata as a phenolic alkaloid.

4) M. Tomita, T. Ibuka Y. Inubushi, K. Takeda, *Tetr. Lett.*, 3605 (1964); M. Tomita, T. Ibuka, Y. Inubushi, K. Takeda, *Chem. Pharm. Bull.*, **13**, 695 (1965); M. Tomita, T. Ibuka, Y. Inubushi, K. Takeda, *ibid.*, **13**, 704 (1965).
5) M. Tomita, T. Ibuka, Y. Inubushi, Y. Watanabe, M. Matsui, *Tetr. Lett.*, 2937 (1964); M. Tomita, T. Ibuka, Y. Inubushi, Y. Watanabe, M. Matsui, *Chem. Pharm. Bull.*, **13**, 538 (1965).
6) M. Tomita M. Kozuka, *Tetr. Lett.*, 6229 (1966).

10.31 BIOSYNTHESIS OF MORPHINE ALKALOIDS

norlaudanosoline **1**   (−)-reticuline **2**   **3**

(+)salutaridine **4**   salutaridinol-I **5**   thebaine **6**

codeinone **7**   codeine **8**   morphine **9**

1) A number of alkaloids such as morphine **9**, codeine **8** or thebaine **6**, generally isolated from Papaveraceae, Ranunculaceae, Berberidaceae, Menispermaceae and Magnoliaceae etc., are known to be produced through several reactions including methylation and phenol oxidative coupling of the key compound, norlaudanosoline **1**.

2) **1** could be formed by the Mannich type condensation of 3,4-dihydroxytyramine and a C_6-C_2 unit, equivalent with 3,4-dihydroxyphenylpyruvic acid probably derived from DOPA.

3) The radioactive salutaridine **4** and salutaridinol-I (one of the stereoisomers) **5** are effectively incorporated into thebaine.[1,2]

4) The role of codeinone **7** as an intermediate in the conversion of thebaine to codeine and morphine was also proved by $^{14}CO_2$ feeding experiments.[3–4]

   The biosyntheses of narcotine **12**, protopine **14**, and chelidonine **16** are proceeded through (+)-reticuline **10**[5] as shown.

1) D. H. R. Barton, G. W. Kirby, W. Steglich, G. M. Thomas, A. R. Battersby, T. A. Dobson, H. Ramuz, *J. Chem. Soc.*, 2423 (1965).
2) D. H. R. Barton, D. S. Bhakuni, R. James, G. W. Kirby, *J. Chem. Soc.* C, 128 (1967).
3) H. Rapoport, F. R. Stermitz, D. R. Baker, *J. Am. Chem. Soc.*, **82**, 2765 (1960).
4) G. Blaschke, H. I. Parker, H. Rapoport, *J. Am. Chem. Soc.*, **89**, 1540 (1967).
5) A. R. Battersby, R. J. Francis, M. Hirst, R. Southgate, J. Staunton, *Chem. Commun.*, 602 (1967).

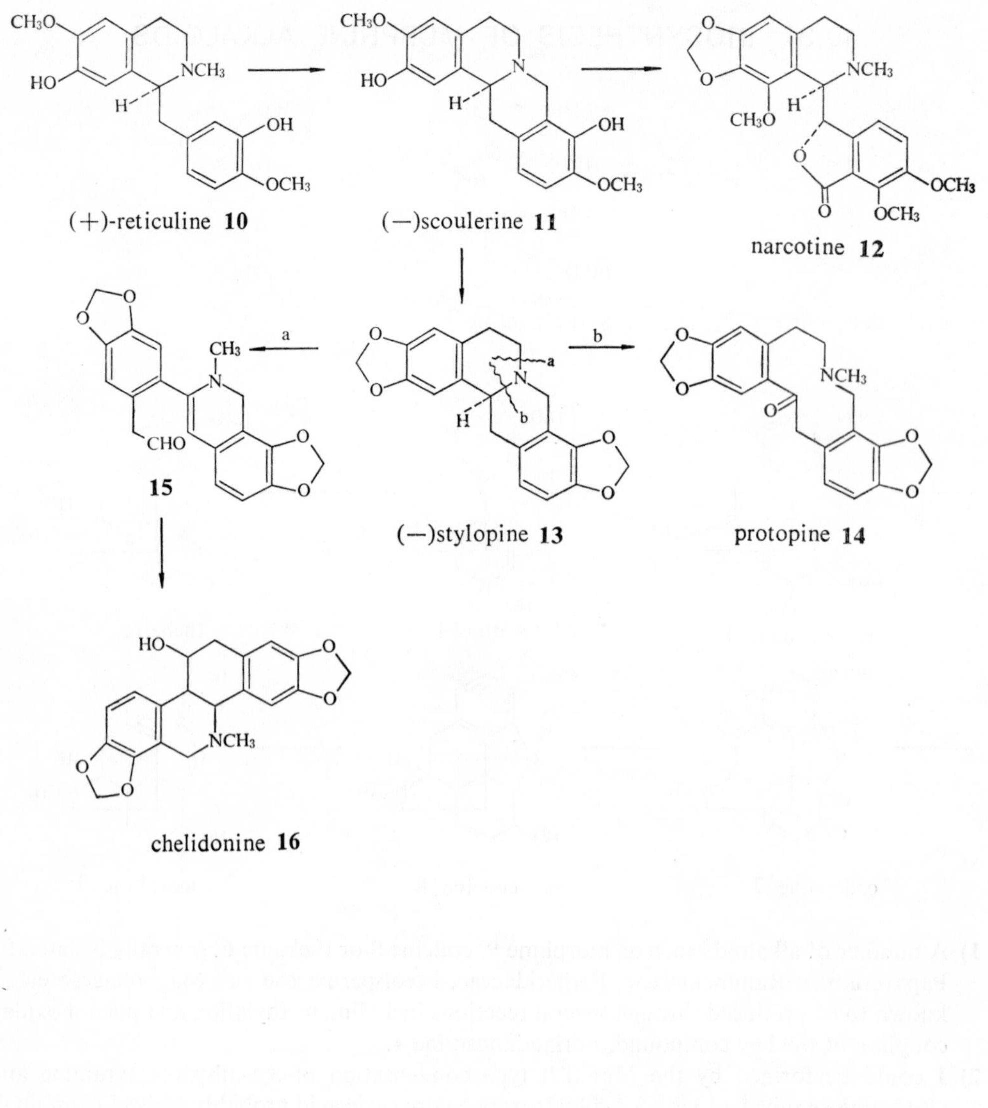

(+)-reticuline **10**   (−)scoulerine **11**

narcotine **12**

15   (−)stylopine **13**   protopine **14**

chelidonine **16**

10.32 STRUCTURE OF ACUTUMINE AND RELATED ALKALOIDS

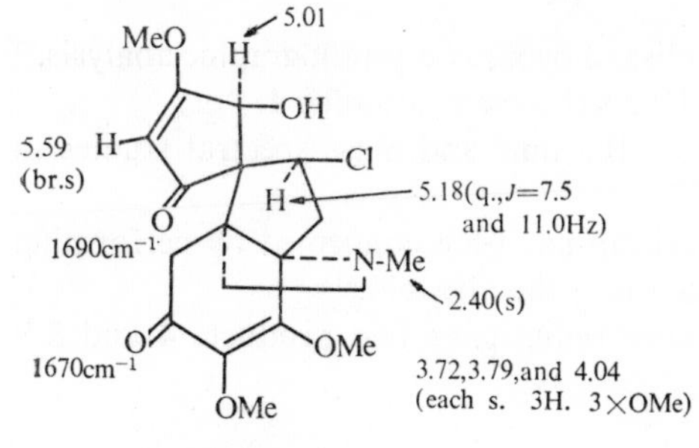

MeO — 5.01

H

—OH

5.59
(br.s) H

Cl

H — 5.18(q.,J=7.5
and 11.0Hz)

1690cm^{-1}

—N-Me

2.40(s)

1670cm^{-1} OMe

OMe 3.72,3.79,and 4.04
(each s. 3H. 3×OMe)

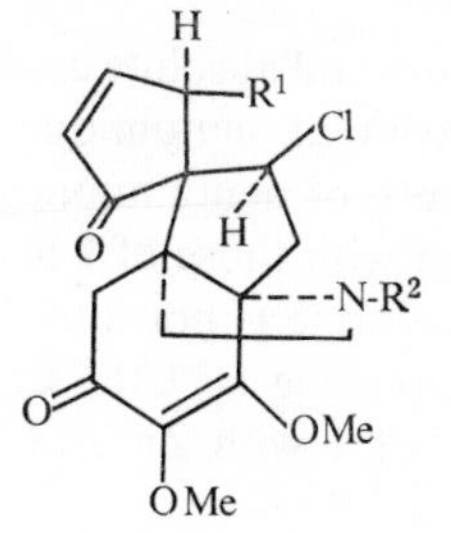

acutumine **1**

m.p : 238—240°
pKa : 5.3(50%EtOH)
$[\alpha]_D$: —206°(c=0.69)
UV : 245(19000) 270(1000)

H

R^1

Cl

H

N-R^2

OMe

O Me

acutumidine **2**(R^1=OH R^3=H)
acutuminine[2] **3**(R^1=H R^2=Me)

1) The chemical shifts values are those in pyridine-d$_6$[1]

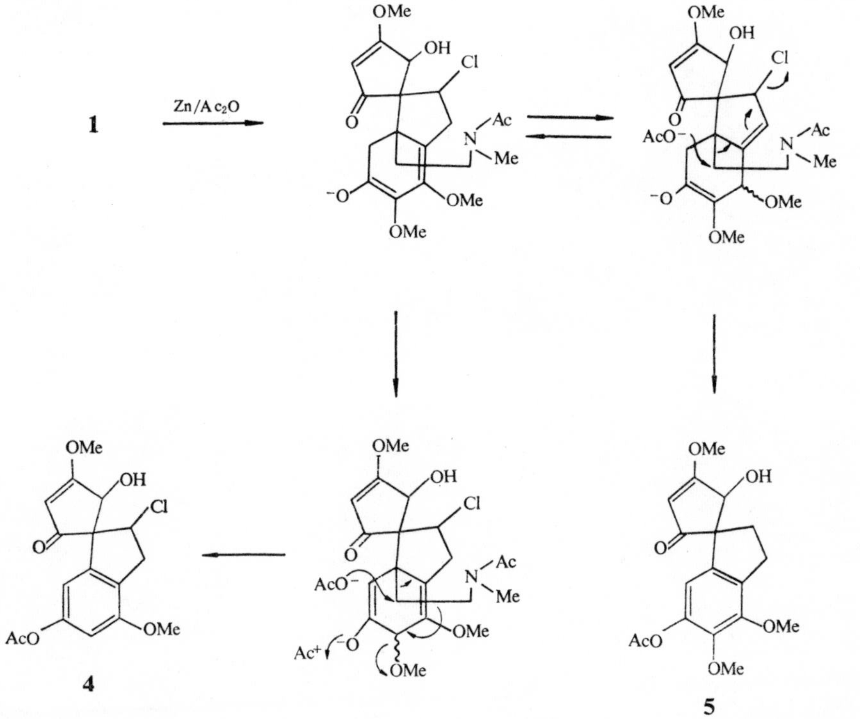

1 $\xrightarrow{\text{Zn/A c}_2\text{O}}$

OMe
OH
Cl
O
N—Ac
Me
-O— OMe
OMe

OMe
OH
Cl
AcO$^-$
N—Ac
Me
-O— OMe
OMe

OMe
OH
Cl
O
AcO$^-$
N—Ac
Me
Ac$^+$ O OMe
OMe

OMe
OH
Cl
O
AcO OMe
OMe

5

OMe
OH
Cl
O
AcO OMe

4

1) M. Tomita, Y. Okamoto, T. Kikuchi, K. Osaki, M. Nishikawa, K. Kamiya, Y. Sasaki, K. Matoba, K. Goto, *Chem. Pharm. Bull.*, **19**, 770 (1971); M. Tomita, Y. Okamoto, T. Kikuchi, K. Osaki, M. Nishikawa, K. Kamiya, Y. Sasaki, and K. Matoba, *Tetr. Lett.*, 2425 (1967).
2) Y. Okamoto, E. Yuge, Y. Nagai, R. Kutsuta, A. Kishimoto, Y. Kobayashi, T. Kikuchi, M. Tomita, *Tetr. Lett.*, 1933 (1969).

Remarks

1. Source. *Sinomenium acutum* REHD. *et* WILS and *Menispermum dauricum* DC (Menisperm-aceae).[1]

2. The structure and absolute configuration were established by x-ray crystallographic analysis.[3]

3. *N*-Methylation of acutumidine **2** with CH_2O and HCO_2H gave acutumine **1**.

4. The structure of acutuminine **3** was proposed from the nmr and mass spectral studies in comparison with those of **1** and **2**.[2]

5. That the chlorine is not introduced during the work-up has been confirmed by performing the work-up using $H^{36}Cl$: ^{36}Cl was not incorporated into the alkaloid.[4]

6. Treatment of **1** with Zn and acetic anhydride under reflux gave two products **4** and **5**.[1]

3) M. Tomita, Y. Okamoto, T. Kikuchi, K. Osaki, M. Nishikawa, K. Kamiya, Y. Sasaki, K. Matoba, K. Goto, *Tetr. Lett.*, 2421 (1967); *J. Chem. Soc.*, B, 652 (1968).

4) D. H. R. Barton, *Chemistry in Britain*, **3**, 330 (1967).

10.33 BIOSYNTHESIS OF ERYTHRINA ALKALOIDS

As an interesting example of isoquinoline alkaloids, the *Erythrina* alkaloids have first been assumed to be formed by the oxidative coupling of the C_6-C_2-N-C_2-C_6 type intermediate **1** which is plausibly derived from two molecules of DOPA through its decarboxylation as shown below.[1] In fact, the participation of tyrosine to the biosynthesis of α- and β-erythroidine has been proved by tracer experiments.[2]

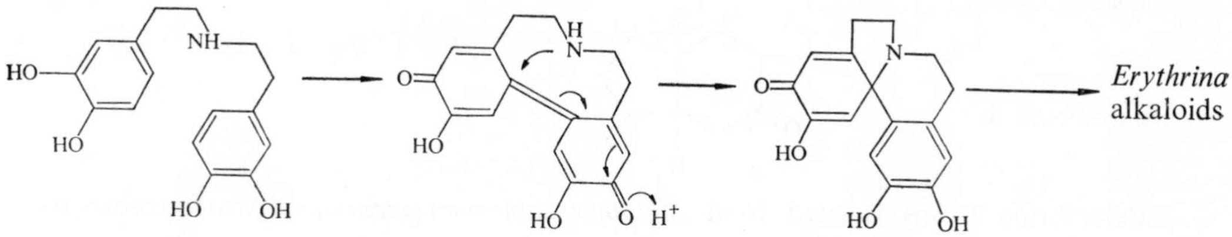

1   **2**   **3**   *Erythrina* alkaloids

Recent results have shown that erythraline **9** is produced by the phenol oxidative coupling and following transformation of the benzylisoquinoline **4**.[3,4]

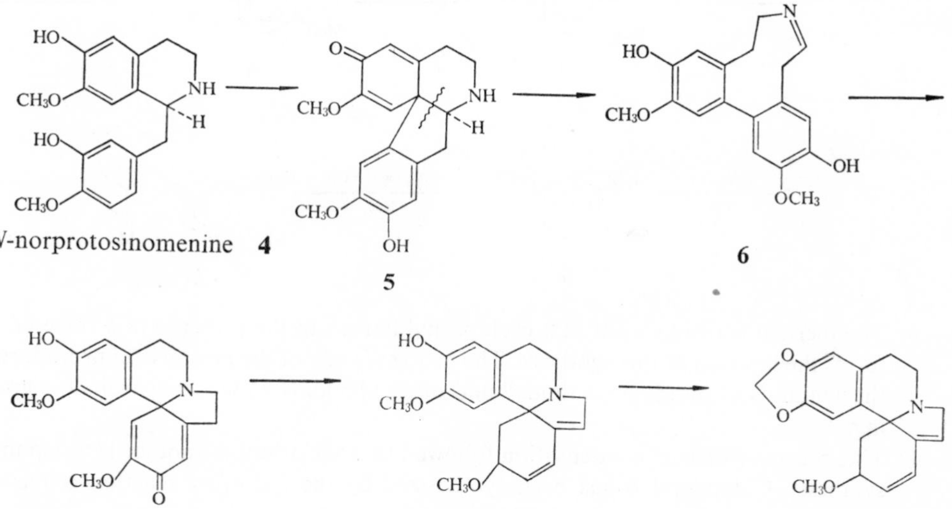

(S)-*N*-norprotosinomenine **4**   **5**   **6**

7   erythaline **8**   **9**

1) E. Leete, *Biogenesis of Natural Compounds* (ed. P. Bernfeld), p. 988, Pergamon Press, (1967).
2) E. Leete, A. Ahmad, *J. Am. Chem. Soc.*, **88**, 4722 (1966).
3) D. H. R. Barton, R. B. Boar, D. A. Widdowson, *J. Chem. Soc. C*, 1213 (1970).
4) D. H. R. Barton, C. J. Potter, D. A. Widdowson, *J. Chem. Soc. Perkin 1*, 346 (1974).

10.34 SYNTHESIS OF MESEMBRINE

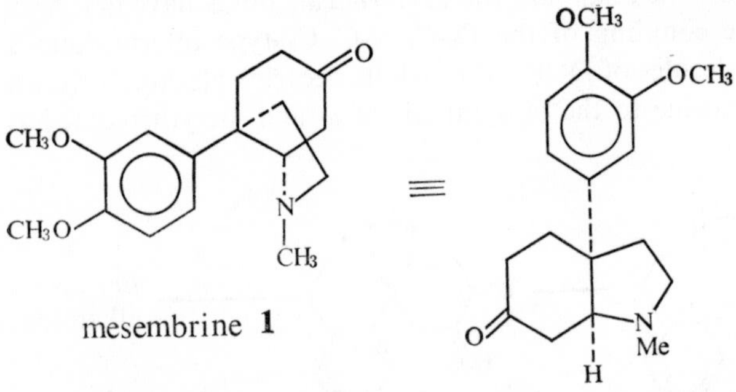

mesembrine **1**

Mesembrine **1** was isolated from *Mesembryanthemum tortuosum* (Aizoaceteae).[1] Its structure has been determined.[2]

Route I[3]

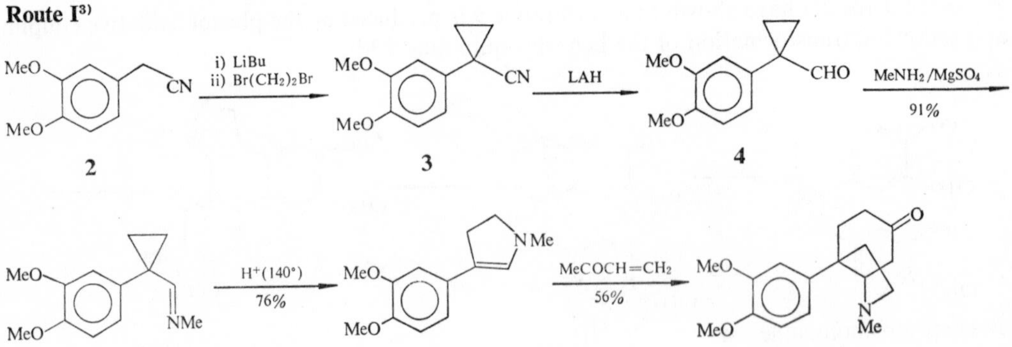

5 ⟶ 6: The thermal rearrangement of a cyclopropyl imine[4] in the presence of a trace of acid forms the key step of this synthesis. The process is one of the most useful procedures in the synthesis of pyridine alkaloids[5] and other alkaloids with octahydroindole groupings.[6]

6 ⟶ 1: This process (Michael condensation followed by annelation) is another key step in the synthesis. Compound **6** has been synthesized by the following reaction sequence.[7]

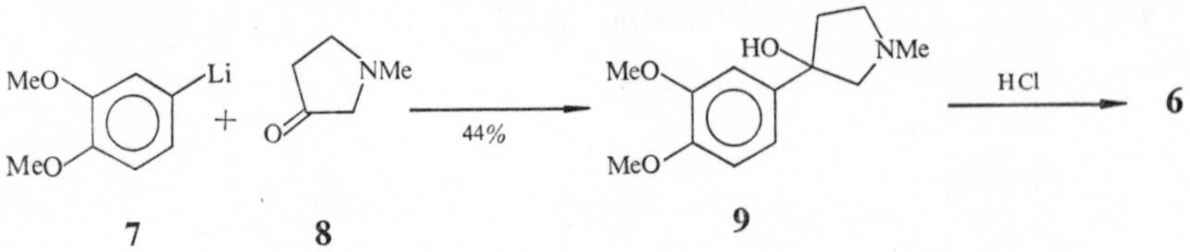

1) K. Bodendorf, W. Krieger, *Arch. Pharm.*, **290**, 441 (1957).
2) P. Coggon, D. S. Farrier, P. W. Jeffs, A. T. McPhail, *J. Chem. Soc.* B, 1267 (1970).
3) R. V. Stevens, M. P. Wentland, *J. Am. Chem. Soc.*, **90**, 5580 (1968).
4) J. B. Cloke, L. H. Baer, J. M. Robbins, G. E. Smith, *J. Am. Chem. Soc.*, **67**, 2155 (1945).
5) R. V. Stevens, M. P. Wentland, *J. Am. Chem. Soc.*, **90**, 5576 (1968).
6) R. V. Stevens, L. E. DuPree, Jr., *Chem. Commun.*, 1585 (1970).
7) T. J. Curphey, H. L. Kim, *Tetr. Lett.*, 1441 (1968).

Route II[8)]

10 → **11** → **12**

13 → **14** → **1**

12 ⟶ 13: This is the key step in the synthesis. The hydrolysis with 65% H_2SO_4 at 140° for 5 min gave **13** as the only product.

Route III[9)]

This is an interesting asymmetric synthesis using L-proline pyrrolidide.[10)] The final product, (+)-mesembrine, is enantiomeric with the natural product.

15 → **16** → **17**

18 → **19**

20 → **21** →

22

$19 \longrightarrow 20$: This process is a modification of Darzen's method.

$21 \longrightarrow 22$: Methylvinylketone attacks the enamine from the less-hindered side, producing partially optically active **22**.

8) T. Oh-ishi, H. Kugita, *Tetr. Lett.*, 5445 (1968); *Chem. Pharm. Bull.*, **18**, 291, 299 (1970).
9) S. Yamada, G. Otani, *Tetr. Lett.*, 1133 (1971).
10) S. Yamada, G. Otani, *Tetr. Lett.*, 4237 (1969).

10.35 STRUCTURE OF LYCORINE

lycorine **1** (—2Ac)
diacetyllycorine **2**

3

1) The values of the chemical shifts shown are those of diacetyllycorine **2** (because of the insolubility of lycorine in the usual organic solvents).[1]
2) The absolute configuration was deduced from the ord spectrum of the ketone **3**.
3) Proof of the structure was provided by x-ray crystallographic analysis of dihydrolycorine hydrobromide.[2]
4) Detailed discussion of ord experiments with this type of alkaloid has been provided by Kotera and Kuriyama.[3]

Remarks

Lycorine **1** is the most abundant alkaloid in Amaryllidaceae plants. Various chemical degradation reactions have been carried out.

4

Narcissidine **4** is one of the trioxygenated alkaloids of this group and the structure was unambiguously established by x-ray crystallographic analysis.[4] A different structure had errously been assigned on the basis of the consumption of one mole of periodate.

1) K. Kotera, Y. Hamada, K. Tori, K. Aono, K. Kuriyama, *Tetr. Lett.*, 2009 (1966).
2) M. Shiro, T. Sato, H. Koyama, *Chem. Ind.*, 1229 (1966).
3) K. Kotera, Y. Hamada, R. Mitsui, *Tetr. Lett.*, 6273 (1966); K. Kuriyama, T. Iwata, M. Moriyama, K. Kotera, Y. Hamada, R. Mitsui, K. Takeda, *J. Chem. Soc.* B, 46 (1967); K. Kotera, Y. Hamada, R. Mitsui, *Tetr.*, **24**, 2463 (1968).
4) J. C. Clardy, W. C. Wildman, F. M. Hauser, *J. Am. Chem. Soc.*, **92**, 1781 (1970).

10.36 SYNTHESIS OF LYCORANE

α-lycorane **1** (BC/CD=*trans/cis*)
β-lycorane **2** (BC/CD=*trans/trans*)
γ-lycorane **3** (BC/CD=*cis/cis*)
δ-lycorane **4** (BC/CD=*cis/trans*)

lycorine **5** (R=OH)
caranine **6** (R=H)

1) Lycorane is the trivial name of the parent skeleton of lycorine and related alkaloids.
2) Four stereoisomeric lycoranes **1, 2, 3** and **4** have been obtained[1] from lycorine **5** and caranine **6** as degradation products.

Synthesis of α-lycorane[2]

Synthesis of β-lycorane[2]

1) K. Takeda, K. Kotera, S. Mizukami, M. Kobayashi, *Chem. Pharm. Bull.*, **8**, 483 (1960); K. Kotera, *Tetr.*, **12**, 240, 248 (1961).
2) R. K. Hill J. A. Joule, L. J. Loeffler, *J. Am. Chem. Soc.*, **84**, 4951 (1962).

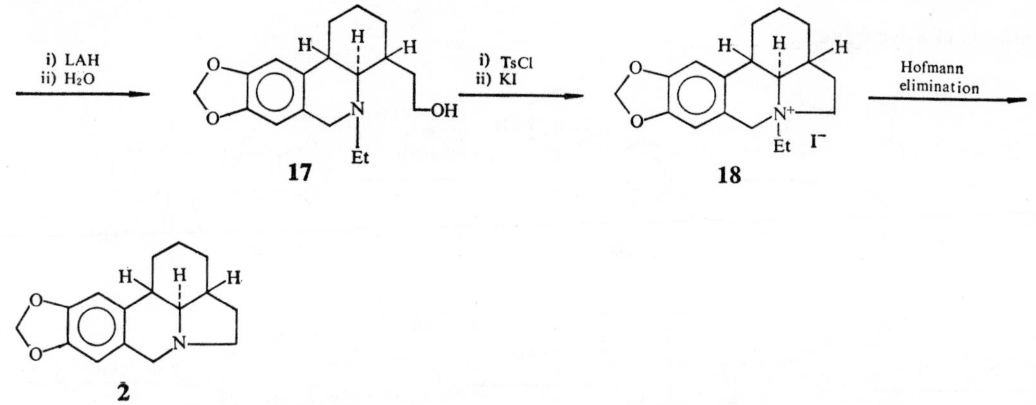

i) LAH
ii) H₂O →

17

i) TsCl
ii) KI →

18

Hofmann
elimination →

2

7 ⟶ 13: Cycloaddition of the nitrostyrene **7** with 1-acetoxybutadiene was carried out in toluene under reflux, yielding **13** as the sole product.

13 ⟶ 14: **14** was obtained as the result of O ⟶ N acyl migration during the hydrogenation and work-up.

Synthesis of γ-lycorane[3]

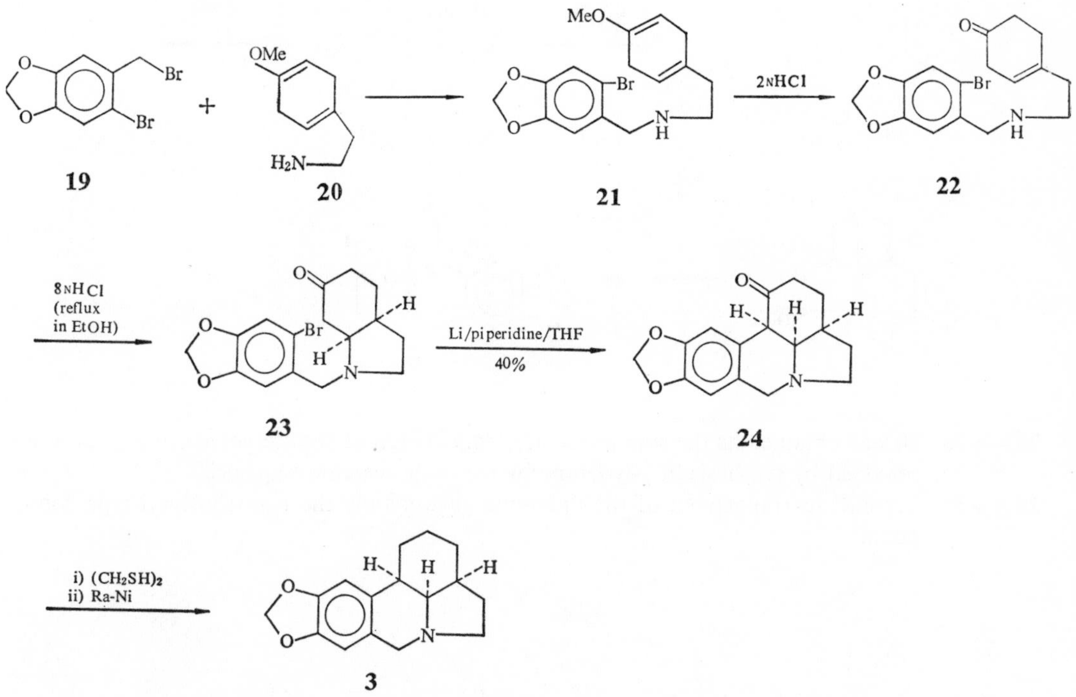

19 + **20** →

21

2 N HCl →

22

8 N HCl
(reflux
in EtOH) →

23

Li/piperidine/THF
40% →

24

i) (CH₂SH)₂
ii) Ra-Ni →

3

19,20 ⟶ 21: Condensation of **19** and **20** gave the monobenzyl **21** and dibenzyl products in a ratio of 3:2.

3) N. Ueda, T. Kokuyama, T. Sakan, *Bull. Chem. Soc. Japan*, **39**, 2012 (1966).

Synthesis of δ-lycorane[4]

25 ⟶ 26: **26** was obtained as the sole product. Confirmation of the stereostructure of **26** was provided by synthesis of γ-lycorane by the same reaction sequence.

29 ⟶ 30: Schmidt rearrangement of the indanone gave mainly the isocarbostyryl-type compound.

4) H. Irie, Y. Nishitani, M. Sugita, K. Tamoto, S. Uyeo, *J. Chem. Soc. Perkin* I, 588 (1972); H. Irie, Y. Nishitani, M. Sugita, S. Uyeo. *Chem. Commun.*, 1313 (1970).

10.37 SYNTHESIS OF CRINANE BY PHOTOCYCLIZATION
OF AN ENAMIDE

Crinane **1**

Crinane is the basic ring system of crinine and the related alkaloids occurring in *Amaryllidaceae* plants.

Remarks[1]

The photolysis of **4** is the key step of this crinane-synthesis, the reaction being controlled by the conrotatory electrocyclic reaction of the six π-electron system followed by suprafacial [1,5]sigmatropic hydrogen rearrangement. The basic investigation of this reaction was reported in connection with the synthesis of 3,4-dihydrocarbostyryl derivatives.[2]

4 ⟶ 5: **4** was irradiated with a low-pressure mercury lamp at room temperature for five hours.

1) I. Ninomiya, T. Naito, T. Kiguchi, *Chem. Commun.* 1669 (1970).
2) I. Ninomiya, T. Naito, T. Kiguchi, *Chem. Commun.*, 1662 (1970); I. Ninomiya, T. Naito, T. Mori, *Tetr. Lett.*, 3643 (1969); I. Ninomiya, T. Naito, T. Kiguchi, *ibid.*, 4451 (1970).

10.38   SYNTHESIS OF OXOCRININE AND OXOMARITIDINE BY ANODIC OXIDATION

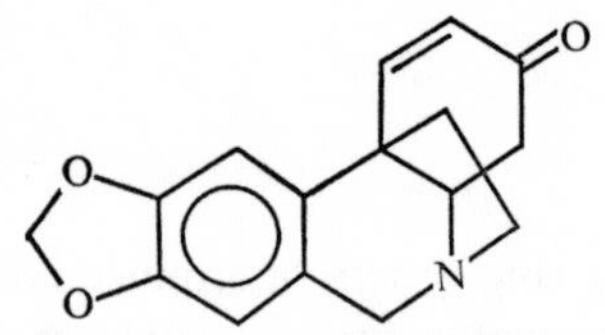

oxocrinine **1** 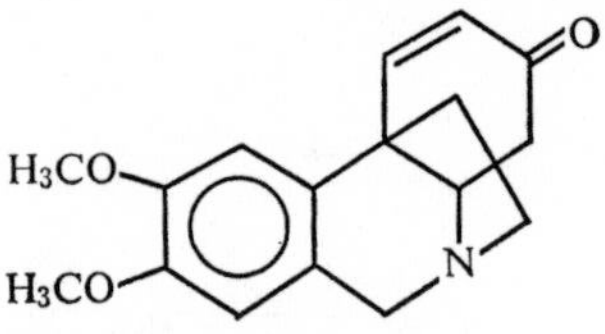   oxomaritidine **2**

The reactions shown below were all carried out using an H-shaped glass cell and platinum electrodes at room temperature in acetonitrile.[1]

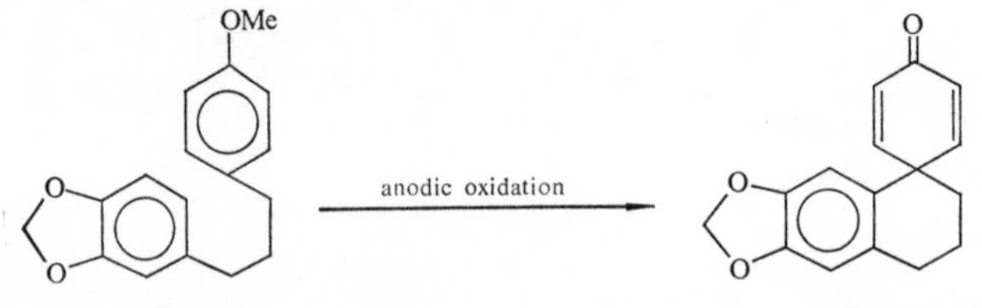

anodic oxidation

3   **4**

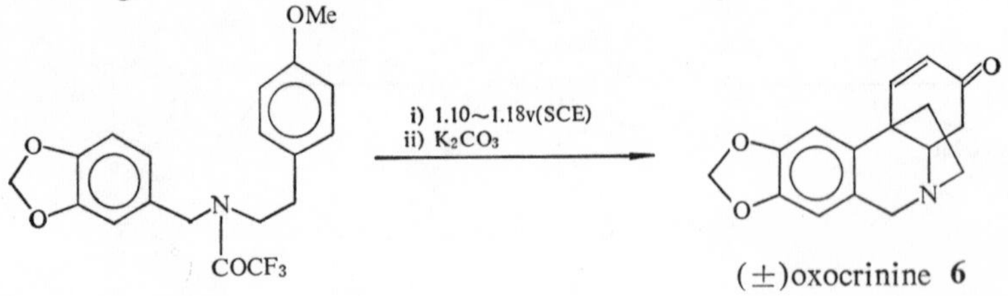

i) 1.10~1.18v(SCE)
ii) K₂CO₃

5   (±)oxocrinine **6**

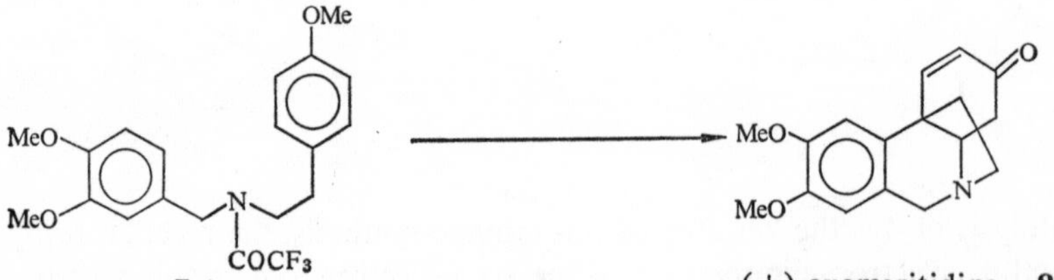

7   (±)-oxomaritidine **8**

1) E. Kotani, N. Takauchi, S. Tobinaga, *Chem. Commun.*, 550 (1973).

10.39   SYNTHESIS OF GALANTHAMINE BY PHENOL OXIDATIVE COUPLING

galanthamine **1**

2 ⟶ 3: The presence of a bromine atom at the para position in relation to the phenolic hydroxy group inhibits para-para coupling[1-3] producing **3** in reasonably good yield. The original reaction was carried out by Barton.[4]

1) T. Kametani, K. Yamaki, H. Yagi, K. Fukumoto, *Chem. Commun.*, 425 (1969); *J. Chem. Soc.* C, 2602 (1969).
2) R. A. Abramovitch, S. Takahashi, *Chem. Ind.*, 1039 (1963).
3) B. Frank, J. Lubs, G. Dienkelmann, *Angew. Chem.*, **79**, 989 (1967).
4) D. H. R. Barton, G. W. Kirby, *J. Chem. Soc.*, 806 (1962).

10.40   SYNTHESIS AND CORRELATION OF HAEMANTHAMINE, HAEMANTHIDINE AND TAZETTINE

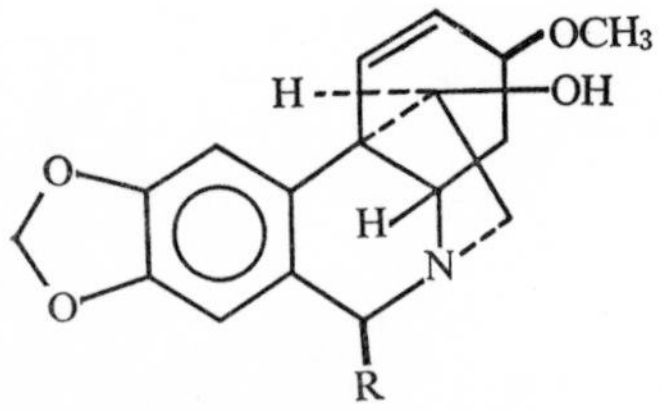

R=H, haemanthamine **1**
R=OH, haemanthidine **2**

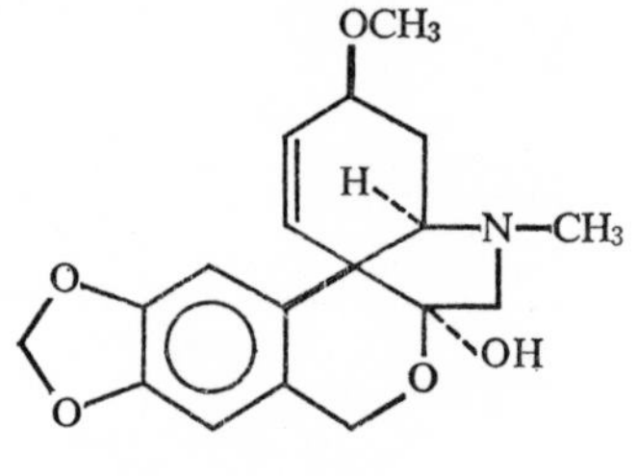

tazettine **3**

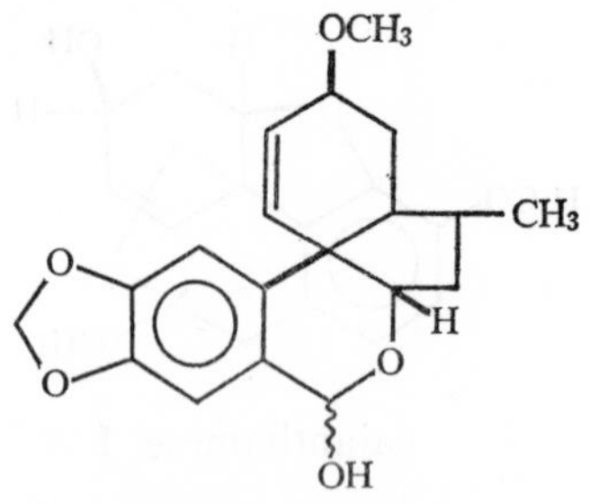

pretazettine **4**

Synthesis

The value of 2-pyrroline-4,5-dione as a dienophile in diene synthesis reactions was elegantly demonstrated[1] in the synthesis of 5,10-ethanophenanthridine alkaloids such as haemanthamine **1**, haemanthidine **2**, and tazettine **3**, which are found in Amaryllidaceae plants.

Route I [1]

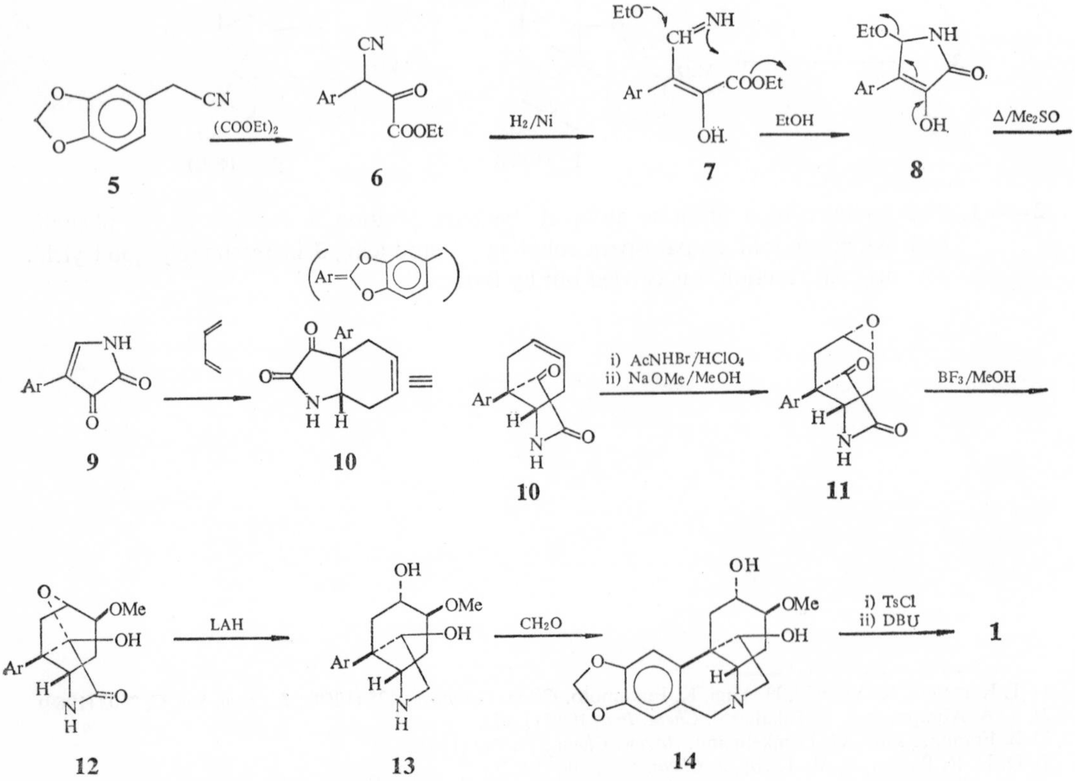

13 → (i) HCOOH/Ac₂O, ii) OH⁻, iii) Ac₂O/py) → **15** → POCl₃ → **16** → MeOH → **17**

16 → H₂O → **19** → MeI/OH⁻ → **20**

17 → ⁻OH → **18**

20 → i) TsCl, ii) DBU → **3**

18 → i) TsCl, ii) DBU, iii) H⁺ → **2**

6 → 7: Partial hydrogenation of **6** with Raney nickel in ether gave the imino-ester **7** which, on treatment with ethanol, was cyclized to **8**.

12 → 13: Reduction gave the key intermediate **13** as a sole product. Thus, the hemiketal ring in **12** provided steric control for reduction of the five-membered ring ketone.

13 → 14: Pictet-Spengler cyclization.

15 → 16: Bischler-Napieralski cyclization yields **16** which, without purification, was converted to **17** and **19** on treatment with methanol and water, respectively.

19 → 20: This type of rearrangement (internal Cannizzaro hydride transfer) occurs under mild alkaline conditions.

Route II

An alternative synthesis of **2** and **3** was achieved by utilizing the cycloaddition of arylmaleic anhydride and butadiene.[2,3]

21 → i) butadiene, ii) MeOH → **22** → i) SOCl₂, ii) NaN₃(Δ) → **23** → H⁺ → **24**

Ar = 1,3-benzodioxol-5-yl

1) Y. Tsuda, K. Isobe, A. Ukai, *Chem. Common.*, 1554 (1971); Y. Tsuda, K. Isobe, *ibid.*, 1555 (1971).
2) J. B. Hendrickson, T. L. Bogard, M. E. Fisch, *J. Am. Chem. Soc.*, **92**, 5538 (1970).
3) J. B. Hendrickson, C. Foote, N. Yoshimura, *Chem. Commun.*, **165** (1965).

27 ⟶ 28: Normal esterification of the hydroxyacid formed by hydrolysis resulted in relactonization: the preparation of the phenacyl ester follows.

29 ⟶ 2: Acidic disiamylborane reduces the 11-keto group stereospecifically because of the presence of the MsO group. Intramolecular Cannizzaro reaction to yield a tazettine skeleton, which occurs under basic conditions, was prevented in this reduction. The acetylation is to prevent the Cannizzaro reaction occurring during the subsequent reaction.

4) R. J. Highet, P. F. Highet, *Tetr. Lett.*, 4099 (1966); *J. Org. Chem.*, **33**, 3105 (1968); W. I. Taylor, S. Uyeo, H. Yajima, *J. Chem. Soc.*, 2962 (1955); T. Ikeda, W. I. Taylor, Y. Tsuda, S. Uyeo, H. Yajima, *ibid.*, 4749 (1956); T. Ikeda, W. I. Taylor, Y. Tsuda, S. Uyeo, *Chem. Ind.*, 411 (1956); R. J. Highet, W. C. Wildman, *ibid.*, 1159 (1955); H. Irie, Y. Tsuda, S. Uyeo, *J. Chem. Soc.*, 1446 (1959); Y. Tsuda, S. Uyeo, *ibid.*, 1159 (1961).
5) W. C. Wildman, D. T. Bailey, *J. Am. Chem. Soc.*, **89**, 5514 (1967); W. C. Wildman, D. T. Bailey, *J. Org. Chem.*, **33**, 3749 (1968).

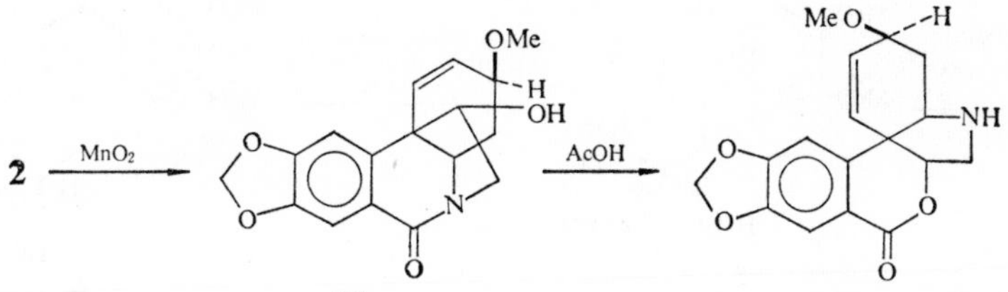

$$2 \xrightarrow{\text{MnO}_2} \mathbf{31} \xrightarrow{\text{AcOH}} \mathbf{32}$$

Remarks

Tazettine **3** is one of the most widely-distributed alkaloids occurring in Amaryllidaceae plants. Its chemistry was extensively studied nearly 15 years ago,[4] culminating in determination of the structure **3**. The compound had long been considered as a genuine alkaloid present in plants. However, in 1967, the alkaloid pretazettine **4** was isolated and tazettine was suggested to be an artifact[5] derived from it. In fact, pretazettine **4** is easily converted to tazettine **3** by treatment with 0.1N sodium hydroxide or basic alumina. **4** was also correlated with haemanthidine **2** by the sequence shown.

10.41   SYNTHESIS OF EMETINE

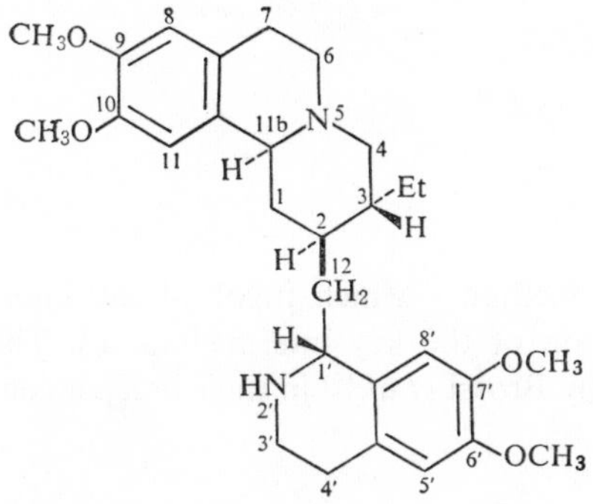

1

Route I[1]

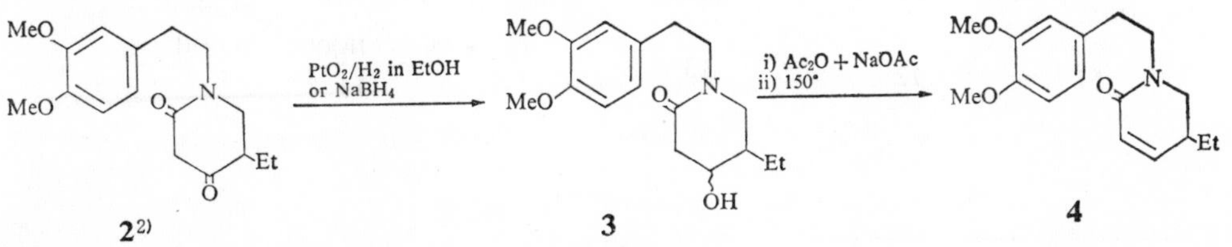

$$\mathbf{2}^{2)} \xrightarrow[\text{or NaBH}_4]{\text{PtO}_2/\text{H}_2 \text{ in EtOH}} \mathbf{3} \xrightarrow[\text{ii) } 150^\circ]{\text{i) Ac}_2\text{O}+\text{NaOAc}} \mathbf{4}$$

---

1) A. R. Battersby, J. C. Turner, *J. Chem. Soc.*, 717 (1960).

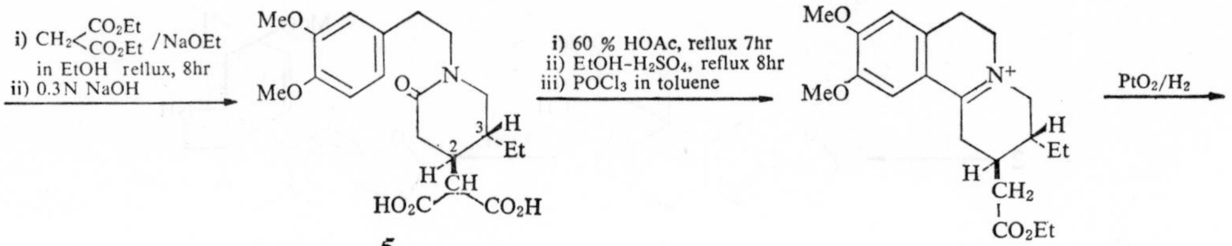

i) $CH_2 \!<\! \begin{smallmatrix} CO_2Et \\ CO_2Et \end{smallmatrix}$ /NaOEt
in EtOH reflux, 8hr
ii) 0.3 N NaOH

i) 60 % HOAc, reflux 7hr
ii) EtOH–H_2SO_4, reflux 8hr
iii) $POCl_3$ in toluene

PtO_2/H_2

5

76%
single product (mp 151.5~2.5°)

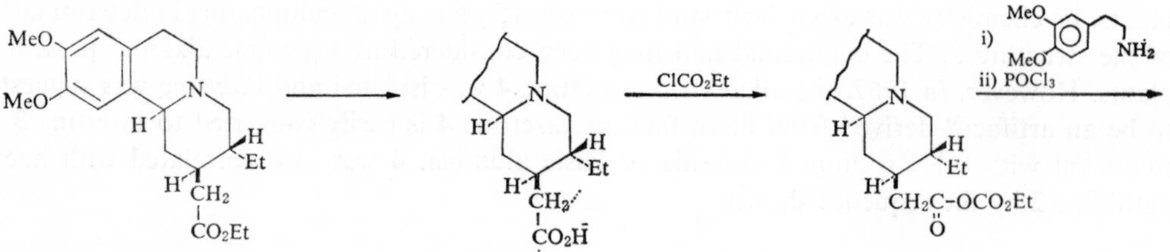

6

7

mp 66~66.5°
(picrate, mp 165~6°)

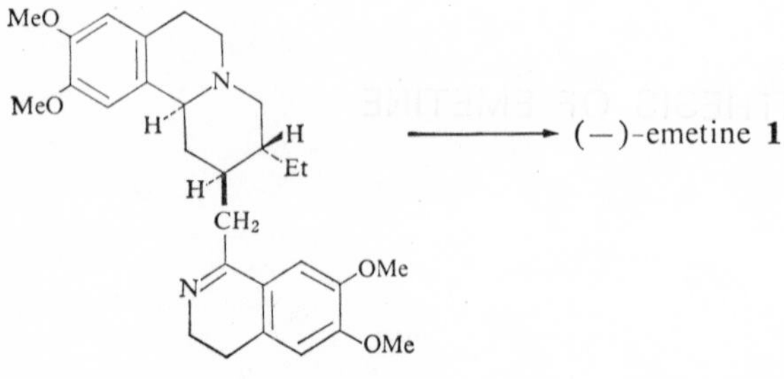

$ClCO_2Et$

8

i) MeO–MeO–$CH_2CH_2NH_2$
ii) $POCl_3$

9

10

$\longrightarrow$ (−)-emetine **1**

16 $\longrightarrow$ 2: The keto-lactam **2** was synthesized by Ban's method,[2] which involved the intermolecular Mannich reaction for the preparation of the key intermediate **14**. The
compound **2** was also used as an intermediate by Brossi et al.[16] in their independent
synthesis of emetine.

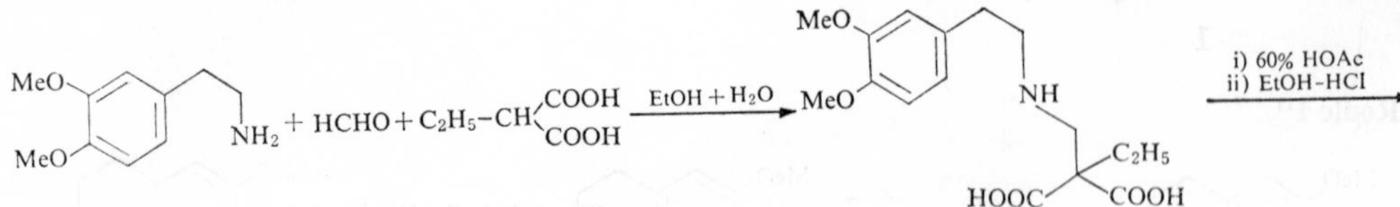

MeO–MeO–$CH_2CH_2NH_2$ + $HCHO$ + C_2H_5–$CH\!<\!\begin{smallmatrix} COOH \\ COOH \end{smallmatrix}$

$EtOH + H_2O$

i) 60% HOAc
ii) EtOH–HCl

11   **12**   **13**

14

2) Y. Ban, *Chem. Pharm. Bull.*, **3**, 53 (1955). *cf.* S. Sugasawa, T. Fujii, *Chem. Pharm. Bull.*, **3**, 47 (1955); S. Sugasawa, T. Fujii, *Chem. Pharm. Bull.*, **6**, 587 (1958); T. Fujii, *Chem. Pharm. Bull.*, **6**, 591 (1958).

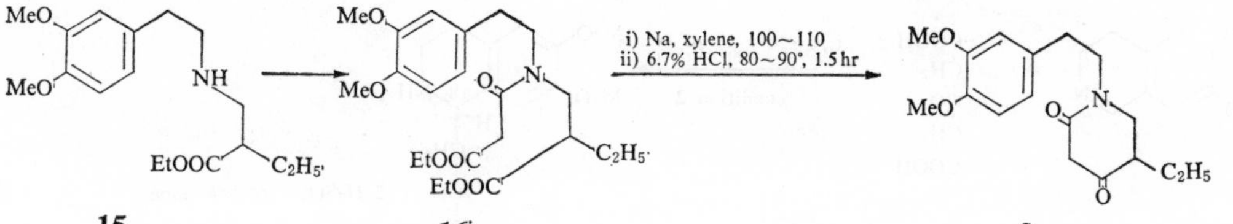

15   **16**   **2**

$\longrightarrow$ $\longrightarrow$ $\longrightarrow$ ($\pm$) rubremetinium
bromide[6]

4 $\longrightarrow$ 5: Michael reaction is known to be reversible and give the thermodynamically more stable
product. Thus, the C_2, C_3-*trans* diacid **5** would be expected to be obtained as a major
product in this reaction. In fact, the single crystalline acid **5** was obtained in 76%
yield. The 2,3-*trans* structure of **5** was rigorously proved as shown below.

5 $\longrightarrow$ 7: The stereostructure of **7** was established by the direct comparison of the ir spectrum
in solution with that of the optically active ester **17** of proven stereochemistry[3]
which was derived from the natural protoemetine.

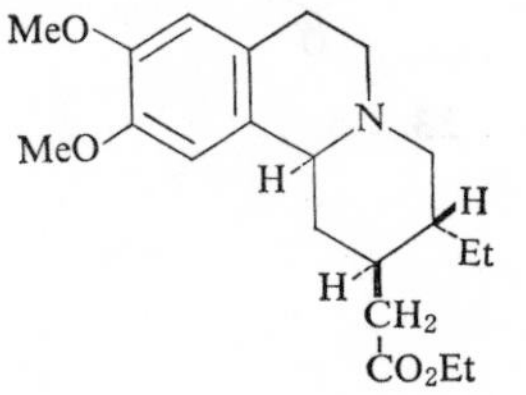

17

10 $\longrightarrow$ 1: The cyclization product **10** was resolved as its dibenzoyltartrate and the (+)-base
so obtained was identical with natural (+)-O-methylpsychotrine. Catalytic hydro-
genation of (+)**10** afforded (−)-emetine **1**, establishing the first stereospecific total
synthesis of the latter.

Route II[3]

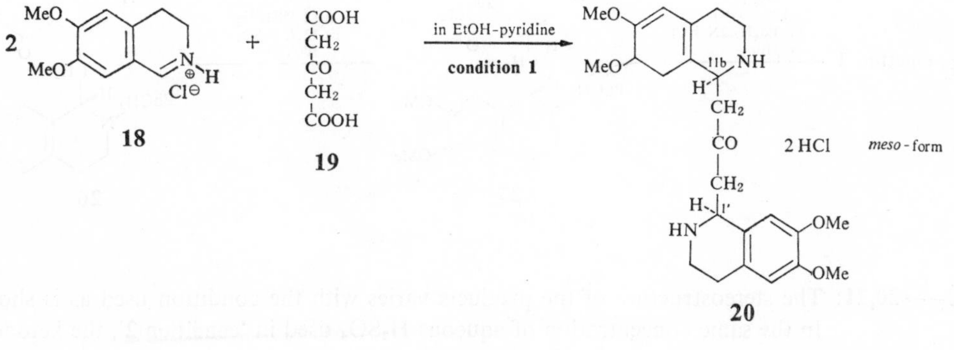

18   **19**   **20**

3) J. H. Chapman, P. G. Holton, A. C. Ritchie, T. Walker, G. B. Webb, K. D. E. Whiting, *J. Chem. Soc.*, 2471
(1962); D. E. Clark, P. G. Holton, R. F. K. Meredith, A. C. Ritchie, T. Walker, K. D. E. Whiting, *J. Chem.
Soc.*, 2479 (1962).

18 → 20,21: The stereostructure of the products varies with the condition used as is shown. In the same concentration of aqueous H₂SO₄ used in 'condition 2', the ketone **20** was inverted, in good yield, into the insoluble sulfate **21**, which may be the reason why only the latter was accumulated in this condition.

Stereostructure of **20** and **21**

The reduction of the dibutyryl derivative of **20** (mp 136°) with NaBH₄ gave two stereoisomeric alcohols. On the other hand, the same derivative of **21** (mp 175°)

afforded only a single alcohol, showing that the parent ketone was the meso-form in the former case and the racemic form in the latter case. It was thus established that the asymmetric center in the ketone **21** have the same relative configuration as that at C_1, and C_{11b} in emetine.

22 + 23 ⟶ 24 + 25: The hydrobromide **25** which had been correlated with **21** of known relative configuration at C_1, and C_{11b} was converted to ($\pm$)-emetine through several steps, which established the stereospecific total synthesis including the configuration at $C_{1'}$.

Addendum

A number of studies aimed for the synthesis of emetine and the related compounds have been reported. These may be roughly classified into two types, type A[4-13] and type B.[14-20] The former involves the ester **28** and the latter, the ketone **29** as a key intermediate.

28

29

4) R. P. Evstigneeva, N. A. Preobrazhensky, *Tetr.*, **4**, 223 (1958), and references cited therein.
5) M. Pailer and G. Beier, *Monatsh.*, **88**, 830 (1957).
6) A. R. Battersby, H. T. Openshaw, *J. Chem. Soc.*, 67 (1949).
7) M. Barash, J. M. Osbond, J. C. Wickens, *J. Chem. Soc.*, 3530 (1959).
8) E. E. van Tamelen, G. P. Schiemenz, H. L. Arons, *Tetr. Lett.*, 1005 (1963); E. E. van Tamelen, C. Placeway, G. P. Schiemenz, I. G. Wright, *J. Am. Chem. Soc.*, **91**, 7359 (1969).
9) E. E. van Tamelen, P. E. Aldrich, *J. Am. Chem. Soc.*, **79**, 4817 (1957); E. E. van Tamelen, J. B. Hester Jr., *ibid.*, **81**, 507 (1959); *ibid.*, **81**, 6214 (1959).
10) A. Grussner, E. Jaeger, J. Hellerbach, O. Schnider, *Helv. Chim. Acta*, **42**, 2431 (1959).
11) A. W. Burgstahler, Z. L. Bithos, *J. Am. Chem. Soc.*, **81**, 504 (1959); *ibid.*, **82**, 5466 (1960).
12) W. Mayer, R. Bachmann, F. Kraus, *Chem. Ber.*, **88**, 316 (1955); W. Mayer, L. Keller, *ibid.*, **92**, 213 (1959).
13) E. E. van Tamelen, M. Shamma, A. W. Burgstahler, J. Wolinsky, R. Tamm, P. E. Aldrich, *J. Am. Chem. Soc.* **80**, 5006 (1958).
14) A. R. Battersby, H. T. Openshaw, H. C. S. Wood, *J. Chem. Soc.*, 2463 (1953).
15) A. R. Battersby, B. J. T. Harper, *J. Chem. Soc.*, 1748 (1959).
16) A. Brossi, M. Baumann, L. H. Chopard-dit-Jean, J. Würsch, F. Schneider, O. Schnider, *Helv. Chim. Acta*, **42**, 772 (1959); A. Brossi, M. Baumann, O. Schnider, *ibid.*, **42**, 1515 (1959); A. Brossi, O. Schnider, *ibid.*, **45**, 1899, (1962).
17) H. T. Openshaw, N. Whittaker, *J. Chem. Soc.*, 1449 (1963); *ibid.*, 1461 (1963).
18) C. Schöpf, J. Thesing, *Angew. Chem.*, **63**, 377 (1951).
19) D. Beke, C. Szántay, *Chem. Ber.*, **95**, 2132 (1962); C. Szántay, J. Rohály, *ibid.*, **98**, 558 (1965); C. Szántay, L. Töke, P. Kolonits, *J. Org. Chem.*, **31**, 1447 (1966).
20) N. L. Allinger, H. M. Blatter, *J. Am. Chem. Soc.*, **83**, 994 (1961).

10.42 PHOTOCHEMISTRY OF COLCHICINE

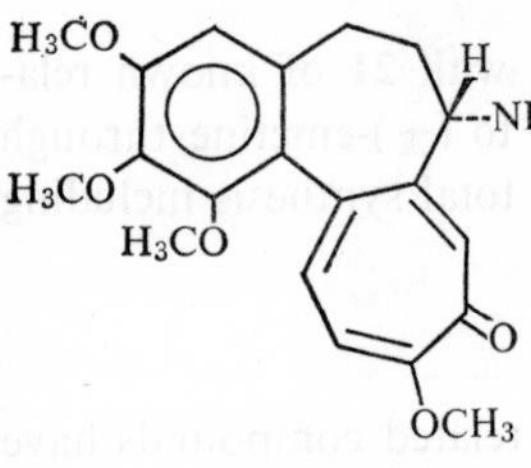

colchicine 1

The photochemical changes of colchicine **1** have been known for over 100 years, but their nature was unknown until the early 1960's.[1-4] This work was the first photochemical study of a troponoid and also a first study of photochemical electrocyclic reactions.

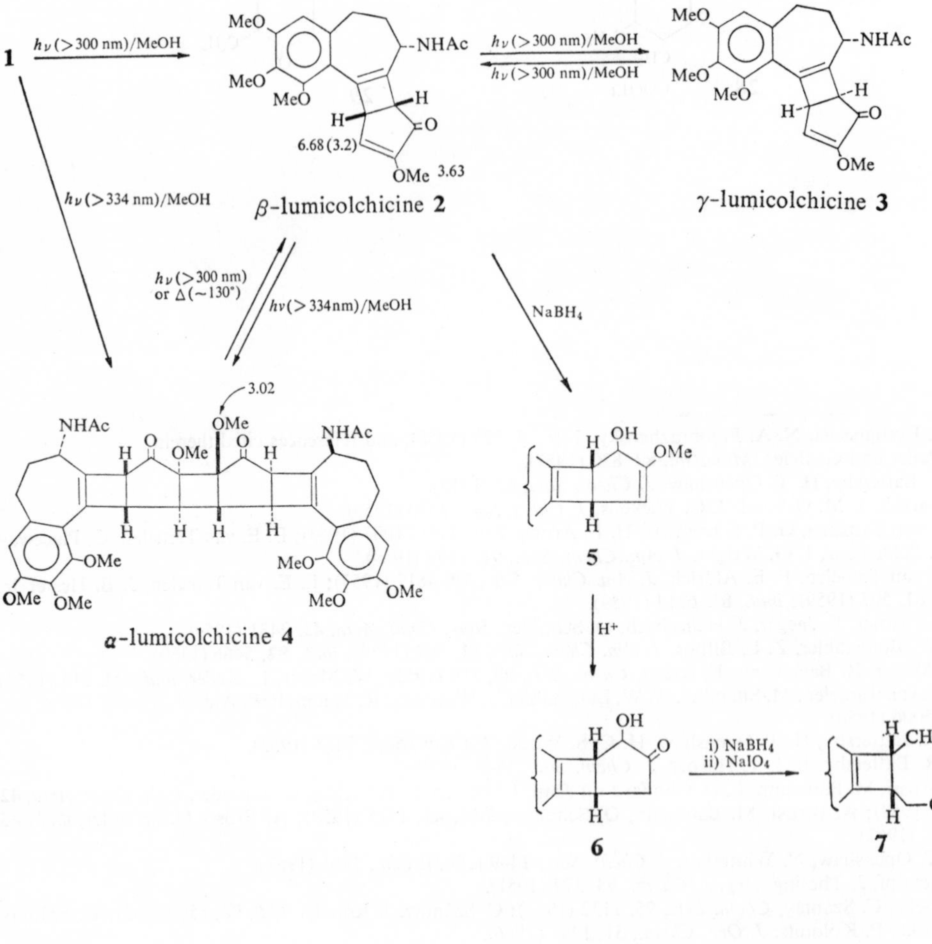

β-lumicolchicine **2**

γ-lumicolchicine **3**

α-lumicolchicine **4**

5

6

7

1) Exposure of **1** to sunlight yielded mixtures of **2**, **3** and **4** in various ratios.[1]
2) The methoxy methyl signal in **4** appears at much higher field than that in **2**, showing that both methoxy groups are diamagnetically shielded.
3) Compound **3** was also degraded to a dialdehyde by a sequence similar to **2** ⟶ **7**.[2]

4) The bathochromic shifts observed in **8** and **9** (22 nm and 26 nm, respectively) were attributed to the arylcyclobutene ring strain. The ε values increase with concentration, and this increase is larger in **9** than in **8**, indicating inter- and intramolecular hydrogen bonding in **9** and **8**, respectively.[3-5]
5) The dilution shift of the nmr signals of NH and OH in **10** showed that the NH proton is intramolecularly hydrogen bonded, whereas the OH proton is intermolecularly hydrogen bonded. Similar measurements with **11** showed that the OH proton, but not the NH proton, is intramolecularly hydrogen bonded.[4]

1) R. Grewe, *Naturwiss.*, **33**, 187 (1946); R. Grewe, W. Wurf, *Chem. Ber.*, **84**, 621 (1951); F. Santavy, *Coll. Czech. Chem. Commun.*, **16**, 655 (1951); G. O. Schcuk, H. J. Kuhu, O. A. Newnuller, *Tetr. Lett.*, 16 (1961).
2) E. J. Forbes, *J. Chem. Soc.*, 3864 (1955).
3) P. D. Gardner, R. L. Brandon, G. R. Haynes, *J. Am. Chem. Soc.*, **79**, 6334 (1957).
4) O. L. Chapman, G. H. Smith, R. W. King, *J. Am. Chem. Soc.*, **85**, 803, 806 (1963); *ibid.*, **83**, 3914 (1961).
5) W. F. Forbes, J. F. Templeton, *Chem. Ind.*, 77 (1957).

6) While **4** reverted to **2** on heating, **13** was stable to heat. Thus, a correct molecular weight was only obtained for **13**. The thermal instability of **4** was rationalized in terms of the stability of the intermediate biradical **14**.

14

Remarks

It is interesting that *β*- and *γ*-lumicolchicine were found in meadow saffron (*Colchicum autumunale*) together with colchicine.[6] Isocolchicine **15**, an isomer of **1**, was found to undergo similar types of electrocyclic reactions to give **16** and **17**.[7]

isocolchicine **15**   **16**   **17**

These fascinating isomerizations have stimulated interest in the photochemistry of troponoids, and many other types of photoreaction of these compounds are now known.[8]

6) F. Santavy, *Coll. Czech. Chem. Commun.*, **15**, 552 (1950).

7) W. G. Dauben, D. A. Cox, *J. Am. Chem. Soc.*, **85**, 2130 (1963); O. L. Chapman, H. G. Smith, P. A. Barks, *ibid.*, **85**, 3171 (1963).

8) E.g., O. L. Chapman, in *Advances in Photochemistry*, vol. 1, p. 330, Interscience, 1963.

10.43 SYNTHESIS OF COLCHICINE

Route I (Eschenmoser synthesis)[1]

1) J. Schreiber, W. Leimgruber, M. Pesaro, P. Schudel, T. Threlfall, A. Eschenmoser, *Helv. Chim. Acta*, **44**, 540 (1961).

1 ⟶ 2: **1** is obtained from purpurogallin.

2 ⟶ 3: The free 1-hydroxyl group plays an important role on this step, since the reaction proceeds as follows:

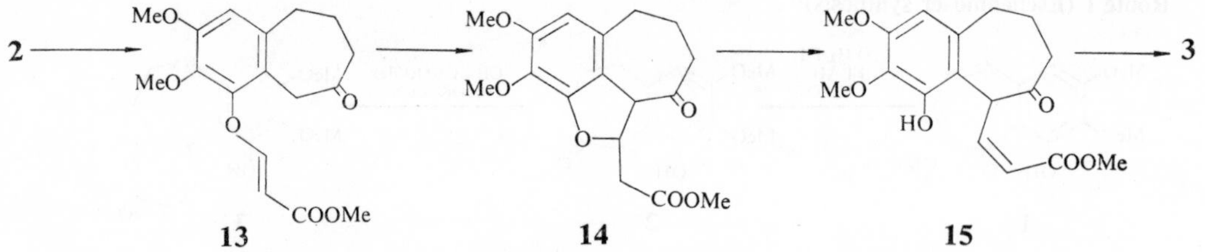

2 ⟶

13	**14**	**15**

6 ⟶ 7: Oxidation with OsO_4 is accompanied by decarboxylation as shown below:

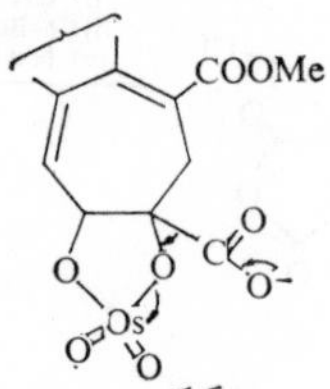

16

10 ⟶ 11: Methylation with CH_2N_2 gave a mixture of two isomeric methyl ethers, but only one isomer could be brominated successfully.

11 ⟶ 12: The tropone ring was also ammonolyzed. Subsequent treatment with alkali reversed this process.

12: Racemic **12** was resolved with camphor-10-sulfonic acid.[2] Methylation with CH_2N_2 gave a mixture of two methyl ethers, which could be separated.

Route II (van Tamelen synthesis)[3]

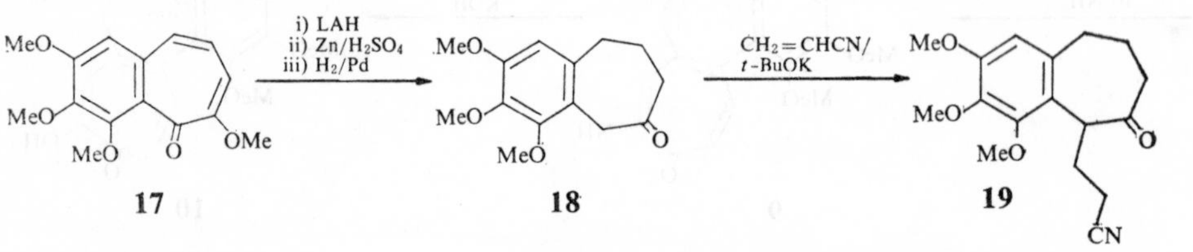

i) LAH
ii) Zn/H_2SO_4
iii) H_2/Pd

$CH_2=CHCN/$
$t-BuOK$

17	**18**	**19**

1) $Zn/BrCH_2COOMe$
ii) NaOH

i) DCC
ii) CH_2N_2

Na/NH_3

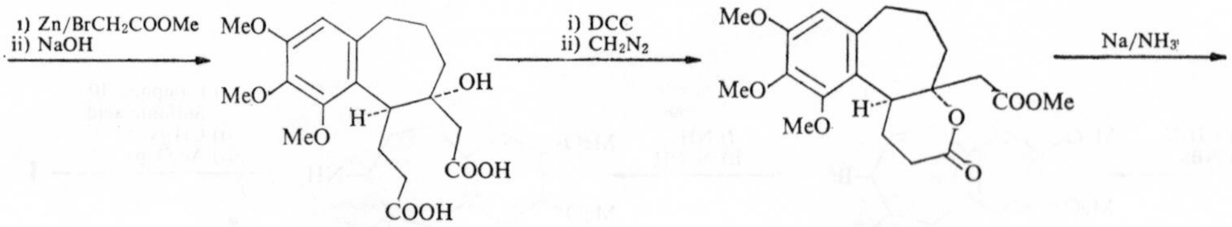

20	**21**

2) H. Corrodi, E. Hardegger, *Helv. Chim. Acta*, **40**, 193 (1957).
3) E. E. van Tamelen, F. A. Spencer, O. S. Allen, R. L. Orvis, *Tetr.*, **14**, 8 (1961).

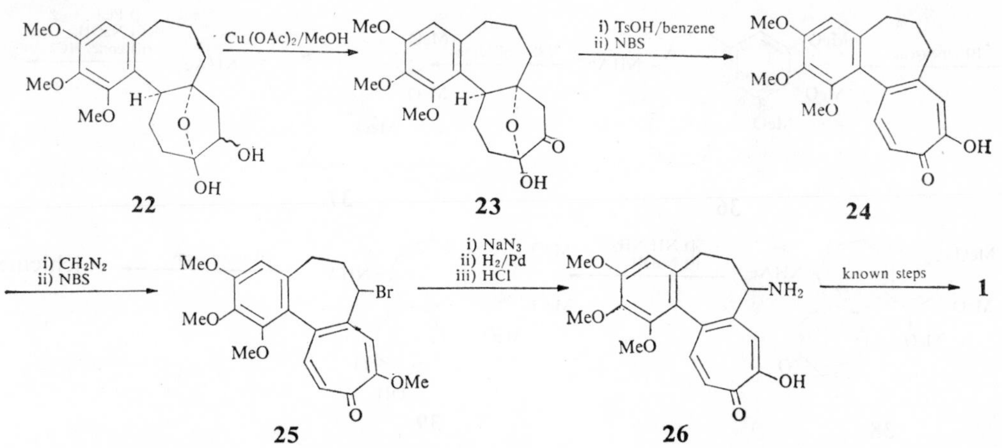

The chemical reaction schemes for compounds 22 through 35 are shown.

22 → 23: Cu(OAc)₂/MeOH

23 → 24: i) TsOH/benzene ii) NBS

24 → 25: i) CH₂N₂ ii) NBS

25 → 26: i) NaN₃ ii) H₂/Pd iii) HCl

26 → 1: known steps

18 ⟶ 19: Only this procedure, among a number of attempts, was found to give a product having an alkyl substituent at C-7.

19 ⟶ 20: The Reformastky reaction gave a mixture of two diastereoisomeric cyano esters, which could be separated. One of them could be hydrolyzed to **20**.

20 ⟶ 21: Protection of the tertiary hydroxyl group by formation of a lactone ring is one of the key steps of this synthesis.

21 ⟶ 22: **21** could be cyclized by acyloin condensation, but an isomeric lactone obtained from the *trans* isomer of **20** did not give any acyloin product.

Route III (Sankyo synthesis)[4]

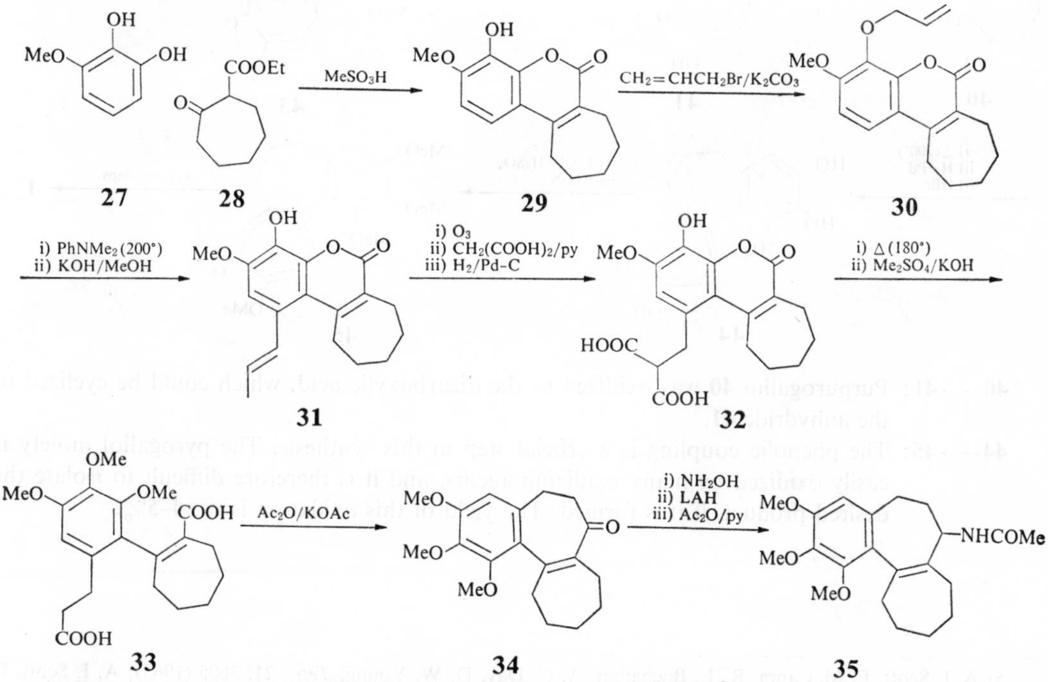

27 + 28 → 29: MeSO₃H

29 → 30: CH₂=CHCH₂Br/K₂CO₃

30 → 31: i) PhNMe₂(200°) ii) KOH/MeOH

31 → 32: i) O₃ ii) CH₂(COOH)₂/py iii) H₂/Pd-C

32 → 33: i) Δ(180°) ii) Me₂SO₄/KOH

33 → 34: Ac₂O/KOAc

34 → 35: i) NH₂OH ii) LAH iii) Ac₂O/py

4) G. Sunagawa, T. Nakamura, J. Nakazawa, *Chem. Pharm. Bull.*, **10**, 291 (1962); T. Nakamura, *ibid.*, **10**, 299 (1962).

36   **37**

38   **39**

37 ⟶ 38: Treatment of the intermediate tropylium salt gave a ditropyl ether, which was disproportionated to the tropone **39** with conc. acid.

Route IV (Scott synthesis)[5]

40   **41**   **43**

44   **45**

40 ⟶ 41: Purpurogallin **40** was oxidized to the dicarboxylic acid, which could be cyclized to the anhydride **41**.

44 ⟶ 45: The phenolic coupling is a crucial step in this synthesis. The pyrogallol moiety is easily oxidized by many oxidizing agents, and it is therefore difficult to isolate the desired product, if it is formed. The yield of this oxidation is *ca.* 4–5%.

5) A. I. Scott, F. McCapra, R. L. Buchanan, A. C. Day, D. W. Young, *Tetr.*, **21**, 3605 (1965); A. I. Scott, E. McCapra, J. Nabney, D. W. Young, A. C. Day, A. J. Baker, T. A. Davidson, *J. Am. Chem. Soc.*, **85**, 3040 (1963).

6) R. B. Woodward, *The Harvey Lectures*, 31 (1963).

Route V (Woodward synthesis)[6]

46 → CSCl₂/Et₃N → **47** → i) NBS/hν ii) Ph₃P → **48** → i) NaOMe ii) 49 →

50 → i) N₂H₄/H₂O₂ ii) LAH iii) MnO₂ → **51** →

i) Ph₃P=CHCH=CHCOOMe ii) NaOH iii) I₂ → **52**

HClO₄ → **53** → i) N₂H₄/H₂O₂ ii) (diphenyl)Li iii) CO₂ iv) CH₂N₂ → **54**

i) NaH ii) AcOH/H₂SO₄/H₂O → **55** → i) HCOOEt ii) Ts-S(CH₂)₃S-Ts/KOAc → **56**

Hg(OAc)₂/HClO₄ → **57** → Ac₂O/py → **58**

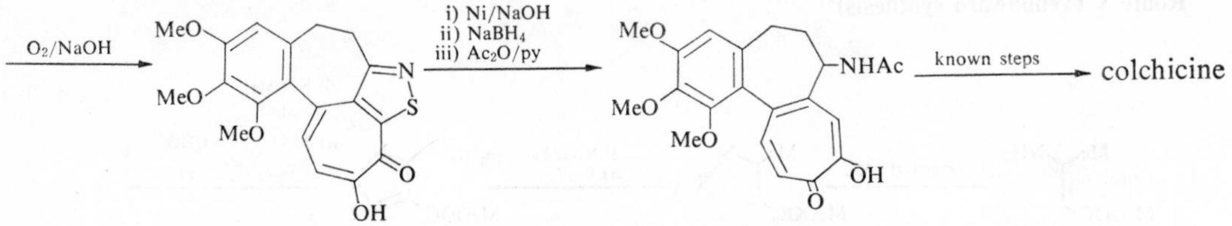

59   **60**

50 ⟶ 51: Reduction of the double bond could not be carried out by catalytic hydrogenation because of the presence of a sulfur atom in the molecule.

53 ⟶ 54: The position next to sulfur in the isothiazole ring is acidic and gives an anion with *o*-biphenyllithium.

55 ⟶ 56: The reaction mechanism is as folllows:

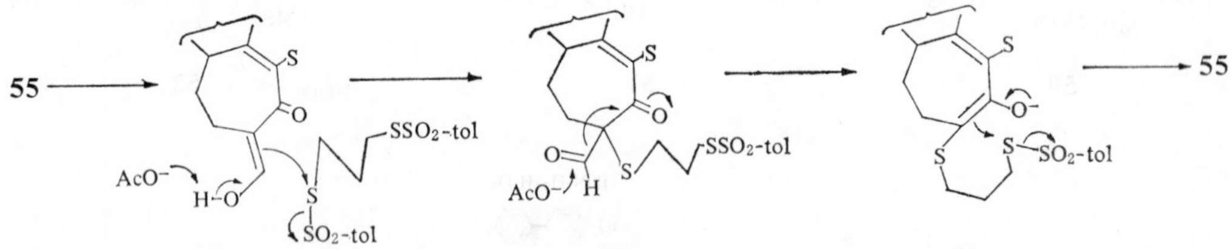

61   **62**   **63**

59 ⟶ 60: Alkaline conditions were used because of the increased stability of the tropolone ring towards reduction in an alkaline medium. NaBH₄ was used to reduce the C=N double bond.

Remarks

Colchicine, an alkaloid isolated from the autumn crocus *Colchicum autumnale* L., has been widely used in recent years by biological workers, who have utilized its effect on cell division. It also shows some anti-cancer activity. The structure was deduced by Dewar in 1945.[7]

7) M. J. S. Dewar, *Nature*, **155**, 141 (1945).

10.44 BIOSYNTHESIS OF COLCHICINE

The biosynthesis of colchicine has attracted much attention, especially as regards the sources of the carbon atoms and the genesis of the tropolone ring. After extensive efforts, specifically labeled phenylalanine **1** and tyrosine **2** were shown to be incorporated, and a possible biogenetic route was proposed on the basis of the labeling positions in the resulting colchicine.[1-3]

9 colchicine (R₁=H, R₂=Ac)
10 demecolcine (R₁=H, R₂=Mc)[4]
11 speciosine (R₁=Me, R₂=CH₂C₆H₄OH)

9 colchicine $(R_1=H, R_2=Ac)$
10 demecolcine $(R_1=H, R_2=Mc)$[4]
11 speciosine $(R_1=Me, R_2=CH_2C_6H_4OH)$

1) ¹⁴C labeling is indicated by ●, ○, * and ■.
2) Phenylalanine loses its amino group but retains its carboxyl group in the process of condensation.

1) For structural studies, see J. W. Cook, J. D. London, *The Alkaloids* (ed. R. H. F. Manske, H. L. Holmes), vol. II, Acedemic Press, 1952.
2) A. R. Battersby, R. Binks, J. J. Reynolds, D. A. Yeowell, *J. Chem. Soc.*, 4257 (1964); E. Leete, *Tetr. Lett.*, 333 (1965); A. R. Battersby, R. B. Herbert, E. McDonald, R. Ramage, J. H. Clements, *Chem. Commun.*, 603 (1966).
3) A. C. Baker, A. R. Battersby, E. McDonald, R. Ramage, J. H. Clements, *Chem. Commun.*, 390 (1967).
4) F. Santavy, T. Reichstein, *Helv. Chim. Acta*, 36 1319 (1953).

3) The reaction **5**⟶**6** is a phenolic oxidative coupling process.

4) Androcymbine **12**, an alkaloid closely related to **5**, has been isolated from *Colchicum* sp.[5]

androcymbine **12**

5) The group shown as OX in **6** represents a good leaving group such as phosphate. The process **7**⟶**8** is a homoallyl-cyclopropylcarbinyl rearrangement.

5) A. R. Battersby, R. B. Herbert, L. Pijewska, F. Santavy, *Chem. Commun.*, 228 (1965).

10.45 BIOSYNTHESIS OF QUINOLINE ALKALOIDS IN RUTACEAEOUS PLANTS

1

5.94×10^7 dpm/mmol

[1—^{14}C] acetate
1.47×10^9 dpm/mmol

8 acetate units

evocarpine **2**

mevalonate

3

4

dictamnine **5** (R = H)
skimmianine **6** (R = MeO)

1) Compound **1** (^{3}H-labeled on the benzene ring) was 0.54% incorporated into **6**. It was thought to be derived from shikimic acid.[1]

2) Sodium [1-^{14}C] acetate was 0.0025% incorporated into **5**, whereas [2-^{14}C] acetate was 0.0076% incorporated into **6**. Sodium ^{14}C-formate (3.01×10^9 dpm/mmol) was 0.011% incorporated into **6**.

Remarks

Compounds **5** and **6** were proved to be derived from anthranilic acid.[1-2] Methionine is the source of methyoxyl groups. The furan portion in **5** and **6** was shown to be derived from an isopentenyl unit *via* the mevalonate pathway.[3,4] The structure of **2**, isolated from *Evodia rutecarpa*, implies the biogenesis shown.[5]

1) M. Matsuo, Y. Kasida, M. Yamazaki, *Biochem. Biophys, Res. Commun.*, **23**, 679 (1966).
2) I. Moncovic, I. D. Spenser, A. O. Plunkett, *Can. J. Chem.*, **45**, 1935 (1967).
3) J. F. Collins, M. F. Grundon, *Chem. Commun.*, 621 (1969); M. F. Grundon, K. J. James, *ibid.*, 1311 (1971).
4) A. O. Colonna, E. G. Gros, *Chem. Commun.*, 674 (1970).
5) R. Tschesche, W. Serner, *Tetr.*, **23**, 1873 (1967); M. Yamazaki, A. Ikuta, T. Mori, T. Kawana, *Tetr. Lett.*, 3317 (1967).

10.46 SYNTHESIS OF CAMPTOTHECIN

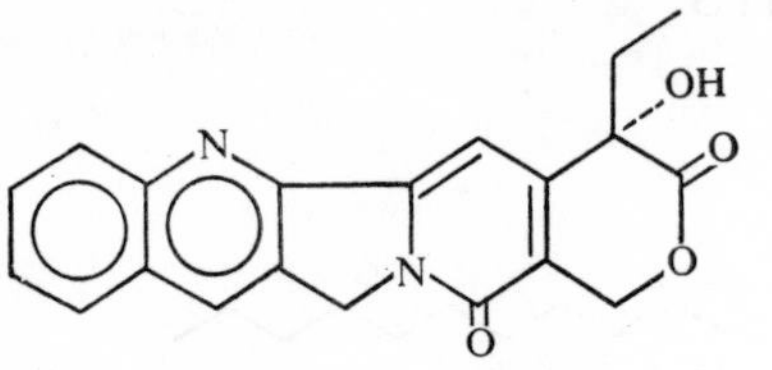

camptothecin **1**

Camptothecin **1** was isolated from *Camptotheca acuminata* (Nyssaceae) and has anti-leukemic and antitumor properties. Its structure was elucidated as shown in 1966.[1] Several syntheses have been achieved recently.

Route I[2]

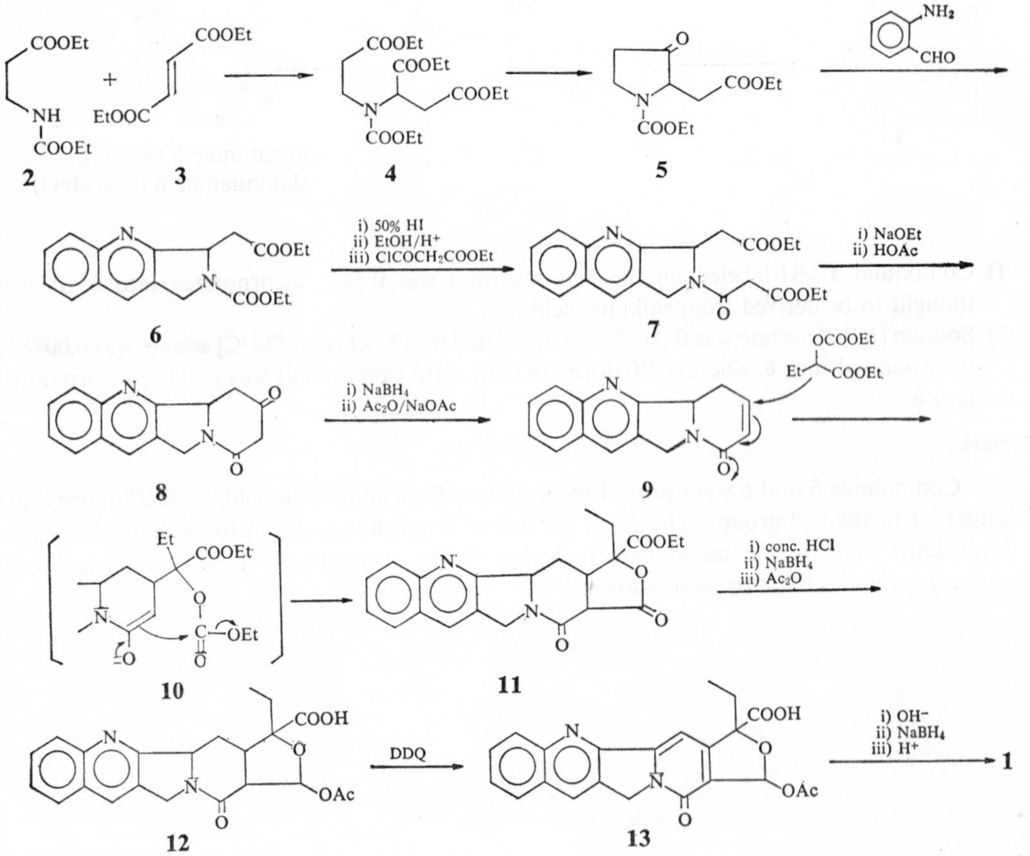

9 → 11: This is the key step of the synthesis, probably proceeding through **10** (Michael-type addition and lactonic annelation).

1) M. Wall, M. C. Wani, C. E. Cook, K. H. Palmer, A. T. McPhali, G. A. Sim, *J. Am. Chem. Soc.*, **88**, 3888, (1966).
2) G. Stork, A. G. Schultz, *J. Am. Chem. Soc.*, **93**, 4074 (1971).

Route II[3]

16 ⟶ 17: Nucleophilic addition of enamines to allenes was developed by the same authors[4] and is the key step in the synthesis.

19 ⟶ 20: Dieckman condensation led to the hydrolysis of one of the ester groups in 20.

20 ⟶ 21: Friedlander condensation.

24 ⟶ 1: The oxidation was observed under different conditions, such as air oxidation of 24 in CH_2Cl_2, of the anion of 24 in DMSO/DMF, etc. H_2O_2 oxidation of the anion is preferred.[5]

Route III[6]

3) R. Volkmann, S. Danishefsky, J. Eggler, D. M. Solomon, *J. Am. Chem. Soc.*, **93**, 5576 (1971).
4) S. Danishefsky, S. J. Etheredge, R. Volkmann, J. Eggler, J. Quick, *J. Am. Chem. Soc.*, **93**, 5575 (1971).
5) G. Buchi, K. E. Matsumoto, H. Nishimura, *J. Am. Chem. Soc.*, **93**, 3299 (1971).
6) M. C. Wani, H. F. Campbell, G. A. Brine, J. A. Kepler, M. E. Wall, S. G. Levine, *J. Am. Chem. Soc.*, **94**, 3631 (1972).

Treatment of **27** with hydrogen cyanide in the presence of a catalytic amount of KCN gave a diastereoisomeric mixture of product, which has all the carbon atoms necesssary to construct camptothecin. Applying the same synthetic scheme, a D, E ring analog of **1** was synthesized.[7]

Route IV[8]

36 ⟶ 37: The auto-oxidative indole-quinoline rearrangement probably proceeds as follows.

This is believed to follow the biogenetic pathway.[9] The first biomimetic conversion was carried out[10] using ajmalicine **47** as the starting material.

41 ⟶ 1: Air oxidation catalysed by $CuCl_2$ was carried out in DMF at room temperature in the presence of a trace of Me_2NH.

7) M. E. Wall, H. F. Campbell, M. C. Wani, S. G. Levine, *J. Am. Chem. Soc.*, **94**, 3632 (1972).
8) J. Warneke, E. Winterfeldt, *Chem. Ber.*, **105**, 2120 (1972); M. Boch, T. Korth, J. M. Nelke, D. Pile, H. Ranunz, E. Winterfeldt, *ibid.*, **105**, 2126 (1972).
9) E. Wenkert, K. G. Dave, R. G. Lewis, P. W. Sprague, *J. Am. Chem. Soc.*, **89**, 6741 (1967).
10) E. Winterfeldt, *Ann. Chem.*, **745**, 23 (1971); E. Winterfeldt, H. Radunz, *Chem. Commun.*, 374 (1971).

10.47 BIOSYNTHESIS OF QUININE

tryptamine **1**

secologanin **2**

vincoside **3** (R=H)

corynantheal **4** (R=H)

cinchonaminal **5**

6

7

8

quinine **9** (R=OMe)
cinchonidine **10** (R=H)

1) Compound **3** with aromatic ^{3}H labeling was 0.008% incorporated into **9** and **10** when fed to *Cinchona ledgeriana*.[1]

2) Compound **4** with aromatic ^{3}H labeling was 0.007% and 0.04% incorporated, respectively, into **9** and **10**.[2]

Remarks

The biosynthesis of quinine, a Cinchona alkaloid, proceeds through a route including the transformation of indole to quinoline, as shown.[3] In regard with the source of the non-tryptamine derived portion, however, the participation of a monoterpene glucoside, secologanin, from the mevalonate pathway has recently been demonstrated.[4–6] The sequence was established by tracer experiments and the isolation of key compounds.[2]

1) A. R. Battersby, A. R. Burnett, P. G. Parsons, *Chem. Commun.*, 1282 (1968); *J. Chem. Soc.* C, 1193 (1969).
2) A. R. Battersby, R. J. Parry, *Chem. Commun.*, 30, 31 (1971).
3) N. Kowanko. E. Leete, *J. Am. Chem. Soc.*, **84**, 4919 (1962).
4) E. Leete, J. N. Wemple, *J. Am. Chem. Soc.*, **88**, 4743 (1966).
5) A. R. Battersby, R. T. Brown, R. S. Kapil, J. A. Knight, J. A. Martin, A. O. Plunkett, *Chem. Commun.*, 888 (1966).
6) H. Inoue, S. Ueda, Y. Takeda, *Tetr. Lett.*, 407 (1969).

10.48   CONVERSION OF QUINOLINE ALKALOIDS TO INDOLE ALKALOIDS

Indole and quinoline alkaloids coexist in plants of *Chinchona* sp. (Quinine tree, Rubiaceae) and a close biogenetic relationship has been established between them by tracer experiments.[1] The two types of alkaloids have also been correlated chemically.[2-5]

The conversion of **9** and **8**, together with that of **1** to **13**, provided the first chemical proof of the C-15 stereochemistry of natural indole alkaloids.[5,6]

i) H_2O_2
ii) SO_2

i) $POCl_3/CHCl_3$
ii) H_2O

dihydrocinchonine
1

2

3

Ni/H_2

$-OH^-$

CH_2COONa

t-BuOLi/Ph_2CO

CH_2COONa

4

5

5a

CH_2COONa

EtOH/HCl

CH_2COOEt

LAH

CH_2CH_2OH

5b

6

dihydrocinchonamine
7

1) A. R. Battersby, R. J. Parry, *Chem. Commun.*, 30, 31 (1971).
2) E. Ochiai, M. Ishikawa, *Chem. Pharm. Bull.*, 6, 208 (1958).
3) E. Ochiai, M. Ishikawa, Y. Oka, *Annual Report of ITSUU Laboratory*, 12, 29 (1962).
4) Y. K. Sawa, H. Matsumura, *Tetr.*, 26, 2919, 2923 (1970).
5) E. Ochiai, M. Ishikawa, *Chem. Pharm. Bull.*, 7, 386 (1959); *Tetr.*, 7, 228 (1959).
6) E. Wenkert, N. V. Bringi, *J. Am. Chem. Soc.*, 80, 3484 (1958); *ibid.*, 81, 1474 (1959).

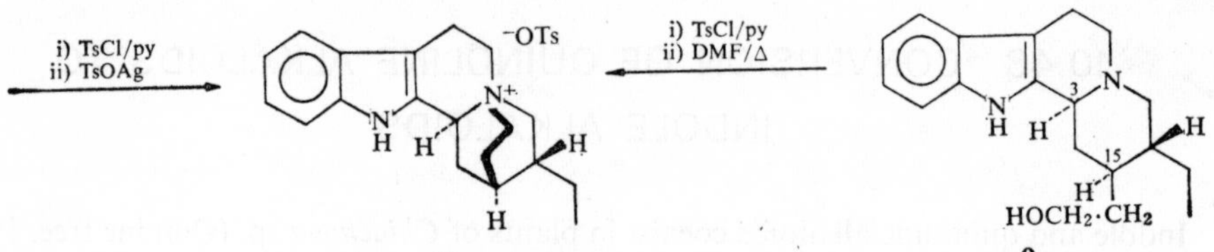

i) TsCl/py
ii) TsOAg →

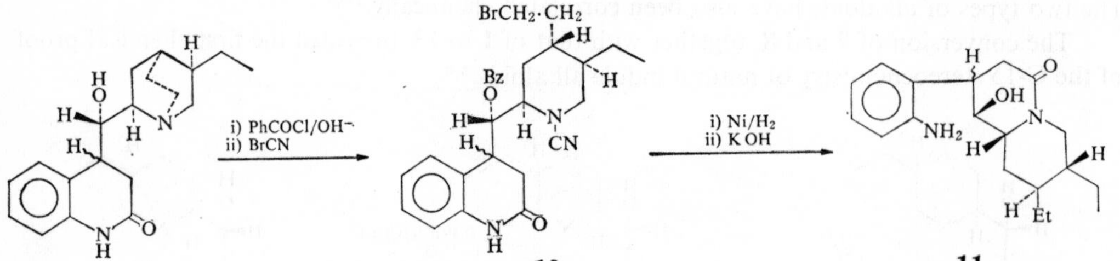

⁻OTs

8

i) TsCl/py
ii) DMF/Δ ←

i) TsCl/py
ii) DMF/Δ

HOCH₂·CH₂

9

i) PhCOCl/OH⁻
ii) BrCN →

4

BrCH₂·CH₂

Bz

CN

10

i) Ni/H₂
ii) KOH →

NH₂

OH

H   Et

11

i) LAH
ii) t-BuOLi/
Ph₂CO⁻ →

H   Et

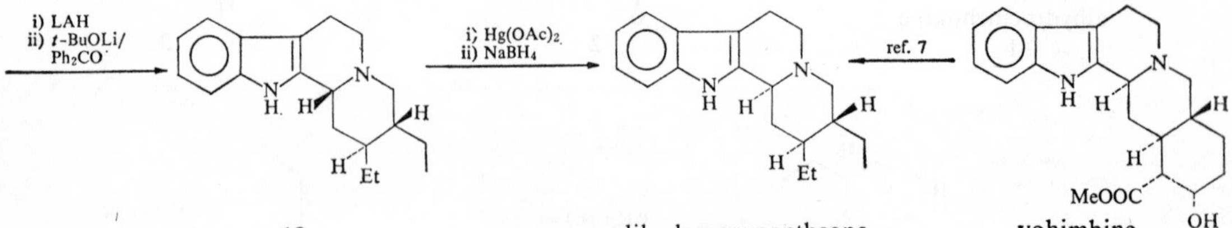

12

i) Hg(OAc)₂
ii) NaBH₄ →

H   Et

dihydrocorynantheane
13

ref. 7 ←

MeOOC   OH

yohimbine
14

7) Y. Ban, O. Yonemitsu, *Tetr.*, **20**, 2877 (1964).

10.49 BIOGENETIC CLASSIFICATION OF INDOLE ALKALOIDS

Condensation of tryptamine (or tryptophan) with secologanin, a monoterpene glucoside, gives rise to a nitrogeneous glucoside, vincoside,[2,3)] from which a great variety of indole alkaloids (about 700) are formed in living plants. The fundamental importance of dehydrosecodine for the skeletal rearrangement has been well discussed.[4)] All these indole alkaloids can be classified into five classes from the structure of the carbon skeleton of the non-tryptamine unit.[1)]

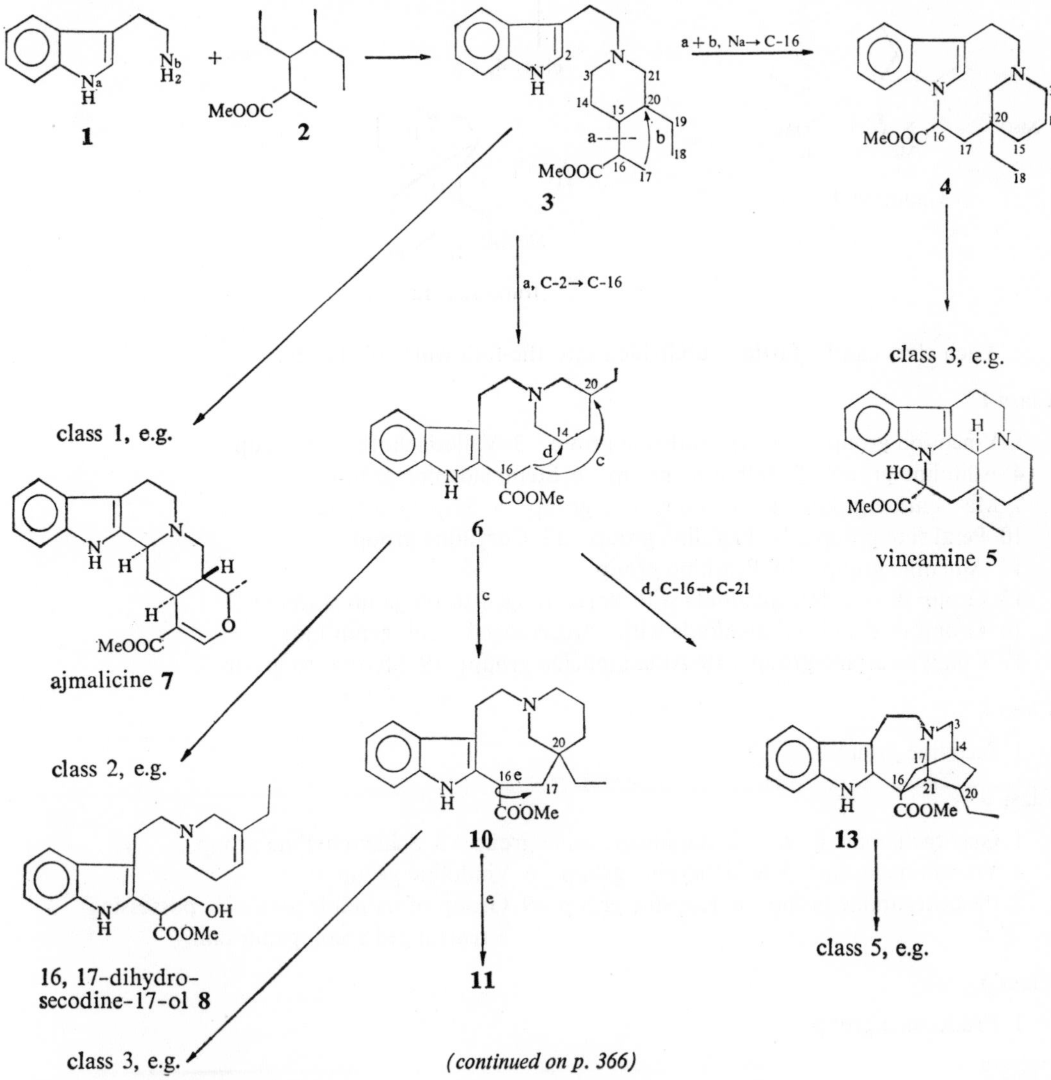

class 1, e.g.

ajmalicine 7

class 2, e.g.

16, 17-dihydro-
secodine-17-ol 8

class 3, e.g.

class 3, e.g.

vincamine 5

class 5, e.g.

(continued on p. 366)

1) I. Kompis, M. Hesse, H. Schnid, *Lloydia*, **34**, 269 (1971).
2) A. R. Battersby, A. R. Burnett, P. G. Parsons, *J. Chem. Soc.*, C, 1193 (1969).
3) W. P. Blackstock, R. T. Brown, G. K. Lee, *Chem. Commun.*, 910 (1971).
4) For example, A. I. Scott, *Accounts Chem. Res.*, **3**, 151 (1970); A. I. Scott, C. C. Wei, *J. Am. Chem. Soc.*, **94** 8263–8267 (1972); J. P. Kutney, J. F. Beck, C. Ehret, G. Poulton, R. S. Sood, N. D. Westcott, *Bioorganic Chem.*, **1**, 194 (1971).

(continued from p. 365)

class 5, e.g.

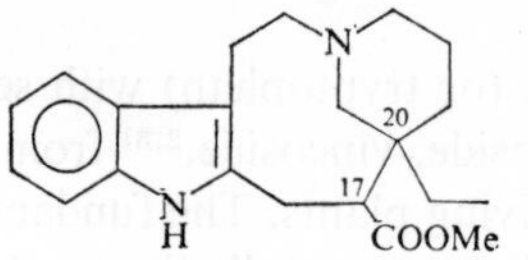

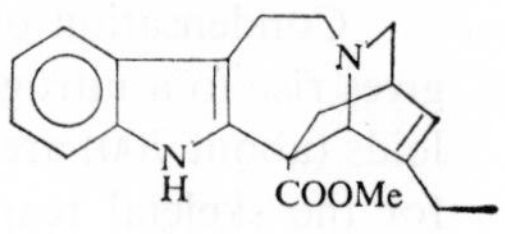

catharanthine **14**

11

class 3, e.g.

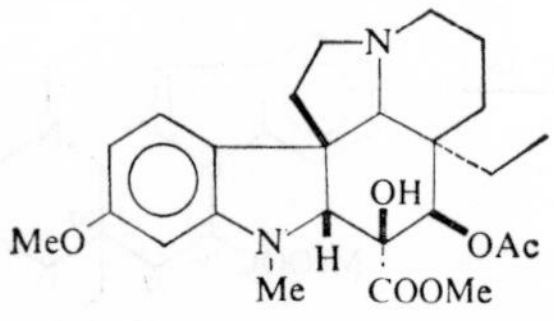

vindoline **9**

class 4, e.g.

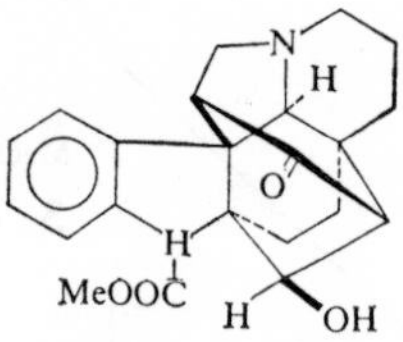

fruticosine **12**

Each class can be further subdivided into the following subclasses:

Class 1

1 Vincoside group   2 Corynantheine group   3 Vallesiachotamine group
4 Adifoline group   5 Talbotine group   6 Stemmadenine group
7 Mavacurine group   8 Cinchonamine group   9 Sarpagine group
10 Peraksine group   11 Picraline group   12 Corymine group
13 Ajmaline group   14 Perakine group
15 Group of oxindole alkaloids with unrearranged secologanin skeleton.
16 Group of Ψ-indoxyl alkaloids with unrearranged secologanin unit.
17 Condylocarpine group   18 Akuammicine group   19 Strychnine group

Class 2

1 Secodine group

Class 3

1 Quebrachamine group   2 Aspidospermine group   3 Schizophylline group
4 Vincamine group   5 Schizozygine group   6 Vindoline group
7 Pleiocarpamine group   8 Kopsine group   9 Group of oxindole alkaloids possessing
                                                a rearranged secologanin unit

Class 4

1 Fruticosine group

Class 5

1 Catharanthine group
2 Group of Ψ-indoxyl alkaloids possessing a rearranged secologanin unit

10.50   SECODINE AND DEHYDROSECODINE

secodine **1**

dehydrosecodine **2**

Though it has not yet been found in nature or synthesized, dehydrosecodine **2** is widely accepted as the key intermediate in indole alkaloid biosynthesis.[1] Its di- and tetrahydro analogues, secodine **1** and dihydrosecodine **3**, have been characterized[3] and the corresponding dimers, presecamine and tetrahydropresecamine **4**, were isolated from natural sources.[3]

3   **4**   **5**

3 ⟶ 4: This is a Diels-Alder reaction which occurs at 0°C over two days in the absence of solvent. The retro-Diels-Alder reaction takes place when presecamine is subjected to short-path distillation, yielding **1**.[3]

4 ⟶ 5: This change occurs during the isolation of the alkaloids, yielding secamine[2] and tetrahydrosecamine **5**[2] from presecamine and **4**, respectively.[3]

Remarks

Scott et al.[4] reported the conversion of tabersonine **6** (aspidosperma) into catharanthine **7** (iboga) and pseudocatharanthine **8**, and of stemmadenine **9** (strychnos) into **6**, **7** and **8** under reflux in acetic acid. The results were taken as a biomimetic in vitro conversion which proceeded via the intermediate **2**. The validity of these experiments, however, was questioned by Smith and his co-workers,[5] who failed to reproduce the results. Several reports concerning the above

1) A. I. Scott, *Accounts Chem. Res.*, **3**, 151 (1970); A. R. Battersby, A. K. Bhatnagar, *Chem. Commun.*, 193 (1970); A. R. Battersby, *Chem. Soc. Specialist Periodical Rep*, **1**, 31 (1971); J. P. Kutney, J. F. Beck, C. Ehret, G. A. Poulton, R. S. Sood, N. D. Wescott, *Bioorg. Chem.*, **1**, 194 (1971); J. P. Kutney, J. F. Beck, N. J. Eggers, H. W. Hanssen, R. S. Sood, N. W. Westcott, *J. Am. Chem. Soc.*, **93**, 7322 (1971).
2) D. A. Evans, G. F. Smith, G. N. Smith, K. S. J. Stapleford, *Chem. Commun.*, 859 (1968).
3) G. A. Cordell, G. F. Smith, G. N. Smith, *Chem. Commun.*, 189 (1970); *ibid.*, 191 (1970).
4) A. A. Qureshi, A. I. Scott, *Chem. Commun.*, 945 (1968).
5) R. T. Brown, J. S. Hill, G. F. Smith, K. S. J. Stapleford, J. Poisson, M. Muquet, N. Kunesch, *Chem. Commun.*, 1475 (1969); R. T. Brown, J. S. Hill, G. F. Smith, K. S. J. Stapleford, *Tetr.*, **27**, 5127 (1971).

and related reactions have appeared from the two groups, continuing to support their respective earlier positions.[6,7]

6) A. I. Scott, P. C. Cherry, *J. Am. Chem. Soc.*, **91**, 5872 (1969); A. I. Scott, *ibid.*, **94**, 8262 (1972); A. I. Scott, C. C. Wei, *ibid.*, **94**, 8263, 8264, 8266 (1972).

7) M. Muquet, N. Kunesch, J. Poisson, *Tetr.*, **28**, 1363 (1972); R. T. Brown, G. F. Smith, J. Poisson, N. Kunesch, *J. Am. Chem. Soc.*, **95**, 5778 (1973).

10.51 CONFIGURATION OF VINCOSIDE AND STRICTOSIDINE

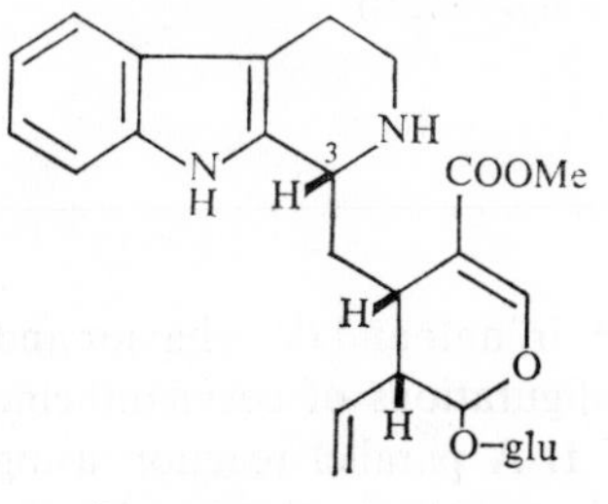
NH
COOMe
N
H
H
3
H
O
H
O–glu

vincoside **1**

Vincoside **1**[1,2] is a nitrogeneous glucoside of major importance as an active precursor for indole alkaloid biosynthesis. The 3-H configuration of **1** was formerly considered to be α. The following transformation,[1] however, has reversed this conclusion.

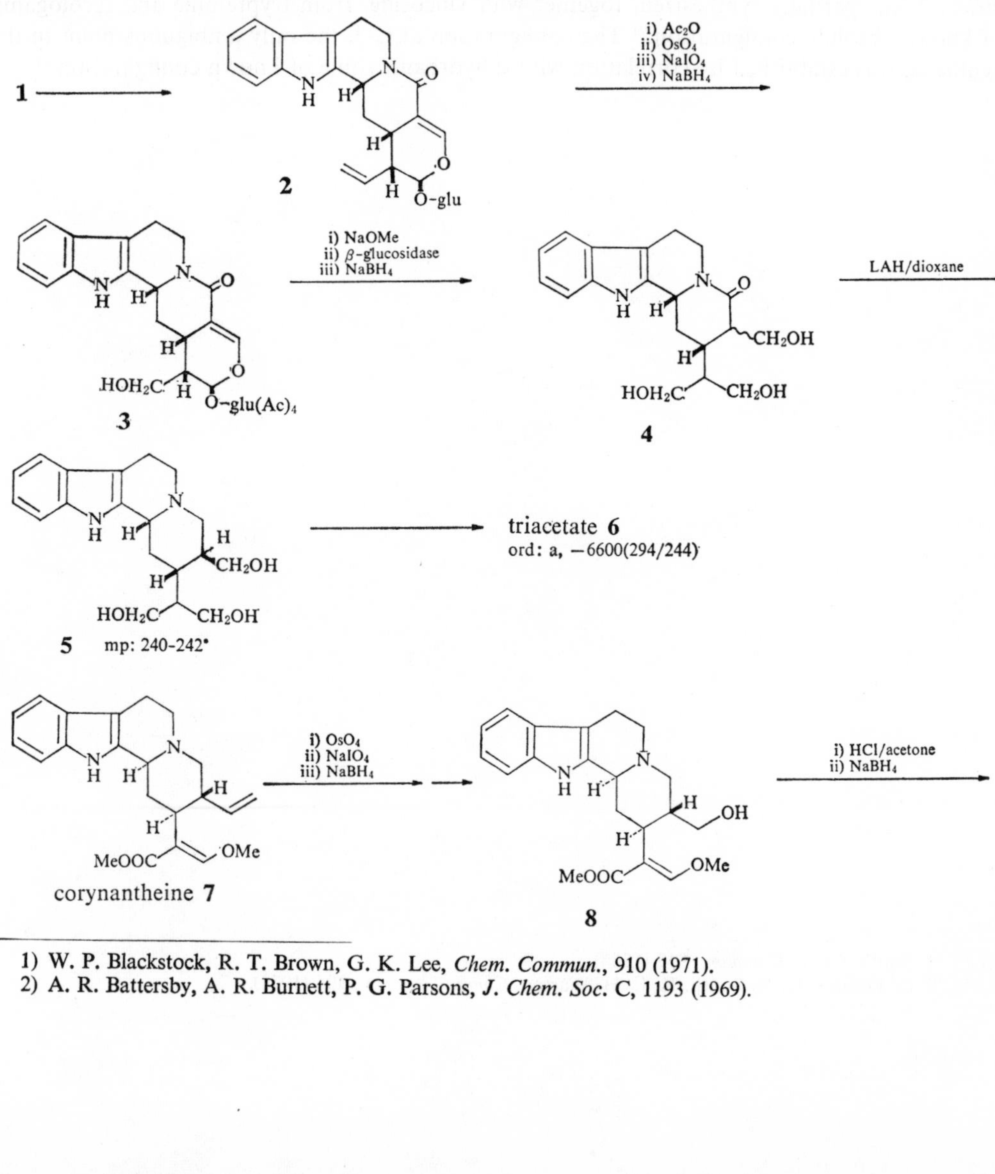
1
i) Ac2O
ii) OsO4
iii) NaIO4
iv) NaBH4
2
O–glu
i) NaOMe
ii) β-glucosidase
iii) NaBH4
3
HOH2C
O–glu(Ac)4
4
CH2OH
HOH2C
CH2OH
LAH/dioxane
5
mp: 240-242°
triacetate 6
ord: a, −6600(294/244)
corynantheine 7
MeOOC
OMe
i) OsO4
ii) NaIO4
iii) NaBH4
8
OH
i) HCl/acetone
ii) NaBH4

1) W. P. Blackstock, R. T. Brown, G. K. Lee, *Chem. Commun.*, 910 (1971).
2) A. R. Battersby, A. R. Burnett, P. G. Parsons, *J. Chem. Soc.* C, 1193 (1969).

triacetate **10**
ord: a, +6300(298/244)

9   mp: 240–242°

The triols **5** and **9** are enantiomeric, as shown by identical mp, ir, nmr and tlc behavior and antipodal optical properties. Since the relative and absolute configurations of corynantheine **7** are known, this correlation proves the 3β-H configuration of **1**. A parallel reaction using strictosamide, the C-3 epimer of **2**, showed that no C-3 epimerization took place under the reaction conditions employed above.

Remarks

Strictosidine (=isovincoside, 3α-H epimer of **1**), which was first isolated from *Rhazia stricta*,[3] was partially synthesized, together with vincoside, from tryptamine and secologanin of known absolute configuration.[2] The configuration at C-3, the only ambiguous point in the synthesis, was established by correlation with dihydroantirrhine of known configuration.[4]

3) G. N. Smith, *Chem. Commun.*, 912 (1968).
4) K. T. D. DeSilva, G. N. Smith, K. E. H. Warren, *Chem. Commun.*, 905 (1971).

10.52 SYNTHESIS OF RESERPINE

reserpine **1** [1]

1) R. B. Woodward, F. E. Bader, H. Bickel, A. J. Frey, R. W. Kierstead, *Tetr.* **2**, 1 (1958).

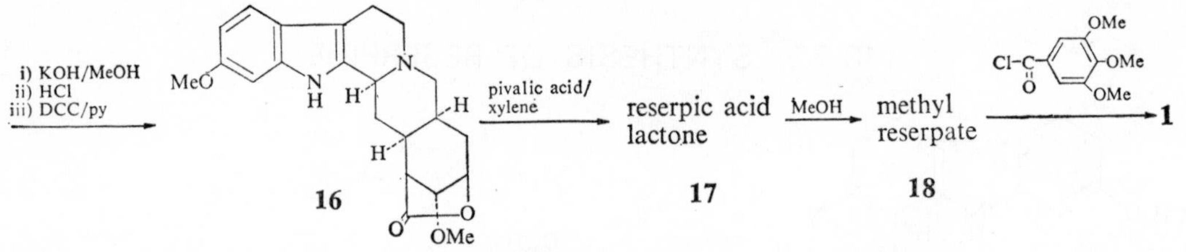

2 ⟶ 3: Approach of NaBH₄ is only sterically easy from the less hindered outer or convex surface of the *cis*-decaline derivative **2**, possessing a cage-like structure. The structure of **3** was confirmed by the conversion into the six-membered lactone **19**.

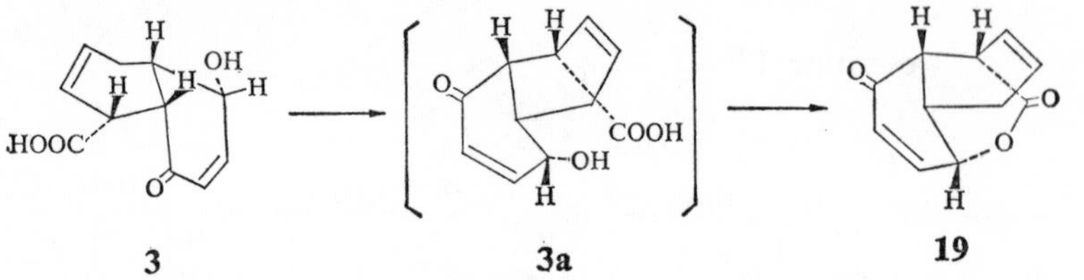

5 ⟶ 6: Formation of the unsaturated ether **6** from **5** may occur as follows.

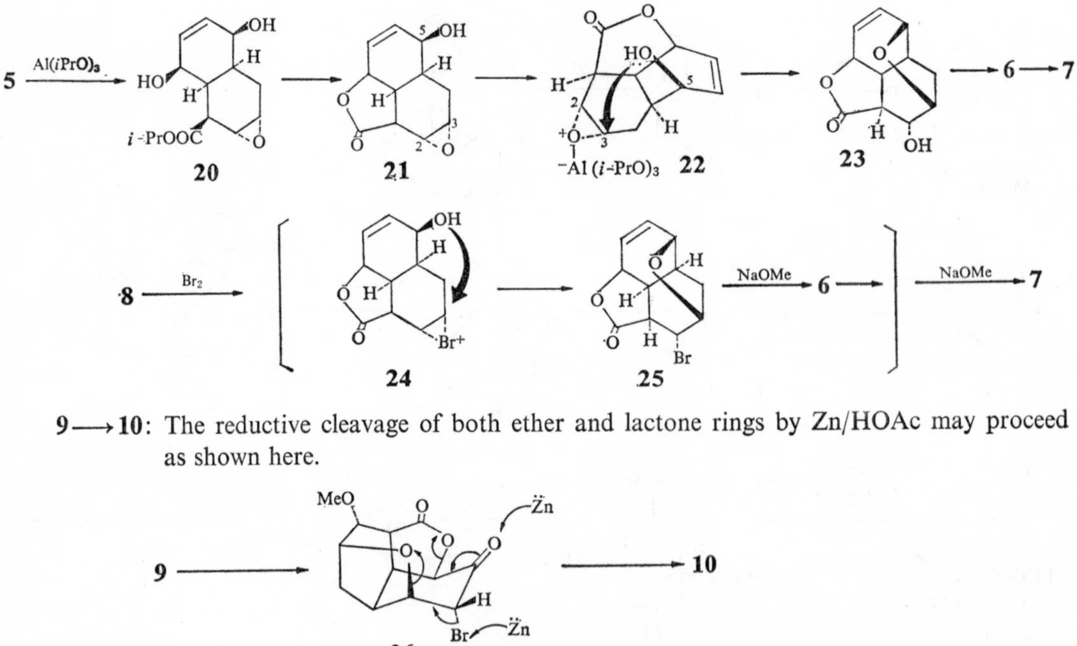

9 ⟶ 10: The reductive cleavage of both ether and lactone rings by Zn/HOAc may proceed as shown here.

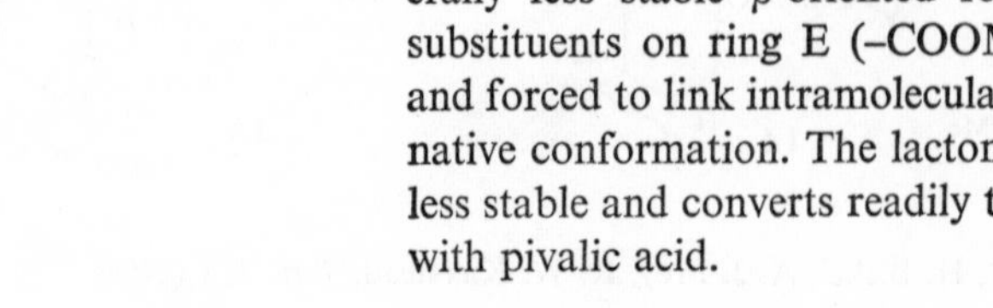

15 ⟶ 16 ⟶ 17 ⟶ 18 ⟶ 1: It is necessary to convert the α-oriented 3-H in **15** to the generally less stable β-oriented form. Thus, the *cis*-diequatorial substituents on ring E (–COOMe, –OCOMe) were hydrolyzed and forced to link intramolecularly to the lactone **16** in an alternative conformation. The lactone **16** is now thermodynamically less stable and converts readily to the more stable **17** on heating with pivalic acid.

2) L. Dorfman, A. Furlenmeier, C. F. Huebner, R. Lucas, H. B. MacPhillamy, J. M. Muller, E. Schilitter, R. Schwyzer, A. F. Andre, *Helv. Chim. Acta*, **37**, 59 (1954).

10.53 SYNTHESIS OF YOHIMBINE

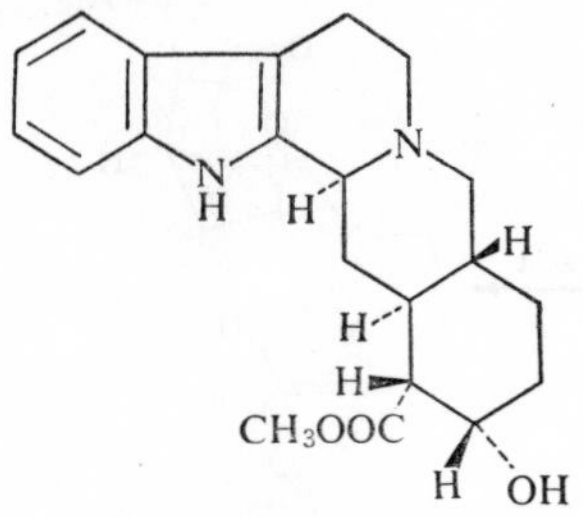

yohimbine **1**

Route I[1-6]

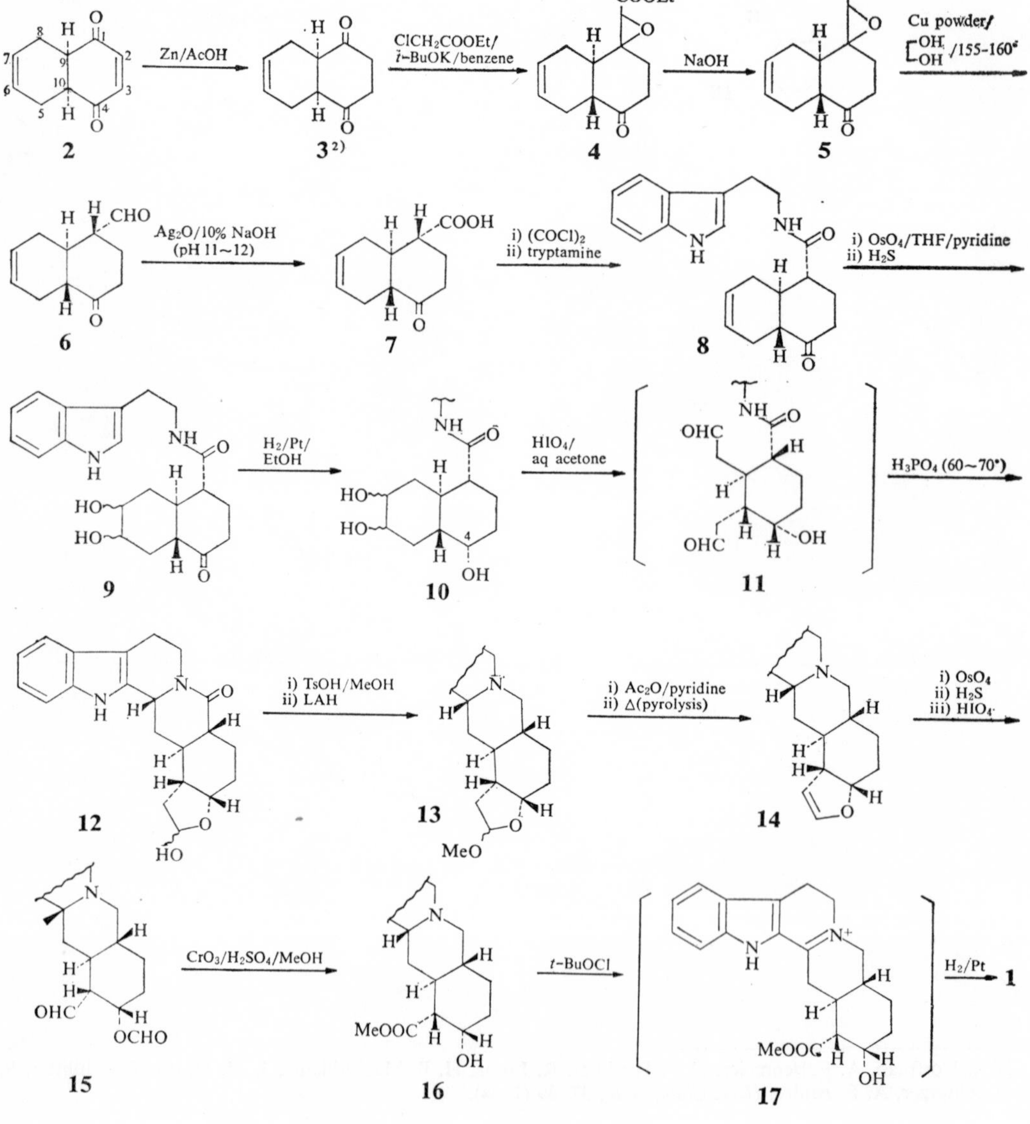

1) The conversion of *cis* **3** to *trans*-decalone **4** took place under basic conditions (the Darzens reaction).

2) The Darzens products **4** and **5** were obtained as a mixture of α- and β-epoxy compounds.

3) The compound **10** having a C-4 axial hydroxy group was obtained selectively due to the well-known approach of hydrogen-bearing catalysts from the less-hindered side of the decalone system.

4) **11 $\longrightarrow$ 12**: The configuration of 3-H is expected to be β (D/E ring junction, *cis*) because the attack of the indole center should be perpendicular to the plane of the π-system.[3,4] This unique Pictet-Spengler type cyclization involving amide nitrogen is interesting.[5]

5) **15 $\longrightarrow$ 16**: The mechanism of the one-step conversion of an aldehyde to an ester is as follows;

Route II[7,8]

1) E. E. van Tamelen, M. Shamma, A. W. Burgstahler, J. Wolinsky, R. Tamm, P. E. Aldrich, *J, Am. Chem. Soc.*, **91**, 7315 (1969).
2) K. Alder, G. Stein, *Ann. Chem.*, **501**, 247 (1933).
3) cf. A. H. Jackson, P. Smith, *Tetr.*, **24**, 403 (1968).
4) A. W. Burgstahler, Z. J. Bithos, *J. Am. Chem. Soc.*, **82**, 5466 (1960).
5) For a precedent in the benzenoid series, see B. Belleau, *J. Am. Chem. Soc.*, **75**, 5767 (1953).
6) W. O. Godtfredson, S. Vandegal, *Acta. Chem. Scand.*, **10**, 1414 (1956).
7) C. Szántay, K. Honty, L. Töke, *Tetr. Lett.*, 4871 (1971).
8) C. Szántay, L. Töke, K. Honty, G. Kalaus, *J. Org. Chem.*, **32**, 423 (1967).

1) The catalytic hydrogenation of **21** affords the diester **24** as a main product. This was subjected to Dieckmann condensation using NaH in toluene to give the desired keto-ester **25** in only 10–15% yield. The main product was found to be the isomer **26**, as expected.[9] Although NaBH$_4$ reduction of **25** affords **1**, the overall yield was unsatisfactory.[10]

24	**25**	**26**

2) **21 ⟶ 22**: This type of directed Dieckmann condensation is very rare in the literature. Its availability in synthetic organic chemistry has now been proved by the present work.

3) **22 ⟶ 23**: No product with a D/E *cis* ring junction could be detected.

Route III[11,12]

1) Birch reduction of **29** afforded the expected *trans*-fused keto ester **30** stereospecifically.[13]
2) **30 ⟶ 31**: A substantial amount of ester with equatorial 17-OH was also produced (~30%) along with the desired **31** (40%).

9) E. E. van Tamelen, M. Shamma, A. W. Burgstahler, J. Wolinsky, R. Tamm, P. E. Aldrich, *J. Am. Chem. Soc.*, **91**, 7315 (1969).
10) L. Töke, K. Honty, C. Szántay, *Chem. Ber.*, **102**, 3248 (1969).
11) G. Stork, R. N. Guthikonda, *J. Am. Chem. Soc.*, **94**, 5109 (1972).
12) cf. I. N. Nazarov, S. I. Zayalov, *Zh. Obshch. Khim.*, **23**, 1703 (1953).
13) G. Stork, S. D. Darling, *J. Am. Chem. Soc.*, **86**, 1961 (1964).
14) T. Fehr, P. A. Stadler, A. Hofmann, *Helv. Chim. Acta*, **53**, 2197 (1970).
15) L. Bartlett, N. J. Dastoor, J. Hrbek, Jr., W. Klyne, H. Schmid, G. Snatzke, *Helv. Chim. Acta*, **54**, 1238 (1971).

3) 33⟶1: Initially, the kinetically controlled product, Ψ-yohimbine **34**, was expected to be produced[1] (*cf.* Route I, comment **4**) and at 120° in the presence of EDTA this would be oxidized to the iminium salt **17**.[15] The reduction of such iminium salts is known to afford the C/D *trans* quinolizidine system.

34   17

10.54 CHEMICAL CORRELATION OF INDOLE ALKALOIDS WITH OXINDOLE ALKALOIDS

On the basis of biogenetic considerations,[1] attempts were made to convert indole alkaloids to the corresponding oxidoles.[2-4] The conversion involves the oxidation of indole to 3-substituted indolenine, followed by rearrangement to an oxindole.

Through the chloroindolenine

1   2   2a

3   4

1) A. I. Scott, *Accounts Chem. Res.*, **3**, 151 (1970).
2) N. Finch, W. I. Taylor, *J. Am. Chem. Soc.*, **84**, 3871 (1962).
3) N. Finch, C. W. Gemenden, I. H-C. Hsu, W. I. Taylor, *J. Am. Chem. Soc.*, **85**, 1520 (1963).
4) H. Zinnes, J. Shavel, Jr., *J. Org. Chem.*, **31**, 1765 (1966).

The following conversions have also been accomplished similarly: ajmalicine⟶mitra-phyllene, corynantheine⟶corynoxeine, dihydrocorynantheine⟶rhynchophyllene. In each case, this represents the conversion of an indole alkaloid to an oxindole alkaloid.

Through the acetoxyindolenine

5   6   7

Example of the reverse conversions[5]

pteropodine **8** (7R)   **10**   **12**
isopteropodine **9** (7S)   **11**

tetrahydroalstonine **13**   akuammigine **14**

---

5) N. Aimi, E. Yamanaka, J. Endo, S. Sakai, J. Haginiwa, *Tetr.*, **29**, 2015 (1973).

10.55 SYNTHESIS OF CORYNANTHEINE

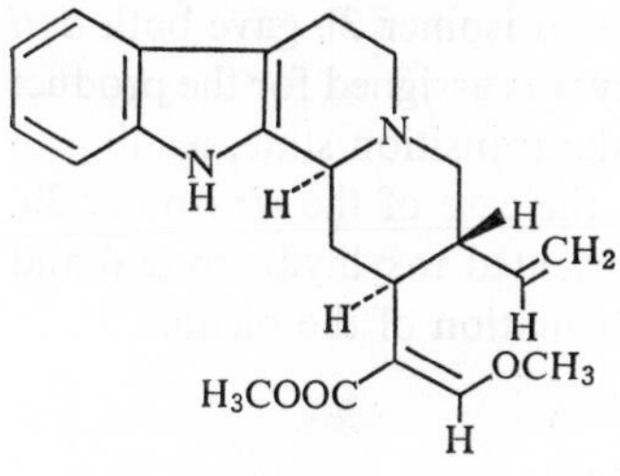

corynantheine **1**

Route I[1]

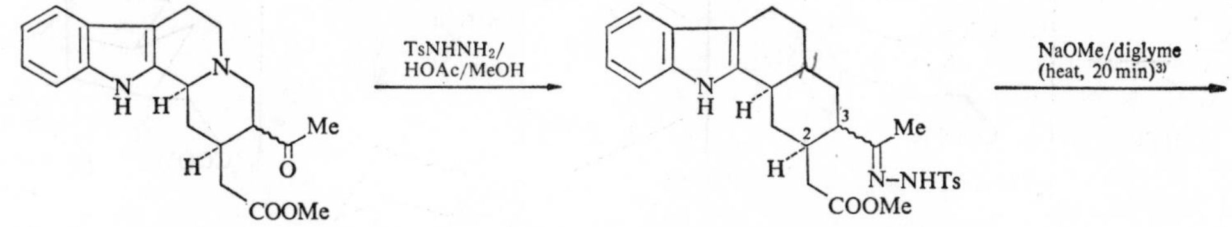

TsNHNH₂/ HOAc/MeOH →

2
(*cis + trans*)[2]

3a (C-2/C-3 *trans*)
3b (C-2/C-3 *cis*)

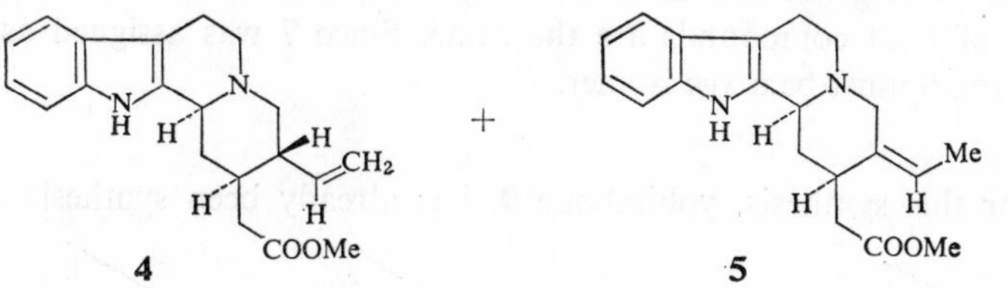

NaOMe/diglyme (heat, 20 min)[3] →

4   +   **5**

i) Ph₃C⁻Na⁺/HCOOMe[4].
ii) CH₂N₂

4 ────────→ **1**

5 ──Ph₃C⁻Na⁺/HCOOMe──→

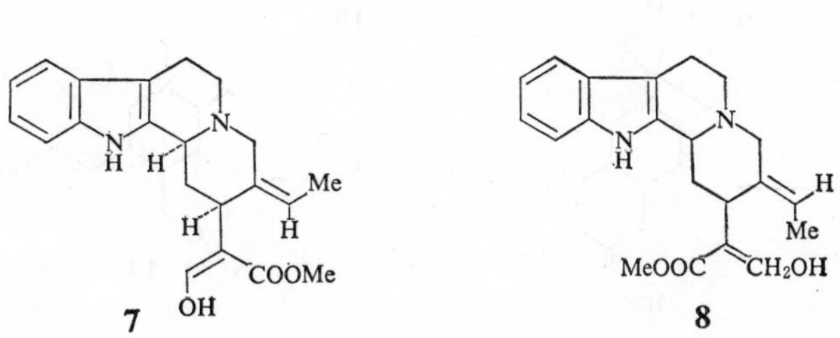

7   **8**

1) E. E. van Tamelen, I. G. Wright, *J. Am. Chem. Soc.*, **91**, 7349 (1969).
2) E. E. van Tamelen, C. Placeway, G. P. Schiemenz, I. G. Wright, *J. Am. Chem. Soc.*, **91**, 7359 (1969).
3) J. W. Powell, M. C. Whiting, *Tetr.*, **12**, 168 (1961).
4) E. E. van Tamelen, J. B. Hester, Jr., *J. Am. Chem. Soc.*, **81**, 3805 (1959).

2—→3: The keto ester **2**, which was used as a key intermediate in the synthesis of ajmalicine, was chosen as a starting material.

3a—→4+5: The *trans*-tosylhydrazone **3a** afforded the single ethylidene isomer **5** (*anti*) as a main product in addition to the desired **4**, while the *cis* isomer **3b** gave both *anti* and *syn* isomers of **5** among others. The *anti* geometry was assigned for the product from **3a** because severe A-strain[6] was expected in the transition state, leading to formation of the *syn* isomer. On the other hand, in the case of the *cis* isomer **3b**, the main steric interactions are due to the axially oriented tosylhydrazone **6** and are relieved by rotation in either direction during formation of the olefin.

5—→7: The compound **7** was not *dl*-geissoschizine **8**[5] although the planar structures and C-3/C-5 stereochemistry of both compounds are the same. Since **7** was assigned as an *anti* isomer (see above), **8** must be a *syn* isomer.

Route II[7]

The starting material for this synthesis, yohimbone **9**, has already been synthesized.[10]

5) M. -M. Janot, *Tetr.*, **14**, 113 (1961).
6) F. Johnson, *Chem. Rev.*, **68**, 375 (1968).
7) R. L. Autrey. P. W. Scullard, *J. Am. Chem. Soc.*, **90**, 4917 (1968); *ibid.*, **90**, 4924 (1968). See, however, R. K. Hill, D. A. Cullison, *ibid.*, **95**, 2923 (1973).

10 ⟶ 11 + 12: When the condensation of **10** with TsSMe was tried under the conditions defined by Woodward and Patchett[9] the unwanted **12** was obtained (20%) in addition to **11** (27%). However, the yield was improved (57–63%) when 95% EtOH was used in place of absolute EtOH as a solvent.

13 ⟶ 14: The mechanism of Beckmann fragmentation in this system may be as follows.

Addendum

Syntheses of dihydrocorynantheine, corynantheidine, and related compounds have been reported.[11]

Syntheses of heteroyohimbine alkaloids, ajmalicine,[12,14] 19-epiajmalicine,[14] akuammigine,[13,14] and tetrahydroalstonine[13,14] have also been reported.

8) L. D. Albright, L. A. Mitscher, L. Goldman, *J. Org. Chem.*, **28**, 38 (1963).
9) R. B. Woodward, A. A. Patchett, D. H. R. Barton, D. A. J. Ives, R. B. Kelley, *J. Chem. Soc.*, 1131 (1957).
10) G. A. Swan, *J. Chem. Soc.*, 1534 (1950).
11) E. E. van Tamelen, J. B. Hester, *J. Am. Chem. Soc.*, **81**, 3805 (1959); J. A. Weisbach, J. L. Kirkpatrick, K. R. Williams, E. L. Anderson, N. C. Yim, B. Douglas, *Tetr. Lett.*, 3457 (1965); E. Wenkert, K. G. Dave, F. Haglid, *J. Am. Chem. Soc.*, **87**, 5461 (1965); C. Szántay, M. Bárczai-Beke, *Chem. Ber.*, **102**, 3963 (1969).
12) E. E. van Tamelen, C. Placeway, G. P. Schiemenz, I. G. Wright, *J. Am. Chem. Soc.*, **91**, 7359 (1969).
13) E. Winterfeld, H. Radunz, T. Korth, *Chem. Ber.*, **101**, 3172 (1968); E. Winterfeld, A. J. Gaskell, T. Korth, H-E. Radunz, M. Walkowiak, *ibid.*, **102**, 3558 (1969).
14) M. Uskokovic, C. Reese, H. L. Lee, G. Grethe, J. Gutzwiller, *J. Am. Chem. Soc.*, **93**, 5902 (1071); J. Gutzwiller, G. Pizzolato, M. Uskokovic, *ibid.*, **93**, 5907 (1971).

10.56   CONVERSION OF INDOLE ALKALOIDS TO 2-ACYL
INDOLE ALKALOIDS

2-Acylindole alkaloids are considered to be biosynthesized from indole alkaloids. The following reaction sequence is a synthesis of dihydroburnamicine[1] based on the biogenetic pathway.

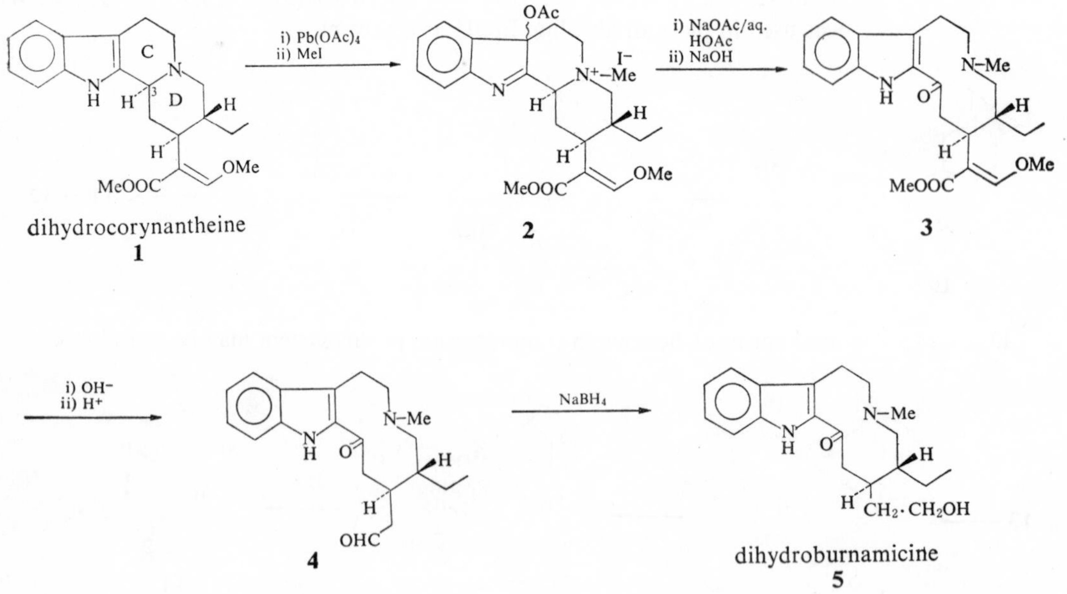

dihydrocorynantheine
1

2

3

4

dihydroburnamicine
5

2 ⟶ 3: This is the first C/D ring cleavage.[1] Other reagents such as BrCN,[2] ROCOCl[3] and (RCO)₂O[4] have been used for the same purpose. The reaction mechanism has been discussed.[5]

1) L. J. Dolby, S. Sakai, *J. Am. Chem. Soc.*, **86**, 5362 (1964); *Tetr.*, **23**, 1 (1967).
2) J. D. Albright, L. Goldman, *J. Am. Chem. Soc.*, **91**, 4317 (1969).
3) S. Sakai, A. Kubo, K. Katano, N. Shinma, K. Sasago, *Yakugaku Zasshi*, **93**, 1165 (1973).
4) G. H. Foster, J. Harley-Mason, W. R. Waterfield, *Chem. Commun.*, 21 (1967); G. C. Crawley, J. Harley-Mason, *ibid.*, 685 (1971).
5) L. J. Dolby, G. W. Gribble, *J. Org. Chem.*, **32**, 1391 (1967).

10.57 SYNTHESIS OF AJMALINE AND ISOAJMALINE

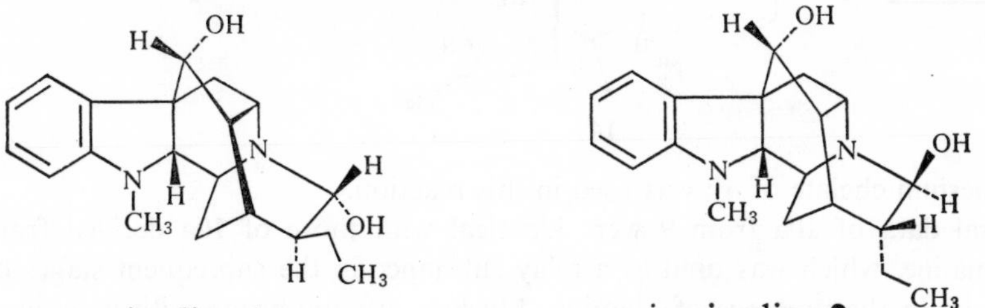

ajmaline **1**   isoajmaline **2**

Route I[1]

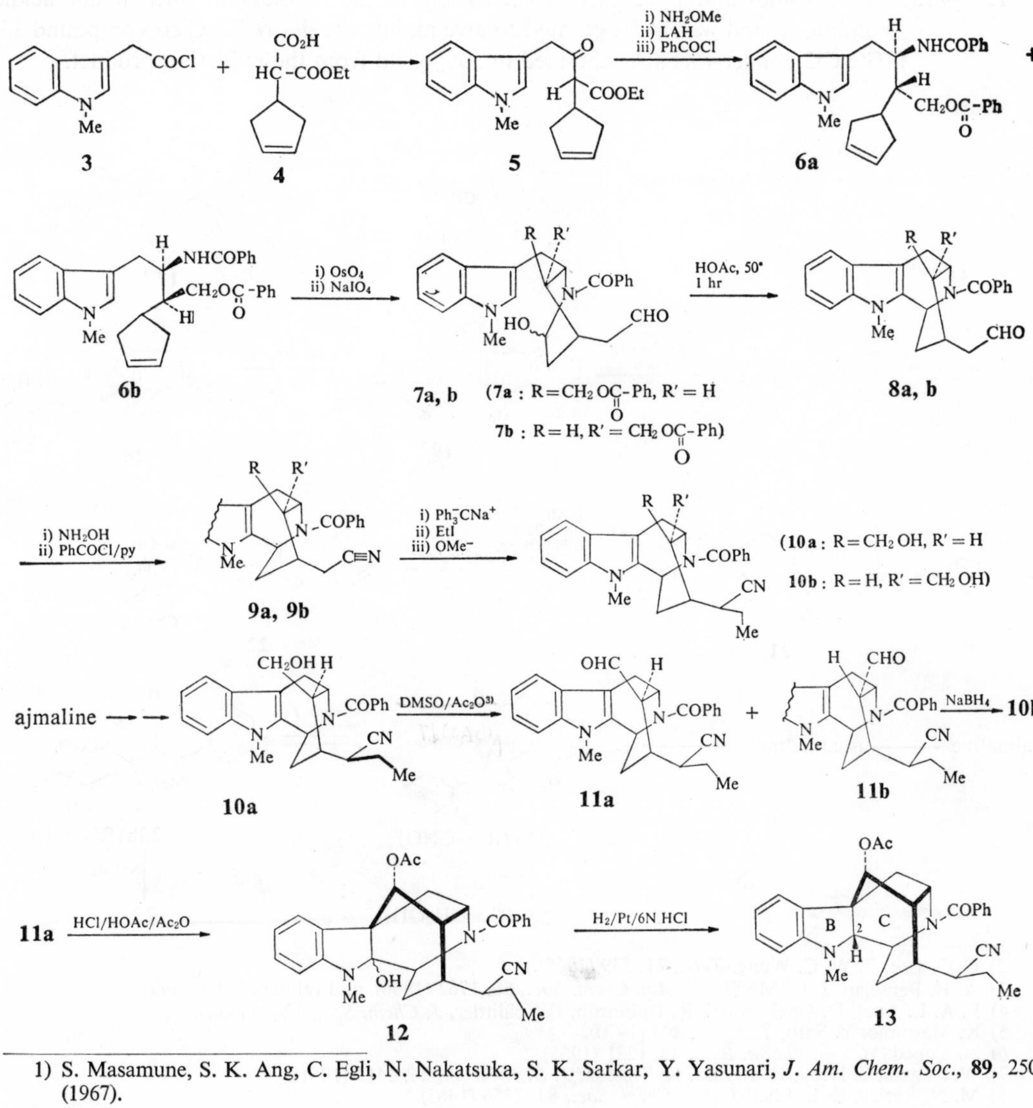

7a : R = CH$_2$OC-Ph, R' = H
          ‖
          O
7b : R = H, R' = CH$_2$OC-Ph)
                    ‖
                    O

(10a : R = CH$_2$OH, R' = H

10b : R = H, R' = CH$_2$OH)

1) S. Masamune, S. K. Ang, C. Egli, N. Nakatsuka, S. K. Sarkar, Y. Yasunari, *J. Am. Chem. Soc.*, **89**, 2506 (1967).

14

$3+4 \longrightarrow 5$: The magnesium chelate of $4^{2)}$ was used in this reaction.

$10 \longrightarrow 11$: The spectral data of **10a** from **9** were identical with those of **10a** derived from natural ajmaline, which was used as a relay substance in the subsequent stages of the synthesis. In the presence of alumina, **11a** was in equilibrium with its epimer **11b** in a 3:7 ratio in favour of **11a**.

$12 \longrightarrow 13$: The carbinol amine **12** existed exclusively in the indolenium form under acidic conditions, and was hydrogenated to give mainly the desired B/C *cis* compound **13**. The C-2 epimer was also obtained in 30% yield from the reduction product.

Route II$^{5)}$

15   **16**   **17$^{6)}$**

18   **19**   **20**

21   **22**

ajmaline $\longrightarrow$ isoajmaline$^{1,4)}$

23a (R′ = CHO)   **23b** (R′ = CHO)

24 (R′ = CH$_2$OH)   **22**

2) C. C. Lee, E. W. C. Wong, *Tetr.*, **21**, 539 (1965).
3) A. H. Fenselau, J. G. Moffatt, *J. Am. Chem. Soc.*, **88**, 1762 (1966), and references therein.
4) F. A. L. Anet, D. Chakravarti, R. Robinson, E. Schlittler, *J. Chem. Soc.*, 1242 (1954).
5) K. Mashimo, Y. Sato, *Tetr.*, **26**, 803 (1970).
6) N. Yoneda, *Chem. Pharm. Bull.*, **13**, 1231 (1965).
7) W. Nagata, S. Hirai, H. Itazaki, K. Takeda, *J. Org. Chem.*, **26**, 2413 (1961).
8) M. N. Rerick, E. L. Eliel, *J. Am. Chem. Soc.*, **84**, 2356 (1962).

25

26   **27**   **28**

$$\xrightarrow{\text{LAH}^{4)}} 2$$

15 ⟶ 17: The keto ester **17**, a key intermediate, was prepared by Yoneda from tryptophan **15**.[6]

19 ⟶ 20: The hydrocyanation reaction proceeded with high stereoselectivity, giving only one product. The stereostructure of the product **20** was clarified by the subsequent reactions (see **21 ⟶ 22**).

21 ⟶ 22: The hydroxy nitrile **22** formed from **21** was found to be identical with that derived from natural isoajmaline, which showed that hydrocyanation took place preferentially to afford the isoajmaline series. For subsequent stages of isoajmaline synthesis, natural **22** was used as a relay substance.

23a ⇌ 23b: In this alumina-induced equilibration reaction, the ratio of **23a** to **23b** was 2:3, while in the ajmaline series, the equilibrium ratio of the corresponding aldehydes **11a, 11b** was in favour of **11a** (7:3).[1]

Route III[10]

29   **30**[11]   **31**   **32**

9) R. F. Borch, *Tetr. Lett.*, 61 (1968).
10) E. E. van Tamelen, L. K. Oliver, *J. Am. Chem. Soc.*, **92**, 2136 (1970).
11) E. E. van Tamelen, L. J. Dolby, R. G. Lawton, *Tetr. Lett.*, 30 (1960).
12) J. C. Collins, W. W. Hess, F. J. Frank, *Tetr. Lett.*, 3363 (1968).

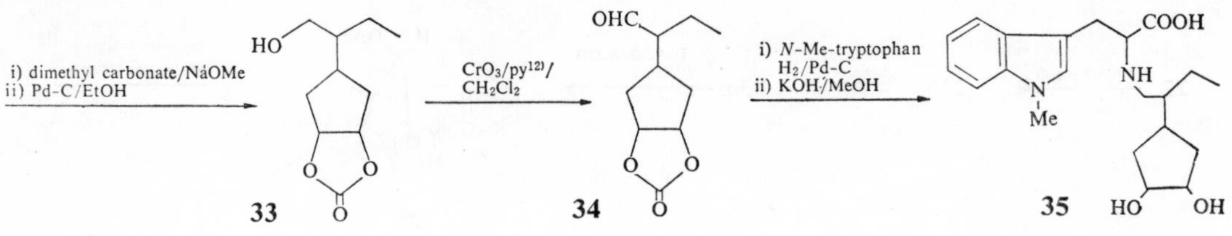

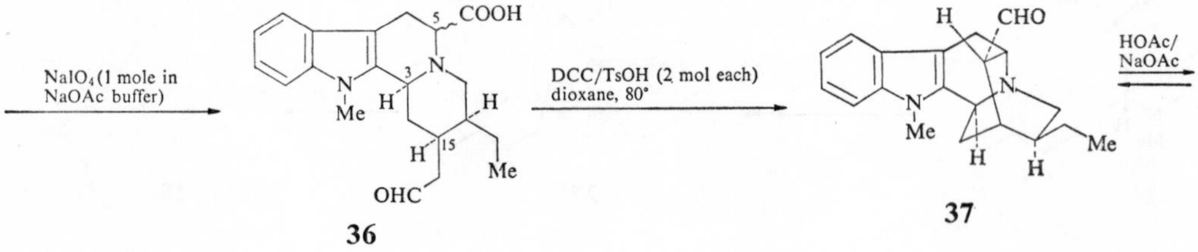

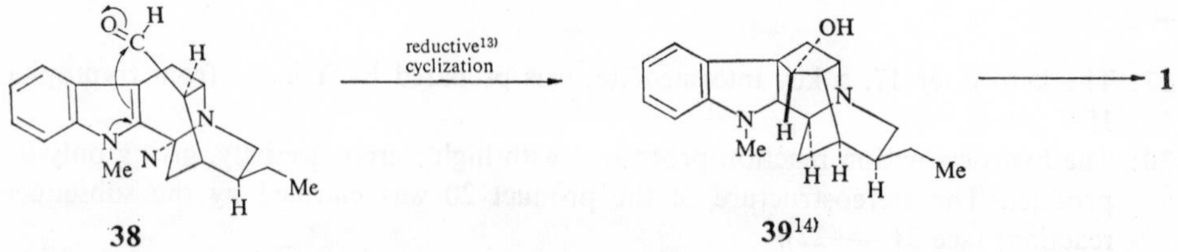

33

34

35

36

37

38

39[14)]

40

41

35 ⟶ 36: The tetracyclic compound **36** was obtained as a mixture of isomers. The desired all-*cis* C-3, C-15 and C-20 material was not separated from stereoisomeric congeners, but the whole mixture was subjected to the next reaction.

36 ⟶ 37: Generation of the iminium salt **40** required for the critical C-5/C-16 bond formation was achieved by a decarboxylative oxidation process, which would not be possible starting from tetrahydrocarboline derivatives. A diastereomeric mixture was used for the reaction.

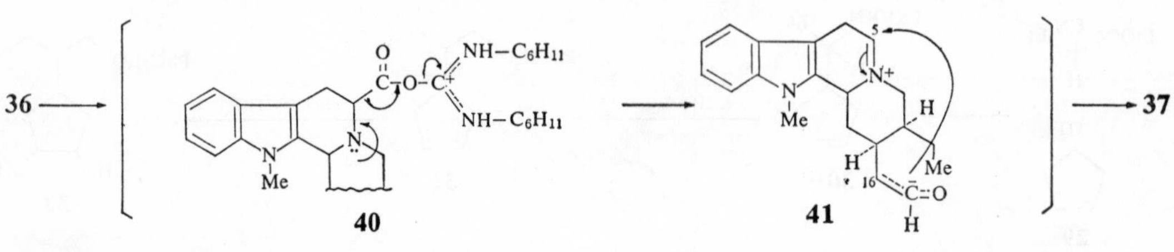

37 was resolved by use of *d*-camphor-10-sulforic acid.

37 ⇌ 38: Treatment of either **37** or **38** with HOAc/NaOAc or Al$_2$O$_3$ in refluxing benzene afforded an equilibrium mixture of **37** (~35%) and **38** (~15%).

39 ⟶ 1: The conversion of deoxyajmaline **39** to **1** has been achieved by the ring opening (using PhCOOCl) and oxidative ring closure sequence of Hobson and McCluskey.[14)]

13) M. F. Bartlett, B. F. Lambert, H. M. Werblood, W. I. Taylor, *J. Am. Chem. Soc.*, **85**, 475 (1963).
14) J. D. Hobson, J. G. McCluskey, *J. Chem. Soc.*, 2015 (1967).

10.58 SYNTHESIS OF RHYNCHOPHYLLINE, ISORHYNCHOPHYLLINE, FORMOSANINE (UNCARINE B), ISOFORMOSANINE (UNCARINE A), MITRAPHYLLINE AND ISOMITRAPHYLLINE

2a (R_1=Me, R_2=H)
3a (R_1=H, R_2=Me)

2b (R_1=Me, R_2=H)
3b (R_1=H, R_2=Me)

(a=normal series; b=iso series)

Route I[1]

4+5——→6: In this intramolecular Mannich reaction, the decarboxylated compound **6′** is frequently obtained. When the reaction mixture was kept at temperatures higher than 60°, the ratio of **6′** in the products increased.

6′

8a,b——→9b: The product **9** was found to be a mixture of geometrical isomers involving the double bond, but the isomer at the spiro-position was not isolated. The iso-structure **9b** was assigned to the product because the compounds in the iso-series usually predominated in basic equilibria.[2]

9b——→10a,b,c: The isomer **10a** was obtained from acidic media and was converted to a mixture of **10a** and **10b** when it was kept with Al_2O_3 in CH_2Cl_2, which suggested that **10a** should be in the normal series.[2] The third product **10c** is assumed to be the stereo-isomer with *cis* substituents at C-15 and C-20.

Route II[3]

1) Y. Ban, M. Seto, T. Oishi, *Tetr. Lett.*, 2113 (1972).
2) cf. J. E. Saxton, *The Alkaloids, Chemistry and Physiology* (ed. R. H. Manske), vol. VIII, p. 51, Academic Press, 1965.
3) Y. Ban, N. Taga, T. Oishi, *Tetr. Lett.*, 187 (1974). Presented at **17**th *Symposium on the Chemistry of Natural Products*, 1973.
4) A. P. Skoldinov, A. P. Arendank, T. M. Godzhello, *J. Org. Chem., USSR*, **6**, 421 (1970).
5) E. Winterfeld, A. J. Gaskell, Y. Korth, H-E Radunz, M. Walkowiak, *Chem. Ber.*, **102**, 3558 (1969).

Me₂N
Me₂N—C–H
16a $\xrightarrow{t\text{-BuO}}$

19

$\xrightarrow{\text{MeOH/HCl}}$

20

$\xrightarrow{\begin{array}{c}\text{aq. dioxane/}\\.\text{H}^+\text{(2 equiv)}\end{array}}$

21

+ **2a** + **2b**

PPA

i) (Me₂N)₂–CH
    t-BuO
.ii) 5% MeOH/HCl
iii) aq. dioxane
iv) PPA

18a $\xrightarrow{\hspace{3cm}}$ 3a + 3b

14b ⟶ 15b or 17b: The reduction of **14b** with Pt or NaBH₄ proceeded quite selectively and afforded **15b** or **17b**, respectively. A similar selectivity has been observed on the reduction of the analogous system in the β-carboline series.[5]

16a ⟶ 19 ⟶ 20: The direct introduction of a carbomethoxy group on ring E using HCOOEt was difficult because **16a** was almost insoluble in inert solvents.[6]

16a ⟶ 2a,b; 18a ⟶ 3a,b: The desired alkaloids were obtained from **16a** or **18a** in a single vessel without isolating any of the intermediates.

Route III[5,7,8]

22 and **24** have been synthesized by a previous route.[5,9]

ajmalicine **22** $\xrightarrow{t\text{-BuClO}}$ **23** $\xrightarrow{\text{HCl/MeOH/H}_2\text{O}}$ **3a, b**[7,8]

24 $\xrightarrow{t\text{-BuClO}}$ **25** $\xrightarrow{\text{HCl/MeOH/H}_2\text{O}}$ **2a, b**[5]

23 ⟶ 3a,b; 25 ⟶ 2a,b: The conversion of indolenines **23** or **25** to the corresponding oxindole derivatives took place readily, irrespective of the configuration at C-3 of the starting materials.[8]

6) H. Bredereck, G. Simchen, S. Rebsdat, W. Kantlehner, P. Horn, R. Wahl, H. Hoffman, P. Grieshaber, *Chem. Ber.*, **101**, 41 (1968).

7) N. Finch, W. I. Taylor, *J. Am. Chem. Soc.*, **84**, 1318 (1962).

8) H. Zinness, J. Shavel, *J. Am. Chem. Soc.*, 1320 (1962); *J. Org. Chem.*, **31**, 765 (1966).

9) E. E. van Tamelen, C. Placeway, G. P. Schiemenz, I. G. Wright, *J. Am. Chem. Soc.*, **91**, 7359 (1969).

10.59   SYNTHESIS OF TUBIFOLINE, TUBIFOLIDINE, CONDYFOLINE, GEISSOSCHIZOLINE, TUBOTAIWINE AND NORFLUOROCURARINE

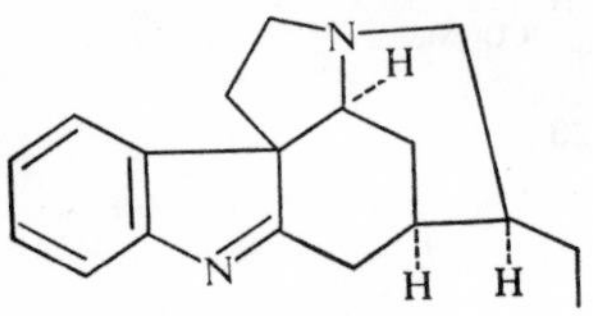

tubifoline **1**[1)] 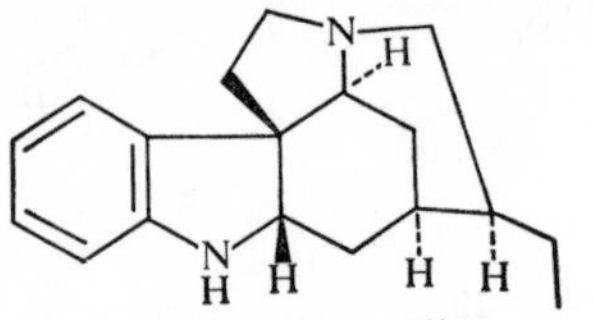   tubifolidine **2**[1)] 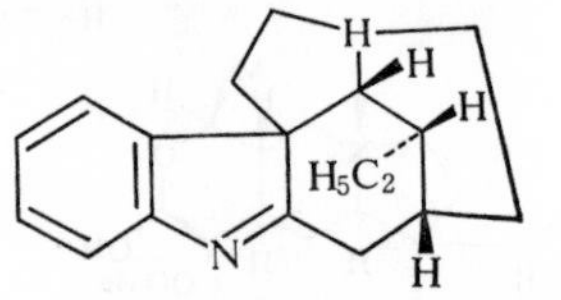   condyfoline **3**[1)]

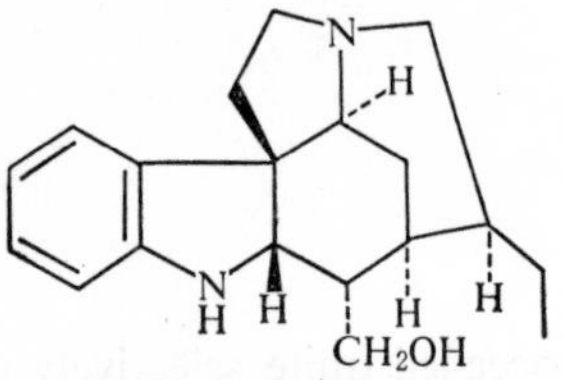

geissoschizoline **4**[2,3)] 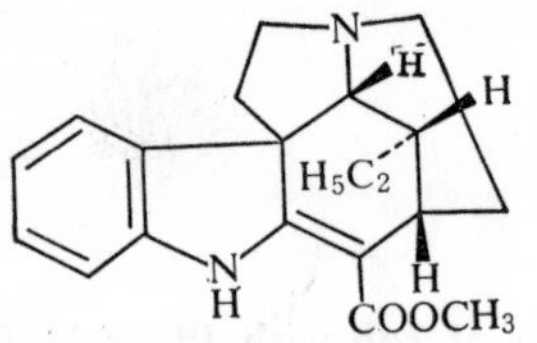   tubotaiwine **5**[4)] 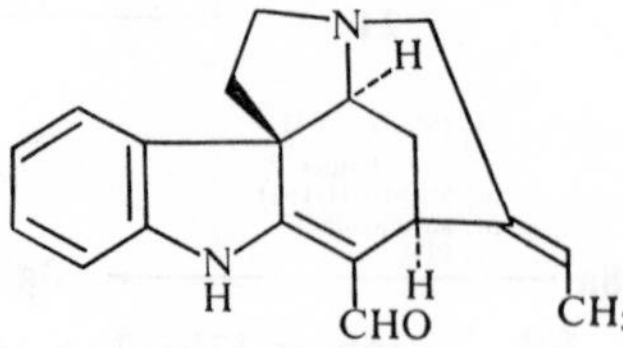   norfluorocurarine **6**[3,5)]

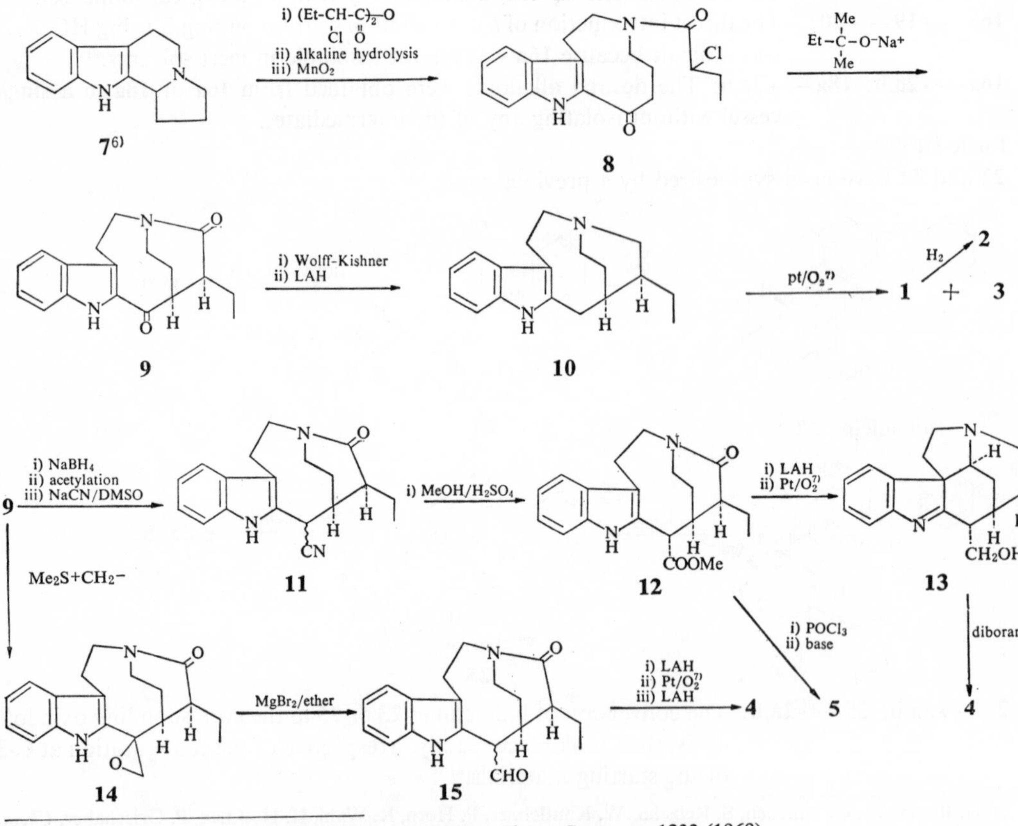

7[6)]

i) (Et-CH-C)$_2$-O
       Cl   O
ii) alkaline hydrolysis
iii) MnO$_2$

8

Me
Et−C−O−Na$^+$
Me

9

i) Wolff-Kishner
ii) LAH

10

pt/O$_2$[7)] $\longrightarrow$ **1** $+$ **3**

H$_2$ **2**

9

i) NaBH$_4$
ii) acetylation
iii) NaCN/DMSO

Me$_2$S$^+$CH$_2^-$

11

i) MeOH/H$_2$SO$_4$

12

i) LAH
ii) Pt/O$_2$[7)]

13

14

MgBr$_2$/ether

15

i) LAH
ii) Pt/O$_2$[7)]
iii) LAH $\longrightarrow$ **4**

i) POCl$_3$
ii) base   **5**

diborane   **4**

1) B. A. Dadson, J. Harley-Mason, G. H. Foster, *Chem. Commun.*, 1233 (1968).
2) B. A. Dadson, J. Harley-Mason, *Chem. Commun.*, 665 (1969).
3) J. Harley-Mason, C. G. Taylor, *Chem. Commun.*, 812 (1970).
4) B. A. Dadson, J. Harley-Mason, *Chem. Commun.*, 665 (1969).

7 → 8: Cleavage of **7** with acid anhydrides constitutes a key step of the present method.

12 → 13: 15 → 4; 20 → 21: The oxidative cyclization proceeds only in the direction indicated. None of the alternative aspidospermatidine skeletons were formed.

17 → 18a + 18b: The *cis* and *trans* isomers (**18a** and **18b**) were obtained in a ratio of 1.4:1 and were characterized by their nmr spectra. The structure **18a** was assigned for the compound whose vinyl proton was strongly deshielded by the nearby carbonyl group.

18a → 19: The ethylene oxide corresponding to **14** was readily obtained but all attempts to convert this compound to the aldehyde were unsuccessful.

19 → 20: Selective reduction of the lactam carbonyl group proved difficult; even if AlH₃ was used, the 19,20-dihydro compound was obtained in addition to **20**.

5) G. C. Crawley, J. Harley-Mason, *Chem. Commun.*, 685 (1971).
6) S. Corsano, S. Algieri, *Ann. Chim.* **50**, 75 (1960).
7) D. Schumann, H. Schmid, *Helv. Chim. Acta*, **46**, 1966 (1963).

10.60 SYNTHESIS OF DASYCARPIDONE, 3-EPIDASYCARPIDONE, ULEINE AND 3-EPIULEINE

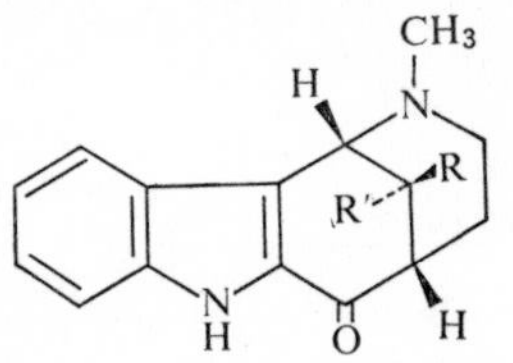

dasycarpidone **1a** (R=H, R′=Et)
3-epidasycarpidone **1b** (R=Et, R′=H)

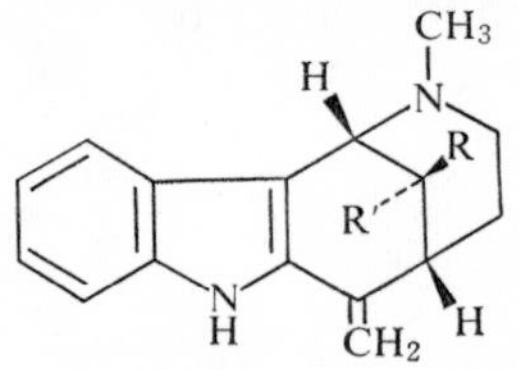

uleine **2a** (R=H, R′=Et)
3-epiuleine **2b** (R=Et, R′=H)

Route I[1]

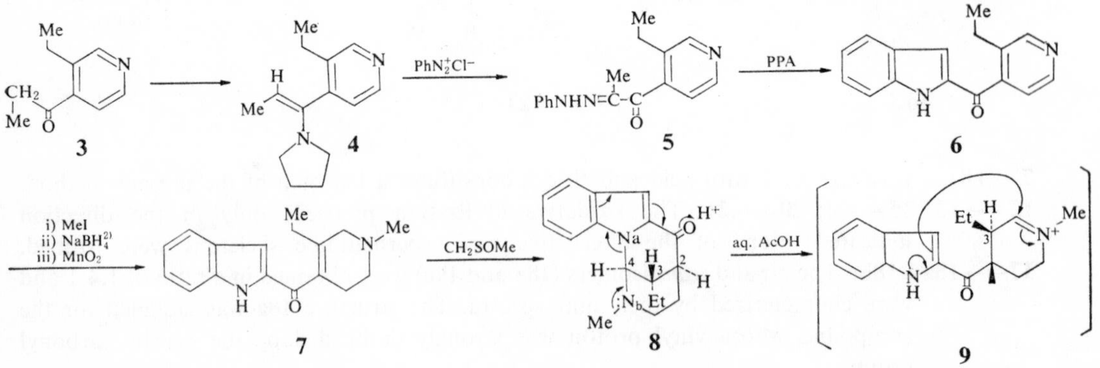

i) MeI[2]
ii) NaBH$_4$[2]
iii) MnO$_2$

$$1a + 1b \text{ (main product)}$$

$$1a \xrightarrow{\text{Mg/Hg/CH}_2\text{I}_2[3]} 2a$$

$$1b \xrightarrow{\text{Ph}_3\overset{+}{\text{P}}\text{-}\overset{-}{\text{CH}}_2} 2b$$

7 ⟶ 8: The mechanism of the formation of one stereoisomer **8** in this conversion may be as follows. The configuration of **8** at C-3 was deduced from the slow rate of quaternization of N$_b$ with methyl iodide.[4]

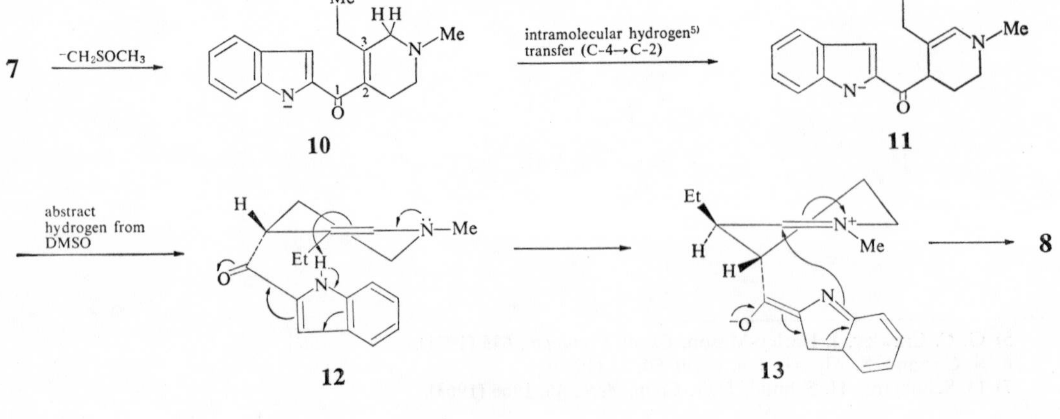

9 ⟶ 1a + 1b: Epimerization at C-3 must occur before ring closure, since acetic acid did not cause isomerization of **1a** or **1b**.

1a ⟶ 2a: The formation of the methylene group using $Mg/Hg/CH_2I_2$, which was used in the conversion of **1b** to **2b**, failed in this case.

Route II[6]

18 ⟶ 19: The amino ester **19** was obtained as a mixture of stereoisomers.

19 ⟶ 1a,b: Epidasycarpidone **1b** was found to be the main product (55%).

Route III[8]

1) A. Jackson, N. D. V. Wilson, A. J. Gaskel, J. A. Joule, *J. Chem. Soc.*, C, 2738 (1969).
2) R. E. Lyle, P. S. Anderson, *Adv. Heterocyclic Chem.*, **6**, 45 (1966),
3) G. Cainelli, F. Bertini, P. Groselli, G. Zubiani, *Tetr. Lett.*, 5153 (1967).
4) M. Shamma, J. A. Weiss, R. J. Shine, *Tetr. Lett.*, 2489 (1967).
5) S. Bank, C. A. Roue, A. Schriesheim, *J. Am. Chem. Soc.*, **85**, 2115 (1963); D. J. Cram, R. T. Uyeda, *ibid.*, **84**, 4358 (1962).
6) L. J. Dolby, H. Biere, *J. Am. Chem. Soc.*, **90**, 2699 (1968).
7) G. A. Youngdale, D. G. Anger, W. C. Anthony, J. P. DeVanzo, M. E. Greig, R. V. Heinzelman, H. H. Keasling, J. Szmuszkovicz, *J. Med. Chem.*, **7**, 415 (1964).
8) G. Büchi, S. J. Gould, F. Naf, *J. Am. Chem. Soc.*, **93**, 2492 (1971).
9) F. Johnson, *Chem. Rev.*, **68**, 375 (1968).
10) P. Karrer, H. Rentshler, *Helv. Chim. Acta*, **27**, 1297 (1955); C. A. Grob, A. Kaiser, E. Renk, *ibid.*, **40**, 2170 (1957).
11) J. H. Chapman, J. Elks, G. H. Phillipps, L. J. Wyman, *J. Chem. Soc.*, 4344 (1956); E. S. Rothman, M. E. Wall, *J. Am. Chem. Soc.*, **79**, 3228 (1957).

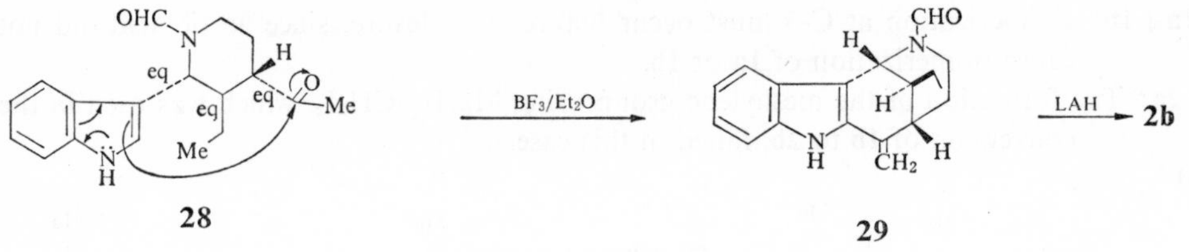

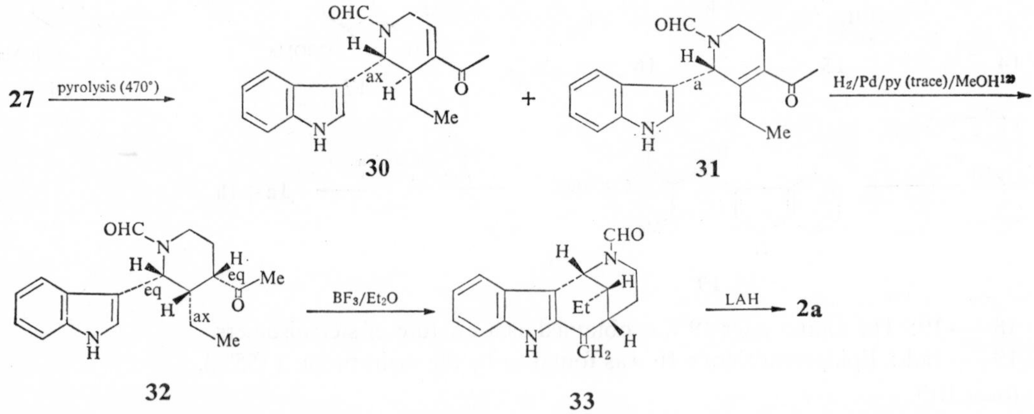

28 → **29** (BF₃/Et₂O), **29** → **2b** (LAH)

27 pyrolysis (470°) → **30** + **31** → H₂/Pd/py (trace)/MeOH[12]

32 → **33** (BF₃/Et₂O), **33** → **2a** (LAH)

23 ⟶ **24**: A chair-like transition state **34** with the large equatorial substituents may be involved in this process.

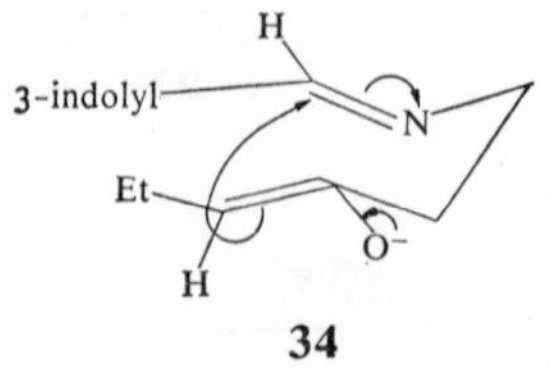

34

24 ⟶ **25**: Formylation was accompanied by conformational inversion of the piperidine ring, which is ascribable to A-strain[9] between the equatorial indole substituent and the amide carbonyl group. Such an effect also occurs in all the subsequent reaction in this synthesis. The N-formyl compounds with axial indole groups were each present as two rotomers due to restricted rotation of the carbon-nitrogen of the amide.

25 ⟶ **26**: The ethynylcarbinol **26** was obtained as a mixture of three conformers but the conformer **26** with the equatorial indole group and the amide carbonyl *trans* to the indole ring was the major product.

27 ⟶ **28**: Epimerization of the acetyl group took place.

28 ⟶ **29**: The mechanism of the cyclization is as follows.

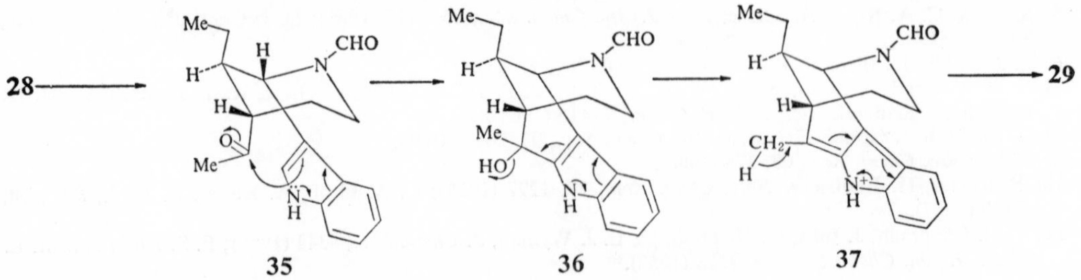

35   **36**   **37**

Route IV[13)]

38 → 40: The feature of this synthesis is the formation of the key intermediate **41** by the reaction of the *N*-oxide activated by *O*-acylation with the indole Grignard reagent **38**.

41·HCl → 42: Catalytic hydrogenation of **41·HCl** proceeded with high stereoselectivity, affording only **42**, while the reduction of the methiodide of **41** gave a mixture of stereoisomers.

42 → 43: Compare Route II.

12) H. J. Ringold, *J. Am. Chem. Soc.*, **82**, 961 (1960).
13) T. Kametani, T. Suzuki, *Chem. Pharm. Bull.*, **19**, 1424 (1971); T. Kametani, T. Suzuki, *J. Org. Chem.*, **36**, 1291 (1971); T. Kametani, T. Suzuki, *J. Chem. Soc.*, C, 1053 (1971).

10.61 SYNTHESIS OF EBURNAMONINE, EBURNAMINE AND VINCAMINE

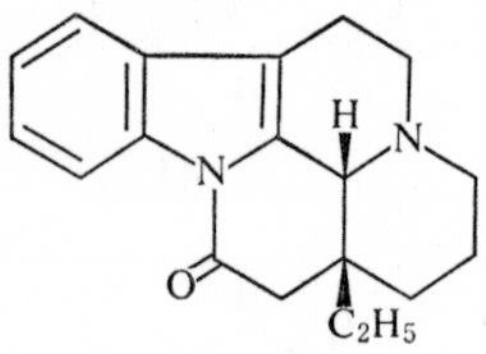

eburnamonine **1** 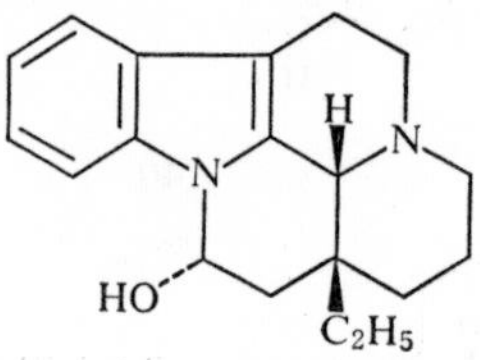   eburnamine **2** 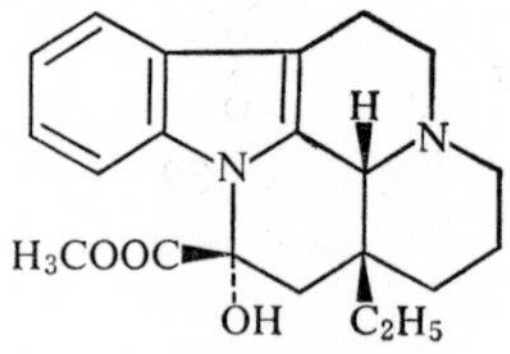   vincamine **3**

Route I[1)]

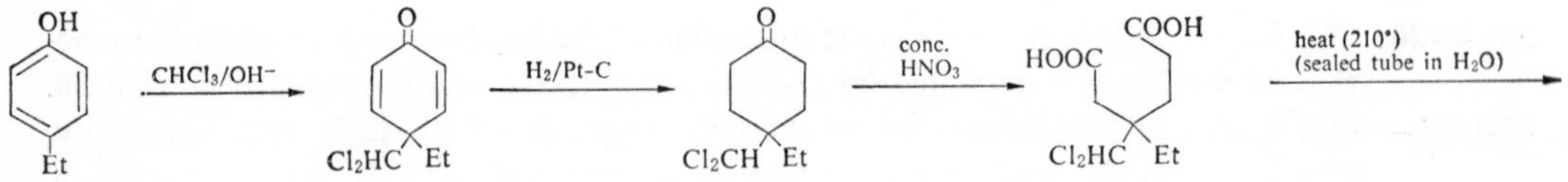

4   **5**   **6**   **7**

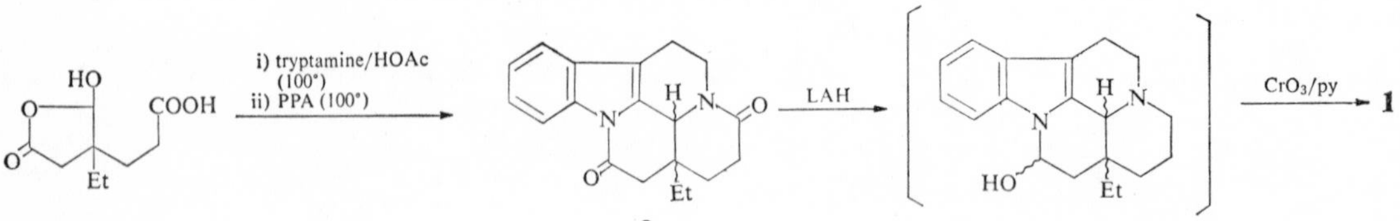

8   **9**

4 ⟶ 5: The quaternary carbon atom carrying the functional group necessary for the subsequent reaction was formed using an "abnormal" Reimer-Tiemann reaction at an early stage of the total synthesis.

8 ⟶ 9 ⟶ 1: The reaction is not stereospecific and thus, although the gross structure of **1** was established by the present synthesis, the stereostructure of **1** remained unknown.

Route II[2)]

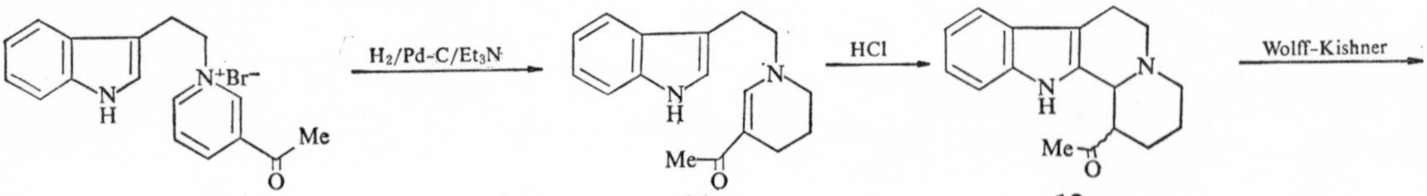

10   **11**   **12**

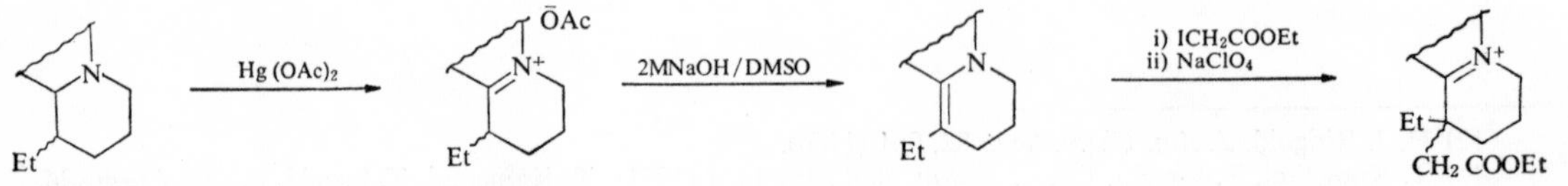

13   **14**   **15**   **16**

17

18

17, 18 — H₂ or NaBH₄ →

isoeburnamonine **19**

16 —reduction→ [20 + 21] —base→ 19 + 1

20

21

10 ——→ 11 ——→ 12: A general method[3] for alkaloid synthesis involving the reaction of a vinylogous amide system such as **11**, prepared by partial hydrogenation of β-acyl pyridinium salts, was successfully applied for the present synthesis.

13 ——→ 14: Direct treatment of **12** with Hg(OAc)₂ afforded mainly starting material along with a small amount of **11**, which might be attributable to the formation of a complex such as **22**. During the work-up, **22** is expected to yield partly **11** via a retro-Pictet-Spengler process.

22

17,18 ——→ 19: Hydrogenation as well as NaBH₄ reduction of **17** and **18** was expected to afford the *trans* D/E system stereoselectively, since approach of the catalyst or NaBH₄ would be controlled by the bulky angular ethyl group. In fact, only isoeburnamonine **19** was obtained. On the basis of these data, the *cis* D/E ring juncture in eburnamonine was proposed. This assignment was confirmed by independent evidence.[4]

16 ——→ 20 + 21: While the NaBH₄ reduction of **16** afforded a *ca.* 1:1 mixture of 20 and 21, catalytic hydrogenation yielded 20 predominantly.

1) M. F. Bartlet, W. I. Taylor, *J. Am. Chem. Soc.*, **82**, 5941 (1960).
2) E. Wenkert, B. Wickberg, *J. Am. Chem. Soc.*, **87**, 1580 (1965).
3) E. Wenkert, *Accounts Chem. Res.*, **1**, 78 (1968).
4) J. Mokry, M. Shamma, H. E. Soyster, *Tetr. Lett.*, 999 (1963).

Route III[5,6]

28 ⟶ 29a,b: The carbinolamide-lactol **28** was obtained as two separable isomers and either isomer afforded a mixture of D/E *cis* **29a** and *trans* **29b** lactams on HOAc treatment.

5) K. H. Gibson, J. E. Saxton, *Chem. Commun.*, 799 (1969).
6) K. H. Gibson, J. E. Saxton, *Chem. Commun.*, 1490 (1969).
7) J. R. Parikh, W. von E. Doering, *J. Am. Chem. Soc.*, **89**, 5505 (1967).

Route IV[8,9)]

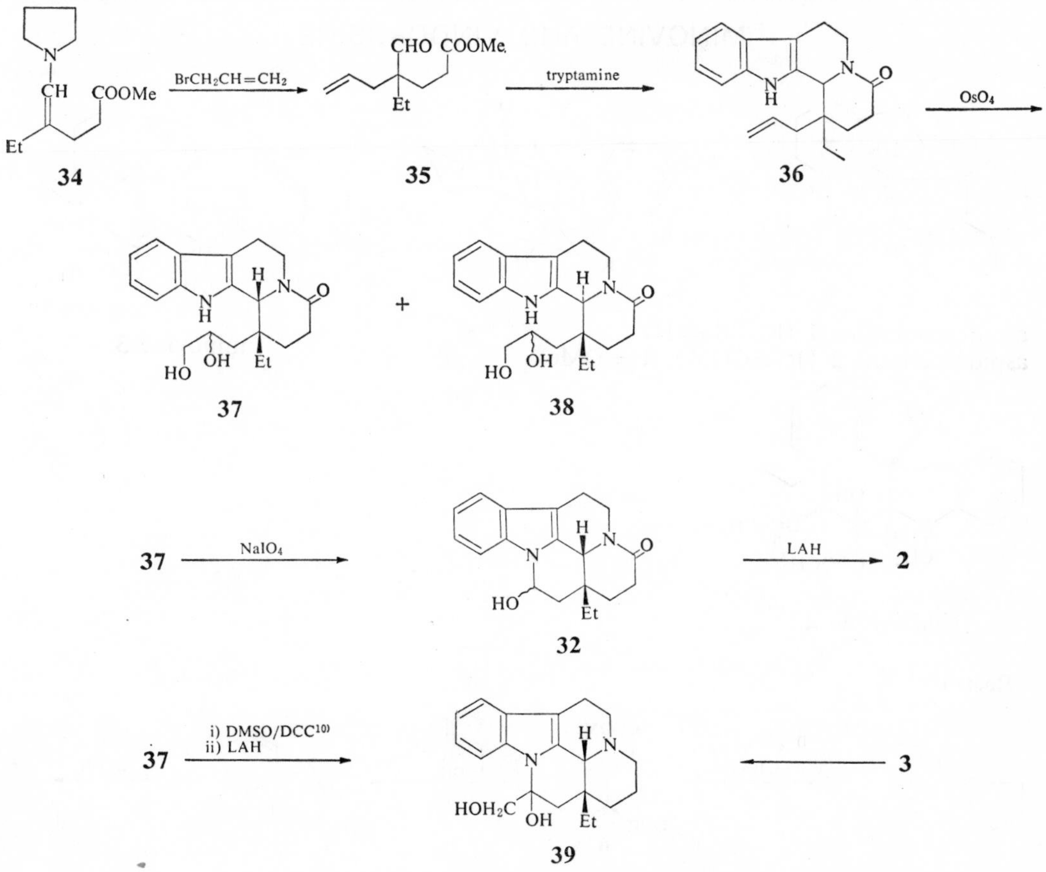

34   35   36

37   +   38

32

39

36 ⟶ 37 + 38: The diol lactam was obtained as a mixtuer of *cis* 37 and *trans* 38 compounds in the ratio 1:6.

37 ⟶ 39:   The diol 39 was identified as vincaminol, the product obtained by LAH reduction of the alkaloid vincamine 3.

8) J. E. D. Barton, J. Harley-Mason, *Chem. Commun.*, 298 (1965).
9) L. Castedo, J. Harley-Mason, T. J. Leeney, *Chem. Commun.*, 1186 (1968).
10) J. D. Albright, L. Goldman, *J. Org. Chem.*, 30, 1107 (1965).

10.62 SYNTHESIS OF ASPIDOSPERMIDINE, ASPIDOSPERMINE
MINOVINE AND VINDOROSINE

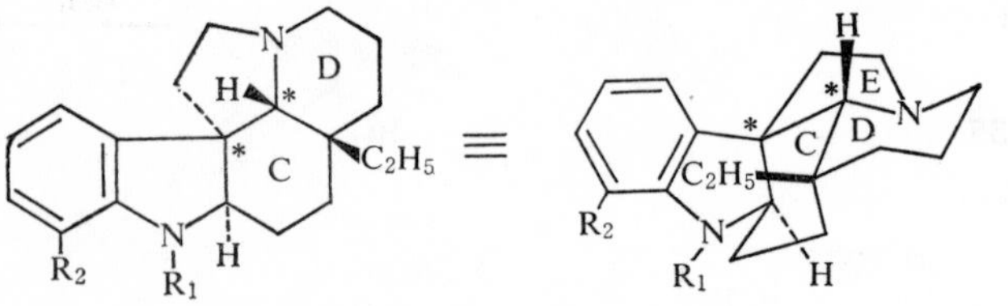

aspidospermidine **1** ($R_1=R_2=H$)
aspidospermine **2** ($R_1=COMe$, $R_2=OMe$)

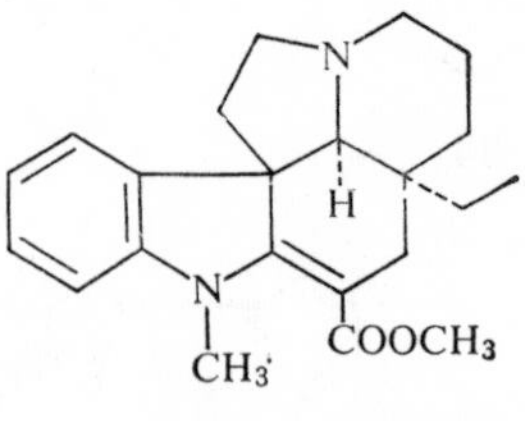

minovine **3**

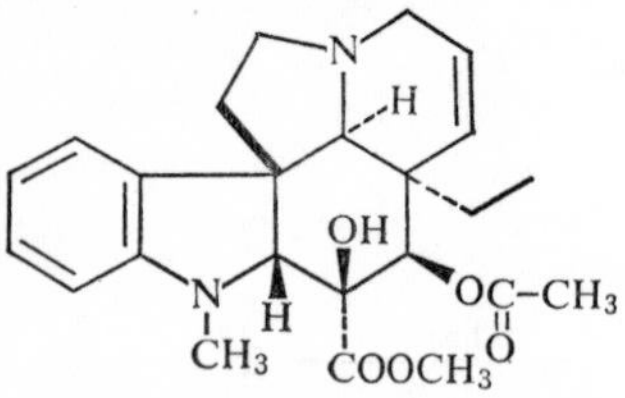

vindorosine **4**

Route I[1)]

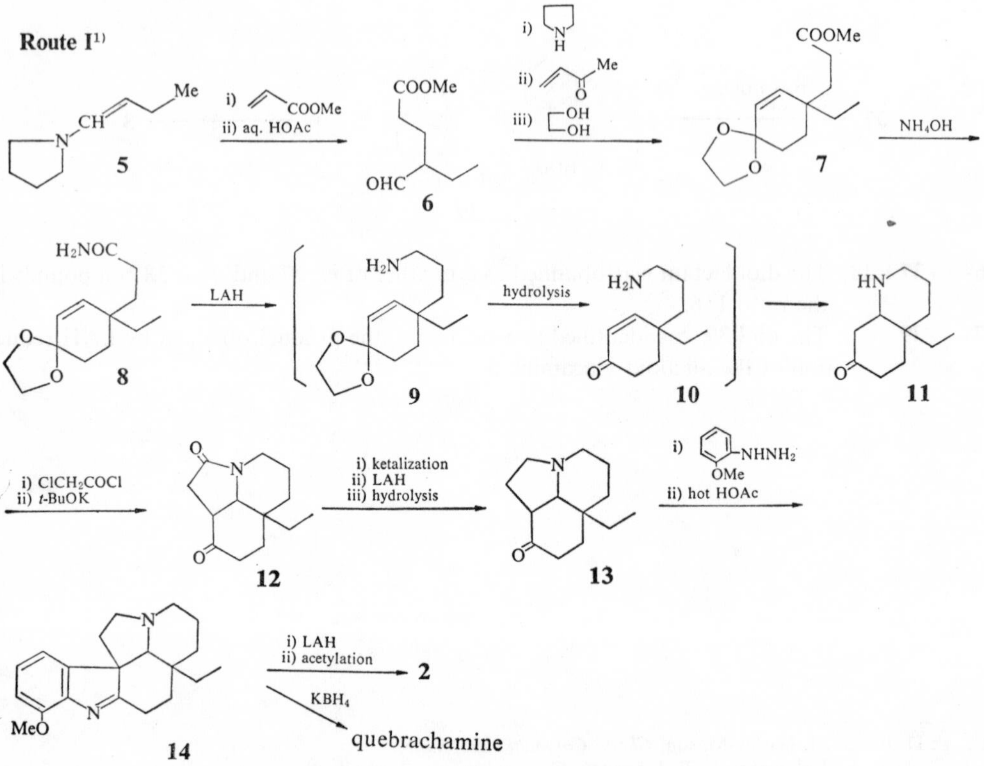

1) G. Stork, J. E. Dolfini, *J. Am. Chem. Soc.*, **85**, 2872 (1963).

A different synthesis of the precursor **13** has also been published.[2] Compound **2** is then obtained in the same way as shown on the previous page.

11,11a ⟶ 2: Although the planar structure of **11**, **12** and **13** were the same as those of **11a**, **12a** and **13a**, respectively, their physicochemical properties were different, indicating that the pairs of compounds were diastereomeric. The stereostructures initially proposed for these compounds were proved to be incorrect by careful study and consideration of model systems, and the following stereostructures were finally assigned.[4]

13,13a ⟶ 2: The stereostructure of the tricyclic ketone **13** was intially considered as **13a** (C/D, C/E, and D/E ring junctions, all *trans*),[5] but **13** yielded aspidospermine **2** (bearing *cis* C/D and C/E ring junctions) via Fisher indole cyclization. To account for this result, a mechanism involving fission and recombination at the two centers marked by asterisks to give eventually a stable aspidospermine skeleton was proposed.[1,5] The validity of this mechanism was later confirmed, since the *o*-methoxyphenylhydrazone of **13a**, which was proved to have the structure

2) Y. Ban, Y. Sato, I. Inoue, M. Nagai, T. Oishi, M. Terashima, O. Yonemitsu, Y. Kanaoka, *Tetr. Lett.*, 2261 (1965).

3) H. A. Bruson, T. W. Walker, *J. Am. Chem. Soc.*, **64**, 2850 (1942).

4) Y. Ban, M. Akagi, T. Oishi, *Tetr. Lett.*, 2057 (1969); M. Akagi, T. Oishi, Y. Ban, *ibid.*, 2063 (1969); Y. Ban, I. Iijima, I. Inoue, M. Akagi, T. Oishi, *ibid.*, 2067 (1969).

5) G. Stork, *Special Lectures presented at the 3rd International Symposium on the Chemistry of National Products held in Kyoto*, p. 131, Butterworths, 1964.

shown, afforded **2** on refluxing in HOAc for a long time, although the yield was poor. Also, the stereoisomer **2a** of **2** was obtained[6] by reductive Fischer indole cyclization using formic acid.[7] These reactions are illustrated below:

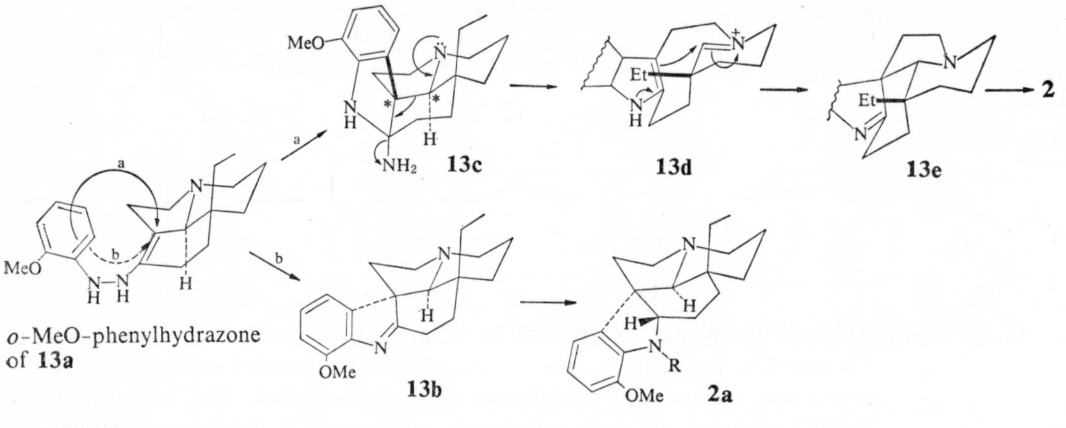

o-MeO-phenylhydrazone of **13a**

Route II[8]

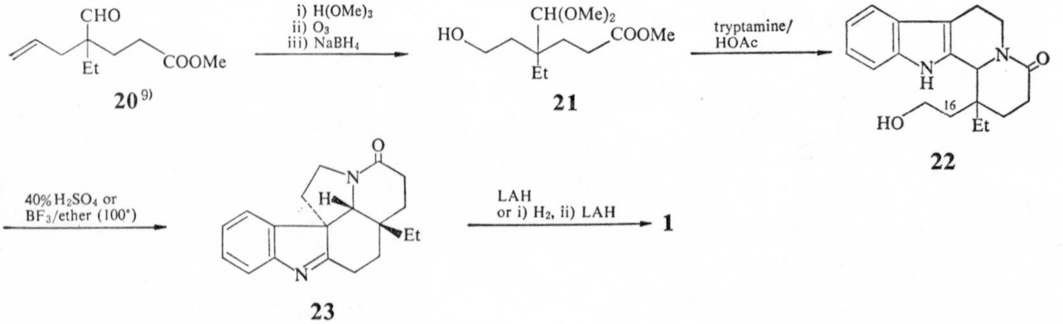

22 ⟶ 23: The process appeared to be entirely stereospecific and no indication of the presence of isomeric products was found. The rearrangement is believed to be initiated by the formation of a carbonium ion at C-16 in **22**, as shown below.

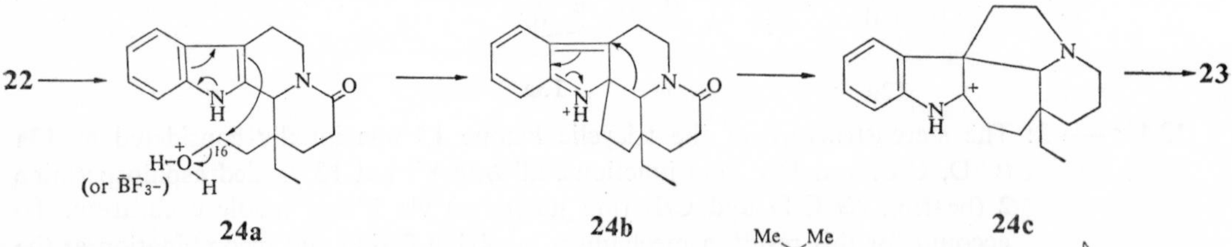

Route III[10]

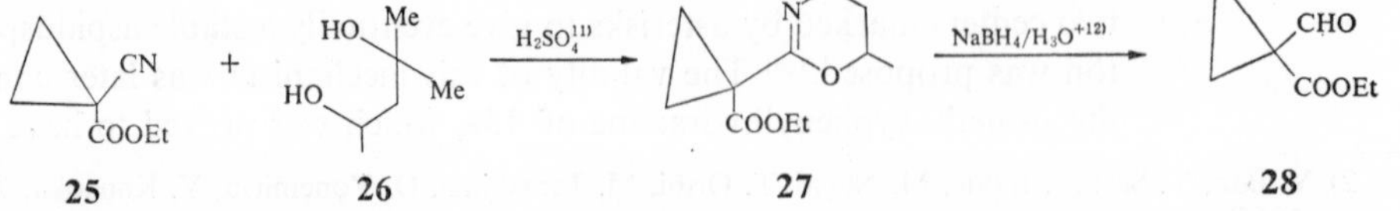

6) Y. Ban, I. Iijima, *Tetr. Lett.*, 2523 (1969).
7) Y. Ban, T. Oishi, Y. Kishio, I. Iijima, *Chem. Pharm. Bull., Japan*, **15**, 531 (1967).
8) J. Harley-Mason, M. Kaplan, *Chem. Commun.*, 915 (1967).
9) J. E. D. Barton, J. Harley-Mason, *Chem. Commun.*, 298 (1965).
10) R. V. Stevens, J. M. Fitzpatrick, M. Kaplan, R. L. Zimmerman, *Chem. Commun.*, 857 (1971).

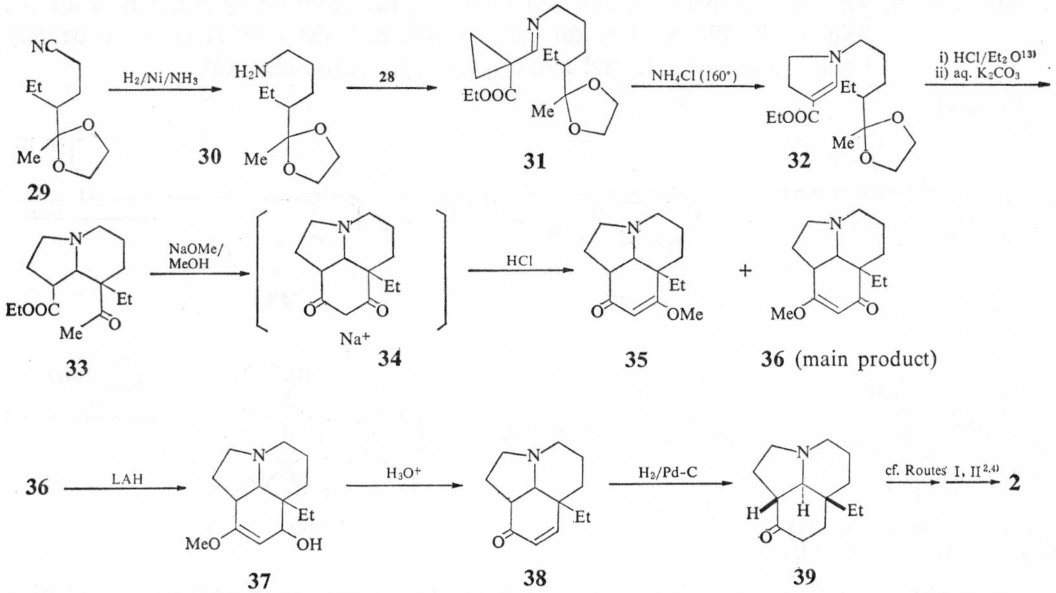

38 ⟶ 39: The tricyclic ketone **39** was found to be identical with **13a**, confirming the all-*trans* stereostructure.

Route IV[14]

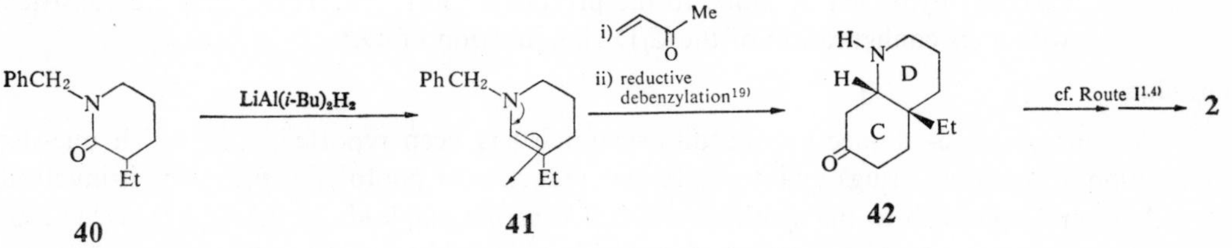

41 ⟶ 42: The amino ketone **42** was found to be identical with **11**, having a *cis* C/D ring junction, which clarified the stereochemical course of methyl vinyl ketone annelation in this system.

Route V[15]

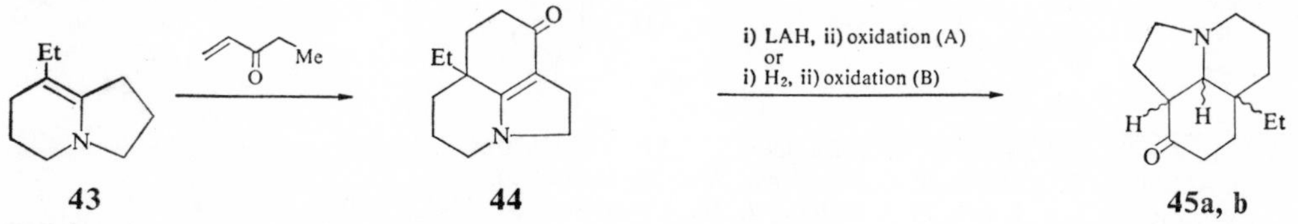

11) L. W. Jones, A. W. Scott, *J. Am. Chem. Soc.*, **44**, 407 (1922).
12) J. M. Fitzpatrick, G. R. Malone, I. R. Politzer, H. W. Adickes, A. I. Meyers, *Organic Preparations and Procedures*, **1**, 193 (1969).
13) E. Wenkert, K. G. Dave, R. V. Stevens, *J. Am. Chem. Soc.*, **90**, 6177 (1968).
14) R. V. Stevens, R. K. Mehra, R. L. Zimmerman, *Chem. Commun.*, 877 (1969).
15) M. E. Kuehne, C. Bayha, *Tetr. Lett.*, 1311 (1966).

44 $\longrightarrow$ 45a,b: The stereoisomers **45a** and **45b** were obtained from **44** by methods A and B, respectively. Physicochemical data for **45a** and **45b** were reported to be very similar to those for **13** and **13a**, respectively, but not identical.

Route VI[16]

46   **47**   **48**

49a (R$_1$=COOMe, R=H)
49b (R$_1$=H, R$_2$=COOMe)

49b $\xrightarrow{\text{10\% Pd-C/MeOH/HCl}}$

50b

46 $\longrightarrow$ 47: Generally, electrophiles are more likely to attack at the β-position than the α-position of indole. Modification of the reactivity of the indole nucleus by initial lithiation is noteworthy.

48 $\longrightarrow$ 49a,b: The presence of a one-proton triplet at δ4.28 (3-H of **49a**) (100 MHz nmr spectrometer) indicates that ring C must be in a half-boat conformation with the ester carbonyl hydrogen bounded to the protonated nitrogen. These data are consistent with a *cis* configuration of the C/D ring junction of **49**.[4]

Addendum

A synthesis of 5-oxo-deethylvincadifformamide has been reported,[19] in which one-step formation of the B/C/D rings by intramolecular reductive or photolytic cyclization is involved.

A general approach to the synthesis of *Aspidosperma* alkaloids including hexacyclic aspidofractinine-type compounds, starting with readily obtainable β-spiro-oxindole derivatives has been developed.[20]

Route VII[21]

51   **52**   **53**   **54**

16) F. E. Ziegler, E. B. Spitzner, *J. Am. Chem. Soc.*, **92**, 3492 (1970).
17) D. A. Shirley, P. A. Roussel, *J. Am. Chem. Soc.*, **75**, 375 (1953).
18) F. E. Ziegler, J. A. Kloek, P. A. Zoretic, *J. Am. Chem. Soc.*, **91**, 2342 (1969).
19) H-P Husson, C. Thal, P. Portier, E. Wenkert, *Chem. Commun.*, 480 (1970); *J. Org. Chem.*, **35**, 442 (1970).
20) T. Oishi, M. Nagai, Y. Ban, *Tetr. Lett.*, 491 (1968); Y. Ban, T. Ohnuma, M. Nagai, Y. Sendo, T. Oishi, *ibid.*, 5023 (1972); Y. Ban, Y. Sendo, M. Nagai, T. Oishi, *ibid.*, 5027 (1972); T. Ohnuma, T. Oishi, Y. Ban, *Chem. Commun.*, 301 (1973).
21) G. Büchi, K. E. Matsumoto, H. Nishimura, *J. Am. Chem. Soc.*, **93**, 3299 (1971).
22) J. M. Conia, *Bull. Soc. Chem. Fr.*, 690 (1954); R. B. Woodward, A. A. Patchett, D. H. R. Barton, D. A. Ives, R. B. Kelly, *J. Chem. Soc.*, 1131 (1957).
23) U. T. Bhalerao, J. J. Plattner, H. Rapoport, *J. Am. Chem. Soc.*, **92**, 3429 (1970).

54 ⟶ 55: Attempts to cyclize 53 failed but the *N*-acetylated compound 54 afforded the desired indoline 55 and the indole 56 in yields of 38% and 20%, respectively. The indoline 55 was also obtained from 56 by BF₃-etherate treatment but only in 8% yield, which showed that electrophilic substitution occurred directly at the β-position to form 55.

55 ⟶ 57: The stereostructure of the hydrolysis product was assigned as 57, which, by inspection of models, was found to be the most stable.

61 ⟶ 62: The oxidation of 61 afforded the β-hydroxy ketone 62 stereoselectively, which may be rationalized by assuming intermediates 64 and 65.

62 ⟶ 63: Reduction of 62 gave a mixture of diols separable by tlc. Acetylation of the slower moving diol yielded 4.

Syntheses of quebrachamine[18,24] 14,15-dehydroquebrachamine,[25] vincadine[24] and their conversion into aspidospermidine 1,[24] tabersonine,[25] and vincadifformine[24] by transannular cyclization have also been reported.

24) J. P. Kutney, K. K. Chan, A. Failli, J. M. Fromson, C. Gletsos, V. R. Nelson, *J. Am. Chem. Soc.*, **90**, 3891 (1968).

25) F. E. Ziegler, G. B. Bennett, *J. Am. Chem. Soc.*, **93**, 5930 (1971).

10.63   ABSOLUTE CONFIGURATION OF IBOGA ALKALOIDS

The following correlation[1,3] established the absolute configuration of (+)-catharanthine
1 and other iboga alkaloids. The correctness of the assignment was further proved by x-ray
analysis of (+)-coronaridine hydrobromide. Since (+)-ibogamine **7** and (+)-epiibogamine **6**
had been derived from (+)-coronaridine **5** and (+)-dihydrocatharanthine **4**, respectively,[3]
both the alkaloids belong to the same absolute ocnfigurational series.

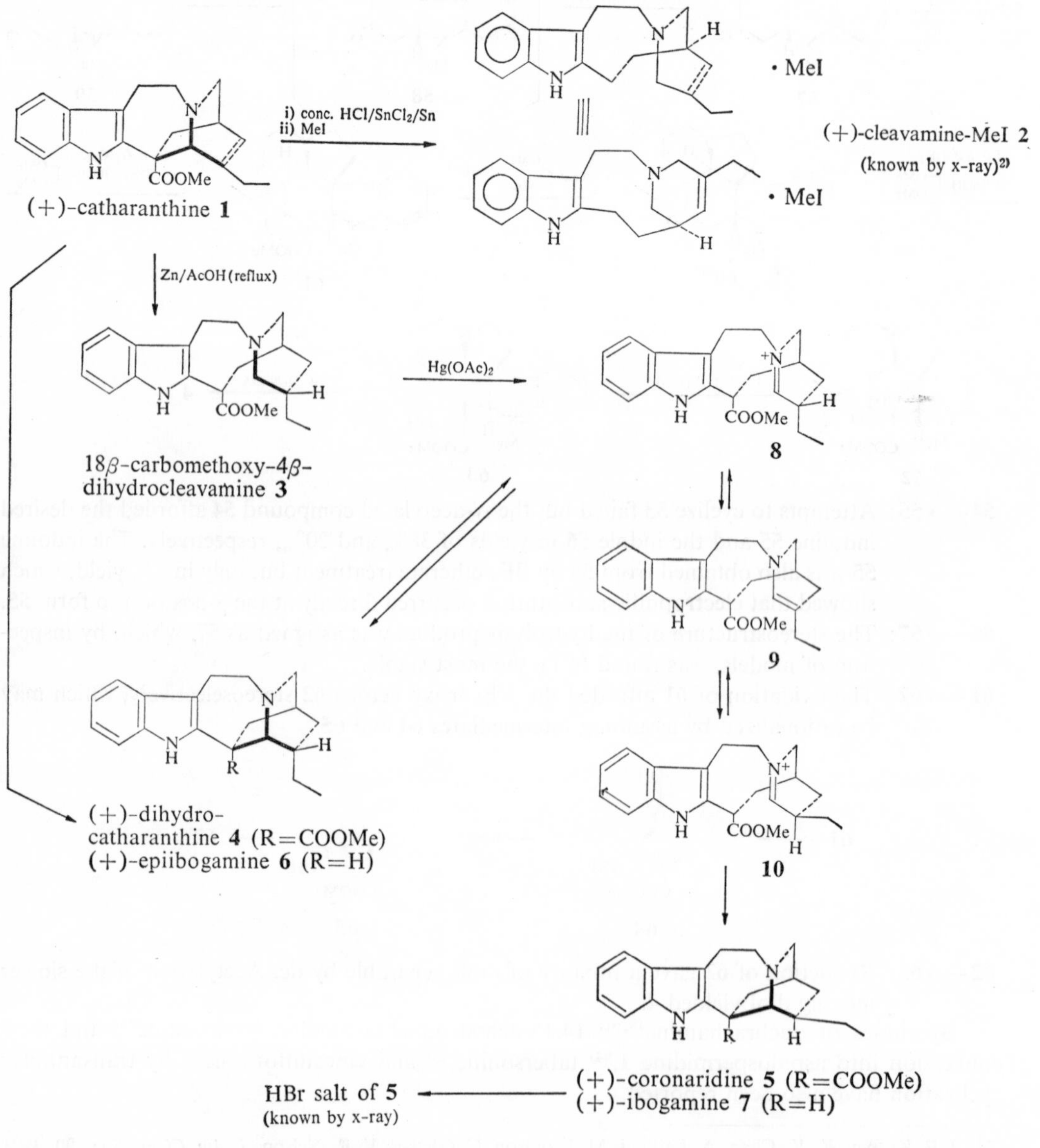

i) conc. HCl/SnCl₂/Sn
ii) MeI

(+)-catharanthine **1**

· MeI

· MeI

(+)-cleavamine-MeI **2**
(known by x-ray)[2]

Zn/AcOH(reflux)

18β-carbomethoxy-4β-
dihydrocleavamine **3**

Hg(OAc)₂

8

9

10

(+)-dihydro-
catharanthine **4** (R=COOMe)
(+)-epiibogamine **6** (R=H)

HBr salt of **5**
(known by x-ray)

(+)-coronaridine **5** (R=COOMe)
(+)-ibogamine **7** (R=H)

1) J. P. Kutney, K. Fuji, A. M. Treasurywala, J. Fayos, J. Clardy, A. I. Scott, C. C. Wei, *J. Am. Chem. Soc.*,
95, 5407 (1973).

Independent chiroptical work by Blaha *et al.*[4] revealed that there are two groups of natural iboga alkaloids, the absolute configurations of which were antipodal. (+)-Catharanthine (whose absolute configuration is known), (+)-dihydrocatharanthine and (+)-epiibogamine belong to one group. The other group contains (−)-coronaridine, (−)-conopharyngine, (−)-isovoacristine, (−)-voacangine, (−)-ibogamine, (−)-ibogaine, (−)-tabernanthine, (−)-iboxygaine and (−)-heyneanine.

	R^1	R^2	R^3	R^4	
(−)-coronaridine	COOMe	H	H	H	16S, 20S
(−)-conopharyngine	COOMe	OMe	OMe	H	16S, 20S
(−)-isovoacristine	COOMe	OMe	H	OH	16S, 20R
(−)-isovoacangine	COOMe	OMe	H	H	16S, 20S
(−)-voacangine	COOMe	H	OMe	H	16S, 20S
(−)-ibogamine	H	H	H	H	16R, 20S
(−)-ibogaine	H	H	OMe	H	16R, 20S
(−)-tabernanthine	H	OMe	H	H	16R, 20S
(−)-iboxygaine	H	H	OMe	OH	16R, 20R
(−)-heyneanine	COOMe	H	H	OH	16S, 20R

	R^1	R^2	R^3	
(+)-catharanthine	COOMe	H	H	$\Delta^{15,20}$, 16R
(+)-dihydrocatharanthine	COOMe	H	H	16R, 20S
(+)-epiibogamine	H	H	H	16S, 20S

2) J. P. Kutney, J. Trotter, T. Tabata, A. Kerigan, N. Camerman, *Chem. Ind.*, 648 (1963); N. Camerman, J. Trotter, *Acta Cryst.*, **17**, 384 (1964).
3) M. Gorman, N. Neuss and N. J. Cone, *J. Am. Chem. Soc.*, **87**, 93 (1965).
4) K. Blaha, Z. Koblicova and J. Trojanek, *Tetr. Lett.*, 2763 (1972); *Idem, Coll. Czech. Chem. Comm.*, **39**, 2258 (1974).

10.64   SYNTHESIS OF IBOGAMINE, IBOGAINE, EPIIBOGAMINE, EPIIBOGAINE AND CORONARIDINE

ibogamine **1a** (R=H)   epiibogamine **1b** (R=H)   coronaridine **3**
ibogaine   **2a** (R=OMe)   epiibogaine   **2b** (R=OMe)

Route I[1)]

4 → **5** → NaBH$_4$[2)] → **6** → **7** → NaOCl/MeOOC → **8**

i) 6N H$_2$SO$_4$ ii) Ac$_2$O → **9** → i) H$_2$/Pd/HCl ii) OH$^-$ → **10** → **11** → TsOH/AcOH

12 → i) LAH ii) DCC/DMSO → **13** → NaOMe → **14** → Zn/HOAc

15 → **16** → Wolff-Kishner reduction → **1a + 1b**

1) G. Büchi, D. L. Coffen, K. Kocsis, P. E. Sonnet, F. E. Ziegler, *J. Am. Chem. Soc.*, **88**, 3099 (1966).
2) H. Diekmann, G. Englet, K. Wallenfels, *Tetr.*, **20**, 281 (1964); P. S. Anderson, R. E. Lyle, *Tetr. Lett.*, **153** (1964).
3) K. Schenker, J. Druey, *Helv. Chim. Acta*, **42**, 1960, 1971 (1959); M. Saunders, E. H. Gold, *J. Org. Chem.*, **27**, 1439 (1962); T. Agawa, S. I. Miller, *J. Am. Chem. Soc.*, **83**, 449 (1961).

6 ⟶ 7: The configuration of the acetyl group follows from the subsequent transformations, but the position of this group on the isoquinuclidine ring was deduced by analogy with similar Diels-Alder adducts of established structure.[4]

11 ⟶ 12: The alcohol **17** derived from the acetate **11** affords the ether **18** on heating with TsOH. Attempted reduction of **18** with LAH to the desired pentacyclic amino alcohol failed; only the stable compound **19** was obtained.

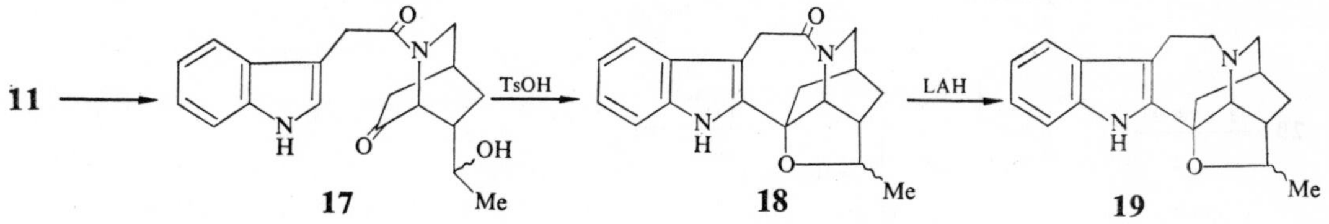

The acetate **11**, on the other hand, yielded the diol monoacetate **12** on the same treatment. Both **18** and **12** are considered to be formed through the hypothetical cations **20a** and **20b**, respectively.

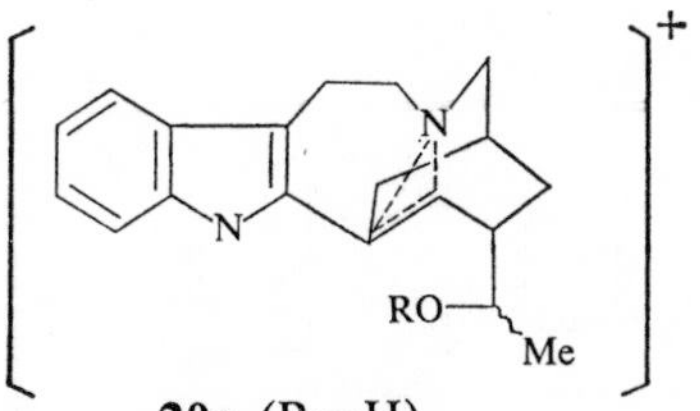

20a (R=H)
20b (R=OMe)

Route II[4]

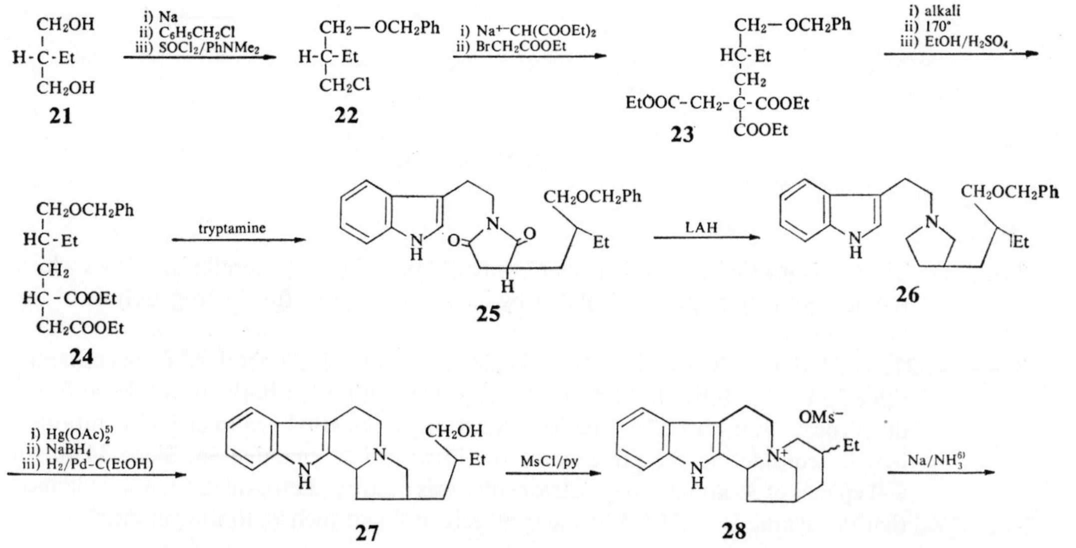

---

4) J. P. Kutney, W. J. Cretrey, P. W. LeQuesne, N. McKague, E. Piers, *J. Am. Chem. Soc.*, **92**, 1712 (1970).
5) E. Wenkert, B. Wickberg, *J. Am. Chem. Soc.*, **84**, 4914 (1962).
6) E. Wenkert, S. Garrett, K. G. Dave, *Can. J. Chem.*, **42**, 489 (1964).

4β-dihydrocleavamine **29**
(+ 4α-dihydrocleavamine **30**)

31

32   **33**   **34**

35
catharanthine

36
18-α-carbomethoxy-4α-
dihydrocleavamine

38

3

37

26 $\longrightarrow$ 27: The five isomeric benzyl ethers of **27** were obtained in 37% overall yield, from which the desired alcohol **27** was isolated by chromatography after hydrogenolysis of the benzyl group.

36 $\longrightarrow$ 3 + 37: Oxidation of **36**, which is the C-4 epimer of synthetic **34** obtained from catharanthine **35** by Zn/HOAc treatment with Hg(OAc)₂, afforded both coronaridine **3** and dihydrocatharanthine **37**.[8] The isomerization of the ethyl group at C-4 is rationalized by considering the presence of the intermediate enamine **38**. Since **34** was a C-4 epimer of **36** and the assymetric center was destroyed on oxidation, it was claimed that the synthesis of **34** led to the synthesis of **3** and thence, to ibogamine.[9]

7) G. Büchi, R. E. Manning, *J. Am. Chem. Soc.*, **88**, 2532 (1966).
8) J. P. Kutney, R. T. Brown, E. Piers, J. R. Hadfield, *J. Am. Chem. Soc.*, **92**, 1708 (1970).
9) M. Gorman, N. Neuss, N. J. Cone, J. A. Deyrup, *J. Am. Chem. Soc.*, **82**, 1142 (1970).

Route III[10,11)]

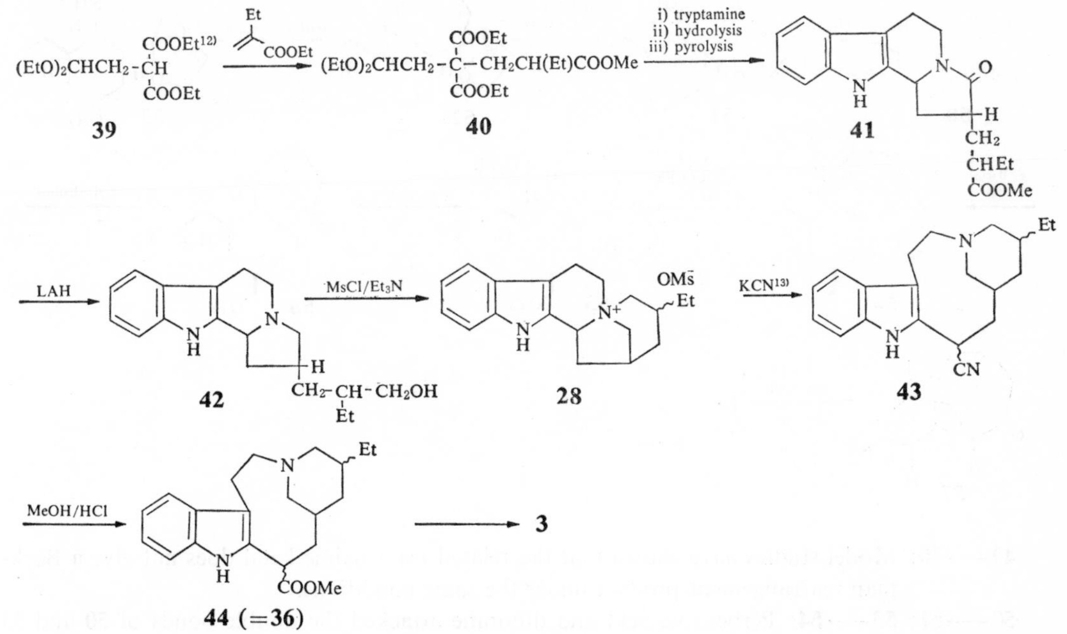

39 → 40 → 41 (i) tryptamine, ii) hydrolysis, iii) pyrolysis)

42 → (MsCl/Et₃N) → 28 → (KCN[13)]) → 43

44 (=36) → 3 (MeOH/HCl)

40 ⟶ 41: Condensation of the acetal **40** with tryptamine afforded the cyclized lactone **41** in one step.

28 ⟶ 43: Although the model compound **45** afforded the unexpected **46**,[13)] **28** yielded **43** successfully.

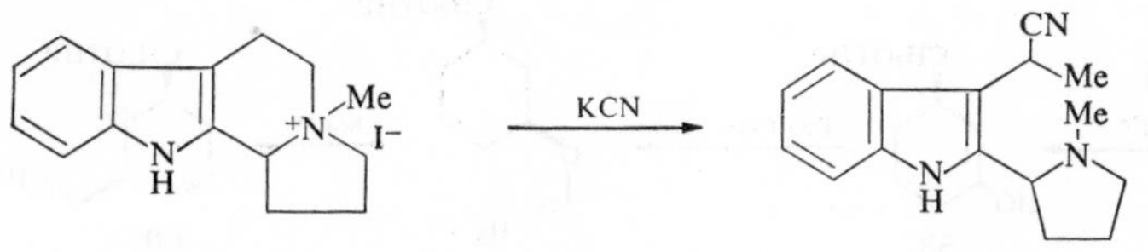

45 → (KCN) → 46

43 ⟶ 44: The compound **44** was obtained as a mixture of the four possible stereoisomers. These were separated easily by chromatography and one was proved to be identical with 18-α-carbomethoxy-4-α-dihydrocleavamine **36**.

Route IV[14,15)]

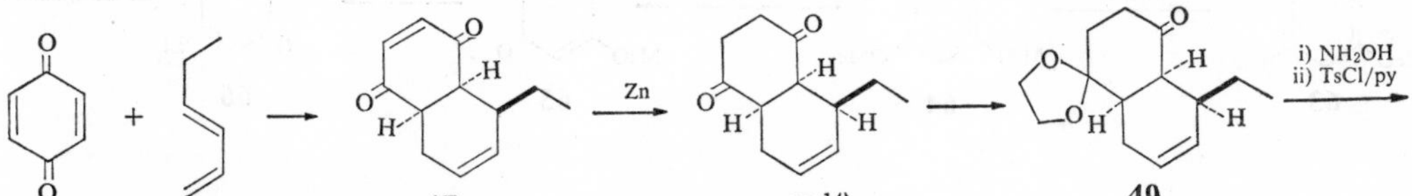

47 → (Zn) → 48[14)] → 49 → (i) NH₂OH, ii) TsCl/py)

10) J. Harley-Mason, Atta-ur-Rahman, J. A. Beisler, *Chem. Commun.*, 743 (1966).
11) J. Harley-Mason, Atta-ur-Rahman, *Chem. Commun.*, 208 (1967).
12) C. Mannich, K. Ritsert, *Chem. Ber.*, **57**, 1116 (1924); Y. Iwakura, M. Sato, Y. Matsuo, *J. Chem. Soc. Japan*, **80**, 502 (1959).
13) G. H. Foster, J. Harley-Mason, W. R. Waterfield, *Chem. Commun.*, 21 (1967).
14) S. I. Sallay, *Tetr. Lett.*, 2443 (1964).
15) S. I. Sallay, *J. Am. Chem. Soc.*, **89**, 6762 (1967).

49 ⟶ 50: Model studies have shown that the related *trans*-oxime ketal does not give a Beckman rearrangement product under the same conditions.

50 ⟶ 51; 53 ⟶ 54: Perbenzoic acid and diborane attacked the double bonds of **50** and **53** from the convex face of the molecules. The processes are well stereo-controlled and provide the first preparative proof that the ethyl side chain of the *iboga* alkaloids has a *cis* configuration with respect to the N_6-function.

Route V[16,17)]

16) W. Nagata, S. Hirai, K. Kawata, T. Aoki, *J. Am. Chem. Soc.*, **89**, 5046; (1967); W. Nagata, S. Hirai, T. Okumura, K. Kawata, *ibid.*, **90**, 1650 (1968).

17) S. Hirai, K. Kawata, W. Nagata, *Chem. Commun.*, 1016 (1968).

18) E. E. van Tamelen, G. T. Hildahl. *J. Am. Chem. Soc.*, **78**, 4405 (1956).

19) A. W. Burgstahler, I. C. Nordin, *J. Am. Chem. Soc.*, **83**, 198 (1961).

66⟶67: The crude alcohol **67** was found to be contaminated with *ca.* 25% of *cis* isomers. This ratio was deduced by gas-liquid chromatography of the corresponding olefinic nitrile **78** derived from the crude alcohol **67**, because separation of the *cis* and *trans* isomers in this way could only be done with **78**. Basic treatment of pure **78a** afforded a 6:5 equilibrium mixture of **78b** and **78a**, respectively, and this proved the *trans* configuration of the latter. Pure *cis* olefinic nitrile was obtained from the alcohol **61**, showing that **61** was also a pure compound.

20) A. J. Birch, P. Hextall, S. Sternbell, *Aust. J. Chem.*, **7**, 256 (1954).
21) J. D. Albright, L. Goldman, *J. Am. Chem. Soc.*, **89**, 2416 (1967).
22) cf. G. Büchi, R. E. Manning, *J. Am. Chem. Soc.*, **88**, 2532 (1966).

62 ⟶ 69 ⟶ 70: The characteristic feature of this synthesis is the formation of the isoquinu-clidine ring from the bridged aziridine derivatives obtained by Pb(OAc)$_4$ oxidation of the suitably oriented unsaturated amine.

71 ⟶ 72: Cyclization of the keto-lactam 71 without rearrangement was effected by refluxing it in benzene solution for 5–10 minutes in the presence of 1.3–1.5 molar eq. TsOH to give 72, while the analogous keto-lactam 11 affords the rearranged alcohol 12 on heating in AcOH.

Route VI[23–5]

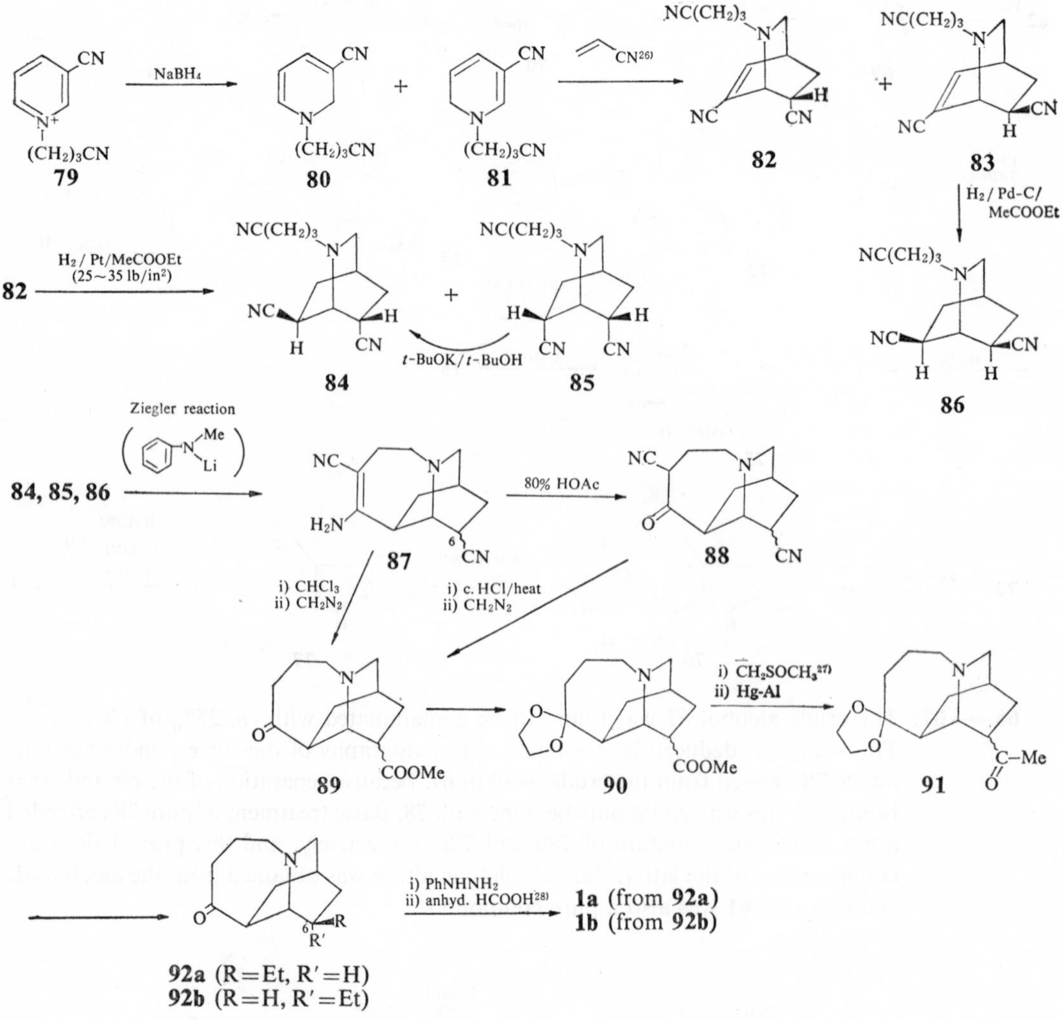

---

23) Y. Ban, T. Oishi, M. Ochiai, T. Wakamatsu, Y. Fujimoto, *Tetr. Lett.*, 6385 (1966).
24) Y. Ban, T. Wakamatsu, Y. Fujimoto, T. Oishi, *Tetr. Lett.*, 3383 (1968).
25) M. Ikezaki, T. Wakamatsu, Y. Ban, *Chem. Commun.*, 88 (1969).
26) K. Schenker, J. Druey, *Helv. Chim. Acta*, **45**, 1344 (1962).
27) E. J. Corey, M. Chaykowsky, *J. Am. Chem. Soc.*, **87**, 1343 (1965).
28) Y. Ban, T. Oishi, Y. Kishio, I. Iijima, *Chem. Pharm. Bull., Japan*, **15**, 531 (1967).

84,85,86 ⟶ 87: The Ziegler cyclizations of the three isomers **84**, **85** and **86** afforded, among others, a mixture of enamines **87** isomeric at the C-6 position. At first, it was thought that the *endo* isomer was predominantly formed but upon reexamination of this reaction, comparable amounts of both isomers were found to be produced.

87,88 ⟶ 89: The nitriles **87** and **88** are convertible to the corresponding acids by refluxing in conc. HCl. Treatment with dilute acids affords only the fragmentation products.

91 ⟶ 92: The isomers of **92** at the C-6 position are separable by chromatography; the slower moving one affords ibogamine **1a** and the other, epiibogamine **1b** on Fischer indole cyclization.

Route VII[29)]

95 ⟶ (96) ⟶ 97: As the keto indole **96** was very unstable, the methanol solution, after filtration of the catalyst, was immediately treated with NH$_2$OH. Four isomeric oxime indoles were detected by tlc, and **97** was isolated in 41% yield.

97 ⟶ 98: In addition to the *cis* amino ester **98** (38%), the corresponding *trans* isomer was also isolated in 25% yield.

The first total synthesis of the Iboga skeleton (deethylibogamine) was achieved by Huffman *et al.*[31)]

A synthesis of deethylibogamine by way of a tricyclic ketone (cf. **56, 92**) has been reported.[32)]
The related alkaloids, velbanamine[33,34,35)] and catharanthine,[33,34)] have also been synthesized starting from the isoquinuclidine derivatives prepared in the above synthetic works.[1,4,8,16,17)]

It has been reported[36)] that corynantheine aldehyde which was synthesized by van Tamelen *et al.*[37)] is converted to catharanthine under biogenetic conditions.

29) P. Rosenmund, W. H. Haase, J. Bauer, *Tetr. Lett.*, 4121 (1969).
30) P. Rosenmund, W. H. Haase, *Chem. Ber.*, **99**, 2504 (1966).
31) J. W. Huffman, C. B. S. Rao, T. Kamiya, *J. Am. Chem. Soc.*, **77**, 2288 (1965).
32) R. L. Augustine, W. G. Pierson, *J. Am. Chem. Soc.*, **34**, 1070 (1968).
33) G. Büchi, P. Kulsa, K. Ogasawara, R. L. Rosati, *J. Am. Chem. Soc.*, **92**, 999 (1970).
34) J. P. Kutney, F. Bylsma, *J. Am. Chem. Soc.*, **92**, 6090, 1970).
35) W. Nagata, S. Hirai, K. Kawata, T. Okumura, *J. Am. Chem. Soc.*, **89**, 5046 (1967); M. Narisada, F. Watanabe, W. Nagata, *Tetr. Lett.*, 3681 (1971).
36) A. A. Qureshi, A. I. Scott, *Chem. Commun.*, 947 (1968); see also, A. I. Scott, *et al.*, *J. Am. Chem. Soc.*, **94**, 8262, 8263, 8264, 8266 (1972).
37) E. E. van Tamelen, I. G. Wright, *J. Am. Chem. Soc.*, **91**, 7349 (1969).

10.65 STRUCTURE AND SYNTHESIS OF CALYCANTHACEOUS ALKALOIDS

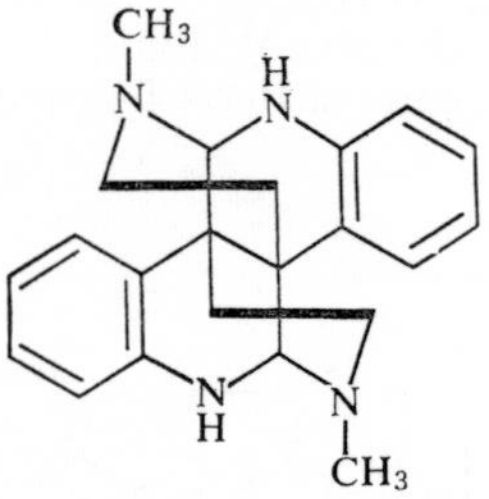

calycanthine, 1 $[\alpha]_D = +684°$

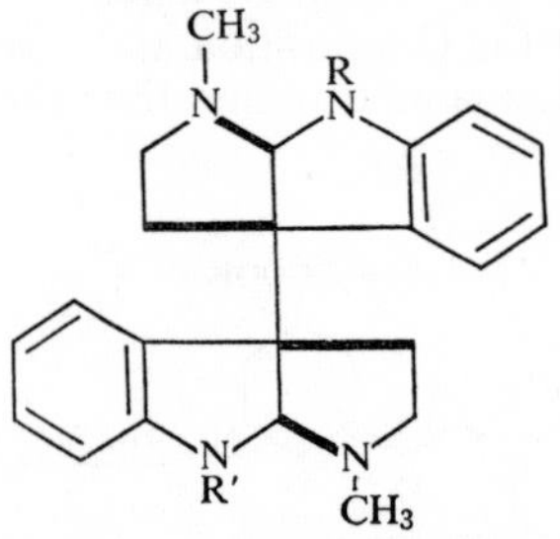

(*rac*-form)

chimonanthine 2 : R, R' = H $[\alpha]_D = -329°$
calycanthidine 3 : R = Me, R' = H $[\alpha]_D = -317°$
folicanthine 4 : R = R' = Me $[\alpha]_D = -364°$

Route I[1]

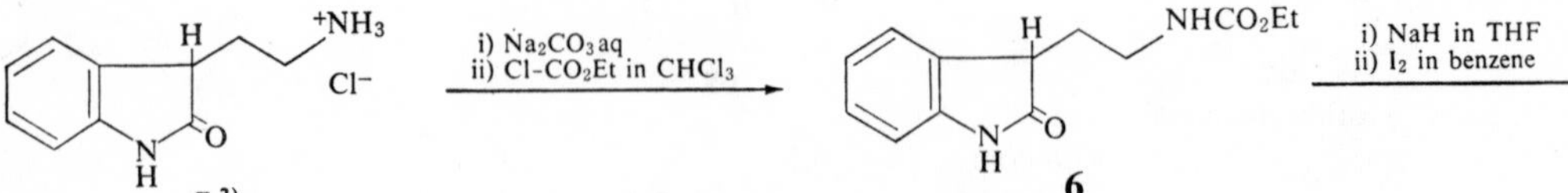

i) Na$_2$CO$_3$ aq
ii) Cl-CO$_2$Et in CHCl$_3$

i) NaH in THF
ii) I$_2$ in benzene

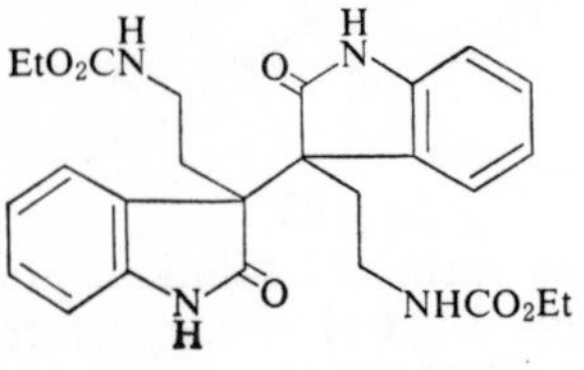

{ 7a, 13%, mp 243~6°
  7b, 3%, mp 214~6°

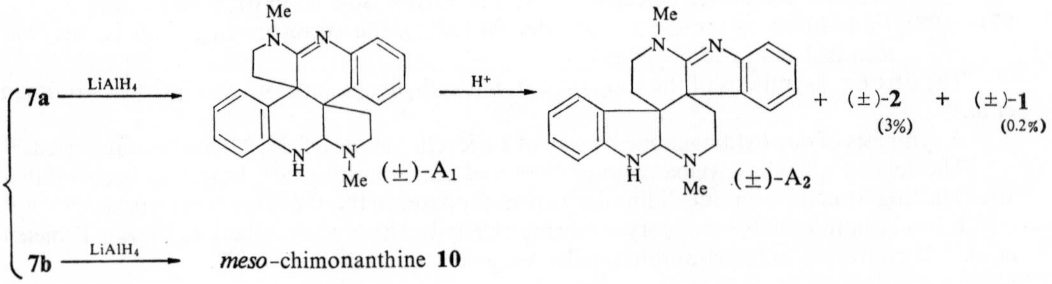

7a $\xrightarrow{\text{LiAlH}_4}$ (±)-A$_1$ $\xrightarrow{\text{H}^+}$ (±)-A$_2$ + (±)-2 + (±)-1
(3%) (0.2%)

7b $\xrightarrow{\text{LiAlH}_4}$ *meso*-chimonanthine 10

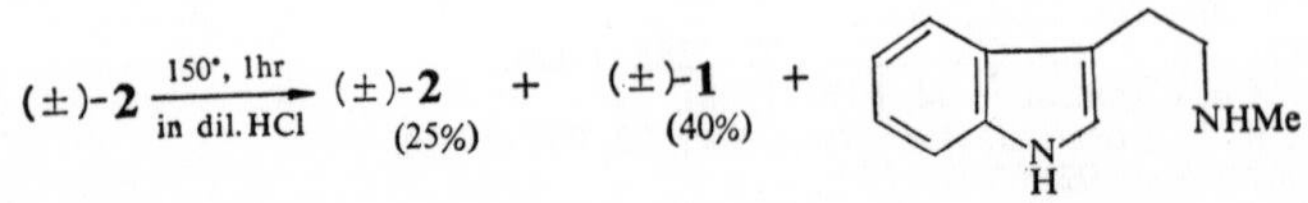

(±)-2 $\xrightarrow[\text{in dil. HCl}]{150°, 1hr}$ (±)-2 + (±)-1 +
(25%) (40%)

8

---

1) J. B. Hendrickson, R. Göschke, R. Rees, *Tetr.*, **20**, 565 (1964); cf., *Proc. Chem. Soc.*, 383 (1962).
2) J. Harley-Mason, R. F. J. Ingleby, *J. Chem. Soc.*, 3639 (1958).

6 ⟶ 7: Initially, oxidation with several salts such as cupric, ferric, or silver ions capable of one-electron reductions as well as electrolytic and *tert*-butyl hydroperoxide oxidations were tested but only the starting material was recovered unchanged.

7 ⟶ 8b: As a precedent for conversion of the oxindole dimer to a compound having caly-canthine skeleton (**9 ⟶ 8a**) had been reported,[3] the transformation of **7** to **8b** was attempted under a various solvolytic condition but **7** was generally recovered.

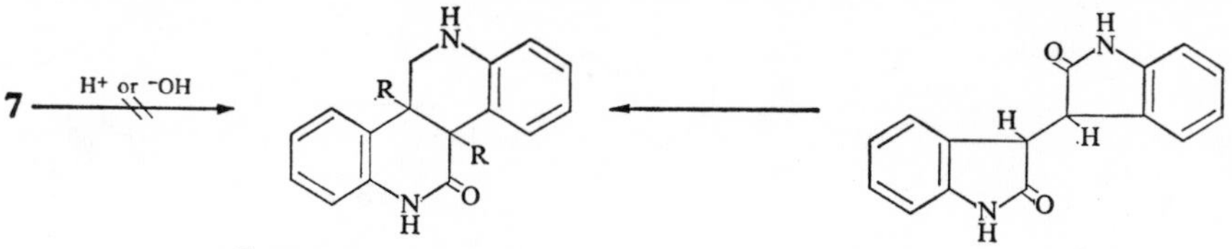

8a, R = H
8b, R = CH$_2$CH$_2$NHCO$_2$Et

9

7b ⟶ 10: The spectral and thin-layer chromatographic data of **10** were very similar but not identical to those of natural (−)-chimonanthine **2**. Therefore, *meso*-structure was proposed for it.

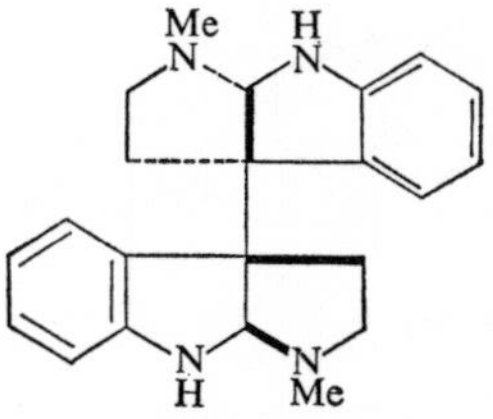

10

Route II[4]

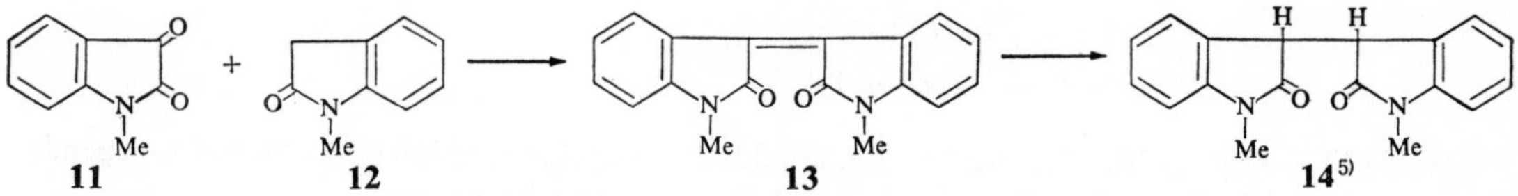

11   **12**   **13**   **14[5]**

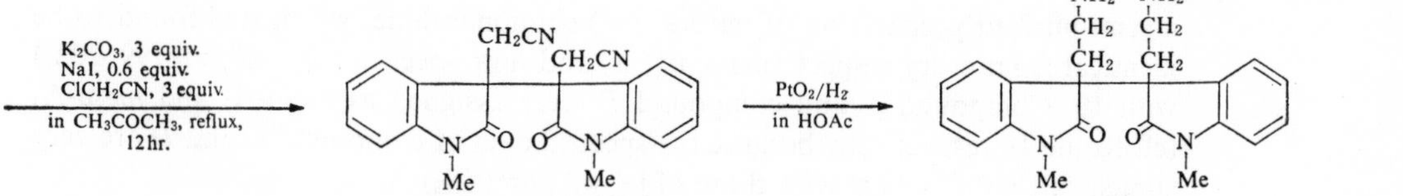

K$_2$CO$_3$, 3 equiv.
NaI, 0.6 equiv.
ClCH$_2$CN, 3 equiv.
in CH$_3$COCH$_3$, reflux,
12 hr.

15 mp 263~265°(47%)

PtO$_2$/H$_2$
in HOAc

16 [6a,b]

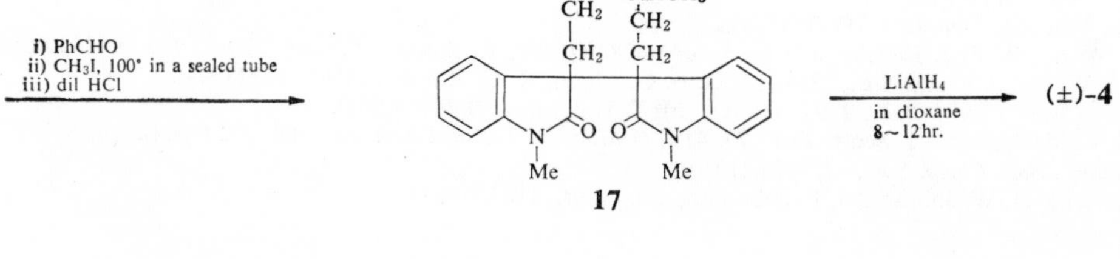

i) PhCHO
ii) CH$_3$I, 100° in a sealed tube
iii) dil HCl

17

LiAlH$_4$
in dioxane
8~12 hr.

(±)-**4**

14 ⟶ 15: The role of NaI in this alkylation reaction is evident since **14** is recovered unchanged in the absence of this salt.

15 ⟶ 16: The use of $NaNH_2$ in liquid NH_3 resulted in the formation of uncharacterized compounds along with a small amount of **15**, which might be ascribed to an unwanted generation of carbanion at the carbon atom adjacent to the cyano group.

Route III[7]

8 ⟶ 18 ⟶ 19 ⟶ 20: The success of β,β'-oxidative coupling reaction of **8** with $FeCl_3$ depends largely on the prior generation of the anion **18** with MeMgI.

20 ⟶ (±)2 + meso-10 + C + D: A small amount of *meso*-chimonanthine was separated from a semi-purified preparation of natural (−)-chimonanthine, which was found to be identical from every respect (mass, uv, ir, and nmr spectra, $[\alpha]_D = 0°$, mp199–202°) with **10**. Compound *C* and compound *D* were assigned as *meso*-A₁ and *meso*-A₂ (cf. Route I), respectively, because the spectral data of compound *C* and *D* are very similar but not identical with those of (±)-A₁ and (±)-A₂.

3) R. B. Woodward, N. C. Yang, T. J. Katz, V. M. Clark, J. Harley-Mason, R. F. J. Ingleby, N. Sheppard, *Proc. Chem. Soc.*, 76 (1970).
4) T. Hino, S. Yamada, *Tetr. Lett.*, 1757 (1963).
5) J. Harley-Mason, R. F. J. Ingleby, *J. Chem. Soc.*, 4782 (1958); R. Stolle, *J. Prakt. Chem.*, **128**, 1 (1930); H. King, J. Wright, *J. Chem. Soc.*, 2314 (1948); F. G. Mann, R. C. Haworth, *ibid.*, 670 (1944).
6) (a) T. Hino, *Chem. Pharm. Bull.*, **9**, 979 (1961); (b), T. Hino, *ibid.*, **9**, 988 (1961).
7) E. S. Hall, F. McCapra, A. I. Scott, *Tetr.*, **23**, 4131 (1967); cf., *J. Am. Chem. Soc.*, **86**, 302 (1964).
8) G. Champetier, *Bull. Chem. Soc.*, **47**, 1131 (1930).
9) M. G. Reinecke, H. W. Johnson, J. F. Sebastian, *Tetr. Lett.*, 1183 (1963).

10.66 STRUCTURE AND BIOMIMETIC SYNTHESIS OF ALSTONIA DIMERIC ALKALOIDS

The structure of villalstonine **1**, a major dimeric indole alkaloid of *Alstonia* spp. (Apocynaceae), was determined through extensive degradative work,[1] which included the isolation of the known monomeric alkaloid, pleiocarpamine **2**, on treatment of **1** with perchloric acid, and thermal degradation of villamine **3**, an isomerization product of **1**, to both halves, pleio-carpamine **2** and macroline **4**.

An independent x-ray crystallographic study of **1** by the direct method led Nordman and Kumra to the same structure.[2]

Synthesis of villalstonine 1 and alstonisidine 5

A biomimetic synthesis of **1** has been attained as shown starting from pleiocarpamine **2** and macroline **4**.

1) M. Hesse, H. Hürzeler, C. W. Gemenden, B. S. Joshi, W. I. Taylor, H. Schmid, *Helv. Chim. Acta*, **48**, 689 (1965); M. Hesse, F. Bodmer, C. W. Gemenden, B. S. Joshi, W. I. Taylor, H. Schmid, *ibid.*, **49**, 1173 (1966).
2) C. E. Nordman, S. K. Kumra, *J. Am. Chem. Soc.*, **87**, 2059 (1965).

Synthesis of another dimeric *Alstonia* alkaloid, alstonisidine **5**, was achieved by the condensation of the two monomeric components, macroline **4** and quebrachidine[3) **6**.

Isolation of the intermediate **7** and a model study of Michael condensation between an indoline and an α,β-unsaturated ketone proved that alstonisidine has the structure **5** instead of the formerly proposed structure **8**.[4]

Two other dimeric alkaloids, macralstonine **9** and macralstonidine **10**, have been isolated and determined to have the structures shown below.[5,6] A partial synthesis of **9** has been reported.[7]

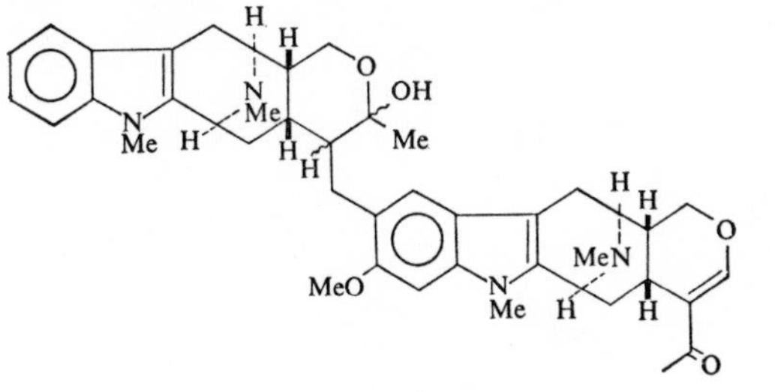

9

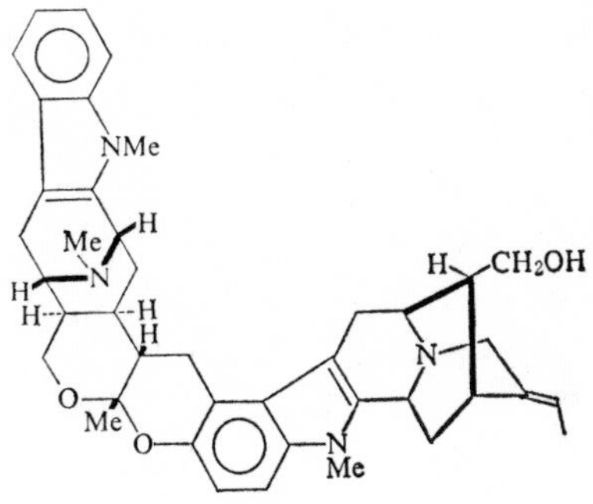

10

3) D. E. Burke, J. M. Cook, P. W. LeQuesne, *J. Am. Chem. Soc.*, **95**, 546 (1973).
4) J. M. Cook, P. W. LeQuesne, *J. Org. Chem.*, **36**, 582 (1971).
5) T. Kishi, M. Hesse, W. Vetter, C. W. Gemenden, W. I. Taylor, H. Schmid, *Helv. Chim. Acta*, **49**, 946 (1965).
6) E. E. Waldner, M. Hesse, W. I. Taylor, H. Schmid, *Helv. Chim. Acta*, **50**, 1926 (1967).
7) D. E. Burke, C. A. DeMarkey, P. W. LeQuesne, J. M. Cook, *Chem. Commun.*, 1346 (1972).

CHAPTER **11**

Non-alkaloidal Nitrogen Compounds

11.1 INTRODUCTION

This chapter includes many biologically important classes of natural products such as antibiotics, vitamins, marine products, animal toxins, fungal metabolites, etc. Owing to the space limitation, emphasis is given to compounds having unusual structures. The common amino acids, polypeptides, proteins and nucleic acids are not included. Certain aminosugars and nucleosides are included in Chapter 8. The order of arrangement of topics is as follows: nitro, azo, amino, amido, heterocyclic compounds with increasing ring size and complexity.

423

11.2   STRUCTURE OF AUREOTHIN

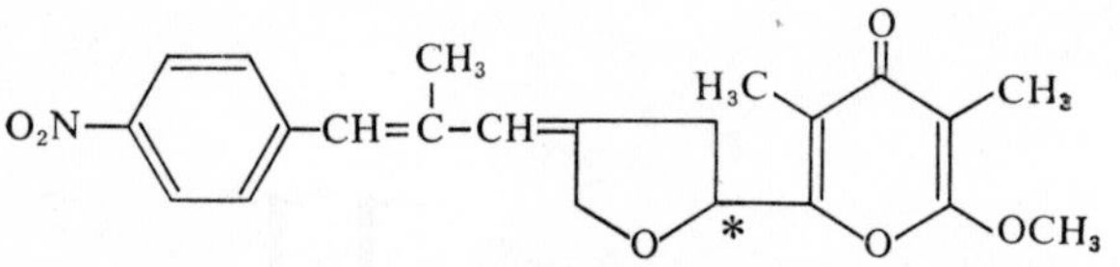

aureothin **1**[1)]

α_D (CHCl$_3$): +51°
uv (EtOH): 257 (4.39), 346 (4.27)
ir (CHCl$_3$): 1668, 1590 (γ-pyrone) 1505, 1332 (NO$_2$)

1) The absolute configuration of **1** at the carbon indicated by * is unknown.

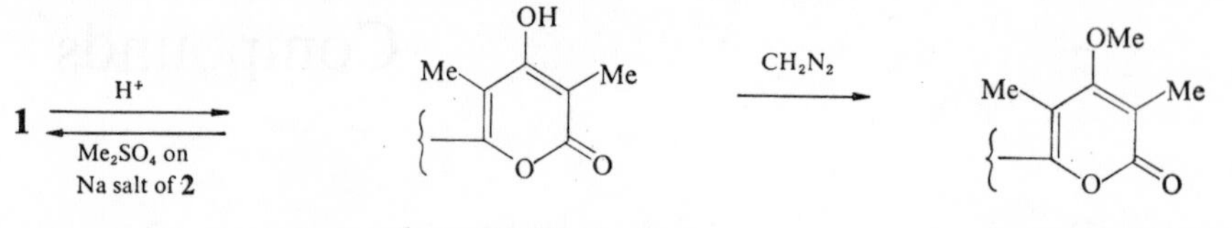

desmethylisoaureothin **2**

isoaureothin **3**

uv:250(4.17),335(4.19)
ir(CHCl$_3$):1700

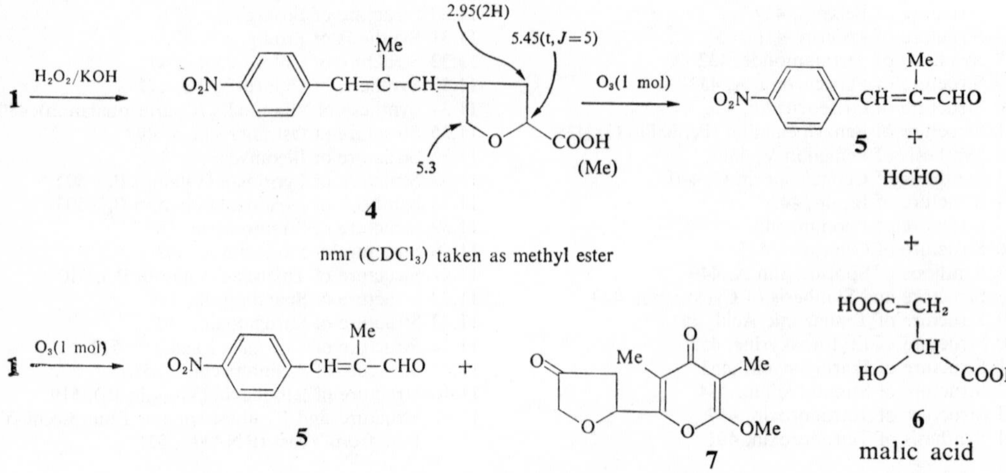

4

nmr (CDCl$_3$) taken as methyl ester

5 +

HCHO

+

HOOC—CH$_2$
CH
HO   COOH

6

malic acid

5

7

uv:263,248
ir:1765(5-membered C=O ring)
1668,1595(γ-pyrone)

1 $\xrightarrow{\text{1.5 N KOH}}$ MeCH$_2$COCH$_2$Me

Remarks

Aureothin **1** is a by-product of the antibiotic aureothricin[2)] and is a yellow toxin containing a novel nitro group. It was isolated from *Streptomyces thioluteus* U.

1) Y. Hirata, H. Nakata, K. Yamada, K. Okuhara, T. Naito, *Tetr.*, **14**, 252 (1961).
2) K. Maeda, *J. Antibiotics* (Japan.), Ser. A**6**, 137 (1953).

11.3 STRUCTURE OF ELAIOMYCIN

6.83(J=9,1.5,1.5)

α_D(EtOH): +38.4
UV: 237.5 (ε 11,000)

elaiomycin 1[1]

1) Dehydration of 1 yields a compound with a uv peak at 260 nm, suggesting the presence of OH vicinal to the azoxy group in 1.
2) 1 gives a positive iodoform test, indicating the presence of MeCHOH–.
3) Rearrangement of 1 to 5 supports structure 1a rather than 1b for 1. This is also supported by the uv data: 1b would be expected to absorb at a longer wavelength than 237.5 nm.
4) The D-threo configuration of 3 was determined by synthesis of the antipode of 3 from L-threonine.[2]
5) The *cis* configuration of the double bond was deduced from the nmr spectrum.[3]
6) The *cis* configuration of the azoxy group was indicated by the cd spectrum.[4]

Remarks

Elaiomycin 1 was isolated from *Streptomyces hepaticus*.[5] It possesses marked activity against tubercle bacilli, and is the first example of a natural product containing an aliphatic α,β-unsaturated azoxy group.

A similar antibacterial azoxy compound, LL-BH 872α, was isolated from *Str. hinnulinus*.

LL-BH872 α 6

1) C. L. Stevens, B. T. Gillis, J. C. French, T. H. Haskell, *J. Am. Chem. Soc.*, **78**, 3229 (1956); *ibid.*, **80**, 6088 (1958).
2) C. L. Stevens, B. T. Gillis, T. H. Haskell, *J. Am. Chem. Soc.*, **81**, 1435 (1959).
3) W. J. McGahren, M. P. Kunstmann, *J. Am. Chem. Soc.*, **91**, 2808 (1969).
4) W. J. McGahren, M. P. Kunstmann, *J. Am. Chem. Soc.*, **92**, 1587 (1970).
5) T. H. Haskel, A. Ryder, Q. R. Bartz, *Antibiot. Chemoth.*, **4**, 141 (1954).

11.4 STRUCTURE OF ANISOMYCIN

anisomycin **1**[1,2]

uv: 224 (ϵ 10800), 277 (ϵ 1600)
ir: 3546 (OH), 1730, 1242 (ester), 1608 (arom.)
nmr in CDCl$_3$/D$_2$O

1) The position of the conjugate ketone in **5** indicates the location of the original acetoxy group in **1**.

Stereochemistry and absolute configuration

The first stereochemical studies of **1** were carried out by chemical methods[2] and lead to the proposal of an all-*trans* relationship for the substituents on the pyrrolidine ring. Recent x-ray studies[3] on *N*-acetylbromoanisomycin have shown the structure **1**.

The absolute configuration[4] was determined by the sequence of reactions shown on the facing page. The ir spectra of **8** and **10** were identical, but the α_D values were opposite.

1) K. Butler, *J. Org. Chem.*, **23**, 2136 (1968).
2) J. J. Beereboom, K. Butler, E. C. Pennington, I. A. Solomons, *J. Org. Chem.*, **30**, 2334 (1965).
3) J. P. Schaefer, P. J. Wheatley, *J. Org. Chem.*, **33**, 166 (1968).
4) C. M. Wong, *Canad. J. Chem.*, **46**, 1101 (1968).
5) B. A. Sobin, F. W. Fanner Jr., *J. Am. Chem. Soc.*, **76**, 4053 (1954).

L-tyrosine **6**

7

8 a_D: +9.7°

9

10 a_D: −11°

Remarks

The antibiotic anisomycin **1** has been isolated from cultures of various *Streptomyces* species.[5] It possesses activity against certain pathogenic protozoa, notably *Trichomonas vaginalis* and *Endamoeba nistolytica*.

11.5   STRUCTURE OF MITOMYCIN A

mitomycin A **1**[1]

purple compound
uv (MeOH): 218 (ϵ 17400),
  320 (ϵ 10400), 520 (ϵ 1400)
ir: 3030 (aziridine CH)

1) The nmr spectrum of **1** indicates that the OMe group is at 9a instead of the 9-position.

2) The C=O band of the *N*-Ac derivative (1712 cm^{-1}) and the uv spectrum of the *N*-*p*-iodobenzoyl derivative suggest an aziridine group.

3) The uv spectrum indicates that OMe is para to the nitrogen on the quinone.

4) In the diacetate of **2**, ir peaks at 1550 (–NHCOMe) and 1738 (–OCOMe) indicate the presence of –OH and –NH$_2$.

5) The formation of CO$_2$ and NH$_3$ on hydrolysis of **2** with 6 N HCl suggests an –OCONH$_2$ group.

1) J. S. Webb, D. B. Cosulich, J. H. Mowat, J. B. Patrick, R. W. Broschard, W. E. Meyer, R. P. Williams, C. F. Wolf, W. Fulmor, C. Pidacks, J. E. Lancaster, *J. Am. Chem. Soc.*, **84**, 3185 (1962); *ibid.*, **84**, 3187 (1962).

2) T. Hata, Y. Sano, R. Sugawara, A. Matsumae, K. Kanamori, T. Shima, T. Hoshi, *J. Antibiotics* (Japan), Ser., A**9**, 141 (1956).

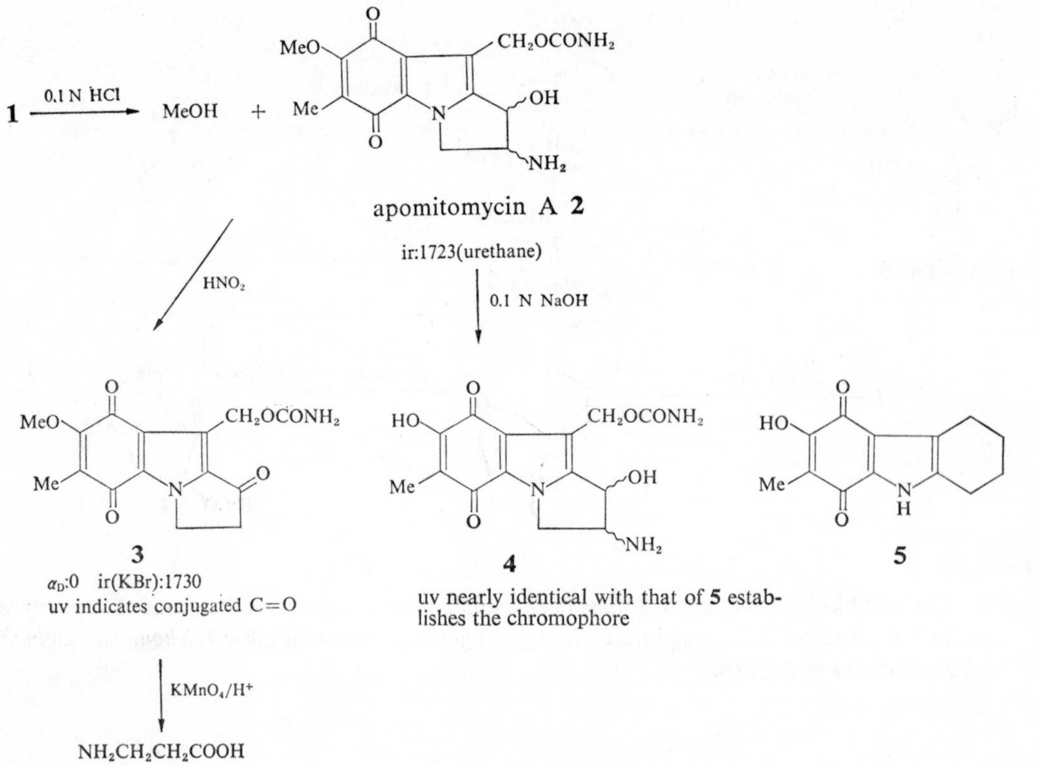
0.1 N HCl
MeOH
MeO
Me
CH2OCONH2
OH
NH2
apomitomycin A 2
ir:1723(urethane)
HNO2
0.1 N NaOH
3
αD:0 ir(KBr):1730
uv indicates conjugated C=O
4
uv nearly identical with that of 5 establishes the chromophore
5
HO
KMnO4/H+
NH2CH2CH2COOH

Remarks

Mitomycin A **1**, one of several mitomycin antibiotics, is a purple pigment isolated from *Streptomyces verticillatus*,[3] *Str. caespitosus*. It possesses strong activity against gram-positive and gram-negative bacteria, as well as against Ehrlich's ascites tumor cells, and was the first naturally occurring aziridine derivative to be identified. The stereochemistry of **1** was established by x-ray analysis of the *N*-4′-bromobenzenesulfonyl derivative.[4]

---

3) D. V. Lefemine, M. Dann, F. Barbatschi, W. K. Hausmann, V. Zbinovsky, P. Monnikendam, J. Adams, N. Bohonos, *J. Am. Chem. Soc.*, **84**, 3184 (1962).
4) A. Tulinsky, *J. Am. Chem. Soc.*, **84**, 3188 (1962).

11.6 STRUCTURE OF BETANIN

betanin 1[1)]

uv:536–538(ε 60,500)
nmr in CF_3COOH

1) The β-glycoside linkage was determined by nmr, and the presence of glucose at C-5 from the formation of 8.

2) The S-configuration at C-2 was deduced from the formation of 4. The S-configuration at C-15 was similarly deduced from the formation of L-aspartic acid 6. However, cyclo-DOPA 4 also yields small amounts of 6, and hence conclusive evidence was obtained using indicaxanthin 9.

1) H. Wyler, A. S. Dreiding, *Helv. Chim. Acta*, **45**, 638 (1962); T. J. Mabry, H. Wyler, G. Sassu, M. Mercier, I. Parikh, A. S. Dreiding, *ibid.*, **45**, 640 (1962); H. Wyler, T. J. Mabry, A. S. Dreiding, *ibid.*, **46**, 1745 (1963); M. E. Wilcox, H. Wyler, T. J. Mabry, A. S. Dreiding, *ibid.*, **48**, 252 (1965); M. E. Wilcox, H. Wyler, A. S. Dreiding, *ibid.*, **48**, 1134 (1965).

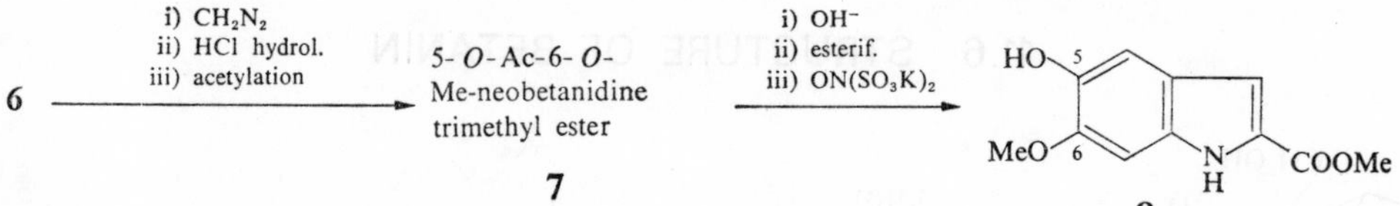

i) CH₂N₂
ii) HCl hydrol.   5- *O*-Ac-6- *O*-
6 ───────────────▶   Me-neobetanidine
iii) acetylation   trimethyl ester

7

i) OH⁻
ii) esterif.
iii) ON(SO₃K)₂
───────────────▶

HO─⁵
MeO─⁶ ... N─H ... COOMe

8

Remarks

Betanin **1** is the violet-red pigment of beet, *Beta vulgaris* var. *rubra*. Indicaxanthin **9** has been isolated from the fruits of *Opuntia ficus-indica*. Indicaxanthin **9** and betanidine **2** can be interconverted, and this reaction can be used for the preparation of their stereoisomers.

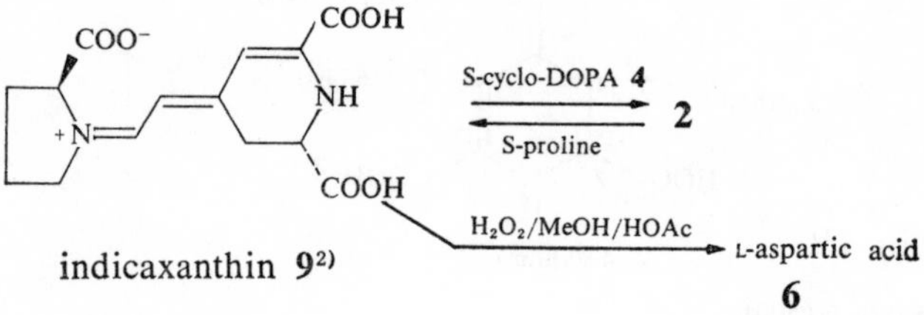

COO⁻   COOH

⁺N ═══   NH

COOH

indicaxanthin **9**[2)]

S-cyclo-DOPA **4**
⇌   **2**
S-proline

H₂O₂/MeOH/HOAc
───────────────▶ ʟ-aspartic acid
6

2) M. Piattelli, L. Minale, G. Prota, *Tetr.*, **20**, 2325 (1964).

11.7 STRUCTURE OF PEDERIN

3.85(ddd, $J=8,5,3$)

1.85(ddd, $J=12,10,5$)
2.10(ddd, $J=0.2,4,3$)

3.75(dd, $J=10,4$)

pederin **1**

nmr: 3.28, 3.30, 3.33, 3.37(s) (OMe),
0.96, 0.87(s) (*t*-ME),
1.18, 1.02(d) (*sec*-Me)

1) One MeO group (*) is easily hydrolyzable by heating in aqueous solution, yielding pseudo-pederin **2**.

Pb(OAc)$_4$/benzene **2** Ba(OMe)$_2$/MeOH(oxidative)

Al$_2$O$_3$

3 **4**

5.30(dd, $J=9.5,7.5$)

7.50(d, $J=9.5$)

i) CH$_2$N$_2$
ii) 1 N HCl **6**

5

9.16(s)

5.71(t, $J=4$)

0.91, 0.99(s)

2.05(d, $J=4$)

6

i) O$_3$
ii) 2N HCl

4.16(AB<u>X</u>) 1.85(<u>AB</u>X)

3.4-3.7(3H)

2.33(s)

1.03, 1.17(s)

3.35, 3.38(s)

7

Remarks

Pederin **1**[1] is one of the toxic principles isolated from blister beetles, *Paederus fuscipes* Curtis. Other minor components, pseudopederin **2**[2] and pederone (**1**, 4-oxo),[3] were later isolated and the structures of the three compounds determined.[3,4] However, correction of the position of the hydroxyl group in **1** (C-4 instead of C-3) was later made by examination of the nmr spectrum of **1**.[5] The stereochemistry and absolute configuration of pederin **1** were established by x-ray analysis.[6]

1) M. Pavan, G. Bo, *Mem. Soc. Entomol. Ital.*, **31**, 67 (1952): *Phys. Comp. Oecol.*, **3**, 307 (1953).
2) A. Quilico, C. Cardani, D. Ghiringhelli, M. Pavan, *Chim. Ind. (Milan)*, **43**, 1434 (1961).
3) C. Cardani, D. Ghiringhelli, A. Quilico, A. Selva, *Tetr. Lett.*, 4023 (1967).
4) C. Cardani, D. Ghiringhelli, R. Mondelli, A. Quilico, *Tetr. Lett.*, 2537 (1965).
5) T. Matsumoto, M. Yanagiya, S. Maeno, S. Yasuda, *Tetr. Lett.*, 6297 (1968).
6) A. Furusaki, T. Watanabe, T. Matsumoto, M. Yanagiya, *Tetr. Lett.*, 6301 (1968).

11.8 SYNTHESIS OF SERRATAMOLIDE

serratamolide 1[1]

The key step of this synthesis is the spontaneous rearrangement of the β-hydroxy acid imide to the amido ester.

Remarks

Serratamolide 1 is an antibiotic produced by *Serratia marcescens*. It was isolated and the structure determined by Wasserman.[2] Hydrolysis of 1 with NaOH yields serratamic acid, $Me(CH_2)_6CH(OH)CH_2CONHCH(CH_2OH)COOH$, in 88% yield. It was shown that the hydroxyl group of the serine moiety is free by reduction followed by hydrolysis, yielding alanine and 3-hydroxydecanoic acid.

Depsipeptides containing an α-hydroxy acid moiety, such as sporidesmolide 6[3] and beauvericin 7[4] have been synthesized by a high dilution method.

sporidesmolide I 6   beauvericin 7

1) M. M. Shemyakin, Y. A. Ovchinnikov, V. K. Antonov, Z. A. Kiryushkin, V. T. Ivanov, V. I. Shchelokov, A. M. Shkrob, *Tetr. Lett.*, 47 (1964); V. K. Antonov, V. I. Shchelokov, M. M. Shemyakin, I. I. Tovarova, O. A. Kiseleva, *Antibiotiki*, 10, 387 (1965).
2) H. H. Wasserman, J. J. Keggi, J. E. McKeon, *J. Am. Chem. Soc.*, 83, 4107 (1961); *ibid.*, 84, 2978 (1962).
3) M. M. Shemyakin, Y. A. Ovchinnikov, V. T. Ivanov, A. A. Kirushkin, *Tetr.*, 19, 995 (1963).
4) Y. A. Ovchinnikov, V. T. Ivanov, I. I. Mikhaleva, *Tetr. Lett.*, 159 (1971).

11.9 STRUCTURE OF FERRICHROME A

ferrichrome A **1**[1]

1) Acid hydrolysis of **1** gives ornithine (0.3–0.6 mole), glycine and serine.

2) Acid hydrolysis of **2** gives glycine (1 mole) and serine (2 mole).

3) Reductive hydrolysis of **2** with HI gives ornithine (3 mole).

4) The existence of the δ-N-hydroxyornithine moiety is inferred firstly from 1) and 3) above; secondly, the hydrolysis of **2** with 12N HCl yields, after electrophoresis, a spot which gives a red color characteristic of the hydroxylamino grouping, when sprayed with tetrazolium reagent; and thirdly the spot was eluted and reduced with H_2/PtO_2 to give ornithine.

5) The absolute configuration of ornithine as L was established by quantitative growth tests with *E. coli*.

6) **1** shows no dissociation constant in the pH range 2.5–9.5 and hence no amino acid end group in the molecule.

7) HIO_4 oxidation of **2** gives 3 moles of **3**, indicating the presence of 3 moles of N-*trans*-β-methylglutaconylhydroxylamino groups since one N-acylhydroxylamino group consumes 1 mole of HIO_4.[2]

8) The presence of the double bond at the α,β-position rather than the β,γ-position in the methylglutaconyl group is suggested from the uv spectrum; **1** [440 nm (ϵ 3740)], on catalytic hydrogenation, yields hexahydro **1** [425 nm (ϵ 2895)], the uv of which is identical with that of ferrichrome having N-acetylhydroxylamino groupings.[2]

1) T. Emery, J. B. Neilands, *J. Am. Chem. Soc.*, **83**, 1626 (1961).
2) T. Emery, J. B. Neilands, *Nature*, **184**, 1632 (1959); *J. Am. Chem. Soc.*, **82**, 3658 (1960).

9) The amino acid sequence was determined by the identification of pyruvamide **5** as its 2,4-dinitrophenylhydrazone and alanylglycine **7**.[3]

10) The structure and absolute configuration of **1** were confirmed by x-ray analysis; the shape of the iron chelate is of the left-hand propeller type.[4]

Remarks

Ferrichrome A **1** is a metabolic product of *Ustilago sphaerogena*, the smut fungus,[2,5] and is related in structure to several substances which are growth factors for certain microorganisms. The molecule is unique in having a ferric atom bound by three hydroxamate rings.

Ferrichrome,[5] ferrichrysin,[6] ferrichrosin,[6] ferrirhodin,[7] and ferrirubin[7] are structurally related to ferrichrome A. Ferrichrome has been synthesized.[9] Ferrioxamine A$_1$, A$_2$, B, C, D$_1$,

3) S. Rogers, J. B. Neilands, *Biochemistry*, **3**, 1850 (1964).
4) A. Zalkin, J. D. Forrester, D. H. Templeton, *J. Am. Chem. Soc.*, **88**, 1810 (1966).
5) J. B. Neilands, *J. Am. Chem. Soc.*, **74**, 4846 (1952).
6) W. Keller-Schierlein, A. Deér, *Helv. Chim. Acta*, **46**, 1907 (1963).
7) W. Keller-Schierlein, *Helv. Chim. Acta*, **46**, 1920 (1963).
8) H. Bickel, E. Gäumann, G. Nussberger, P. Reusser, E. Vischer, W. Voser, A. Wettstein, H. Zähner, *Helv. Chim. Acta*, **43**, 2105 (1960); H. Bickel, P. Mertens, V. Prelog, J. Seibl, A. Walser, *Tetr. Suppl.*, **8**, 171 (1966).
9) W. Keller-Schierlein, B. Maurer, *Helv. Chim. Acta*, **52**, 388, 603 (1969).

D_2, E, and G,[10-13] which are metabolites of *Streptomyces pilosus*, are structurally related to ferrimycin A **8**,[8] an antibiotic produced by *Streptomyces griseoflavus*. A metabolite of *Neurospora crassa*, coprogen, has the unique structure **9**.[14]

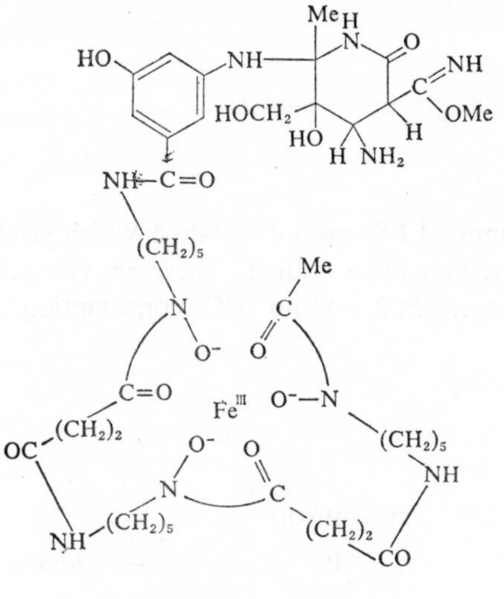

ferrimycin A **8**[8]

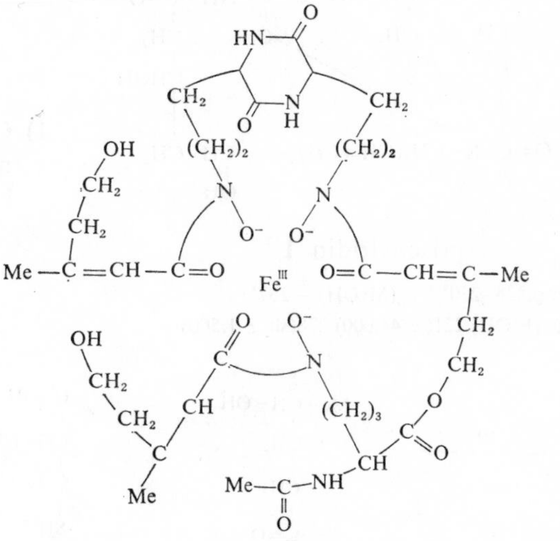

coprogen **9**[14]

10) H. Bickel, G. E. Hall, W. Keller-Schierlein, V. Prelog, E. Vischer, A. Wettstein, *Helv. Chim. Acta*, **43**, 2118 (1960); *ibid.*, 2129 (1960).
11) W. Keller-Schierlein, V. Prelog, *Helv. Chim. Acta*, **44**, 709 (1961); *ibid.*, **44**, 1981 (1961); *ibid.*, **45**, 590 (1962).
12) V. Prelog, A. Walser, *Helv. Chim. Acta*, **45**, 631 (1962).
13) W. Keller-Schierlein, P. Mertens, V. Prelog, A. Walser, *Helv. Chim. Acta*, **48**, 710 (1965).
14) W. Keller-Schierlein, H. Diekmann, *Helv. Chim. Acta*, **53**, 2035 (1970).

11.10   STRUCTURE OF GRISEOVIRIDIN

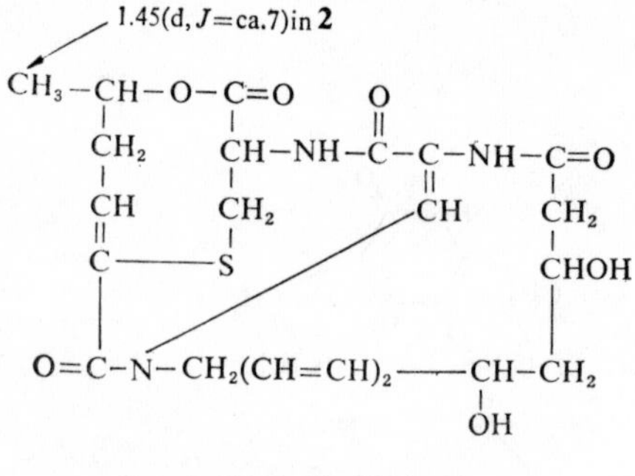

1.45(d, *J*=ca.7)in **2**

CH₃–CH–O–C=O   O
     CH₂ ||
    CH₂   CH–NH–C–C–NH–C=O
    CH   CH₂ || CH₂
    || CH   CHOH
    C–––S
O=C–N–CH₂(CH=CH)₂––––CH–CH₂
                OH

griseoviridin **1**[1)]

mp:228–230°   α_D(MeOH):−237°
uv(EtOH):221(ε 44,000),278(sh, ε 1,500)

1) Compound **1** forms a diacetate **2**, which yields a maleic anhydride adduct. There are two active hydrogens in **2**, and no HIO₄ consumption.

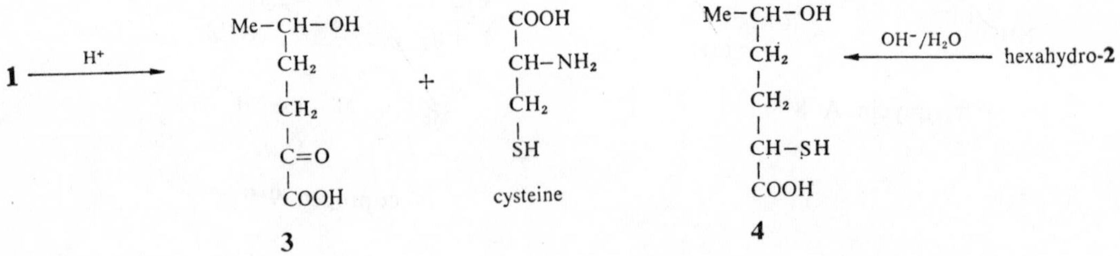

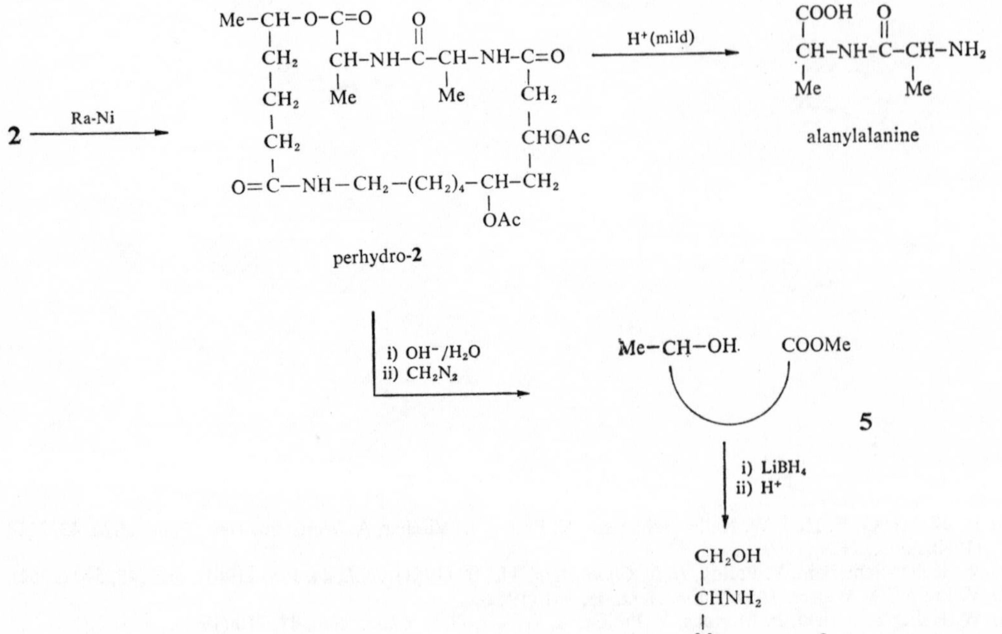

1 $\xrightarrow{H^+}$

Me–CH–OH
 CH₂
 CH₂
 C=O
 COOH
3

+

COOH
CH–NH₂
CH₂
SH
cysteine

Me–CH–OH
 CH₂
 CH₂
 CH–SH
 COOH
4 $\xleftarrow{OH^-/H_2O}$ hexahydro-**2**

2 $\xrightarrow{Ra-Ni}$

Me–CH–O–C=O   O
 CH₂ CH–NH–C–CH–NH–C=O
 CH₂ Me Me CH₂
 CH₂   CHOAc
O=C–NH–CH₂–(CH₂)₄–CH–CH₂
           OAc

perhydro-**2**

$\xrightarrow{H^+(mild)}$

COOH   O
CH–NH–C–CH–NH₂
Me   Me
alanylalanine

i) OH⁻/H₂O
ii) CH₂N₂ $\longrightarrow$

Me–CH–OH   COOMe
5

i) LiBH₄
ii) H⁺

CH₂OH
CHNH₂
Me   **6**

Evidence for the partial structure 1b

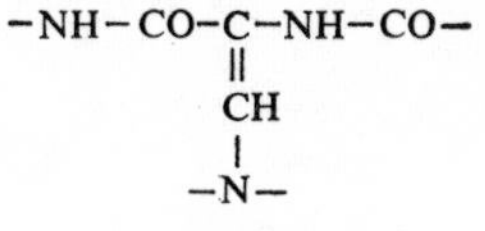

```
-NH-CO-C-NH-CO-
         ‖
         CH
         |
        -N-

      1b
```

1) An ir band at 1625 cm^{-1} indicates C=C.
2) An nmr singlet at 8.1 ppm (1 H) indicates C=CH–X.
3) Hydrolysis with OH$^-$/H$_2$O yields HCOOH and HOOCCH$_2$NH$_2$.
4) The absence of basicity and the presence of the amido C=O bands at the same intensity in the ir spectra of both 2 and perhydro-2 indicate the absence of an imino ether linkage such as

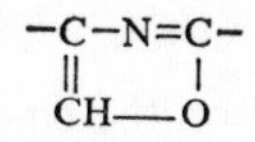

```
 -C-N=C-
  ‖   |
  CH—O
```

5) 2 has only two active hydrogens and hence has two –NH–CO– groups and one –N–CO– group.

Evidence for the partial structure 1c

The following reactions indicated the partial structure 1c.

```
O=C-N-CH₂-(CH=CH)₂-CH-CH₂-CH-CH₂C=O    MnO₂    O=C-N-CH₂-(CH=CH)₂-C-CH₂-CH-CH₂-C=O
              |        |                              ‖              |
              OH       OH                             O              OH
      1c                                        7  uv:277(ε 20,100)
```

```
  acetyl.                         H₂/Pd-C
 ────────►   monoacetate   ──────────►   O=C-N-(CH₂)₄-C-CH₂CH-CH₂-C=O    0.023 N NaOH
                  8                                   ‖     |              ──────────►
                                                      O     OAc
                                                         9
```

```
                                         OH⁻
O=C-N-(CH₂)₄-C-CH=CH-CH₂-C=O    ⇌    O=C-N-(CH₂)₄-C=CH-CH=CH-C=O
             ‖                   H⁺                  |
             O                                       O⁻

  10  uv:255                            11  uv:385(ε 23,000)
```

```
               reductive hydrolysis
perhydro-2   ──────────────────────►   H₂N-(CH₂)₁₀-COOH

                         12
```

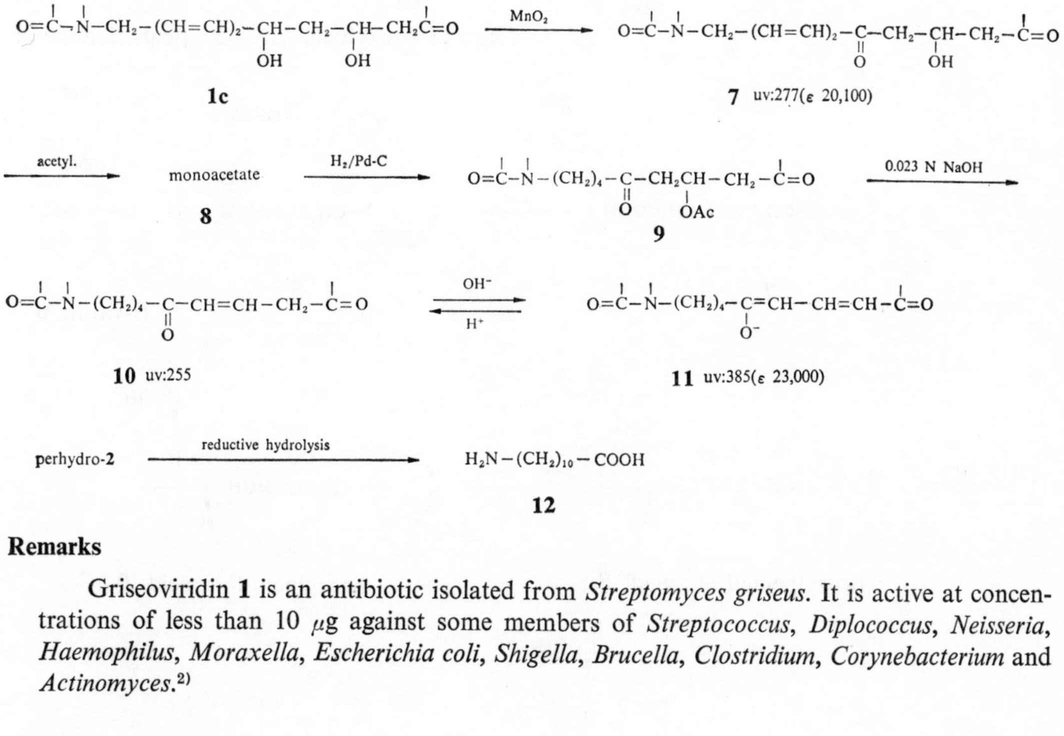

Remarks

Griseoviridin 1 is an antibiotic isolated from *Streptomyces griseus*. It is active at concentrations of less than 10 μg against some members of *Streptococcus*, *Diplococcus*, *Neisseria*, *Haemophilus*, *Moraxella*, *Escherichia coli*, *Shigella*, *Brucella*, *Clostridium*, *Corynebacterium* and *Actinomyces*.[2]

1) M. C. Fallona, T. C. McMorris, P. deMayo, T. Money, A. Stoessl, *J. Am. Chem. Soc.*, **84**, 4162 (1962); M. C. Fallona, P. deMayo, T. C. McMorris, T. Money, A. Stoessl, *Can. J. Chem.*, **42**, 371 (1964); M. C. Fallona P. deMayo, A. Stoessl, *ibid.*, **42**, 394 (1964).
2) T. Korzybski *et al., Antibiotics* (English ed.), vol. 1, p. 351–356, Pergamon, 1967.

11.11 STRUCTURE OF BENZYLPENICILLIN (PENICILLIN G)

penicillin G 1[1]

1) The β-lactam structure is assumed on the basis of the ir spectrum of **10**.
2) The configuration at C-3 was determined by the isolation of D-valine **6**.
3) The structure and stereochemistry were determined by x-ray analysis.

methyl benzylpenaldate **3**   penicillamine **4**

2

7   **5**

D-valine **6**

benzylpenicilloic acid **8**   benzylpenilloic acid **9**

desthiobenzylpenicillin **10**   benzylpenillic acid **11**

1) *Chemistry of Penicillins* (ed. H. T. Clarke, J. R. Johnson, R. Robinson), Princeton Univ. Press, 1949.
2) A. Fleming, *Brit. J. Exp. Path.*, **10**, 226 (1929).

Remarks

Penicillin, the first antibiotic, was discovered in 1928 by Fleming.[2] It is formed by *Penicillium notatum*, and is active against gram-positive bacteria. It has a very low toxicity to man.

The structure was determined jointly by English and American scientists during World War II. This work was not published in the journals, but is reported in ref. 1.

1 was the first known example of a compound with a β-lactam structure to be isolated from natural sources. Other naturally occurring penicillins differ from **1** only in the substitution of the side chain (benzyl group): penicillin F has an $MeCH_2CH=CHCH_2-$ group in place of the benzyl group in **1**; penicillin X has $p-HOC_6H_4CH_2-$; penicillin K has $Me(CH_2)_6-$; and flavacidin has $MeCH=CHCH_2CH_2-$.

11.12   SYNTHESIS OF PENICILLIN V

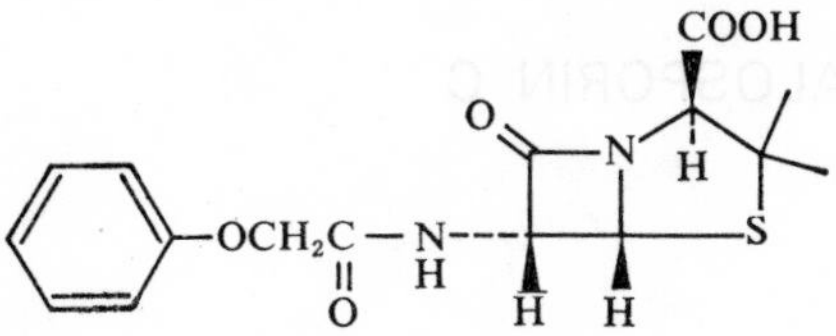

penicillin V **1**[1]

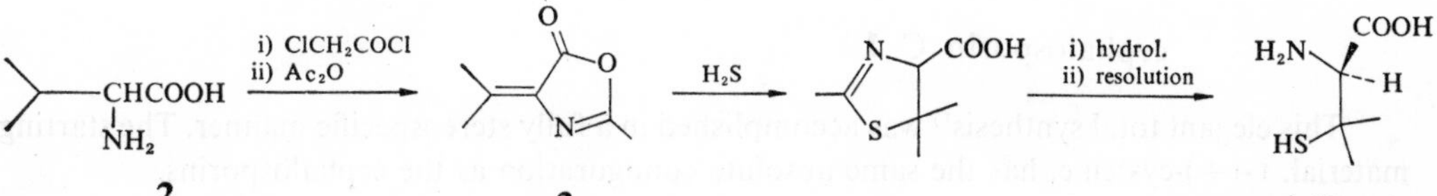

D-penicillamine **5**[2]

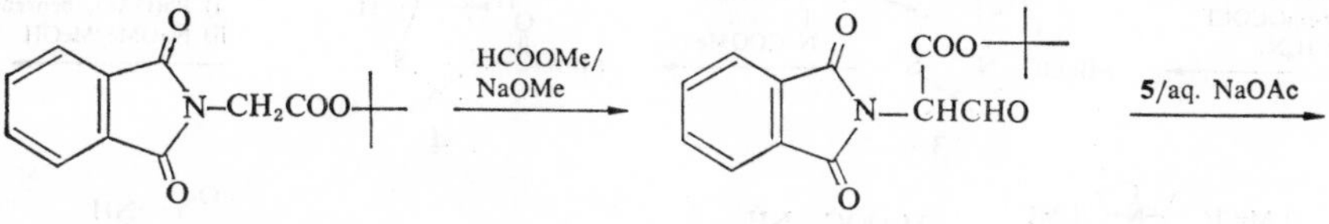

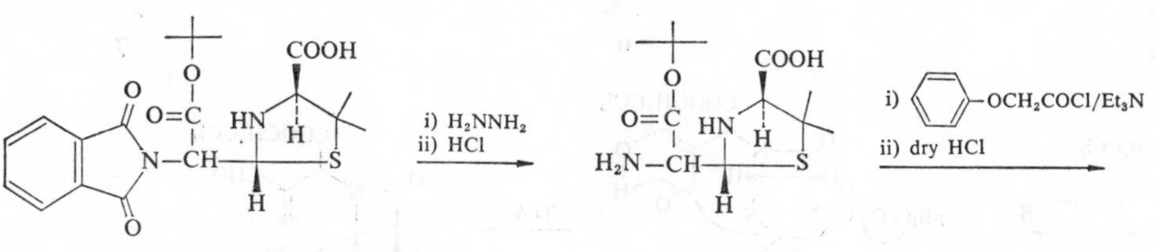

8 (α-D-form)   **9**

---

1) J. C. Sheehan, D. A. Johnson, *J. Am. Chem. Soc.*, **76**, 158 (1954); J. C. Sheehan, K. R. Henery-Logan, *ibid.*, **79**, 1262 (1957); *ibid.*, **81**, 3089 (1959).

2) B. E. Leach, J. H. Hunter, *Biochemical Preparations*, **3**, 111 (1953).

$$10$$

7 ⟶ 8: A mixture of α and γ stereoisomers is produced. The α isomer is predominant.

10 ⟶ 1: The key step in this synthesis is the closure of the β-lactam ring under very mild conditions, using DCC. The yield is 10–12%. D,L-Penicillin V obtained from D,L-**5** has a bioactivity 51.4% of that of D-**1**.

11.13   SYNTHESIS OF CEPHALOSPORIN C

cephalosporin C **1**

This elegant total synthesis[1] was accomplished in a fully stereospecific manner. The starting material, L-(+)-cysteine, has the same absolute configuration as the cephalosporins.

1) R. B. Woodward: K. Heusler, J. Gosteli, P. Naegeli, W. Oppolzer, R. Ramage, S. Ranganathan, H. Vorbrüggen, *J. Am. Chem. Soc.*, **88**, 852 (1966).

i) trichloroethoxycarbonyl-
   D-(−)-α-aminoadipic acid/
   THF/DCC
ii) B₂H₆/THF
iii) Ac₂O/py

$$\xrightarrow{\text{py}} \quad K(5/6)=1/4$$

11   **12**

$$\xrightarrow{\text{Zn/aq/HOAc}} \mathbf{1}$$

$$(R=-CH_2CH_2CH_2CH(NH_3^+)COO^-)$$

3 ⟶ 4: The methylene group (indicated in **3** by an arrow) was far more reactive in a stereo-specific manner than **2**, because rotation about the α,β-C–C bond is impossible. Thus, the dimethyl azodicarboxylate was introduced solely on one side of the ring.

   8: This highly reactive dialdehyde was synthesized from β,β,β-trichloroethyl glyoxylate and malondialdehyde.

12 ⟶ 1: The elegant protecting group for carboxylic acid was removed under mild reductive conditions.

Remarks

An interesting direct chemical correlation between a penicillin **13** and a cephalosporin **15** has been performed.[2]

$$\xrightarrow{\text{P-TSA/xylene}}$$

$$(R=C_6H_5OCH_2-)$$

$$\xleftarrow[\text{dioxane}]{\text{H}_2/\text{Pb-C/}}$$

13   **14**   **15**

2) R. B. Morin, B. G. Jackson, R. A. Mueller, E. R. Lavognino, W. B. Scanlon, S. L. Andrews, *J. Am. Chem. Soc.*, **91**, 1401 (1969).

11.14   STRUCTURE OF BIOTIN

biotin **1**

1) **1** forms a methyl ester with CH_2N_2 and a sulfone with $H_2O_2/AcOH$.[1,6]

2) **2** yields a pyrazine derivative with 9,10-phenanthrenequinone.[2]

3) A comparison of the rates of hydrolysis of four synthetic stereoisomers indicated that the two rings in **1** are *cis* fused.[3,4]

4) The stereochemistry of the third asymmetric carbon and the absolute configuration of **1** were established by x-ray analysis of the bis-*p*-bromoanilide of carboxylated biotin **8**.[5]

Remarks

Biotin **1** is present in animals and yeasts in a bound form, but exists in the free state in plants. It was first isolated from yolks of eggs,[11] and then in a pure form from liver.[12] Biotin

1) F. Kögl, T. J. de Man, *Z. Physiol. Chem.*, **269**, 81 (1941).
2) K. Hofmann, G. W. Kilmer, D. B. Melville, V. du Vigneaud, H. H. Darby, *J. Biol. Chem.*, **145**, 503 (1942).
3) S. A. Harris, R. Mozingo, D. E. Wolf, A. N. Wilson, K. Folkers, *J. Am. Chem. Soc.*, **67**, 2102 (1945).
4) B. R. Baker, W. L. McEwen, W. N. Kinley, *J. Org. Chem.*, **12**, 323 (1947).
5) C. Bonnemere, J. A. Hamilton, L. K. Steinrauf, J. Knappe, *Biochemistry*, **4**, 240 (1965); J. Trotter, J. A. Hamilton, *ibid.*, **5**, 713 (1966).
6) K. Hofmann, D. B. Melville, V. du Vigneaud, *J. Biol. Chem.*, **141**, 207 (1941); F. Kögl, L. Pons, *Z. Physiol. Chem.*, **269**, 61 (1941).
7) D. B. Melville, K. Hofmann, V. du Vigneaud, *Science*, **94**, 308 (1941).
8) K. Hofmann, D. B. Melville, V. du Vigneaud, *J. Am. Chem. Soc.*, **63**, 3237 (1941); *J. Biol. Chem.*, **144**, 513 (1942).
9) V. du Vigneaud, D. B. Melville, K. Folkers, D. E. Wolf, R. Mozingo, J. C. Keresztesy, S. A. Harris, *J. Biol. Chem.*, **146**, 475 (1942).
10) D. B. Melville, A. W. Moyer, K. Hofmann, V. du Vigneaud, *J. Biol. Chem.*, **146**, 487 (1942).
11) F. Kögl, B. Tönnis, *Z. Physiol. Chem.*, **242**, 43 (1936).
12) P. György, R. Kuhn, E. Lederer, *J. Biol. Chem.*, **131**, 745 (1939).
13) J. Knappe, B. Werger, U. Wiegand, *Biochem. Z.*, **337**, 232 (1963).

is a growth factor for rats, yeasts and some microorganisms. True deficiency disease is virtually unknown in man, since biotin can be synthesized by the intestinal bacteria.

The mechanism of carboxylation and transcarboxylation reactions catalysed by biotin enzymes involves the intermediary formation of the carboxylated biotin **8**.[5,13]

8

11.15   STRUCTURE OF PHOMIN

phomin **1**[1]

mp: 218–220°
uv (EtOH): 213 (4.42), 219 (4.32), 258 (inf., 2.69), 264 (inf., 2.48), 267 (inf., 2.32)

1) CrO_3/py oxidation of **1** gave 5-dehydro **1**, with an enedione chromophore.
2) The absolute configuration at C-9 was established by synthesis of **3** from (+)-pulegon **4**.
3) The structure **5** was assigned mainly by nmr analysis. The configuration of **5** was determined by x-ray analysis of deacetyl **5**.
4) The presence of a γ-lactam was indicated by the isolation of **7**.[2]
5) The presence of two *trans* double bonds was deduced from the nmr spectrum.
6) The absolute stereochemistry of **1** was determined by x-ray analysis of the **1**-silver fluoroborate complex.[3]

1) W. Rothweiler, Ch. Tamm, *Experientia*, 750 (1966); *Helv. Chim. Acta*, **53**, 696 (1970).
2) D. C. Aldridge, J. J. Armstrong, R. N. Speake, W. B. Turner, *Chem. Commun.*, 26 (1967); *J. Chem. Soc. C*, 1667 (1967).
3) G. M. McLaughlin, G. A. Sim, *Chem. Commun.*, 1398 (1970).

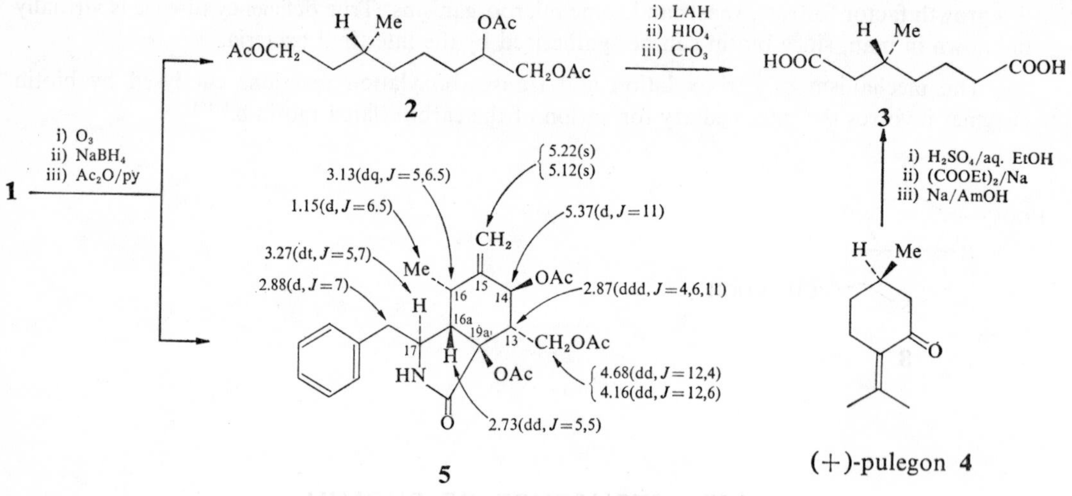

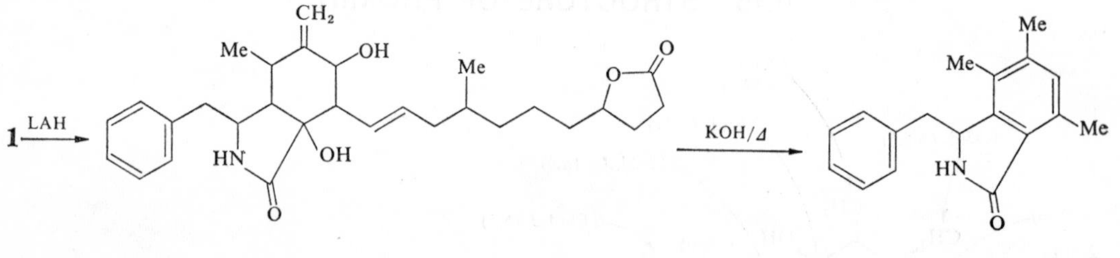

2

3

(+)-pulegon **4**

5

6

7

Remarks

Phomin[1] (cytochalasin B[2]) **1** was isolated from *Phoma* sp., *Helminthosporium dematioideum*. This compound is a cytostatic active metabolite, and is a novel type of macrolide in which the large lactone ring is fused to a highly substituted octahydroisoindole system. Many congeners have been isolated, and are classified into three types: the phomin type includes cytochalasin A and B;[2] the zygosporin A[4] type includes cytochalasin C and D (zygosporin A) and zygosporin D, E, F, and G[4]; and the chaetoglobosin type includes chaetoglobosin A and B.[6]

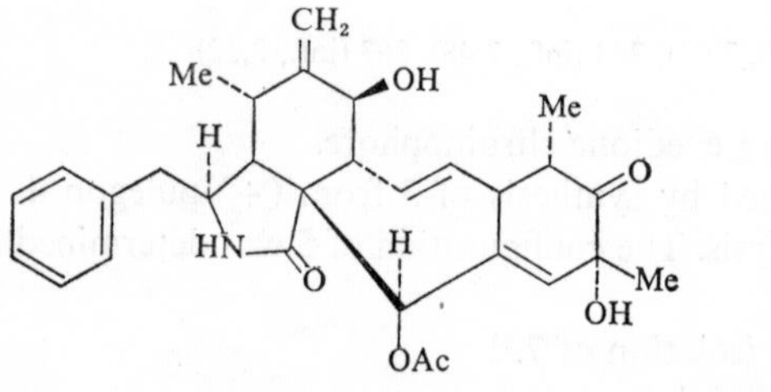

zygosporin A **8**[4]

(from *Zygosporium masonii* and *Metarrhizium anisopliae*)

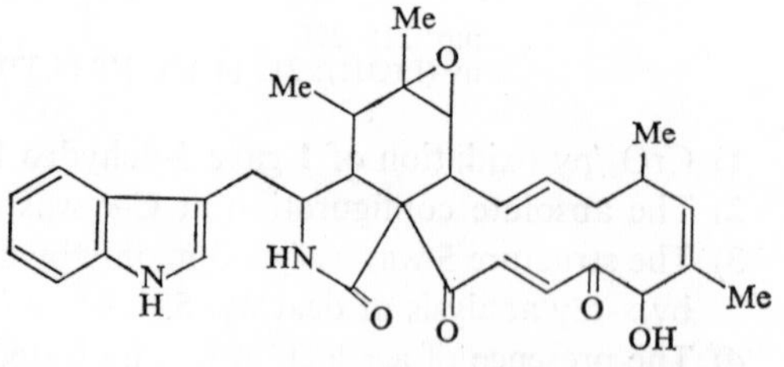

chaetoglobosin A **9**[6]

(from *Chaetomium globosum*)

4) H. Minato, M. Matsumoto, *J. Chem. Soc.* C, 38 (1970); Y. Tsukada, M. Matsumoto, H. Minato, H. Koyama *Chem. Commun.*, 41 (1969).
5) D. C. Aldridge, B. F. Burrows, W. B. Turner, *Chem. Commun.*, 148 (1972).
6) S. Sekita, K. Yoshihira, S. Natori, H. Kuwano, *Tetr. Lett.*, 2109 (1973).

11.16　STRUCTURE OF GLIOTOXIN

gliotoxin **1**

1) The absence of an aromatic ring system and the presence of a 1,3-diene were indicated by the uv and ir spectra.

2) Chemical and stereochemical considerations preclude the attachment of a disulfide bridge at any other positions.

3) The aromatic chromophore in **5** was suggested by uv.

4) The polymeric product **7** showed a striking resemblance to the very insoluble, high melting, dimeric substances produced by the action of ammonia and other bases on dipeptides or 2,5-piperazinediones containing a serine moiety, suggesting the presence of such a moiety in **1**.

5) The sulfur-containing compound **8** was synthesized: its structure suggests the point of attachment of sulfur.

1) Weiddling, *Phytopathology*, **31**, 991 (1941); *Botan. Rev.*, **4**, 475 (1938).
2) M. R. Bell, J. R. Johnson, B. S. Wildi, R. B. Woodward, *J. Am. Chem. Soc.*, **80**, 1001 (1958).
3) A. F. Beecham, F. Fridrichsons, A. McL. Mathieson, *Tetr. Lett.*, 3131 (1966).
4) J. A. Winstead, R. J. Suhadolnik, *J. Am. Chem. Soc.*, **82**, 1644 (1960).
5) A. K. Bose, K. S. Khanchandani, R. Tavares, P. T. Funke, *J. Am. Chem. Soc.*, **90**, 3593 (1968).
6) R. Hodges, J. W. Ronaldson, J. S. Shannon, A. Taylor, E. P. White, *J. Chem. Soc.*, 26 (1964); see also A. Taylor *et al., Chem. Commun.*, 1032 (1967); *ibid.*, 1571 (1968).

Remarks

Gliotoxin **1**, a sulfur-containing antibiotic, has been obtained from various species of *Trichoderma*, *Gladiochladium fimbriatum*, *Aspergillus fungatus* and *Penicillium*.[1] A structure proposed in 1958[2] was later confirmed together with its stereochemistry, by x-ray analysis.[3]

The biosynthesis of gliotoxin has been extensively investigated. D- and L-Phenylalanine, *m*-tyrosine, serine and glycine are all incorporated in **1** in high yield.[4] The positions of isotopically labeled atoms were located by chemical degradation and by mass spectroscopy.[5]

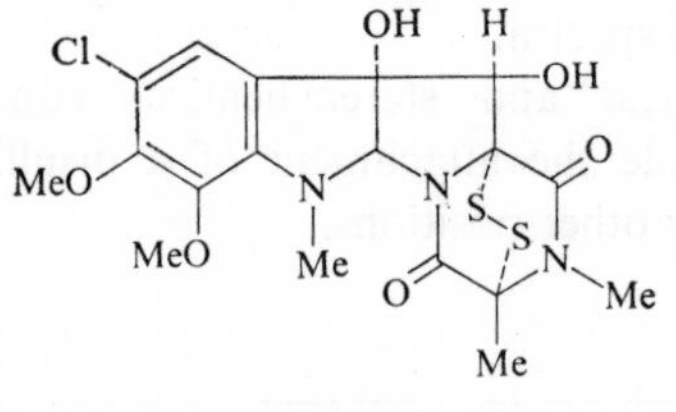

sporidesmin **9**

Analogous toxic, sulfur-containing metabolites, the sporidesmines, are responsible for liver damage and facial eczema in sheep in New Zealand. These compounds were isolated from *Pithomyces chartarum*.[6]

11.17   SYNTHESIS OF SPORIDESMIN A

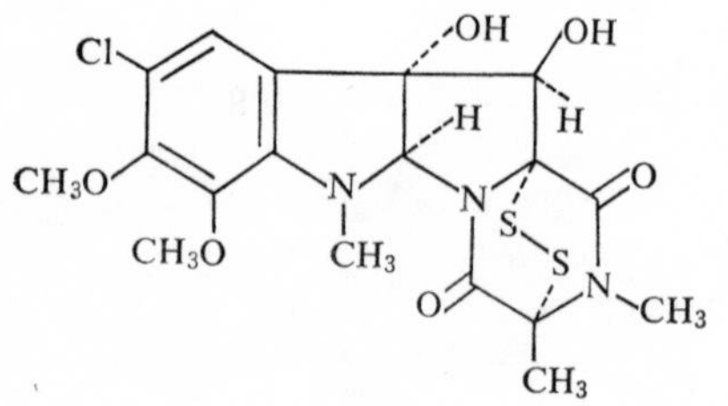

sporidesmin **A 1**[1]

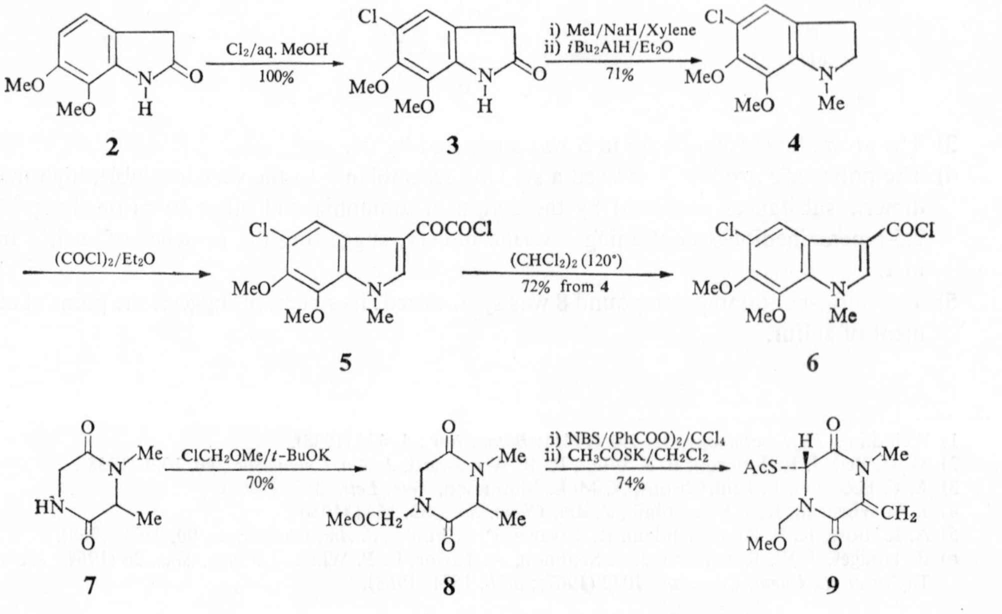

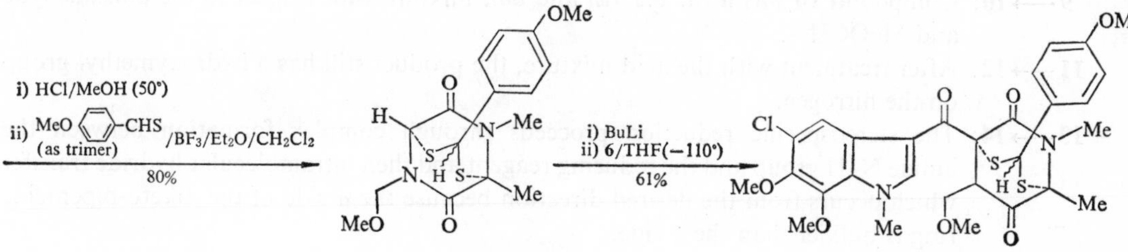

i) HCl/MeOH (50°)

ii) MeO⬡–CHS /BF₃/Et₂O/CH₂Cl₂
   (as trimer)

   80%

10

i) BuLi
ii) 6/THF(−110°)
   61%

11

i) conc. HCl/CF₃COOH (70°)
ii) NaOH
iii) tlc separation
   62%

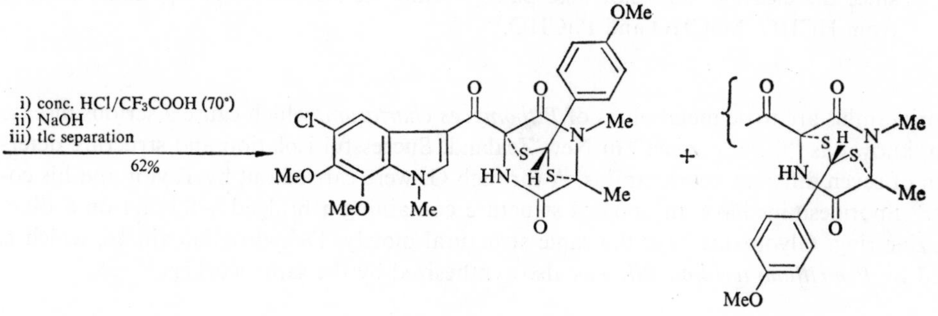

12 (50%)   +   **13** (12%)

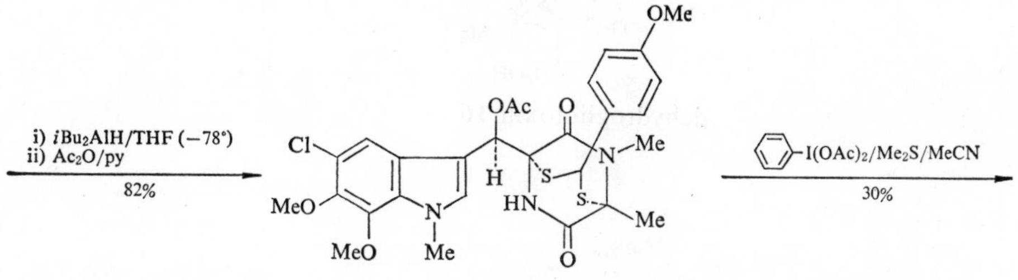

i) iBu₂AlH/THF (−78°)
ii) Ac₂O/py
   82%

14

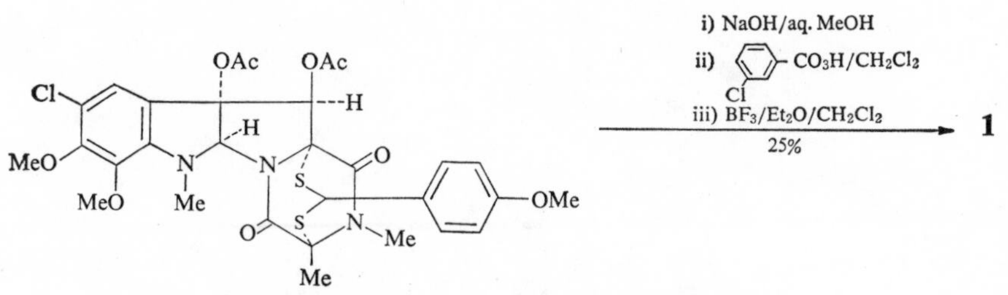

⬡–I(OAc)₂/Me₂S/MeCN
   30%

15

i) NaOH/aq. MeOH
ii) ⬡–CO₃H/CH₂Cl₂
iii) BF₃/Et₂O/CH₂Cl₂
   25%

   1

1) Y. Kishi, S. Nakatsuka, T. Fukuyama, M. Havel, *J. Am. Chem. Soc.*, **95**, 6493 (1973).

9 ⟶ 10: Compound **10** was a *ca.* 1:2 *syn* and *anti* mixture with respect to the anisaldehyde and MeOCH$_2$–.

11 ⟶ 12: After treatment with the acid mixture, the product still has a hydroxymethyl group on the nitrogen.

13 ⟶ 14: The stereospecific reduction proceeds through complex formation between the amide N–H group and the reducing reagent and then intramolecular hydride transfer which occurs from the desired direction because the α side of the diketo-piperazine ring is bulkier than the β side.

14 ⟶ 15: The two new asymmetric centers introduced in this step are the desired ones for steric reasons.

15 ⟶ 1: In this acid cleavage of the sulfoxide, a facile carbon-sulfur bond fission is crucial since the cleavage does not take place in thio acetals (model compounds) derived from HCHO, MeCHO and PhCHO.[2]

Remarks

Sporidesmins are toxic metabolites of *Pithomyces chartarum*, which cause a serious disease in sheep, known as "facial eczema" in New Zealand. Successful isolation and structure determination of seven different sporidesmins, A through G, were carried out by Taylor and his coworkers.[3] Sporidesmins have an unusual structure containing a bridged S–S bond on a diketopiperazine ring. Glyotoxins have the same structural moiety. Dehydrogliotoxin **16**, which is produced by *Penicillium terlikowskii*, was also synthesized by the same workers.[4]

dehydrogliotoxin **16**

---

2) Y. Kishi, T. Fukuyama, S. Nakatsuka, *J. Am. Chem. Soc.*, **95**, 6490 (1973).
3) S. Safe, A. Taylor, *J. Chem. Soc.*, P. I, 472 (1972), and earlier papers.
4) Y. Kishi, T. Fukuyama, S. Nakatsuka, *J. Am. Chem. Soc.*, **95**, 6492 (1973).

11.18 STRUCTURE AND SYNTHESIS OF CYCLOPENIN

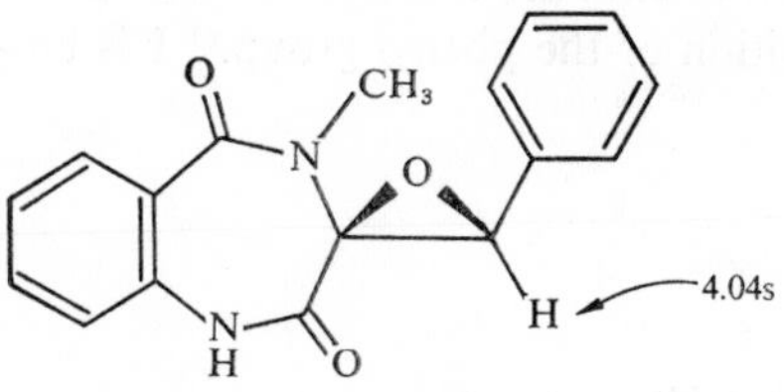

cyclopenin **1**

1) Cyclopenin **1** readily rearranges to **2**, having a different skeleton.[1] The correct structure **1**[2] was finally deduced from **3**: radioactive **3** was obtained from radioactive **1** prepared by *in vivo* incorporation on anthranilic acid (^{14}COOH).

2) An eight-membered epoxydiamide and a seven-membered ketodiamide structure cannot be ruled out completely by the degradation studies. However, final confirmation of the structure **1** was obtained by total synthesis.[3]

3) The stereochemistry was deduced from the nmr of **6**.[3]

4) The absolute configuration is unknown.

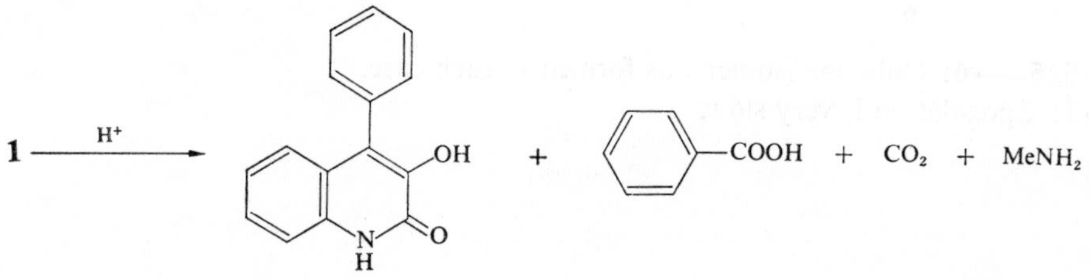

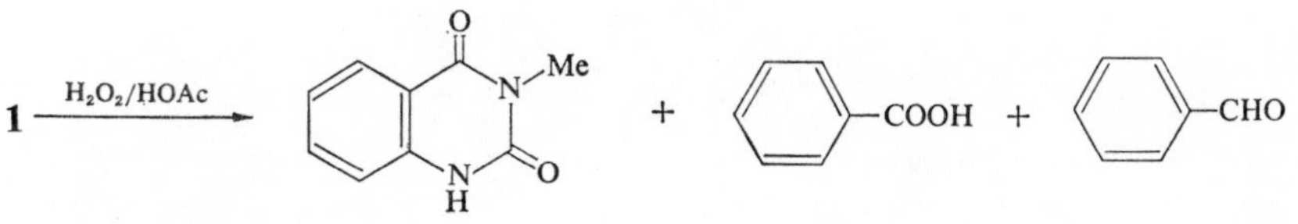

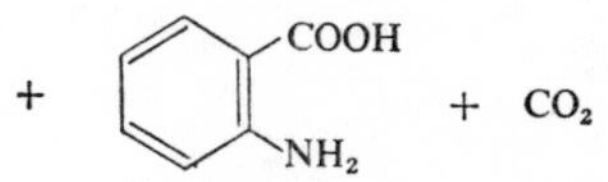

1) Y. S. Mohammed, M. Luckner, *Tetr. Lett.*, 1953 (1963).
2) M. Luckner, Y. S. Mohammed, *Tetr. Lett.*, 1987 (1964).
3) H. Smith, P. Wegfahrt, H. Rapoport, *J. Am. Chem. Soc.*, **90**, 1668 (1968); P. K. Martin, H. Rapoport, H. W. Smith, J. L. Wong, *J. Org. Chem.*, **34**, 1359 (1969).

Remarks

Cyclopenin **1** is a mould metabolite obtained from *Penicillium cyclopium*.[4] Viridicatin **2** is also obtained from the same mould.[5] Cyclopenol and viridicatol are analogs of **1** and **2**, respectively, having an extra hydroxyl group on a *meta* position of the phenyl group.[6] **1** is biosynthesized from anthranilic acid and phenylalanine.[7]

Synthesis

4⟶**5, 5**⟶**6**: Only one isomer was formed in each case.

6⟶**1**: Epoxidation is very slow.

---

4) A. Bracken, A. Pocker, H. Raistrick, *Biochem. J.*, **57**, 587 (1954).

5) M. Luckner, K. Mothes, *Tetr. Lett.*, 1035 (1962).

6) H. Birkinshaw, M. Luckner, Y. S. Mohammed, K. Mothes, C. E. Stickings, *Biochem. J.*, **89**, 196 (1963).

7) M. Luckner, *Eur. J. Biochem.*, **2**, 74 (1967).

11.19 STRUCTURE OF TENUAZONIC ACID

tenuazonic acid **1**[1] (mixture of tautomers **1a** and **1b**)

liquid pK_a: 3.35 (acid)
α_D (CHCl$_3$, c=0.2): −132°
uv (EtOH): 217 (3.72), 277 (4.13)
nmr in CDCl$_3$[2]

1) **1** gives an orange-red color with Fe^{3+} and a green complex with Cu^{2+}.

2) The formation of a mono-2,4-dinitrophenylhydrazone and a positive iodoform test indicate the presence of a –COMe group.

3) The stronger acidity of **1** than **2**, and the formation of metal complexes suggest the presence of –COMe at the 3-position.

L-isoleucine **4**

Remarks

Tenuazonic acid **1** was isolated from *Alternaria tenuis*.[1,2] It has been shown that **1** inhibits the growth of human adenocarcinoma-1 in embryonated eggs.

1 has been synthesized from L-isoleucine[3] by treatment with diketene, followed by methylation with diazomethane and cyclization in the presence of sodium ethoxide.

1) C. E. Stickings, *Biochem. J.*, **72**, 332 (1959).
2) E. A. Kaczka, C. O. Gitterman, E. L. Dulaney, M. C. Smith, D. Hendlin, H. B. Woodruff, K. Folkers, *Biochem. Biophys. Res. Commun.*, **14**, 54 (1964).
3) S. A. Harris, L. V. Fisher, K. Folkers, *J. Med. Chem.*, **8**, 478 (1965).

11.20   STRUCTURE OF ERYTHROSKYRINE

erythroskyrine **1**[1]

mp: 133°
uv (EtOH): 409 (4.45), 260 (3.95)
uv (0.1 N NaOH): 392 (4.78), 260 (4.14)

1) The presence of a polyene system is clear from the blue-violet color reaction with conc. H_2SO_4, from ir bands at 1584, 1550 and 1010 cm^{-1} which disappeared in the ir of **2**, and from the 10 H signal at 6.3–7.1 in the nmr, which was not present in **2**.

2) Proton nmr data for protons at C-20 to C-26 were assigned by the spin-decoupling technique.

3) The uv and ir spectra of **2** are quite similar to those of tenuazonic acid **8**, indicating the presence of the cyclic amide.

4) Compound **6** shows an ms peak at m/e 101 attributable to **7**, and consumes one mole of HIO_4.

1) J. Shoji, S. Shibata, *Chem. Ind.*, 419 (1964); J. Shoji, S. Shibata, U. Sankawa, H. Taguchi, Y. Shibanuma, *Chem. Pharm. Bull.*, **13**, 1240 (1965).
2) R. H. Howard, H. Raistrick, *Biochem. J.*, **56**, 216 (1954).
3) S. Shibata, U. Sankawa, H. Taguchi, K. Yamasaki, *Chem. Pharm. Bull.*, **14**, 474 (1966).

7

tenuazonic acid **8**

Remarks

Erythroskyrine **1** is an orange-red pigment first isolated from *Penicillium islandicum*.[2] It was shown that **1** is biosynthesized from acetate (one mole), malonate (nine moles) and valine (one mole).[3]

11.21 STRUCTURE OF IKARUGAMYCIN

ikarugamycin **1**

α_D: $+390°$ pK_a', (67% EtOH): 5.6
uv (MeOH): 227 (ε 20,700), 327 (ε 17,300)
uv (0.1N NaOH/MeOH): 243 (ε 21,400), 321 (ε 13,300)
nmr in py-d$_6$

1) The β-diketone system in **1** is suggested from pK_a, uv, positive FeCl$_3$ test, and formation of a Cu salt.[1]

2) The five-membered amide system is suggested from the amide ir band at 1695 cm^{-1} in the deoxydecahydro derivative of **1**, obtained by catalytic hydrogenation of **1** followed by reduction with LiBH$_4$.[1]

3) The configuration of the double bonds was assigned from nmr data.

4) The presence of a CH$_2$ group between C-6 and the *cis* double bond is suggested from the nmr data.

5) Structure **2** is assumed form the combination of **9** and **11**.[2]

6) The relative stereochemistry of the asymmetric carbon atoms in the two five-membered carbocyclic rings was determined by synthesis of racemic **8** and **11**.[3]

7) The absolute configuration was determined by synthesis[3] of (+)-**10** from·(+)-erythro-2-ethyl-3-methylsuccinic acid,[4] whose absolute configuration was determined as 2R, 3S, and from **13**, whose absolute configuration was already determined as R.[5]

1) S. Ito, Y. Hirata, *Tetr. Lett.*, 1181 (1972).
2) S. Ito, Y. Hirata, *Tetr. Lett.*, 1185 (1972).
3) S. Ito, Y. Hirata, *Tetr. Lett.*, 2557 (1972).
4) H. Brockmann Jr., D. Müller-Enoch, *Chem. Ber.*, **104**, 3704 (1971).
5) K. Freudenberg, J. Geiger, *Ann. Chem.*, **575**, 145 (1952).
6) K. Jomon, Y. Kuroda, M. Ajisaka, H. Sakai, *J. Antibiotics*, **25**, 271 (1972).

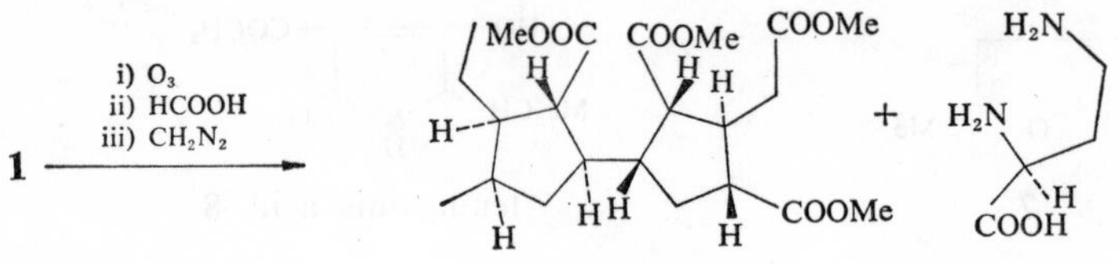

i) O$_3$
ii) HCOOH
iii) CH$_2$N$_2$

1 ⟶

MeOOC   COOMe   COOMe

+   H$_2$N ... H$_2$N ... COOH

2   L-ornithine **3**
(after hydrolysis)

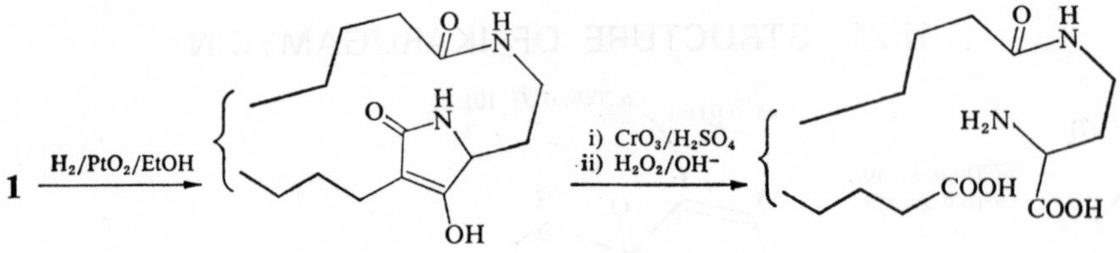

1   H$_2$/PtO$_2$/EtOH ⟶   OH   **4**   i) CrO$_3$/H$_2$SO$_4$ ii) H$_2$O$_2$/OH⁻   H$_2$N ... COOH COOH   **5**

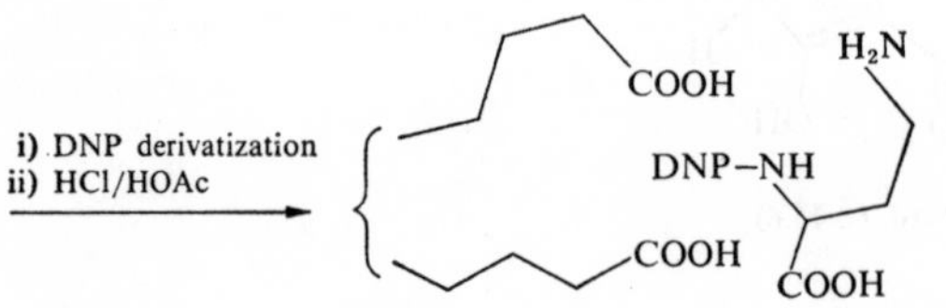

i) DNP derivatization
ii) HCl/HOAc ⟶

COOH   H$_2$N

DNP–NH

COOH   COOH

6   **7**

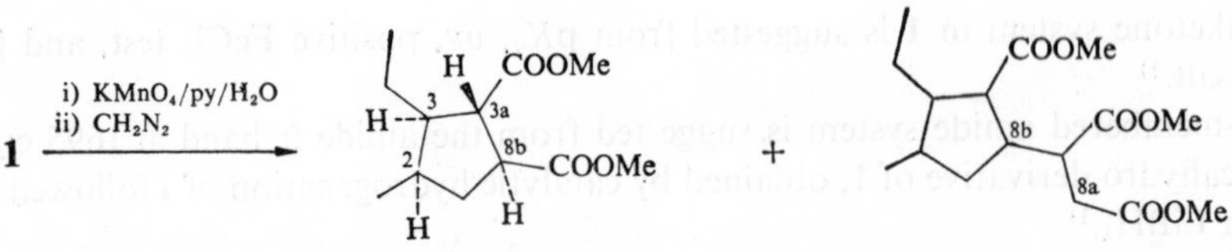

i) KMnO$_4$/py/H$_2$O
ii) CH$_2$N$_2$

1 ⟶

H   COOMe

3   3a

2   8b   8a   COOMe

8   **9**

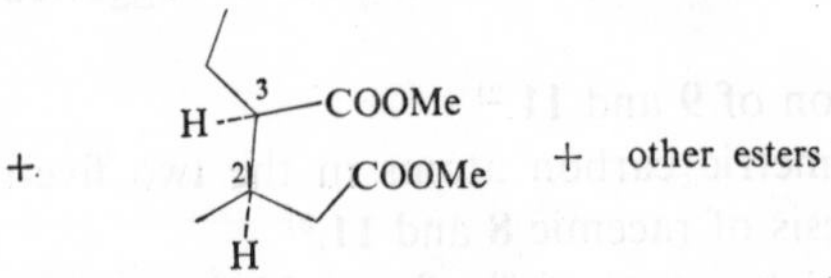

COOMe

COOMe

COOMe

H   3   COOMe

2   COOMe   + other esters

10

i) CrO$_3$/6 N H$_2$SO$_4$
ii) CH$_2$N$_2$

1 ⟶

COOMe COOMe

5a   6

8a   8b   7

MeOOC   COOMe   + other esters

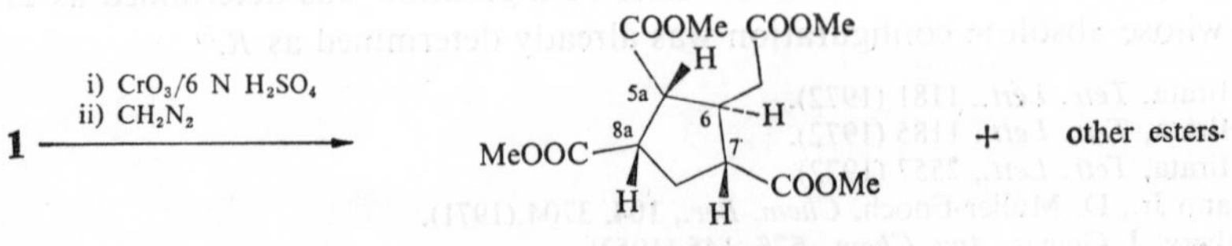

11

i) CrO_3/6 N H_2SO_4
ii) CH_2N_2

1 $\xrightarrow{H_2/PtO_2/EtOH}$ hexahydro **1** $\longrightarrow$ ... + other esters

12　　　　　　　　　　　　　　　　　　**13**

Remarks

Ikarugamycin **1** was isolated from *Streptomyces phoeochromogenes* var. *ikaruganensis*. It is a new antibiotic compound with specific antiprotozoal activity.[6] The unique structure was suggested to be constructed from two hexa-acetate units and L-ornithine, by a process involving intramolecular Diels-Alder reaction to form the *trans-anti-cis* decahydro-*as*-indacene skeleton.[3]

11.22 STRUCTURE OF STREPTOLYDIGIN

streptolydigin **1**[1]

mp: 148°　　α_D (CHCl$_3$) −93°
pK_a (65% aq. MeOH): 5.3
uv (0.01 N H_2SO_4): 357 ($E^{1\%}_{1cm}$ 591), 370 (560)
uv (0.01 N NaOH): 262 (227), 291 (273), 335 (333)

1) The spectral behavior of octahydro **1** is nearly identical to that of the chromophore **2**.
2) Hydrolysis in the presence of dinitrophenylhydrazine gave L-rhodinose **3** as the phenylhydrazone.

2　　　　　　　　　　　**3**

1) K. L. Rinehart Jr., J. R. Beck, W. W. Epstein, L. D. Spicer, *J. Am. Chem. Soc.*, **85**, 4035 (1963); K. L. Rinehart Jr., D. B. Borders, *J. Am. Chem. Soc.*, **85**, 4037 (1963); K. L. Rinehart Jr., J. R. Beck, D. B. Borders, T. H. Kinstle, D. Kraus, *J. Am. Chem. Soc.*, **85**, 4038 (1963).

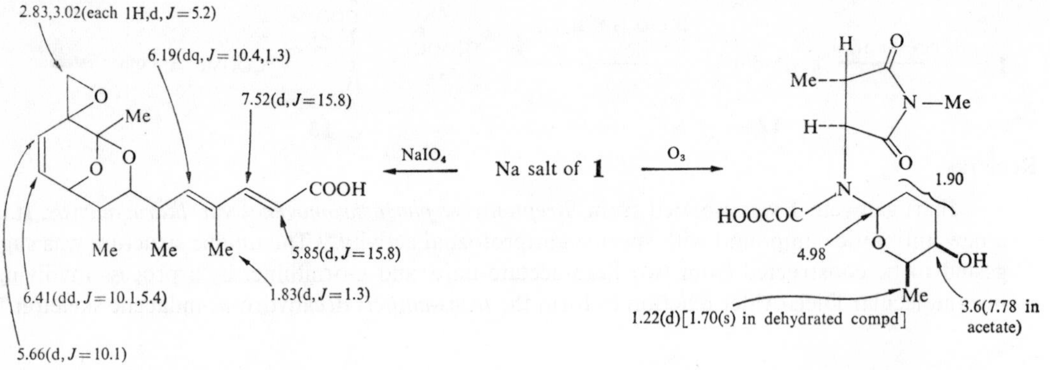

2.83,3.02(each 1H,d, $J=5.2$)

6.19(dq, $J=10.4,1.3$)

7.52(d, $J=15.8$)

NaIO₄ ← Na salt of **1** → O₃

Me---C---COOH

H---C---COOH

5.85(d, $J=15.8$)

6.41(dd, $J=10.1,5.4$)

1.83(d, $J=1.3$)

5.66(d, $J=10.1$)

streptolic acid **4**

H   O

Me--

H--

N —Me

O

1.90

HOOCOC

4.98

O

OH

Me

3.6(7.78 in acetate)

1.22(d)[1.70(s) in dehydrated compd]

ydiginic acid **5**

4N NaOH
(−MeNH₂)
→

H

Me---C---COOH

H---C---COOH

NH₂

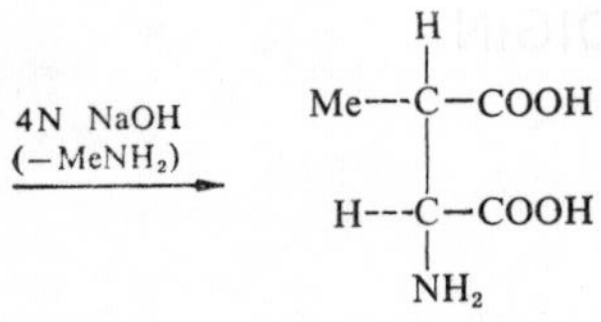

L-threo-β-methyl-aspartic acid

Remarks[2]

Streptolydigin, which was isolated from *Streptomyces lydicus*, is a new antibiotic active against gram-positive bacteria other than micrococci.

---

2) T. E. Eble, C. M. Larg, W. H. DeVries, G. F. Crum, J. W. Shnell, *Antibiotic Ann.*, 893 (1955–56).

11.23 STRUCTURE OF TETRODOTOXIN

tetrodotoxin $\mathbf{1}^{1-3)}$

1) Permanganate oxidation of **1** gave guanidine.[2a,3a]

2) Formation of the 2-aminoquinazoline derivatives **2**[4] and **3**[5] suggests the presence of a perhydroquinazoline nucleus in **1**.

2 ←— aq. NaOH/Δ —— **1** —— conc. H_2SO_4 —→ 3

3) The structures of hydrobromide **4**,[1a] hydrobromide **8**[6] and hydrochloride **9**[3a] were determined by x-ray analysis. The structure of **4** was also suggested chemically from periodate oxidations.[2e,7]

4) The configuration at the C-9 position in **4** differs from that in **7** and **9**. Examination of the nmr spectrum of **4** prepared in D_2O solution showed that epimerization at the 9 position of **4** occurred during the reaction.[1d,2c,7]

5) The structure **1** was suggested for tetrodotoxin on the basis of the structures **4**,[1] **8**[2] and **9**,[3] together with the following data: **1** is not a lactone since it shows no C=O absorption in the ir spectrum (KBr disc) whereas **9** and amorphous **1** hydrochloride show a C=O band

1) (a) K. Tsuda, C. Tamura, R. Tachikawa, K. Sakai, O. Amakasu, M. Kawamura, S. Ikuma, *Chem. Pharm. Bull.*, **11**, 1473 (1963); (b) *ibid.*, **12**, 634 (1964); (c) K. Tsuda, R. Tachikawa, C. Tamura, O. Amakasu, M. Kawamura, S. Ikuma, *ibid.*, **12**, 642 (1964); (d) K. Tsuda, S. Ikuma, M. Kawamura, R. Tachikawa, K. Sakai, O. Amakasu, C. Tamura, *ibid.*, **12**, 1357 (1964).

2) (a) T. Goto, Y. Kishi, S. Takahashi, Y. Hirata, *Tetr. Lett.*, 2105 (1963); (b) *ibid.*, 2115 (1963); (c) *ibid.*, 779 (1964); (d) T. Goto, S. Takahashi, Y. Kishi, Y. Hirata, *ibid.*, 1831 (1964); (e) T. Goto, Y. Kishi, S. Takahashi, Y. Hirata, *Tetr.*, **21**, 2059 (1965); (f) T. Goto, Y. Kishi, S. Takahashi, Y. Hirata, *Nippon Kagaku Zasshi*, **85**, 661 (1964); (g) *ibid.*, 667 (1964).

3) (a) R. B. Woodward, *Pure Appl. Chem.*, **9**, 49 (1964); (b) R. B. Woodward, J. Z. Gougoutas, *J. Am. Chem. Soc.*, **86**, 5030 (1964).

4) M. Kawamura, *Chem. Pharm. Bull.*, **8**, 262 (1960); K. Tsuda, S. Ikuma, M. Kawamura, R. Tachikawa, I. Baba, T. Miyadera, *ibid.*, **10**, 247, 856, 865 (1962); T. Goto, Y. Kishi, Y. Hirata, *Bull. Chem. Soc. Japan*, **35**, 1045 (1962).

5) T. Goto, Y. Kishi, Y. Hirata, *Bull. Chem. Soc. Japan*, **35**, 1244 (1962); Y. Kishi, H. Taguchi, T. Goto, Y. Hirata, *Nippon Kagaku Zasshi*, **85**, 564 (1964),

6) Y. Tomiie, A. Furusaki, K. Kasami, N. Yasuoka, K. Miyake, M. Haisa, I. Nitta, *Tetr. Lett.*, 2101 (1963); A. Furusaki, Y. Tomiie, I. Nitta, *Bull. Chem. Soc. Japan*, **43**, 3325 (1970).

7) Y. Kishi, T. Goto, Y. Hirata, *Nippon Kagaku Zasshi*, **85**, 572 (1964).

at 1750 cm^{-1}; the solvent effect on the pK_a (8.8 in water, 9.4 in 50% EtOH) of **1** indicates that dissociation occurs from an acidic group (OH) and not from the guanidine group; the nmr of **1** shows $J_{4,4a} = 10$ Hz, indicating that 4-H and 4a-H are *trans*-diaxially oriented; finally, there is no measurable uv absorption.

6) **1** and **10** are interconvertible on treatment with acid.[1,2]
7) The **11** hydroiodide in the anhydro series was analyzed by x-ray crystallography.[1]
8) Compound **4** and the anhydro series all show $J_{4,4a} = ca.$ 0 Hz in the nmr.[1,2]
9) Compound **21** is obtained from each of the acetates in the anhydro series on treatment with conc. ammonia. $J_{4,4a} = 10$ Hz in the nmr indicates that **21** belongs to the normal series.[1c,2d,3b]
10) The absolute configuration of **1** was determined by x-ray analysis.[6]

$$1 \xrightarrow[\text{ii) MeOH}]{\text{i) py/Ac}_2\text{O}} 11 \xrightarrow{\text{dil. OH}^-} 10 \xrightarrow{5\% \text{ HCl}} 1$$

$$1 \underset{5\% \text{ HCl}}{\overset{\text{i) HCOOH}}{\underset{\longleftarrow}{\xrightarrow{\text{ii) MeOH}}}}} 12$$

$$1 \xrightarrow{\text{Ac}_2\text{O}/p\text{-TSA}} 13 \xrightarrow{\text{Ac}_2\text{O/py}} 14 \xrightarrow{\text{aq. NH}_3} 10$$

$$1 \xrightarrow{\text{HCl/acetone}} 15 \xrightarrow{\text{Me}_2\text{SO}_4/\text{MeI}} 16$$

$$1 \xrightarrow{\text{Ac}_2\text{O/py}} 17 + 18$$

$$1 \underset{5\% \text{ HCl}}{\overset{\text{ROH/HCl}}{\underset{\longleftarrow}{\longrightarrow}}} 19 \quad (\text{R}=\text{Me or Et})$$

$$1 \xrightarrow{\text{H}_2/\text{PtO}_2/\text{HCl}} 20 \xleftarrow{\text{H}_2/\text{PtO}_2/\text{HCl}} 10$$

$$\mathbf{10, 11, 12, 13, 14, 19} \xrightarrow{\text{NH}_3} 21 \xrightarrow{10\% \text{ HCl}} 1$$

normal series ($J_{4,4a}=10$ Hz)
(R=OH) (R=OMe or OEt)[1]
(R=H)[1] (R=NH$_2$)[1,2]

anhydroepi series ($J_{4,4a}=0$ Hz)

Compound (anion)	R$_1$	R$_2$	R$_3$	R$_4$	R$_5$	R$_6$	Ref.
10	H	H	H	H	H	—*	1,2
11	Ac	Ac	H	H	H	—	1,2
12 (HCOO⁻)	CHO	H	H	H	H	H	1
13 (p-TsO⁻)	Ac	Ac	Ac	H	H	Ac	2
14 (p-TsO⁻)	Ac	Ac	Ac	Ac	H	Ac	2
15	>CMe$_2$	>CMe$_2$	H	H	H	—	3
16 (MeSO$_4$⁻)	>CMe$_2$	>CMe$_2$	H	H	H	Me	3
17	Ac	Ac	Ac	Ac	Ac	Ac	3

*(—=negative charge)

$$18^{2g,3)}$$

Remarks

Tetrodotoxin is found in *Spheroides rubripes* and other *Spheroides,* and in *Taricha torosa.* It is one of the most toxic of low molecular weight poisons, being found in the ovaries and liver of puffer fish. Poisoning due to the toxin has long been a serious problem in Japan, where the puffer fish is highly prized as a food item.

Since the toxin is only soluble in acids, its purification was difficult,[8] but pure toxin was eventually obtained via the picrate.[8d] Determination of the molecular formula of the toxin was also extremely difficult, and the correct composition was determined only at a fairly late stage in the investigation. Several derivatives of the toxin as well as tetrodotoxin hydrobromide itself[9] were subjected to x-ray analysis in attempts to determine the structure. Dimeric forms of **21** and **1** have been suggested[1c] but evidence supporting the monomeric formula has accumulated.[2,3b,9]

Tarichatoxin, obtained from California salamander, *Taricha torosa,* is identical with **1.**[10]

8) (a) A. Yoboo, *Nippon Kagaku Zasshi,* **71,** 590 (1950); (b) K. Tsuda, M. Kawamura, *Yakugaku Zasshi,* **72,** 187, 771 (1952); (c) Y. Hirata, H. Kakisawa, Y. Okumura, *Nippon Kagaku Zasshi,* **80,** 1483 (1959); (d) T Goto, S. Takahashi, Y. Kishi, Y. Hirata, *ibid.,* **85,** 508 (1964).
9) A. Furusaki, Y. Tomiie, I. Nitta, *Bull. Chem. Soc. Japan,* **43,** 3332 (1970).
10) H. S. Mosher, F. A. Fuhrman, H. D. Buchwald, H. G. Fischer, *Science,* **144,** 1100 (1964).

11.24 SYNTHESIS OF TETRODOTOXIN

tetrodotoxin **1**[1)]

1) Y. Kishi, F. Nakatsubo, M. Aratani, T. Goto, S. Inoue, H. Kakoi, S. Sugiura, *Tetr. Lett.*, 5127 (1970); Y. Kishi, F. Nakatsubo, M. Aratani, T. Goto, S. Inoue, H. Kakoi, *ibid.*, 5129 (1970); Y. Kishi, M. Aratani, T. Fukuyama, F. Nakatsubo, T. Goto, S. Inoue, H. Tanino, S. Sugiura, H. Kakoi, *J. Am. Chem. Soc.*, **94**, 9217 (1972); Y. Kishi, T. Fukuyama, M. Aratani, F. Nakatsubo, T. Goto, S. Inoue, H. Tanino, S. Sugiura, H. Kakoi, *ibid.*, **94**, 9129 (1972).

12 $\xrightarrow{\Delta}$ **13** $\xrightarrow{m\text{-Cl-PBA/CH}_2\text{Cl}_2/\text{K}_2\text{CO}_3}$ **14**

(12: EtO, EtO, AcHN, OAc, CH$_2$OAc, H, H, O)

(13: EtO, AcHN, O, H)

(14: EtO, AcHN, OAc, CH$_2$OAc, H, H, O, O)

$\xrightarrow[\text{70\% from 11}]{\text{AcOH}}$ **15** $\xrightarrow[\text{quant.}]{m\text{-Cl-PBA/CH}_2\text{Cl}_2}$ **16** $\xrightarrow[\text{quant.}]{\substack{\text{i) KOAc/HOAc} \\ \text{ii) Ac}_2\text{O/CSA}}}$ **17**

(15: AcO, NHAc, O, H)

(16: AcO, OAc, AcNH, CH$_2$OAc, O, H, H, H)

(17: AcO, H, OAc, CH$_2$OAc, AcNH, AcO, AcO, H, O, H, O)

$\xrightarrow[\text{80\%}]{\Delta}$ **18** $\xrightarrow[\text{70\%}]{\substack{\text{i) OsO}_4\text{/THF/py} \\ \text{ii) acetone/CSA}}}$ **19** $\xrightarrow[\text{93\%}]{\substack{\text{i) Et}_3\text{O}^+\text{BF}_4^-\text{/} \\ \text{CH}_2\text{Cl}_2\text{/Na}_2\text{CO}_3 \\ \text{ii) aq. HOAc}}}$ **20**

(18: O, H, H)

(19: O, O, OAc, CH$_2$OAc, AcNH, AcO, AcO, H, O, H, H, H)

(20: O, O, OAc, CH$_2$OAc, NH$_2$, AcO, AcO, H, O, H, H, H)

$\xrightarrow{\text{NCBr/NaHCO}_3}$ **21** NCNH— $\xrightarrow{\text{H}_2\text{S}/\Delta}$ **22** NH$_2$CNH— (∥S) $\xrightarrow[]{\substack{\text{i) Et}_3\text{O}^+\text{BF}_4^-\text{/CH}_2\text{Cl}_2 \\ \text{ii) Ac}_2\text{O/py}}}$ **23** AcNHC=N— (| SEt) $\xrightarrow[\text{50\% from 20}]{\text{MeCONH}_2/\Delta}$ **24** AcHNC=N— (| NHAc)

(EtS)$_2$C=NAc (alternative route) → **20**

$\xrightarrow[\text{60\%}]{\text{BF}_3\text{/CF}_3\text{COOH/CH}_2\text{Cl}_2}$ **25** $\xrightarrow[\text{15\%}]{\substack{\text{i) aq. CF}_3\text{COOH} \\ \text{ii) HIO}_4\text{/aq. MeOH} \\ \text{iii) NH}_4\text{OH}}}$ **1**

(25: HO, HO, OAc, CH$_2$OAc, AcNHC=N, AcHN, AcO, H, O, H, O, H)

3 $\longrightarrow$ **4**: Without the Lewis acid the yield of this Diels-Alder reaction was only 30%.

9 $\longrightarrow$ **10**: Epoxidation of the olefin, which has exceptionally poor reactivity, can be effected at elevated temperature in the presence of a radical inhibitor such as 4,4′-thiobis-(6-t-butyl-3-methylphenol), which prevents radical decomposition of the peracid at high temperature.[2]

10 $\longrightarrow$ **11**: The acetyl group at the 11 position was hydrolyzed with CF$_3$COOH.

13 $\longrightarrow$ **14**: The epoxide is extremely acid-labile and in the absence of K$_2$CO$_3$, the C-10 ketone having an m-Cl-PhCOO group at C-9 was produced.

2) Y. Kishi, M. Aratani, H. Tanino, T. Fukuyama, T. Goto, S. Inoue, S. Sugiura, H. Kakoi, *Chem. Commun.*, 64 (1972).

14 ⟶ 15: The reaction can be considered to take the following course.

26   **27**   **28**   **15**

15 ⟶ 16: Extremely high selectivity of the migrating bond could be attributed to the effect of the ether linkage at the 12-position.

20 ⟶ 21: Usual methods to convert amines to the corresponding guanidines were not applicable because the lactone group, as well as the acyl group, is exceptionally labile to bases.

Remarks

Tetrodotoxin, which is found in puffer fish, *Spheroides*, is one of the most toxic poisons having a low molecular weight. This elegant total synthesis was accomplished in a fully stereospecific manner. The following alternate route for the last few steps was also developed.[1]

29   **30**

11.25 STRUCTURE OF SAXITOXIN

4.27(q, $J=11,9$)
4.05(q, $J=11,5$)
3.87(dq, $J=9,5,1$)
2.37(m)
4.77(d, $J=1$)
3.85(d, $J=10$)
3.57(d, $J=10$)

saxitoxin $1^{1)}$ (incorrect)

saxitoxin **1b** (correct)

1) The coupling constant $J_{4,6}=1$ Hz indicates *cis* ring fusion.

2) The formation of 140 mole % of total guanidine (**2**+**3**+**4**) indicates the presence of two guanidine residues and no N–N bonding.

3) The N–C–N–CH$_2$CH$_2$–C–R (R=alkyl or aryl but not an ester or amide) group must be present since **5** contains two deuterium atoms, whereas **6** does not, and since the mild reduction of **1** afforded **7**, with which the CH$_2$ was no longer exchangeable with D$_2$O, and H$_2$O$_2$ oxidation no longer occurs.

partial structure **1a**

15% H$_2$O$_2$(pH4.6)

guanidine +
2 0.8 mol.

3 0.15 mol.   **4** 0.3 mol.

$H_2N-\overset{NH}{\overset{\|}{C}}-NHCH_2CH_2COOH$

D$_2$O(r.t.)

deuterium exchanged **1**
no signal at δ 2.37

H$_2$O$_2$/H$_2$O

$H_2N-\overset{NH}{\overset{\|}{C}}-NHCH_2CD_2COOH$
5

D$_2$O$_2$/D$_2$O

$H_2N-\overset{NH}{\overset{\|}{C}}-NHCH_2CH_2COOH$
6

1

H$_2$/PtO$_2$ or NaBH$_4$ dihydro-**1**

7

1) J. C. Wong, M. S. Brown, K. Matsumoto, R. Oesterlin, H. Rapoport, *J. Am. Chem. Soc.*, **93**, 4633 (1971); W. Schuett, H. Rapoport, *ibid.*, **84**, 2266 (1962); J. L. Wong, R. Oesterlin, H. Rapoport, *ibid.*, **93**, 7344 (1971).
2) E. J. Scuantz, J. M. Lynch, G. Vayvada, K. Matsumoto, H. Rapoport, *Biochemistry*, **5**, 1191 (1966).
3) E. J. Schanz, V. E. Ghararossin, H. K. Schnoes, F. M. Strong, J. P. Springer, J. O. Pezzanite, J. Clardy, *J. Am. Chem. Soc.*, **97**, 1238 (1975).
4) J. Bordner, W. E. Thiessen, H. A. Bates, H. Rapoport, *J. Am. Chem. Soc.*, **97**, 6008 (1975).

4) The structure **3** and the partial structure in **3)** suggest the partial structure **1a**, in which the propionyl residue must be attached at C-4, C-5 or C-6, (nmr analysis indicates C-5); extra nitrogen atoms must be at the NH_3 level, since no N–N or N–O bonds are present; and the extra carbon atom must be at the CO_2 oxidation level, hence –$CONH_2$.

5) Definitive placement of –$CONH_2$ follows from a pK_a study:

$pK_a(H_2O)$ 8.24 $\longrightarrow pK_a(20\% \text{ EtOH})$ 8.50 $\longrightarrow pK_a(50\% \text{ EtOH})$ 9.05

the pK_a values are attributable to $-OH \rightleftharpoons -O^- + H^+$ but not $-N^+H \rightleftharpoons -N + H^+$
and hence

6) Recent X-ray analysis[3,4] of saxitoxin derivatives has indicated that the correct structure is **1b**; the pK_a may be attributable to one of the two guanidines.[4] Considering the spatial arrangement, one guanidine could have a pK_a at least three units below the other because of their proximity.

Remarks

Paralytic shellfish poisoning is a severe form of food intoxication. The toxic principle responsible is saxitoxin, a powerful neurotoxin produced by the dinoflagellate *Gonyaulax catenella* and accumulated in some otherwise edible species of shellfish.[2]

(Note added: the structure **1** may not be correct; the C=O may be attached at C-6, and the –$CONH_2$ group as in –$OCONH_2$.)

11.26   STRUCTURE OF AEROTHIONIN

aerothionin **1**[1)]

uv (MeOH): 284 (ε 12,660)
nmr in CD_3COCD_3

1) The molecular weight of **1** was determined by osmometry.
2) **1** yields a diacetate.
3) The value of α_D (acetone) of 252° indicates that **1** is not a *meso* form, but is D or L.
4) The absolute configuration of **1** has not been determined.

1) E. Fattorusso, L. Minale, G. Sodano, *Chem. Commun.*, 752 (1970).

Remarks

Aerothionin **1** occurs in *Aplysina aerophoba* and *Verongia thiona*. It is the major bromo compound in the sponge *A. aerophoba*, and is probably biosynthesized from dibromothyrosine through a benzene epoxide intermediate. The following bromo compounds, which show antibiotic activity, have also been isolated from the above sponges.

8[2]   aeroplysinin-1 **9**[3]   aeroplysinin-2 **10**[4]

2) C. M. Sharma, P. R. Burkholder, *Tetr. Lett.*, 4147 (1967).
3) E. K. Fattorusso, L. Minale, G. Sodano, *Chem. Commun.*, 751 (1970).
4) L. Minale, G. Sodano, W. R. Chan, A. M. Chen, *Chem. Commun.*, 674 (1972).

11.27   STRUCTURE OF LEUCOGENENOL

leucogenenol **1**[1]

pK_a: 3.8 (enol)
uv: ε 9,000 at 220, ε 420 at 266
nmr in acetone-d_6

1) Acetylation of an enol methyl ether formed from **1** by reaction with CH_2N_2 yields a tetra-acetate which in turn affords bis-2,4-DNP–hydrazone, suggesting the presence of C=C–OH, four OH and two C=O groups.

2) The presence of a C=C–CH=N group is indicated by nmr (singlet at 8.1), ir (1634 cm^{-1}), uv (ϵ=9,000 at 220) and by the formation of NH_2CH_2CHO on hydrolysis.

3) **2** gives an α,β-unsaturated ketone on treatment with Ac$_2$O/NaOAc, indicating the presence of OH β to C=O.

4) COOH in **2** is equatorial on the basis of energy considerations.

5) $J_{2,3}$=4 Hz in **2** suggests a di-equatorial conformation of the two protons.

6) The *tert*-OH and Me groups in **2** are in a 1,3-diaxial relationship, since the Me signal shifts from 1.35 to 1.30 on acetylation.

7) The enolizable α-diketone in **3** consumes one mole of KOH.

8) **3** yields a dibenzoate which in turn forms a bis-2,4-DNP–hydrazone, indicating the presence of two OH and two C=O groups.

9) The *tert*-OH and Me groups in **3** are in a 1,3-diaxial relationship, since the Me signal shifts from 1.25 to 1.20 on acetylation.

1) F. A. H. Rice, *J. Chem. Soc.* C, 2599 (1971).

3-methylglutaric acid **10**

Remarks

Leucogenenol **1** is a metabolic product of *Penicillium gilmanii*.[2] When injected into animals, it produces a neutrophilia without a concurrent febrile response. It also stimulates the production of precursor cells to peripheral blood cells,[3] and in general stimulates the regeneration of myeloid and lymphoid tissues.[4] It occurs in normal bovine and human liver[5] and hence could be a compound that normally plays a role in the regulation of the number and type of blood cells in the body.

2) F. A. H. Rice, *Proc. Soc. Expt. Biol. Med.*, **123**, 189 (1966).
3) F. A. H. Rice, J. H. Darden, *J. Infectious Diseases*, **118**, 289 (1968).
4) F. A. H. Rice, J. Lepick, J. H. Darden, *Radiation Res.*, **36**, 144 (1968).
5) F. A. H. Rice, B. Shaikh, *Biochem. J.*, **116**, 709 (1970).

11.28 STRUCTURE OF DIBROMOPHAKELLIN

dibromophakellin **1**[1]

—6.88(s)

pK_a (20% methyl cellosolve): ca. 7.7
uv (MeOH): 233 (ε 8,877), 281 (ε 8,813)

1) The pK_a of **1** is abnormally low.
2) The uv spectrum suggested the presence of a pyrrole ring carrying a carbonyl group in the α position
3) **1** has three D_2O-exchangeable protons.
4) The nmr spectrum of **1** indicated the partial structure $-CH_2CH_2CH_2N-C=O$, a 1,2-disubstituted pyrrole ring and a disubstituted guanidine.

1 $\xrightarrow{\text{catalytic hydrogenation}}$

4.0~5.0(4H,D$_2$O exchangeable)
7.45(J=3 and 1.8)
6.43(J=4 and 3)
6.86(J=4 and 1.8)
6.18(s)
2.1(2H, br.q)
2.5~2.15(2H, m)
3.70(q, J=18 and 8)
3.56(q, J=18 and 8)

phakellin **2**

Remarks

Dibromophakellin was isolated from the marine sponge *Phakellia flabellata,* found on the Great Barrier Reef. The structure was confirmed by x-ray analysis of mono-acetyl-**1**. 4-Bromophakellin is also produced by the same sponge.

The following bromopyrroles are also found in the sponge, *Agelas oroides*, in the bay of Naples.[2] Oroidin **6** appears to be closely related biogenetically to dibromophakellin **1**.

3 (R=COOH)
4 (R=CN)
5 (R=CONH$_2$)

oroidin **6**[3]

1) G. M. Sharma, P. R. Burkholder, *Chem. Commun.*, 151 (1971).
2) S. Forenza, L. Minale, R. Riccio, *Chem. Commun.*, 1129 (1971).
3) E. E. Garcia, L. E. Benjamin, R. I. Fryer, *Chem. Commun.*, 78 (1973).

11.29   STRUCTURE AND SYNTHESIS OF PYRROLNITRIN

3.25 or 3.35(d)

3.25 or 3.35(d)

pyrrolnitrin **1**

1) The ir band at 3480 cm^{-1} and positive Ehrlich reaction indicate the presence of a pyrrole ring.
2) The presence of two α-protons and no β-protons in the pyrrole ring was suggested by nmr.

uv:252(ε 2,500)
ir:1530,1375(NO$_2$),1600(arom.)

Structure[1]

$\xleftarrow{\text{CrO}_3 \text{ or O}_3}$ **1** $\xrightarrow{\text{KMnO}_4}$

2

uv(EtOH):310(2870)

3

Remarks

   Pyrrolnitrin is an antibiotic isolated from bacterial cells of *Pseudomonas pyrocinia*[2] and is the first antibiotic substance discovered that contains nitro and chlorine in the molecule.

Synthesis[3]

Several other routes for the synthesis of pyrrolnitrin have been developed.[3]

1) H. Imanaka, M. Kousaka, G. Tamura, K. Arima, *J. Antibiotics* (Japan) Ser. A, **18**, 207 (1965).
2) K. Arima, H. Imanaka, M. Kousaka, A. Fukuda, G. Tamura, *J. Antibiotics* (Japan) Ser A, **18**, 201 (1965).
3) H. Nakano, S. Umio, K. Kariyone, K. Tanaka, T. Kishimoto, H. Noguchi, I. Ueda, H. Nakamura, Y. Morimoto, *Tetr. Lett.*, 737 (1966); *Yakugaku Zasshi*, **86**, 159 (1966).

11.30 STRUCTURE OF PRODIGIOSIN

3.92(s)   2.35(t)   0.84(t)

OCH₃   CH₃

N H   N H   N   CH₃

1.26(m)

1.65(s)

mp: 174–175° (as Zn salt), 238–240° (as perchlorate)
uv (EtOH): 538 (ε 110,000), 354 (ε 1,600), 290 (ε 9,700).
270 (ε 9,700) (Zn salt)

prodigiosin **1**

1 $\xrightarrow{\text{soda-lime (heat)}}$

Me

N H   Me

2

3.93(s)

OMe

N H   N H   CHO

9.43(s)

3

$\xrightarrow{\text{H}_2\text{O}_2/\text{OH}^-}$

N H   CONH₂

4

CHO   OMe

N H   N H

3b

1) The structure of **2** was confirmed by synthesis.[1]
2) A mutant strain produces **3**, which is converted to **1** by another mutant strain.[2]
3) The structure of **3** was suggested by the following evidence, but **3b** was not excluded.[3] Firstly, the α,α′-linkage of the pyrrole rings was indicated by the isolation of **4**, and secondly, the strong uv absorption indicates conjugation of both pyrrole chromophores with the formyl group; the uv also excludes a structure having the –CHO group at C–3.
4) Structure **3** was confirmed by synthesis.[4]
5) The condensation of **2** and **3** under the conditions of dipyrrylmethane synthesis gave **1**.[3]

Remarks

Prodigiosin is a red pigment of *Serratia marcescens*, a widely distributed, non-pathogenic bacterium often found in soil and water. Prodigiosin itself has considerable antibiotic and antifungal activity, but high toxicity precludes its use as a therapeutic agent.

Several prodigiosin analogs were isolated.[5] Metacycloprodigiosin **5**[6] is one of the more interesting because **5** contains an unusual *meta*-bridged pyrrole moiety.

OMe

N H   N   N H

5

1) F. Wrede, O. H. Hetche, *Chem. Ber.*, **62B**, 2678 (1929); F. Wrede, *Z. Hyg. Infektionskrankh.*, **111**, 531 (1930); *Z. Physiol. Chem.*, **210**, 125 (1932); F. Wrede, A. Rothhaas, *Z. Physiol. Chem.*, **226**, 95 (1934).
2) U. V. Santer, H. J. Vogel, *Federation Proc.*, **15**, 1131 (1956); *Biochim. Biophys. Acta*, **19**, 578 (1956).
3) H. H. Wasserman, J. E. McKeon, L. Smith, P. Forgione, *J. Am. Chem. Soc.*, **82**, 506 (1960); *Tetr. Suppl.*, **8**, 647 (1966).
4) H. Rapoport, K. G. Holden, *J. Am. Chem. Soc.*, **82**, 5510 (1960): *ibid.*, **84**, 635 (1962).
5) W. R. Hearn, R. E. Worthington, R. C. Burgus, R. P. Williams, *Biochem. Biophys. Res. Commun.*, **17**, 517 (1964), H. H. Wasserman, G. C. Rodgers, D. D. Keith, *Chem. Commun.*, 825 (1966).
6) H. H. Wasserman, G. C. Rodgers, D. D. Keith, *J. Am. Chem. Soc.*, **91**, 1263 (1969); H. H. Wasserman, D. D. Keith, J. Nadelson, *ibid.*, **91**, 1264 (1969).

11.31 SYNTHESIS OF PRODIGIOSIN

In the first step of this synthesis the formation of the unwanted **2** is rationalized by considering the initial step to this compound as involving Michael addition of a carbanion rather than of the amide anion, which produces the desired **4**.

prodigiosin **1**[1]

i) Na/xylene/benzene/Δ
ii) H_2O
iii) CH_2N_2/MeOH/Et_2O

2 (14%)   +   **3** (1%)

4 (11%)

4 $\xrightarrow[59\%]{\text{conc.}H_2SO_4}$ **5** $\xrightarrow[72\%]{200°}$ **6** $\xrightarrow[13\%]{\Delta^1\text{-pyrroline/EtOH} \ (150°)}$

7 $\xrightarrow[82\%]{5\%\text{Pd-C/}p\text{-cymene/}\Delta}$ **8** $\xrightarrow[32\%]{\substack{\text{i) } NH_2NH_2/\Delta \\ \text{ii) TsCl/py} \\ \text{iii) diethylene glycol/}Na_2CO_3(170°)}}$ **9**

9 + **10** $\xrightarrow[55\%]{\text{conc.HCl/MeOH(r.t.)}}$ **1**

4 ⟶ 5: Mild alkali hydrolyzes only the α-ester of **4**, but conc. sulfuric acid gives **5**.

6 ⟶ 7 ⟶ 8: The crucial step in this synthesis was to establish a method for the synthesis of a 2,2'-bipyrrole.

Remarks

Synthetic **9** was found to be identical with naturally occurring prodigiosin precursor[2] and therefore the structures of the precursor as well as of **1**[3] were established. Biogenetic[2] and chemical[3] methods to reconstitute **1** from **9** and **10** are known.

Synthesis of the prodigiosin analog, metacycloprodigiosin was also reported.[4]

1) H. Rapoport, K. G. Holden, *J. Am. Chem. Soc.*, **82**, 5510 (1960); *ibid.*, **84**, 635 (1962).
2) U. V. Santer, H. J. Vogel, *Federation Proc.*, **15**, 1131 (1956); *Biochim. Biophys. Acta*, **19**, 578 (1956).
3) H. H. Wasserman, J. E. McKeon, L. Smith, P. Forgione, *J. Am. Chem. Soc.*, **82**, 506 (1960); *Tetr. Suppl.*, **8**, 647 (1966).
4) H. H. Wasserman, D. D. Keith, J. Nadelson, *J. Am. Chem. Soc.*, **91**, 1264 (1969).

11.32   STRUCTURE OF CHLOROPHYLL *a*

chlorophyll *a* **1**

mp: 117–120°
nmr in $CDCl_3/CD_3OD$ (3%)

1) The cyclic tetrapyrrole structure of **1** was indicated by the porphyrin chemistry and the oxidation products **3**, **4** and **5**.[2] This was confirmed by the synthesis of **7**, and also of **6**[3] and **8**.[4]

2) It was shown that phytol is attached at the carboxyl group in **2**.[5,6]

3) The presence of a dihydropyrrole ring in **1** is clearly indicated by the formation of **5**, whose structure, including stereochemistry, was deduced.[7]

4) The position of the cyclic *β*-keto ester group was proven by the reaction sequence leading to **12**,[8,9] the structure of which was confirmed by synthesis.[10]

5) The presence of a vinyl group is clear from the formation of a cyclopropane derivative[11] **13** and its position was indicated from the structure of **15**,[12,13] which was confirmed by synthesis.

1) G. L. Closs, J. J. Katz, F. C. Pennington, H. R. Thomas, H. H. Strain, *J. Am. Chem. Soc.*, **85**, 3809 (1963).
2) H. Fischer, H. Wedenroth, *Ann. Chem.*, **537**, 170 (1939); *ibid.*, **545**, 140 (1940).
3) H. Fischer, H. Grosselfinger, G. Stangler, *Ann. Chem.*, **461**, 221 (1928); H. Fischer, H. K. Weichmann, K. Zeile, *Ann. Chem.*, **475**, 241 (1929).
4) H. Fischer, H. Helerlerger, *Ann. Chem.*, **480**, 235 (1930); H. Fischer, W. Lautsch, *Ann. Chem.*, **528**, 265 (1937).
5) J. B. Conant, J. F. Hyde, *J. Am. Chem. Soc.*, **51**, 3668 (1929).
6) J. B. Conant, E. M. Dietz, C. F. Bailey, S. E. Kamerling, *J. Am. Chem. Soc.*, **53**, 2382 (1931).
7) G. E. Ficken, R. B. Johns, R. P. Linstead, *J. Chem. Soc.*, 2272 (1956).
8) H. Fischer, O. Süs, *Ann. Chem.*, **482**, 225 (1930).
9) H. Fischer, O. Moldenhauer, O. Süs, *Ann. Chem.*, **485**, 1 (1930); *ibid.*, **486**, 107 (1931).
10) H. Fischer, H. Berg, A. Schormüller, *Ann. Chem.*, **480**, 109 (1930).
11) H. Fischer, H. Medick, *Ann. Chem.*, **517**, 245 (1935).
12) H. Fischer, J. Riedmair, *Ann. Chem.*, **505**, 87 (1933).
13) H. Fischer, W. Röse, *Ann. Chem.*, **519**, 1 (1935).

$1^{14)}$ $\xrightarrow{\text{HCl}}$ [structure] $+$ Mg^{2+} $+$ phytol

pheophorbide *a* **2**

2 $\xrightarrow{\text{CrO}_3}$ [structure] **3** $+$ [structure] **4** $+$ [structure] **5**

$1^{14)}$ $\xrightarrow{\text{degradation}}$ [structure]

	R_1	R_2	R_3
6	COOH	H	CH_2CH_2COOH
7	H	H	CH_2CH_2COOH
8	H	Me	CH_2CH_2COOH
9	H	H	CH_2Me

2 $\xrightarrow{\text{HI/HOAc}}$ [structure] **10** $\underset{}{\overset{\text{OH}^-}{\rightleftarrows}}$ [structure] **11**

14) R. Willstätter., A. Stoll, *Untersuchungen uber Chlorophyll*, Springer-Verlag, 1931.

rhodoporphyrin **12**

14   **15**

Stereochemistry and absolute configuration

1) The *trans* configuration of –CH₂CH₂COO–phytyl and –Me, and the absolute configuration of **1** were established by the synthesis of **16** from α-santonin **18**.[15]

2) The relative configuration of the –COOMe and –CH₂CH₂COO–phytyl groups in **1** was deduced to be more stable as *trans* by correlation of the stereochemistry of compound **19** to **23** with their respective cd and nmr spectra.[16]

5   **16**   **17**   α-santonin
                                                                        18

15) I. Fleming, *Nature*, **216**, 151 (1967).
16) H. Wolf, H. Brockmann Jr., H. Biere, H. H. Inhoffen, *Ann. Chem.*, **704**, 208 (1967).

19   **20** (62%)   **21** (38%)

22 (11%)   **23** (89%)

Remarks

Chlorophyll is the pigment of green vegetation, and plays an important role in the photosynthesis of carbohydrates. Higher plants usually contain chlorophylls *a* and *b* in the ratio 3:1.[17,18]

chlorophyll *b* **24**[19]   bacteriochlorophyll **25**[20]

17) A. Winterstein, K. Schön, *Z. Physiol. Chem.*, **230**, 139 (1934).
18) A. Winterstein, G. Stein, *Z. Physiol. Chem.*, **220**, 263 (1933).
19) A. Winterstein, S. Breitner, *Ann. Chem.*, **510**, 183 (1934).
20) H. Fischer, W. Lautsch, K. H. Lin, *Ann. Chem.*, **534**, 1 (1938).

11.33   SYNTHESIS OF CHLOROPHYLL *a*

chlorophyll *a* **1**

2,3 ⟶ 4: A single product is obtained.

8,9 ⟶ 10: It is difficult to form the Schiff base from **8** and the corresponding aldehyde, but the thioaldehyde **9** forms **10** readily.

10 ⟶ 11: Only phlorin 11 is formed because of the steric effects of the γ-substituent.

12 ⟶ 13: Steric hindrance of the γ-substituent causes isomerization of 12 to a phlorin having an acrylic ester side chain at the γ-position, which is then oxidized to a porphyrin 13.

13 ⟶ 14: The purpurin 14 is readily formed by release of the steric compression, and this remarkable reaction represents the first interconversion of a porphyrin and a chlorin ever observed.

$h\nu$ (vis)/O$_2$ →

i) KOH/MeOH
ii) dil.NaOH →

17

18

i) opt. resol.
ii) CH$_2$N$_2$
iii) HCN/NEt$_3$ →

i) Zn/HOAc
ii) CH$_2$N$_2$ →

19

20

HCl/MeOH →

MeONa →

i) phytylation
ii) Mg^{2+} → **1**

chlorin e_6 trimethyl ester
21

methyl phaeophorbide *a*
22

17 ⟶ 18: This reaction confirms the *trans* configuration of the H atoms at C-7 and C-8, since inversion at C-7 would involve the passage of the C-7 propionic ester side chain through a molecular plane which is very crowded due to the substituents in the γ-position. Optical resolution of **18** was carried out with quinine.

21 ⟶ 1: This transformation was previously carried out by Willstätter and Fischer[2] with the degradation products of natural chlorophyll. Optically active phytol has been synthesized.[3]

1) R. B. Woodward, *Angew. Chem.*, **72**, 651 (1960); *Pure Appl. Chem.*, **2**, 383 (1961); R. B. Woodward, W. A. Ayer, J. M. Beaton, F. Bickelhaupt, R. Bonnett, P. Buchschacher, G. L. Closs, H. Dutler, J. Hannah, F. P. Hauck, S. Ito, A. Langemann, E. Le Goff, W. Leimgruber, W. Lwowski, J. Sauer, Z. Valenta, H. Volz, *J. Am. Chem. Soc.*, **82**, 3800 (1960).

2) C. A. Stall, E. Wiedemann, *Fortschr. Chem. Forsch.*, **2**, 538 (1942); R. Willstätter, A. Stall, *Ann. Chem.*, **380** 148 (1911); H. Fischer, A. Stern, *Ann. Chem.*, **519**, 244 (1935).

3) J. W. K. Burrell, L. M. Jackman, B. C. L. Weedon, *Proc. Chem. Soc.*, 263 (1959).

11.34   SYNTHESIS OF VITAMIN B_{12}
(CYANOCOBALAMIN)

vitamin B_{12} **1**

$\alpha_{656\cdot3}: -59° \pm 9°$
uv: 278 ($E_{1cm}^{1\%}$ 115), 361 (204), 550 (64)

Vitamin B_{12} occurs in practically all animal tissues. Man requires an external supply, and deficiency of this vitamin causes pernicious anaemia. The normal requirement for the adult human may be about 1 μg a day.

The effectiveness of liver as a specific agent in the dietary treatment of pernicious anaemia was described by Minot and Murphy[1] in 1926. In 1948 it was isolated by a Merck group[2] and by a Glaxo group.[3] It is now commercially produced by fermentation, employing *Streptomyces griseus*.[4] Although the structure of the vitamin was extensively studied chemically and the nucleotide part and β-hydroxypropylamine were characterized,[5] the structure of the nucleus, which contains cobalt, was not completely solved until 1956, when the complete structure was finally elucidated by the x-ray diffraction studies of Hodgkin *et al.*[6] The vitamin has one of

1) G. R. Minot, W. P. Murphy, *J. Am. Med. Assoc*, **87**, 470 (1926).
2) E. L. Rickes, N. G. Brink, F. R. Koniuszy, T. R. Wood, K. Folkers, *Science*, **107**, 396 (1948).
3) E. L. Smith, *Nature*, **161**, 638 (1948); E. L. Smith, L. F. J. Parker, *Biochem. J.*, **43**, viii (1948).
4) E. L. Rickes, N. G. Brink, E. R. Koniuszy, T. R. Wood, K. Folkers, *Science*, **108**, 634 (1948).
5) E. Chargaff, C. Levine, C. Green, J. Kveam, *Experientia*, **6**, 229 (1950); G. Cooley, M. T. Davies, B. Ellis, V. Petrow, B. Sturgeon, *J. Pharm. Pharmacol.*, **5**, 257 (1953); K. Folkers *et al.*, *J. Am. Chem. Soc.*, **71**, 2951 (1949); *ibid.*, **72**, 1866, 2820, 4442 (1950); *ibid.*, **74**, 4521 (1953); A. R. Todd *et al.*, *J. Am. Chem. Soc.*, 2845 (1950); *ibid.*, 3061 (1953); *ibid.*, 1148, 1158, 1168 (1957).
6) D. C. Hodgkin, J. Pickworth, J. H. Robertson, N. K. Trueblood, R. J. Prosen, J. G. White, *Nature*, **176**, 325 (1955); D. C. Hodgkin, J. Kamper, M. Mackay, J. Pickworth, K. N. Trueblood, J. G. White, *ibid.*, **178**, 64 (1956).

the most complex and beautifully constructed molecular structures known in organic chemistry.

In 1960, conversion of naturally derived cobyric acid **63** to vitamin B$_{12}$ was carried out by Friedrich *et al.*[7] Then, the synthesis of cobyric acid **63** became an objective as a means for the total synthesis of the vitamin. In the cobyric acid molecule there are nine centers of chirality, each requiring correct orientation of the side chains, and thus presenting a staggering stereo-chemical problem. Moreover, cobyric acid contains chains terminating in six instances in primary amide groups, while a seventh chain terminates in a free carboxyl group. Thus the seventh chain terminus must be differentiated in some way from all the others.

In 1972 an extremely elegant and laborious total synthesis of cobyric acid, and hence of vitamin B$_{12}$ itself, was achieved by a collaborative effort of the Woodward group at Harvard and the Eschenmoser group at ETH (Route I).[8] It is a real pyramid in the field of organic synthesis. The Eschenmoser group also announced an entirely different route (Route IV)[9] for the total synthesis of cobyric acid by applying the Woodward-Hoffmann rules,[10] one of the most significant by-products of the Woodward-Eschenmoser synthesis of the vitamin (Route I).

Route I—the Woodward-Eschenmoser synthesis[8]

(Star mark circles in the formulas indicate a mixture of stereoisomers at that center.)

7) W. Friedrich, G. Gross, K. Bernhauer, P. Zeller, *Helv. Chim. Acta*, **43**, 704 (1960).
8) R. B. Woodward, *Pure Appl. Chem.*, **33**, 145–177 (1973).
9) Y. Yamada, D. Miljkovic, P. Wehrli, B. Golding, P. Löliger, R. Keese, K. Müller, A. Eschenmoser, *Angrew. Chem.*, **81**, 301 (1969); A. Eschenmoser, *Pure Appl. Chem.*, **25** (Supple.), 69 (1971).
10) R. B. Woodward, R. Hoffmann, *The Conservation of Orbital Symmetry,* Academic Press, 1970.

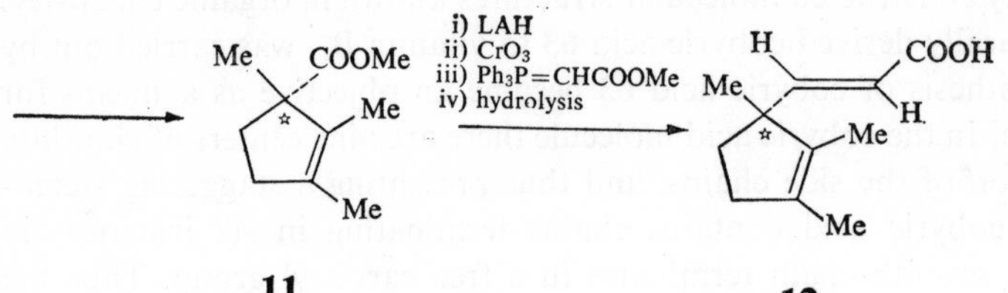

11 12

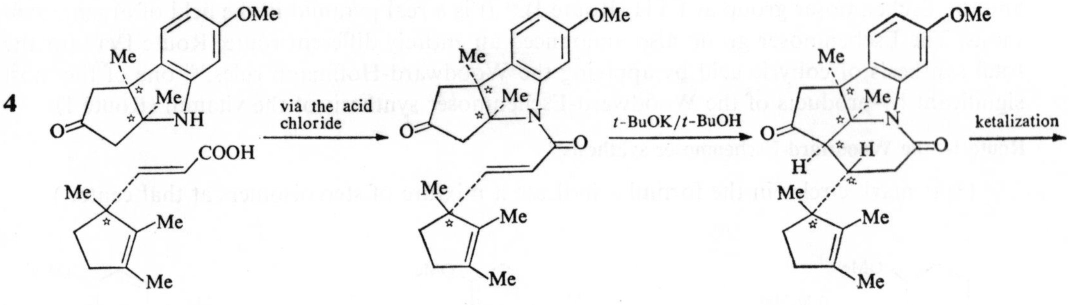

12 13 14

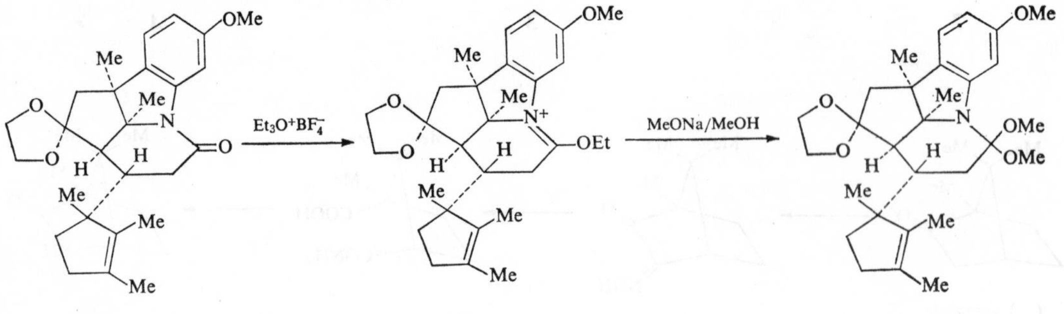

15 16 17

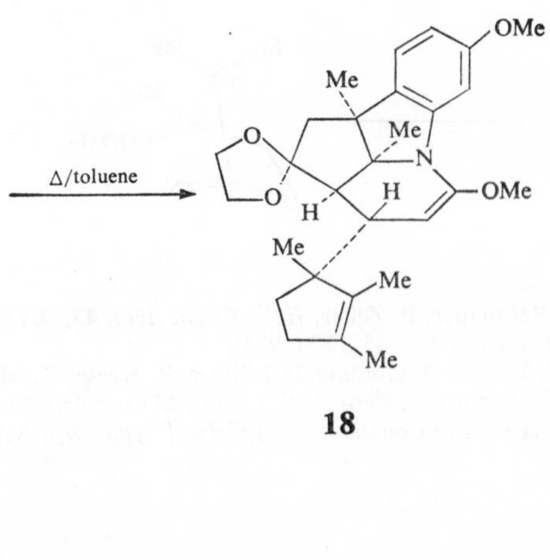

18

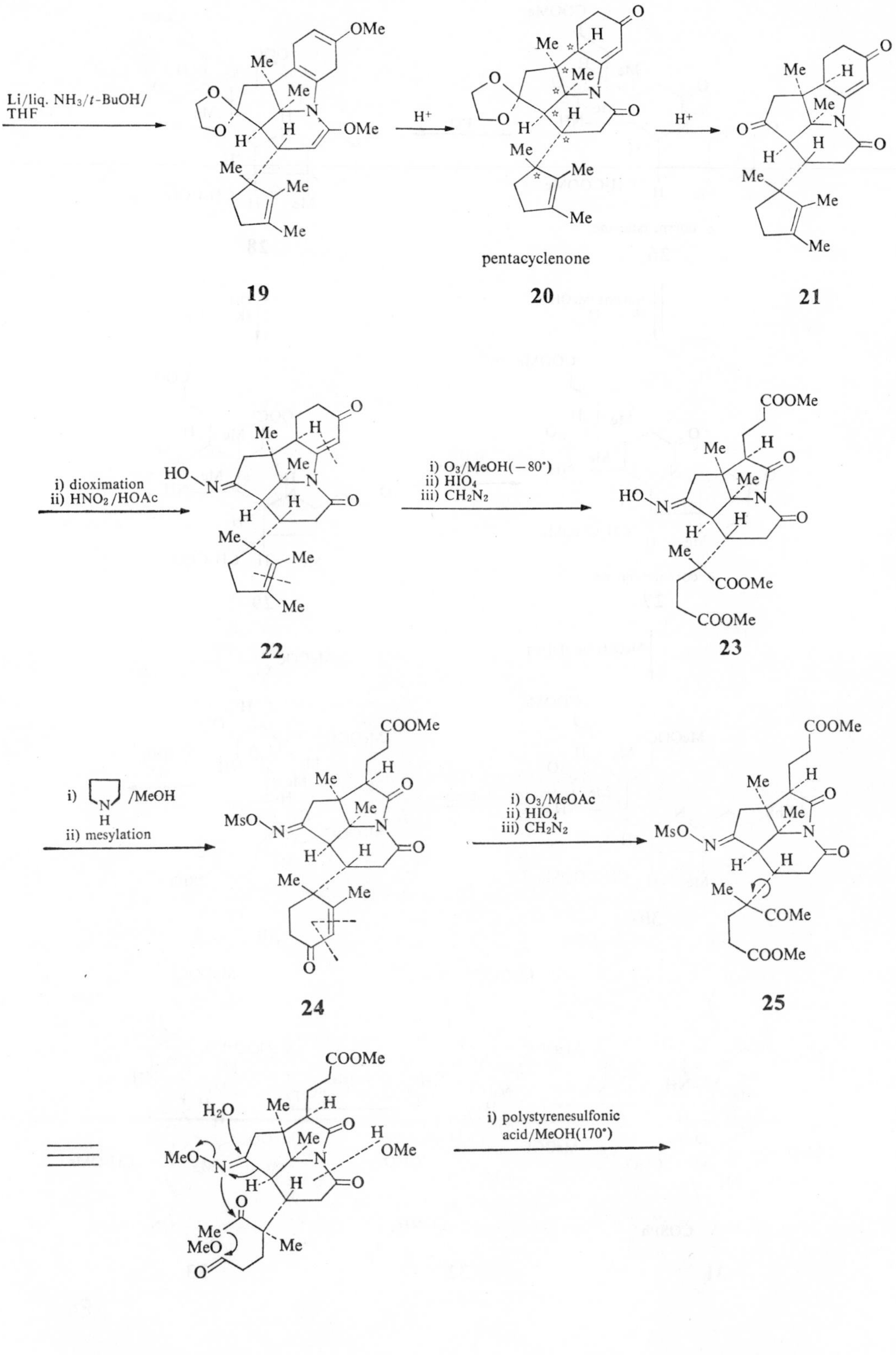

Li/liq. NH$_3$/t-BuOH/
THF

19

H$^+$

20

pentacyclenone

H$^+$

21

i) dioximation
ii) HNO$_2$/HOAc

22

i) O$_3$/MeOH($-80°$)
ii) HIO$_4$
iii) CH$_2$N$_2$

23

i) [pyrrolidine]/MeOH
ii) mesylation

24

i) O$_3$/MeOAc
ii) HIO$_4$
iii) CH$_2$N$_2$

25

i) polystyrenesulfonic
acid/MeOH(170°)

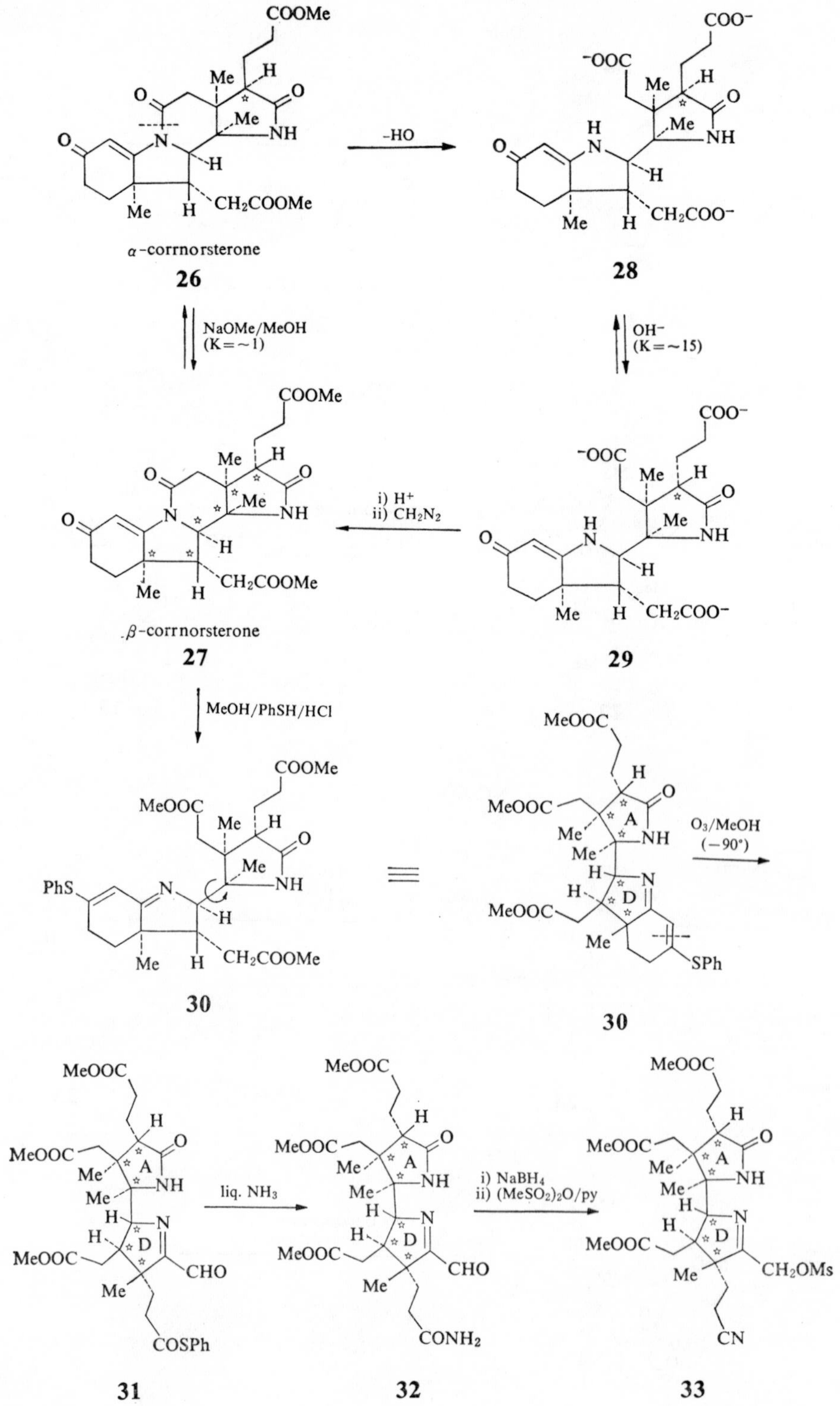
α-corrnorsterone
26
-HO
28
NaOMe/MeOH
(K=~1)
OH⁻
(K=~15)
β-corrnorsterone
27
i) H⁺
ii) CH₂N₂
29
MeOH/PhSH/HCl
30
O₃/MeOH
(−90°)
30
liq. NH₃
i) NaBH₄
ii) (MeSO₂)₂O/py
31
32
33

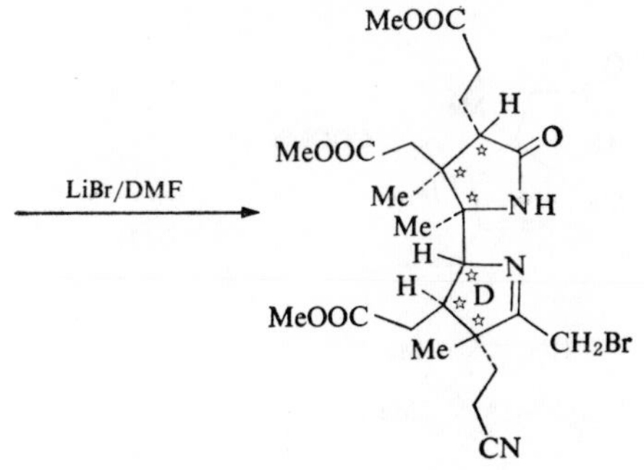

34

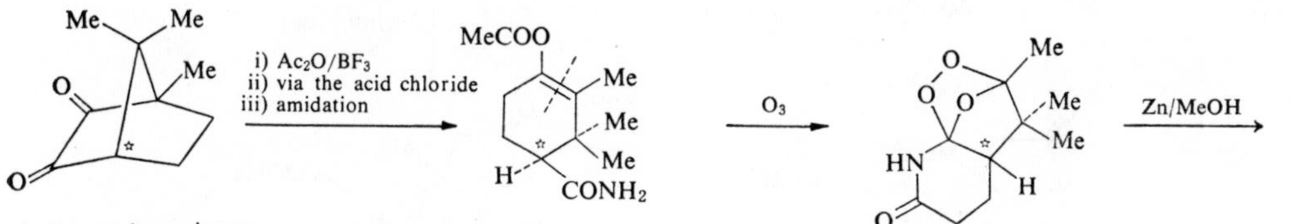

(+)-camphorquinone

35   **36**   **37**

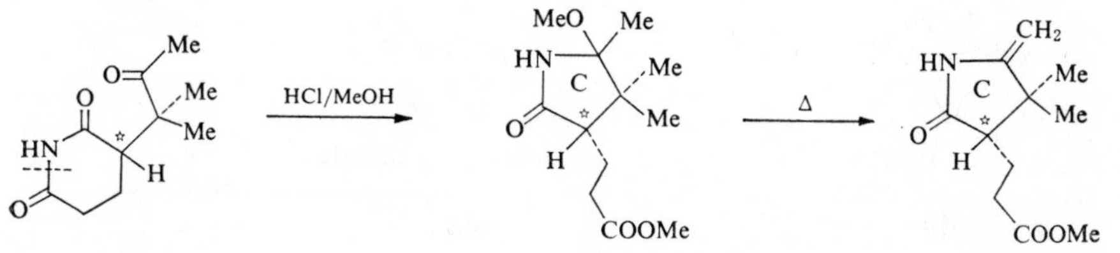

38   **39**   **40**

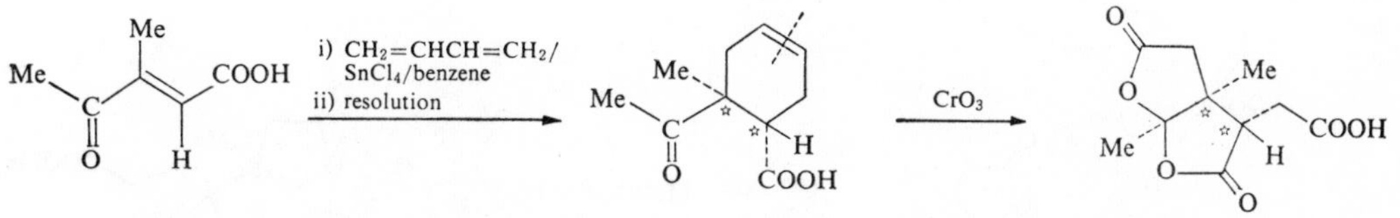

41   **42**   **43**

$\alpha_D : \sim 85.7°$

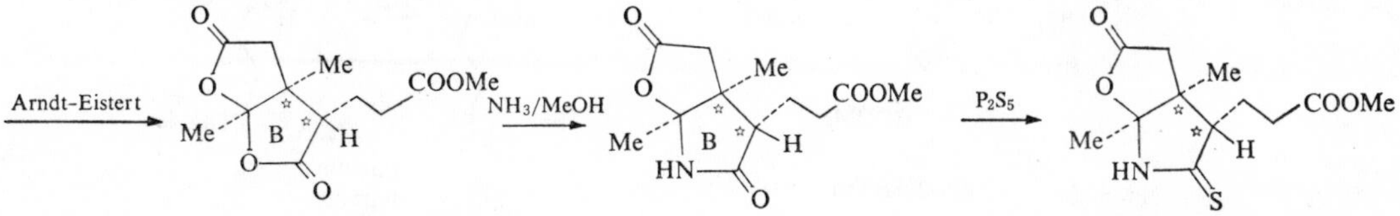

44   **45**   **46**

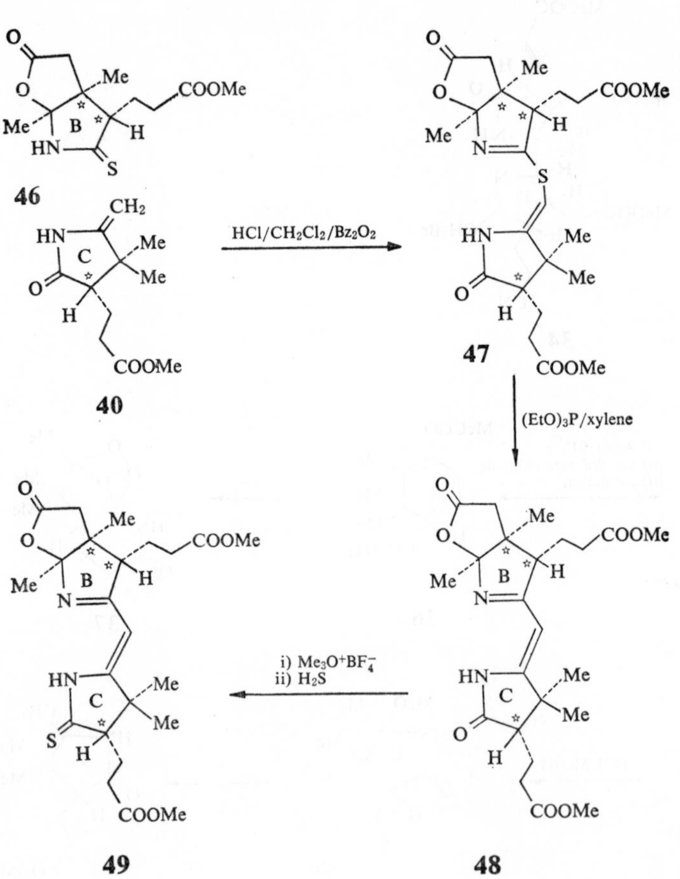

46 40 47 48 49

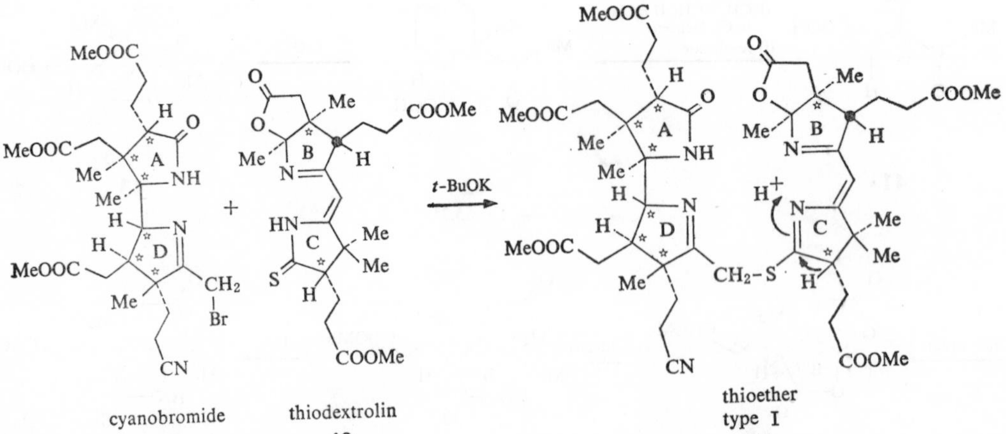

cyanobromide **34** thiodextrolin **49** thioether type I **50**

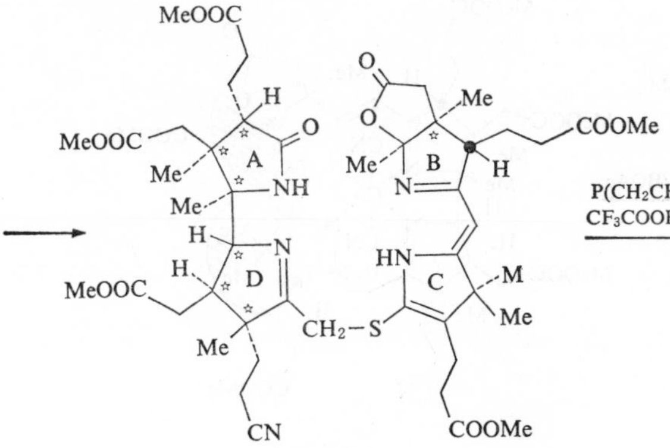

thioether
type II
51

P(CH₂CH₂CN)₃/
CF₃COOH/MeNO₂

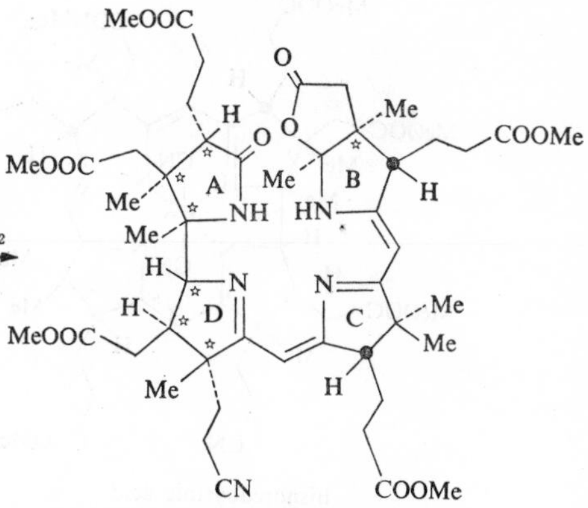

cyanocorrigenolide
52

P₂S₅/toluene/
γ-picoline

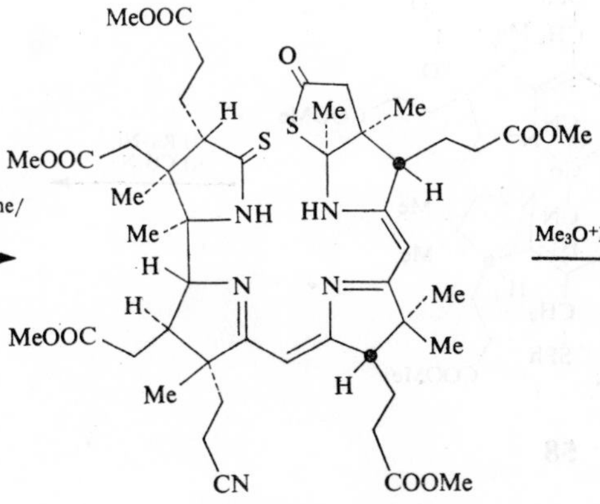

dithiocyanocorrigenolide
53

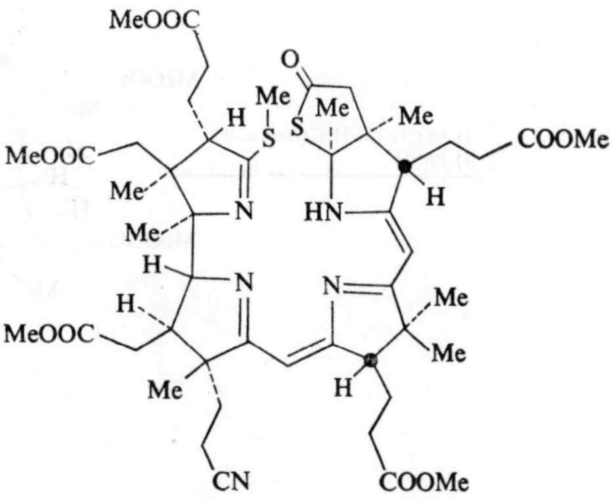

Me₃O⁺BF₄⁻

S-methyldithiocyanocorrigenolide
54

i) Me₂NH/MeOH
ii) CoCl₂/THF

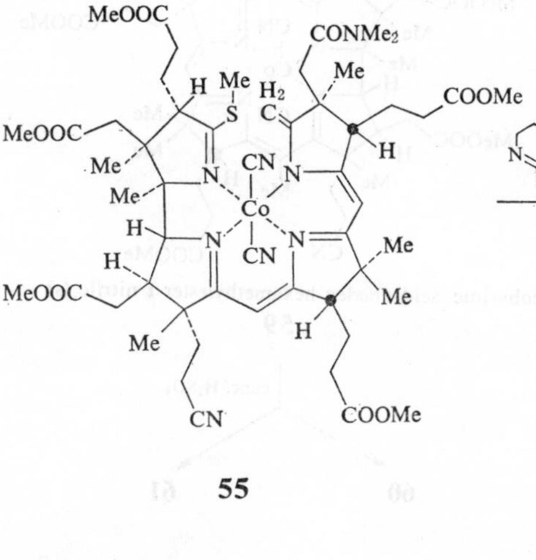

/Me₂NCOMe

55

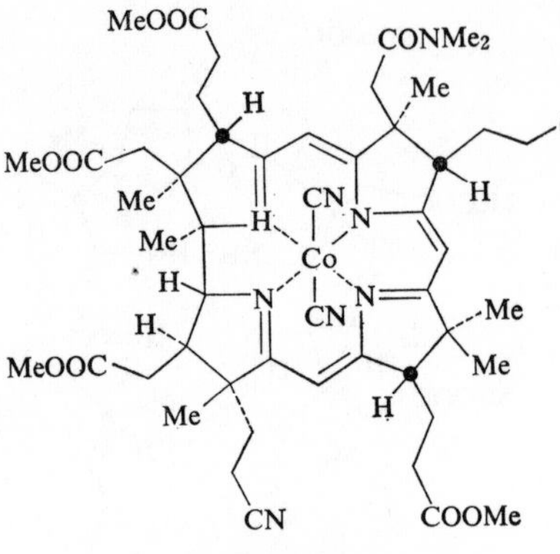

bisnorcobyrinic acid
a b d e g pentamethyl ester
c dimethylamide f nitrile (s)

56

I_2/HOAc

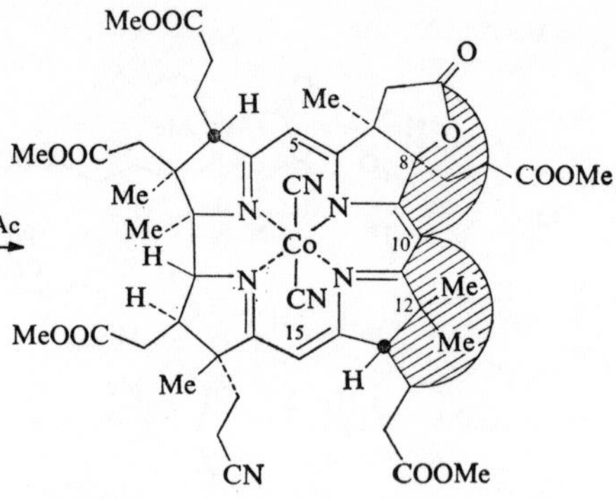

57

i) $PhCH_2OCH_2Cl$/sulfolane
ii) PhSH

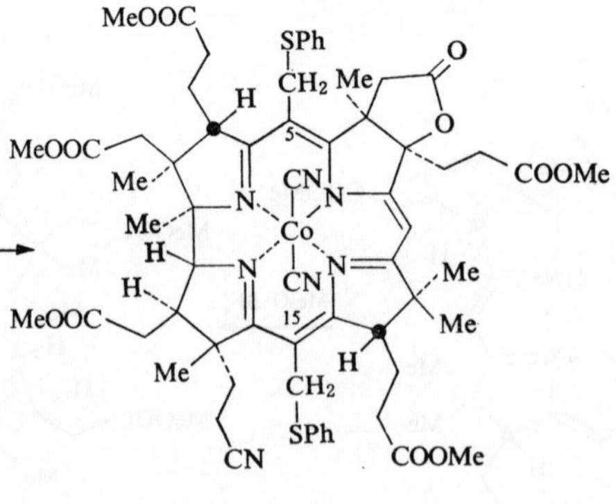

58

i) Ra-Ni
ii) CH_2N_2

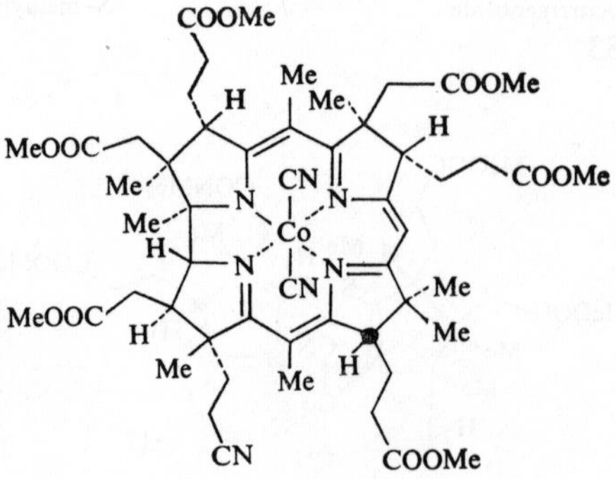

cobyrinic acid abcdeg hexamethylester f nitrile (s)

59

conc. H_2SO_4

60 **61**

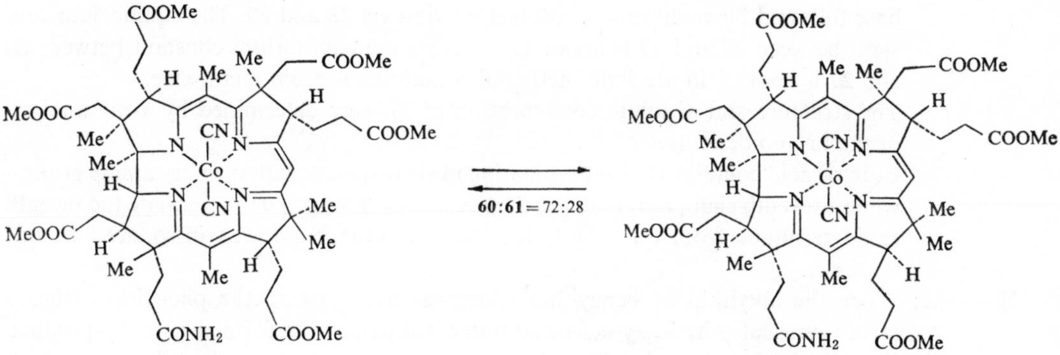

cobyrinic acid abcdeg hexamethylester f amide
60

neocobyrinic acid abcdeg hexamethylester f amide
61

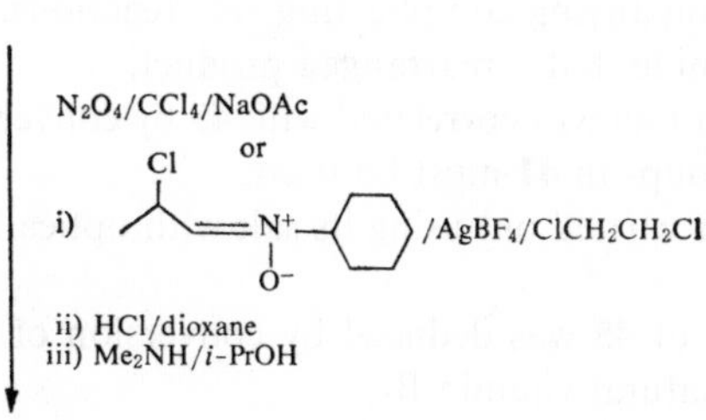

60:61 = 72:28

N$_2$O$_4$/CCl$_4$/NaOAc
or
i) [structure] /AgBF$_4$/ClCH$_2$CH$_2$Cl
ii) HCl/dioxane
iii) Me$_2$NH/i-PrOH

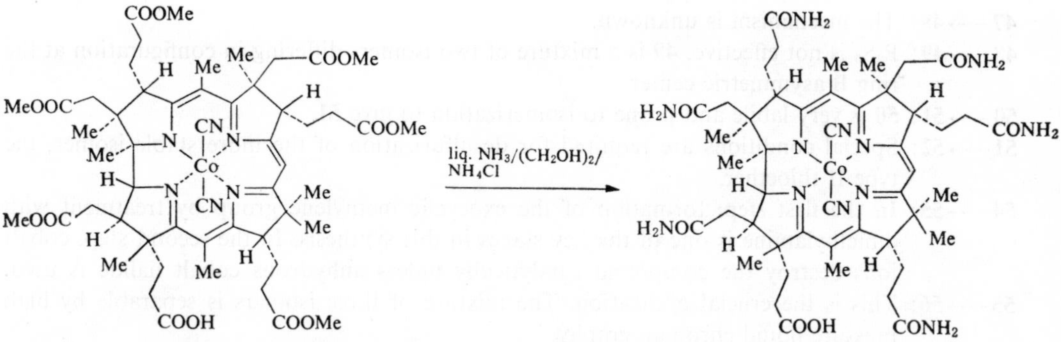

cobyrinic acid abcdeg hexamethylester
62

liq. NH$_3$/(CH$_2$OH)$_2$/NH$_4$Cl

cobyric acid
63

3 ⟶ 4: Optical resolution of **4** was effected using the urea derivative obtained by the action of α-phenylethylisothiocyanate on **4**. The absolute configuration of **4** was determined by correlation of **4** with camphor.

11 ⟶ 12: The nmr of **12** shows that **12** has a *trans* double bond, which determines the stereochemistry of **14** at the newly created asymmetric carbon atoms.

23 ⟶ 24: Treatment of **23** with hydroxide ions, instead of pyrrolidine acetate, gave an isomeric cyclization product.

25 ⟶ 26: This is the crucial Beckmann rearrangement; very drastic conditions are necessary.

26 ⟶ 27 ⟶ 30: The lactam ring in **26** is difficult to open, but the lactam ring in **27** is very susceptible to acid or base. **26** could be isomerized to **27** by treatment with strong base followed by acidification and methylation via **28** and **29**. The equilibrium constant between **28** and **29** is about 15, whereas the equilibrium constant between **26** and **27** is about 1 in absolute methanol containing sodium methoxide.

The structure and absolute configuration of **27** were determined by x-ray analysis of the bromo derivative of **27**.

Cobyric acid contains chains terminating in six instances with primary amide groups, while a seventh chain terminates in a free carboxyl group. To differentiate the seventh chain terminus from all others, the carbon-sulfur bond was introduced in this sequence.

30 ⟶ 31: When the alkylthio or benzylthio ether was used instead the phenylthio ether a rather unusual β-hydroxy-α,β-unsaturated sulfoxide was formed as a by-product.

31 ⟶ 32: The attack of nitrogen bases on thio esters is very much easier than the corresponding attack upon the analogous oxygen esters.

32 ⟶ 33: Methanesulfonic anhydride is superior to methanesulfonyl chloride since the reaction proceeds without accompanying complicating side reactions.

36 ⟶ 37: Here **37** is not a direct ozonide, but a rearranged product.

40: The absolute configuration of **40** was correlated with **45** by conversion of **45** into **40**.

41 ⟶ 42: The acetyl and carboxyl groups in **41** must be *trans*.

42: Optical resolution of **42** was carried out using its salt with optically active α-phenylethylamine.

45: The absolute configuration of **45** was deduced by conversion of **45** to a derivative which was prepared from natural vitamin B_{12}.

40,46 ⟶ 47: In this reaction, **46** is oxidized first to its disulfide, the S–S bond of which is then attacked by **40** to form **47**.

47 ⟶ 48: The mechanism is unknown.

48 ⟶ 49: P_2S_5 is not effective. **49** is a mixture of two isomers differing in configuration at the ring B asymmetric center.

50 ⟶ 51: **50** is very labile and prone to isomerization to give **51**.

51 ⟶ 52: Special conditions are required for desulfurization of the more stable isomer, the type II thioether.

54 ⟶ 55: In the first step, formation of the exocyclic methylene group by treatment with dimethylamine is one of the key stages in this synthesis. In the second step, cobalt ions destroy the compound catalytically unless anhydrous cobalt halide is used.

55 ⟶ 56: This is the crucial cyclization. The mixture of three isomers is separable by high pressure liquid chromatography.

56 ⟶ 57: This reaction does not take place with the corresponding ester. The newly formed lactone group created a situation in which C-10 was so crowded as to be virtually inaccessible to attack by the reagent in the next step.

57 ⟶ 58: Probable intermediates are those having benzyloxy groups and chloromethyl groups. Yields are rather low.

58 ⟶ 59: Raney nickel reduces the lactone groups as well as cleaving the C–S bonds. The mixture of nitriles obtained from **58** was separated into a number of fractions by liquid–liquid chromatography under high pressure. One of the fractions consisted of just the two substances **59** (normal and neo) which are isomeric at the C-13 position. Separation of these isomers is not necessary since in the next step equilibration occurs at C-13.

59 ⟶ 60 and 61: These compounds are readily separable by high pressure liquid chromatography.

60 ⟶ 62: Deamination using nitrous acid or nitrous ester is accompanied by extremely ready nitrosation of the C-10 position. The second method developed by the Eschenmoser group is an extremely elegant method for selective transformation, as outlined below, but the first approach, specially developed by the Woodward group, is simpler and easier.

62 ⟶ 63: The action of ammonium chloride is very important, since without the addition of ammonium chloride this reaction produced pseudocobyric acid, which could not be converted into **63**.

Route II—the synthesis of Friedrich *et al.*[7]

Friedrich *et al.*[7] developed the following route from **64** to **1**.

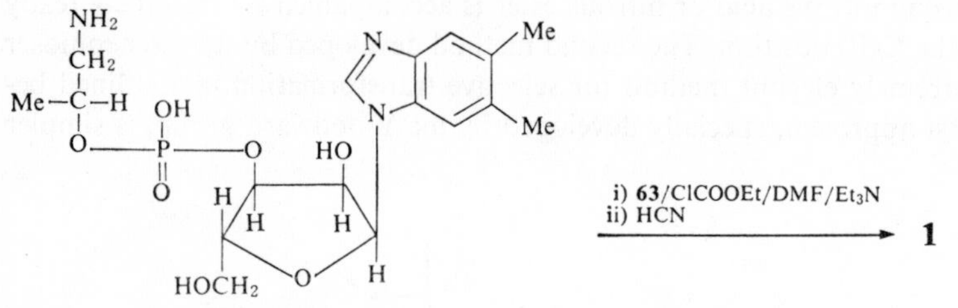

i) **63**/ClCOOEt/DMF/Et₃N
ii) HCN

66

Route III—Modification of Route I by the Eschenmoser group[8]

In practice, this method is somewhat superior to the Woodward-Eschenmoser synthesis (Route I) in that it is relatively easier to reproduce, even though it is a very complicated sequence.

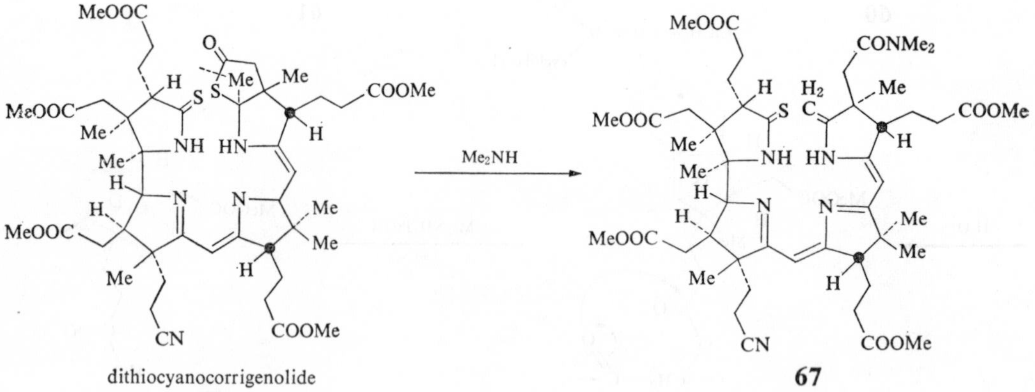

dithiocyanocorrigenolide
53

67

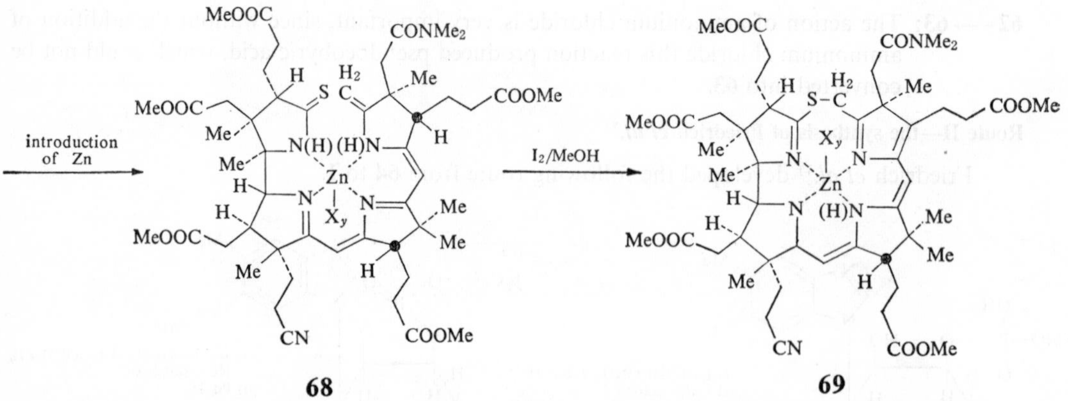

introduction of Zn

68

I₂/MeOH

69

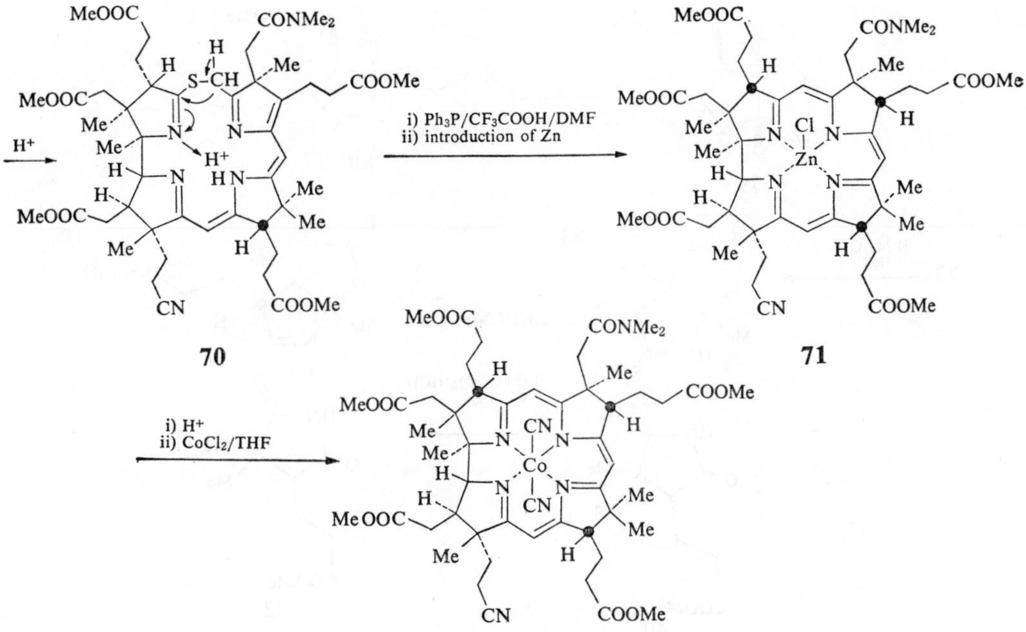

70

i) Ph₃P/CF₃COOH/DMF
ii) introduction of Zn

71

i) H⁺
ii) CoCl₂/THF

bisnorcobyrinic acid
abdeg pentamethyl ester
c dimethylamide f nitrile (s)

56

67 ⟶ 68: The structure of the zinc complex was not defined completely.
70 ⟶ 71: Zinc is introduced simply for purification of the product.

Route IV—the Eschenmoser synthesis[9]

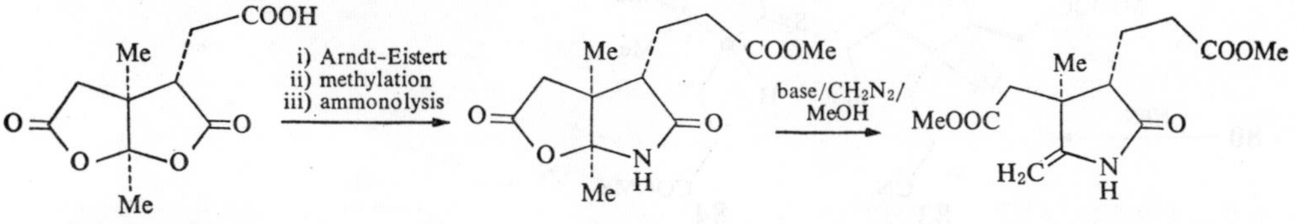

(+)-C₁₀-acid
72

i) Arndt-Eistert
ii) methylation
iii) ammonolysis

73

base/CH₂N₂/
MeOH

74

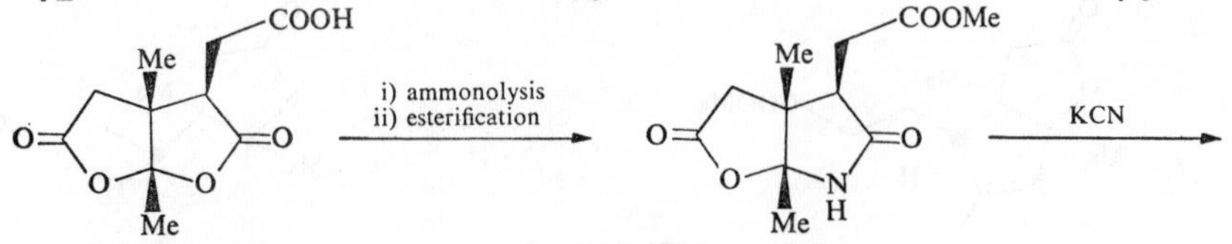

(−)-C₁₀-acid
75

i) ammonolysis
ii) esterification

76

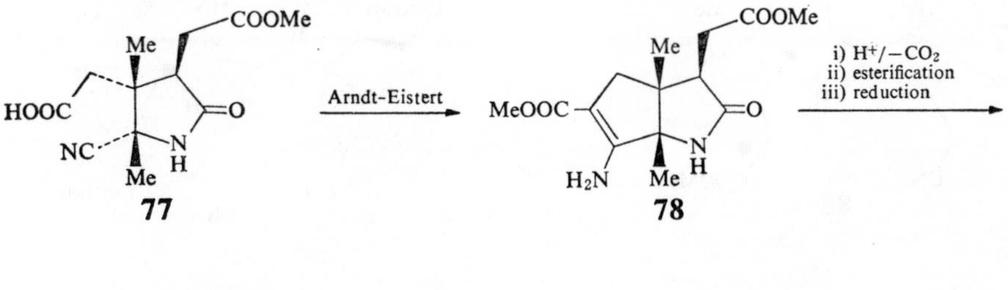

KCN

77

Arndt-Eistert

78

i) H⁺/−CO₂
ii) esterification
iii) reduction

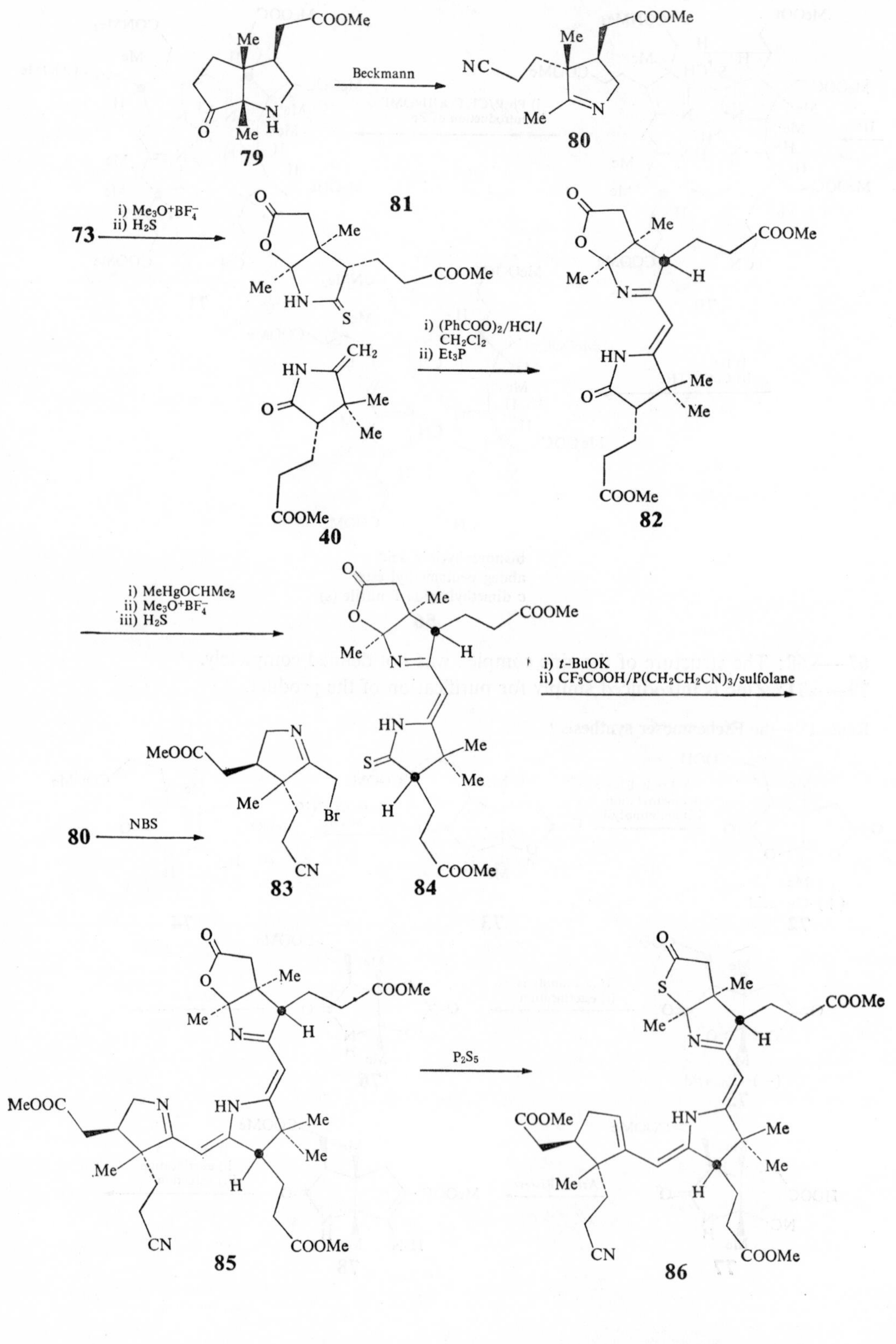

COOMe
Me
O
Me
N
H
79
Beckmann
NC
COOMe
Me
Me
N
80
73
i) Me3O+BF4−
ii) H2S
81
O
O
Me
Me
HN
S
COOMe
HN
CH2
O
Me
Me
COOMe
40
i) (PhCOO)2/HCl/
CH2Cl2
ii) Et3P
82
i) MeHgOCHMe2
ii) Me3O+BF4−
iii) H2S
i) t-BuOK
ii) CF3COOH/P(CH2CH2CN)3/sulfolane
80
NBS
MeOOC
Br
CN
83
S
H
84
MeOOC
CN
COOMe
85
P2S5
S
COOMe
86

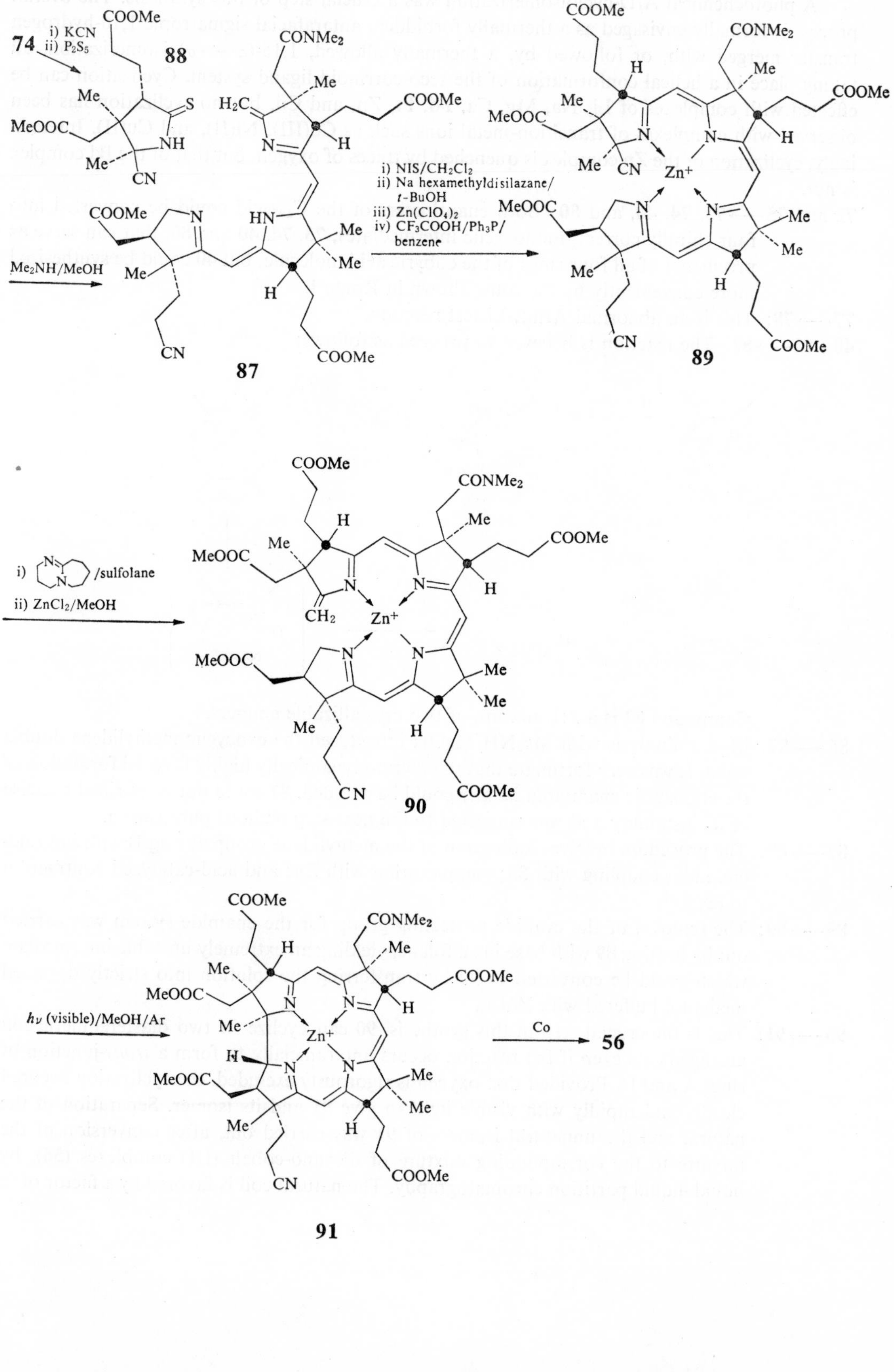
74
i) KCN
ii) P₂S₅
88
Me₂NH/MeOH
87
i) NIS/CH₂Cl₂
ii) Na hexamethyldisilazane/t-BuOH
iii) Zn(ClO₄)₂
iv) CF₃COOH/Ph₃P/benzene
89
/sulfolane
ii) ZnCl₂/MeOH
90
hν (visible)/MeOH/Ar
Co
56
91

A photochemical A/D-cycloisomerization was a crucial step of this synthesis. The overall process is formally envisaged as a thermally forbidden, antarafacial sigmatropic 1,16-hydrogen transfer merged with, or followed by, a thermally allowed, 1,15-($\pi \longrightarrow \sigma$) isomerization, all taking place in a helical conformation of the seco-corrinoid ligand system. Cyclization can be effected with complexes of Li, Na, Mg, Ca, Pd, Pt, Zn, and Cd, but no cyclization has been observed with complexes of transition-metal ions such as Co(III), Ni(II), and Cu(II). Interestingly, cyclization of the Zn complex is quenched by traces of oxygen, but that of the Pd complex is not.

72 and 75 $\longrightarrow$ 73, 74, 40, and 80: Both enantiomers of the C_{10}-acid could be converted into four chirally correct, monocyclic intermediates, **73, 74, 40** and **80**, that can serve as precursors of all four rings of the cobyric acid molecule, but **40** could be synthesized more conveniently by the route shown in Route I.

77 $\longrightarrow$ 78: This is an abnormal Arndt-Eistert reaction.

40 + 81 $\longrightarrow$ 82: The reaction is believed to proceed as follows:

Compound **82** is a 2:1 mixture of two crystallizable epimers.

86 $\longrightarrow$ 87: Mild aminolysis with $Me_2NH/MeOH$ introduced the exocyclic methylidene double bond. It was very fortunate that the thermodynamically highly favored formation of the endocyclic enaminoid isomer could be avoided. **87** could not be purified because of its instability and was subjected to the next step without purification.

87 $\longrightarrow$ 89: The procedure involves iodination of the methylidene group at ring B with iodosuccinimide; coupling with **88**; complexation with Zn; and acid-catalyzed contraction to **89**.

89 $\longrightarrow$ 90: The removal of the cyanide protecting group for the enamide system was carried out by heating **89** with base in sulfolane, yielding an extremely unstable intermediate which could be converted to **90** by transferring the solution into strictly degassed methanol buffered with $ZnCl_2$.

90 $\longrightarrow$ 91: This is the crucial step in this synthesis. **90** can cyclize in two diastereomeric coil arrangements even if the reaction occurs antarafacially to form a *trans*-junction of rings A and D. Provided that oxygen is rigorously excluded, the cyclization occured cleanly and rapidly with visible light to give **91** and its isomer. Separation of the natural and the unnatural isomers of **91** was carried out, after conversion of the mixture to the corresponding mixture of dicyano-cobalt (III) complexes (**56**), by liquid–liquid partition chromatography. The natural coil is favored by a factor of 2.

11.35 STRUCTURE OF OSTREOGRYCIN A

ostreogrycin A **1**[1]
mp:203-205°
α_D: −285°
uv(EtOH):228(4.51),272(4.00)
nmr in CDCl$_3$

1) Reductive ozonolysis of **1** yields methylglyoxal and 2,4-dimethylpent-2-enal.

2) **1** gives a negative iodoform test, suggesting the absence in **1** of the MeCO– group which is present in **6**.

3) The presence of Δ^2-pyrroline–2-carboxylic acid residue in **1** is suggested by the formation of β-alanine on ozonolysis followed by oxidation with H$_2$O$_2$ and hydrolysis with 6 N HCl.

4) Acid hydrolysis of **1** and **2** yields trace amounts of glycine and serine, possibly produced from the imidazole moiety.

5) **2** is not a single compound, but consists of at least five components.

6) The uv absorption of **2** in alkaline media at 298 nm (4.31) suggests the presence of an enolizable carbonyl group β to the imidazole moiety.

7) Reduction of **2** with LiBH$_4$ followed by acid hydrolysis yields prolinol and the γ-lactone **5**, indicating an ester linkage at * and an amide linkage at **.

8) Compound **3** consumes no HIO$_4$, and hence the two OH groups are not in a vicinal position.

2 X=H or OH

3

1) G. R. Delpierre, F. W. Eastwood, G. E. Gream, D. G. I. Kingston, P. S. Sarin, Lord Todd, D. H. Williams, *Tetr. Lett.*, 369 (1966); *J. Chem. Soc.* C, 1653 (1966); D. G. I. Kingston, Lord Todd, D. H. Williams, *J. Chem. Soc.* C, 1669 (1966).

2

dil OH⁻   6 N HCl

3

HCl/HCOOH

$HOOC-CH-Me$
$|$
NH_2 + **5** + **7**

8

H_2N ... O   **6**

(+)-form **4**

+

6 N HCl

HOOC—NH

5 (*dl*-form)

partially racemic
7

H_2N ... O ... O

9

Remarks

Ostreogrycin complex, which was isolated from *Streptomyces ostreogriseus*,[2] consists of at least six components. These fall structurally into two distinct groups; the A and G group[3] and the B series. Although these compounds individually have rather low activity, they show strong synergism,[4] combinations of each of the two groups showing marked activity against gram-positive bacteria. Other similar antibiotics have been reported, e.g. staphylomycin,[5] vernamycin,[6] etamycin[7] and mikamycin.[8] A structurally similar antimucobacterial antibiotic, pyridomycin **12**,[9] is produced by *Str. pyridomyceticus*.

etamycin **10** (R=H)
ostreogrycin B **11** (R=NMe₂)

pyridomycin **12**

2) S. Ball, B. Boothroyd, K. A. Lees, A. H. Raper, E. L. Smith, *Biochem. J.*, **68**, 24 (1958).
3) F. W. Eastwook, B. K. Snell, Sir A. Todd., *J. Chem. Soc.*, 2286 (1960).
4) J. Bessell, K. H. Fantes, W. Hewitt, P. W. Muggleton, J. P. R. Toothill, *Biochem. J.*, **68**, 24 (1958).
5) P. de Somer, P. van Dijk, *Antibiotics Chemoth.*, **5**, 632 (1955); H. Vanderhaeghe, P. van Dijk, G. Parmentier, P. de Somer, *ibid.*, **7**, 606 (1957); H. Vanderhaeghe, G. Parmentier, *J. Am. Chem. Soc.*, **82**, 4414 (1960).
6) M. Bodanszky, M. A. Ondetti, *Antimicrobial Agents Chemoth.*, **13**, 360 (1963).
7) J. C. Sheehau, H. G. Zachau, W. B. Lawson, *J. Am. Chem. Soc.*, **80**, 3349 (1958); R. B. Arnold, A. W. Johnson, A. B. Mauger, *J. Chem. Soc.*, 4466 (1958).
8) M. Arai, S. Nakamura, Y. Sakagami, K. Fukuhara, H. Yonehara, *J. Antibiotics* (Japan), Ser. A, **9**, 193 (1956); M. Arai, K. Karasawa, S. Nakamura, H. Yonehara, H. Umezawa, *ibid.*, **11**, 14 (1958); K. Watanabe, *ibid.*, **14**, 14 (1961).
9) H. Ogawara, K. Maeda, G. Koyama, H. Naganawa, H. Umezawa, *Chem. Pharm. Bull.*, **16**, 679 (1968); G. Koyama, Y. Iitaka, K. Maeda, H. Umezawa, *Tetr. Lett.*, 3587 (1967).

11.36 STRUCTURE OF BLEOMYCIN A₂

bleomycin A₂ **1**[9]

1) Acid hydrolysis (6N HCl/105°/20 hr) of **1** gave the amine **2** and six amino acids, **3–8**, the structures of which were proved by comparison with authentic samples, by synthesis[1,4,5,7] or by x-ray analysis.[2,11]

2) The nmr spectrum shows that **1** contains 1 mole each of **2–8**.

3) Structure **12** is established from the evidence that **4** is an *N*-terminal amino acid (DNP method) and that partial hydrolysis affords the peptide **3–2**.[10]

4) Transformations **13 ⟶ 14** and **15 ⟶ 17** involve *N ⟶ O* acyl migration; DNP derivatization followed by acid hydrolysis gave DNP-**5** from **14** but not from **13**.[10]

5) A series of reactions leading to **5** and **16** from **14** indicates an amide bond between **5** and **6**.[10]

6) **7** and **8** are produced from **17** by competitive β-eliminations.[7]

1) T. Takita, Y. Muraoka, K. Maeda, H. Umezawa, *J. Antibiotics* (Japan), **21**, 79 (1968).
2) G. Koyama, H. Nakamura, Y. Muraoka, T. Takita, K. Maeda, H. Umezawa, Y. Iitaka, *Tetr. Lett.*, 4635 (1968).
3) T. Takita, K. Maeda, H. Umezawa, S. Omoto, S. Umezawa, *J. Antibiotics* (Japan), **22**, 237 (1969).
4) Y. Muraoka, T. Takita, K. Maeda, H. Umezawa, *J. Antibiotics* (Japan), **23**, 252 (1970).
5) T. Takita, T. Yoshioka, Y. Muraoka, K. Maeda, H. Umezawa, *J. Antibiotics* (Japan), **24**, 795 (1971).
6) Y. Muraoka, T. Takita, K. Maeda, H. Umezawa, *J. Antibiotics* (Japan), **25**, 185 (1972).
7) T. Yoshioka, Y. Muraoka, T. Takita, K. Maeda, H. Umezawa, *J. Antibiotics* (Japan), **25**, 625 (1972).
8) S. Omoto, T. Takita, K. Maeda, H. Umezawa, S. Umezawa, *J. Antibiotics* (Japan), **25**, 752 (1972).
9) T. Takita, Y. Muraoka, T. Yoshioka, A. Fujii, K. Maeda, H. Umezawa, *J. Antibiotics* (Japan), **25**, 755 (1972).
10) T. Takita, Y. Muraoka, K. Maeda, H. Umezawa, *Proc. Eighth Symposium on Peptide Chem.*, 179 (1970).
11) G. Koyama, H. Nakamura, Y. Muroaka, T. Takita, K. Maeda, H. Umezawa, Y. Iitaka, *J. Antibiotics* (Japan), **26**, 109 (1973).

7) Structure **17** was suggested by the nmr of a triol derived from **17** by methylation and reduction,[4] and was proved by synthesis.[7]

8) Methylation (MeOH/H[+]) of **15** followed by reduction with LiBH$_4$ and acid hydrolysis gave reduced **17**, having a carboxyl group, which is stable even on drastic acid hydrolysis (6N HCl/105°), indicating that in **15** the COOH group on the pyrimidine forms a peptide bond with **16**.[10]

9) NBS oxidation of **1** gave the tetrapeptide **11**, whose N-terminal amino acid is **5**, indicating that **5** is linked to **4**.[6]

10) Amberlyst-**15** catalyzed methanolysis of **1** gave 1 mole each of L-gulose **9** and 3-O-carbamoyl-D-mannose **10**.[3]

11) Mild acid hydrolysis (0.3N H$_2$SO$_4$, 80°/6 hr) of **1** gave a disaccharide, which on reduction with NaBH$_4$ followed by methanolysis, gave D-sorbitol and the methyl glycoside of **10**.[8]

12) The nmr of the disaccharide indicates that the C-2 position of **9** is attached to **10**.[8]

13) Hudson's rules suggest the α-glycoside linkage in the disaccharide.[8]

14) Treatment of **1** with 0.1N NaOH at room temperature gave the disaccharide accompanied by increased uv absorption at 290 nm, indicating the formation by β-elimination of **18** of a fragment **19** that contains a dehydrohistidine moiety.[5,8]

15) **19** shows an olefinic proton in its nmr spectrum and on catalytic hydrogenation followed by hydrolysis gave histidine.[8]

16) The α-glycoside linkage between **18** and **19** is suggested from the nmr of **1**.[8]

17) Methylation of **1** with MeI/Et$_3$N to give the quaternary ammonium salt of **1**, which on acid hydrolysis gave β-aminoalanine α-betaine amide, indicated that the α-amino group of **8** is free in **1** and that the carboxyl group forms an amide linkage.[9]

18) Total carbon numbers, and hence the molecular formula, were determined from the ^{1}H-decoupled ^{13}C-PFT nmr spectrum of bleomycin A$'_2$-b.[9]

19) The β-lactam in **7** is suggested from the molecular formula and pK_a measurements.[9]

20) **1** shows a strong ir band at 1650 cm^{-1} with a shoulder at 1720 cm^{-1}; although a 1720 cm^{-1} band seems too low for the β-lactam,[9] there is a similar instance in the case of pachystermine.[12]

12) T. Kikuchi, S. Uyeo, *Chem. Pharm. Bull.*, **15**, 549 (1967).
13) H. Umezawa, K. Maeda, T. Takeuchi, Y. Okami, *J. Antibiotics* (Japan), **19A**, 200 (1966).
14) K. Maeda, H. Kosaka, K. Yagishita, H. Umezawa, *J. Antibiotics* (Japanese), **9A**, 82 (1956); K. Ikekawa, F. Iwami, H. Hiranaka, H. Umezawa, *J. Antibiotics* (Japan), **17A**, 194 (1964).

Remarks

Bleomycin **1** is a unique peptide antibiotic, which exhibits strong inhibition of tumors.[13] It is obtained from *Streptomyces verticillus*. It forms a stable copper chelate, from which it is isolated by treatment with 8-hydroxyquinoline. Bleomycin complex contains several congeners (A$_1$, DM-A$_2$, A$'_2$-a, A$'_2$-b, B$_2$, B$_4$, A$_5$, A$_6$ etc.) which are modified at the amine **2** moiety by other amines.

Phleomycins,[14] produced by *Streptomyces verticillus*, is identical with dihydrobleomycin having a

moiety.[9]

11.37   STRUCTURE OF PYRIDOXOL (VITAMIN B$_6$)

pyridoxol 1[1]

1) Compound **1** has an Me group and 3 active hydrogens.

2) The uv spectrum of **1** is similar to that of β-hydroxypyridine.

3) Compound **1**, but not **2**, gives a positive FeCl$_3$ test and Gibbs phenol test, indicating the presence of phenolic OH, the position *para* to which is unsubstituted.

4) Compound **3** gives a negative FeSO$_4$ test, indicating the absence of a carboxyl group at the α position.

5) The structure of **3** was confirmed by synthesis.[2]

Remarks

Pyridoxol **1** is widely distributed in the plant and animal kingdoms, and was first isolated from rice bran.[4] Deficiency of the vitamin causes a specific dermatitis in rats, and growth is also inhibited. The daily requirement for man may be 1–2 mg. Pyridoxal **5** and pyridoxamine **6** are also found in microorganisms.[5] Codecarboxylase is identical with pyridoxal phosphate **7**.

1) E. T. Stiller, J. C. Keresztesy, J. R. Stevens, *J. Am. Chem. Soc.*, **61**, 1237 (1939).
2) S. A. Harris, E. T. Stiller, K. Folkers, *J. Am. Chem. Soc.*, **61**, 1242 (1939); R. Kuhn, G. Wendt, K. Westphal, *Chem. Ber.*, **72**, 310 (1939); R. Kuhn, K. Westphal, G. Wendt, O. Westphal, *Naturwiss.*, **27**, 469 (1939).
3) R. Kuhn, H. Andersag, K. Westphal, G. Wendt, *Chem. Ber.*, **72**, 309 (1939); R. Kuhn, G. Wendt, K. Westphal, *ibid.*, **72**, 310 (1939).
4) P. György, *J. Am. Chem. Soc.*, **60**, 983 (1938); S. Lepkovsky, *J. Biol. Chem.*, **124**, 125 (1938); *Science*, **87**, 169 (1938).
5) E. E. Snell, B. M. Guirard, R. J. Williams, *J. Biol. Chem.*, **143**, 519 (1942).

11.38   SYNTHESIS OF PYRIDOXOL (VITAMIN B₆)

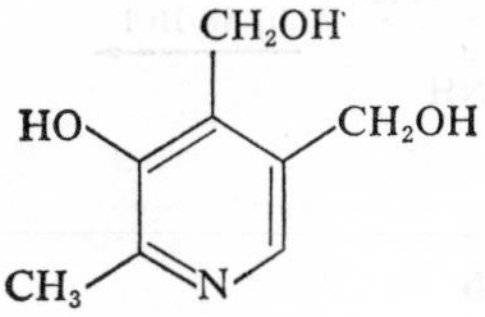

pyridoxol **1**

Route I[1]

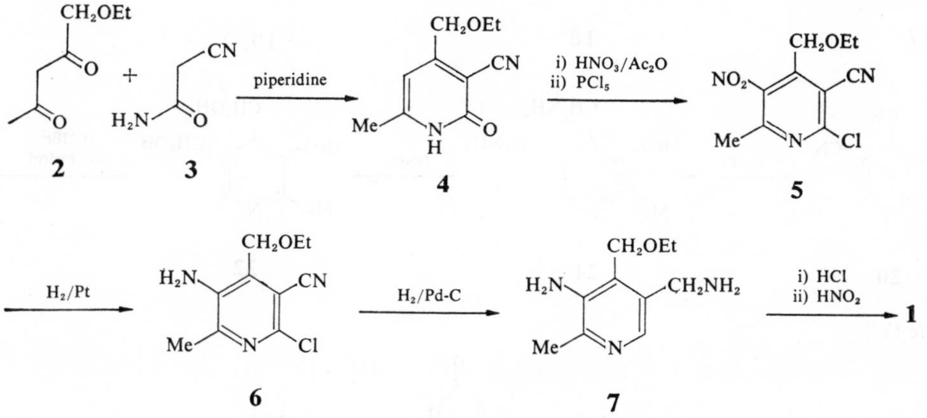

Route II[2]

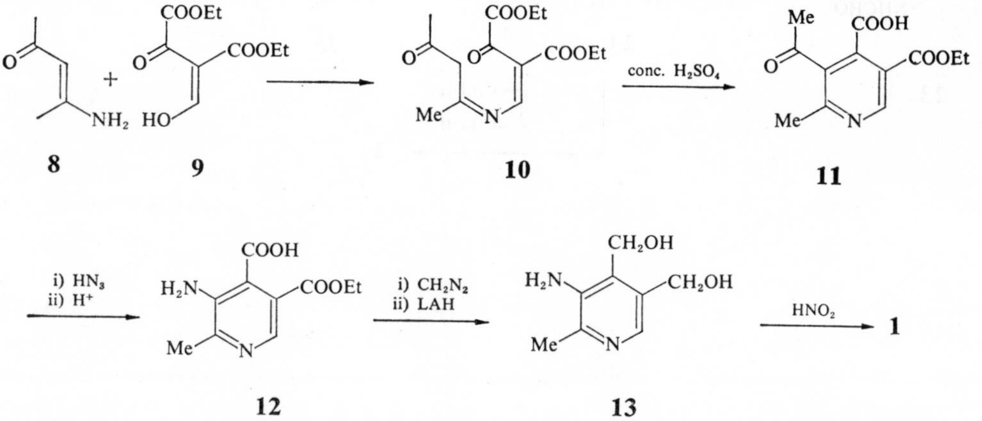

1) S. A. Harris, E. T. Stiller, K. Folkers, *J. Am. Chem. Soc.*, **61**, 1242 (1939); S. A. Harris, K. Folkers, *J. Am. Chem. Soc.*, **61**, 1245, 3307 (1939).
2) R. G. Jones, E. C. Kornfield, *J. Am. Chem. Soc.*, **73**, 107 (1951); R. G. Jones, *J. Am. Chem. Soc.*, **73**, 5244 (1951).

Route III[3,4]

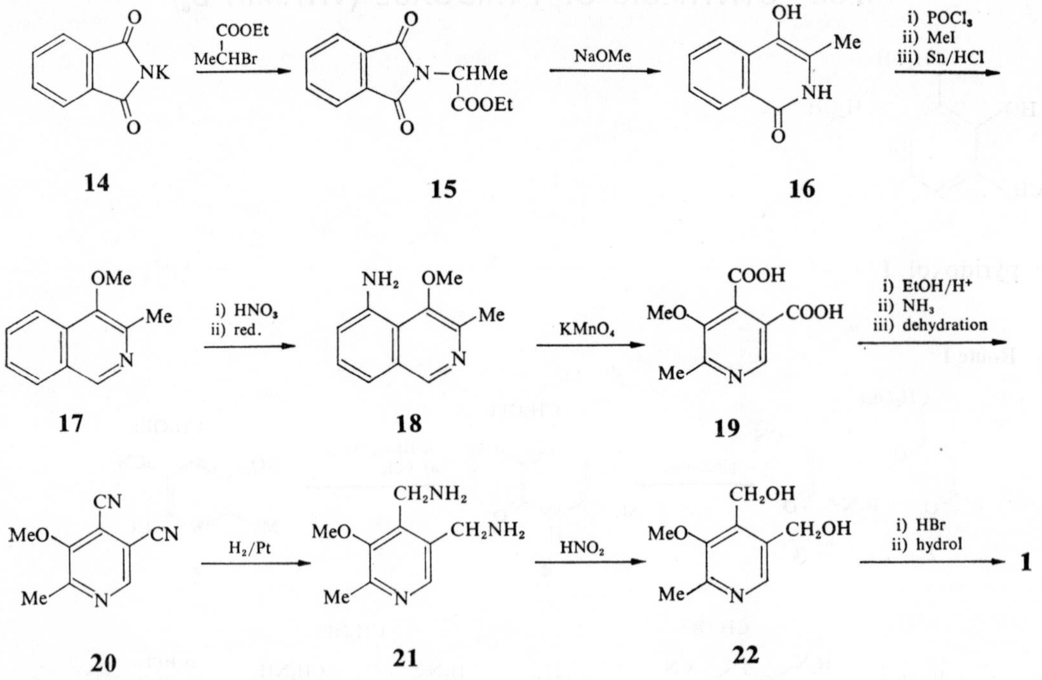

14   **15**   **16**

17   **18**   **19**

20   **21**   **22**

Route IV[5]

23   **24**   **25**

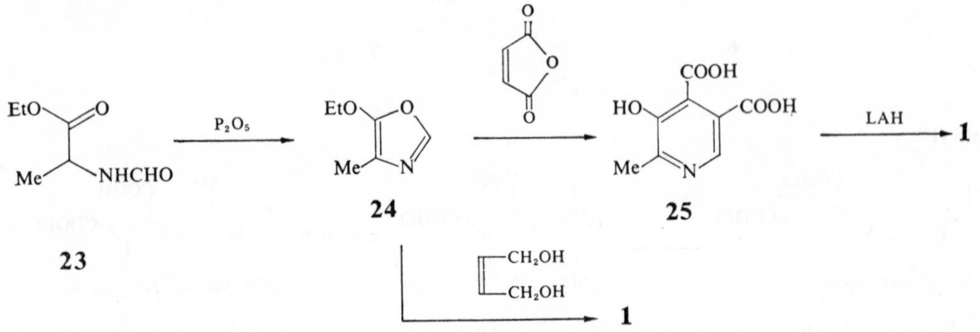

3) R. Kuhn, K. Westphal, G. Wendt, O. Westphal, *Naturwiss.*, **27**, 469 (1939).
4) A. Itiba K. Miki, *Sci. Pap. Inst. Phys. Chem. Res. (Tokyo)*, **36**, 173 (1939).
5) E. E. Harris, R. A. Firestone, K. Pfister III, R. R. Boettcher, F. J. Cross, R. B. Currie, M. Monaco, E. R. Peterson, W. Reuter, *J. Org. Chem.*, **27**, 2705 (1962); R. A. Firestone, E. E. Harris. W. Reuter, *Tetr.*, **23**, 943 (1967).

Route V[6)]

Route VI[7)]

6) A. Cohen, J. W. Haworth, E. G. Hughes, *J. Chem. Soc.*, 4374 (1952).
7) N. Elming, N. Clauson-Kaas, *Acta Chim. Scand.*, **9**, 23 (1955).

11.39   STRUCTURE OF STREPTONIGRIN

pKa: 6.3
uv (MeOH): 245 ($E^{1\%}_{1cm}$ 760), 390 ($E^{1\%}_{1cm}$ 344)

streptonigrin **1**[1]

1) The molecular weight was determined by ms of hexamethyldihydro **1**, obtained from **1** by catalytic reduction followed by methylation with Me_2SO_4.
2) The presence of a primary amino, a carboxyl and a methyl group attached to aromatic rings was suggested by nmr, ms and other physical data.
3) The structures **4, 5** (as the decarboxylated product) and **6** were confirmed by synthesis.
4) The formation of the benzofuran **8** indicated the relative positions of the amino and methoxyphenyl groups.
5) The position of the phenolic hydroxyl group was deduced on the basis of the formation of **8**, which does not contain deuterium, from **2** by trideuteriomethylation followed by deamination.
6) The position of the amino group on the quinonoid ring was deduced from the structure **11**.

$$ \mathbf{1} \xrightarrow{H_2O_2/OH^-} \mathbf{2} \xrightarrow{KMnO_4/OH^-} \mathbf{3} \longrightarrow \mathbf{4, 5, 6} $$

___

1) K. V. Rao, K. Biemann, R. B. Woodward, *J. Am. Chem. Soc.*, **85**, 2532 (1963).
2) K. V. Rao, W. P. Cullen, *Antibiot. Ann.*, 950 (1959–60).

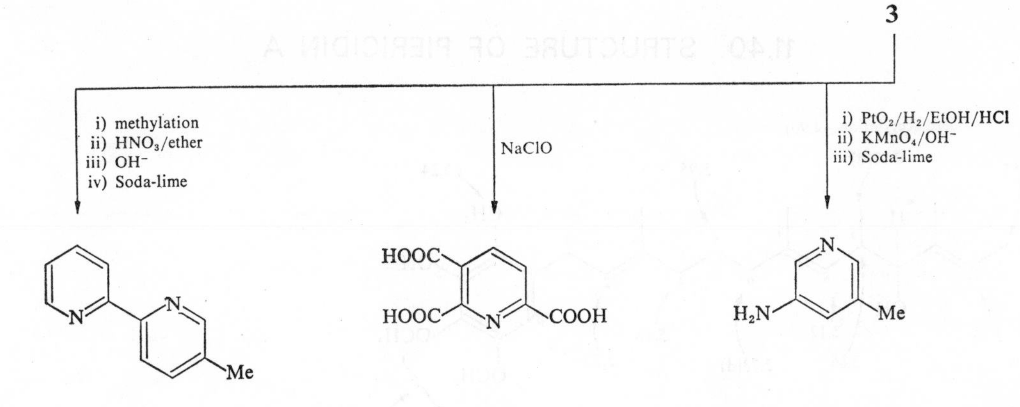
3
i) methylation
ii) HNO_3/ether
iii) OH^-
iv) Soda-lime
NaClO
i) PtO_2/H_2/EtOH/HCl
ii) $KMnO_4/OH^-$
iii) Soda-lime
N
Me
HOOC
HOOC
COOH
H_2N

4 5 6

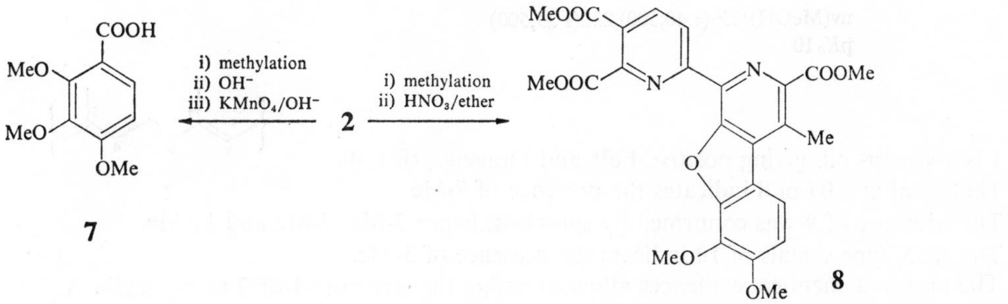
COOH
MeO
MeO
OMe
7
i) methylation
ii) OH^-
iii) $KMnO_4/OH^-$
2
i) methylation
ii) HNO_3/ether
MeOOC
MeOOC
N
COOMe
Me
O
MeO
OMe
8

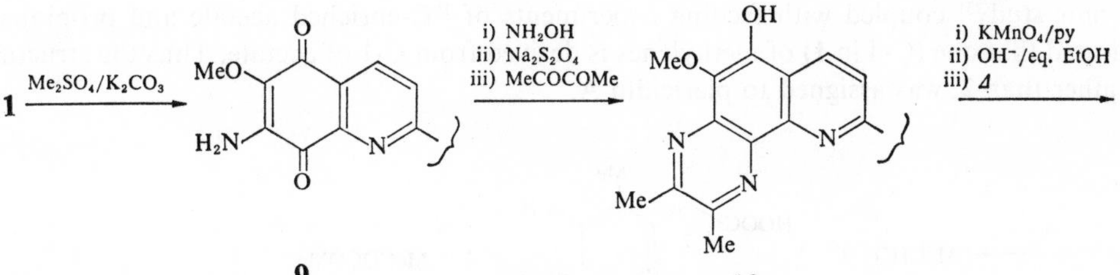
1
Me_2SO_4/K_2CO_3
O
MeO
H_2N
N
i) NH_2OH
ii) $Na_2S_2O_4$
iii) MeCOCOMe
OH
MeO
N
Me
Me
i) $KMnO_4$/py
ii) OH^-/eq. EtOH
iii) Δ

9 10

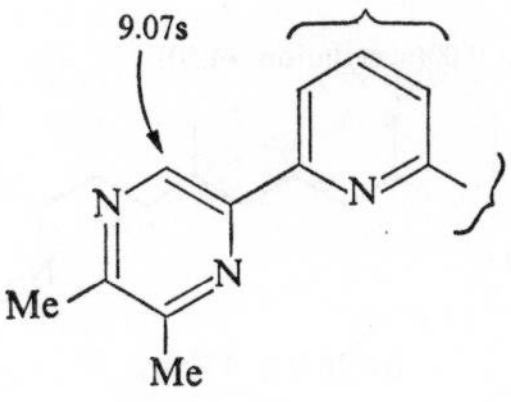
ABC pattern
9.07s
N
Me
Me

11

Remarks

Streptonigrin **1** is an antibiotic produced as a metabolite of *Streptomyces flocculus*[2] and exhibits striking activity against a variety of animal tumors. The name originated from its dark brown color.

11.40 STRUCTURE OF PIERICIDIN A

piericidin A **1**[1-3)]

uv(MeOH):239(ε 40,500),232(ε 39,500)

pK_a10

2

1) **1** is a viscous oil, giving positive Fe^{3+} and Dragendorff tests.
2) The signal at 1.09 in **8** indicates the presence of 7-Me.
3) The structure of **9** was confirmed by synthesis, hence 3-Me, 7-Me and 11-Me.
4) The ABX type signals of **10** indicate the presence of 3-Me.
5) The nmr and chemical evidences allow to assign the structure **1** or **2** to piericidin A.
6) The biosynthetic study[4)] using ^{14}C-precursors disclosed that piericidin A is formed *via* a polyketide intermediate which is consisted of five propionates and four acetates.
7) ^{13}C nmr study[5)] coupled with feeding experiments of ^{13}C-enriched acetate and propionate indicated that one (C-4 in **1**) of methylenes is derived from C-1 of acetate. Thus the structure **1**, rather than **2**, was assigned to piericidin A.

3   uv:267(ε 4300)

1) N. Takahashi, A. Suzuki, S. Tamura, *J. Am. Chem. Soc.*, **87**, 2066 (1965).
2) S. Tamura, N. Takahashi, S. Miyamoto, R. Mori, A. Suzuki, J. Nagatsu, *Agr. Biol. Chem.* (Tokyo), **27**, 576, 583 (1963).
3) N. Takahashi, A. Suzuki, S. Tamura, *Agr. Biol. Chem.* (Tokyo), **27**, 798 (1963); *ibid.*, **30**, 1, 13, 18 (1966).
4) Y. Kimura, N. Takahashi, S. Tamura, *Agr. Biol. Chem.* (Tokyo), **30**, 1507 (1969).
5) S. Yoshida, S. Shiraishi, K. Fujita, N. Takahashi, *Tetr. Lett.*, 1863 (1975).

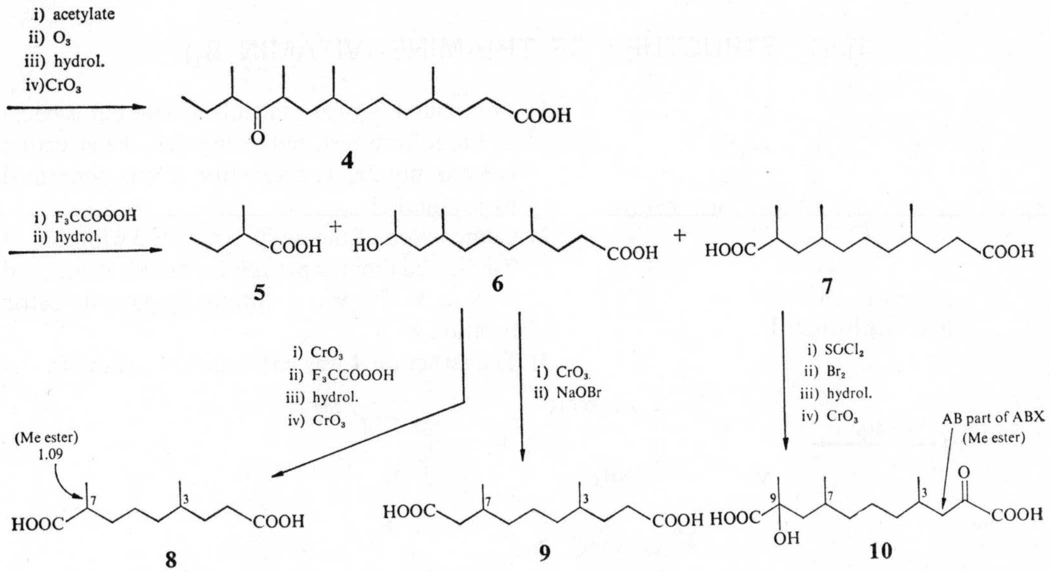

i) acetylate
ii) O$_3$
iii) hydrol.
iv) CrO$_3$

⟶

4

(structure: branched chain with ketone O and COOH)

i) F$_3$CCOOOH
ii) hydrol.

⟶

5   +   HO—...**6**   +   HOOC—...**7**

(structures 5, 6, 7 with COOH groups)

i) CrO$_3$
ii) F$_3$CCOOOH
iii) hydrol.
iv) CrO$_3$

i) CrO$_3$
ii) NaOBr

i) SOCl$_2$
ii) Br$_2$
iii) hydrol.
iv) CrO$_3$

AB part of ABX
(Me ester)

(Me ester)
1.09

HOOC—7—3—COOH
8

HOOC—7—3—COOH   HOOC—9—7—3—COOH
9   OH   **10**

Remarks

Piericidin A is an insecticidal metabolite having a fully substituted pyridine nucleus. It was isolated from *Streptomyces mobaraensis*. Recently sixteen piericidin homologues were isolated from *Streptomyces pactum* and their structures were established.[6]

6) N. Takahashi, Proceedings of the first International Congress of IAMS, **5**, 472 (1975).

11.41 STRUCTURE OF THIAMINE (VITAMIN B$_1$)

thiamine chloride
hydrochloride 1

1) Compound 3 gives *p*-nitrobenzoate but a negative iodoform test, indicating that the structure is 3 and not 3a. The structure 3 was confirmed by synthesis.[2]

2) Comparison of the uv spectrum of 3 with that of 7 indicated that the pyrimidine moiety is attached to N in 3. This was confirmed by potentiometric titration.[3]

3) The structure 4 was confirmed by synthesis.

$$1 \xrightarrow{\text{Na}_2\text{SO}_3^{1)}} \quad 2^{5)} \quad + \quad 3$$

$$1 \xrightarrow{\text{Na/liq. NH}_3} \quad 4 \xrightarrow{\text{HBr}} \quad 5^{6)}$$

$$3 \xrightarrow{\text{Na}_2\text{SO}_3}$$

$$3 \xrightarrow{\text{HNO}_3} \quad 6^{7)} \qquad 3a \qquad 7$$

Remarks

Thiamine occurs in all living cells as its pyrophosphate ester, cocarboxylase. It was first isolated from rice pericarp.[8] The daily requirement of thiamine in man is 1–3 mg: polyneuritis or beriberi is associated with thiamine deficiency. Plants can synthesize the vitamin.

1) R. R. Williams, *J. Am. Chem. Soc.*, **57**, 229 (1935); R. R. Williams, E. R. Buchman, A. E. Ruehle, *J. Am. Chem. Soc.*, **57**, 1093 (1935); R. R. Williams, R. E. Waterman, J. C. Keresztesy, E. R. Buchman, *J. Am. Chem. Soc.*, **57**, 536 (1935).
2) H. T. Clarke, S. Gurin, *J. Am. Chem. Soc.*, **57**, 1876 (1935).
3) R. R. Williams, A. E. Ruehle, *J. Am. Chem. Soc.*, **57**, 1856 (1935).
4) R. Grewe, *Z. Physiol. Chem.*, **237**, 98 (1935).
5) R. R. Williams, *J. Am. Chem. Soc.*, **58**, 1063 (1936).
6) R. Grewe, *Z. Physiol. Chem.*, **242**, 89 (1936).
7) E. R. Buchman, R. R. Williams, J. C. Keresztesy, *J. Am. Chem. Soc.*, **57**, 1849 (1935).
8) B. C. P. Jansen, W. F. Donath, *Chem. Weekblad*, **23**, 201 (1926); *Proc. Koninkl. Ned. Akad. Wetenschap.*, **29**, 1390 (1926).

11.42 STRUCTURE OF SPARSOMYCIN

sparsomycin $\mathbf{1}^{1,2)}$

1) Compound **1** has no basicity, indicating the presence of an amide linkage.
2) The nmr spectrum does not show an Me-CH signal, so that the Me group in 2-aminopropanol **8** is formed during desulfurization.
3) The chemical shift of Me at 2.30 suggests MeS–, not MeSO–.

4) Compound **7** was identified by synthesis. The tribenzoate derivative of D-2-amino-propanol **9** was identified by synthesis of the antipode from L-2-aminopropanol.

Remarks

Sparsomycin **1** has a unique structure containing a C–SO–C–S–C moiety. It shows broad-spectrum antibacterial and antifungal activities, and also has strong antitumor activity. It is produced by *Streptomyces sparsogenes*, which also produces another antibiotic, tubercidine.[3]

1) P. F. Wiley, F. A. MacKeller, *J. Am. Chem. Soc.*, **92**, 417 (1970).
2) A. D. Argoudelis, R. R. Herr, *Antimicrobial Agents and Chemotherapy*, 780 (1962).
3) S. P. Owen *et al.*, *Antimicrobial Agents and Chemotherapy*, 772 (1962).

11.43   STRUCTURE OF SURUGATOXIN

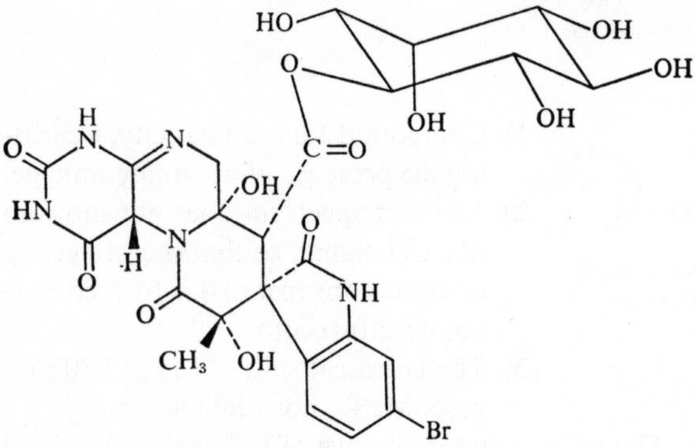

surugatoxin **1**

mp: > 300°

uv (H_2O): 276 (ε 15,000)

Remarks[1]

Intoxication occurred from ingestion of the carnivorous pastropod, *Babylonia japonica*, captured in Suruga Bay, Japan, from which surugatoxin **1** was isolated. The patients complained of visual defects, including amblyopia and mydriasis, with thirst, numbness of lips, speech disorders, constipation and dysuria. The pure toxin evoked mydriasis in mice at a minimum dose of 0.05 μg/g body weight. The molecular structure including the absolute configuration of the toxin were determined by x-ray analysis of single crystals. Surugatoxin is the first marine toxin so far discovered that contains indole and pteridine skeletons in the structure.

1) T. Kosuge, H. Zenda, A. Ochiai, N. Masaki, M. Noguchi, S. Kimura, H. Narita, *Tetr. Lett.*, 2545 (1972).

11.44   STRUCTURE OF *CYPRIDINA* LUCIFERIN

Cypridina luciferin **1**[1)]

mp (dihydrobromide): 254°
pK_a (H_2O): <2, 8.35, >11
uv (MeOH): 218 (ε 27,500), 270 (ε 17,000), 310 (ε 11,500), 435 (ε 9,000)
nmr in DMSO-d_6

1) The molecular formula of **3** was deduced from elemental analysis of **3** and high resolution ms of **4**.

2) The presence of monosubstituted guanidine was indicated by the positive Sakaguchi test and pK_a' above 11; **4** has an amino group since the pK_a' is at 9.6 in 50% ethanol.

3) Hydroetioluciferin **5**, obtained from **3** by catalytic hydrogenation, shows a uv spectrum typical of a β-aminomethylindole and gives indole on heating.

4) Acid hydrolysis of **1** gives arginine (0.62 mole) and glycine (0.96 mole).

5) **3** gives an orange color when treated with HNO_2 followed by β-naphthol, indicating the presence of a primary aromatic amino group in **3**.

6) An alternative structure having the alkylguanidine side chain at C-6 is eliminated since an m/e 141 peak ($C_{10}H_7N$) is present in ms spectrum of **4**.

7) An alternative structure having the amino and alkylguanidine groups at C-3 and C-2, respectively, is eliminated since the nitrogen atom in the arginine moiety is different from that in the glycine moiety, as shown in **4**).

8) Hydroluciferin **6**, obtained from **1** by catalytic hydrogenation, afforded L-isoleucine (0.50 mole) and D-alloisoleucine (0.48 mole) on acid hydrolysis, indicating C-α in the isoleucine moiety in **1** is not asymmetric, whereas C-β is asymmetric and its absolute configuration is as shown in **1**.

---

1) Y. Kishi, T. Goto, Y. Hirata, O. Shimomura, F. H. Johnson, *Tetr. Lett.*, 3427 (1966); *Bioluminescence in Progress* (ed. F. H. Johnson, Y. Haneda), p. 89, Princeton University Press, 1966; Y. Kishi, T. Goto, S. Eguchi, Y. Hirata, E. Watanabe, T. Aoyama, *Tetr. Lett.*, 3437 (1966).

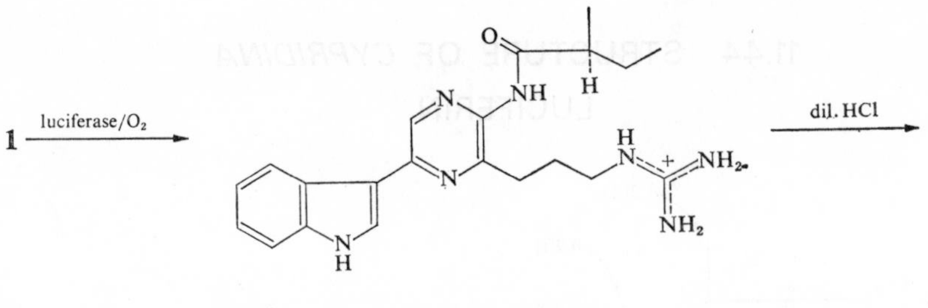

oxyluciferin **2**

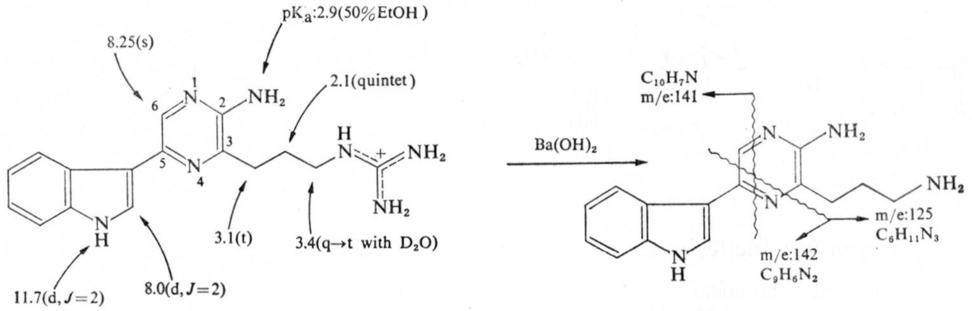

etioluciferin **3**   **4**

Remarks[1,2]

Cypridina luciferin **1**[2] is a bioluminescent substance isolated from *Cypridina hilgendorfii*, and affords light in the presence of luciferase and molecular oxygen, giving oxyluciferin **2**. The structure **1**, **2**, **3**, and **4** have been confirmed by synthesis.[3] The chemical transformations for the production of light are assumed to be as shown below.[4]

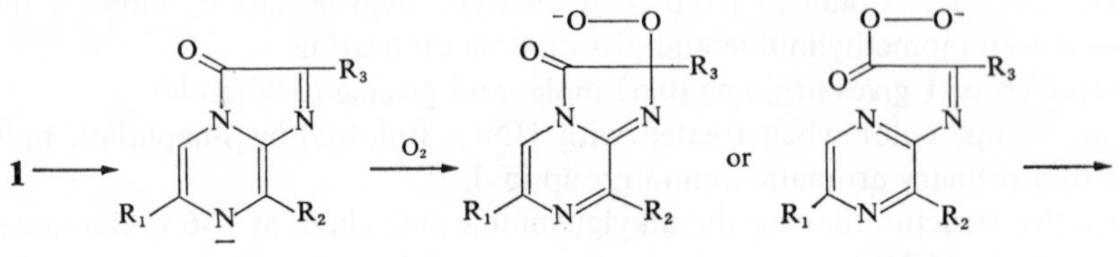

2) O. Shimomura, T. Goto, Y. Hirata, *Bull. Chem. Soc. Japan*, **30**, 929 (1957).
3) Y. Kishi, T. Goto, S. Inoue, S. Sugiura, H. Kishimoto, *Tetr. Lett.*, 3445 (1966).
4) F. McCapra, Y. Chang, *Chem. Commun.*, 1011 (1967); T. Goto, *Pure Appl. Chem.*, **17**, 421 (1968); T. Goto, S. Inoue, S. Sugiura, K. Nishikawa, M. Isobe, Y. Abe, *Tetr. Lett.*, 4035 (1968); H. Stone, *Biochem. Biophys. Res. Commun.*, **31**, 386 (1968).

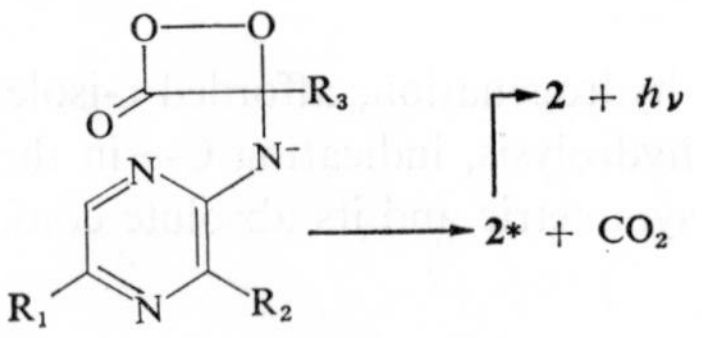

Aequorea, Renilla, and squid bioluminescences have been shown to involve similar types of bioluminescent substances. For structures of other luciferins, see ref. 8.

9 (R=H) chromophore obtained from *Aequorea*[5]   **11** *Renilla* luciferin[7]
10 (R=SO₃) chromophore obtained from a squid,
      Watasenid[6]

Mechanism of firefly bioluminescence, which involves firefly luciferin **12**, has also been extensively studied.[9]

firefly luciferin **12**

5) O. Shimomura, F. H. Johnson, *Biochem.*, **11**, 1602 (1972); *Tetr. Lett.*, 2963 (1973); Y. Kishi, H. Tanino, T. Goto, *Tetr. Lett.*, 2747 (1972).
6) T. Goto, H. Iio, S. Inoue, H. Kakoi, *Tetr. Lett.*, 2321 (1974).
7) K. Hori, M. J. Cormier, *Proc. Nat. Acad. Sci. U.S.A.*, **70**, 120 (1973).
8) T. Goto, Y. Kishi, *Angew. Chem. Intern. Ed. in Engl.*, **7**, 407 (1968).
9) W. D. McElroy, M. DeLuca, *Chemiluminescence and Bioluminescence* (ed. M. J. Cormier, D. M. Hercules, J. Lee), p. 285–311, Plenum Press, 1973.

11.45   STRUCTURE OF ACTINOMYCIN C$_3$

3.75~3.70 or 2.67

3.62

~2.2

1.13 or 0.91 (d, $J = 6.5$)

7.37 ($J = 7.8$)

7.64 ($J = 7.8$)

2.56

ca 4.0 or 3.9

4.79 and 3.63 or 4.72 and 3.63 (AB, $J = 17.8$)

2.75~2.60

0.96 or 0.76

6.03 or 5.98 ($J \cong 7.5$)

4.62 or 4.51 ($J = 2$)

1.27 (d, $J = 6.0$)

5.21 or 5.15 (dq, $J = 2$ and 6)

2.24

CH$_3$ HN   NH$_2$ O   N   CH$_3$—N   CH$_3$

actinomycin   C$_3$ **1**[1,2]

weak base
uv (cyclohexane): 241, 424, 446
nmr in CDCl$_3$[3]

1) Acid hydrolysis of **1** gave 2 moles each of L-threonine, sarcosine, L-proline, L-*N*-methyl-valine and D-alloisoleucine.[4] The amino acid sequence was determined by hydrazine cleavage.[5]

2) The quinonoid structure of **1** is indicated by reductive acetylation.[6]

3) Acid hydrolysis of **1** gave **2**, **3**, 2,5-dihydroxy-*p*-toluquinone and 3-hydroxy-1,8-dimethyl-2-phenoxazone;[7] structure **3** was proved by synthesis.[8]

4) Treatment of **1** with methanolic alkali gave, as a result of opening the lactone rings, actinomycin acid which could be oxidized with HIO$_4$ to a product containing no threonine moiety.[6]

5) The chromophore of actinomycin **9** was synthesized by oxidative coupling of **8**.[9]

1) H. Brockmann, G. Bohnsack, B. Franck, H. Gröne, H. Muxfeldt, C. H. Süling, *Angew. Chem.*, **68**, 70 (1956),
2) H. Brockmann, *Fortschr. Chem. Org. Naturstoffe* (ed. L. Zechmeister), **18**, 1 (1960); *Pure Appl. Chem.*, **2**, 405 (1960).
3) H. Arison, K. Hoogsteen, *Biochemistry*, **9**, 3976 (1970).
4) H. Brockmann, B. Franck, *Angew. Chem.*, **68**, 70 (1956).
5) H. Brockmann, G. Bohnsack, C. H. Süling, *Angew. Chem.*, **68**, 66 (1956).
6) H. Brockmann, B. Franck, *Angew. Chem.*, **68**, 68 (1956).
7) H. Brockmann, H. Gröne, *Angew. Chem.*, **68**, 66 (1956).
8) H. Brockmann, H. Muxfeldt, *Angew. Chem.*, **68**, 67 (1956).
9) H. Brockmann, H. Muxfeldt, *Angew. Chem.*, **68**, 69 (1956).

6) The smaller extinction of **1** than **9** in their uv absorptions is attributable to the distortion of the blocking peptide residues.[9]

7) The production of 2 moles of **4** indicates a symmetrical structure for the peptide moieties.[10]

desaminoactinomycin **2**   actinocinin **3**

actinocin **5**   **6**

depeptidoactinomycin
7[12,13]

10) H. Brockmann, P. Boldt, *Naturwiss.*, **50**, 19 (1963).
11) H. Brockmann, H. Muxfeldt, *Chem. Ber.*, **91**, 1242 (1958).
12) H. Brockmann, K. Vohwinkel, *Chem. Ber.*, **89**, 1373 (1956).
13) H. Brockmann, H. Maxfeldt, *Chem. Ber.*, **87**, 1379 (1956); *ibid.*, **89**, 1397 (1956).

$$\text{8} \xrightarrow{\underset{90\%}{O_2/(NH_4)_2CO_3/pH\ 9}} \text{9}$$

Remarks

The actinomycins are mixtures of yellow-red crystalline antibiotics produced by *Actinomyces antibioticus*[14] and various species of *Streptomyces*.[15] They are highly toxic and have some anti-cancer effect.[16] The main components of the actinomycins are: C_1 (=D),[17] C_2, C_3, X_2, $X_0\delta$, $X_0\beta$, $X_0\gamma$, $X_1\alpha$, etc.; each of which is composed of one phenoxazone chromophore and two pentapeptide-lactone rings. Variation of the peptide moieties in each congener is listed in the following table.[2] Other variants can be produced by the addition of appropriate amino acids to the culture broth.[2] Total syntheses of actinomycin C_1, C_2 and C_3 have been achieved.[18-20]

Actinomycin C_3	Alloisoleu–Pro	Alloisoleu–Pro
C_2[21]	Val–Pro	Alloisoleu–Pro
C_1[21,17]	Val–Pro	Val–Pro
X_2	Val–L-γ-Oxoproline	Val–Pro
$X_0\delta$	Val–L-Allohydroxyproline	Val–Pro
$X_0\beta$	Val–L-Hydroxyproline	Val–Pro
$X_0\gamma$	Val–Sar	Val–Pro
$X_1\alpha$	Val–Sar	Val–L-Hydroxyproline

14) S. A. Waksman, H. B. Woodruff, *Proc. Soc. Exp. Biol. Med.*, **45**, 609 (1940); *J. Bacteriol.*, **42**, 231 (1941).
15) C. E. Dalgliesh, A. W. Johnson, A. R. Todd, L. Vining, *J. Chem. Soc.*, 2946 (1950); H. Brockmann N. Grubhofer, W. Kass, H. Kalbe, *Chem. Ber.*, **84**, 260 (1951).
16) C. Hackmann, *Z. Krebsforsch.*, **58**, 607 (1952).
17) E. Bullock. A. W. Johnson, *J. Chem. Soc.*, 1602, 3280 (1957)
18) H. Brockmann, H. Lackner, *Naturwiss.*, **51**, 384, 407, 435 (1964); *Chem. Ber.*, **100**, 353 (1967).
19) H. Brockmann, H. Lackner, *Tetr. Lett.*, 3517, 3527 (1964).
20) J. Meienhofer, *J. Am. Chem. Soc.*, **92**, 3771 (1970).
21) H. Brockmann, P. Boldt, H. S. Petras, *Naturwiss.*, **47**, 62 (1960).

11.46   STRUCTURE OF RIBOFLAVIN (VITAMIN B₂)

riboflavin **1**

1) The following evidence indicated the presence of a ribityl group at N-9[1]: **1** forms a tetra-*O*-acetate and a diacetonide; **1** consumes Pb(OAc)₄, yielding 1 mole of HCHO; **2** yields no acetate, and is not oxidized by Pb(OAc)₄; **2** contains N–Me which is not present in **1**[2]; the configuration of the tetraol was determined by synthesis.[3]

2) Comparison of the uv spectra indicates that **1** and **2** have the same chromophore.

3) **2** must contain a cyclic ureide, since it yields **3** and urea with the consumption of two moles of water.[4]

4) The position of COOH in **3** was assumed on the basis of the ease of decarboxylation of **3**.[4,5]

5) The structure of **4** was confirmed by synthesis.[6]

lumiflavin **2**[7]

5[2]   **4**[5]

1) R. Kuhn, H. Rudy, T. Wagner-Jauregg, *Chem. Ber.*, **66**, 1950 (1933); R. Kuhn, T. Wagner-Jauregg, *ibid.*, **66**, 1577 (1933); R. Kuhn, H. Rudy, F. Weygand, *ibid.*, **68**, 625 (1935).
2) R. Kuhn, H. Rudy, *Chem. Ber.*, **67**, 1298 (1934).
3) P. Karrer, K. Schöpp, F. Benz, *Helv. Chim. Acta*, **18**, 426 (1935).
4) R. Kuhn, H. Rudy, *Chem. Ber.*, **67**, 892 (1934).
5) R. Kuhn, T. Wagner-Jauregg, *Chem. Ber.*, **66**, 1577 (1933).
6) R. Kuhn, K. Reinemund, *Chem. Ber.*, **67**, 1932 (1934); R. Kuhn, K. Reinemund, F. Weygand, *Chem. Ber.*, **67**, 1460 (1934).

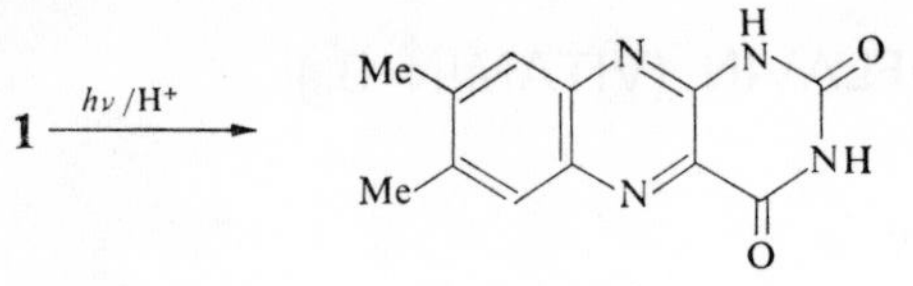

$$1 \xrightarrow{h\nu / H^+}$$

lumichrome 6[8]

Remarks

Riboflavin 1 is widely distributed in both the plant and animal kingdoms. It was first isolated from whey. In man, 2–3 mg of riboflavin are required per day: deficiency results in chellosis. It is a growth factor in most of the animal kingdom.

Riboflavin is a component of flavin mononucleotide (FMN) and also of the coenzyme flavin adenine dinucleotide (FAD) 7, which is the prosthetic group of flavoproteins. FAD was isolated from yeast in 1938.[9]

L-Lyxoflavin has been isolated from human myocardium, and possesses vitamin B_2 activity.[10]

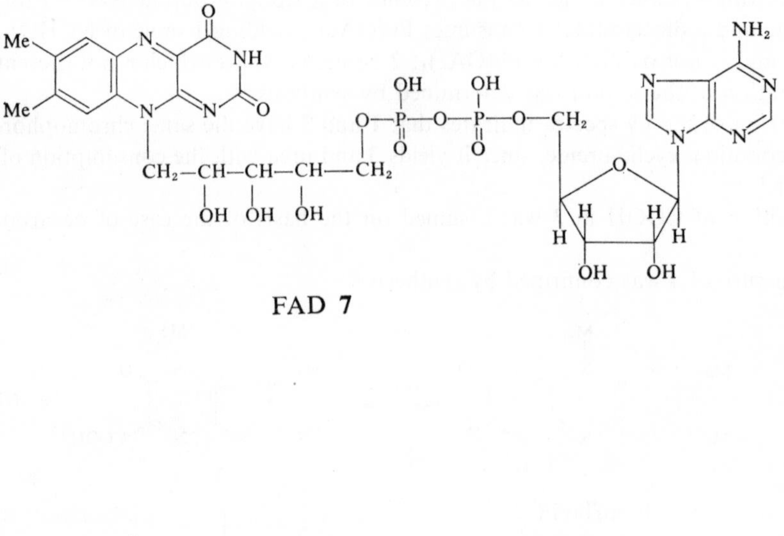

FAD 7

7) R. Kuhn, H. Rudy, T. Wagner-Jauregg, *Chem. Ber.*, **66**, 1950 (1933); O. Warburg, W. Christian, *Biochem. Z.*, **266**, 377 (1933).
8) P. Karrer, H. Salomon, K. Schöpp, E. Schlittler, H. Fritzsche, *Helv. Chim. Acta*, **17**, 1010 (1934).
9) O. Warburg, W. Christian, *Biochem. Z.*, **298**, 150 (1938).
10) D. Heyl, E. C. Chase, F. Koniuszy, K. Folkers, *J. Am. Chem. Soc.*, **73**, 3826 (1951); E. S. Pallares, H. M. Garza, *Arch. Biochem.*, **22**, 63 (1949).

11.47 STRUCTURE AND SYNTHESIS OF
THE FLUORESCENT Y BASE
FROM YEAST tRNA^Phe

Y base **1**[1]

uv (10% MeOH): 235 (ε 23,500), 263 (ε 4,500), 313 (ε 3,500)
uv (pH 2.1): 235 (ε 22,800), 285 (ε 4,615)
uv (pH 9.4): 236 (ε 23,500), 264 (ε 4,400), 303 (ε 4,350)

1) The side chain was established by comparison of the nmr and mass spectra of **1** with those of the model compound **3**.

2) The nucleus was established by comparison of the uv spectrum with that of synthetic **4**.

3) The methyl group at C-11 was deduced by comparison of the nmr spectra of **1** and synthetic **5**.

uv (10% MeOH): 237 (ε 26,000), 265 (ε 4,700), 315 (ε 3,850)
uv (pH 2.1): 237 (ε 23,600), 292 (ε 5,850)

---

1) K. Nakanishi, N. Furutachi, M. Funamizu, D. Grunberger, I. B. Weinstein, *J. Am. Chem. Soc.*, **92**, 7617 (1970).

Absolute Configuration[2]

Natural **1** $\xrightarrow[\text{ii) CH}_2\text{N}_2]{\text{i) O}_3\text{/EtOAc}}$ (structure with NH—COOMe, MeOOC—C—CH$_2$CH$_2$COOMe, H, S-configuration) $\xrightarrow[\text{ii) CH}_2\text{N}_2]{\text{i) ClCOOCH}_3\text{/MgO}}$ (S)(+)-glutamic acid

Δ_ε-0.14(232nm)
Δ_ε-0.76(206nm)

10

Remarks

tRNA's contain numerous odd bases, one of which, Y base in phenylalanine tRNA of yeast, has attracted great interest because of its important biochemical role and intense fluorescence. The structure of Y base was determined with 300 μg of the base obtained from 400 mg of tRNA^{Phe}. Biosynthetically, Y base may be regarded as a guanine modified by a glutamic acid residue at N-1 and a C$_2$ unit at the C-2 amino group, both of which are cyclized between C-10 and C-11. Originally ribose is presumed to have been attached at N-9.

The congeners **11**[3] and **12**[4] were isolated from various types of liver tRNA^{Phe} and from brewer's yeast tRNA^{Phe}, respectively.

11[3]   **12**[4]

Synthesis[2]

(structure **6**: NH$_2$—C(COOMe)H—CH$_2$—CH$_2$—I) $\xrightarrow{\text{ClCOOMe/NEt}_3}$ (structure **7**: MeOOCNH—C(COOMe)H—CH$_2$—CH$_2$—I) $\xrightarrow[\text{NaH/dioxane/benzene}]{\text{MeCOCH}_2\text{COOCH}_2\text{Ph/}}$

(structure **8**: MeOOCNH—C(COOMe)H—CH$_2$—CH$_2$—CH—COOCH$_2$Ph—COMe) $\xrightarrow[\text{iii) heat at 55-60°}(-\text{CO}_2)]{\substack{\text{i) Pd-C/H}_2\text{/MeOH} \\ \text{ii) Br}_2\text{/MeOH/CHCl}_3}}$ (structure **9**: MeOOCNH—C(COOMe)H—CH$_2$—CH$_2$—CH—Br—COMe) $\xrightarrow[\text{(pH4.5-6)}]{\substack{\text{(guanine structure)} \\ \text{/aq.DMSO}}}$ dl-**1**

---

2) M. Funamizu, A. Terahara, A. M. Feinberg, K. Nakanishi, *J. Am. Chem. Soc.*, **93**, 6706 (1971).
3) K. Nakanishi, S. Blobstein, M. Funamizu, N. Furutachi, G. Van Lear, D. Grunberger, K. W. Lanks, I. B. Weinstein, *Nature, New Biol.*, **234**, 107 (1971).
4) H. Kasai, M. Goto, S. Takemura, T. Goto, S. Matsuura, *Tetr. Lett.*, 2725 (1971).

CHAPTER **12**

Aspects of
Natural Products Photochemistry

12.1   INTRODUCTION

Photochemistry is one of the most rapidly expanding fields in organic chemistry. Until the early 1950's, light-induced organic reactions were not well understood because of the unexpected and diversified reactivities displayed by excited molecules. However, this situation has changed dramatically in subsequent years owing to developments in theory and in experimental techniques. Natural products such as santonin, a molecule which offers the possibility of numerous structural and stereochemical variations, have provided excellent models for experimental and theoretical chemists to conduct studies aimed at clarifying the chemistry of excited-state molecules.

In this chapter, several topics from natural products photochemistry have been selected. They are described mainly with a view to clarifying some of the underlying photochemical principles.

12.2   PHOTOCHEMISTRY OF CROSS CONJUGATED KETONES

Ever since α-santonin was found to undergo light-induced rearrangement,[1] the photochemistry of cross conjugated ketones has drawn great attention among organic chemists. However, it was only a decade ago that correct structures were elucidated for various photoproducts. Following the discovery of this unique rearrangement, analogous photoisomerizations have been reported for numerous cyclohexadiene derivatives.[2-4]

Extensive effort has been devoted to mechanistic studies and generalization of the facile photorearrangements, and it is now widely accepted that photoisomerizations take place via an n, π^* triplet state and show strong solvent dependence.

General isomerization paths[4]

1) For old history of santonin, see J. L. Simonsen, D. H. R. Barton, *The Terpenes*, Vol. III, p. 292, Cambridge Univ. Press, 1952.
2) O. L. Chapman, *Advan. Photochem.* (ed. W. A. Noyes Jr., G. S. Hammond, J. N. Pitts Jr.), **1**, p. 323–436 Wiley-Interscience, 1963.
3) K. Schaffner, *Advan. Photochem.* (ed. W. A. Noyes Jr., G. S. Hammond, J. N. Pitts Jr.), **4**, p. 81–112, Wiley-Interscience, 1966.
4) P. J. Kropp, *Organic Photochemistry* (ed. O. L. Chapman), **1**, p. 1–86, Marcel Dekker, 1967.

1) The lumiketone **3** is the primary photoproduct in neutral media, e.g. dioxane; however, **3** is photochemically very labile and it is usually obtained in the form of secondary photoproducts.

2) Isomerizations in acidic media, e.g. 45% aq. AcOH, lead to two kinds of hydroxy ketones, **5** and **6**.

3) The stability of the intermediate cation **4** determines the reaction path in acidic media; thus, when R_1 is more electron releasing, e.g. $R_1 = Me$ and $R_2 = H$, the major product is the ketone **5**, whereas when R_2 is more electron releasing, the ketone **6** is formed preferentially.

4) These reactions proceed via n,π^* triplets.

Photoisomerizations in neutral media

$\xrightarrow{hv}$

solvent (isolation yield)
dioxane (42%)[5]
ethanol (13%,[6] 25%[5])

Santonin **7**   Lumisantonin **8**

$\xrightarrow{hv}$

(steroid)

9a $R_1 = R_2 = H$ **10a** $R_1 = R_2 = H$ dioxane (11%[7,8,9]) 60% on irradiation with a low pressure lamp at 2537Å[7]

9b $R_1 = Me, R_2 = H$ **10b** $R_1 = Me, R_2 = H$ dioxane (4%)[10]

9c $R_1 = H, R_2 = Me$ **10c** $R_1 = H, R_2 = Me$ dioxane (60–70%)[11]

9d $R_1 = R_2 = Me$ **10d** $R_1 = R_2 = Me$ dioxane (60%)[8,11]

$\xrightarrow{hv}$

ethanol (11%)[12,13]

11 (steroid)   **12**

$\xrightarrow{hv}$

dioxane (3%)[7]

13 (steroid derived)   **14**

5) D. Arigoni, H. Bosshard, H. Bruderer, G. Büchi, O. Jeger, L. J. Krebaum, *Helv. Chim. Acta*, **40**, 1732 (1957).
6) D. H. R. Barton, P. de Mayo, M. Shafiq, *J. Chem. Soc.*, 140 (1958).
7) H. Dutler, C. Ganter, H. Ryf, E. C. Utzinger, K. Weinberg, K. Schaffner, D. Arigoni, O. Jeger, *Helv. Chim. Acta*, **45**, 2346 (1962).
8) L. Ruzicka, O. Jeger, *Chem. Abstr.*, **55**, 26041, 22388 (1961).
9) H. Dutler, H. Bosshard, O. Jeger, *Helv. Chim. Acta*, **40**, 494 (1957).
10) C. Ganter, F. Greuter, D. Kägi, K. Schaffner, O. Jeger, *Helv. Chim. Acta*, **47**, 627 (1964).
11) K. Weinberg, E. C. Utzinger, D. Arigoni, O. Jeger, *Helv. Chim. Acta*, **43**, 236 (1960).

1) The photoproducts (lumiketones) shown are also photochemically labile.

2) Photochemical destruction of the 4-methyl lumiketones **8** and **9c** appears to be slower than their formation.

3) Further transformation of **10b** on irradiation with a low pressure lamp (2537 Å) is slow compared to its rate of formation, because of the smaller ε value of **10b** than of **9b** at 254 nm.

Photoisomerizations in acidic media

1) A methyl substituent effect is observed; thus, the 4-methyl dienones **7** and **9c** yielded only the hydroazulenone derivatives **15** and **18**, whereas the non-substituted dienone **9a** afforded the two expected hydroxy ketones (differences in reactivity are due to the stability of intermediate cations, such as **4**).

2) Lumiketones such as **3** are not intermediates since irradiation of them does not yield the product hydroxyketones.

12) K. Tsuda, E. Ohki, J. Suzuki, H. Shimizu, *Chem. Pharm. Bull.*, **9**, 131 (1961).
13) D. H. R. Barton, J. F. McGhie, M. Rosenberg, *J. Chem., Soc.*, 1215 (1961).
14) P. J. Kropp, W. F. Erman, *J. Am. Chem. Soc.*, **85**, 2456 (1963).
15) D. H. R. Barton, W. C. Taylor, *J. Chem. Soc.*, 2500 (1958); L. J. Lorenc, M. Miljkovic, K. Schaffner, O. Jeger, *Helv. Chim. Acta*, **49**, 1183 (1966).

3) The formation of these hydroxy ketones suggests that zwitterion intermediates such as **2** may be involved in the photorearrangements (see refs. 3 and 4 for dienones substituted at 2-C).

4) As the structure of the minor product from **19** is not yet known, the effect of the 11-keto group is not clear.

Abnormal photoisomerizations

1) Conversion of **21** to the "lumiketone" does not occur because of steric strain.

2) Photoirradiation of **24** is accompanied by loss of OH and OAc, a general occurrence with allylic OR.

3) The mechanism proposed for the **27a** ⟶ **28a** conversion, i.e. initial cleavage of the 9,10-bond followed by rearrangement,[15] has since been criticized because compounds of type **28a** may also be formed from the more common lumiketones.[15]

Mechanistic studies

Mechanistic studies have been carried out primarily by Zimmerman et al.[18] The currently accepted pathways for these photoisomerizations are as follows:

16) G. Bozzato, H. P. Throndsen, K. Schaffner, O. Jeger, *Helv. Cheim. Acta*, **86**, 2073 (1964).

17) C. Ganter, R. Warszawski, H. Wehrli, K. Schaffner, O. Jeger, *Helv. Chim. Acta*, **46**, 320 (1963).

18) H. E. Zimmerman, J. S. Swenton, *J. Am. Chem. Soc.*, **86**, 1436 (1964); H. E. Zimmerman, R. G. Lewis, J. J. McCullough, A. Padwa, S. W. Stanley, M. Semmelhack, *ibid.*, **88**, 1965 (1966); H. E. Zimmerman, D. S. Crumrine, *ibid.*, **90**, 5612 (1968); H. E. Zimmerman, D. S. Crumrine, D. Dopp, P. S. Huyffer, *ibid.*, **91**, 434 (1969): H. E. Zimmerman, G. Jones II, *ibid.*, **92**, 2753 (1970).

1) Kinetic studies and measurements of the phosphorescence spectrum of **30**, which is very similar to that of benzophenone and has a short life time of *ca.* 0.5 msec, showed the reactive excited state to be an *n*, π^* triplet.

2) 3,5-Bond formation is permitted in view of the symmetry properties of the lowest excited state.

3) This electron demotion is an electronic transition from the singly occupied antibonding molecular orbital of **32** to the low energy, singly occupied oxygen non-bonding orbital.

4) Evidence for a zwitterion (such as **33**) in a similar photoreaction has been reported by trapping.[19]

---

19) D. I. Schuster, Kou-Chang Liu, *J. Am. Chem. Soc.*, **93**, 6711 (1971); see also D. J. Patel, D. I. Schuster, *ibid.*, **89**, 184 (1967); *ibid.*, **90**, 5137 (1968); D. I. Schuster, V. Y. Abraitys, *Chem. Commun.*, 419 (1969); D. I. Schuster, W. V. Curran, *J. Org. Chem.*, **35**, 4192 (1970).

12.3 PHOTOCHEMISTRY OF THE BICYCLO[3.1.0] HEXANE-2-ONE SYSTEM

Bicyclo[3.1.0]hexane-2-ones, e.g. **1**, give rise to a large number of photochemical transformations.[1-3] The primary processes encountered in bicyclo[3.1.0]hexanones are: 1,5-bond cleavage followed by 1,2-bond cleavage, leading to a diene-ketone without skeletal rearrangement, **1**⟶**4**; 5,6-bond cleavage leading to a diene-ketene with rearrangement, **5**⟶**7**; and rearrangement to a cyclopropanone (formally a 1,3-suprafacial sigmatropic rearrangement) followed by thermal rearrangement, **5**⟶**8**⟶**10/11**.[3]

1) 1,5-bond cleavage should be favored by the presence of substituents on C-1 and C-5, and 5,6-bond cleavage should be favored by C-6 substituents.
2) In addition to substituents, geometric constraints and differences in experimental conditions can also influence the initial bond cleavage.
3) Phenolic products such as **10** and **11** are formed via a cyclopropane intermediate such as **8**.

1) P. J. Kropp, *Organic Photochemistry*, (ed. O. L. Chapman), vol. 1, p. 1–86, Marcel Dekker, 1967.
2) K. Shaffner, *Advances in Organic Photochemistry* (ed. W. A. Noyes Jr., G. S. Hammond, J. N. Pitts Jr.), vol. 4, p. 81–112, Wiley, 1966.
3) O. L. Chapman, *XXIIIrd International Congress of Pure and Applied Chemistry*, vol. 1 (special lecture), p. 311–333, Butterworth, 1971.

Umbellulone[4,5]

umbellulone 12 hv/neat MeOH   13   14   15

MeOH/−190°   MeOOC   16

1) Irradiation of **12** at −190° produces **13** and **14** but no thymol **15**. On warming to −70°, the ketene **13** disappears with concurrent increase in thymol.[5]

2) Diene-ketene **13** can be trapped by irradiation at −190° of a methanol solution of **12**, and then warming.[5]

3) Although methanol does not trap a measurable amount of ketene **13** at room temperature, it is reasonable to conclude that this failure to trap **13** is due to rapid, irreversible cyclization to **14**.[3]

Lumisantonin[1−3,5,6]

18 (blue species)

hv

hv

17   19 Δ / hv mazdasantonin 20   EtOOC   21

1) Lumisantonin **17** gives photosantonic acid derivatives such as **21**, when irradiated in the presence of nucleophiles. In contrast, mazdasantonin **20** is formed when nucleophiles are not present.[5]

2) Ketene **19** was detected as a primary photoproduct by ir measurements at −190°.

3) An unknown blue species **18** was also observed during irradiation at −190°, but disappearance of the color caused no change in the ir spectrum.[5]

4) Ketene **19** disappeared with concurrent formation of mazdasantonin **20** on warming to room temperature (in the absence of nucleophiles).[5]

5) If low-temperature irradiation is carried out in EPA glass, warming of ketene **19**, affords the ester **21** rather than **20**.[3]

---

4) J. W. Wheeler, R. H. Eastman, *J. Am. Chem. Soc.*, **81**, 236 (1959).

5) L. L. Barber, O. L. Chapman, J. D. Lassila, *J. Am. Chem. Soc.*, **90**, 5933 (1968).

Steroids

1) Phenolic compounds and related rearranged products such as **29** and **30** are formed via cyclopropanone intermediates,[7] **24** and **33**.
2) Such intermediates rearrange thermally to spiro-ketones, which are further photochemically converted to spiro-cyclopropane derivatives, **27** and **36** (see Photochemistry of cross conjugated ketones).[9]
3) See refs. 1,2 and 3 for similar photochemical rearrangements of other steroid derivatives.

6) M. H. Fisch, J. H. Richards, *J. Am. Chem. Soc.*, **90**, 1553 (1968).
7) C. Ganter, F. Greuter, D. Kägi, K. Schaffner, O. Jeger, *Helv. Chim. Acta*, **47**, 627 (1964).
8) K. Weinberg, E. C. Utzinger, D. Arigoni, O. Jeger, *Helv. Chim. Acta*, **43**, 236 (1960).
9) L. L. Barber, O. L. Chapman, J. D. Lassila, *J. Am. Chem. Soc.*, **91**, 3664 (1969).

12.4   PHOTOCHEMISTRY OF α,β-UNSATURATED KETONES

Cyclic α,β-unsaturated ketones such as **1** undergo photochemical rearrangements formally similar to those of cross-conjugated dienones to afford cyclopropyl ketones such as **3** in polar solvents (e.g. t-BuOH, EtOH).[1-4] On the other hand, such enones isomerize photochemically to β,γ-unsaturated ketones **4** in non-polar solvents (e.g. cyclohexane, benzene) via n, π^* triplets.[4-6]

The detailed course of the cyclic enone ⟶ cyclopropyl ketone rearrangement has not been determined unambiguously. It has been shown by Schaffner and his associates[4,7] that the overall reaction is formally a concerted $(\sigma 2a + \pi 2a)$[8] cycloaddition, and that it proceeds with retention at C-1 and with inversion at C-10. These stereospecific reactions can be rationalized on the basis of a step-wise mechanism $(\mathbf{1} \longrightarrow \mathbf{2} \longrightarrow \mathbf{3})$ or a photochemically allowed concerted $(\sigma 2a + \pi 2a)$ mechanism.[4,5] Kinetic and solvent dependency studies of these rearrangements so far suggest that the reactive excited state involved is a π, π^* triplet.[4,5]

General isomerization paths

1) The isomerizations $(\mathbf{1} \longrightarrow \mathbf{3}$ and $\mathbf{1} \longrightarrow \mathbf{4})$ are triplet state reactions.[5-7]
2) $^3(n, \pi^*)$ and $^3(\pi, \pi^*)$ are almost degenerate (phosphorescence, phosphorescence excitation, and photoselection techniques).[4,9,10]
3) The enones adopt a considerably distorted non-planar geometry in the $^3(\pi, \pi^*)$ state.
4) The isomerizations $(\mathbf{1} \longrightarrow \mathbf{3})$ are stereospecific (concerted bond migration is required).

1) O. L. Chapman, *Advances Photochemistry* (ed. W. A. Noyes Jr., G. S. Hammond, J. N. Pitts Jr.) vol. 1, p. 323–436, Wiley, 1963.
2) P. J. Kropp. *Organic Photochemistry*, (ed. O. L. Chapman) vol. 1, p. 1–86, Marcel Dekker, 1967.
3) N. J. Turro, *Molecular Photochemistry*, p. 169, Benjamin, 1967.
4) K. Schaffner, *XXIIIrd International Congress of Pure and Applied Chemistry*, vol. 1 (special lecture), p. 405–417), Butterworth, 1971; see also: W. G. Dauben, W. A. Spitzer, M. S. Kellog, *J. Am. Chem. Soc.*, **93**, 3674 (1971); H. E. Zimmerman, J. W. Wilson, *J. Am. Chem. Soc.*, **86**, 4036 (1964); H. E. Zimmerman, K. G. Hancock, *ibid.*, **90**, 3749 (1968).
5) S. Kuwata, K. Schaffner, *Helv. Chim. Acta*, **52**, 173 (1969).
6) N. Furutachi, Y. Nakadaira, K. Nakanishi, *J. Am. Chem. Soc.*, **91**, 1028 (1969).
7) D. Bellus, D. R. Kearns, K. Schaffner, *Helv. Chim. Acta*, **52**, 971 (1969).
8) R. B. Woodward, R. Hoffmann, *Angew. Chem.*, **81**, 797 (1969).
9) D. R. Kearns, G. Marsh, K. Schaffner, *J. Chem. Phys.*, **49**, 3316 (1968).
10) G. Marsh, D. R. Kearns, K. Schaffner, *Helv. Chim. Acta*, **51**, 1890 (1968).

Solvent dependency

Since the π, π^* singlet state is stabilized in hydroxylic solvents, e.g. t-BuOH, by solvation due to the strong ionic character of the excited state ($\overset{+}{C}$–C=C–$\overset{-}{O}$), the excitation energy is less than that in non-polar solvents, e.g. cyclohexane or benzene. In contrast, since non-bonding electrons in the ground state molecule are stabilized by H-bonding with the solvent –OH group, a larger excitation energy (n, π^*) is required in order to break the H-bonding ($\overset{\delta+}{C}=\overset{\delta-}{O} \ldots$ H–O–R $\xrightarrow{n,\pi^*} \overset{\delta-}{C}=\overset{\delta+}{O} \ldots$ H–O–R). Moreover, the singlet-triplet splittings of n, π^* states are smaller than those of π, π^* states. Accordingly, if the energy difference between the n, π^* and π, π^* singlet states is small, the lowest excited triplet can be the π, π^* state, particularly in hydroxylic solvents.

| in polar solvents (hydroxylic solv.) | in non polar solvents (non-hydroxylic solv.) |

Some examples are given below.

Steroids

1) No isomerization occurs when R = Me, OAc, or halogen in **5**.
2) Steroids such as **8** give no cyclopropyl ketone even in hydroxylic solvents (nature of the excited state differs from that of **5**).

α,β-Unsaturated ketones undergo unique H-abstractions in addition to the normal skeletal rearrangements mentioned above (see following examples).[12,13]

---

11) J. Pfister, H. Wehri, K. Schaffner, *Helv. Chim. Acta*, **50**, 166 (1967).
12) T. Kobayashi, M. Kurono, H. Sato, K. Nakanishi, *J. Am. Chem. Soc.*, **94**, 2863 (1972).
13) W. Hertz, M. G. Nair, *J. Am. Chem. Soc.*, **89**, 5474 (1967).

Taxinines[12]

Photoirradiations of taxinines yielded very interesting isomerized photoproducts via π, π^* triplet states.

1) Transannular H-abstractions ($\mathbf{10} \longrightarrow \mathbf{11}$) can be regarded as following a concerted ($\sigma^2 s + \pi^2 s$) route (a C-11, C-12 biradical is also conceivable but less likely).

2) Molecular models indicate that initial transfer of 3-H to the carbonyl oxygen would be impossible in view of the direction of the carbonyl group (see **12**).

3) The isopropylidene group greatly hindered formation of the transannular bond by pulling apart the 3-H and 11-ene.

4) Kinetic studies and solvent dependency reveal that all isomerizations proceed via π, π^* triplet states.

Levopimaric acid-*para*-benzoquinone adduct[13]

1) **19a** gave exclusively **20a**.

2) **19b** gave only 20% of **20b**, and 60% of **21**.

3) The phosphorescence spectrum of **19a** exhibited a highly structured spectrum characteristic of emission from an n, π^* triplet, whereas the entirely featureless phosphorescence spectra of **19b** and **21** were indicative of π, π^* triplets.

4) Further irradiation of **21** gave **22**.

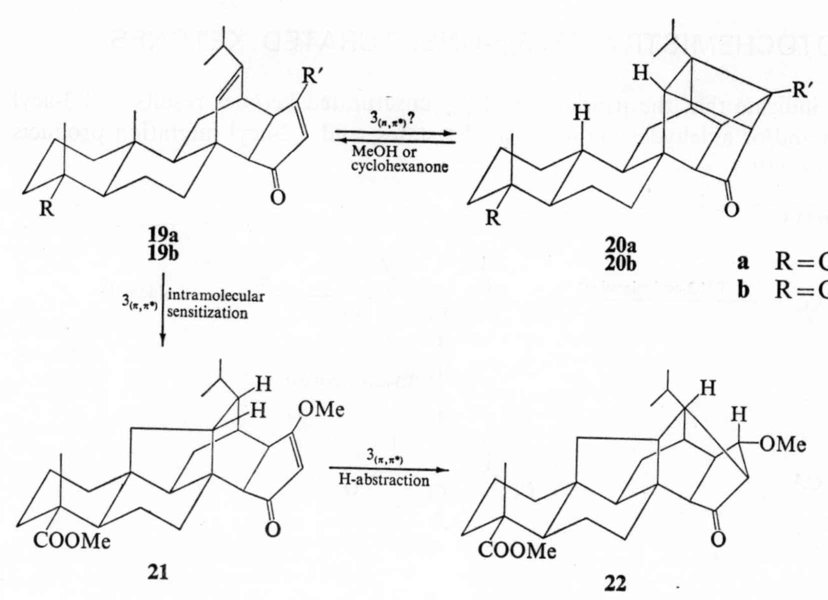

a   R=CH₂OH, R′=H

b   R=COOMe, R′=OMe

12.5 PHOTOCHEMISTRY OF β,γ-UNSATURATED KETONES

Recent results indicate that the irradiation of β,γ-unsaturated ketones results in 1,3-acyl migration products and/or aldehydes via n, π^* singlet states, and 1,2-acyl migration products via π,π^*-triplet states.[1-12]

General reaction pathways

1) 1,3-Acyl migrations and aldehyde formation are n,π^*-singlet reactions.[5,7]

2) 1,2-Acyl migrations are π, π^* triplet reactions.[8,9,11]

3) 1,3-Acyl migrations are stereospecific reactions,[2,5] whereas 1,2-acyl migrations proceed nonstereospecifically.[10,11]

4) Irradiation of β,γ-enones gives a photoequilibrium mixture with corresponding 1,3-acyl migration products, the photoequilibrium ratio being controlled by the relative ε values of the starting ketone and its 1,3-acyl migration products.[5]

5) Conformational and conjugative effects on the reaction center are important factors in such photochemical isomerizations.[5,11]

6) A strong interaction between $-C=O$ and $-C=C-$ favors 1,3-acyl migrations, whereas weak interaction favors 1,2-acyl migrations.[5,11,12]

1) J. R. Williams, H. Ziffer, *Chem. Commun.*, 194, 496 (1967); *Tetr.*, **24**, 6728 (1968).
2) N. Furutachi, Y. Nakadaira, K. Nakanishi, *J. Am. Chem. Soc.*, **91**, 1028 (1969).
3) K. Kojima, K. Sasaki, K. Tanabe, *Tetr. Lett.*, 1925 (1969).
4) K. Kojima, K. Sasaki, K. Tanabe, *Tetr. Lett.*, 3399 (1969).
5) H. Sato, N. Furutachi, K. Nakanishi, *J. Am. Chem. Soc.*, **94**, 2150 (1972).
6) K. G. Hancock, R. O. Grider, *Tetr. Lett.*, 4281 (1971).
7) N. Furutachi, J. Hayashi, H. Sato, K. Nakanishi, *Tetr. Lett.*, 1061 (1972).
8) R. S. Givens, W. F. Oettle, R. L. Coffin, R. G. Carlson, *J. Am. Chem. Soc.*, **93**, 3957 (1971) and references cited therein.
9) R. S. Givens, W. F. Oettle, *J. Am. Chem. Soc.*, **93**, 3963 (1971); and references cited therein.
10) S. Domb, K. Schaffner, *Helv. Chim. Acta*, **53**, 677 (1970).
11) H. Sato, K. Nakanishi, J. Hayashi, Y. Nakadaira, *Tetr.*, **29**, 275 (1973).
12) J. R. Williams, G. M. Sarkisian, *Chem. Commun.*, 1564 (1971).

Aldehyde formation[3-5]

1,3-Acyl migrations

1) Aldehydes **7** and **10** could not be isolated but their formation can be inferred from the isolation of oxetanes **8** and **11**.[4] (See also ref. 5).

2) The sharp contrast observed in the reaction modes of oxosteroids such as **6** and **9**, or **12** and **14**, can be explained as follows: in the case of **6** and **9** the expected 1,3-acyl migration compound **16** would have larger ε value (strong orbital interaction between $-C=O$ and $C=C-$) than the starting enones **6**, **9** because of the chair conformation of ring A. On the other hand, 1,3-acyl migration products **13** and **15** have ε values smaller than those of the starting ketones because ring B adopts a boat conformation in which only a weak interaction between $-C=O$ and $-C=C-$ is expected.[5]

1,2-Acyl migration

19 (R = H$_2$)[3,11]
20 (R = O)[10]

21 (R = H$_2$)[3,11]
22 (R = O)[10]

β,γ-Unsaturated aldehyde[13-17]

23

24

1) Photochemical decarbonylation of C-10 aldehydes occurs with intramolecular transfer of the aldehydic hydrogen to C-10.

2) The reaction can be rationalized by assuming that the reactions either proceed in a concerted manner or through an intimate "radical pair".

13) J. Iriarte, J. Hill, K. Schaffner, O. Jeger, *Proc. Chem. Soc.*, 114 (1963).
14) M. Akhtar, *Tetr. Lett.*, 4727 (1965).
15) K. Schaffner, *Chimia*, **19**, 575 (1965).
16) J. Hill, J. Iriarte, K. Schaffner, O. Jeger, *Helv. Chim. Acta*, **49**, 292 (1966).
17) See also, E. Baggiolini, H. P. Hamlov, K. Schaffner, *J. Am. Chem. Soc.*, **92**, 4906 (1970).

12.6 PHOTOCHEMISTRY OF EPOXY KETONES

Epoxy-ketones have been found to undergo facile photochemical rearrangements to β-diketones.[1] These photochemical reactions can be rationalized on the following grounds. Namely, since the three-membered ring possesses a somewhat delocalized electronic character,[2] it can interact with an adjacent carbonyl group under certain geometrical conditions, the interaction being maximal when the plane of the three-membered ring lies parallel to the carbonyl π orbitals.

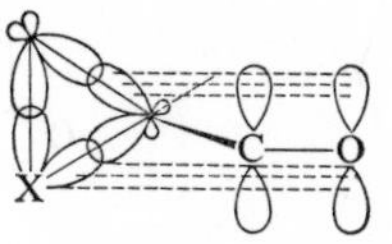

X=C,O,N

α,β-Epoxy-ketones

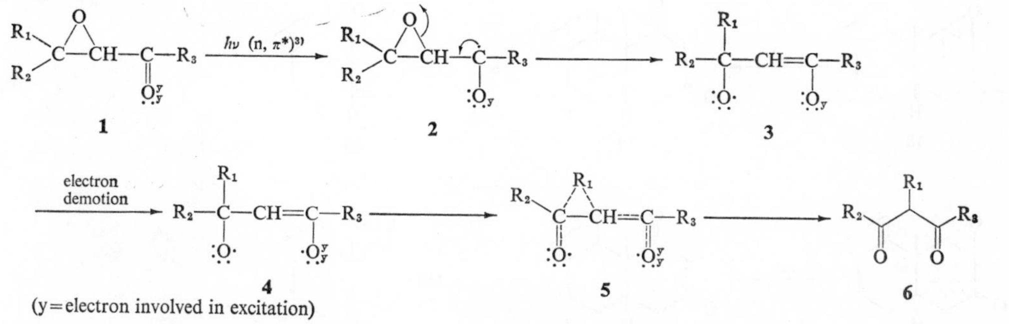

(y=electron involved in excitation)

1) The transformations shown here are initiated by n, π^* excitations.

2) Since the electron density at the carbonyl carbon in the excited state is enhanced relative to that in the ground state, substituents α to the carbonyl group are readily expelled from the n, π^* excited state, probably as odd-electron species.

3) Negative-quenching experiments show the probable n, π^* singlet nature of these reactions.

Pulegonoxide (7, 8)[4]

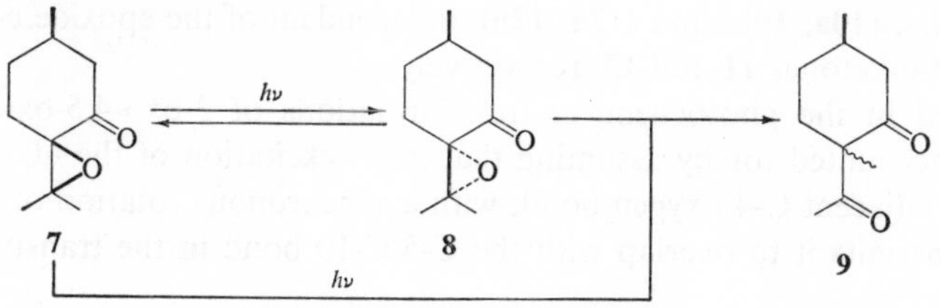

1) A. Padwa, *Organic Photochemistry* (ed. O. L. Chapman), vol. 1, p. 91–126, Marcel Dekker, 1967.
2) R. Parker, N. Issac, *Chem. Rev.*, **59**, 737 (1959).
3) H. E. Zimmerman, B. R. Cowley, C. Y. Tseng, J. W. Wilson, *J. Am. Chem. Soc.*, **86**, 947 (1964).
4) C. K. Johnson, B. Dominy, W. Reusch, *J. Am. Chem. Soc.*, **85**, 3894 (1963).

Steroids

1) Each of the two isomeric pairs, (**10a, 10b**) and (**12a, 12b**), independent of the epoxide configuration, gave the single 1,3-diketones **11** and **13**, respectively.

2) The stereospecificity observed in the photochemical transformations of 3-oxo-4,5-oxido-steroids (**14** and **16**) can be accounted for by assuming that n, π^* excitation of the $-C=O$ group induces rupture of the adjacent C-4 oxygen bond, with a synchronous rotation of the unfilled orbital at C-4; this permits it to overlap with the C-5/C-10 bond in the transition state, as shown in **24**.[7,8]

3) The stereospecificity observed in irradiation of 3-oxo-1,2-oxidosteroids **18, 21** can be rationalized in terms of the occurrence of a concerted H-shift and 1,10-bond migration, which leads to carbonyl formation at C-1.

5) C. Lehmann, K. Schaffner, O. Jeger, *Helv. Chim. Acta*, **45**, 1031 (1962).
6) H. Wehrli, C. Lehmann, K. Schaffner, O. Jeger, *Helv. Chim. Acta*, **47**, 1336 (1964).
7) H. Wehrli, C. Lehmann, P. Keller, J. J. Bonte, K. Schaffner, O. Jeger, *Helv. Chim. Acta*, **49**, 2218 (1966).
8) H. Wehrli, C. Lehmann, T. Iizuka, K. Schaffner, O. Jeger, *Helv. Chim. Acta*, **50**, 2403 (1967).
9) O. Jeger, K. Schaffner, H. Wehrli, *Pure Appl. Chem.*, **9**, 555 (1964).

α,β-Unsaturated γ,δ epoxy-ketones[10,11]

1) Irradiation of the n, π^* absorption band of **25** in ethanol at $-77°$ exclusively afforded the rearranged ene-dione **28**.

2) Both direct irradiation (n, π^*) and sensitized iradiation of **25** in ethanol at 24° gave identical mixtures of photoproducts, the ratio of **28**, **32**, and **34** being 1 : 0.05 : 0.1.

3) Selective π, π^* excitation of **25** at 24° yielded a somewhat less specific product mixture of **28**, **32** and **34**, i.e. 1; 0.05: 0.5.

4) The π, π^* excitation of isomer **29** in ethanol at 24° resulted in a mixture of **34** and **32** in the ratio 1 : 0.6.

5) Separate photolyses of photoproducts **28**, **32**, and **34** showed that these compounds are not photochemically interconvertible under the above conditions.

6) The stereospecificity observed in photoinduced transformations of **25** (at $-77°$) or **29** (at room temperature) can be accounted for by assuming that 9,10-bond migration from the intermediate **26** (or **30**) is a concerted process (see **27**, **31**, and **33**).

10) M. Debono, R. M. Molly, D. Bauer, T. Iizuka, K. Schaffner, O. Jeger. *J. Am. Chem. Soc.*, **92**, 420 (1970).

11) K. Schaffner, *XXIIIrd International Congress of Pure and Applied Chemistry*, vol. 1 (special lecture), p. 405-417, Butterworth, 1971.

12.7 PHOTOCHEMISTRY OF SATURATED KETONES

Photochemical reactions of ketones and aldehydes are initiated by formally forbidden n,π^* excitations. Two kinds of reaction modes, α-cleavage and H-abstraction,[1] have so far been reported.

α-Cleavage

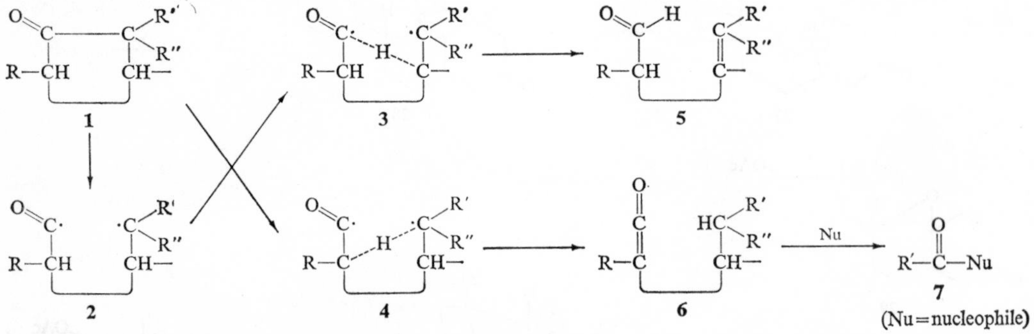

1) As shown in **8**, the singly occupied oxygen orbital can overlap with the C-1/C-2 bond in the excited state, thus giving rise to α-cleavage.[2,3]

2) The resulting diradical **2** undergoes hydrogen abstractions via two kinds of six-membered transition states, e.g. **3** and **4**, resulting in the formation of an aldehyde **5** and a ketene: the ketene **6** can be trapped as the corresponding carboxylic acid derivative **7** in the presence of a nucleophile.[1]

3) In the presense of oxygen, the aldehyde **5** is further oxidized to the carboxylic acid.

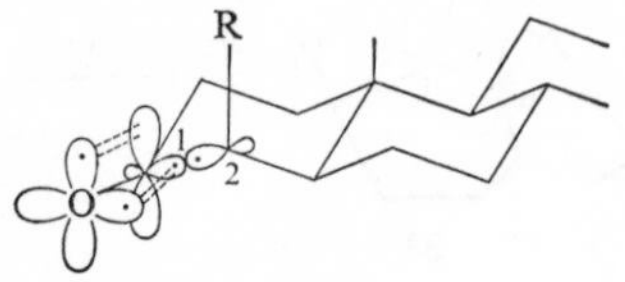

8

Some examples of α-cleavage are shown below.

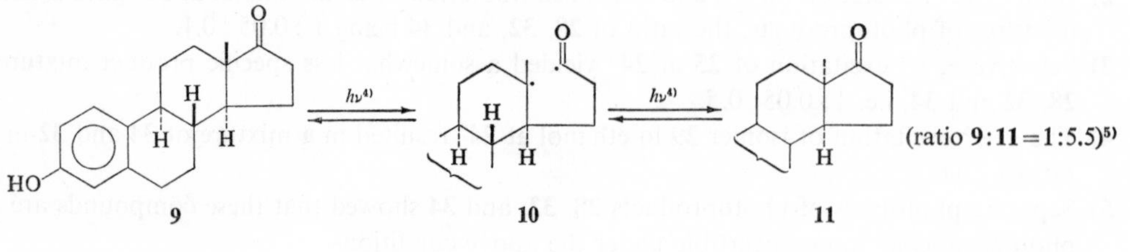

(ratio **9:11**=1:5.5)[5]

1) N. J. Turro, *Energy Transfer and Organic Photochemistry* (ed. A. A. Lamola, N. J. Turro), p. 133, Wiley, 1969.
2) H. E. Zimmerman, *Advances in Photochemistry* (ed. W. A. Noyes, G. S. Hammond, J. N. Pitts Jr.), p. 183–208, Wiley, 1963.
3) D. I. Schuster, G. R. Underwood, T. P. Knudsen, *J. Am. Chem. Soc.*, **93**, 379 (1971).

4) A. Butenandt, A. Wolff, P. Karlson, *Chem. Ber.*, **74**, 1308 (1941); A Butenandt, W. Friedrich, L. Poschmann, *ibid.*, **75**, 1931 (1942); A Butenandt, L. Poschmann, *ibid.*, **77**, 394 (1944).
5) H. Wehrli, K. Schaffner, *Helv. Chim. Acta*, **45**, 385 (1962).
6) P. Bladon, W. McMeekin, I. A. Williams, *J. Chem. Soc.*, 5727 (1963).
7) J. Iriate, K. Schaffner, O. Jeger, *Helv. Chim. Acta*, **47**, 1255 (1964).
8) G. Quinkert, B. Wegemund, F. Hombury, G. Cimbollek, *Chem. Ber.*, **97**, 958 (1962); D. Arigoni, D. H. R. Barton, R. Bernasconi, C. Djerassi, J. M. Mills, R. E. Wolff, *Proc. Chem. Soc.*, 306 (1959); *J. Chem. Soc.*, 1900 (1960).

γ-Hydrogen abstraction

Almost all saturated ketones possessing γ-hydrogen atoms undergo an intramolecular hydrogen transfer. A cyclic six-membered transition state is involved in these reactions, which proceed both by n, π^* singlet and triplet states.[9-12]

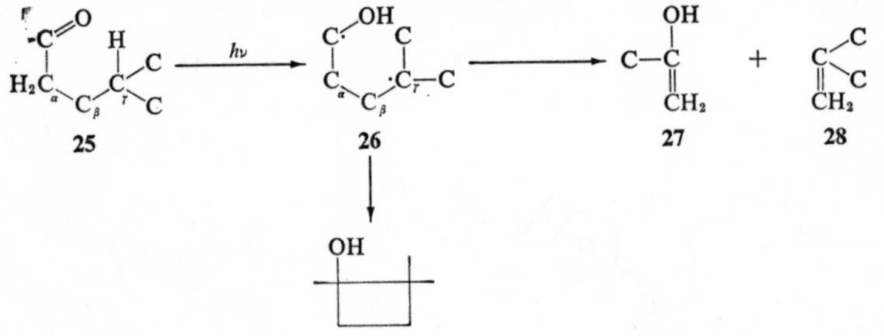

Some examples of γ-hydrogen abstraction are shown below.

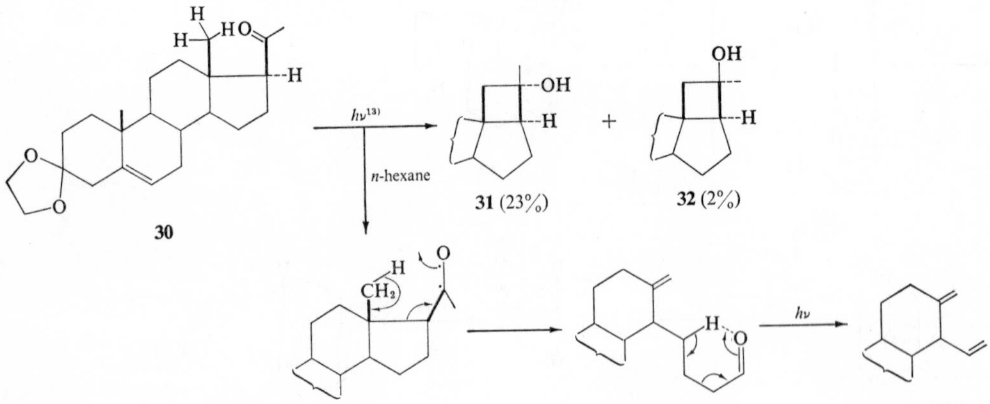

9) P. J. Wagner, G. S. Hammond, *J. Am. Chem. Soc.*, **88**, 1245 (1966); *ibid.*, **87**, 4009 (1965).
10) D. R. Coulson, N. C. Yang, *J. Am. Chem. Soc.*, **88**, 4511 (1966).
11) R. Srinivasan, *J, Am. Chem. Soc.*, **81**, 5061 (1959).
12) I. Orban, K. Schaffner, O. Jeger, *J. Am. Chem. Soc.*, **85**, 3034 (1963).
13) P. Buchschacher, M. Cereghetti, H. Wehrli, K. Schaffner, O. Jeger, *Helv. Chim. Acta*, **42**, 2122 (1959); H. Wehrli, M. Cereghetti, K. Schaffner, O. Jeger, *ibid.*, **43**, 354 (1960); N. C. Yang, D. D. Yang, *Tetr. Lett.*, 10 (1960).

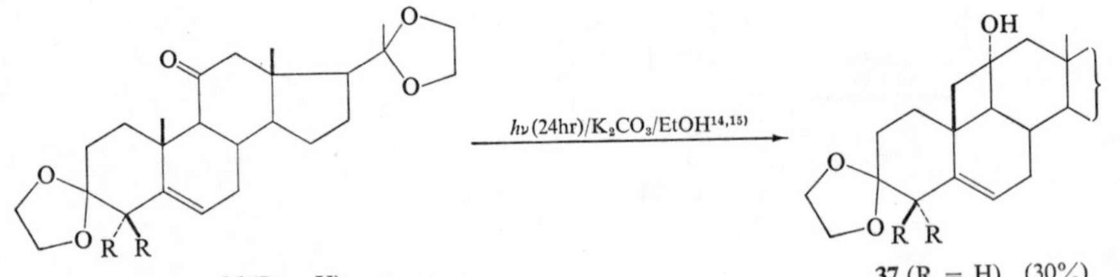

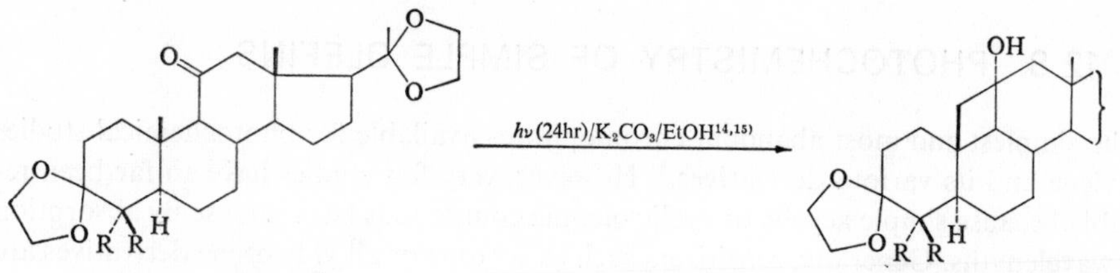

40 (R = H)
42 (R = Me)

$h\nu$ (24hr)/K_2CO_3/EtOH[14,15]

41 (R = H) (40%)
43 (R = Me) (65%)

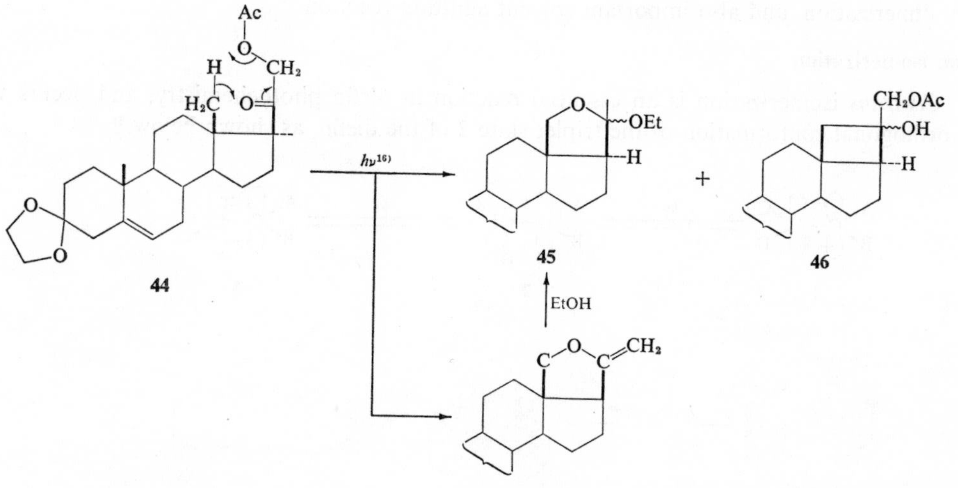

44

$h\nu$[16]

45 + 46

EtOH

47

1) As exemplified by ketones **36**, **38**, **40**, and **42**, the introduction of a 5-ene decreases the photoproduct yield, whereas the introduction of a 4,4-dimethyl group increases the yield.

2) These results can be rationalized as follows. Introduction of the 5-ene flattens ring B and increases the distance between the 10-Me and the carbonyl; on the other hand, the 4β-Me in **38** and **42** pushes the 10-Me towards the 11-carbonyl, thus leading to an increase in yield.

14) M. S. Heller, H. Wehrli, K. Chaffner, O. Jeger, *Helv. Chim. Acta*, **45**, 1261 (1962).
15) J. Iriate, K. Schaffner, O. Jeger, *Helv. Chim. Acta*, **46**, 1599 (1963).
16) H. Wehrli, M. Cereghetti, K. Schaffner, O. Jeger, *Helv. Chim. Acta*, **43**, 367 (1960).

12.8 PHOTOCHEMISTRY OF SIMPLE OLEFINS

One of the simplest and most abandant chromophores available for photochemical studies is that of ethylene and its various derivatives.[1] However, very few studies have so far been reported, probably because simple acyclic or cyclic olefinic compounds have intense uv absorption at rather low wavelengths. Generally, sensitizers such as acetone or alkyl benzene derivatives are employed in studying the photochemical behavior of olefins. Irradiation of olefins under these conditions results in *cis-trans* isomerization, addition to the sensitizers via hydrogen abstraction, dimerization, and also important solvent addition reactions.

Cis-trans isomerization

Cis-trans isomerization is an essential reaction in olefin photochemistry, and occurs via the orthogonal conformation of the triplet state **2** of the olefin, as shown below.[2]

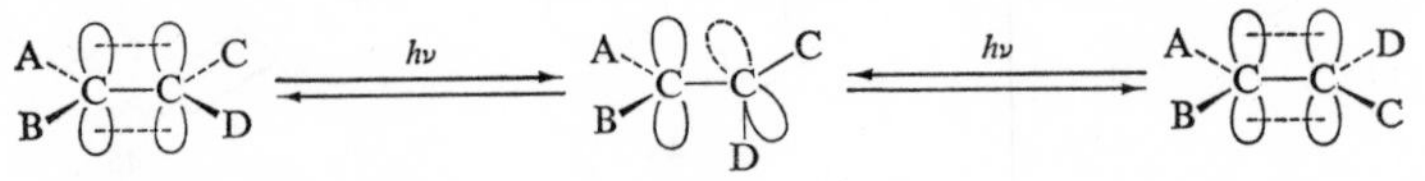

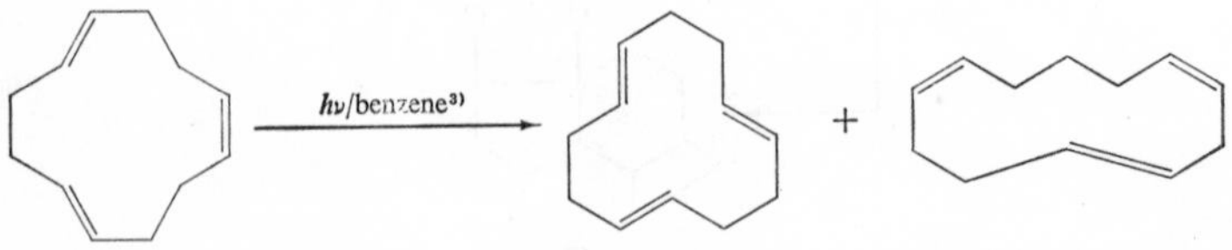

Addition to sensitizers

In the example below, this reaction proceeds via hydrogen abstraction from the substrate olefin by the sensitizer, acetone.[1] In this case, acetone does not behave as a sensitizer because the triplet energy of acetone is not sufficient to excite an unstrained olefin such as cyclohexene (acetone is generally an excellent sensitizer for strained 3-, 4- and 5-membered cyclic olefins).

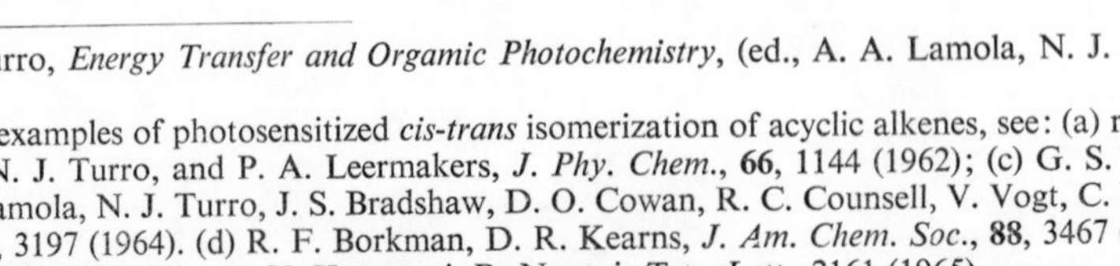

1) N. J. Turro, *Energy Transfer and Organic Photochemistry*, (ed., A. A. Lamola, N. J. Turro) Wiley, p. 133, 1969.
2) For the examples of photosensitized *cis-trans* isomerization of acyclic alkenes, see: (a) ref. 1); (b) G. S. Hammond, N. J. Turro, and P. A. Leermakers, *J. Phy. Chem.*, **66**, 1144 (1962); (c) G. S. Hammond, J. Saltiel, A. A. Lamola, N. J. Turro, J. S. Bradshaw, D. O. Cowan, R. C. Counsell, V. Vogt, C. Dalton, *J. Am. Chem. Soc.*, **86**, 3197 (1964). (d) R. F. Borkman, D. R. Kearns, *J. Am. Chem. Soc.*, **88**, 3467 (1966).
3) H. Nozaki, Y. Nishikawa, Y. Kamatani, R. Noyori, *Tetr. Lett.*, 2161 (1965).
4) P. de Mayo, J. B. Stothers, W. Templeton, *Can. J. Chem.*, **39**, 488 (1961).
5) J. S. Bradshaw, *J. Org. Chem.*, **31**, 237 (1966).

Dimerization

Most small-ring olefins (3-, 4- and 5-membered cyclic olefins) which have relatively low triplet energy are sensitized by acetone to give dimeric photo-products.

$$11 \xrightarrow{h\nu/\text{acetone}^{6)}} 12 \;+\; 13$$

$$14 \xrightarrow{h\nu/\text{acetone}^{7)}} 15 \;+\; 16\ (\text{CH}_2\text{COMe})$$

$$17 \xrightarrow{h\nu/\text{acetone}^{8)}} 18 \;+\; \text{others}$$

$$19 \xrightarrow{\text{C}_6\text{H}_5\text{COCH}_3{}^{9)}} 20$$

These dimers are formed via a triplet biradical intermediate such as **22**.

$$21 \xrightarrow{h\nu/\text{sensitizer}} \left[22 \;+\; 21 \right] \longrightarrow 23$$

Solvent addition in a protic medium

Irradiation of alkyl substituted 6- or 7-membered cyclic compounds gives an isomerized product and a solvent addition product in a protic medium such as methanol or ethanol in the presence of alkyl-substituted benzene as a sensitizer.[10,11]

$$24 \xrightarrow{h\nu/\text{sensitizer}^{10)}} [\pi, \pi^* \text{ triplet}] \longrightarrow 25$$

$$\xrightarrow{\text{H}^+} 26 \xrightarrow{\text{HOR}} 27 \;+\; 28$$

6) H. H. Stechl, *Angew. Chem.*, **75**, 1176 (1963).
7) R. Srinivasan, K. A. Hill, *J. Am. Chem. Soc.*, **88**, 3765 (1966).
8) H. D. Scharf, F. Korte, *Ber.*, **97**, 2425 (1964).
9) H. D. Schaff, F. Korte, *Tetr. Lett.*, 821 (1963).
10) P. J. Kropp, H. J. Krauss, *J. Am. Chem. Soc.*, **89**, 5199 (1967); P. J. Kropp, *ibid*, **88**, 4091 (1966).
11) J. A. Marshall, R. D. Carrol, *J. Am. Chem. Soc.*, **88**, 4092 (1966).

1) Triplet energy transfer occurs from the sensitizer (benzene, toluene, or xylene, Et = 81–84 kcal/mole) to the cycloalkane ($n \geqslant 6$, Et = *ca.* 70–82 kcal/mole), resulting in *cis-trans* isomerization.
2) The resulting *trans* isomer **25** ($n=6, 7$) is a highly strained compound, and rapid protonation occurs to release the strain energy in protic medium.
3) Smaller ring olefins (n = 5, 4, 3) cannot form the orthogonally oriented π, π^* triplet for structural reasons, and show radical characteristics. Such an excited olefin abstracts hydrogen atoms to give a reduction product or a dimer.
4) Acyclic, exocyclic, and larger ring compounds which can readily form both an orthogonally oriented triplet and a relatively unstrained *trans* isomer show no tendency to undergo double bond isomerization (endo to exo), solvent addition or reduction.
5) The solvent approaches from the less-hindered side of the cation intermediate **26**.[12]

Some examples are shown below.

29 **30** **31**

32 **33** **34** **35** **36**

37 **38** **39** **40**

41 **42** (R = Me, OMe, or OEt)

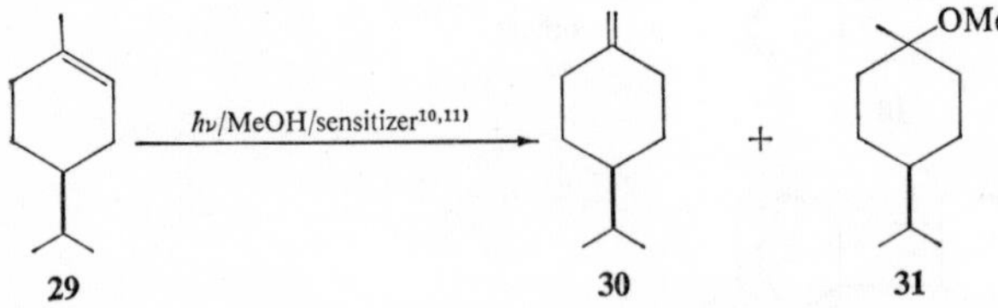

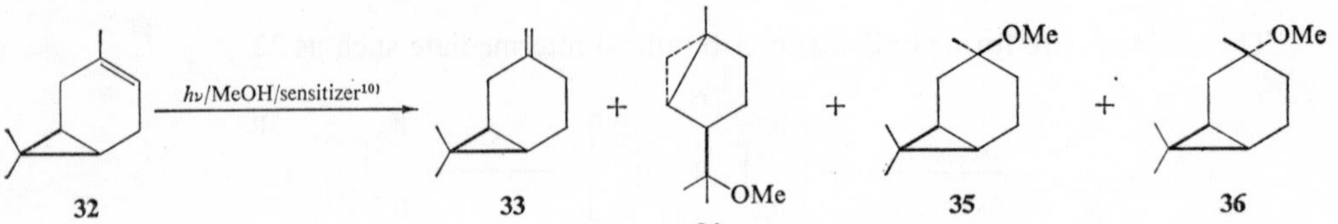

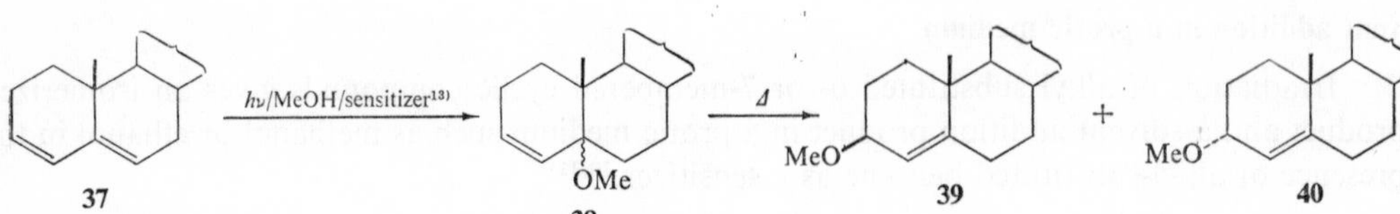

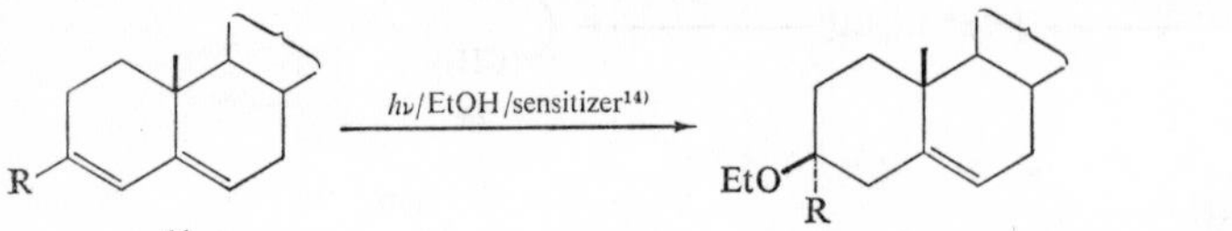

12) J. A. Marshall, M. J. Wurth, *J. Am. Chem. Soc.*, **89**, 6788 (1967).
13) J. Pusset, R. Bengelmans, *Tetr. Lett.*, 3249 (1967).
14) G. Bauslaugh, G. Just, E. L. Ruff, *Can. J. Chem.*, **44**, 2837 (1966).

12.9 PHOTOCHEMISTRY OF CYCLIC DIENES AND RELATED COMPOUNDS

Considerable efforts have been devoted to the clarification of photochemical transformations occurring in the ergosterol/vitamin D series because of their important physiological activities.[1] However the mechanism of these reactions was not understood until Woodward and Hoffmann[2] formulated their selection rules for "electrocyclic reactions" in 1965.

The Woodward-Hoffmann rules[2]

Woodward and Hoffmann predict that the stereochemistry of polyene cyclizations involving σ-bond formation between the terminal atoms of a polyene system is determined by the symmetry of the molecular orbital. Similar relations are also true for ring-opening reactions which give open-chain polyenes.

highest occupied orbital **1**   **2**

highest occupied orbital **5**   **6**

lowest vacant orbital **3**   **4**

lowest vacant orbital **7**   **8**

1) The highest occupied orbital (H.O.) is responsible for thermal reactions, whereas the lowest vacant orbital (L.V.) is responsible for photochemical reactions.

2) Since p orbitals having the same sign have to overlap, two terminal bonds must rotate specifically according to the orbital symmetry, namely, **conrotatory** (two bonds rotate in the same direction) or **disrotatory** (two bonds rotate in the opposite direction).

Photochemistry of the vitamin D series[1,3-5]

ergosterol **9**   precalciferol **10**

1) L. F. Fieser, M. Fieser, *Steroids*, ch. 4, McGraw-Hill, 1959.
2) R. B. Woodward, R. Hoffmann, *J. Am. Chem. Soc.*, **87**, 395 (1965); *Angew. Chem.*, **81**, 797 (1969).
3) G. J. Fonken, *Organic Photochemistry* (ed. O. L. Chapman), p. 197–244, Marcel Dekker, 1967.
4) R. Srinivasan, *Advances in Photochemistry* (ed. W. A. Noyes Jr., G. S. Hammond, J. N. Pitts Jr), p. 113–142, Wiley, 1966.
5) E. Havinga, R. J. de Kock, M. P. Rappoldt, *Tetr.*, **11**, 276 (1960); E. Havinga, J. Schaltmann, *ibid.*, **12**, 146 (1961); G. M. Sanders, E. Havinga, *Rec. Trav. Chim.*, **83**, 665 (1964).

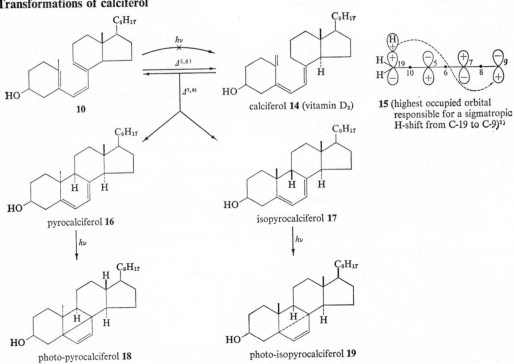

1) Photochemical electrocyclic reactions, such as the formation of *anti*-isomers **9** and **13**, follow the Woodward-Hoffmann rules.[2]

2) Excited tachysterol can rotate around the 6,7-bond to give lumisterol **13** by bonding between C-9 and C-10 (see **12**).[5]

3) In the formation of **13**, the approach of C-10 to C-9 from the β-side is subject to considerable steric hindrance by the 10-Me.[5]

Transformations of calciferol

1) All photochemical and thermal reactions follow the Woodward-Hoffmann rules.[2]

2) Reaction **10 ⇌ 14** is an antarafacial sigmatropic reaction (see **15**).[2]

6) M. Akhtar, C. J. Gibbons, *J. Chem. Soc.*, 5964 (1965).
7) W. G. Dauben, G. J. Fonken, *J. Am. Chem. Soc.*, **81**, 4060 (1959).
8) K. Dimorth, *Chem. Ber.*, **70**, 1631 (1937).

3) Photochemical reaction of dienes **16** (or **17**) cannot yield the corresponding trienes such as **10** for stereochemical reasons (orbital symmetry requirements do not permit arrangements such as **10**).

4) Since **16** and **17** adopt a chair conformation in ring B, cyclobutenes **18** and **19** can be readily formed via photochemical disrotatry ring closures).

Other transformations

AcO ... COOMe **20** $\xrightarrow[\text{conrotatory}]{h\nu^{9)}}$ **21** $\xrightarrow[\substack{\text{antarafacial} \\ \text{H-shift}}]{\Delta^{9)}}$ AcO ... **22**

C_8H_{17} **23** $\underset{\text{conrotatory}}{\overset{h\nu^{10)}}{\rightleftarrows}}$ C_8H_{17} **24** $\overset{\Delta^{10)}}{\rightleftarrows}$ C_8H_{17} **25**

C_9H_{17} AcO **25** $\xrightarrow[\text{stereospecific reaction}]{h\nu^{11,12)}}$ C_9H_{17} AcO **26**

Cyclic transoid dienes gives a bicyclobutane derivative such as **28**, and the detail mechanism was recently explained on the basis of the relaxed singlet excited state concept.[13]

C_8H_{17} **27** $\xrightarrow{h\nu^{14,15)}}$ **28** $\xrightarrow{\text{EtOH}}$ H OEt **29** + CH_2OEt **30**

9) R. L. Autrey, D. H. R. Barton, A. K. Ganguly, W. H. Reusch, *J. Chem. Soc.*, 3313 (1961); See also W. G. Dauben, M. S. Kellog, *J. Am. Chem. Soc.*, **93**, 3805 (1971).
10) W. R. Roth, B. Peltzer, *Angew. Chem. Intern. Ed.*, **3**, 440 (1964).
11) D. H. R. Barton, A. S. Kende, *J. Chem. Soc.*, 688 (1958).
12) D. H. R. Barton, R. Bernasconi, J. Klein, *J. Chem. Soc.*, 511 (1960).
13) W. G. Dauben, J. S. Ritscher, *J. Am. Chem. Soc.*, **93**, 2925 (1970).
14) W. G. Dauben, J. A. Ross, *J. Am. Chem. Soc.*, **81**, 6521 (1959).
15) W. G. Dauben, F. G. Willey, *Tetr. Lett.*, 893 (1962).

12.10 PHOTOCHEMISTRY OF β-IONONE AND RELATED COMPOUNDS

The photochemistry of polyenic compounds related to ionones has attracted much attention because of their structural relation to physiologically active compounds such as vitamin A and retinal. However, the photochemical behavior is still not fully understood.[1,2]

Although 11-*cis* vitamin A has been synthesized by selective hydrogenation of the corresponding acetylenic compounds,[3] the 7-*cis* isomers have not been prepared so far due to the large steric hindrance in the planar form[4] (to be strain-free the 7-ene should be close to orthogonal with respect to the 5-ene). The photochemical conversion of an all-*trans* isomer to its hindered *cis* isomer presents an intriguing synthetic and theoretical subject.

β-Ionol 1[1,2]

1   **2**

1) The hindered 7-*cis* isomer **2** is readily obtained under the photochemical conditions described.

2) The triplet excitation energy of the hindered 7-*cis* isomer **2** is probably higher than that of the nearly planar isomer **1** because the diene group is not conjugated. The photostationary ratio in this system is therefore dependent on the triplet energies of sensitizers employed.

Sensitizer (E_t, kcal/mole)	**1** (%)	**2** (%)
acetonaphthone (59.3)	2	98
benzophenone (68.5)	25	75
acetophenone (73.6)	40	60
xanthone (74.2)	75	25

β-Ionone 3[1,2]

5   **4**   **3**   **6**

1) Direct irradiation of **3** results in a mixture of **4** (minor product) and α-pyran **5** (major product).[5]

1) M. Mousseron, *Advances in Photochemistry* (ed. W. A. Noyes Jr., G. S. Hammond, J. N. Pitts Jr.), vol. 4, p. 195–224), Wiley, 1966.
2) R. S. Liu, *XXIIIrd International Congress of Pure and Applied Chemistry*, vol. 1 (special lecture), p. 335–350, Butterworth, 1971.
3) For a review of the synthesis of vitamin A and isomers, see O. Isler, R. Ruegg, U. Schieter, J. Wursch, *Vitamins and Hormones*, **18**, 295 (1960).
4) G. Wald, P. K. Brown, R. Hubbard, W. Oroshnik, *Proc. Nat. Acad. Sci.*, **41**, 438 (1955).
5) G. Büchi, N. C. Yang, *J. Am. Chem. Soc.*, **79**, 2318 (1957); P. de Mayo, J. B. Stothers, R. W. Yip, *Can. J. Chem.*, **39**, 2135 (1961); M. Mousseron-Canet, M. Mousseron, P. Legendre, J. Wylde, *Bull. Soc. Chim. Fr.*, 279 (1963).

2) Sensitized irradiation of **3** first yields the 7-*cis* isomer **6** which undergoes rapid thermal cyclization to the α-pyran **5**.[2,6]

3) Formation of **4** could be a singlet reaction (sigmatropic H-shift).

β-Ionylidene 11[1,2]

1) Irradiation of an isomeric mixture of **6** and **7** in benzene in the presence of a sensitizer results in a photostationary mixture of **6**, **7**, **8** and **9**.

2) The photostationary ratios are dependent on the triplet energies of the sensitizer.

Sensitizer (E$_t$, kcal/mole)	6(%)	7(%)	8(%)	9(%)
benzanthrone (47)	<1	<1	35	65
fluorenone (51)	12	16	30	41

3) The triplet energies of **6** and **7** can be assumed to be similar to those of conjugated trienes[7] (43–48 kcal/mole); on the other hand, those of the hindered ketones **8** and **9** are similar to those of dienes (52–60 kcal/mole)[7,8] due to their nonplanarity. Hence specific energy transfer to **6** and **7** can be carried out preferentially by employing low-energy sensitizers such as benzanthrone.

Vitamin A[1,2]

1) The triplet energy of all-*trans* vitamin A is ca. 40 kcal/mole.[9]

2) Irradiation of **10** in the presence of deuterated 9,10-dibromoanthracene (E$_t$ 40.2) in CDCl$_3$, gives only the dimer **11**.[2] This was consistent with the production of cyclobutane dimers via a biacetyl-sensitized reaction of retinyl acetate.[10]

3) The dimer resulting from direct irradiation differs from that obtained by sensitized irradiation.[2]

6) E. N. Marvell, presented at the 26th ACS Northwest Regional Meeting, 1971.
7) D. F. Evans, *J. Chem. Soc.*, 1735 (1960).
8) R. E. Kellogg, W. T. Simpson, *J. Am. Chem. Soc.*, **87**, 4230 (1965).
9) A. Sykes, T. G. Truscott, *Chem. Commun.*, 929 (1969); *Trans. Faraday Soc.*, **67**, 679 (1971).
10) C. Giannotti, *Can. J. Chem.*, **46**, 3025 (1968); for direct irradiation, see M. Mousseron-Canet, D. Lerner, J. Mani, *Bull. Soc. Chim. Fr.*, 4639 (1968).

12.11 DYE-SENSITIZED PHOTOOXYGENATIONS

Since Kautsky discovered the reactivity of singlet oxygen,[1] dye-sensitized photooxygenation of organic compounds has been extensively investigated from both synthetic and mechanistic viewpoints.[2-4]

Molecular oxygen has a unique electronic structure in that the electronic character of ground state oxygen is triplet.[5] On the other hand, two kinds of singlet oxygen can be generated photochemically; one of these is known to react with olefinic double bonds.[2-4]

Electronic structure of the oxygen molecule[6,7]

			vacant orbital	
$\pi 2p$				
$\pi 2p$				
σsp				
σsp				
σsp				
1s, 1s				
	ground state ($^3\Sigma^-$g)	$^1\Delta$g singlet state	$^1\Sigma^+$g singlet state	

1) Two possible singlet excited oxygens are formed photochemically; these are $^1\Delta$g and $^1\Sigma^+$g singlets.

2) $^1\Sigma^+$g singlet oxygen has a cationic character because the vacant orbital behaves as an electron pair acceptor.

3) $^1\Delta$g singlet oxygen has a radical character, because two electrons occupy the highest orbitals independently, as shown above.

4) $^1\Delta$g singlet oxygen is mostly responsible for usual photooxygenation reactions.

5) $^1\Delta$g singlet oxygen can be prepared chemically without photosensitization.[4]

General reaction modes

In the case of cyclic or cisoid dienes, the reaction proceeds concertedly via a cyclic six-membered transition state,[2-4] in agreement with orbital symmetry predictions.[8-9]

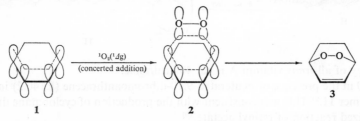

$$\xrightarrow[\text{(concerted addition)}]{^1O_2(^1\Delta g)}$$

1   2   3

1) H. Kautsky, *Trans. Faraday Soc.*, **35**, 216 (1939).
2) K. Gollnick, *Advances in Photochemistry*, (ed., W. A. Noyes Jr., G. S. Hammond, N. J. Pitts Jr.) vol. 6, p. 1–122, Wiley, 1968.
3) D. R. Kearns, *Chem. Reviews*, **71**, 395 (1971).
4) C. S. Foote, *Accounts of Chemical Research*, **1**, 104 (1968).
5) R. S. Mulliken, *Rev. Mod. Phys.*, **4**, 1 (1932).
6) G. Herzberg, *Spectra of Diatomic Molecules*, Van Nostrand, 1950.
7) J. S. Griffin, *Oxygen in Animal Organisms*, (ed. F. Dickens, E. Niel), p. 481, Pergamon Press, 1964.
8) A. U. Khan, D. R. Kearns, *Adv. Chem. Ser.*, **77**, 143 (1968).
9) D. R. Kearns, *J. Am. Chem. Soc.*, **91**, 6554 (1969).

With mono-olefins, the reaction is more complex, and follows the general scheme shown below (see **1** to **7**).[10-16]

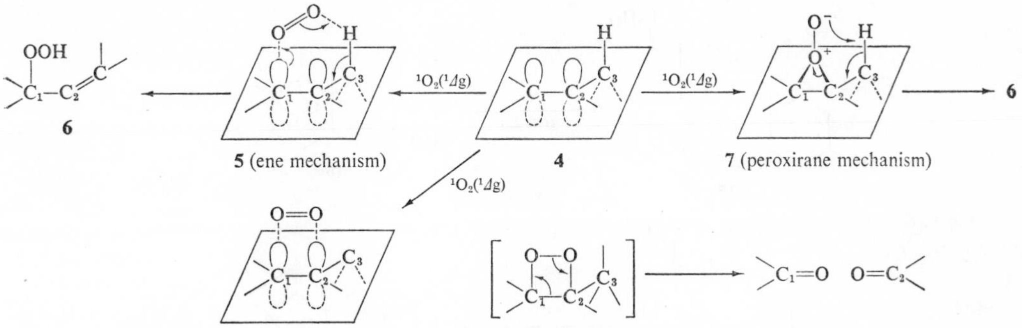

5 (ene mechanism)   **4**   **7** (peroxirane mechanism)

6

8   **9**

1) The hydrogen parallel to the π-orbital of a double bond is abstracted by $^1\Delta g$ singlet oxygen with concomitant double bond migration (see **4** and **6**).[2-4]

2) Solvent effects and Markownikoff-type directing effects are not observed.[4,12]

3) No correlation between relative rates of photooxygenation and rates of reaction with radicals is observed.[16]

4) The isotope effect appears to be small.[17]

5) The facts described above suggest that photooxygenation of a monoolefin usually proceeds in a concerted manner.[2-4]

6) Two possible mechanisms have so far been proposed in order to account for the facts described; a six-membered cyclic mechanism (**4**$\longrightarrow$**5**$\longrightarrow$**6**),[10-14] and a peroxirane mechanism (**4**$\longrightarrow$**7**$\longrightarrow$**6**).[15,16]

7) Photooxygenation of an electron-rich double bond such as occurs in enamines and styrenes results in the formation of an unstable dioxetane **9**, which readily decomposes to ketones.[17-19]

Dienes

Some examples of the photooxygenation of dienes are shown below.

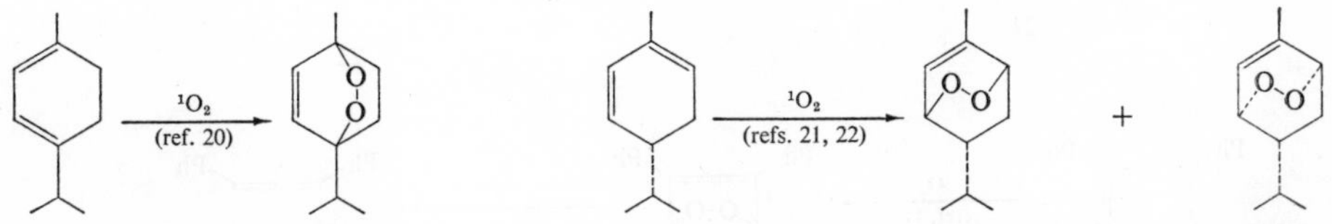

10   **11**   **12**   **13**   **14**

10) A. Nickon, W. L. Mendelson, *J. Am. Chem. Soc.*, **87**, 3921 (1965); *Can. J. Chem.*, **43**, 1419 (1965).
11) A. Nickon, J. F. Bagli, *J. Am. Chem. Soc.*, **83**, 1498 (1961).
12) K. Gollnick, G. O. Schenk, *Pure and Appl. Chem.*, **9**, 507 (1964).
13) C. S. Foote, *Science*, **162**, 963 (1968).
14) K. Gollnick, D. Haisch, G. Schade, *J. Am. Chem. Soc.*, **94**, 1747 (1972).
15) W. Fenical, D. R. Kearns, P. Radlick, *J. Am. Chem. Soc.*, **91**, 7771 (1969).
16) K. R. Kopeky, H. H. Reich, *Can. J. Chem.*, **43**, 2265 (1965).
17) F. A. Litt, A. Nickon, *Oxidation of Organic Compounds III, Adv. Chem. Ser*, **77**, 143 (1968).
18) C. S. Foote, *Pure and Appl. Chem.*, **27**, 635 (1971).
19) K. R. Kopecky, C. Mumford, *Can. J. Chem.*, **47**, 709 (1969).
20) G. O. Schenk, K. Ziegler, *Naturwiss.*, **32**, 157 (1944).
21) A. Schönberg, *Preparative Organische Photochemie*, Springer, 1958.
22) G. O. Schenk, K. G. Kinkel, H.-J. Mertens, *Am. Chem.*, **584**, 125 (1953).

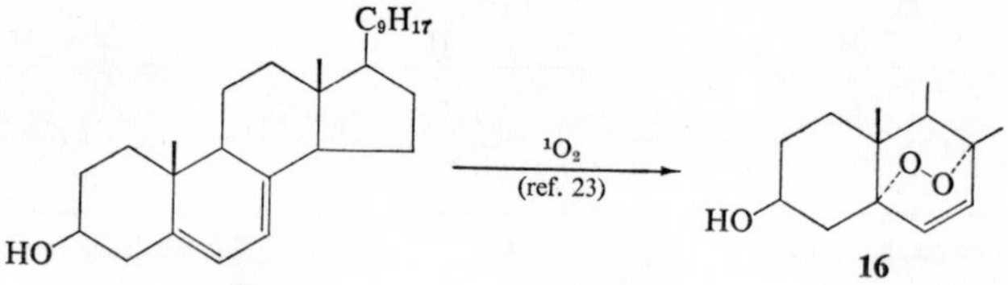

15 16

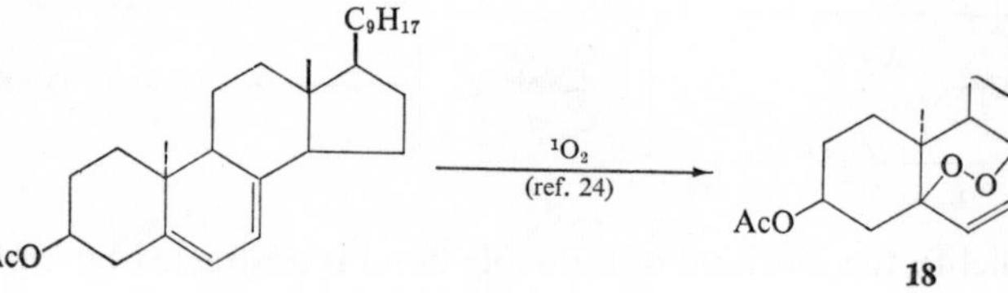

17 18

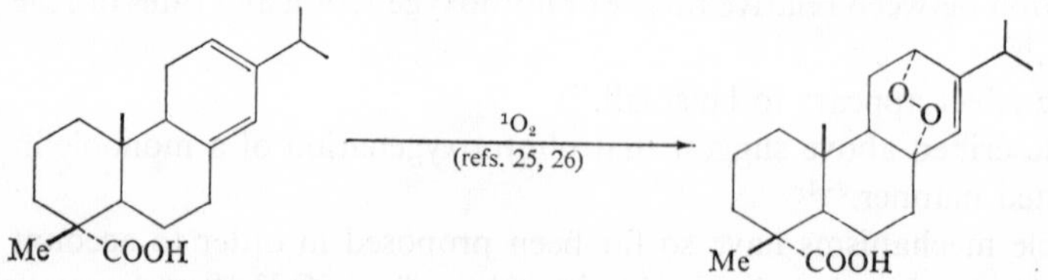

19 20

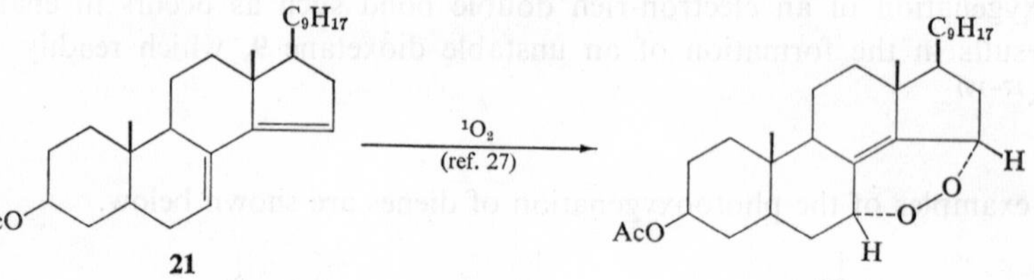

21 22

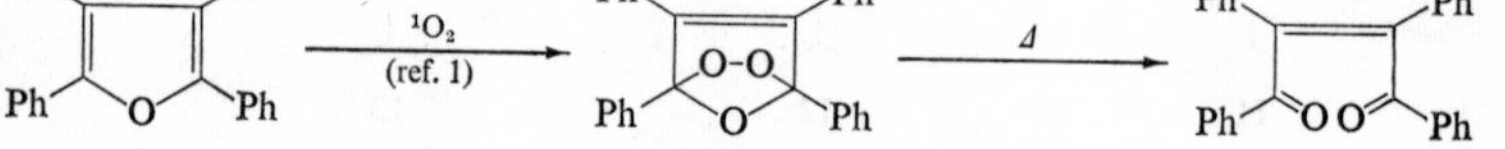

23 24 25

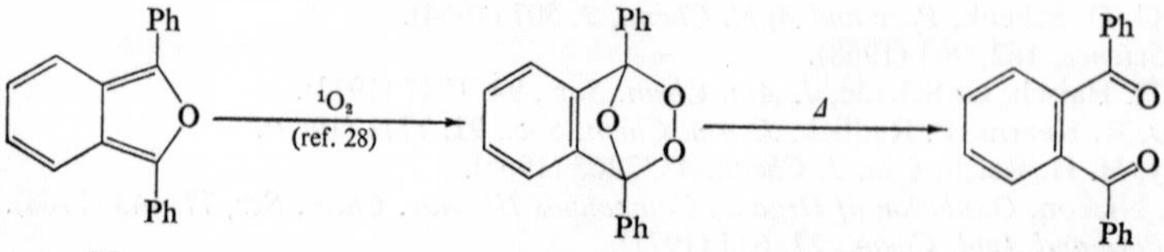

26 27 28

29 (X = H, Ph, Me)   30

31 (X = H, Ph)   32

Mono-olefins

Various examples of the photooxygenation of mono-olefins are given below.

33   34   35   36   37

38   39   40

1) 8-H cannot be abstracted since the conformer **40** is very unstable.

2) Conformers **38** and **39** are involved in the photooxygenation of **33**.[31]

94%   <1%
42   41   43

3) Oxygen approaches from the less hindered side (α-side) of **41**.

4) **43** is a minor product, since 4-H is oriented less favorably (quasi-axial) for oxidation as compared to the 2-Me protons.

23) L. F. Fieser, *The Chemistry of Natural Products Related to Phenanthrene*, Reinhold, 1936.
24) P. Blandon, *J. Chem. Soc.*, 2176 (1955).
25) R. N. Moore, R. V. Lawrence, *J. Am. Chem. Soc.*, **80**, 1438 (1958).
26) R. N. Moore, R. V. Lawrence, *J. Am. Chem. Soc.*, **81**, 458 (1959).
27) D. H. R. Barton, G. F. Laws, *J. Chem. Soc.*, 52 (1954).
28) C. Dufraisse, S. Ecary, *Compt. Rend.*, **223**, 735 (1946).
29) C. Dufraisse, R. Horclois, *Bull. Soc. Chim. France*, 3, 1880 (1936).
30) C. Dufraisse, R. Horclois, *Bull. Soc. Chim. France*, 3, 1894 (1936).
31) (a) G. Ohloff, G. Uhde, *Helv. Chim. Acta*, **48**, 10 (1965); (b) Klein, W. Rojahn, *Tetr.* **21**, 2173 (1965).
32) G. O. Schenk, H. Eggert, W. Denk, *Ann. Chem.*, **584**, 177 (1953).

44a ⇌ **44b**

i) ¹O₂ (ref. 33) ii) reduction

45 (50%)   +   **46** (27%)   +   **47** (23%)

5) In **44b**, the shielding of 7-Me renders the quasi-axial hydrogens (a'-H) nonreactive.

6) Conformer **44a** is reactive to photooxygenation, and singlet oxygen approaches from the less hindered side (α side).

48 → i) ¹O₂ ii) reduction (ref. 34) → **49**

50 → i) ¹O₂ ii) reduction (ref. 34) → **51**

52 → hν/O₂/sens. (ref. 11) → **53**

7) 8-H is strongly shielded by 10- and 13-Me's in **52**, so that oxygen approaches from the less hindered side (α side) to abstract 5-H.

33) K. Gollnick, S. Shroeter, G. Chloff, G. Schade, G. O. Schenk, *Ann. Chem.*, **687**, 14 (1965); cf. K. Gollnick, G. Schade, *Tetr. Lett.*, 2335 (1966).
34) R. A. Bell, R. E. Ireland, L. N. Mander, *J. Org. Chem.*, **31**, 2536 (1966).

8) The *cis*-fused compound **54** is unreactive to photooxygenation, since conformers **55a** and **55b** both lack unshielded axial (or quasi-axial) hydrogens.[2]

9) *Tert*-OOH groups isomerize thermally to more stable *sec*-OOH compounds (**57 ⟶ 60**, **59 ⟶ 61**).

10) 6-H (β) is hindered by the 10-Me group and hence oxygen abstracts the 14α-H in **62**.

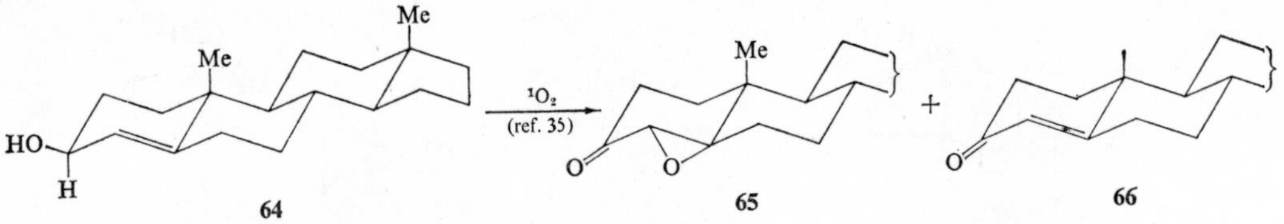

64   65   +   66

11) The epoxy-ketone **65** is assumed to be formed from $^1\Delta$g singlet oxygen.

12) The product ratio **65/66** varies with the E_t values of the sensitizers employed; i.e., higher energy sensitizers give lower product ratios.[36]

13) The keto-compound **66** may be produced by $^1\Sigma_g^+$ singlet oxygen.[36]

β,γ-Unsaturated ketones

Mechanistic studies with β,γ-unsaturated ketones suggest that photooxygenation occurs via the cyclic "ene" mechanism (see **5**).

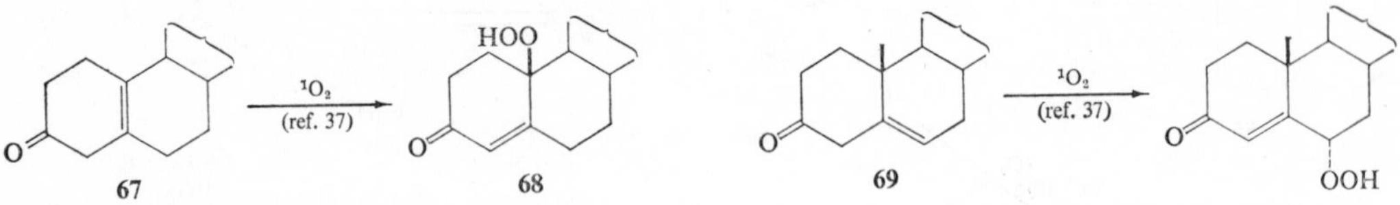

67   68   69   70

Carotenoids and related terpenoids

Photooxygenation of carotenoids and related terpenoids affords ketene derivatives in addition to normal oxygenation products. This unique ketene formation is attributed to the non-planarity of the 5,6- and 7,8-double bonds, resulting from steric crowding of the methyl groups at C-1 and C-5. This unique photooxidation has been applied to syntheses of naturally occurring ketenic carotenoids.

1) Photooxygenations of β-ionol **71** to form **73** and **75**, and of dehydro β-ionone **76** to form **79** and **80**, are shown below.[38]

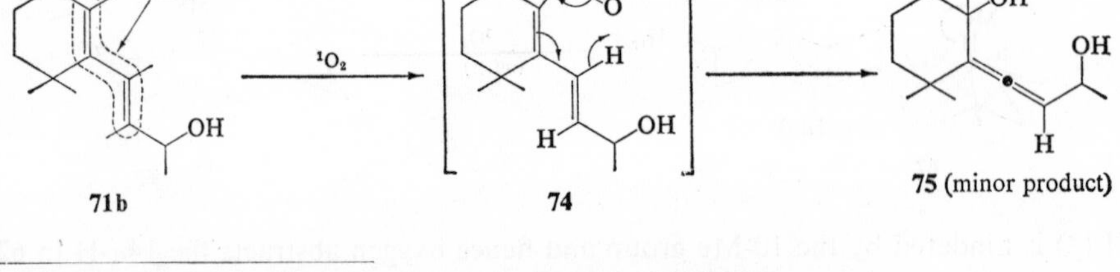

71a

72   73 (major product)

non-planar

71b   74   75 (minor product)

35) A. Nickon, W. L. Mendelson, *J. Am. Chem. Soc.*, **85**, 1894 (1963).
36) D. R. Kearns, R. A. Hollins, A. U. Khan, R. W. Chembers, P. Radlick, *J. Am. Chem. Soc.*, **89**, 5455 (1967).
37) N. Furutachi, Y. Nakadaira, K. Nakanishi, *Chem. Commun.*, 1625 (1968).
38) S. Isoe, S. B. Hyeon, H. Ichikawa, S. Katsumura, T. Sakan, *Tetr. Lett.*, 5561 (1968).

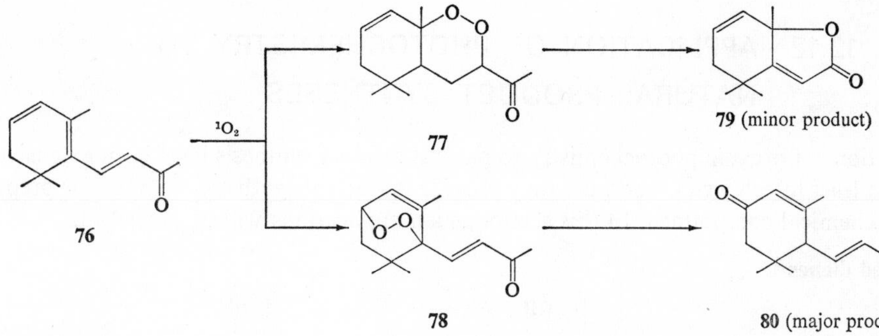

2) A biogenetic-type synthesis of grasshopper ketone **96** and loliolide **92**[39] based on these photo-oxygenation reactions is shown below.

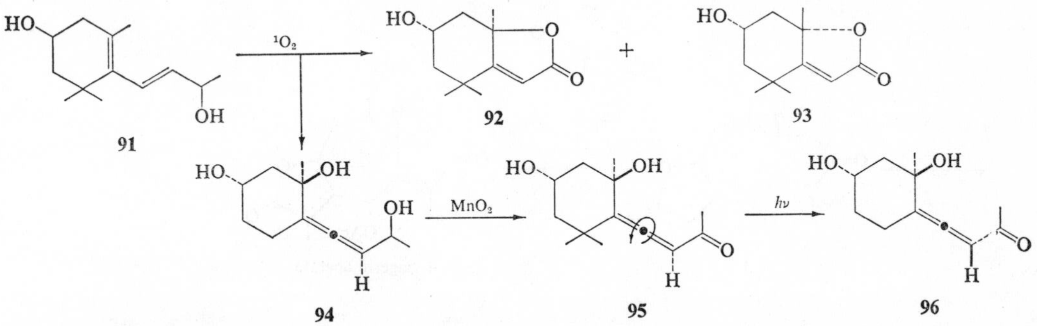

3) β-Carotene itself can be photooxygenated to yield **100**.[40,41]

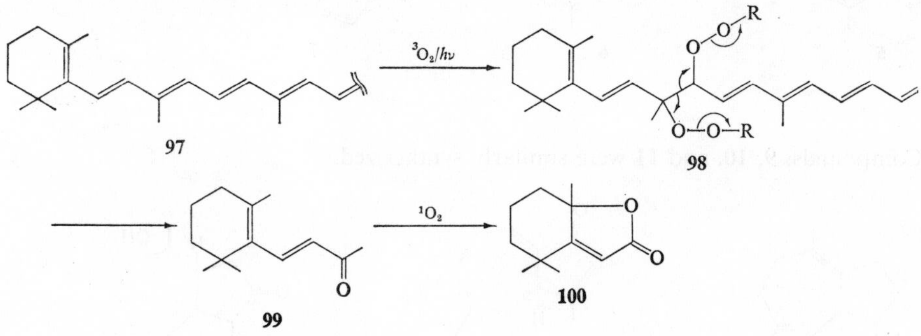

39) S. Isoe, S. Katsumura, S. B. Hyeon, T. Sakan, *Tetr. Lett.*, 1089 (1971).
40) S. Isoe, S. B. Hyeon, T. Sakan, *Tetr. Lett.*, 279 (1969).
41) S. Isoe, S. B. Hyeon, S. Katsumura, T. Sakan, *Tetr. Lett.*, 2517 (1972).

12.12   APPLICATION OF PHOTOCHEMISTRY TO NATURAL PRODUCT SYNTHESES

Applications of organic photochemistry to natural product synthesis have been extensively studied for at least fifteen years,[1] because they provide important synthetic methods for preparing unusual chemical compounds. In this section, such applications will be described.

Cross-conjugated dienes

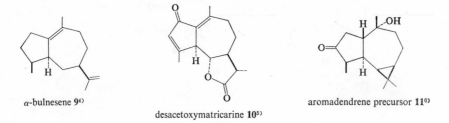

Compounds, **9, 10,** and **11** were similarly synthesized.

α-bulnesene **9**[4]

desacetoxymatricarine **10**[5]

aromadendrene precursor **11**[6]

1) related reviews: (a) K. Schaffner, *Fortschr. Chem. Org. Naturstoffe,* **22,** 1 (1964). (b) P. de Mayo, *Adv. Org. Chem.,* **2,** 367 (1960); P. G. Sammes, *Quart Rev.,* **24,** 37 (1970).
2) D. H. R. Barton, J. T. Pinhey, R. J. Wells, *J. Chem. Soc.,* 2518 (1964).
3) J. A. Marshall, P. C. Johnson, *Chem. Comm.,* 391 (1968).
4) E. Piers, K. F. Cheng, *Chem. Comm.,* 562 (1969).
5) H. White, S. Eguchi, J. N. Marx, *Tetr.,* **25,** 2099 (1969).
6) J. Streith, A. Blind, *Bull. Soc. Chim. France,* 2133 (1968).

Cycloadditions of α,β-unsaturated ketones with olefins

Irradiation of α,β-unsaturated ketones in the presence of olefins yields cyclobutane derivatives which are potentially useful in natural product synthesis. The mechanism has not been established in detail, but the triplet enones were found to be reactive species.[7]

1) *Cyclohexenones*: Cyclohexenones give cyclobutane derivatives photochemically in the presence of olefins.

Orientation in the cycloaddition of cyclohexenone to olefins can be accounted for on the basis of a complex **13** between the excited triplet enones (n, π^*) and electron-rich olefins in which the olefin serves as the electron donor and the excited ketone acts as the electron acceptor.[8] Cycloaddition proceeds in a two-step mechanism via the biradical intermediate **14**, and yields *cis*- and *trans*-fused cyclobutanes **15**.[8] A π, π^* triplet mechanism has also been proposed for this reaction.[8]

Applications of such reactions are shown below.

7) N. J. Turro, *Energy Transfer and Organic Photochemistry*, (ed. A. A. Lamoda and N. J. Turro), p. 133, Wiley, 1969.
8) (a) E. J. Corey, J. D. Bass, R. LeMahieu, R. B. Mitra, *J. Am. Chem. Soc.*, **86**, 5570 (1964); (b) G. J. Fonken, *Organic Photochemistry*, (ed. O. L. Chapman), vol. 1, p. 197, Marcel Dekker, 1967.
9) E. J. Corey, R. B. Mitra, H. Uda, *J. Am. Chem. Soc.*, **86**, 485 (1964).
10) E. J. Corey, S. Nozoe, *J. Am. Chem. Soc.*, **86**, 1652 (1964).
11) K. Wiesner, I. Jirkovsky, M. Fishman, C. A. J. Williams, *Tetr. Lett.*, 1523 (1967), see also K. Wiesner, V. Musil, K. J. Wiesner, *Tetr. Lett.*, 5643 (1968).

24   28   29   atisine[12] and veatchine[13]

30   31

32   33 (ormosia alkaloid)

2) *Cyclopentenones*: Cyclopentenone triplet molecules add similarly to olefins to give cyclo-butane derivatives (only *cis*-fused cyclobutanone).

34   35   36   37   α-bourbonene 38

39   40   41   prostanic acid 42

R¹= (CH₂)₆COOCH₃
R²= (CH₂)₄CH₃

3) *1,3-Diketones and their enol acetates*: Cycloaddition of these compounds is extremely im-portant in synthetic reactions.

12) R. W. Guthrie, Z. Valenta, K. Wiesner, *Tetr. Lett.*, 4645 (1966).
13) K. Wiesner, S. Uyeo, A. Philipp, Z. Valenta, *Tetr. Lett.*, 6279 (1968).
14) N. R. Hunter, G. A. MacAlpine, H. J. Liu, Z. Valenta, *Can. J. Chem.*, **48**, 1436 (1970).
15) J. D. White, D. N. Gupta, *J. Am. Chem. Soc.*, **90**, 6171 (1968).
16) J. F. Bagli, T. Bogri, *Tetr. Lett.*, 1639 (1969).
17) H. Hikino, P. de Mayo, *J. Am. Chem. Soc.*, **86**, 3582 (1964).

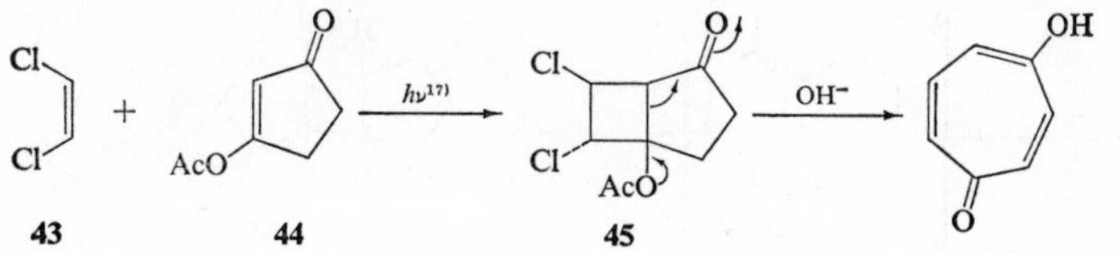

43 44 45 troplolone 46

Other tropolones have been similarly synthesized.[18)]

47 48 49 50 51

52 53 54 55

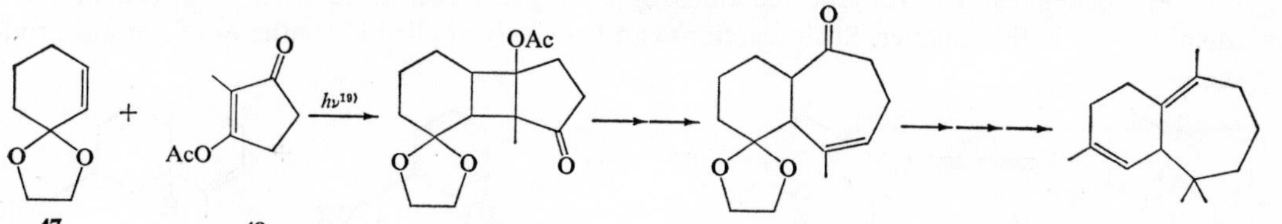

4) *Intramolecular cycloadditions*

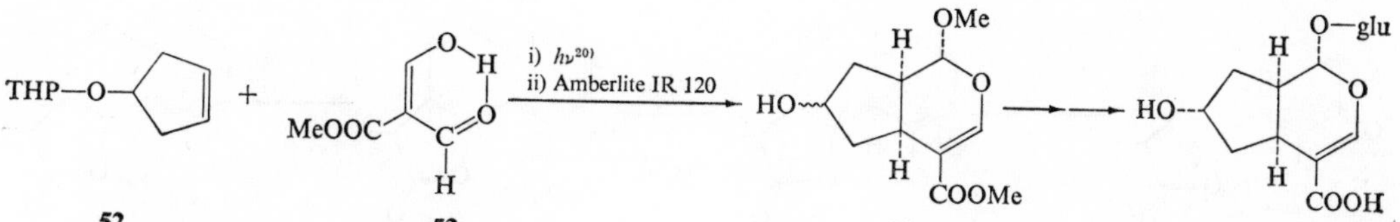

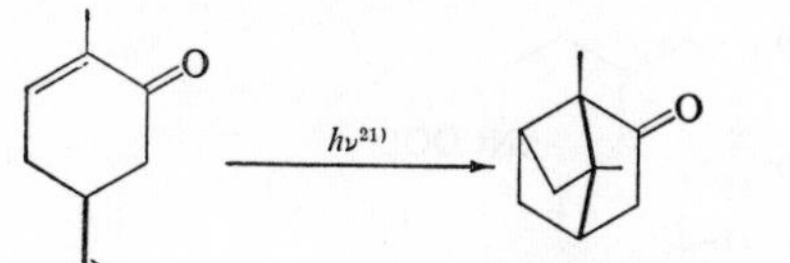

carvone camphor 57

56

58 59 copaene 60

Cycloadditions of conjugated dienes with olefins

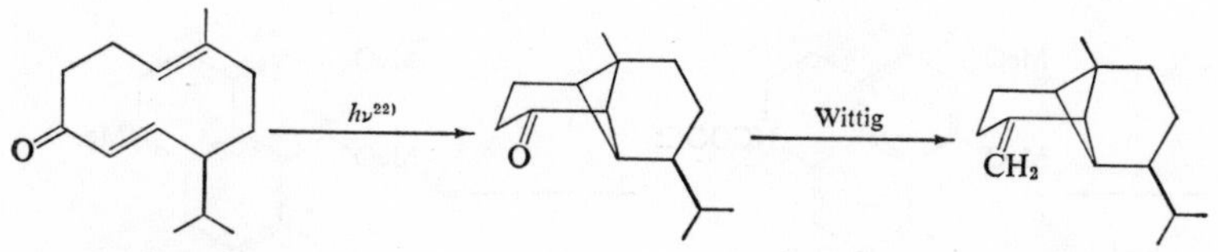

Butadienes add photochemically to olefins to give cyclobutane derivatives via a biradical intermediate formed from the triplet state.[23)]

18) G. L. Lange, P. de Mayo, *Chem. Comm.*, 704 (1967).
19) B. D. Challand, G. Kornis, G. L. Lange, P. de Mayo, *Chem. Comm.*, 704 (1967); see also B. D. Challand, P. de Mayo, *Chem. Comm.*, 982 (1968).
20) G. Büchi, J. A. Carlson, J. E. Powell, L. -F. Tietze, *J. Am. Chem. Soc.*, **92**, 2165 (1970).
21) G. Büchi and I. M. Goldman, *J. Am. Chem. Soc.*, **79**, 4741 (1957).
22) C. H. Heathcock, R. M. Badger, *Chem. Comm.*, 1510 (1968).
23) R. S. H. Liu, N. J. Turro, G. S. Hammond, *J. Am. Chem. Soc.*, **87**, 3406 (1965).

61   62   (−)-β-bourbonene **63**

Electrocyclic reactions

Photochemical electrocyclic reactions of conjugated enes have been described in more detail earlier in this chapter. Such reactions can be widely applied in syntheses of natural products.

64   **65**   occidentalol **66**   **67**

68   **69**   **70**   dihydrocostunolide **71**

Cyclizations of styrene-type aromatic compounds are important in alkaloid syntheses.

72   **73**

74   **75**

24) K. Yoshihira, Y. Ohta, T. Sakai, Y. Hirose, *Tetr. Lett.*, 2263 (1969).
25) A. G. Hortmann, *Tetr. Lett.*, 5785 (1968).
26) E. J. Corey, A. G. Hortmann, *J. Am. Chem. Soc.*, **85**, 4033 (1963).
27) M. P. Cava, S. C. Havlicek, A. Lindert, R. J. Spangler, *Tetr. Lett.*, 2937 (1966).
28) N. C. Yang, A. Shani, G. R. Lenz, *J. Am. Chem. Soc.*, **88**, 5369 (1966).
29) M. P. Cava, S. C. Havlicek, *Tetr. Lett.*, 2625 (1967).
30) (a) G. R. Lenz, N. C. Yang, *Chem. Commun.*, 1136 (1967); (b) S. M. Kupchan, H. C. Wormser, *J. Org. Chem.*, 30, 3792 (1965); (c) I. Ninomiya, T. Naito, T. Kiguchi, *Tetr. Lett.*, 4451 (1970); (d) D. H. Hey, G. H. Jones, M. J. Perkins, *J. Chem. Soc.*, C 116 (1971).

76 77 78

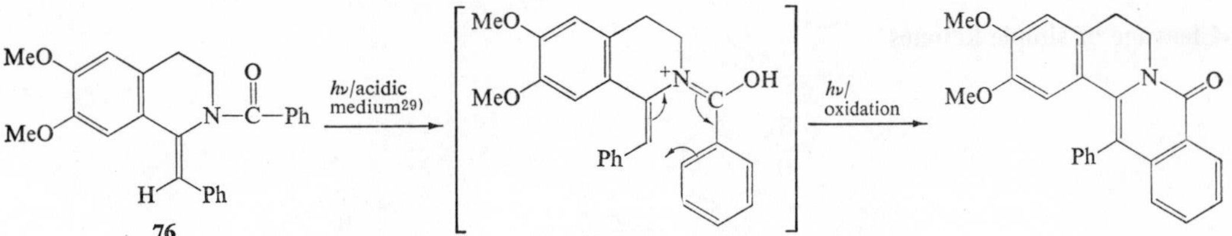

Under acidic conditions, amides, such as **75** can cyclize on irradiation via the protonated intermediate 77. Many similar reaction have been reported.[30]

Photochemical cyclization of benzophenone-type aromatic compounds may be useful for the synthesis of gibberellins or indole alkaloids.

79 80

81 82 83

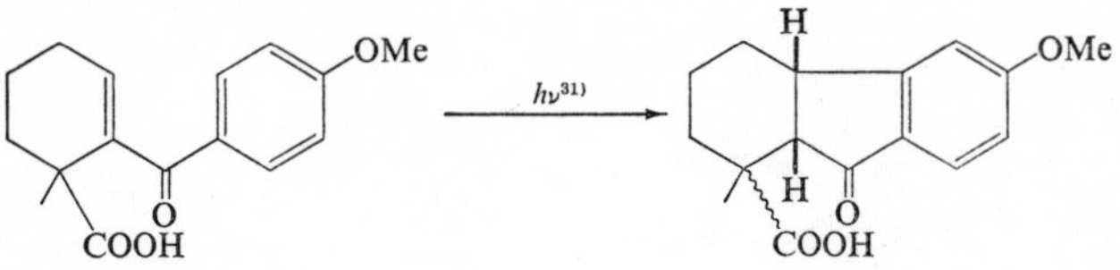

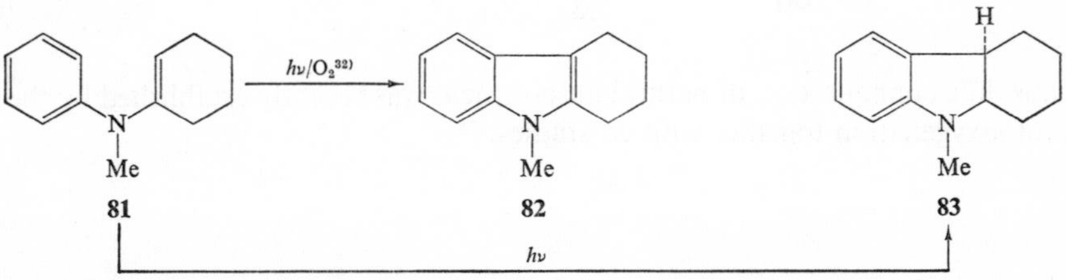

Photochemical Fries rearrangements[33]

The general photo Fries rearrangement is shown below.

84 85 86

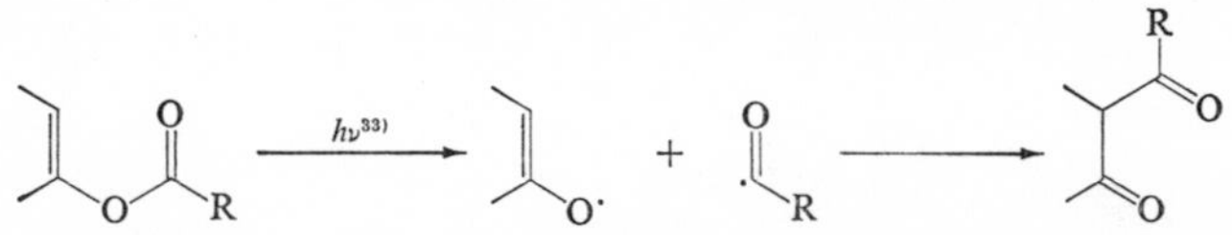

This reaction was applied for the synthesis of griseofulvin.[34]

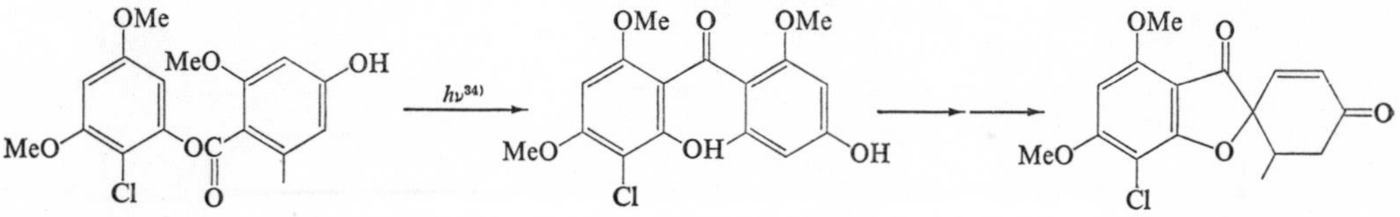

87 88 griseofulvin **89**

31) L. M. Jackman, E. F. M. Stephenson, H. C. Yick, *Tetr. Lett.*, 3325 (1970).
32) O. L. Chapman, G. L. Eian, A. Bloom, J. Clardy, *J. Am. Chem. Soc.*, **93**, 2918 (1971).
33) D. Bellus, P. Hrdlovic, *Chem. Rev.*, **67**, 599 (1967).
34) D. Taub, C. H. Kuo, H. L. Slates, N. L. Wendler, *Tetr.*, **19**, 1 (1963).

α-Cleavage of simple ketones

The photochemical primary product is the corresponding aldehyde, which is further oxidized by oxygen to the acid.

Dye-sensitized photooxygenation

The absolute configuration of natural abscisic acid was recently established by the application of photooxygenation together with cd studies.[37]

---

35) G. Quinkert, H. G. Heine, *Tetr. Lett.*, 1659 (1963); K. Habaguchi, M. Watanabe, Y. Nakadaira, K. Nakanishi, A. K. Kiang, *Tetr. Lett.*, 3731 (1968).

36) J. W. Cornforth, B. V. Milborrow, G. Ryback, *Nature*, **206**, 715 (1965); see also M. Mousseron-Canet, J. C. Mani, J. P. Dalle, J. L. Olive, *Bull. Soc. Chim. France*, 2874 (1966).

37) M. Koreeda, G. Weiss, K. Nakanishi, *J. Am. Chem. Soc.*, **95**, 239 (1973).

INDEX